High Speed Turbomachines and Electrical Drives Conference (HSTED-2020)

E3S Web of Conferences Volume 178 (2020)

Prague, Czech Republic
14 – 15 May 2020

Editor:

Alexander Drozdov

ISBN: 978-1-7138-1439-9

Printed from e-media with permission by:

Curran Associates, Inc.
57 Morehouse Lane
Red Hook, NY 12571

Some format issues inherent in the e-media version may also appear in this print version.

For additional information, please contact EDP Sciences – Web of Conferences
at the address below.

EDP Sciences – Web of Conferences
17, Avenue du Hoggar
Parc d'Activité de Courtabœuf
BP 112
F-91944 Les Ulis Cedex A
France

Phone: +33 (0) 1 69 18 75 75
Fax: +33 (0) 1 69 28 84 91

contact-edps@webofconferences.org

Additional copies of this publication are available from:

Curran Associates, Inc.
57 Morehouse Lane
Red Hook, NY 12571 USA
Phone: 845-758-0400
Fax: 845-758-2633
Email: curran@proceedings.com
Web: www.proceedings.com

TABLE OF CONTENTS

HIGH SPEED TURBOMACHINES AND ELECTRICAL DRIVES CONFERENCE 2020 (HSTED-2020)

Preface

We are delighted to present the proceedings of the High Speed Turbomachines and Electrical Drives Conference 2020, which was held in an online format on May 14-15, 2020.

This conference was organized by Business Communication Center (Czech Republic) in cooperation with University of Florence (Italy) and Research Laboratory "Gas Dynamics of Turbomachines" (Peter the Great St. Petersburg Polytechnic University, Russia). The conference aims to discuss the current state of the industry in the development of gas turbines, steam turbines, bearing for high speed machinery, materials for high speed machinery, high speed electrical drives, power electronics, rotor dynamics, compressors and their systems, hydraulic machinery and systems, energy efficiency, electrical engineering.

We would like to thank the keynote speakers, members of the scientific committee and all participants for their active position and desire to continue research despite the unexpected coronavirus crisis.

Dr. Alexander Drozdov, Senior Researcher of the Research Laboratory "Gas Dynamics of Turbomachines", Peter the Great St. Petersburg Polytechnic University

Prague (online), May 14-15, 2020

edp sciences

E3S Web of Conferences

Statement of Peer review

In submitting conference proceedings to E3S Web of Conferences, I certify to the Publisher that I adhere to the **Policy on Publishing Integrity** of the journal in order to safeguard good scientific practice in publishing.

1. All articles have been subjected to peer review administered by the proceedings editors.
2. Reviews have been conducted by expert referees, who have been requested to provide unbiased and constructive comments aimed, whenever possible, at improving the work.
3. Proceedings editors have taken all reasonable steps to ensure the quality of the materials they publish and their decision to accept or reject a paper for publication has been based only on the merits of the work and the relevance to the journal.

Title, date and place of the conference

" High Speed Turbomachines and Electrical Drives Conferense-2020", may 14-15, 2020, Prague, Czech Republic

Proceedings editor(s):

Alexander Drozdov

Date and editor's signature

30.06.2020

17, avenue du Hoggar - PA de Courtabœuf – BP 112 - 91944 Les Ulis Cedex A (France)
Tél. : 33 (0)1 69 18 75 75 - Fax : 33(0)1 69 07 45 17 – www.edpsciences.org

Modeling and analysis of vector control systems for asynchronous motor

Viktor Meshcheryakov[1,*], *Tatyana* Sinyukova[1], *Alexey* Sinyukov[1], *Andrei* Boikov[1] and *Rustem* Mukhametzhanov[2]

[1]Electric Drive Department, Lipetsk State Technical University, Lipetsk, Russia
[2]Kazan State Power Engineering University, Kazan, Russia

Abstract. The study aims at the analysis of vector control asynchronous electric drive systems. For comparison and evaluation, mathematical models of systems are implemented in the environment of simulation modeling Matlab Simulink. The evaluation criteria selected were: complexity of implementation, energy efficiency of the inverter, accuracy of speed maintenance, torque ripple, reaction speed of the system to disturbances from the side of the drive mechanism, impact on the supply network. Vector control systems by ensuring the maintenance of accuracy of the moment in the entire range of speed control are more widespread. The study of vector systems, the formation of the stator voltage vector in which is carried out using pulse-width modulation. The signal organization during the study was carried out by several methods. At the initial stage, the signal was formed due to relay-vector control in a closed loop for monitoring the instantaneous values of current errors without forced modulation; at the next stage of the study, the signal was generated using sinusoidal pulse-width modulation based on a comparison of control signals with some reference vector, the final stage became a spatial-vector modulation method.

1 General information

The basis of vector control of an asynchronous electric drive are differential equations that allow you to describe an electric machine quite accurately in both statics and dynamics. The control of the electromagnetic torque is achieved by controlling the amplitude and instantaneous phase of the stator current or its voltage, which allows achieving complete control of the machine in dynamics and obtaining the desired transients, the quality of which is much higher than in scalar control systems. This fact has ensured the widespread use of vector systems, including in high-precision electric drives [1-15]. The stator voltage vector is formed using pulse-width modulation. At the same time, on the basis of a voltage inverter, pulse-width modulation signal can be obtained in the following ways:

- relay-vector formation of pulse-width modulation in a closed loop for monitoring the instantaneous values of current errors without forced modulation;
- sinusoidal pulse-width modulation based on a comparison of the control signals with some reference signal;
- space-vector modulation method.

1.1 Relay-vector control

In the relay-vector control systems, the reference vector is the rotor flux linkage vector (Fig. 1). The system has two control channels: a speed control channel and a rotor flux link control channel. As feedback channels, signals from current sensors and a speed sensor (in a real system represented by an encoder) are used. The rotor flux linkage value is calculated indirectly. Relay-vector PWM signal is generated by the relay current regulator (Current Regulator). Sequence control of power switching transistors made IGBT- current regulator.

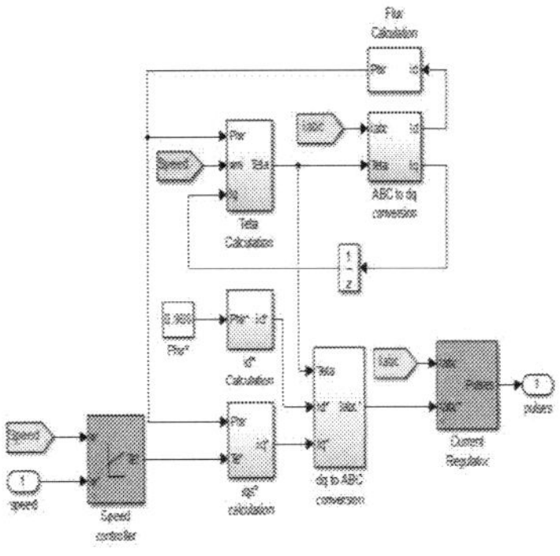

Fig. 1. Relay-vector control.

1.2 Vector control with sinusoidal PWM

The main drawback of standard sinusoidal PWM modulation is the underutilization of the DC link voltage by about 14%. This leads to a decrease in the motor torque, an increase in current consumption when working with the rated torque, as well as the

* Corresponding author: mesherek@yandex.ru

non-optimal use of high-voltage elements of the power circuit. One way to solve this problem is premodulation with the third harmonic of the line voltage. The idea of the work is that a third harmonic premodulation signal is added to the signal of the fundamental frequency generator, which allows to realize the maximum amplitude of the output voltage. The functional diagram of the generator that implements a sinusoidal PWM with premodulation is shown in Figure 2.

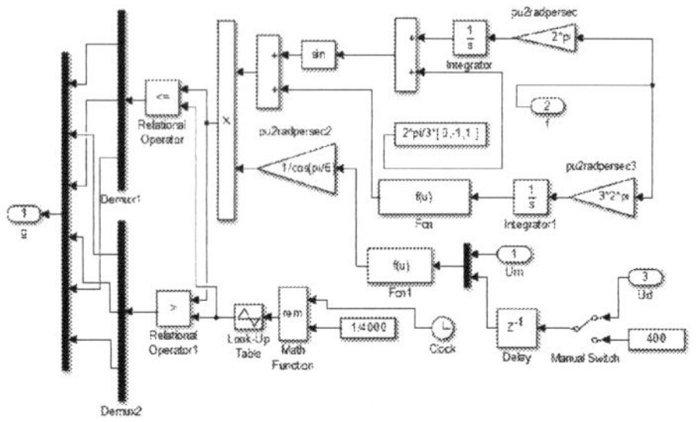

Fig. 2. Pre-modulated Sine-PWM Generator.

1.3 Vector control with space-vector PWM

The spatial-vector PWM of the voltage vector is implemented using basic vectors in a three-phase coordinate system, with each basic vector characterized by a certain state of the inverter keys (Fig. 3). The speed reference comes to classical PI controller, whose output signals are proportional to the required frequency and voltage magnitude. Based on these signals, a three-phase voltage generating unit generates sinusoids of a given frequency and amplitude. In the sector calculation unit, the sector number in the coordinate system αβ is determined.

The calculation of the duration of the inclusion of basic and zero vectors is carried out by the corresponding unit, the conversion of the relative durations into real ones is carried out by the signal of the ramp generator, which determines the carrier frequency of the pulse-width modulation of the inverter. The power driver is controlled directly by the pulse shaper.

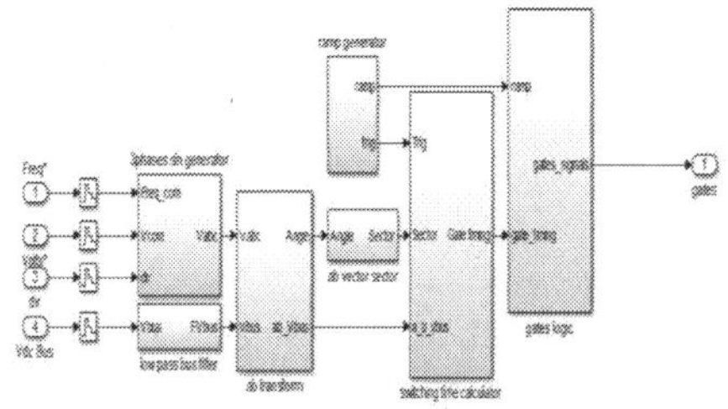

Fig. 3. Control system with space-vector PWM.

2 Simulation results

2.1 Vector system (relay-vector control)

The characteristics obtained during modeling are presented in Figures 4-6.

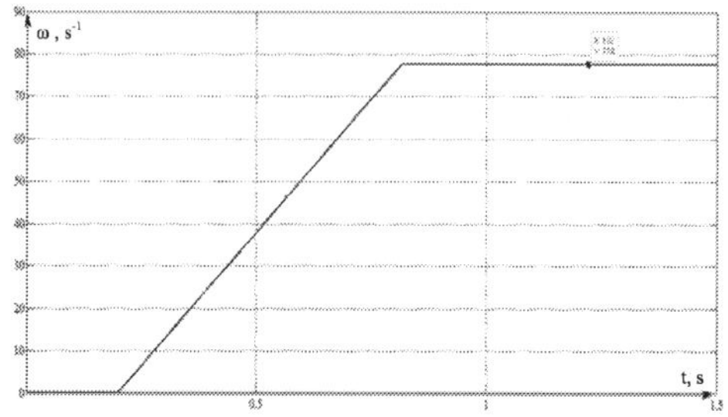

Fig. 4. Transition process of the torque.

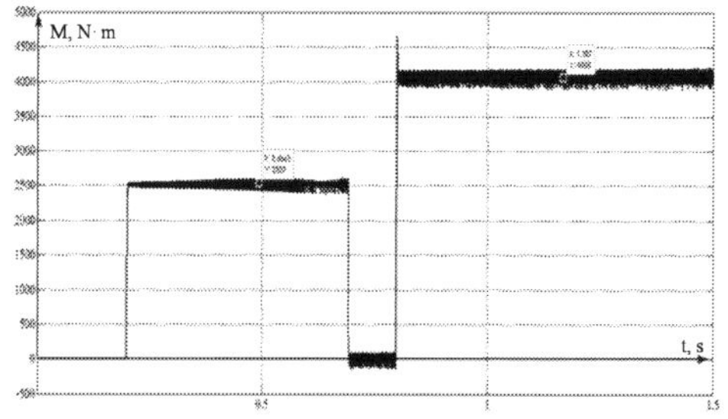

Fig. 5. Transient process of the stator current.

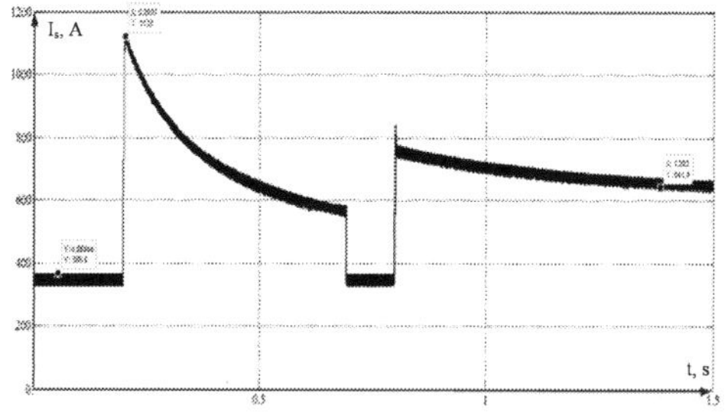

Fig. 6. Transient process of the speed.

As can be seen from the oscillogram of the stator current, at a time when there is no speed reference, a voltage is applied to the stator windings, causing a magnetizing current of 366 A to flow, which is necessary to create a nominal rotor flux linkage. At a time of 0.2 s, a speed reference is received through the intensity controller. During acceleration, a constant dynamic torque of 2500 Nm is maintained. At a moment of time of 0.8 s, a load surge occurs, while there is no subsidence of speed, and the torque is formed with a slight overshoot. So, the relay-vector control system allows for full control of the asynchronous motor in dynamics.

https://doi.org/10.1051/e3sconf/202017801001

2.2 Vector system (control with sinusoidal PWM)

The characteristics obtained during modeling are presented in Figures 7-9.

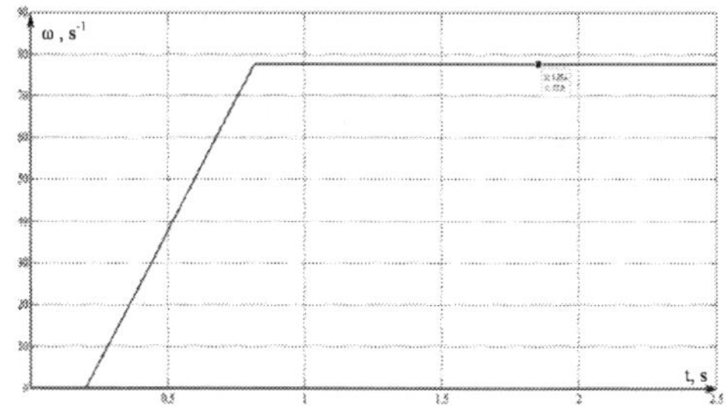

Fig. 7. Transition process of the torque.

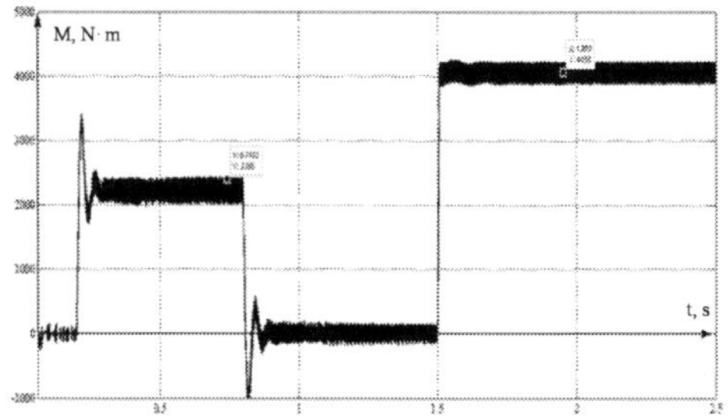

Fig. 8. Transient process of the stator current.

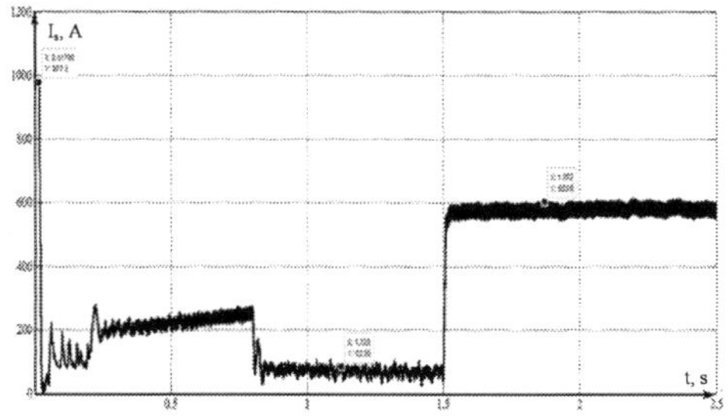

Fig. 9. Transient process of the speed.

At the initial torque, a magnetization current of 977 A is formed, not exceeding twice the nominal value, necessary to create the nominal rotor flux. In the process of acceleration is maintained constant dynamic torque 2300 Nm. At a moment of time of 1.5 s, a load surge occurs, the torque is formed practically without overshoot. So, a vector control system with a sinusoidal pulse-width modulation algorithm allows to obtain a good quality of transients.

2.3 Vector system (control with space-vector PWM)

The characteristics obtained during modeling are presented in Figures 10-12.

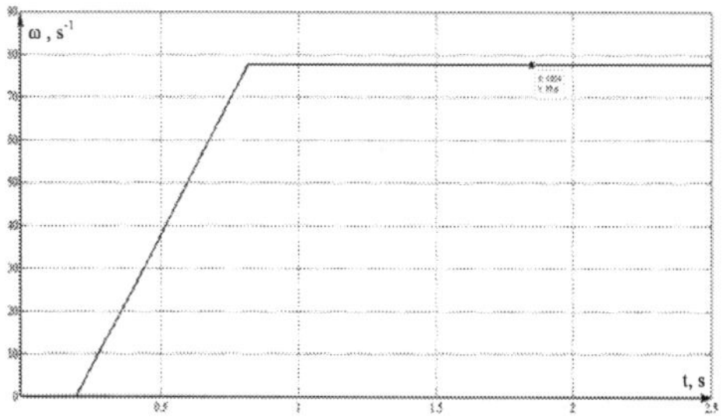

Fig. 10. Transition process of the torque.

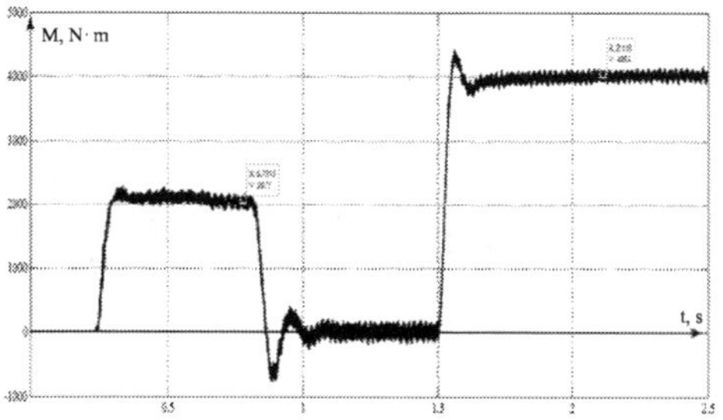

Fig. 11. Transient process of the stator current.

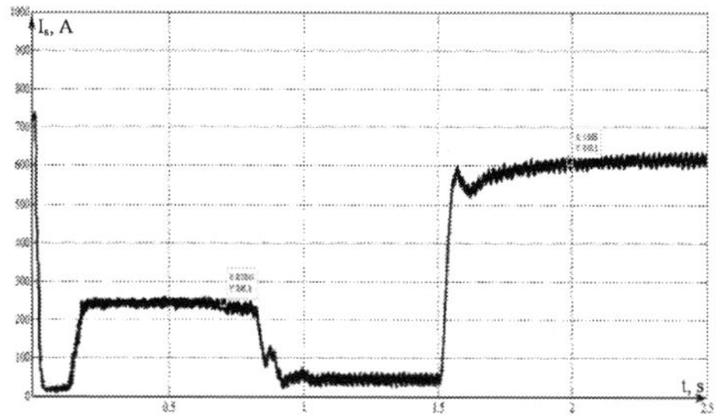

Fig. 12. Transient process of the speed.

The rated motor flux is formed by a stator current surge of 735 A; during startup, the current value is 245 A. During acceleration, a constant dynamic torque of 2000 N · m is maintained. At a moment of time of 1.5 s, a load surge occurs, the torque is formed with permissible overshoot. So, the spatial-vector control system allows you to implement the desired transients.

3 Comparison of different types of PWM

3.1 Torque ripple

The oscillograms of the electromagnetic motor torque during steady state operation at rated load for different types of PWM type shown in Figures 13- 15.

https://doi.org/10.1051/e3sconf/202017801001

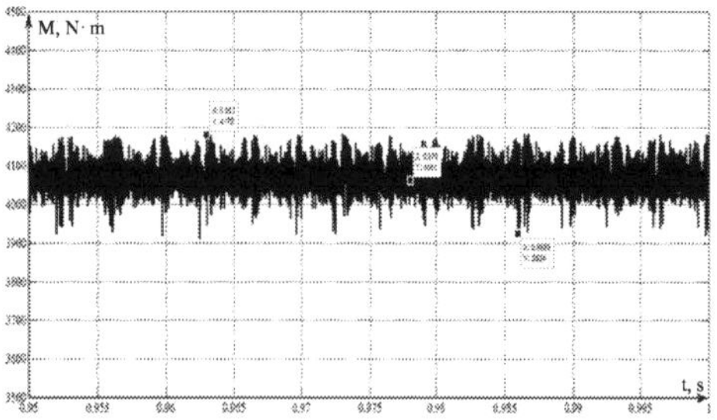

Fig. 13. Transient process for relay-vector pulse-width modulation.

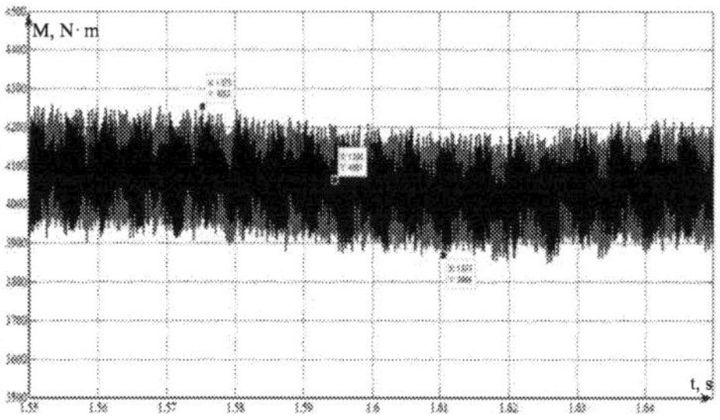

Fig. 14. Transient for sinusoidal pulse width modulation.

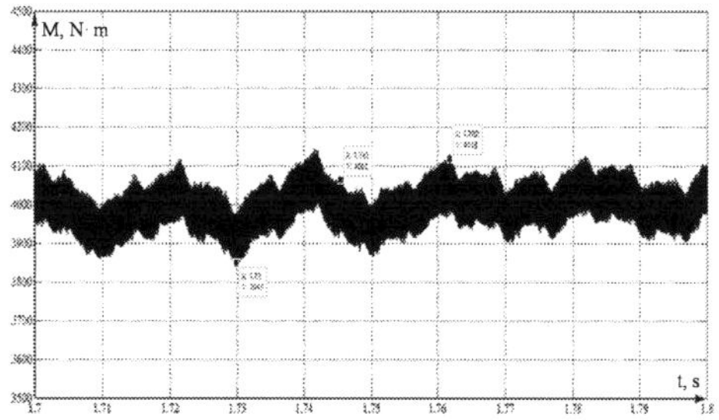

Fig. 15. Transient for spatially vector pulse width modulation.

The average amplitude of the pulsations of the electromagnetic torque for the relay-vector and space-vector pulse-width modulation lies in the range of 120 - 130 N · m, for sinusoidal pulse-width modulation in the range of 150 - 160 N · m. So, sinusoidal pulse-width modulation provides a slightly worse quality of the electromagnetic torque, however, the ripple value is still small and does not exceed 7% of the nominal value of the torque.

3.2 System response time to perturbation

Oscillograms of the formation of the electromagnetic trque of the engine when the rated load is drawn during idling for various types of pulse-width modulation are shown in Figures 16, 17 and 18.

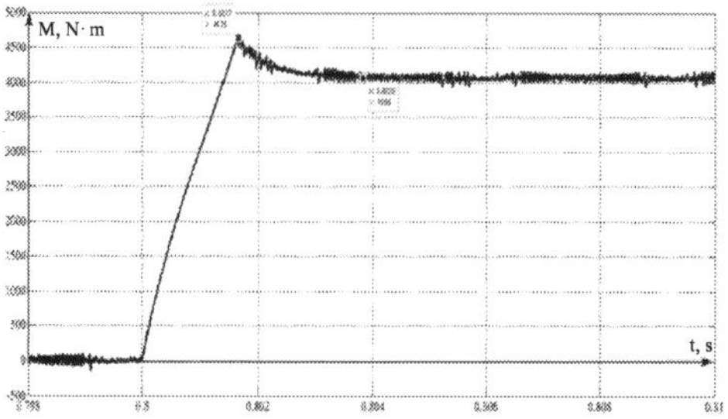

Fig. 16. Transient process for relay-vector pulse-width modulation during load transfer.

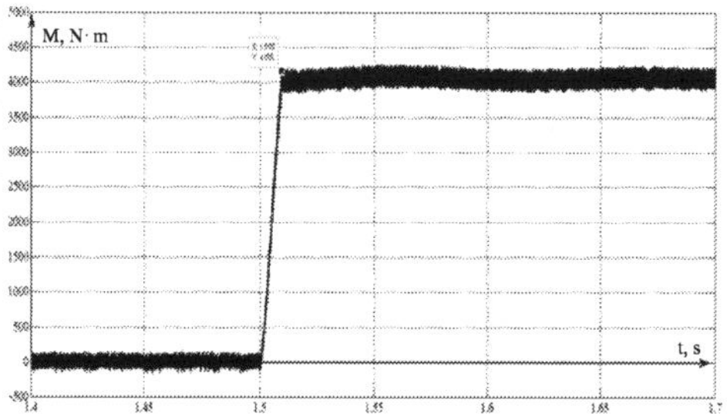

Fig. 17. Transient for sinusoidal pulse width modulation during load transfer.

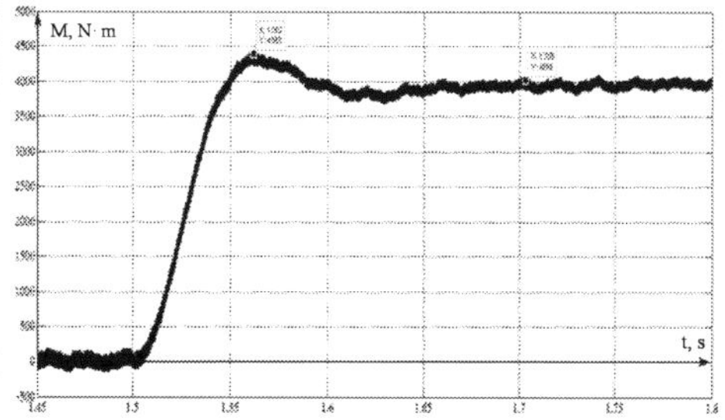

Fig. 18. Transient for spatially vector pulse-width modulation during load transfer.

It is seen from the graphs that the system with ladder-vector control has an order of magnitude higher speed torque control channel that is connected to the lack of a fixed switching frequency keys.

4 Conclusion

Vector control systems as a whole make it possible to achieve complete control of blood pressure in both static and dynamic modes. Due to the presence of two control channels, it is possible to separately or jointly control the flow and speed of the machine, which in turn provides complete control over the current and electromagnetic torque of the machine. Since the system is focused on a high-performance microprocessor system, it is very flexible and allows implementing various control laws, methods for modulating phase voltage, and increasing the energy efficiency of the system as a whole. These positive qualities ensured its wide distribution in a wide range of industrial electric drives.

Acknowledgements

The paper was powered by research grant RFBR 19-48-480001 «Development, investigation and optimization of energy-saving electrical and electrically driven automated systems for plasma, electrometal slag and induction technologies and units».

References

[1] Y.I. Gracheva, N.A. Alimova, Calculating Methods and Comparative Analysis of Losses of Active and Electric Energy in Low Voltage Devices, International Ural Conference on Electrical Power Engineering (UralCon), 361–367 (2019)

[2] Y.I. Gracheva, O.V. Naumov, Estimation of Power Losses in Electric Devices of the Electrotechnical Complex, International Conference on Industrial Engineering, Applications and Manufacturing (ICIEAM), 6 (2019)

[3] V.N. Meshcheryakov, V.V. Danilov, Sh.R. Khasanov, S. Valtchev, Minimization of the stator current in induction motor with defined load on the shaft by maintaining optimum absolute slip, Kazan, SES 2019, E3S Web of Conferences, 01036 (2019)

[4] V.N. Meshcheryakov, D.V. Lastochkin, Z.M. Shakurova, S. Valtchev, Kazan, SES 2019. E3S Web of Conferences, 01037 (2019)

[5] A.M. Abakumov, D.G. Randin, Research of Dual-Mass Oscillation System with Linear Motor, International Conference on Industrial Engineering, Applications and Manufacturing (ICIEM), Sochi, Russia, 1–5 (25-27 March 2019)

[6] T.V. Sinykova, E.V. Sentsov, A.V. Sinyukov, Neural Network Speed Observers, Proceedings 2019 1st International Conference on Control Systems, Mathematical Modelling, Automation and Energy Efficiency (SUMMA), Lipetsk, Russia, **288** (2019)

[7] T.V. Sinykova, V.E. Gladyshev, A.V. Sinyukov, Methods for Reducing Electromechanical Oscillations in Conveyor Control Systems, Proceedings 2019 1st International Conference on Control Systems, Mathematical Modelling, Automation and Energy Efficiency (SUMMA), Lipetsk, Russia, **435** (2019)

[8] R.T.Y. Thien, Y. Kim, Decentralized Formation Flight via PID and Integral Sliding Mode Control, Aerospace Science and Technology. Elsevier Science Publishing Company, Inc 81, 322–332 (2018)

[9] Z. Hu, Y.V. Bodyanskiy, O.K. Tyshchenko, A Multidimensional Adaptive Growing Neuro-Fuzzy System and Its Online Learning Procedure, Advances in Intelligent System and Computing **689**, 186–203 (2018)

[10] A. Pugachev, Efficiency increasing of induction motor scalar control systems, International Conference on Industrial Engineering, Applications and Manufacturing (ICIEM), St. Petersburg, Russia, 1–5 (16-19 May 2017)

Experience in multistage steam-driven ejectors

Ilia Murmanskii[1,*], *Konstantin* Aronson[1], *Nickolay* Zhelonkin[1], *Alexander* Ryabchikov[1], *Adriano* Milazzo[2], and *Yuri* Brodov[1]

[1]Ural Federal University, Turbines & Engines Dep., 620002, Mira, 19, Ekaterinburg, Russia
[2]University of Florence, Industrial Engineering Dep., 50121, Piazza San Marco,4, Florence, Italy

Abstract. The article presents some trends of multistage steam-driven ejectors improvement and examples of its implementation. Three relevant directions were picked out: increasing of reliability, profitability and efficiency. Regarding the reliability increasing, existing serial ejector designs of various manufactures were analyzed. A number of new technical solutions were designed. For profitability raising two approaches were considered: decreasing of primary steam parameters and effect of high ejector productivity to the vacuum deviation in the condenser. Efficiency increasing comprises jet devices improvement. For these reasons a design method was refined. In this case geometrical parameters of existing serial ejectors were summarized, CFD research was performed. The improved design method along with new design solutions are implemented in several ejectors successfully. New ejectors are set at thermal power plants and have shown a high technical-economical effect.

1 Introduction

In recent years a significant interest in jet pumps (ejectors) has appeared over the world. On the one hand, this is due to the use of these devices in various energy conversion plants, such as refrigerators, solar plants, fuel cells, chemical reactors, oil refineries, etc. On the other hand, a large number of existing jet pumps are operating on chemical reagents, gases and other environmentally hazardous substances, so it is advisable to convert them to a water steam. The main trends in the study of ejectors can be traced on the example of a review [1], which presents an analysis of 48 works published only in 2018 in the Energy journal (Elsevier). A large number of studies are devoted to modern computational and experimental work aimed at complex gas-dynamic and heat and mass processes study.

Under these conditions, the modernization of multi-stage steam-jet ejectors of steam turbines, the main technical solutions and designs of which were developed in the 50s and 80s of the last century, is an urgent and promising task. The improvement of these devices is necessary in view of the increased requirements for the reliability, output and efficiency of the ejectors of condensing plants. Research aimed at multi-stage steam-jet ejectors improving is carried out in several areas: increasing of reliability, efficiency (calculation methods refining and design changes), cost-effectiveness, and developing new designs. Further on are the options for implementing research in these areas, the work of "Turbines and Engines" department of Ural Federal University being an example.

2 Materials and methods

2.1 Reliability improvement

An analysis of the damage and the main disadvantages of serial multi-stage steam-jet ejectors was carried out over a 25-year period of operation of more than 500 turbines with a capacity of 100 to 500 MW based on the study of repair documentation for thermal power plants, equipment failure reports and the authors' experience in repairing and upgrading more than 100 different ejectors. The analysis found out that about 13% of turbine shutdowns are associated with a malfunction of the condensing unit. In turn, 23% of condensation unit failures are due to the main ejectors malfunctions. Moreover, almost every failure of the main ejector leads to a shutdown of the turbine [2–5].

In many ways, the causes of ejector failures are related to their wear over a long term of operation. This is manifested in various defects [6–7]. For example, jet apparatus damage is a crucial defect caused by erosive wear of the nozzle throat (Fig. 1), mechanical damage to the nozzle exit edge, etc.

* Corresponding author: i.b.murmansky@urfu.ru

 https://doi.org/10.1051/e3sconf/202017801002

Fig. 1. Damaged nozzle - view from the outlet edge.

Various damages of et apparatus were recorded by the authors in most of the investigated serial ejectors. This can be caused both by a long service life and improper technical parameters of the driving steam. Part of the mechanical damage appears during routine inspections and repairs of the ejector.

A significant number of defects are also associated with the wear of the housing elements and the tube system. When manufacturing a multi-stage ejector in a single casing, the partitions separating the intercoolers are eroded over time in the zone of connection with the tube plate. In this case, the overflow occurs of the steam-air mixture to the zone of reduced pressure. A significant number of defects are associated with materials used for the manufacture of steam jet ejectors. The heat transfer tubes installed by most manufacturers are made of brass. Over time a significant portion of the heat transfer surface is plugged due to damage.

To increase reliability, new designs of multi-stage ejectors are being developed. At the same time, intermediate coolers are made remote to prevent steam-air mixture overflow and to simplify a repair. Stainless steel profile-twisted heat transfer tubes are installed. So one can maintain the efficiency of the cooler and at the same time increase its reliability characteristics. Ejector jet devices are configured to suit specific operating conditions by changing the axial position of the nozzle. A change is also being implemented in existing schemes for intercoolers draining [5].

2.2 Efficiency improvement

This direction is associated with a reduction in the thermal power plants upkeep costs both directly on the ejector, and on the condensation unit, and on the turbine unit as a whole.

The efficiency of the ejector can be achieved by reducing the motivating steam flow rate, its parameters and by changing its source. In this case, the key role is played by the characteristics of a particular operation object. The experience of Ural Federal University in the reduction of costs and parameters of ejector motivating steam is based on the analysis of layout solutions for specific TPPs. For example, the situations are rather often when the source parameters of the motivating steam significantly exceed the required ones. This may be due to the modernization carried out at the TPP, to a change in the layout, to ejector replacement and other factors. In this case, it is necessary not only to reduce possible losses (for example, from steam throttling to reduced pressure), but also to provide the required steam parameters at the ejector inlet.

When a high-potential steam source with parameters greater than those required is connected to the ejector, for example, from the steam cooling reducer, steam temperature often significantly exceeds the saturation temperature corresponding to motivating steam pressure in the ejector. The tests of the ejectors at various temperatures of motivating steam made it possible to determine the permissible value of motivating steam superheating temperature as $t_{ms} = t_{sat} + 100$ °C, where t_{sat} is the saturation temperature of motivating steam at its pressure at the nozzle inlet. A significant increase in steam temperature leads to a sharp decrease in the ejector output. This can be explained by the overload of the II and III stages of the ejector, since most of the heat exchange surface of the ejector coolers serves to reduce the steam superheat, and not to condense it; the intercooler is not able to condense the steam entering it. Non-condensed steam enters the next stage, overloading it.

From the point of view of ejectors improving the steam turbine units cost-effectiveness is in fuel burns reducing when the vacuum deviates in the condenser. When analyzing the deviation of the vacuum in the condenser from the standard values, the following factors are taken into account: heat exchange surface pollution, reduction of the heat exchange surface due to plugged tubes, increased air suction. The feasibility study is carried out individually for each power unit. For example, we present one of the power units with a turbine of the K-200-130 LMZ type, for which the feasibility of using a highly efficient ejector was substantiated. The power unit under consideration operates under conditions when the proportion of plugged tubes in the condenser reaches 21%. At the same time, a ball cleaning system is regularly used to reduce condenser pollution, i.e. the deterioration of the heat transfer process can be neglected. In this case, we determine the deviation of the vacuum in the condenser by its standard characteristics, as well as by the condenser verification calculation assuming that heat transfer surface is reduced by 21%. The deviation about 0,5% of the actual vacuum value from the standard one calculated for the full heat transfer surface was found. We consider the deviation is caused by increased air suction to the turbine, which allows us to evaluate the tightness of the vacuum system and the effect of air suction on the condenser steam pressure, taking into account different temperatures of circulation water at various seasons. These data make it possible to justify both the replacement of the tubes in the condenser and the need to modernize the ejectors.

Due to the significant depletion of turbines service life, serial ejectors, even in the design state, are not always able to maintain the standard vacuum in the condenser. This is confirmed by both manufacturers' and

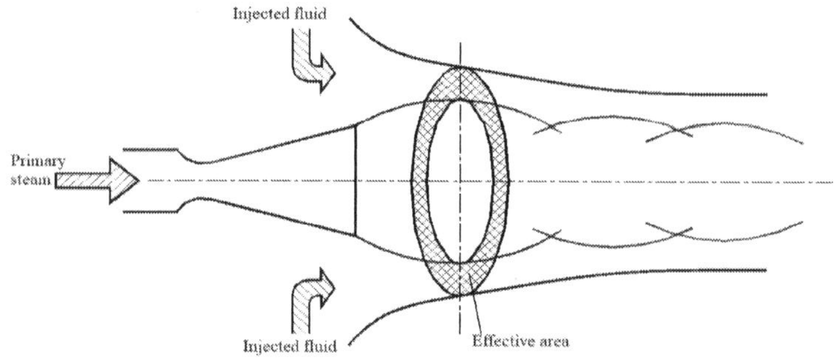

Fig. 2. Effective cross section in the mixing chamber of jet apparatus.

experimental performances. In this case, to increase the efficiency of the condensing unit, a strategy can be applied to increase the efficiency of the ejector or to develop new ejectors for increased parameters.

3 Results

3.1 The development of ejectors

The efficiency increasing is based on ongoing theoretical, numerical, laboratory and experimental studies [8-12]. The analysis and refinement of existing calculation techniques were carried out. The development of a refined technique is based on the results of a generalization both of serial ejectors geometric characteristics and of industrial test data and numerical calculations analysis [8]. The generalization of geometric characteristics is based on data on jet apparatuses of 24 types and sizes of serial ejectors. Geometric parameters were analyzed, such as the position of the effective cross section in which the injected flow reaches the sonic velocity (determined by the parameter μ, Fig. 2), the ratio of the cross-sectional areas of the mixing chamber cylindrical part and the critical section of the nozzle ("the main geometric parameter of the ejector" — F), the relation of nozzles critical sections diameters by stages.

The main refinements developed by the authors of the technique are as follows:

• As a result of numerical simulation, the shape of the working stream jet leaving the nozzle was specified. It is assumed that the pressure in the outlet section of the nozzle is equal to the pressure in the receiving chamber (a designed mode of ejector operation). In this case, the jet cross section must have a constant diameter up to the mixing chamber cylindrical part inlet.

• Based on a generalization of the results of industrial tests and on comparison with the obtained geometric characteristics, the recommended values of the coefficient μ (which determines the position of the effective cross section in which the injected flow reaches the sonic velocity) are specified. The choice of coefficient μ is based on the revealed relationship between the coefficient μ and the main geometric parameter of the ejector.

• Based on data generalization of the compression ratios from the first stage of the ejector to the last one, a strategy for changing the critical nozzles diameters is determined for the developed multi-stage ejectors. It has been established that each manufacturer is characterized by a specific strategy for increasing or decreasing the diameters of the critical nozzle sections from the first stage of the ejector to the last one. On this basis a strategy was chosen — as for the condenser turbine ejector it was decided to increase the diameters of the nozzle critical sections from the first stage to the last one, and on the contrary, for cogeneration turbine ejectors to reduce them.

• An algorithm has been developed and tested for the ejector calculation — i.e. construction of calculated performance of the ejector for dry air and for steam-air mixture. Based on the developed algorithm, the performance of the first stage of the ejector can be calculated. The performances of the subsequent stages can be calculated after determining the amount of steam condensed in the intermediate heat exchangers.

The refined designed method is basing on all these studies.

Using the refined method, new multi-stage steam-jet ejectors of the EPO-3-80 and EPO-3-120 types were designed (Fig. 3). The features of these ejectors are shown in a number of new patented technical solutions incorporated in the design [11,12]:

• coolers are made external and vertical; the diameters of the cooler casings are unified, the cooler tubes are U-shaped and made of stainless steel, the pipes between the nozzle and diffuser are also located vertically to reduce corrosion wear;

• the I stage jet apparatus is divided into two parallel ones. This decision was made to increase the apparatus reliability and ease its manufacture;

• the nozzle fixing unit provides for moving the nozzles along the axis of the jet apparatus. This is necessary for adjusting the ejector operation under various operating conditions of a specific steam turbine unit. When setting up the ejector, the nozzles could be installed in various positions, dependent of their operation parameters. The design of the nozzle axial displacement unit is shown in Fig. 4.

https://doi.org/10.1051/e3sconf/202017801002

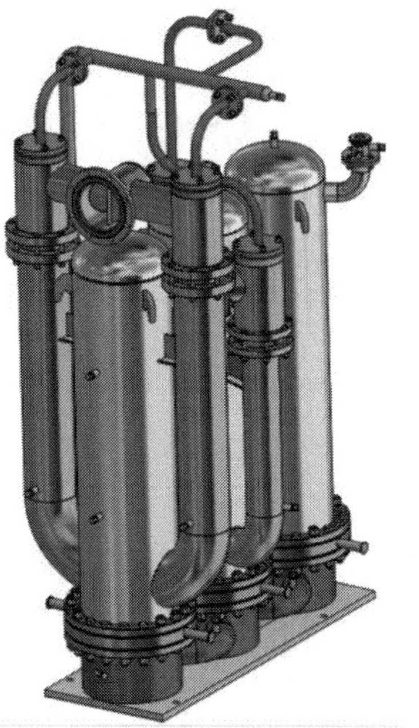

Fig. 3. 3D-model of EPO-3-120 ejector with the first stage dualized jet device and external coolers.

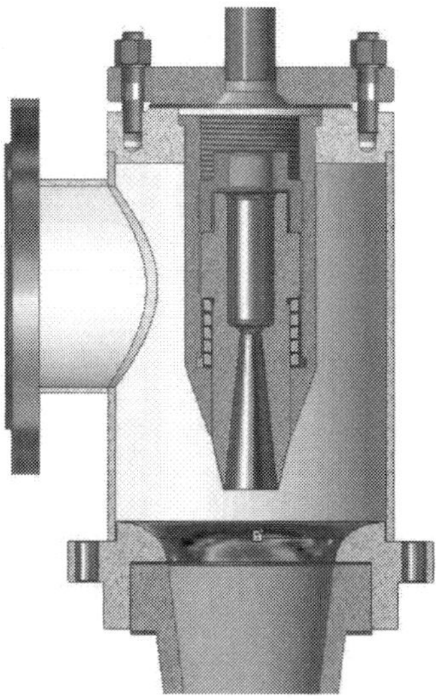

Fig. 4. Nozzle fixing unit with threaded installation.

Tests and tuning of new ejectors of the EPO-3-80 and EPO-3-120 type, developed in accordance with the above technical recommendations, were carried out under operating conditions at thermal power plants. To conduct experimental studies of the new ejector, the test procedure based on the recommendations [13] has been refined. A new expanded scheme has been developed for measuring the performance of the ejector, including also a number of new parameters, along with the pressure of the steam-air mixture in the receiving chambers of the jet devices. The determination is provided of pressures and temperatures behind the diffusers of the jet devices at each stage, along with the temperatures in the receiving chambers, temperatures of the drains and cooling condensate at the inlet and outlet of each intercooler and also the cooling condensate flow rate. The developed measurement scheme made it possible to fix the gas-dynamic resistance of the intercoolers.

To study the effect of cooling condensate temperature on the ejector operation, tests were carried out at higher temperatures of the circulating water at condenser inlet and, accordingly, of the main condensate at the inlet to the intermediate ejector coolers.

Cooling condensate temperature varied from t_{cc} = 10 — 12 °C to t_{cc} = 32 — 35 °C. The tests were conducted at a greater motivating steam pressure P_{ms} = 0,81 and P_{ms} = 0,86 MPa.

Fig. 5 presents a comparison of the first stage performance obtained at various t_{cc}.

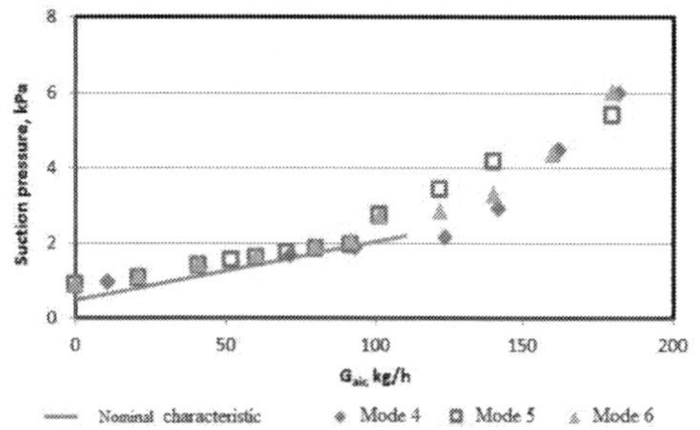

Fig. 5. First stage performance of EPO-3-80 ejector: Mode No. 4 – P_{ms} = 0.7 MPa, nozzle position is adjusted, t_{cc} = 10 — 12 °C; Mode No. 5 - P_{ms} = 0.81 MPa, nozzle position is adjusted, t_{cc} = 32 — 35 °C; Mode No. 6 - P_{ms} = 0.86 MPa, nozzle position is adjusted, t_{cc} = 32 — 35 °C.

In all investigated modes, the ejector output is about 140 kg/h. At a higher t_{cc} the 1st stage suction pressure within the air flow range from 90 kg/h to 140 kg/h is slightly increased in comparison with the mode with t_{cc} = 10...12 °C. Throughout the rest of the stage characteristics, the effect of t_{cc} is absent.

To justify the replacement of serial ejectors with new ones, an analysis and comparison of their performances was carried out. Fig. 6 shows the performance of the first stages of the new EPO-3-80 ejector and serial one EP-3-700 (replaced by a new EPO-3-80 ejector). The performances were obtained at the recommended for each ejector steam pressure — ΔP_{ms} = 0.7 MPa for EPO-3-80 and ΔP_{ms} = 0.5 MPa for EP-3-700.

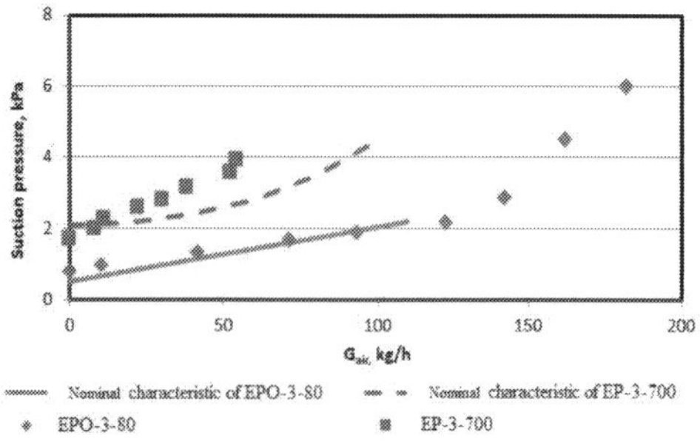

Fig. 6. Operating performances comparison for EPO-3-80 and EP-3-700 ejectors.

From the graph presented, it can be seen that operation and certified performances of the first stage of EP-3-700 ejector are located above the corresponding performances of EPO-3-80. The starting points of the EP-3-700 performances are 0.9 kPa higher than that of EPO-3-80 performances. The inclination angle of the of the EPO-3-80 performances is much smaller, which allows maintaining a deep vacuum in the turbine condenser at high air suction into the vacuum system. The output of the EP-3-700 ejector for "dry" air is 54 kg/h, the output of EPO-3-80 is -140 kg/h.

4 Discussion

The installation of EPO-3-80 ejector on K-200-130 LMZ turbine made it possible to maintain a deep vacuum in the turbine condenser under conditions of increased (by 6 times relative to standard) air suction. At the same time, an increase in the efficiency of the turbine unit was achieved. Fig. 7 presents a retrospective (according to monthly average data) of the deviation of standard value of turbine condenser vacuum from the actual one.

It can be seen that the excess of the standard vacuum value over the actual one during the turbine operation in 2016 (with the two serial ejectors EP-3-700 employed) ranged from 0.8 to 1.9%. When the EPO-3-80 was turned on and the both standard serial ejectors were turned off, the vacuum deviation decreased on average by $\Delta V = 0.5\%$. Data for February 2017 are not available, because the turbine was in a stand-by mode.

According to [14], a change in condenser pressure by 1 kPa causes a change in the power of the K-200-130 LMZ turbine by 1.9 MW. The effect of the new ejector operation is proposed to be determined by the additional generation of electricity. This effect amounted to 5 million rubles per year.

EPO-3-80 and EPO-3-120 ejectors have been already installed on 8 turbines with a capacity of 65 to 200 MW at various thermal power plants. When developing the designs of these ejectors, the listed technical solutions were implemented, which made it possible to achieve a significant technical and economic effect.

5 Conclusion

1. The main trends in the research and development of multi-stage steam-jet ejectors are formulated. These trends are associated with an increase in three key parameters: efficiency, profitability and reliability.

2. To increase the efficiency, the development of modern calculation techniques is carried out. The techniques are developed on the basis of numerical simulation of gas-dynamic processes, as well as a synthesis of operating experience and the characteristics of serial ejectors of various manufacturers. A refined technique has been developed to improve the flow part of jet devices.

3. The efficiency improving is achieved by several approaches. The efficiency of ejector itself is achieved by choosing the optimal parameters of the ejector motivating steam. The efficiency of the condensing unit and the entire steam turbine is increased by bringing the

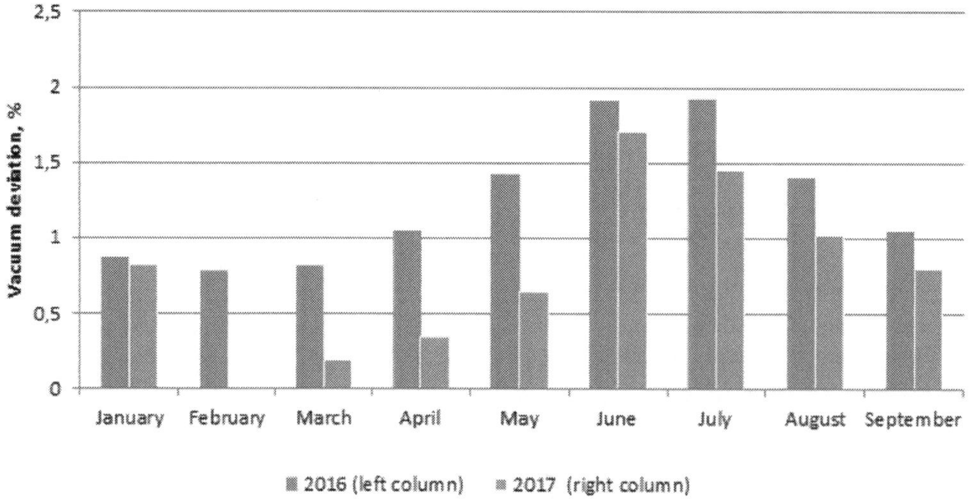

Fig. 7. Retrospective deviation of the actual value of the condenser vacuum from the standard one.

vacuum in the condenser to standard values due to ejector output rise.

4. To improve the reliability characteristics of ejectors, a number of new technical solutions have been developed based on the generalized experience gained from serial ejectors operation of various manufacturers.

5. EPO-3-80 and EPO-3-120 ejectors are now installed on 8 turbines with a capacity of 65 to 200 MW at various thermal power plants and permit to achieve a high technical and economic effect.

References

[1] G. Besagni, Ejectors on the cutting edge: The past, the present and the perspective, Energy, 998–1003 (1 March 2019)

[2] G.G. Shklover, O.O. Milman, *Research and calculation of steam turbines condensing units* (M.: Energoatomizdat, 240, 1985)

[3] Yu.M. Brodov, K.E. Aronson, A.Yu. Ryabchikov, B.E. Murmansky, I.B. Murmansky, N.V. Zhelonkin, Reliability of turbines steam jet ejectors, Scientific and Technical Journal Reliability and Safety in Power Industry **2**, 33, 60–64 (2016)

[4] K.E. Aronson, A.Yu. Ryabchikov, I.B. Murmanskii, et al., Ejectors of power plants turbine units efficiency and reliability increasing, Journal of Physics Conference **891**, 012249 (2017)

[5] Yu.M. Brodov, K.E.Aronson, I.B. Murmansky, S.I. Khaet, Reliability of steam jet ejectors for steam turbine units at thermal power plants, Power engineer **12**, 40–41 (2016)

[6] Yu.M. Brodov, K.E.Aronson, B.E. Murmansky, I.B. Murmansky, M.A. Nierenstein, V.B. Novoselov, P.N. Plotnikov, A.Yu. Ryabchikov, *Reliability of steam turbine equipment: a training manual* (Ekaterinburg: UrFU Publishing House, 144, 2017)

[7] A.Y. Ryabchikov, K.É. Aronson, Y.M. Brodov, B.E. Murmanskii, I.B. Murmanskii, N.V. Zhelonkin, S.I. Khaet Increasing the reliability of steam-jet ejectors in power plant turbines, Power Technology and Engineering **50**, 5, 546–548 (2017)

[8] K.É. Aronson, A.Y. Ryabchikov, Y.M. Brodov, D.V. Brezgin, N.V. Zhelonkin, I.B. Murmanskii, Analysis of experimental characteristics of multistage steam-jet electors of steam turbines, Thermal Engineering **64**, 2, 104–110 (2017)

[9] K.É. Aronson, A.Y. Ryabchikov, Y.M. Brodov, N.V. Zhelonkin, I.B. Murmanskii, Functioning efficiency of intermediate coolers of multistage steam-jet ejectors of steam turbines, Thermal Engineering **64**, 3, 170–175 (2017)

[10] I.B. Murmansky, K.E. Aronson, Yu.M. Brodov, L.G. Galperin, A.Yu. Ryabchikov, D.V. Brezgin, Investigation of the effect of pressure increasing in condensing heat-exchanger, Materials of the International Conference Modern problems of thermal physics and energy, Moscow, 9-11 Oct. 2017, in 2 vol., V. 1., M.: MEI Publishing House, 260–262 (2017)

[11] Yu.M. Brodov, V.K. Kuptsov, A.Yu. Ryabchikov, K.E. Aronson, I.B. Murmansky, N.V. Zhelonkin, D.V. Brezgin, Utility Model Patent no. 170935 Russian Federation, MPK F04F5/00, Three-stage steam jet ejector, Applicant and patent holder: Ural federal university named after the First President of Russia B.N. Yeltsin, no. 2016119824. Appl. 23.05.2016, published. 15.05.2017, Bull. 14, 9 (2017)

[12] Yu.M. Brodov, V.K. Kuptsov, A.Yu. Ryabchikov, K.E. Aronson, I.B. Murmansky, N.V. Zhelonkin, D.V. Brezgin, S.I. Khaet, Patent for invention no. 2645635 Russian Federation, MPK F04F5/30, Three-stage steam jet ejector, Applicant and patent holder: Ural federal university named after the First President of Russia B.N. Yeltsin, no. 2016126736, Appl. 04.07.2016, published. 26.02.2018, Bull. 6, 4 (2018)

[13] RD 34.30.302-87, Guidelines for testing and operating of turbine condensing units steam jet ejectors for thermal and nuclear power plants, M.: Joint Stock Co United energy system of Russia, 34 (1987)

[14] N.A. Natanson, *Typical net energy characteristic of a K-200-130 LMZ turbine unit* (Specialized Center for Scientific and Technical Information ORGRES, Moscow, 31, 1972)

Heat pump in a new modular configuration to recover low-grade heat emissions at enterprises

Altai Sh. Alimgazin[1], *Saule G.* Alimgazina[2], and *Mikhail G.* Zhumagulov[1,*]

[1]L.N. Gumilyov Eurasian National University, Dept. of Power engineering, 2 Satpayev str, 010003 Nur-Sultan, Kazakhstan
[2]Startup company BMHU LLP, 13 Kazhymukan str, 010008 Nur-Sultan, Kazakhstan

Abstract. This paper reviews application of energy-saving heat pump technologies using non-traditional and renewable energy sources (NRES) to improve the efficiency of industrial equipment, heat supply systems in public sector and in agroindustrial industry in the Republic of Kazakhstan. The heat pump technology to transform low-grade heat of secondary low-temperature power resources into high-grade heat energy suitable for practical use is not a regular upgrade of traditional power resources but introduction a new, advanced, high-efficient and ecofriendly heat generation method. Introducing Modular Heat Pumps (MHP) at metallurgic and other industrial enterprises (power industry, petrochemsitry, machinery and etc.) as well as agroindustrial complex, housing and public utilities, and state-funded organizations of the Republic of Kazakhstan is expected to get a significant economic and ecological effects, foremost, by replacing traditional heat sources (solid and liquid fuel boilers, electric boilers), decrease greenhouse gases emissions into the atmosphere, and improving ecological situation. MHP application will allow to get 2-4 times saving budget funds annually allotted for heat supply (procurement various fuel types, repair current boiler equipment, ecological payments and etc.) in comparison with the existing traditional boilers.

1 Introduction

One of perspective ways in energy saving and energy efficiency enhancement of the existing equipment at the Republic of Kazakhstan's enterprises is recycling low-temperature waste heat of process water-supply systems (WSS) at power, ferrous and non-ferrous metallurgy, and oil and chemistry [1 -14, 16-18, 20-23].

Introducing MHP at metallurgic and other industrial enterprises (power industry, petrochemsitry, machinery and etc.) as well as agroindustrial complex, housing and public utilities, and state-funded organizations of the Republic of Kazakhstan is expected to get a significant economic and ecological effects, foremost, by replacing traditional heat sources (solid and liquid fuel boilers, electric boilers), decrease greenhouse gases emissions into the atmosphere [2, 3, 5-7, 12-14, 20-23].

Application heat pump technologies at the mentioned above enterprises will allow to recover considerable volume of irrevocable lost heat into the heat supply system at enterprises without additional combustion of ordinary fuel that will allow to get a significant economic effect, refuse from sometimes from purchased energy outside, thereby reducing heat combustion of entertainment.

It should be noted that prime cost of the own heat pump heat from twice to four times will be less than purchased heat, i.e.after start heat pump plant up a real opportunity to provide an existing necessity in heat energy is appeared.

As the known experience of the world leaders in a regular use of heat pump technologies (Finland, Sweden, Japan, China, Russia and others) shows a number of projects have already implemented using HP at Power Plant [2- 4, 9 -12, 14].

1.1 Examples

As an illustrative example of the HP application efficiency at stations let us give the data on some objects in the PRC [9-14].

Chinese power plant in Shentu

The efficiency of modern coal power plants with steam turbines is from 40% to 45%, which means that about 55% - 60% of the fuel in the form of energy that enters the power station is discharged through condensers into the atmosphere.

The key element of heating system is the heat pump which extracts a low-grade heat from cooling water after the condenser and transfers it to the hot water heating circuit (see Fig. 1.2).

After pass through the heat pump the cooling water cools down from approximately 30°C to 26°C and also enters the cooling towers. As a result of the fact that the water temperature has become lower it will be less evaporated into the atmosphere. Consequently, the consumption of additional water per cycle will be decreased.

Hot water is heated to 75 °C in heat pumps, the final heating to 130 °C occurs in network heaters.

* Corresponding author: mixailgng@gmail.com

Fig. 1. Power Plant in Shentu.

Fig. 2. TPP Assembly view.

As a result of the work carried out (2011-2014), the efficiency of station itself was significantly increased and environmental situation in the city and region as a whole was significantly improved. It is estimated that for 1 MW of heat energy supplied from the Shentu station approximately 0.3 MW is taken from cooling water after the condenser.

Another example of project execution with HPP

Six HPP with 30 MW capacity each from Shuangliang were installed at the power station in Yangjin (China) in 2010.

They recycle the heat energy of cooling towers (temperature 30–40°C), and produce hot water with parameters 70–90°C for additional heat supply to the whole district of the city utilizing the heat previously discharged by the cooling towers.

HPP introduction in this project annually allowed to:

• sell additionally heat energy to the amount more than $ 5 million (and tariffs for heat energy in this city are lower than in Kazakhstan);

• save 500 000 tons of water (it is not evaporated in cooling towers);

• save 49 300 tons of coal;

• reduce CO_2 emissions by 98 600 tons;

• reduce nitrogen oxide emissions by 730 tons.

The payback period for introduction HPP and use a low-temperature waste heat for heating the enterprise was less than two years.

However, there are no for a while any precedents in practical use of heat pumps (hereinafter - HP) at power enterprises (power stations, regional hydroelectric power stations) in the Republic of Kazakhstan. There are single examples of heat pumps application using waste heat at only one metallurgical enterprise Kazzinc JSC (Ust-Kamenogorsk, 2000) [15, 20, 22].

2 The research Object and research methods

Starting from 2000, the scientists from Kazakhstan and Russia (MEI, Moscow, Energiya CJSC, Novosibirsk) have been carrying out joint works developing application methods and technologies for HP using waste heat of technical water-supply systems (TWSS) at industrial enterprises (metallurgic complexes, power station, oil processing plants and etc.) to improve heat-power equipment efficiency of the aforesaid enterprises [15, 20].

During 2005-2019 Kazakhstani scientists have been conducting research on the prospects to use heat pumps (HP) using waste heat of circulating and direct-flow water cycles of a number of real industrial facilities – TETS-2 JSC in Astana, Ekibastuz TPP-2 LLP, Petropavlovsk TPP-2, SevKazEnergo JSC, Ust-Kamenogorsk TPP LLP, Sogrinsk TPP LLP, Aluminum of Kazakhstan JSC, Arselor Mittal Temirtau JSC, Pavlodar oil chemistry Plant LLP, PS KSP Steel LLP (Pavlodar).

The important precondition to use new generation MHP at mentioned above enterprises is quite large volumes of heat discharged with circulating water into cooling towers, cooling ponds, and rivers (Irtysh and etc.) [19, 21, 22].

Together with Russian scientists (Energiya CJSC, Novosibirsk) based on grants of the Ministry of Education and Science of the Republic of Kazakhstan (2012 – 2015) some research activities have been carried out [15, 20] according to results of which for the first time ever a multistage application technology was developed for the new generation HP operating at a more effective thermodynamic Lorenz cycle compared with reverse Carnot cycle and providing coolant delivery into the heat system with temperature up to 85°C (but not 55-62°C as other types of HP produced worldwide).

This unique technology was patented both in Russia and in the Republic of Kazakhstan and is ready to be practically implemented [12,15].

The said technology having passed a multilevel selection in the Ministry of Energy, Ministry of Education and Science of the Republic of Kazakhstan in 2016-2017 was represented at the Sphere National booth (Kazakhstan) during the International exhibition EXPO-2017 among the 25 best scientific projects of the national scientists [21].

During some research activities carried out in 2010-2016 based on scientific and research activities supported by the national and business contracts for commercial purposes a certain result was presented, i.e.

a sample of a new product that is a modular heat plant (MHP) [19, 21] (see Fig.3).

Fig. 3. General assembly of modular heat pump (MHP).

MHP being a separate stage of multistage heat pump is a metallic heat-insulated container with supporting frame. The guards are made as three-layed panels consisting of profile sheets with insulants of mineral-wool boards.

All the necessary main and ancillary equipment and instrumentation will be installed inside MHP with only 4 outside pipelines for input a low-grade coolant (waste water from circulation water supply system) and output of hot water with temperature of 75-80°C into heat supply system of enterprise.

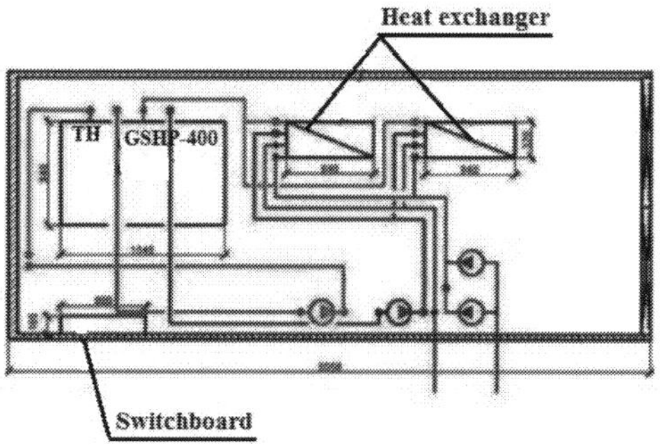

Fig. 4. Demonstrates a pilot MHP with heat capacity 400 kW.

Supply of various modular heat pump capacities (from 200 kW to 2000 kW) is planning to customers. The proposed container-type MHP [19, 21] is designed as a standard element of enterprise heat circuit ready for practical implementation at any enterprise and is planned to be used both at industrial enterprises and state-funded organizations, housing and public utilities where various alternative energy sources are available.

MHP advantages:

- a singular patented technology and multi-stage MTU application scheme ready to be practically implemented at enterprises using alternative energy sources in various regions of the country;
- adaptability of the proposed technology to severe climate conditions of the Republic of Kazakhstan (up to -40°C);
- portable and versatile (can be used in different industrial sectors where circulation water supply system is available in process cycles);
- in comparison with other existing world heat plant technologies the proposed project provides more effective power (SOR) and economical characteristics (production cost for 1 Gcal of heat, specific fuel costs when replacing traditional heat sources.
- a lower market price of our products compared to peers;

The market price for a unit of production and (or) services of this competitor can be estimated as follows.

The cost of MTU proposed (heat power up to 2 MW) is significantly lower than the cost of heat pumps (HP) of the equal power to foreign manufacturing companies (Germany, Sweden and etc.). The approximate cost of MTU heat power equal to 1 Gcal/hour is about 200-220 thousand US dollars (in comparison with foreign heat pumps of the equal heat power are about 350-440 thousand US dollars);

- high-quality warranty and after-sales service of MTU installed at enterprises;
- 80% Kazakhstani content of the project (equipment is in modular configuration, technology, design, a complex of installation and commissioning works, highly qualified scientific and engineering personnel).

In 2017-2020, under the grant funding of the Science Fund JSC of the Ministry of Education and Science of the Republic of Kazakhstan, the scientists of the L.N. Gumilyov Eurasian National University and specialists from the start-up company BMTU LLP (Astana) have implemented a "Project oriented to commercialize modular heat pump to recover a low-potential waste heat obtained from industrial water supply systems at enterprises" at the pilot facility - pipe rolling production of a large metallurgical enterprise in the country – Pavlodar Subsidiary(hereinafter referred to as PS) KSP Steel LLP (see Fig.6-7) [19, 22].

This article describes commercialization aspects of a new MHP with heat capacity 400 kW for low-temperature heat waste recovery from technological processes at enterprises (enterprise waste water, water from technical water-supply system (TWSS) and etc.) to reduce emissions of hazardous substances into the atmosphere and in parallel get a high-grade heat at a certain metallurgic enterprise – PS KSP Steel LLP).

This survey explores the following issues:

- development MTU application diagram as a typical element of heat supply system at pipe rolling production (hereinafter referred to as PRP) of PS KSP Steel LLP.

Fig. 5. General view of pipe rolling production at PS KSP Steel LLP and ventilator cooling tower of circulation water supply system.

- conducting a series of pilot tests of new equipment to determine optimal operating conditions for new generation heat pump technology when MTU prototype is included in the theat circuit of PRP KSP Steel LLP;

- analysis of energy, environmental and economic efficiency of MTU application to recover waste heat of technical water supply system (TWS) of PRP and heat generation for heating and hot water supply of the enterprise.

At the moment, the existing power supply diagram of PS KSP Steel LLP is based on purchased heat and electric energy from TPP -3 (Pavlodar) (steam, hot water). The enterprise annually spends great funds to purchase heat power (about 46 Gcal/hour) and electric power (about 6.5 mln USD (data of March, 2015)).

Moreover, the current fan cooling towers used in circulation water supply system (CWSS) at the enterprise emit large volumes of heat energy into the atmosphere (according to our assessments, about 280-600 Gcal/hour), and costs enterprise spent to purchase electric power for towers operation are substantial.

However, physical heat of water circulation systems emitted into the atmosphere through the towers is irrevocably lost which is economically inadvisable.

Besides, getting substantial volumes of own heat with temperature up to 60-85°C using MHP will give a chance to save large volumes of power purchased at the moment at TPP-3 (Pavlodar) to provide continuous tower fans operation.

The following table shows results of economical effect estimation from MHP implementation for hot water supply at KSP Steel LLP Pipe Rolling Plant[22].

Table 1. Comparison of annual operational costs at Pipe Rolling Production for hot water supply from various heat resources (heat load - 2 Gcal/hour).

No	Heat source	Cost per annum operating costs, th. KZT (th. USD)	Heat Cost, KZT/Gcal (USD/ Gcal)
1	Central heating (from TPP-3 of Pavlodar)	72 180 (195)	3 313.8 (9)
2	Heat pump boiler room based on MHP	13 160 (35)	751.14 (2)

Thus, the results of preliminary survey give the reason to conclude that when introducing energy-saving heat pump technologies based on MHP using alternative heat sources a deeper cooling of circulating water is achieved at the KSP Steel LLP and costs to purchase heat and power from CHPP-3 in Pavlodar are reduced at the enterprise.

The source of project profitability when introducing MHP at enterprises are:

1. Remarkable economical effect, i.e. difference between heat cost the consumers purchased at CHPP and the heat generated using heat pumps (2-4 times).

2. Considerable enterprise OPEX reduction using circulating and in-line water-supply system.

3. Reduction of enterprise greenhouse emissions and ecological payments decrease consequently.

Based on the results of pilot research on MHP operation in real production conditions an analysis of power, environmental and economic efficiency of its application to recover waste heat from recycling water supply system at enterprises of the country will be carried out, specific recommendations for the further commercialization of this "green" technology at various facilities in the Republic of Kazakhstan be proposed.

In years ahead MHP of different capacities introduction is expected. It will be introduced at various enterprises of ferrous and non-ferrous metallurgy, power industry, oil and chemical industry and etc. with existing TWS (cooling towers, cooling ponds, water reservoirs.).

In particular, during 2018-2019 a research “Improving energy efficiency of metallurgical equipment at the Aksu Ferroalloy Plant (AFP) which is branch of Kazchrome TNK JSC by introducing power saving heat pump technologies recovering waste heat from circulating water supply system (Aksu, Pavlodar region)” has begun [23].

3 Conclusion

Thus, the following conclusions can be made.

1. Heat pump technology to transform low-grade natural heat of secondary low-temperature energy resources into high-grade heat energy for practical use is not another upgrade of traditional energy resources but implementation of a new, advanced, high-efficient and ecofriendly method of heat generation.

2. Application a technology and new generation MHP adapted to operation in cold climate conditions of Kazakhstan (up to -45°C) using discharged water heat from service water system (SWS) as a low-grade source, are perspective to introduce at existing 57 TPPs, several GRES, and a serie of ferrous and non-ferrous metallurgy, oil and chemistry industry, machinery plants and etc. that will allow to improve efficiency of heat-power equipment thereof, save budget funds, boost heat generating capacities while not increasing consumption of primary fuel (coal, residual fuel oil), reduce greenhouse emissions into the atmosphere, and improve ecological situation.

3. It is the first time when MHP is recommended to use to recover waste heat from cooling towers of real enterprise PS KSP Steel LLP (Pavlodar) in the Republic of Kazakhstan.

References

[1] D. Ray, D. McMichael, *Heat pumps* (Energoizdat, 1982)

[2] A.P. Burdukov, Yu.M. Petin, Heat pumps for Russia: technology of use of geothermal and waste heat by industrial enterprises, Equipment, Developments, Technologies **7**, 07, 27–32 (2007)

[3] D.N. Devyanin, S.I. Pisshikov, Yu.N. Sokolov, Development and test of a laboratory stand for approbation of heat pump plants for power industry at CHPP-28 Mosenergo OJSC, News of heat supply **1**, 86–93 (2000)

[4] G. Heinrich, et al., *Heat pump installations for heating and hot water supply* (Stroyizdat, 1985)

[5] Yu.M. Petin, The decade of heat pump production experience at Energia CJSC, Energy policy **3**, 28–33 (2001)

[6] D.G. Zakirov, Utilization of secondary energy resources and the use of renewable energy sources using heat pumps is the main way of burning the energy intensity of production, Thermal engineering, 5 (2002)

[7] E. Bubyalis, K. Marcinauskas, P. Shkema, Opportunities and prospects for the use of heat pumps in the production of low-grade heat. Indastrial power engineering **22**, 3 (2000)

[8] A.B. Baranenko, L.S. Timofeevsky, A.G. Dolotov, A.I. Popiov, *Absorption Heat Converters: monograph* (Saint Petersburg, 2005)

[9] Jian Sun, Lin Fu, Fangtian Sun, Shigang Zhang, Experimental study on a project with CHP system basing on absorption cycles, Applied Thermal Engineering **73**, 1, 732–738 (2014)

[10] S. Jian, F. Lin, Zh. Shigang, Experimental study of heat exchanger basing on absorption cycle for CHP system. Applied Thermal Engineering, 102, 1280–1286 (2016)

[11] H. Tianle, X. Xiaoyun, J. Yi, Simulation research on a variable-lift absorption cycle and its application in waste heat recovery of combined heat and power system, Energy 140, 912–921 (2017)

[12] Yu.M. Petin, S.N. Shamanayev, Ye.V. Oparin, B.S. Golodnikov, Ways of hot water supply and ways of heating thereby, RF Patent no. 2454608 (2012)

[13] X. Xiaoyun, J. Yi, Absorption heat exchangers for long-distance heat transportation, Energy **141**, 2242–2250 (2017)

[14] H.S. Zhang, H.B. Zhao, Z.L. Li, Performance analysis of the coal-fired power plant with combined heat and power (CHP) based on absorption heat pumps, Journal of the Power Institute **89**, 70–80 (2016)

[15] A.Sh. Alimgazin, Yu.M. Petin, S.G. Alimgazina, A.Ye. Bakhtiayrov, Heat pump plant for heating and hot water supply, KZ Patent no. 1843 (2015)

[16] I.A. Sultanguzin, T. Shomova, P. Shomov, V. Dostovalov, The use of heat pumps in gas processing plants, Abstracts and reports of the XV Minsk International Forum on Heat and Mass Transfer **3**, 420–424 (2016)

[17] I.A. Sultanguzin, T. Shomova, P. Shomov, The use of heat pumps in gas processing enterprises: experience and prospects, Plumbing, Heating, Air conditioning, 5, 48–51 (2015)

[18] A. Fedyukhin, I. Sultanguzin, S.Kurzanov, R. Belov, A. Bakulin, T. Shomova, *The use of applied software to solve the problems of industrial heat power: a training manual* (Moscow: MEI Publishing House, 2016)

[19] A.Sh. Alimgazin, S.G. Alimgazina, K.S. Omarov, A.Ye. Bakhtiyarov, Heat pump for head and cold supply, KZ Patent no. 2289 (2017).

[20] A.Sh. Alimgazin, S.G. Alimgazina, Y.M. Petin, Application of new generation heat pump technologies using alternative energy sources to generate additional heat energy a the heat power plants-2 in Astana city, IX Minsk International Seminar Heat Pipes, Heat Pumps, Refrigerators, Power Sourses, Minsk, Belarus, 230–235 (2015)

[21] A.Sh. Alimgazin, A.N. Berguzinov, Application of Green Heating technology for power supply of the Energy Research Center facility of the international exhibition EXPO-2017, Herald of S.Toraighyrov PSU, a series of Power **1**, 21–28 (2017)

[22] A.Sh. Alimgazin, S.G. Alimgazina, A.S. Sharypov, S.Ye. Bakhtiyarova, Commercialization of the modular heat pump for recovering of low-grade waste heat of technical water supply system at enterprises, Volume of the IV International Conference ICITE Industrial technologies and engineering, Shymkent (2017)

[23] A.Sh. Alimgazin, S.G. AlimgazinA, Opportunity to use power-saving heat pumps using alternative sources of power at Aksu Ferroalloy Plant – branch of TNK Kazchrome JSC. Herald S.Toraighyrov PSU, Energy series **4**, 35–38 (2019)

Assessment of energy efficiency of application heat-insulating paint for the needs of district heat supply systems

Mikhail Pavlov[1,*], *Denis* Karpov[1], *Irina* Akhmetova[2] and *Nikolay* Monarkin[1]

[1]Vologda State University, Institute of civil engineering, 160000 Vologda Lenina st. 15, Russia
[2]Kazan State Power Engineering University, Kazan, Russia

Abstract. Energy conservation is a priority for most of the nation states. Its relevance in modern world energy industry is growing every year. This problem acquires particular importance when a country is characterized by a lack of fuel and energy resources or adverse climatic conditions. The most effective way to reduce heat losses from heat network pipelines to environment is the use of heat insulation. This paper presents the results of application heat-insulating paint on the surface of main pipeline. The infrared thermometry devices (pyrometer and thermal imager) were used to evaluate the energy efficiency of modern high-tech insulation. The thermograms from thermal imager and the readings from pyrometer were analyzed, it was established that the absolute difference in temperatures of an isolated pipeline section and an unpainted section are 5-10 °C. The studies have also shown that when a 250×6 pipeline is coated with a 1 mm thick heat-insulating paint, it will save 126.1 m3 of natural gas per 1 running meter per year. In monetary terms it will be approximately 610 rubles or about $9 (for the first half of 2020). The payback period of this energy-saving measure should not exceed six months. Increasing the thickness of liquid heat insulation layer is not economically feasible. The equivalent fuel economy is approximately 65% when the heat-insulating paint is of 1 mm thickness, and it will increase by only 8% for a thickness of 1.5 mm.

1 Introduction

The existing regulations of the Russian Federation enact to use special materials as heat insulation for pipelines. For pipelines with temperature of substances contained in them in the range from 20 to 300 °C, for all laying methods, except for channelless, one should use heat-insulating materials and products with density of not more than 200 kg/m3, coefficient of thermal conductivity in dry state not more than 0.06 W/(m•K) at an average temperature of 25 °C. More than 30 types of heat insulation materials meet these requirements, including reinforced concrete, bitumen perlite, gas silicate, polyurethane foam and others. Due to wide market of offers and high competition, modern heat insulation materials should posses not only good heat insulation properties, but also should be characterized by incombustibility, water repellency, sound insulation, long service life, ease of installation, sanitariness, corrosion protection, wide temperature range of application [1, 2].

The main type of insulation used in Russia is mineral wool products, whose share in the total volume of production and consumption is more than 65%. Glass wool materials ac-count for 8%, expanded polystyrene and other foams account for 20%. The share of heat-insulating cellular concrete in the total volume of produced heat insulation materials does not exceed 3%, the share of expanded perlite, vermiculite and products based on them does not exceed 2-3% (for expanded product). The structure of production volume of heat insulation in Russia is close to that of the advanced world countries, where fibrous heat insulation also account for 60-80 % of the total output of heat-insulating materials [3, 4].

Currently, heat-insulating paint is actively used in practice, the thickness of which on the surface of heat-conducting material can be from 1.5 to 3.5 m. Its popularity is explained by ultra-low thermal conductivity λ (Figure 1), high adhesion to metal, ease of application to pipeline surface, short payback period (up to two years). The low thermal conductivity of ultra-thin liquid heat insulation is based on the properties of discharged gas contained in various types of substance (ceramic, silicone) microspheres with a diameter up to 100 ı m. Aside from low coefficient of thermal conductivity, it also have high heat-reflecting characteristics [5, 6].

Due to its high heat insulation qualities, liquid insulation has found its wide application in construction and heat power engineering. The studies of ultra-thin liquid-ceramic heat insulation on pipelines of heating networks showed that, for a given coefficient of thermal conductivity, heat-insulating paint is the most energy-efficient heat-insulating material. For example, when Nobasil LSP liquid heat insulation of 0.4 mm thickness is used as heat insulation for a tank with high-temperature pitch, the heat losses reduce by a factor of 25, compared to that of non-insulated tank. When

* Corresponding author: pavlov_kaftgv@mail.ru

TEMP-COAT is used, the heat losses are reduced by a factor of 28. The payback period of the energy-saving measures for the project will be 0.22 of a whole year (about three months).

It is known that capital investments aimed at production and transportation of thermal energy from a heat source to the final consumer are directly related to the amount of fuel consumed for heating needs. For Russia, this issue is particularly acute, due to the harsh climatic conditions in which most of the country is located. According to the Ministry of Regional Development of the Russian Federation, the average cost of heating residential buildings in Russia is five to seven times higher that for Germany and other countries of the European Union. However, now Russia, being one of the leading world countries in energy reserves, is far below the economically developed countries while concerning the rational use of fuel and energy resources. The purpose of this work is to assess the energy efficiency of the use of modern high-tech insulation using the example of heat-insulating paint com-pared to that of traditional heat-insulating materials.

using the thermal imager, processing of thermal images, their qualitative and quantitative analysis were carried out according to methods described in textbook [7].

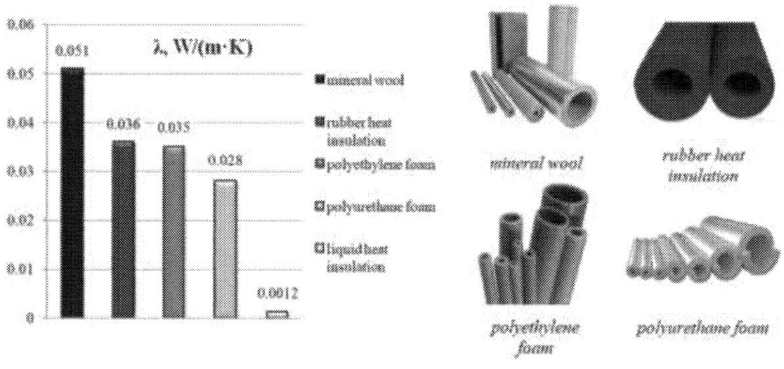

Fig. 1. Thermal conductivity of heat insulation materials of pipelines.

2 Materials and methods

Consider the results of application of liquid heat insulation by the example of insulation the supply main pipeline (steel of St3 grade, 250×6) of the district heating system (nominal thermal power of the boiler house is 18 Gcal/h). The Teplomett Standard heat insulation paint was used as a heat insulation coating with the declared coefficient of thermal conductivity $\lambda_{ins} = 0.003\ \mathrm{W/(m \cdot K)}$.

Figure 2 shows the district heating boiler room (a); the pipeline section, on which a layer of heat insulation paint of variable thickness is applied (b); and the layout of heat insulation along the pipeline length (c).

The quality control of heat insulation on the surface of examined pipeline was carried out in a non-destructive way. The devices for non-contact temperature measurements were used: the Testo 845 laser infrared thermometer, and the Testo 875-2 thermal imager. Ther-mography of the object of thermal control

3 Results and discussion

Table 1 shows temperatures of pipeline surface measured using the Testo 845 infrared thermometer, the sections numbers are shown in Figure 2.

Table 1. Averaged surface temperature of the main pipeline

	The number of pipeline section (Figure 2)			
	№1	№2	№3	№4
t, °C	68.4	53.2	59.8	61.0

According to the heating and domestic temperature schedule of boiler room, the temperature of network water in supply and return pipelines of the centralized heat supply system at the time of heat measurements were $\tau_1 = 77$ °C and $\tau_2 = 55$ °C, respectively, for the outdoor temperature $t_{out} = 0$ °C. The internal air temperature in the boiler room was $t_{in} = 24$ °C.

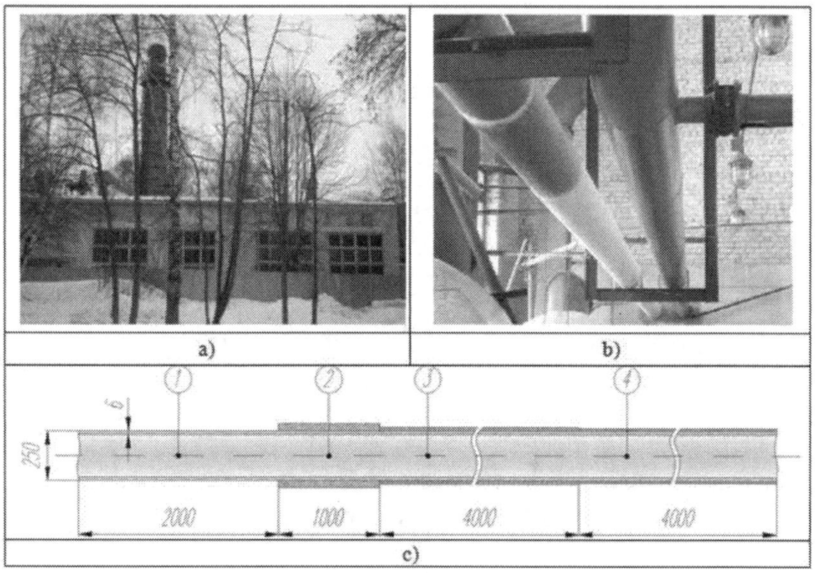

Fig. 2. a) The district heating boiler room in Vologda, Russia; b) the main pipeline; c) the layout of heat insulation along the pipeline length: 1 – unpainted pipeline; 2 – painted pipeline, δ_{ins} = 2.2 mm; 3 – the same, δ_{ins} = 1.8 mm; 4 – the same, δ_{ins} = 1.5 mm.

The linear heat loss to environment from an isolated pipeline can be calculated using the heat transfer equation:

$$q_l = \frac{\pi(\tau_1 - t_{int})}{\frac{1}{\alpha_1 d_1} + \frac{1}{2\lambda}\ln\frac{d_2}{d_1} + \frac{1}{2\lambda_{ins}}\ln\frac{d_{ins}}{d_2} + \frac{1}{\alpha_2 d_{ins}}} \quad (1)$$

where α_1 and α_2 are the coefficients of convective heat exchange (heat transfer), which were calculated based on the theory of similarity and dimensionless numbers (criteria), they are equal to 5232 and 4.77 W/(m^2·K), respectively; λ is the coefficient of thermal conductivity of steel, for St3 grade it is equal to 50.2 W/(m·K); d_1 and d_2 are the inner and outer diameters of steel pipeline, respectively, m; $d_{ins} = d_2 + 2\delta_{ins}$ is the outer diameter of heat insulation (heat insulation paint), m.

The results of solving the heat transfer equation (1) were used to construct the plot showing relationship between linear heat loss to the environment and the thickness of heat insulation layer (Figure 2).

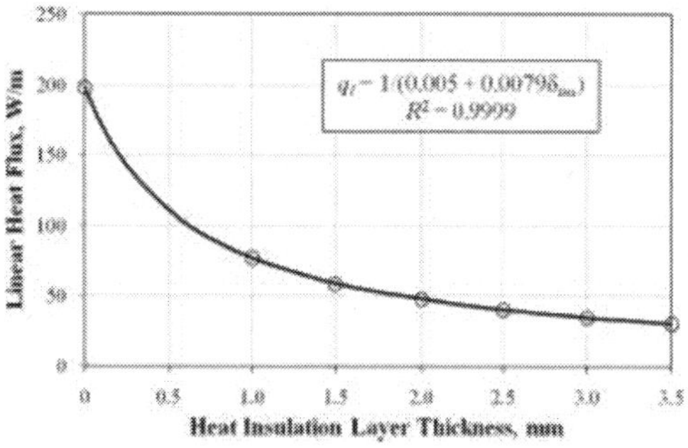

Fig. 3. Relationship between linear heat flux and the thickness of heat insulation layer.

Figure 3 shows that under the considered conditions a layer of heat-insulating paint with a thickness of allows reducing the heat loss to the environment by about 60%.

Table 2 presents the results of thermal imaging of the examined pipeline with tempera-ture markers (Figure 2).

Thermal imaging gives a more complete picture of temperature state of a thermal control object than point pyrometry by infrared thermometer. For example, frame 1 clearly shows how the surface temperature of pipeline changes before and after heat insulation. Analyzing the temperature field of pipeline in frames 2, 3, and 4, we can conclude that heat insulation has an uneven coating thickness. At the points of thermogram where the temperature is higher, the layer of heat insulation paint is smaller and vice versa. Despite the conclusions made, the readings obtained by the Testo 875-2 thermal imager are generally comparable with the results of infrared diagnostics by the Testo 845 pyrometer. The absolute difference in temperatures of an isolated pipeline section

Table 2. Thermographic studies of the examined pipeline.

Frame number	Photograph	Thermal image (thermogram)	Description
1.		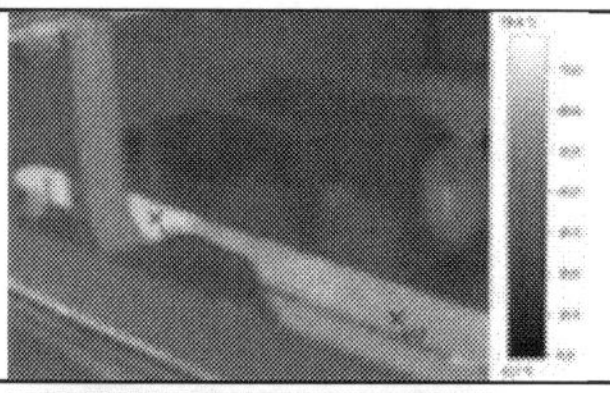	**M1**: 69.3 °C **M2**: 56.3 °C
2.			**M1**: 60.1 °C **M2**: 66.7 °C
3.		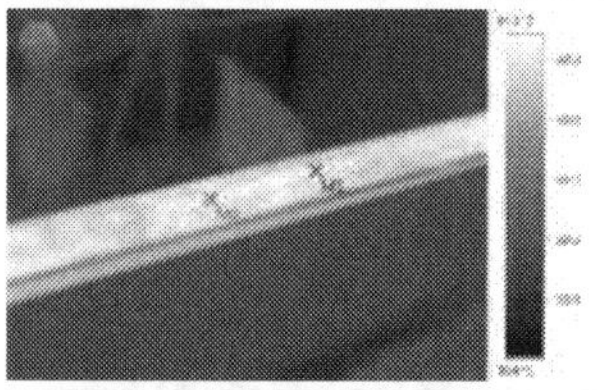	**M1**: 59.4 °C **M2**: 63.7 °C
4.		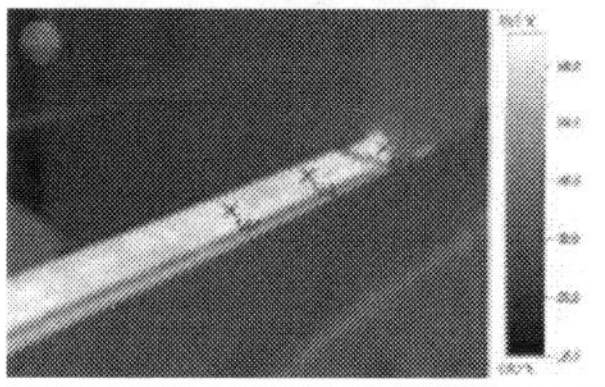	**M1**: 60.3 °C **M2**: 62.2 °C **M3**: 64.7 °C

and an unpainted section are in the range 5-10 °C. It should also be noted that for this object of thermal control the results of non-contact and contact (using the TK-5 thermometer) thermometry diverge on average on 8-12 °C. The readings of the contact thermometer turned out to be less than the values obtained by infrared measuring devices. This is confirmed by tests on liquid heat insulation of the Teplomett Standard brand on the surface of hot water supply boiler. These tests were carried out by the energy audit company CKT CJSC in Novosibirsk (Russia) using the SDS HotFind-D thermal imager and the IT-5-T Pt100-2 surface temperature meter, and showed that the deviation in temperatures was also approximately 8-10 °C. The question on reasons for discrepancy in temperature readings obtained on the surface of the studied object of thermal control using non-contact and contact thermometry remains open.

Tests of liquid heat insulation on the surface of main pipeline were carried out in a closed room of boiler house (Figure 2), where the temperature of internal air is much higher than the outdoor air temperature registered during the heating season. However, even under such conditions, there is a possibility to achieve substantial savings in fuel and energy resources, which is clearly confirmed by the graph presented in Figure 3.

Figure 4 presents a histogram of annual consumption of equivalent fuel as a function of thickness of heat-insulating paint. The paint covers the surface of a 250×6 steel feed pipeline, the considered length is a running meter. The pipeline is air laying in Vologda (Russia). The parameters of boiler house are as follows: heating and domestic temperature schedule; design temperatures of network water $\tau_1/\tau_2 = 130/70°C$; break point of temperature graph $\tau'_1 = 70$ °C at outdoor temperature $t'_{out} = -0.925$ °C.

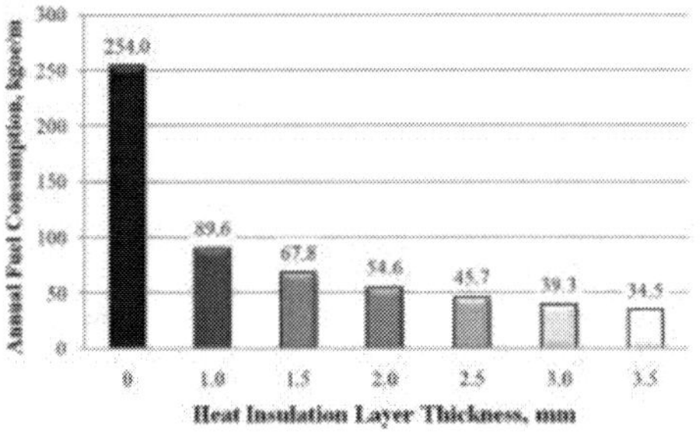

Fig. 4. Annual specific consumption of equivalent fuel.

4 Conclusions

The presented study shows that the annual economic benefit of application of liquid heat insulation is as obvious as reduction in the cost of fuel and energy resources for production of thermal energy (Figure 4). For example, heat insulation of a pipeline with a standard size of 250×6 by heat-insulating paint with a 1 mm layer thickness will save 126.1 m^3 of natural gas from 1 running meter of pipeline per year. In monetary terms it will be approximately 610 rubles or about $9 (for the first half of 2020 for the conditions of Vologda, Russia). As calculations showed, a significant economic effect can be achieved for a minimum thickness of heat insulation. For a thickness of heat-insulating paint of $\delta_{ins} = 1\,\text{mm}$, the economy of equivalent fuel is approximately 65%, for a thickness of $\delta_{ins} = 1.5\,\text{mm}$ it will increase by only 8%. In any case, for the declared coefficient of material thermal conductivity and the current cost of heat-insulating paint, the payback period of energy-saving measure to heat the pipelines of the district heating system should not exceed six months, which is quite cost-effective for the conditions of the Russian Federation.

References

[1] J. Zhao, G. Wang, Ch. Wang, Ch. B. Park, Ultra-lightweight, super thermal-insulation and strong PP/CNT microcellular foams, Composites Science and Technology 191, 108084 (2020).

[2] E.G. de Moraes, L. Sangiacomo, N.P. Stochero, S. Arcaro, L.R. Barbosa, A. Lenzi, C. Siligardi, A.P. Novaes de Oliveira, Innovative thermal and acoustic insulation foam by using recycled ceramic shell and expandable styrofoam (EPS) wastes, Waste Management 89, 336-344 (2019).

[3] B. Kiss, Cl.G. Manchon, L. Neij, The role of policy instruments in supporting the development of mineral wool insulation in Germany, Sweden and the United Kingdom, Journal of Cleaner Production 48, 187-199 (2013).

[4] St. Lang, M. Gerschitzka, D. Bauer, H. Drück, Thermal Conductivity of Vacuum Insulation Materials for Thermal Energy Stores in Solar Thermal Systems, Energy Procedia 91, 172-181 (2016).

[5] D. Bozsaky, Laboratory Tests with Liquid Nano-ceramic Heat insulation Coating, Procedia Engineering 123, 68-75 (2015).

[6] B. Abu-Jdayil, A.-H. Mourad, W. Hittini, M. Hassan, S. Hameedi, Traditional, state-of-the-art and renewable thermal building insulation materials: An overview, Construction and Building Materials 214, 709-735 (2019).

[7] A.A. Sinitsyn, D.F. Karpov, M.V. Pavlov, Basic Thermal Imaging Diagnostics of Heat-Consuming Construction Facilities (VoSU, 160, 2014).

Integrated modernization of the gas-and-air system of a turbocharged diesel engine (21/21)

Leonid Plotnikov[1,*], *Nikolay* Kochev[1], and *Leonid* Osipov[1]

[1]Ural Federal University named after the first President of Russia B.N. Yeltsin, Yekaterinburg, Russia

Abstract. Improving the exploitative and environmental performance of piston engines (PICE) is an urgent task for many engineers and scientists. The article presents the results of the upgrade of a gas-and-air system of a diesel PICE, carried out through changing the turbocharging system's configuration and modernizing the design of the admittance collector. The authors present a review of studies on the given subject and a description of the object of the research. The study was conducted on the basis of bench tests at a manufacturing plant and mathematical modeling using ACTUS program. The results of experimental studies on the main indicators of a basic and upgraded PICEs are presented. The gas exchange processes in the PICE under examination were studied in detail using mathematical modeling. For the given diesel PICE, improvement of the gas-and-air system leads to a growth in charging efficiency by 2.45-3.92%, a decrease in scavenging factor by 3.11-6.31% and a reduction of specific fuel consumption up to 3.33%. In the conclusion, new directions for increasing the efficiency of the given PICE are offered.

1 Introduction

Today, piston internal combustion engines (PICE) are some of the most convenient and mobile sources of energy. They can operate on various types of fuel (gasoline, gas, biofuels, alcohol, hydrogen, etc.). Improving the performance, efficiency and environmental friendliness of ICE is the main goal in the development of this industry. Their performance depends not only on improving processes in the engine cylinder, but also on the quality of the cylinder filled with a working substance and the degree to which said cylinder is clean of end gases [1, 2]. These processes determine the amount of the working medium at the start of combustion and its physical and chemical properties, thus having a direct influence on the exploitative indicators of power systems based on PICEs.

Many engineers and scientists have perfected the designs of the admittance and final systems of PICEs. It is possible to single out works where mathematical modeling is used to evaluate the influence of the spiral canal in the cylinder head on PICE's exploitative and ecological indicators [3-5]. It is demonstrated that the use of a spiral canal results in a growth in engine power by up to 6.05% and the reduction of NOx emission up to 40.55% compared to the basic system. Other experts have explored the possibility of developing an admittance manifold in order to grow the performance of PICEs through numerical simulation in different programs [6-8]. It has been stated that a refined configuration of admittance and final systems can grow the airflow rate in the system by 10.5-25.2%, which also leads to increased PICE power [9-11]. Thus, the authors assess the influence of the GaAS's configuration on the processes taking place inside the cylinder. It is notable that studies using mathematical modeling to examine heat and mass transfer processesin the admittance and final systems have often been confirmed by the results of experiments [12, 13]. Still, it is important to note that experimental studies are carried out under steady-state conditions. It is obvious that the PICE working processes are periodic with a high cyclicity. Vibrations, high temperatures, external turbulence, etc. should be considered in research as well. Modern PICE research should be carried out in conditions as close as possible to reality. The turbocharging system (TCS) is actively used in PICE in order to grow their exploitative performance [14]. Effective application of TCS in diesel PICE has revealed a number of problems and the tasks for further improvement of gas-and-air systems (GaAS). Modern requirements for the exploitative and ecological indices of PICEs with TCS are closely related to the strict requirements for gas-dynamics and heat exchange in GaAS [15-17]. Many studies are aimed at designing GaAS s with optimum geometrical forms and without sources for the possible appearance of stagnant zones [18-20]. Still, the maximum effect is achieved by the integrated upgrade of the admittance and final systems, TCS adjustment and engine development [21, 22]. Such upgrade results in a growth in PICE power by up to 28% and the reduction of specific fuel consumption (SFC) b_e by 2.54% on average, compared to a basic PICE.

This work describes the results of bench tests and numerical modeling as applied to a universal diesel ICE of dimension 210/210 (cylinder diameter – 210 mm, piston stroke – 210 mm). During the study, a

* Corresponding author: leonplot@mail.ru

comprehensive upgrade of the GaAS was carried out by changing the TCS and fine-tuning the admittance collector. It is important to obtain reliable information about these processes.

2 Description of the study subject

As a study subject, the authors chose a diesel PICE produced by the Ural Diesel-Motor Plant (Russian Federation, Yekaterinburg). It is an eight-cylinder 4-stroke engine with a V-type cylinder arrangement at an angle of 90°, with direct fuel injection, a closed-type liquid cooling system, gas turbine charging, aftercooling, and a dry-sump lubrication system. The engine's power rating is 980 kW at a crankshaft speed of 1500 rpm and a maximum torque of 1500 N•m. The factory designation of the unit is 8UDMP-21. In order to define the directions for improving diesel engine 8UDMP-21, the authors have conducted bench tests at a manufacturing plant and mathematical modeling of working processes using ACTUS program, developed by ABB Turbo Systems. Let us take a brief look at the basic schematic of the GaAS in the PICE (Figure 1). Arrows indicate the direction of gas movement.

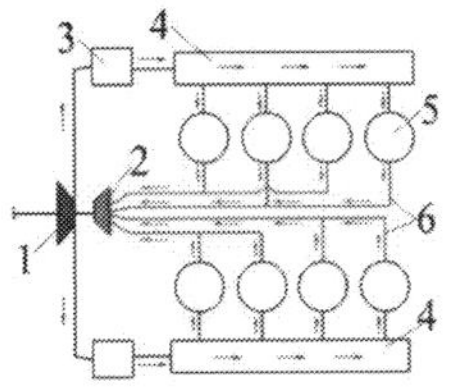

Fig. 1. Schematic of the gas-and-air system of 8UDMP-21: 1 – compressor; 2 – turbine; 3 – aftercooler; 4 – admittance collector; 5 – diesel cylinders; 6 – final collector.

In figure 1, the basic GaAS has one turbocharger produced by LLC "Special turbocharger design bureau" (Russia). Air, compressed in the centrifugal compressor, goes through the pipelines to the admittance collectors on the left and right sides of the V-type diesel. The turbine case has four inlets. In the given configuration, the final hoods from two cylinders form a single pipeline in order to reduce the pressure wave energy. It should be noted that the 8UDMP-21 has a traditional layout.

3 Results of the studies and analysis

As a result of studies on the examined diesel engine, the following basic characteristics were obtained (Figure 2). The minimum effective b_e is about 212 g/(kW·h), and the maximum about 250 g/(kW·h). These values of b_e are high: it is necessary to reduce them in order to grow the engine's competitiveness. Actual air consumption through the PICE is 0.182 to 0.802 kg/s in the idle mode and nominal mode, respectively. Along with this, the compressor pressure ratio ranges from 1.08 to 2.38, while the air ratio α ranges from 1.62 to 2.23. The minimum values of α for the initial diesel PICE are close to engine smoking, which was observed visually during the tests, which is undesirable under operational conditions. Obviously, upgrade of the TCS is necessary.

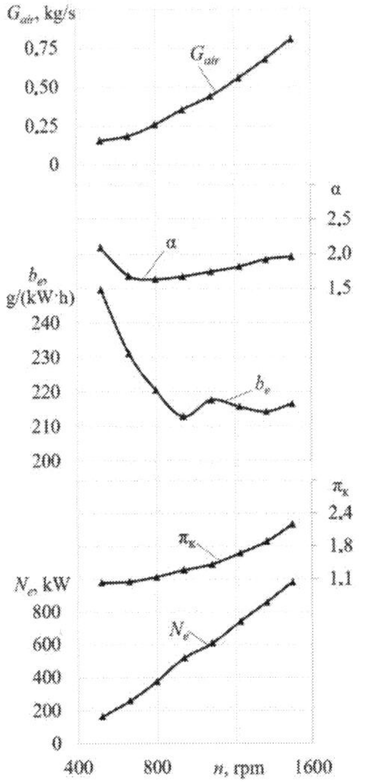

Fig. 2. Basic characteristics of the 8UDMP-21: G_{air} – actual air consumption through the cylinders; α – air ratio; b_e – effective SFC; π_K – compressor pressure ratio; N_e – effective power.

Modeling of the PICE working process was carried out in the ACTUS program. A mathematical model of the 8UDMP-21 was developed. It included the main systems of this PICE. The program has a block structure. Each block has its own physical properties and mathematical description. A set of interconnected blocks defines a particular system or engine element. The simulation results of the base PICE were correlated with the test results. Differences in the main indicators did not exceed 5.22%. The program simulates the working processes of PICEs with a TCS with fairly high accuracy. This is confirmed by a large number of studies comparing the results of modeling and bench tests of PICEs. Some examples can be found in [22, 23].

The results of modeling the gas dynamics in the GaAS are presented in Figure 3. At the moment the admittance port is opened, the pressure in the final collector is higher than in the admittance one (this area is circled in Figure 3). This points to the fact that products get into the admittance collector in the process of inlet combustion, which, in turn, may lead to a deficiency of oxygen, necessary for complete fuel combustion, and, consequently, to decreased ecological indices and excessive smoke. Previously, this design was used to reduce the size of the PICE.

 https://doi.org/10.1051/e3sconf/202017801005

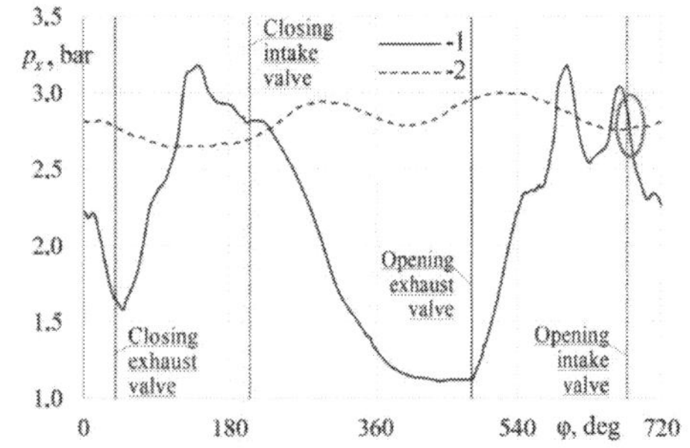

Fig. 3. Estimated dependence of pressures p_x in the admittance port (1) and the final port (2) for the third left cylinder of the 8UDMP-21 on the crankshaft rotation angle φ in the nominal operational mode.

The basic configuration of this engine's admittance collector has strongly marked technical deficiencies in terms of its gas-dynamics (Figure 4). There are fixture elements in the way of the air flow, which cause vortices and reduce the flow section (they are circled in Figure 4b). This can lead to deterioration in the quality of the cylinder filled with a working substance. This will lead to a reduction in charging efficiency, a growth in exhaust smoke, a growth in b_e (reduction in the efficiency of the PICE and TCS).

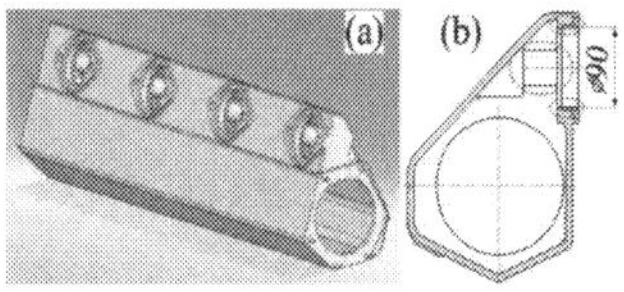

Fig. 4. General view of the admittance collector (a) and its cross section (b) for the basic 8UDMP-21.

The principal directions of gas-dynamic improvement for the GaAS of engine 8UDMP-21 were established, which consist of changing the TCS's configuration (in order to grow the air ratio α and refine carburation and combustion) and upgrading the construction of the admittance collector (with the aim of improving charging efficiency and reducing SFC b_e).

On the basis of literature analysis, the scheme of the TCS was changed. A pulse TCS with two small turbochargers produced by ABB Turbo Systems was used. This design solution allowed for the compressor pressure ratio to be increased, thus increasing the actual air consumption through the PICE. The TCS's response time was also reduced, which should grow PICE acceleration compared to a basic PICE equipped with one large turbocharger. Concurrently, the configuration of the admittance collectors was upgraded, namely, their in-line and cross-cut profiling was performed (Figure 5). The new design is more modern. The key changes are related to the change in the geometry of the admittance ports to each cylinder: this is aimed at minimizing the sources of flow swirling. The proposed design of the admittance collector has smoother surfaces compared to the base collector. At the same time, fasteners on the path of air movement are absent in the new design. It should be noted that the main characteristics of the bore sections were saved in this design. Thus, the proposed design has a reduced hydraulic resistance, and therefore, has refined flow characteristics. This should lead to a growth in charging efficiency and a reduction in fuel consumption for the 8UDMP-21.

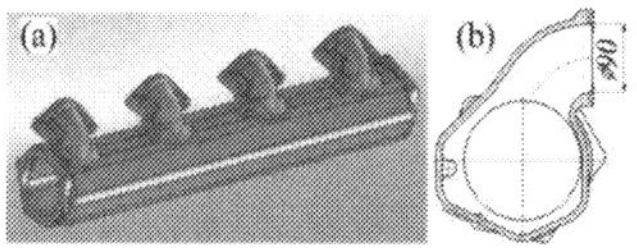

Fig. 5. General view of the upgraded admittance collector (a) and its cross-section (b) for 8UDMP-21.

After selecting a new TCS scheme and applying new admittance collectors, the upgraded 8UDMP-21 was tested (Figure 6).

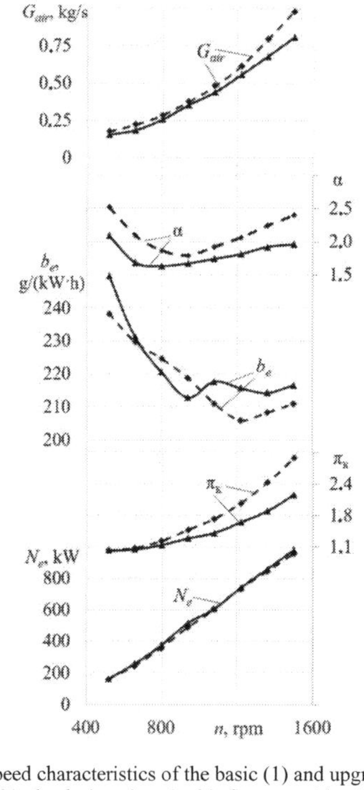

Fig. 6. Speed characteristics of the basic (1) and upgraded (2) 8UDMP-21: the designations in this figure are identical to Figure 2.

The figure shows the combined dependencies for both PICEs. In Figure 6, the N_e of the upgraded diesel PICE in partial modes is slightly higher (by 5-7%) than that of a basic diesel. Presumably, this effect was achieved by increasing the compressor pressure ratio π_K, and, consequently, the actual air consumption, which led to increased charging efficiency, along with refined carburation and combustion. In this case, the compressor pressure ratio ranges from 1.12 to 2.49, while the air ratio α ranges from 1.81 to 2.52. The values of the α coefficient are far from the values characteristic of the exhaust smoke.

The charging efficiency of the upgraded diesel increased by 2.45-3.92%. The growth of charging efficiency occurs in all operating modes of the PICE and TCS. The charging efficiency determines the degree to which the cylinder is filled with air. This should lead to an increase in the efficiency of both the PICE and the turbocharger. The smoke of the gases also decreased. Moreover, the upgrade of the GaAS also led to a reduction in scavenging factor by 3.11-6.31% compared with the base system. This indicates an improvement in exhaust cleaning of the cylinders as well. Reducing the scavenging factor will also increase the efficiency of the TCS, as the gas flow through the final system will grow. The charging efficiency and scavenging factor are key indicators of gas exchange and have a direct relationship with each other.

4 Conclusions

Application of the new configuration of the admittance collector and two turbochargers with refined exploitative and flow characteristics increased the compressor pressure ratio (by about 8.02%) and the charging pressure (on average by 10.5-15.3%). It also resulted in an increased air ratio (by 5.42% on average) in partial modes, which had a positive effect on the processes of carburation and combustion, leading to reduced smoke and emission toxicity. Despite the relatively small growth in the power of the upgraded PICE, the effective SFC b_e was reduced by about 3.11% in some modes, which will allow for a reduction in the overall fuel consumption by the end user of up to 10.51 kg/h. Using a TCS with two small turbochargers also refined the mass and size characteristics of the PICE and increased its response. This grows the competitiveness of the 8UDMP-21.

As an additional potential direction for further upgrade of the 8UDMP-21, we can consider thermomechanical improvement of the final system (a thorough refinement of the final collectors) and studying the application of a constant-pressure TCS in front of the turbine. The directions of improving the GaAS of PICEs considered in the article are universal. The results of this study can be extended to PICEs of other dimensions in order to refine their exploitative indicators.

Acknowledgements

The work has been supported by the Russian Science Foundation (grant No. 18-79-10003).

References

[1] C.R. Ferguson, A.T. Kirkpatrick, *Internal combustion engines: applied thermosciences* (New York: John Wiley & Sons, 2016)

[2] W.W. Pulkrabek, *Engineering Fundamentals of the Internal Combustion Engine* (New Jersey: Prentice Hall, Upper Saddle River, 2005)

[3] A.M. Sadeq, M.A. Bassiony, A.M. Elbashir, S.F. Ahmed, M. Khraisheh, Fuel **255**, UNSP 115769 (2019)

[4] G.X. Wang, W.B. Yu, X.B. Li, Y.P. Su, R. Yang, W.T. Wu, Energy **180**, 1008–1018 (2019)

[5] E.A.A. Silva, A.A.V. Ochoa, J.R. Henriquez, Energy conversion and management **188**, 310–320 (2019)

[6] S.Q. Guo, S. Huang, M.S. Chi, Eurasip J. image and video processing **70** (2018)

[7] G.K. Giannakopoulos, C.E. Frouzakis, K. Boulouchos, P.F. Fischer, A.G. Tomboulides, Int. J. heat and fluid flow **68**, 257–268 (2017)

[8] S.D. Hanriot, J.M. Queiroz, C.B. Maia, J. Brazilian society of mechanical sciences and engineering **41**, 2, UNSP 79 (2019)

[9] L. Kapitza, O. Imberdis, H.P. Bensler, J. Willand, D. Thevenin, Experiments in fluids **48**, 2, 265–280 (2010)

[10] G.J. Micklow, W.D. Gong, Proc. institution of mechanical engineers part d-J. automobile engineering **221**, D11, 1425–1440 (2007)

[11] E. Yilmaz, M. Ichiyanagi, T. Suzuki, Int. J. automotive technology **20**, 5, 1065–1071 (2019)

[12] D.R. Jebamani, T.M.N. Kumar, Thermal science **12**, 1, 25–32 (2008)

[13] M.A. Ceviz, M. Akin, Energy conversion and managemen **51**, 11, 2239-2244 (2010)

[14] N. Watson, M.S. Janota, *Turbocharging the Internal Combustion Engine* (London, Palgrave Macmillan Press, 1982)

[15] L.V. Plotnikov, IOP Conf. Series: J. Physics **899**, 042008 (2017)

[16] L.V. Plotnikov, B.P. Zhilkin, Applied Thermal Engineering **160**, 114123 (2019)

[17] M.W. Bae, Y.J. Ku, H.S. Park, Tran. Korean society of mechanical engineers B **43**, 5, 379–388 (2019)

[18] U.B. Khairuddin, A.W. Costall, J. Engineering for Gas Turbines and Power **140**, 5, 052801 (2018)

[19] Y.M. Brodov, L.V. Plotnikov, B.P. Zhilkin Technical Physics **63**, 3, 319–324 (2018)

[20] G.X. Wang, W.B. Yu, X.B. Li, R. Yang, Fuel **262**, UNSP 116548 (2020)

[21] M.A. Jemni, G. Kantchev, M.S. Abid, Energy **36**, 5, 2701–2715 (2011)

[22] L.V. Plotnikov, S. Bernasconi, Y.M. Brodov, Procedia Engineering **206**, 140–145 (2017)

[23] L.V. Plotnikov, S. Bernasconi, P. Jacoby, Lecture Notes in Mechanical Engineering, 549–558 (2020)

High Speed Electrical Drives – Perspective of VFD Manufacturer

Martin Bruha[1,*], *Kai* Pietiläinen[1], and *Axel* Rauber[1]

[1]ABB Switzerland Ltd, Medium Voltage Drives, 5300 Turgi, Switzerland

Abstract. This paper deals with high-speed electrical drives utilizing power electronic converters (commonly abbreviated as ASD, VFD or VSD). Existing solutions vary mainly on the motor side while the power electronic converter is very similar for all cases. Various advantages as well as technical challenges are discussed and illustrated. At certain stages comparisons between conventional and high-speed drives are made. The paper summarizes the experience of a VFD manufacturer based on state of the art technology in medium voltage and multi-megawatt power range. The authors believe that main complexity around high-speed drives is the motor design while the VFD requires only small adaptations or can sometimes be used directly without any modifications of standard design. The technology readiness is evaluated to be on a medium to high level.

1 Introduction

Variable frequency drives (VFD) gain increasing popularity across many industries. The various applications have very different speed ranges from ultra-low speed machines such as large mills in mining and minerals industry, low speed rolling mills in metals industry, medium speed pumps and fans up to high speed turbomachinery in oil and gas sector. The motor is connected with the load machine either directly or through a gear.

In turbomachinery a competing technology to VFD is a gas turbine or steam turbine. The turbine rotates with high speed and is directly coupled to the load machine (typically compressor) without any gear. Due to environmental aspects, easier controllability, reduced maintenance and other motivations there is a trend to replace such turbines with all electric drive system. In some cases both conventional and high-speed technologies can be used. However, if there are space and weight restrictions high-speed technology might be the only solution. Another case is e.g. electric motor drive as starter/helper with gas turbine as main driver. VFD technology also allows the use of 4-pole machines for higher speeds by increasing stator frequency.

As the authors illustrate, the VFD design for a high-speed application does not significantly differ from a conventional VFD. It is either a standardized product with small adaptations or even completely free of any modifications. The electric motor, on the other hand, is the component with higher complexity. One step from a conventional motor is motor with increased speed and 'standard' bearings. If the speed range requires magnetic bearings the complexity goes again one step higher.

2 Variable speed solutions for turbomachinery

2.1 Motor technology

First, it shall be defined what is a high-speed application, resp. where the high-speed range starts. To authors' knowledge there is no standardized classification in this regard. One can distinguish between conventional speed, semi high-speed and high-speed motors. In this paper conventional speed is considered up to 3'600 rpm (60 Hz / 2-pole machine). Semi high-speed is up to approx. 6'400 – 7'500 rpm [1]. Above 7'500 rpm it is usually specific high-speed design with active magnetic bearings [2-5]. These considerations apply to larger machines in power range of several megawatt. Obviously, for smaller machines it is easier to achieve higher speed.

For turbomachinery applications there are principally two solutions: conventional technology with a gearbox or high-speed direct drive, also called gearless.

Conventional solution in electrically driven turbomachinery is a standard 4-pole electric motor with rated speed 1'500 rpm or 1'800 rpm (or somewhere nearby those speeds) and speed-increasing gear. Due to the gear the motor and compressor speeds are decoupled. Such motor design is derived from a direct on-line (DOL) design while considering variable speed operation and inverter duty.

Direct drive solution utilizes motor with same speed as the driven machine. It eliminates the gear and its mechanical losses. The installation is typically more compact than the conventional approach. That fact makes it attractive in case of space or weight restrictions. The complexity is around the active magnetic bearings and their control.

* Corresponding author: martin.bruha@ch.abb.com

Fig. 1. Typical LCI converter suitable to supply conventional as well as (semi) high-speed motors.

If the speed of turbomachine is not too high (semi high-speed) then a 2-pole turbo-motor can be used. Such motor is usually a round rotor type of synchronous machine derived from a turbo-generator and eventually adopted for higher speed. Maximum speed can reach approx. 7'500 rpm. Such design is normally super-critical meaning that motor is operated above the first critical speed. This brings some challenges in case the controllable speed range is very wide. In lower semi high-speed range (up to approx. 4'000 rpm, eventually higher) 4-pole rotor design with higher supply frequency can be realized. The 4-pole rotor has fundamentally different mechanical behaviour compared to his 2-pole counterpart. Its design is sub-critical providing many benefits with respect to critical speeds. This approach with 4-pole rotor is followed by many leading manufacturers and some machines have already been manufactured [6, 7].

For speeds above 7'000 – 8'000 rpm the solution is usually induction motor with magnetic bearings. Several rotor designs can be used with their pros and cons. Main solutions are variants of solid cylindrical rotor and laminated rotor with special shaft arrangement [8].

2.2 VFD technology

Variable frequency drive (VFD) for high-speed application nowadays does not significantly differ from a conventional VFD. Drive configuration suggested in [2] with so called twin inverter configuration and open winding stator design is principally still possible, but there are meanwhile more straight forward ways. The power hardware is either the same as conventional VFD or it experiences minor modifications. The inverter modulation scheme might be different. As for conventional drives also for high-speed two basic technologies are available: current source inverters (CSI) and voltage source inverters (VSI).

2.2.1 Load-commutated inverter (LCI)

LCI is a specific category of current source inverters. Both line side and motor side converters consist of classical thyristors commutated by external voltage source: grid voltage on the line side and motor voltage on the motor side. This explains the name '*load-commutated*' and also the fact that LCI only works in combination with synchronous machine. LCI is very mature and reliable technology that is available on the market since 1970s. Due to easy power and voltage scalability a very high-power drive can be realized without too much complexity and with low parts count. Although majority of the LCI drive applications have rated output frequency in the range 40 to 60 Hz the drive can reach output frequency up to 100...120 Hz with minimum modifications. In combination with 2-pole turbo-motor a direct drive with up to 7'200 rpm can be realized [1].

First projects semi high-speed LCI driven motors had already been realized in early 1980s and some installations therefore reached over 30 years operational experience demonstrating high reliability and availability.

2.2.2 Voltage source inverter (VSI)

VSI technology became dominant in 1990s with the evolution of self-commutated power electronic switches (first GTO thyristors, later IGCT thyristors and IGBT

Fig. 2. Typical VSI converter (5-L NPC) suitable to supply conventional as well as high-speed motors.

transistors). Nowadays VSI drives represent vast majority of all new electric drive installations. There are several topologies on the market. In medium voltage range the most common are 3-level neutral point clamped (3-L NPC) inverter, 5-level neutral point clamped (5-L NPC) inverter, multiwinding multicell or modular multilevel converter (MMC, M2C). Topology further described is a 5-L NPC inverter which in fact consists of 3-L-NPC modules in H-bridge connection. It is "6kV class" VFD with rated output voltage selectable in the range 5.5 kV to 7.2 kV. The liquid cooled variant offers power range starting from approx. 6 MW up to almost 40 MW as single unit. Higher power is achieved by paralleling multiple units. Standard output frequency is up to 66 Hz in conventional applications. In high-speed configuration more than 250 Hz output frequency can be easily reached (field reference up to 260 Hz, technical feasibility up to at least 400 Hz).

VSI type of converter can supply both induction and synchronous machine and the true high-speed applications typically consist of cage induction machine and active magnetic bearings (AMB).

3 VFD design considerations for high-speed

3.1 Modulation technique and switching frequency

3.1.1 Load-commutated inverter (LCI)

The inverter is load-commutated and the principle is the same as in conventional drive system. Higher output frequency might lead to small adaptations of the snubber circuit and more frequent commutation shall be considered in voltage design, but otherwise the LCI is basically the same regardless if designed for rated motor frequency of 50 Hz or 100 Hz. Semi high-speed synchronous machines with 2-pole round rotor usually inherently have quite low sub-transient reactance which is beneficial for LCI operation. The commutation notches are smaller and the inverter can operate with better (higher) firing angle. This fact in turn means higher power factor and lower torque ripple.

3.1.2 Voltage source inverter (VSI)

High speed applications naturally need higher fundamental output frequency of the inverter. Too high switching frequency would cause high additional losses, higher semiconductor junction temperatures and consecutive power derating of VFD. Example of inverter loadability versus semiconductor switching frequency is shown in Fig. 3. The VFD has certain design margin so that small increase of switching frequency does not cause any power derating. However, above certain frequency the current derating becomes significant eventually leading to uneconomical VFD design.

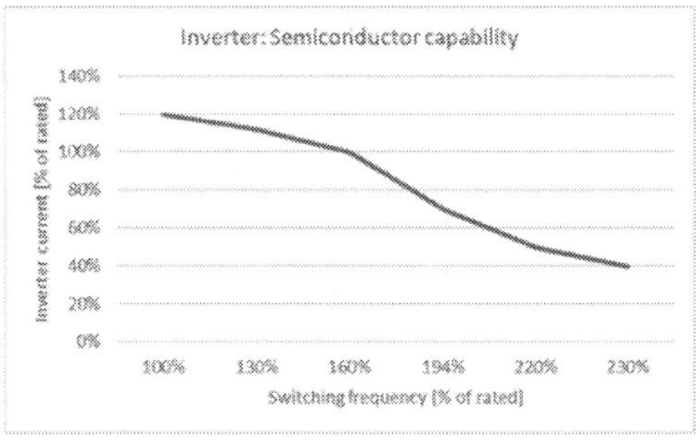

Fig. 3. Inverter capability versus semiconductor switching frequency.

Therefore, different modulation scheme is used than in conventional applications. Optimized pulse pattern (OPP) provides a good balance between good output waveform and reasonable switching frequency. It allows operation with lower ratio between switching frequency and fundamental output frequency. Several OPPs are pre-defined in the software and the VFD changes automatically from one PP to another one at certain motor fundamental speeds. Those speed thresholds are defined project specifically as part of system design. Such flexibility allows optimization between low current harmonics and maximum utilization of VFD.

The output voltage is axis-symmetrical with pulse pattern modulation. Plots in Fig. 4 show real field

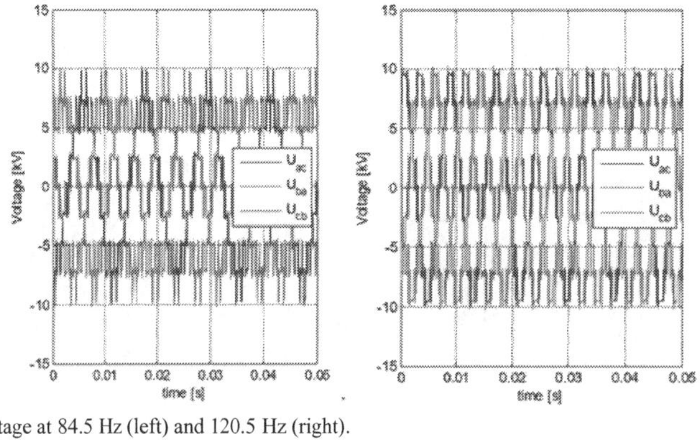

Fig. 4. VFD output voltage at 84.5 Hz (left) and 120.5 Hz (right).

Table 1. Drive system efficiency: Conventional versus high-speed.

Motor type	Induction, conventional	Synchronous, conventional	Induction, high-speed	Induction, high-speed
Shaft power [kW]	12'000	12'000	12'245	12'245
Motor voltage [kV]	6.6	6.6	6.6	6.6
Motor current [A]	1'198	1'090	1'395	1'392
Rated speed [rpm]	1'493 / 1'793	1'500 / 1'800	12'000	12'000
Power factor [pu]	0.92	1.00	0.80	0.80
Gear efficiency [%]	98.00%	98.00%	N/A	N/A
Motor efficiency [%]	97.20%	98.30%	96.00%	96.20%
Switching frequency	Standard	Standard	Increased	Standard
Output choke	No	No	No	Yes
VFD heat loss [kW]	133	121	172	200
VFD efficiency [%]	98.96%	99.04%	98.67%	98.45%
Combined efficiency VFD-Motor-Gear [%]	94.26%	95.41%	94.72%	94.%

measurements for two different motor speeds (measured signals plotted in MATLAB® environment).

3.2 Efficiency and losses

Efficiency is nowadays very important performance criteria as it directly impacts the operational expenditure (OPEX) and return of investment. Proper efficiency assessment shall always be done on system level. Benchmarking of individual system components might be misleading. For example, one VFD design has higher efficiency, but its output waveform contains lot of harmonics that cause additional losses in the motor. It shall be stated if the efficiency includes all auxiliary consumption or not.

VFD with higher output frequency tendentially has somewhat higher losses. As a rule of thumb approx. 1/3 of total losses is scaled up with increased switching frequency. The high-speed motor has somewhat lower efficiency than a conventional one. However, the elimination of gear saves 2 to 3% of throughput power. In Table 1 typical values are provided considering a drive system with 12 MW compressor input power. The drive system consists of (a) conventional induction machine, (b) conventional synchronous machine, (c, d) high-speed induction machine and VFD without and with output reactor (choke).

Gear efficiency of 98% is assumed. Different gear value could potentially shift the results little bit. Input transformer is disregarded in comparison as it would have approx. the same efficiency for each case and is 'neutral' in that sense. VFD losses include auxiliary consumption, such as cooling water pumps, cooling fans, control boards etc. It is obvious that overall efficiency is very similar in all cases. High-speed electrical drives have somewhat higher losses (both VFD and motor) that are offset by the mechanical losses of the speed increasing gear.

3.3 VFD motor friendliness

Motor friendliness is important characteristic of VFD to ensure compatibility with the electric motor. In case the motor is located in hazardous area the compatibility is even more critical and shall be carefully checked.

3.3.1 Load-commutated inverter (LCI)

The advantage of LCI is that the motor voltage is almost sinusoidal with typical commutation notches (see Fig. 5 as example) and the motor insulation system requires only small margin compared to direct on-line motor. The output current and torque contain harmonics that need to be respected in motor design.

LCI configured as full drive, i.e. for continuous operation, typically consists of a 12-pulse inverter in combination with dual star stator winding on the

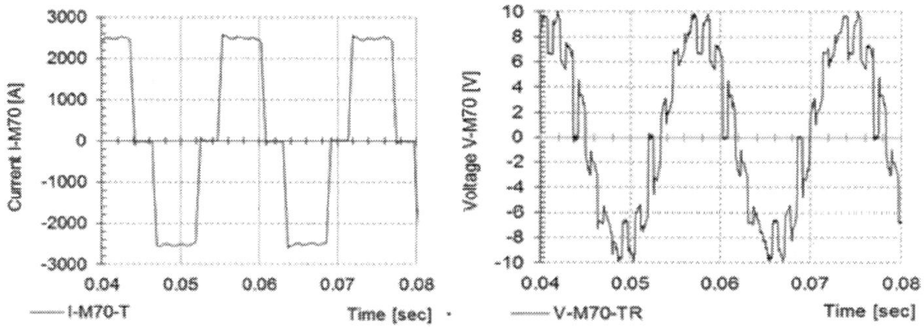

Fig. 5. Current and voltage output waveform of LCI (37'000 kW / 1'800 rpm).

 https://doi.org/10.1051/e3sconf/202017801006

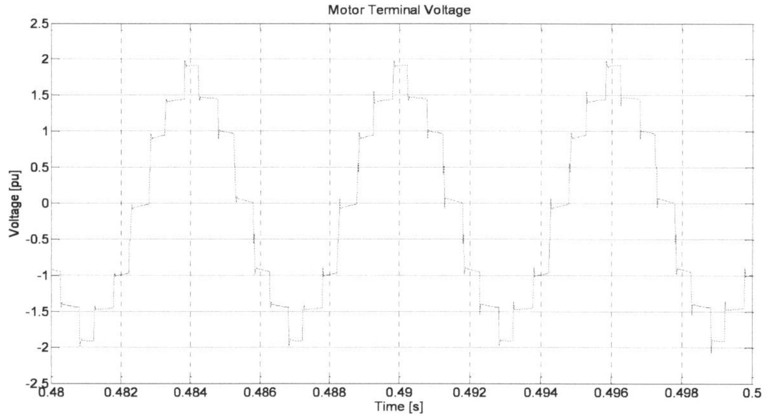

Fig. 6. Motor phase-phase voltage at 166.7 Hz.

machine. The stator windings have relative displacement of 30° electrical (YY30°). Such phase shifting helps to reduce the current and torque harmonics in the machine by up to 3-times compared to 6-pulse inverter and standard three phase motor.

3.3.2 Voltage source inverter (VSI)

If the VFD has a full-blown grounded sine filter at its output, there are no special considerations for the motor apart from mechanical design and cooling supporting variable speed operation [9]. However, sine filter is not common in high-speed and/or high-power range. The physical size of sine filter would be very large at high power. There are also extra losses associated with sine filter. And finally, there is the topic of self-excitation which might be even more challenging for wide speed control range.

For all other topologies (without sine filter at the output) the VFD has certain impact on motor. Generally, the more voltage steps (levels) in the output waveform the closer is the waveform to the ideal sine wave.

It can be divided into two categories: output voltage waveform influencing insulation system of motor and output current waveform impacting thermal design of motor.

For the integrity of motor insulation following VFD parameters are typically asked:

- Peak phase to phase voltage
- Peak phase to ground voltage
- Peak neutral to ground (common mode) voltage
- Voltage rate of rise (dv/dt)

Example of output voltage waveform is displayed in Fig. 6. Note that scaling of y-axis is based on phase to ground nominal voltage.

For the thermal design of motor following information is typically asked either for worst case point or for certain design or envelope points:

- Frequency spectrum of motor current
- Total current harmonic distortion (THDi)

The motor friendliness can be enhanced by using an output choke. The choke is specifically designed for high frequency operation. It is either stand-alone unit or integrated inside VFD panel. The integrated version allows easier handling and installation. Choke is directly water-cooled (WF) sharing the same deionized water with the VFD. The majority of heat losses of the choke are dissipated into water avoiding extra heat load for air-conditioning system of the electric room. Fig. 7 shows an example from the assembly.

Fig. 7. Output choke integrated inside VFD lineup and directly water cooled.

https://doi.org/10.1051/e3sconf/202017801006

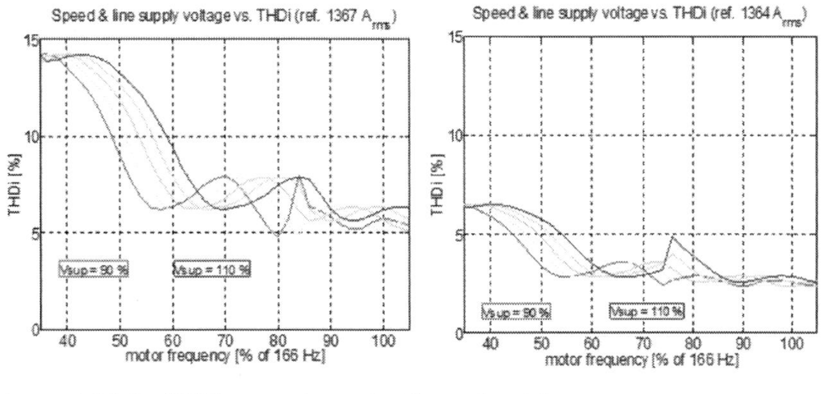

Fig. 8. Motor current distortion (THDi) across motor speed and line supply variation.

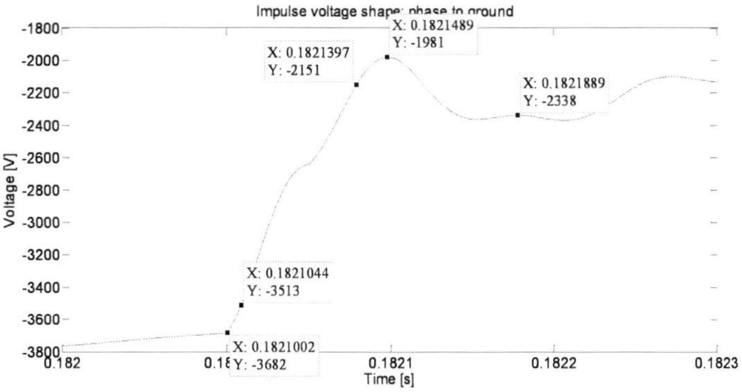

Fig. 9. Voltage impulse waveshape (phase-ground).

The positive impact of the output choke on the motor current distortion (THDi) is depicted in Fig. 8. The graphs show the dependence on the motor fundamental frequency and incoming supply voltage variation. The pulse patterns change at pre-programmed motor speed that can be project-specifically shifted. The supply voltage affects the dc voltage and therefore the modulation index at given speed and in turn also THDi.

Presence of output choke also reduces the rate of rise (dv/dt) of the inverter output voltage. Detail of voltage waveform is depicted in Fig. 9.

Parameters determined from the waveshape as per [10] are shown in Table 2:

Table 2. Parameters of voltage impulse waveshape.

t_{10}	[s]	0.1821044
t_{90}	[s]	0.1821397
t_r	[s]	3.53e-5
U_p	[V]	1701
U_b	[V]	357
U_a	[V]	1344
dv/dt	[V/μs]	38.5

Remark: Simulation results do not consider effects of motor cables and non-linearity of components at high frequencies. Recommended dv/dt for insulation consideration is increased to 120 – 150 V/μs. That is still much lower value compared to standard inverter with typical dv/dt of 1'000 – 3'000 V/μs.

3.3 VFD load friendliness

The VFD-motor drive system shall also be load friendly. VFD driven motor inherently has certain torque ripple that influences torsional behaviour of the entire shaft string [11]. It must be ensured that all shaft components are properly dimensioned for the alternating torque occurring in normal operation as well as transient torque during possible fault conditions. Therefore, torque ripple of the VFD driven motor shall be kept within certain limits. More important than the overall peak-peak torque is however the frequency spectrum of the motor torque [12, 13]. The frequency content of the motor torque defines if torsional modes of the shaft string will be excited or not. This is of utmost importance in modern turbomachinery as the damping values are extremely small and the amplification factors correspondingly large. In other words, even a small excitation torque can cause large alternating torque on the shaft components (e.g. couplings) when the frequency of motor air gap

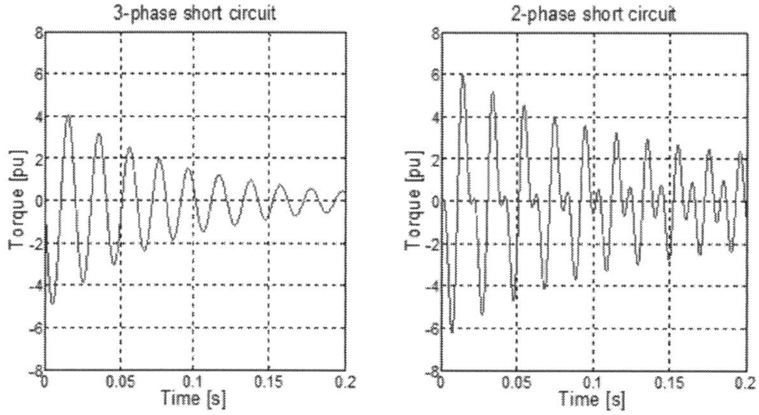

Fig. 10. Motor transient torques: 3-phase short circuit (left), 2-phase short circuit (right).

torque component matches one of the dominant torsional natural frequencies.

The torsional integrity shall be verified by a torsional analysis. Practical visualization of excitation torque components and their interaction with torsional modes over the entire speed range is by means of Campbell diagram.

Protection concept of the VFD also affects the torsional design. Many modern VFDs are fuseless. In case of major issue inside the VFD the reaction is so called protection firing which is seen by the motor as a symmetrical three-phase short circuit. Since the protection is extremely fast a phase to phase short circuit can be disregarded. The three-phase short circuit and corresponding transient torque is less harmful for the shaft system and easier to deal with as seen in Fig. 10.

Comprehensive electro-mechanical analysis of variable speed electric drive systems from torsional perspective is contained in [14]. In case the power electronic based loads are dominant loads and /or in case of island network there is certain risk of sub-synchronous torsional interaction (SSTI) – see [15].

4 Field experience

The area of (semi) high speed electric drives is not fully new and some of the drive systems have in the meanwhile over 30 years of running experience. Tables 3 and Table 4 summarize manufacturer's experience stating the basic technical data and year of delivery. Note that tables show motor rated speed while maximum continuous speed is typically 5 to 10% higher. The reference installations incorporate motors from various manufacturers and include both new installations as well

Table 3. Field experience in high-speed electrical drives – selected LCI references.

Motor technology	Shaft power [kW]	Rated speed [rpm]	Load machine	Year of delivery	Quantity
Synchronous, 2-pole round rotor, brushless excitation, sleeve bearings	9'600	6'000	Pump	1984	18
	8'000	5'521	Pump	1986	18
	13'000	6'400	Compressor	1986	1
	13'000	5'700	Pump	1988	2
	18'000	5'500	Compressor	1992	1
	41'000	3'600	Compressor	1993	5
	16'500	4'225	Compressor	2002	1
	48'000	3'600	Compressor	2005	3
	42'200	3'600	Compressor	2005	1
	13'500	6'000	Compressor	2008	1
	15'000	5'064	Pump	2008	1
	14'500	5'300	Pump	2009	1
	15'000	5'064	Pump	2011	1
	48'000	3'600	Pump	2014	1
	14'200	4'200	Compressor	2015	1
	17'100	5'100	Compressor	2015	1
	11'800	5'050	Pump	2015	1
	8'470	5'400	Pump	2015	12
	8'000	5'521	Pump	2016	2
	11'800	5'050	Pump	2017	1
	8'000	5'521	Pump	2017	2

Table 4. Field experience in high-speed electrical drives – selected VSI references.

Motor technology	Shaft power [kW]	Rated speed [rpm]	Load machine	Year of delivery	Quantity
Induction	11'500	11'200	Compressor	2007	1
Induction	6'600	5'100	Compressor	2008	1
Induction	4'700	14'715	Compressor	2011	1
Induction	12'000	10'002	Test stand	2011	1
Induction	14'821	9'175	Compressor	2011	1
Induction	14'060	9'045	Compressor	2011/12	3
Induction	11'460	6'364	Compressor	2012	2
Induction	13'075	8'371	Compressor	2012	6
Induction	7'678	11'702	Compressor	2013	2
Synchronous	10'000	5'202	Compressor	2013	7
Induction	4'700	14'715	Compressor	2015	1
Induction	5'950	11'905	Compressor	2015	2
Induction	7'412	11'429	Compressor	2016	1
Induction	6'635	10'769	Compressor	2017	1
Induction	7'000	10'000	Compressor	2017	2
Induction	9'500	12'000	Compressor	2018	2
Induction	11'877	9'959	Compressor	2018	3
Induction	8'932	12'600	Compressor	2019	1
Induction	9'068	12'600	Compressor	2019	1

as retrofit projects.

The references from Table 3 and Table 4 are graphically summarized in Fig. 11 as function of shaft power and rotor speed.

5 Conclusion

Use of semi high-speed and high-speed (direct) electric drives in megawatt power range as alternative to mechanical drivers and conventional geared electric drives seems to be a trend in the industry. There is already significant field experience gathered up to now: over 30 years operational experience with semi high-speed drives in speed range 3'600…6'400 rpm and over 10 years operational experience with high-speed drives with maximum speed up to 15'000 rpm or even higher.

In semi high-speed range it is principally possible to use machine with 4-pole rotor and adjust the stator frequency correspondingly. Such motor design is then sub-critical (unlike 2-pole design that is normally super-critical) and allows wide controllable speed range. Utilizing wider speed range enables larger energy savings and lower operational cost / life cycle cost.

The VFD for high-speed application is a derivative of standard VFD requiring minimum adaptations. Sometimes even standard VFD can be used with no modifications required. From this perspective the VFD

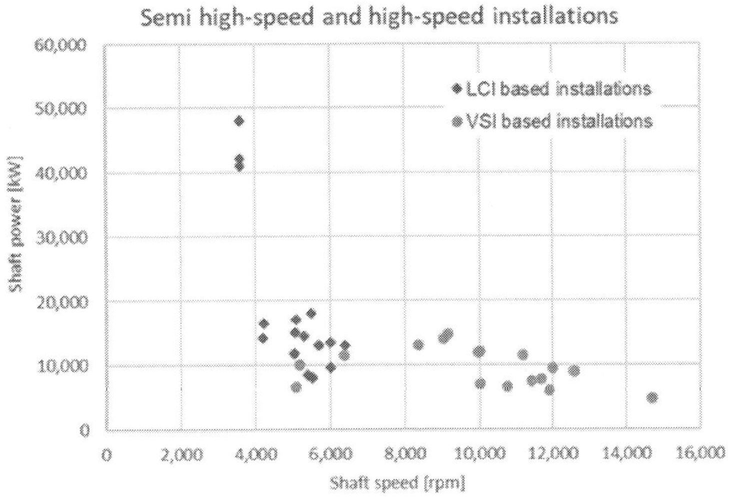

Fig. 11. (Semi) high-speed installations.

technology fully supports the trend of going in the direction of high-speed/gearless solutions. Attention shall be paid to correct system design (e.g. protection, grounding etc).

References

[1] G.J. Neidhöfer, A.G. Troedson, Large converter-fed synchronous motors for high speeds and adjustable speed operation: Design features and experience, IEEE Trans. on Energy Conversion **14**, 3 (September 1999)

[2] M. Ahrens, U. Bikle, R. Gottkehaskamp, H. Prenner, Electrical Design of High-Speed Induction Motors of up to 15 MW and 20000 rpm, IEE Conf. on Power Electronics, Machines and Drives, 487, 381–386 (2002)

[3] A. Binder, T. Schneider, High-speed inverter-fed AC drives, Int. Aegean Conference on Electrical Machines and Power Electronics, 9–16 (2007)

[4] R.R. Moghaddam, High speed operation of electrical machines, a review on technology, benefits and challenges, IEEE Energy Conversion Congress and Exposition, 5539–5546 (2014)

[5] N. Uzherov, A. Smirnov, C.H. Park, J.H. Ahn, J. Heikkinen, J. Pyrhönen, Design aspects of high-speed electrical machines with active magnetic bearings for compressor applications, *IEEE Trans. on Ind. Electronics* **64**, 8427–8436 (2017)

[6] A. Tessarolo, G. Zocco, C. Tonello, Design and testing of a 45-MW 100-Hz quadruple-star synchronous motor for a liquefied natural gas turbo-compressor drive, IEEE Trans. on Industry Applications **47**, 3, 1210 – 1219 (May/June 2011)

[7] T. Holopainen, P. Jörg, O. Liukonnen, Comparison of two- and four-pole VSD motors up to 4000 RPM, 45th Turbomachinery and 32nd Pump Symposia, Houston, Texas (2016)

[8] T. Mauffrey, J-F. Pradurat, L. Durantay, J. Fontini, Comparison of 5 different squirrel cage rotor designs for large high speed induction motors, PCIC Europe 2013, Istanbul, Turkey (2013)

[9] J.K. Steinke, R. Vuolle, H. Prenner, J. Järvinen, New variable speed drive with proven motor friendly performance for medium voltage motors, IEEE International Electric Machines and Drives Conference IEMDC'99, 235–239 (1999)

[10] IEC 60034-18-42: Rotating electrical machines – Part 18-42: Partial discharge resistant electrical insulation systems (Type II) used in rotating electrical machines fed from voltage converters – Qualification tests (2017)

[11] T. Holopainen, P. Jörg, J. Niiranen, D. Andreo, Electric motors and drives in torsional vibration analysis and design, 40th Turbomachinery Symposium, Houston, Texas (2013)

[12] M. Bruha, Z. Peroutka, Torsional vibration in large variable speed drive systems: Origin and mitigation methods, EPE'15 ECCE-Europe – 17th European Conference on Power Electronics and Applications, Geneve, Switzerland (2015)

[13] M. Bruha, et al., Torsional issues related to variable frequency control of elastic drive systems, IECON 2016 – 37th Annual Conference on IEEE Industrial Electronics Society, Florence, Italy, 2981–2987 (2016)

[14] M. Bruha, *Electro-mechanical interaction between electric drive and its mechanical load and control interventions mitigating unwanted vibration phenomena: PhD thesis* (University of West Bohemia, 2018)

[15] P. Jörg, A. Tresch, M. Bruha, A model based approach to optimize controls of a large LCI VSD for minimal grid-side sub-synchronous torsional interaction, PCIC Europe 2013, Istanbul, Turkey, (2013)

Russian and foreign steam generators for NPP power units with wet steam turbines

Mikle Egorov[1,2,*], *Ivan* Kasatkin[1], *Ivan* Kovalenko[2], *Irina* Krectunova[2], *Nataliya* Lavrovskaya[2], and *Nadezhda* Litvinova[2]

[1]Peter the Great St. Petersburg Polytechnic University, Russia, 195251 Saint-Petersburg, Polytechnicheskaya 29
[2]St. Petersburg State University of Aerospace Instrumentation, Russia, 190000 Saint Petersburg, Bolshaya Morskaya st. 67

Abstract. The main aim of the current study is to analyze advantages and shortcomings of horizontal and vertical types of steam generator design. Design solutions and experience of operation of steam generators of horizontal type accepted in Russia and of vertical type applied by Westinghouse, Combustion Engineering, Siemens, Mitsubishi, Doosan were analyzed within the framework of the present study. It was established that steam generator equipment of horizontal type is characterized by disadvantages of design, technological and operational nature. Thus, horizontal steam generators with dimensions permissible for railroad transportation and, for VVER-1200 with reactor vessel diameter equal to 5 m, by water transport as well, have exhausted the possibilities for further significant increase of the per unit electric power. The demonstrated advantages of vertical-type steam generators are as follows: 1) absence of stagnant zones within the second cooling circuit; 2) uniformity of heat absorption efficiency of the heating surface that ensures improved conditions for moisture separation; 3) increased temperature drop with parameters of generated steam elevated by 0.3 – 0.4 MPa. Conclusion was made on the advisability of introduction of steam generators with vertical-type layout in the Russian nuclear power generation.

1 Introduction

Steam generator is one of the most important elements of a nuclear power installation. Its purpose is to produce water steam – the working medium of the power generation cycle. Steam production is associated with the following complex processes [1, 2]:

– Transfer of heat from heat transfer fluid of the NPP primary cooling circuit to the heated water of the secondary cooling circuit;

– Conversion of thus heated water into water steam;

– Circulation of steam-water mixture;

– Separation of water from steam.

A significant part of emergency shutdowns of NPPs is caused, in particular, by failures of steam generators [3, 4].

Steam generator for VVER-1000 power unit (Figure 1) is a large-sized equipment with vessel length of 14.5 m and diameter up to 4.2 – 4.3 m. Russian NPPs equipped with VVER-1000 reactors have four-loop cooling configuration, i.e. each power unit incorporates four steam generators occupying significant space inside the containment shell of the reactor unit. The larger are the dimensions of the containment shell the higher is its cost. Containment shell must be strong enough, i.e. it must be capable to withstand excess pressure of steam-water mixture in case of an accident accompanied with a failure of pressurization in the primary cooling circuit. The larger is the diameter of the shell which is a pressurized vessel, the more difficult it becomes to ensure its strength.

Comparative analysis of contemporary design of steam generators and earlier design configurations presented, for instance, in [5], demonstrated that during the period of development of "large-scale" nuclear power generation industry beginning from the time of its initiation, design layouts of steam generators remain practically unchanged.

The positive side of such conservatism is associated with the fact that the design solutions supported by several decades of practical application in the process of development of different technical fields including power generation, ship building and chemical technologies, allow obtaining in many cases durable and reliable equipment.

The negative side of such conservatism amounts to obtaining equipment with excessive mass and, consequently, increased metal consumption and, in separate cases, equipment which is not sufficiently reliable when unfavorable processes must be dealt with in the process of operation of such equipment.

* Corresponding author: mikhail.yu.egorov@gmail.com

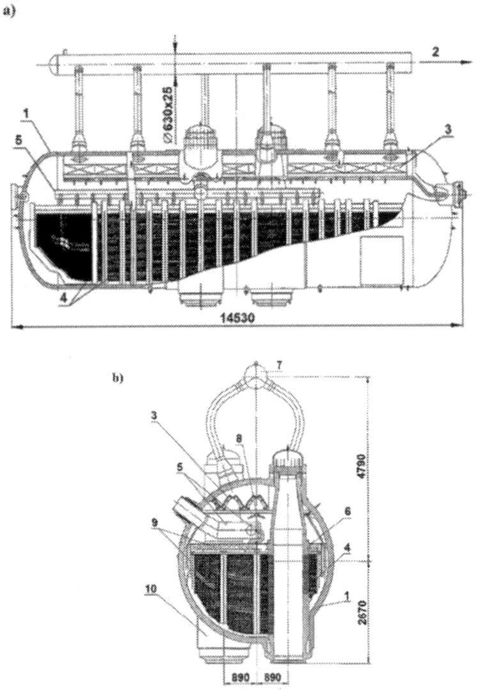

Fig. 1. Horizontal type steam generator for VVER -1000 (Unit 5 Novovoronezh NPP): a) – longitudinal section, b) – cross-section, 1 – vessel; 2 – steam outlet; 3 – chevrons of the separator unit; 4 – heat exchange tube bundle; 5 – supply of feed water; 6 – outlet header; 7 – steam header; 8 – supply of auxiliary feed water; 9 – tube bundle support; 10 – inlet header.

Let us analyze the advantages and drawbacks of design solutions and organization of working processes in steam generators for the purpose of search for the potential for enhancement of per unit power of power units of nuclear power installations equipped with light water nuclear power reactors and enhancement of parameters of the generated steam.

2 Results and Discussion

2.1 Steam generators of horizontal type

Steam generators of horizontal type are used on NPPs with VVER reactors from the moment of initiation of development of nuclear power generation [6].

2.1.1 Advantages of the design

1. The design is relatively straightforward and simple: manufacturing of chevron-type separation devices is simpler as compared with swirl-vane separators.

2. Dimensions allow railroad transportation of the equipment.

3. Absence of thick flat tube sheets. Heating medium headers have cylindrical shape.

The first horizontal steam generators were used in relatively low-power nuclear installations with low parameters of steam which were intended for different applications (not power generation).

Starting from 1964, steam generators of horizontal type were used in power units of the Novovoronezh NPP with per unit electric power of 210, 365 and 440 MW. Operation of this equipment proved to be reliable enough.

2.1.2 Shortcomings of the design revealed in the process of operation of Russian NPPs

Fractures in the header structures took place already during the initial period of operation of VVER-440 power units. In case of fluctuation of water level inevitable on NPP, variable cyclic thermal stresses can develop within the header wall. This problem was successfully resolved by sealing voids between the steam generator partitions protecting the header wall in the region of water level fluctuations from direct contact with water of the secondary cooling circuit. Prior to that, inadequate sealing of the voids resulted in the contact between the relatively cool water with relatively hot header walls.

Horizontal PG-1000 steam generators were introduced during the subsequent period of operation initially of Unit 5 of the Novovoronezh NPP followed by other NPP power units. In the course of transition from power installation with VVER-440 to significantly higher electric power (1000 MW), design of horizontal steam generators reached their limiting thermal hydraulic characteristics.

Besides the above considerations, the external dimensions of steam generators are limited by the requirements of their transportation by railroad transport. Therefore, it was necessary to in-crease specific thermal and steam loads by more than two times and to reduce factual inventories as refers to water level in the secondary cooling circuit and the emergency reserve of this boiler water in the shell in the case of accident of the "feed water supply break" type. These restrictions initiated certain negative consequences – 45 steam generators had to be replaced on the Novovoronezh, South Ukraine, Zaporozhye and Balakovo NPPs [7]. The main cause that prompted the necessity of replacement of steam generators was the development of fractures in the walls of "cold" headers collecting flows of heating medium from heat exchange tubes at the outlet of the equipment. When water levels in the secondary cooling circuit fluctuated in case of transients accompanied with variation of pressure and flow rates of the heat transferring medium, lowering of the level caused by faulty automation or human errors [8], situations developed when tubes in upper levels of the heating surface appeared to be above the level of water, heat transferring medium was not cooled in the tubes and "cold" header was becoming "hot" on certain parts of its surface area. This process was non-stationary because of instability both of the water level, and of the nature of boiling process [9–11] in the vicinity of outer wall of the "cold" header, in connection with which fluctuating thermal stresses developed in the parts of the wall in question.

Designers (OKB "Gidropress") and manufacturers (Machine-Building Plant of Podolsk) of horizontal steam generators undertook significant efforts for enhancing reliability of the equipment including the following:

– Replacement of header materials;

– Increased spacing of partitions between the holes for tubes in the headers within the critical area;

– Rejection of explosive tube expansion, which may cause additional stresses in the material.

Advancement of design of horizontal steam generators produced positive results. Currently, search for design solutions aimed at the enhancement of intensity of heat exchange processes both in the equipment of NPP as a whole and in NPP steam generators [12] is ongoing.

Nevertheless, principal shortcomings of the concept of steam generator of horizontal type and, first of all, impossibility of enhancement of per unit electric power of the equipment in excess of 50 MW, remain. This results in the situation when configurations of reactor facilities with horizontal steam generators requiring significant area under the containment shell of the primary cooling circuit are still used in the design of modern Russian VVER NPPs.

2.2 Steam generators of vertical type

The design concept of steam generators with vertical spatial orientation of vessel for NPP equipped with light water reactors (PWR) is well known. It was implemented on NPPs in different countries beginning from the middle of 1950s by Westinghouse, Combustion Engineering and other companies [13]. Working process in the vertical steam generator is achieved by repetitive natural circulation of steam-water mixture inside the vessel and in the annulus under significant dynamic pressure (due to the height of the equipment) [14–15]. This leads to high rate of circulation of water-steam mixture and to high steam quality in the mixture at the outlet from the bundle of tubes conducting heating medium, i.e. water of the NPP primary cooling circuit. Possibility emerges as the result to achieve in one reactor installation higher electric power equal to at least 500 MW.

Considerable experience of operation of vertical steam generators on foreign NPPs demonstrated both their advantages and shortcomings.

2.2.1 Shortcomings of the design revealed in the process of operation on foreign NPPs

The main disadvantages of design of steam generators of vertical type were manifested in the form of massive damages leading to the development of penetrating fractures in the tubes. The damages took place due to hideout corrosion in the region of flat bottom tube sheets, as well as due to deformation, corrosion and wear in the places where holdups of solid impurities were accumulated in the grids spacing tube bundles. This resulted in significant financial losses due to the replacement of about 300 items of equipment on the operated NPPs [16].

Steam generators of vertical type exist in Russia only as design documents [17]. This obstacle does not allow comprehensive enough and objective comparison of the horizontal and vertical types of the equipment because of the absence of first-hand experience of operation of vertical steam generators.

2.2.2 Advantages of the design

Vertical steam generators, which must be developed taking into consideration the experience of operation on foreign NPPs, including negative experience, possess the following advantages as compared to horizontal steam generators.

1. Improved arrangement of components under the containment shell, which results in the saving of materials and reduction of volumes of construction operations. Four horizontal steam generators installed under the containment shell in the modern VVER-1000 reactor installations occupy four times more area than four and, even more so, than two or three vertical steam generators with higher per unit capacity.

2. Absence of slurry holdups in stagnation zones along the secondary cooling circuit. There are no such zones in vertical steam generators, while they are present in horizontal steam generators in the region of lower tube rows.

3. Possibility to have tubes of the same length and uniformity of open area for penetration of steam-water mixture flow over the cross-section of the equipment due to the bend of platen of heat exchange surface along the involute curve result in the uniformity of heat absorption efficiency of the heating surface and improves operational conditions of the separator.

4. Application of two-stage (chevron and swirl-vane) system for separation of moisture with higher degree of steam drying (it is known that foreign separation devices are used in intermediate separators-superheaters of Russian nuclear power plants [18–19]).

5. "Hot" upper and "cold" lower parts of the vertical cylindrical header of the heating medium do not intersect the level of water in the secondary cooling loop.

6. Design of vertical single-phase pre-boiling steam generator allows organizing the so-called separate section, i.e. the heat exchanger for heating feed water to saturation temperature. This allows increasing average temperature difference in the equipment (Figure 2). Enhancement of parameters of the produced steam by 0.3–0.4 MPa is achieved due to that. It is not possible to organize single-phase pre-boiling section in the horizontal steam generator because feed water is supplied to the upper part of the device and is mixed with boiler water generating steam-water mixture at saturation temperature.

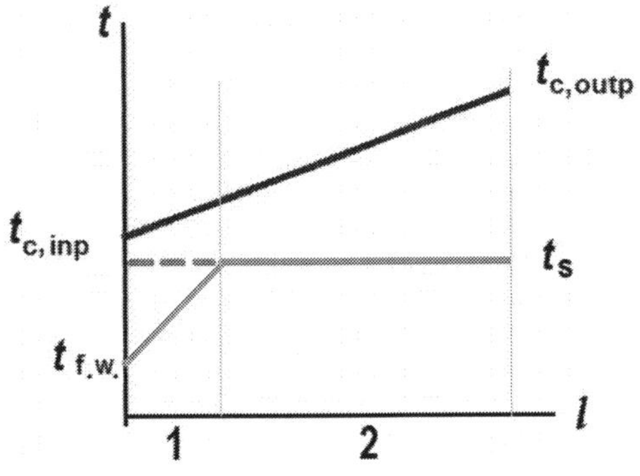

Fig. 2. Layout of distribution of temperatures of the heating medium (coolant) (t_C) and medium of the secondary loop (feed water) ($t_{f.w.}$) in the steam generator: 1 – separate pre-boiling section; 2 – evaporation section; (t_S); - - - – steam generator without separate pre-boiling section.

Because of the thrust height of the circulation loop is much larger than in horizontal steam generators (~10 m versus ~2.5 m), the driving head of the natural circulation also becomes much greater. The velocity of the steam–water mixture and the intensity of its laving of the tube surfaces are also greater, which not only promotes intensification of heat-and-mass transfer in the inter-tubes space but also reduces the likelihood of steaming – stagnation of steam with degradation of heating emission, observed in tight tube bundles, as well as accumulation and settling of products of corrosion on the tubes.

In domestic designs, there is no thick flat, bottom tube board on which sludge can settle and whose materials undergo large temperature stresses because one part of the board is subjected to the hot heat-carrier flowing from the reactor and the other to cold heat-carrier. Since the collector walls are vertical, sludge is not retained on them and is washed off by the intensely circulating steam-water mixture.

The operating temperature conditions of the headers in dynamic regimes change: the evaporation time of the water from the steam generator after the feed water supply stops (both complete evaporation, i.e., emptying, as well as partial evaporation, i.e., to a definite level) is 1.5–1.6 times longer in a vertical steam generator then in a horizontal one. The design of a vertical steam generator is less sensitive to these irregularities, since the changes in the position of the level will be smaller.

2.3 Analysis of operating experience of foreign steam generators

Operational practice indicates that the pipe heating surface and the area of connection of the pipes with the pipe boards are crucial for the reliability of the steam generators [20].

The main events causing breakdowns in the elements of the heat exchange surface of steam generators are denting (due to clogging of the gaps between the tubes and holes in the spacer grids), tube wear, fatigue cracking, fretting (due to vibration), stress-corrosion (due to impurities of the secondary water), pitting, wastage. [21]

Long-term foreign experience in steam generators of nuclear power plants with PWR-type reactors operating has shown that at their design development, at determining the modes of their operation as part of the first and second circuits of nuclear power plants, the following conditions are important [22-24]:

– The material of the heat exchange tubes must be corrosion-resistant in the environment of both circuits;

– The design of the tube spacing devices (grids, strips) should exclude the possibility of accumulation of corrosion products in the secondary circuit and possible vibration wear of the tubes;

– In the second circuit, copper-containing alloys should be completely eliminated as structural materials of the equipment elements of this circuit;

– Since the formation of "sludge" in the second circuit is inevitable, means should be provided for purging volumes ("pockets", slots) where accumulation of "sludge" takes place, as well as the possibility of cleaning these volumes from "sludge" and washing the surfaces of the tube;

– The structural material of the heating coolant collectors should be not only high strength, but also sufficiently ductile (crack-resistant), since during operation of steam generators it is impossible to completely avoid the regime with time-varying loading of the collectors;

Careful observance of the given water-chemical regime of the second circuit must be ensured, and, if possible, in the second circuit should be excluded presence of chlorine ion (for example, together with feed water in case of leaks in the turbine condensers) and oxygen (in the presence of an air suction in these capacitors);

During operation of a nuclear power plant, it is necessary to exclude modes with "dips" in pressure in steam generators, as well as modes with "drops" of a level below the nominal;

During operation of a nuclear power plant, the possibility of water hammer in circuits must be excluded, especially condensation water hammer in the primary circuit, in which during normal operation there are no elastic volumes absorbing the energy of the shock in the reactor loops; the appearance of steam volumes with local boiling of water in the zone of low pressure creates the prerequisites for water hammer when the steam collapses in the zone of higher pressure.

3 Conclusion

The performed analysis of the development of concepts of horizontal and vertical steam generators for NPPs with Russian designed VVER pressurized light water reactors and with foreign PWR reactors allows formulating the following conclusions.

1. Fifty years of experience of operation of hundreds of horizontal and vertical steam generators revealed both positive and negative sides of both types of steam

generator design. Both types of steam generators have deficiencies that are difficult to overcome.

2. A number of shortcomings of design, technological and operational mode nature were revealed in the process of operation of horizontal steam generators. Most part of the revealed drawbacks were removed during modernization of the equipment. Modern horizontal steam generators are characterized by the following non-removable drawbacks:

– Small height and volume of steam space above the evaporation mirror reducing efficiency of gravitational separation and prohibiting installation of swirl-vane separators between the perforated sheet and chevron-type separators thus reducing the capacity of the device [25];

– Complexity of organization in the tube bundle between two vertical headers of separate pre-boiling section cooled down with "cold" feed water;

– Equipment of reactor installation with horizontal steam generators is inferior in terms of compactness as compared with vertical steam generators and requires arrangement of bulky and cost intensive containment shell.

3. Since the design features of horizontal steam generators restrict the potential of enhancement of capacity of the power unit and parameters of generated steam, which negatively affects the financial performance of the reactor installation, development of reliable indigenous vertical steam generator in the nearest future may become one of the priority directions of further development of nuclear machine-building industry.

The following is suggested for the purpose of implementation of vertical steam generators in the Russian nuclear power generation industry based on high-power VVER reactors:

– Conducting testing of large-scale models of vertical steam generators;

– Development of design project and testing the pilot steam generator device in the composition of NPP with VVER-1200 reactors.

Coordination of the above tasks can be performed by the group of experts [26–30] representing design bureaus and research institutes and production facilities manufacturing heat exchange equipment for NPPs.

References

[1] N.G. Rassokhin, *Steam generation equipment of nuclear power plants* (Moscow, Energoatomizdat, 1987)

[2] L.F. Fedorov, V.F. Titov, N.G. Rassokhin, *Steam generators of nuclear power plants* (Saint-Petersburg – Moscow, Energoatomizdat, 1992)

[3] N.B. Trunov, V.V. Denisov, S.A. Kharchenko, B.I. Lukasevich, Thermal Engineering **53**, 1, (2006)

[4] N.B. Trunov, S.B. Ryzhov, S.E. Davidenko, Thermal Engineering **58**, 3 (2011)

[5] B.I. Lukasevich, N.B. Trunov, Yu.G. Dragunov, S.E. Davidenko, *Steam generators of VVER reactors for nuclear power plants* (Moscow, Akademkniga, 2004)

[6] S.M. Dmitriev, D.L. Zverev, O.A. Bykh, Yu.K. Panov, N.M. Sorokin, V.A. Farafonov, *Main equipment of NPP with freestanding thermal neutron reactors* (Moscow, Mashinostroenie, 2013)

[7] N.A. Mahutov, K.V. Frolov, Yu.G. Dragunov, G.S. Vasilchenko, M.M. Gadenin, A.F. Getman, V.P. Gorbatykh, B.N. Dranchenko, V.V. Zatsarinny, G.P. Karzov, B.I. Lukasevich, I.V. Makarenko, L.V. Makarenko, Yu.K. Mikhalev, I.A. Razumovsky, A.N. Salin, A.V. Seleznev, A.V. Fomin, N.V. Shary, *The bearing capacity of steam generators of pressurized water reactors* (Moscow, Nauka, 2003)

[8] V.V. Bazhanov, A.A. Vavilov, S.S. Davydov, P.I. Zalevsky, V.A. Ivanov, Energomashinostroenie **4** (1988)

[9] A.A. Sinitsyn, D.F. Karpov, M.V. Pavlov, *Theory and practice of heat transfer* (Vologda: Vologda State Technical University, 2013)

[10] P.L. Kirillov, A.V. Zhukov, N.I. Loginov, *Reference book on thermohydraulic calculations in nuclear power industry* (Moscow, IzdAT, 2013)

[11] F.F. Tsvetkov, B.A. Grigoriev, *Heat and mass transfer* (Moscow, Moscow Energy Institute, 2011)

[12] Yu.F. Gortyshov, I.A. Popov, V.V. Olimpiev, A.V. Shchelchkov, S.I. Kaskov, *Thermohydraulic efficiency of promising approaches to intensification of heat transfer in the channels of heat transfer equipment Intensification of heat transfer* (Kazan, Center for Innovative Technologies, 2009)

[13] N.I. Kolev, *Multiphase flow dynamics 5: nuclear thermal hydraulics* (Switzerland, Springer International Publishing, 2015)

[14] V. Bezlepkin, I. Kukhtevich, S. Svetlov, S. Alekseev, A. Solodovnikov, A. Molchanov, Y. Krylov, A. Blagoveshchenskij, O. Krektunov, Tyazheloe Mashinostroenie **8** (2004)

[15] A. Blagoveshchenskii, S. Bor, M. Konovich, V. Mityukov, V. Sokolov, B. Shumskii, S. Vygovskii, A. Pinegin, L. Bogachek, V. Bai, N. Ignat'ev, Thermal Engineering **51**, 2 (2004)

[16] N.B. Trunov, S.E. Davidenko, V.V. Denisov, VANT: Obespechenie bezopasnosti AES,Reaktornye ustanovki s VVER **9** (2005)

[17] A.V. Sudakov, V.V. Silin, M.E. Lebedev, B.S. Fokin, B.K. Danilin, M. Ya., *Proc. of the 6th Int. scientific-practical Conf.* (Moscow, Spetskniga Publ., 2012)

[18] V.V. Legkostupova, A.V. Sudakov, St. Petersburg Polytechnic University Journal of Engineering Science and Technology **23**, 3 (2017)

[19] V.V. Legkostupova, A.V. Sudakov, St. Petersburg Polytechnic University Journal of Engineering Science and Technology **4**, 254, (2016)

[20] R. Fournier, M. Thibodeau, C. T. French, Measurement of steam generator or reactor vessel moisture carryover using a non-radioactive chemical tracer, International Conference on Nuclear Engineering **43536** (2009)

[21] B. John, S.P. Dharne, S.G. Ghadge, S.A. Bhardwaj, Evolution of 540 MWth steam generator from 434 MWth (2005)

[22] X. Kong, J. Zhang, Y. Xiao, L. Qian, L. Su, B. Chen, M. Xu, Performance optimization for steam generator level control based on a revised simultaneous perturbation stochastic approximation algorithm, 3rd International Conference on Intelligent Green Building and Smart Grid (IGBSG) IEEE (2018)

[23] S.M. Hoseyni, F. Di Maio, E. Zio, Condition-based probabilistic safety assessment for maintenance decision making regarding a nuclear power plant steam generator undergoing multiple degradation mechanisms, Reliability Engineering & System Safety **191**, 106583 (2019)

[24] G. Yang, V. Pointeau, E. Tevissen, A. Chagnes, A review on clogging of recirculating steam generators in Pressurized-Water Reactors. Progress in Nuclear Energy **97** (2017)

[25] N.B. Trunov, B.I. Lukasevich, D.O. Veselov, Yu.G. Dragunov, Atomic Energy **105**, 3 (2008)

[26] N.D. Agafonova, I.L. Paramonova, Thermal Engineering **60**, 3 (2013)

[27] S. Alekseev, Y. Ilyukhin, V. Kukhtevich, I. Paramonova, S. Svetlov, V. Sidorov, High Temperature **37**, 4 (1999)

[28] N.D. Agafonova, I.L. Paramonova, Thermal Engineering **61**, 8 (2014)

[29] N.D. Agafonova, I.L. Paramonova, Journal of Engineering Physics and Thermophysics **89**, 4 (2016)

[30] M. Gorpinyak, A. Solodov, Thermal Engineering **66**, 6 (2019)

Increasing the efficiency of nuclear power plant equipment at the design stage

Mikle Egorov[1,2,*], *Anastasiya* Ukolova[2], *Ivan* Kovalenko[2], *Irina* Krectunova[2], *Nataliya* Lavrovskaya[2], and *Nadezhda* Litvinova[2]

[1]Peter the Great St. Petersburg Polytechnic University, Russia, 195251 Saint-Petersburg, Polytechnicheskaya 29
[2]St. Petersburg State University of Aerospace Instrumentation, Russia, 190000 Saint Petersburg, Bolshaya Morskaya st. 67

Abstract. It is possible to increase the efficiency of the nuclear power plants equipment in various ways. In particular, one of the most relevant is the active use of computer modeling at different stages of work. The effectiveness the software package used directly affects the quality of the installation equipment. Depending on the stage at which the software package is used, it has various priority properties for the most effective application.

1 Introduction

It is difficult to overestimate the importance nuclear facilities in modern industry. The development of nuclear energy began with the launch on June 27, 1954 in the USSR in Obninsk at the First NPP with a capacity of 5000 kW. Its operation convincingly proved the technical feasibility of converting nuclear energy into electrical energy on an industrial scale [1].

Work with nuclear energy requires the most accurate calculations when designing and working with facilities [2-8], because in the event of an accident the consequences will be catastrophic for people, especially for the economy — the restoration of the surrounding territories and the development of other energy sources after an accident at a nuclear power plant requires huge costs and not always completely possible, which was proved by the experience of the infamous Chernobyl nuclear power plant [9].

It should be noted that at present there is a variety of existing complex technological systems, during the implementation of which the parameters of their functioning and the structure of the technological cycle can change, as well as the complexity of practical tasks that arise when assessing the level of reliability and safety of functioning of potentially dangerous industrial facilities, approaches and new technical decisions in optimizing the structure of production require the development of specific systems at the stage of their design modeling [10, 11].

With the development of computer technology software has also appeared for modeling, designing and evaluating the operation of various nuclear power plants. This significantly increased their efficiency - the automation of many calculations and technological processes allowed minimizing the influence the human factor, as well as accelerating some processes at the stage of modeling and design of installations.

The main life stages of any nuclear power unit (NPU) are design, modeling, analysis work taking into account environmental factors and direct operation. Since the construction of such an installation is an extremely expensive process, at the above stages the installation concept is tested and all its parameters are calculated with high accuracy.

A feature of solving the problems planning and production management is the need to take into account, when solving them, many variables that characterize constantly changing external conditions.

One of the most promising areas of the tasks being solved is simulation, which allows one to obtain qualitative and quantitative estimates of the possible consequences of managed decisions. Simulation methods are the most common means of management theory and operations research in the management of industrial enterprises and organizations.

This is due to the fact that they provide instrumental support for the analysis of functioning in order to improve production and management processes, coordinated and controlled work of all subsystems [12].

2 Materials and methods

Using computer simulation can improve the efficiency of nuclear power plants in the following ways:

– Error reduction through intelligent modeling of piping and equipment elements;

– Reduced costs for the design and testing of both equipment and installations in general;

– Modeling of installations and processes in them, allowing to identify the most vulnerable places and establish ways out of emergency situations.

* Corresponding author: mikhail.yu.egorov@gmail.com

Traditional approaches to the development of production systems and the adoption of design decisions are based, as a rule, on engineering methods for substantiating and calculating the design of technical equipment, their kinematics, the strength of individual components and elements, as well as the technological aspects of the operation of such systems. Such methods do not take into account the high structural and organizational complexity of modern production, which significantly reduces the quality of design and the accuracy of the resulting solutions.

This is unacceptable, given the high cost of implementing such projects. Therefore, a necessary prerequisite for the successful design, research and implementation of technological systems in modern conditions is a strategy in which the main technical decisions at the early stages of development are made, first of all, on the basis of scientific approaches using previous modeling and computer-aided design [13].

Intelligent modeling (for an example see Figure 1) is the most time-consuming stage of work on industrial facilities and equipment. The design of nuclear power plants, in particular, requires the use of non-trivial templates and the consideration of many factors, for which far from all automatic design systems are suitable.

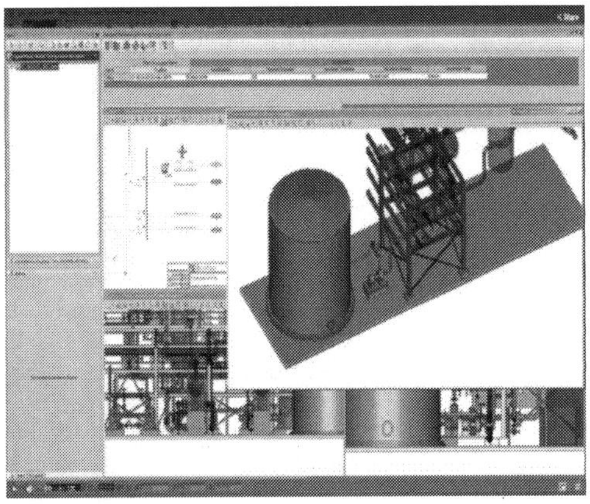

Fig. 1. AXSYS. Engine, CAD for modeling an object and its components.

For example, SolidWorks does not allow modeling complex installations taking into account all physical factors, including many properties of materials. That is why when working with nuclear power plants, software is used to design industrial facilities.

Software tools for the intelligent modeling of industrial facilities are distinguished by large volumes of processed information, in contrast to software systems for the development, for example, of small parts. Also significant differences are powerful graphical tools used when working with three-dimensional models and libraries, which often contain many rare templates and models, which affects the amount of computer memory used by the program.

Modern intelligent modeling programs also allow the analysis of large data sets to also simulate the behavior of equipment in certain conditions, for example, weather or emergency.

Modeling of production processes is carried out by various methods, divided into two classes: analytical and simulation modeling. In this case, simulation models are more optimal, since simulation allows you to work with more complex models of greater dimension, taking into account the greater number of factors influencing the functioning of the object [14].

Similar software tools used when working with nuclear power plants can improve the efficiency of model development, as well as avoid the influence of the human factor at the design stage. However, in the case when the software package is unstable or incomplete (in particular, if the process of connecting the necessary libraries is complicated), the efficiency decreases, which affects the quality of the final equipment.

The main difficulty in creating such a software product can be called ensuring the rational use of computer resources and the stability of the software complex. In modern programming, this is solved through quality testing of the software product and its components, as well as regular code refactoring.

Depending on the required level of abstraction to simulate production systems, various paradigms of simulation modeling can be applied. For example, at the level of global interconnections, an approach to the study of system dynamics is used. To simulate production processes, we apply an approach that explores specific processes separately. Using a mixed approach allows us to study the system at several levels of abstraction [15].

Conceptual design is a mandatory stage in the development of any installation and equipment. The most effective software is one that simplifies work according to standards and allows the use of more libraries with components, Figure 2.

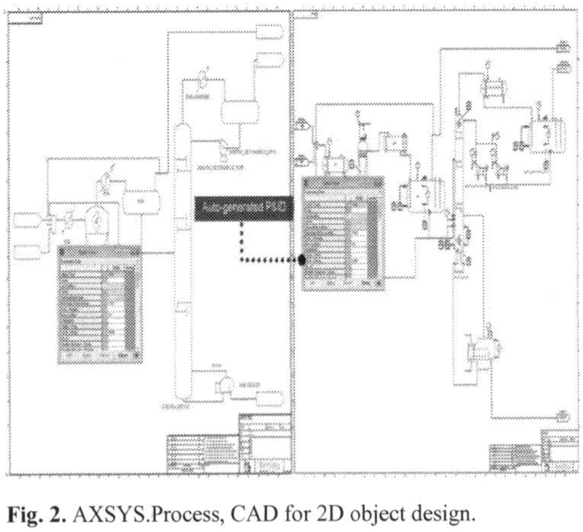

Fig. 2. AXSYS.Process, CAD for 2D object design.

The main purpose of using such programs is to create drawings according to various standards, which requires providing the user with a comfortable interface and easy access to libraries.

A well-constructed model allows you to reduce risks during the operation of the installation, and computer simulation significantly reduces the influence of the human factor.

3 Results and Discussion

One of the most important aspects of the operation of a nuclear power unit is a probabilistic safety analysis, taking into account the probability of failure of technological equipment. This avoids many malfunctions, which significantly affects the overall efficiency of the installation.

When it comes to the nuclear industry, computer simulation is imperative, since real experiments are impossible. The development and improvement of the accuracy of computer simulation programs makes it possible to improve the quality of real installations.

Realistic assessment software is most relevant for nuclear power plants - it allows you to assess the vulnerability of the entire installation or equipment and to design emergency escape routes

This software works with accurate calculations and constantly updated data, but does not require powerful hardware and is fast enough.

It can be assumed that currently relevant areas of information technology Big Data and Data Science will find application, including in realistic assessment programs [16].

Figure 3 shows a comparative diagram of the average requirements of software systems to the hardware of a computer, depending on their area of application.

The diagram shows that the programs for intelligent modeling are most effective on powerful computers, which is primarily due to graphical tools and a large number of libraries used.

These programs, however, do not require a high processor frequency, since they process a relatively small amount of data.

The vast majority of modern software systems for simulation use a graphical interface and a graphical representation of the simulation model. This allows you to visually assess the functioning of the relationships between the objects of the simulation model. In addition, the visual graphical interface simplifies the work with a complex of simulation modeling, which allows even a poorly trained person to develop a simulation model. This is especially important when creating a model by a person who has extensive practical experience working with a real system but does not have modeling experience [17].

Conceptual design requires significantly less hardware resources, however, for the correct and effective work it is necessary to correctly organize the interaction of the template libraries, as well as their compliance with the standards of graphic symbols [18, 19]. In this case, it is important to update the software products when a new GOST or ISO standards are released.

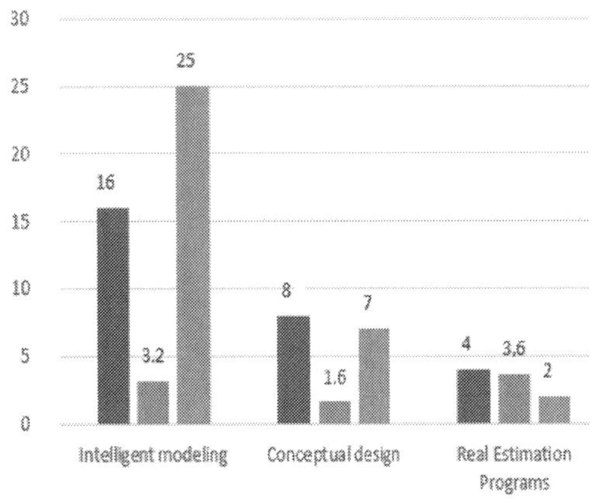

Fig. 3. Diagram of requirements for the hardware of a computer for the operation of various types of software systems: (blue – average power requirements (RAM, GB); orange – average speed requirements (processor frequency, GHz; gray – average memory requirements (hard drive, GB)).

Realistic Evaluation Programs, as it might seem at first glance, have the lowest hardware requirements on average. However, if you pay attention, the most demanding on the processor frequency, due to the need to process a large amount of data in a minimum amount of time. For example, this can be data for the installation itself, in particular, the dynamics of its various indicators in normal and emergency operation modes, and data on external factors - weather conditions, the condition of the rest of the equipment, and others [20-22].

The main goal of realistic assessment programs is to evaluate the efficiency and safety of the installation and equipment both in normal operation and in emergency mode, taking into account all external and internal factors. In such programs, most often, graphical tools are not used and sets of numbers act as output or just brief answers about whether the results are within normal values.

Nevertheless, in the development of realistic assessment programs, the most optimized mathematical and physical calculations play an important role, which allows minimizing or eliminating the risks at all during the operation of the analyzed product [23, 24].

Determining the priority sides of software systems is an extremely important stage of their development when working with industrial nuclear installations, since this is precisely the area where risks of incorrect execution of instructions by a computer are enormous. Good-planned development of software products in the field of nuclear energy is ensured, first of all, due to the high-quality technical specifications and the availability of at least a superficial understanding of the scope of work.

This is necessary because, taking into account the constraints imposed by the economy and business, development is often limited in time and resources, which often leads to a poor-quality product, which, subsequently, reduces the efficiency of plants and equipment, and in some cases seriously increases risks

https://doi.org/10.1051/e3sconf/202017801008

when working with nuclear power on specific equipment.

Depending on the design stage of the nuclear power plant, for maximum efficiency, the software should have some of the following properties:

– Extensive library of objects;

– Powerful graphical tool for visualizing models;

– Ability to work with databases;

– Ability to conduct highly accurate calculations with constant updating of input data;

– Strict adherence to equipment design standards.

4 Conclusion

1. Computer modeling, in particular, intellectual and conceptual modeling, is quite common at the stage of development of directly installations and equipment. Realistic assessment programs, in turn, must be applied with a ready-made model of the entire enterprise using the developed nuclear power plants.

2. Modern nuclear power can't go without partial or full automation at any stage of work.

3. The development of computer simulation is the safest and most promising way to improve the efficiency of equipment and nuclear power plants in general.

References

[1] B.G. Ganchev, L.L. Kalishevsky, R.S. Lemeshev, E.B. Kolosov, L.A. Kuznetsov, N.F. Rekshnia, S.V. Selikhovkin, *Nuclear Power Plants* (Energoatomizdat, Moscow, 1990)

[2] G. Li, Ch. Yan, J. Wang, Annals of Nuclear Energy **73** (2014)

[3] M. Le, J. Kim, J. Kim, H. Do, J. Lee, Nondestructive Testing and Evaluation **33**, 1 (2018)

[4] R. Rafee, H. Rahimzadeh, Iranian Journal of Chemistry and Chemical Engineering **29**, 3 (2010)

[5] E. Narimani, S. Shahhoseini, Applied Thermal Engineering **31**, 2 (2011)

[6] H. Zhang, Q. Liu, B. Qin, H. Bo, Annals of Nuclear Energy **75** (2015)

[7] X. Wang, P. Cheng, J. Xie, Z. Zhou, Journal of Mechanical Science and Technology **29**, 8 (2015)

[8] F. Kavousi, Ya. Behjat, Sh. Shahhosseini, Chemical Engineering Research and Design **91** (2013)

[9] D.S. Pashkevich, E.D. Fedorovich, V.V. Kapustin, Atomic Energy **123**, 6 (2018)

[10] V.S. Smorodin, Mathematical machines and systems **2** (2012)

[11] V.S. Smorodin, A.V. Klimenko, E.I. Sukach, O.A. Shimchik, Problems of Physics, Mathematics and Technics **1**, 26 (2016)

[12] T.V. Mikheeva, Izvestia Altai State University **1**, 61 (2009)

[13] A.R. Rott, S.Ya. Alibekov, A.V. Maryashev, A.V. Medvedev, R.S. Salmanov, Bulletin of Kazan Technological University **1** (2014)

[14] V.G. Martynov, Dynamics of systems, mechanisms and machines **2** (2014)

[15] S.S. Zhuravlev, Problems of Informatics **3** (2009)

[16] *The Big Data Phenomenon* (Age of Quality, 2014)

[17] M.I. Chizhov, Yu.S. Skripchenko, P.Yu. Gusev, Computer Research and Modeling **6**, 2 (2014)

[18] K.N. Malovik, East European Journal of Advanced Technologies **1** (2012)

[19] V.K. Semenov, M.A. Wolman, Global Nuclear Safety **4** (2012)

[20] S.F. Sandu, Geo-Resource Engineering **2** (2003)

[21] A.R. Belozerova, B.F. Melnikov, Izvestiya Vysshikh Uchebnykh Zavedenii. Physics and mathematics **1** (2014)

[22] N.N. Chernyshov, Radioelectronics and Informatics **3** (2013)

[23] I.I. Ovchinnikov, I.G. Ovchinnikov, M.Yu. Bogina, A.V. Matora, Bulletin of Eurasian Science **4** (2013)

[24] Yu.I. Dimitrienko, Yu.V. Yurin, S.V. Europine, News of higher educational institutions, Engineering **6** (2013)

E3S Web of Conferences 178, 01009 (2020) https://doi.org/10.1051/e3sconf/202017801009
HSTED-2020

Hybrid power quality conditioner for three-phase four-wire power systems

Maxim Chernyshov[1], *Valery* Dovgun[1], *Sergei* Temerbaev[1,*], and Zumeyra Shakurova[2]

[1]Siberian Federal University, 79 Svobodny pr., Krasnoyarsk, 660041, Russia
[2]Kazan State Power Engineering University,51 Krasnoselskaya street, Kazan, 420066, Republic of Tatarstan, Russia

Abstract. The article considers a hybrid power quality conditioner (HQPC) for 3-phase 4-wire systems with a distributed modular structure. Some conditioner modules provide compensation for the component currents and voltages that form the negative and zero sequence systems. The open structure of the HQPC, consisting of independent modules, allows compensating for distortions of currents and voltages of the 3-phase network caused by the nonlinear nature and asymmetry of single-phase loads. The compensation characteristics of the proposed conditioner were researched using a model developed in the MatLab environment. The simulation showed that the proposed conditioner can ensure normalization of power quality in 3-phase 4-wire system at various modes of network operation.

1 Introduction

The issue of ensuring the power quality (PQ) in 3-phase 4-wire systems has become particularly important in recent years. The reason is the continuous increase of non-linear loads, such as energy-saving lighting systems, office and computer equipment.

Loads in low-voltage systems have a number of features that differ them from non-linear loads of industrial consumers. In contrast to industrial networks, where high-power loads are concentrated, office and commercial consumers are characterized by low-power single-phase non-linear loads, which current spectrum contains low-frequency 3, 5, and 7th harmonics predominate. Harmonics that are multiples of three in 3-phase networks form a zero-sequence system and are summed in neutral conductors.

Another feature of low-voltage four-wire network is the unequal distribution of single-phase loads. This causes phase current asymmetry, including higher harmonics asymmetry [1].

The widespread use of nonlinear single-phase loads has led to the fact that the asymmetry and non-sinusoidal nature of currents and voltages in 3-phase 4-wire systems have become permanent factors that reduce the efficiency of power supply systems [1, 2]. It should also be taken into account that many consumers in such systems have increased requirements for power quality [3, 4]. Asymmetry and non-sinusoidal nature of currents and voltages can lead to failures of control systems, failures in information transmission and processing systems. Therefore, the problem of creating filter-compensating devices for 3-phase 4-wire systems reducing the influence of the nonlinear nature and asymmetry of single-phase loads requires the new approaches.

Compensating devices for 3-phase 4-wire power systems are considered in [1-9]. In many cases, passive and hybrid compensating devices for 3-phase 4-wire systems are implemented on the basis of transformers with windings connected according to the "zigzag" scheme [1, 5-7]. However, the zigzag transformer also has a low impedance for zero-sequence components created by the external network. This can cause a significant increase the current in the neutral conductor if the network voltage is not symmetrical [5, 6].

The general task of PQ normalizing in 3-phase 4-wire systems can be solved with the help of unified power quality conditioner (UPQC), proposed for the first time in [8]. The UPQC considered in [8, 9] contains both series and parallel active filters. The disadvantage of UPQC is the high cost of the device. In many cases, a structure that includes both a serial and parallel active filter may be redundant

A more economical version of UPQC for 3-phase 4-wire systems is proposed in the article [7]. It consists of two parallel hybrid filters that provide separate compensation for the symmetric components of the negative and zero sequence. The circuit solutions proposed in [7] allow reducing the power of active filters and thereby reducing the cost of the device.

Power quality conditioners, considered in [7 – 9], allow to solve the problems caused by harmonic distortion and unbalanced (asymmetric) load. However, these devices are designed to regulate PQ in network nodes that are connected to a concentrated non-linear load. For 3-phase 4-wire systems, in which the main share of the distorting load is made up of 1-phase low

* Corresponding author: s.a.temerbaev@gmail.com

 https://doi.org/10.1051/e3sconf/202017801009

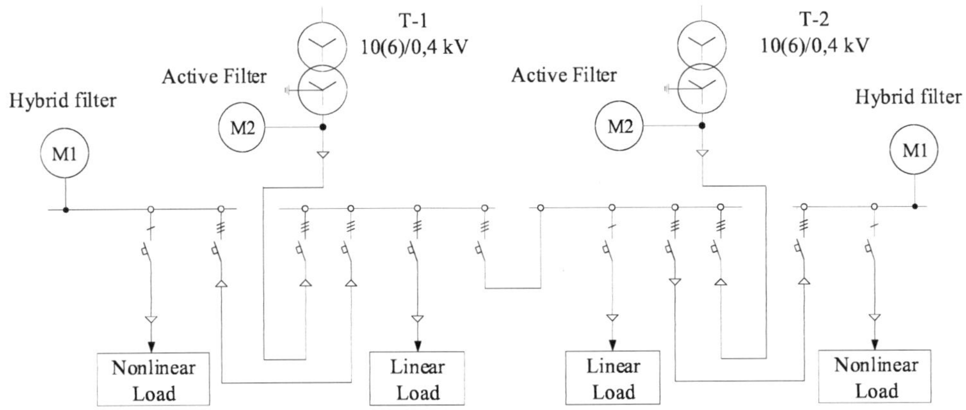

Fig. 1. Structure of the HPQC.

power receivers, it is necessary to use distributed power quality conditioners.

The article considers a hybrid power quality conditioner (HPQC) for 3-phase 4-wire systems with a distributed modular structure. The conditioner modules are installed in the network nodes to which a significant part of the nonlinear load is connected, and provide compensation for the currents and voltages components that form symmetrical systems of negative or zero sequence. A digital system for generating control signals for active filters that are part of the conditioner is considered. A model in the MatLab (Simulink) environment was developed to study the characteristics of HPQC.

2 Configuration of a HPQC

The structure of the proposed HPQC is shown in Fig. 1. The conditioner contains modules of filter compensating devices of two levels. First-level modules (M1), as shown in Fig.1, carry out the attenuation of the component currents and voltages that form a symmetrical system of zero sequence.

Since the majority of single-phase non-linear loads that generate zero-sequence currents are connected to network nodes inside the building, it is advisable to place M1 modules in such nodes.

Second-level modules (M2), as shown in Fig. 2, designed to compensate for components that form a symmetrical negative sequence system. They are installed in nodes that are connected to a load sensitive to reduced PQ, or they are connected to the secondary winding of a step-down transformer. This will reduce the impact of nonsymmetric and non-linear loads on the operating modes of consumers with high requirements for the PQ.

Let's consider the configuration of compensating devices that are part of the conditioner. The M1 module is a hybrid filter formed by a serial connection of a passive filter and a single-phase voltage inverter as shown in Fig. 2. The passive filter can be realized either in the form of resonant circuit tuned to the frequency of the dominant third harmonic, or in the form of a broadband passive filter that has low resistance at higher harmonic frequencies ($k=3,5,\dots$).

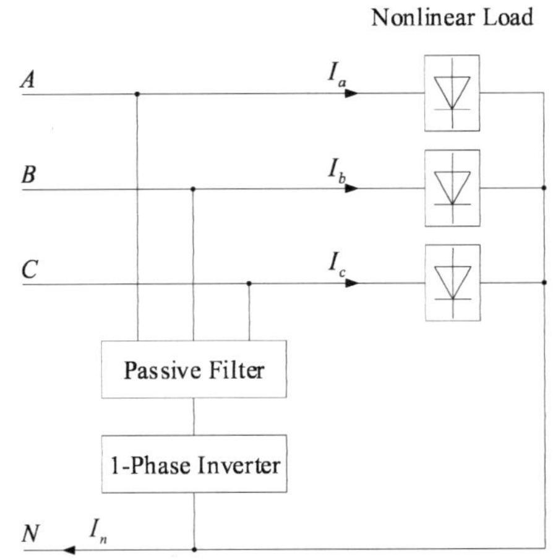

Fig. 2. Hybrid filter structure to reduce neutral wire current.

Single-phase inverter in the diagram in Fig. 2 act as an active filter. The inverter voltage is proportional to the harmonic components of the neutral conductor current:

$$U_{AF} = R_{AF1} I_n^{(1)} + R_{AF2} I_n^{(h)} \quad (1)$$

R_{AF1}, R_{AF2} are the control parameters of the active filter; $I_n^{(1)}$ is a component of the neutral conductor current with the fundamental frequency; $I_n^{(h)}$ are high-frequency components of the neutral current.

It was shown in [10] that at the frequency of the k-th harmonic, the current of the neutral conductor in the circuit as in Fig. 2.

$$\dot{I}_n = \frac{3E_{0C}}{Z_C + Z_{PF} + 3(Z_N + R_{AF})} - \frac{Z_{PF} \dot{J}_{0L}}{Z_C + Z_{PF} + 3(Z_N + R_{AF})} \quad (2)$$

Voltage between neutral points

$$\dot{U}_{Nn} = \frac{3Z_N E_{0C}}{Z_C + Z_{PF} + 3(Z_N + R_{AF})} - \frac{Z_N Z_{PF} \dot{J}_{0L}}{Z_C + Z_{PF} + 3(Z_N + R_{AF})} \quad (3)$$

E_{0C} is a component of the power system voltage that forms a symmetric zero-sequence system; J_{0L} is zero sequence component of load current; Z_C is the power system impedance; Z_{PF} is the passive filter impedance; Z_N is the impedance of the neutral wire.

From the formulas (1, 2), it follows that the action of the active filter is equivalent to turning on the resistance in series with the impedance of the neutral wire. This reduces the current in the neutral conductor and reduces the voltage between the neutral points. The hybrid filter also reduces the components of the neutral conductor current caused by the asymmetry and non-sinusoidal form of the external network voltage.

The M2 module is designed to compensate for the components that make up the negative sequence system. A parallel or serial hybrid filter can implement it. The M2 module in the form of a parallel hybrid filter is shown in Fig. 3.

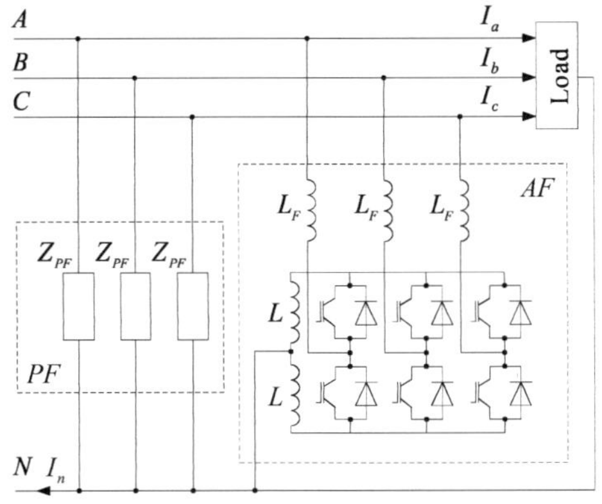

Fig. 3. Structure of a 3-phase hybrid filter.

The passive filter is tuned to the fifth harmonic. The active filter is implemented by a 3-phase 3-wire current inverter controlled by the voltage at the module connection point.

Active filter current J_{AF} in the diagram in Fig. 3 is proportional to the voltage component at the point of common coupling, which forms the negative sequence system: $J_{AF} = G_{AF} U_{PCC}^{(2)}$. Voltage at the connection point of the nonlinear load and the hybrid filter can be expressed as

$$U_{PCC}^{(2)} = \frac{Y_C E_C^{(2)}}{Z_C + Y_{PF} + G_{\text{аф}}} - \frac{J_L^{(2)}}{Z_C + Y_{PF} + G_{AF}} \quad (4)$$

$E_C^{(2)}$ and $J_L^{(2)}$ are the components of voltage and current of the load forming system of negative sequence; Y_{PF} is the conductivity of passive filter; G_{AF} is the parameter of active filter, which has the conductivity dimension.

In accordance with (3), the active filter reduces the components of the negative sequence voltage at the point of common coupling (PCC) created by both the load and the external network. The reduction degree can be adjusted by changing the parameter of the active filter G_{AF}.

The proposed distributed structure of the HPQC is open and easy to modify, including new modules. Failure of one of the modules will not affect the efficiency of the conditioner in other nodes of the network. The distributed nature of the compensating system allows reducing the power of active filters. This reduces the cost of the system and increases its reliability.

3 Digital system for generating control signals for active filters

The proposed HPQC can be used as a multifunctional device that provides suppression of neutral conductor currents, compensation of voltage asymmetry in network nodes, and so on. Its characteristics depend on the methods used to generate control signals for active filters. In this case, it is advisable to use a strategy for generating control signals based on the use of instantaneous values of symmetrical components of voltages and currents of a 3-phase network [11-16].

The digital system for generating control signals for active filters consists of two parts. The first part (shown in Fig. 4) provides control of a 1-phase inverter that suppresses the current of the neutral conductor. It includes a frequency separation digital filter (FSDF), which generates signals proportional to the spectral components of the neutral current having the frequency of the main harmonic $i_N^{(1)}$ and high-frequency harmonics $i_N^{(h)}$, as well as a block for generating control signals for a 1-phase inverter.

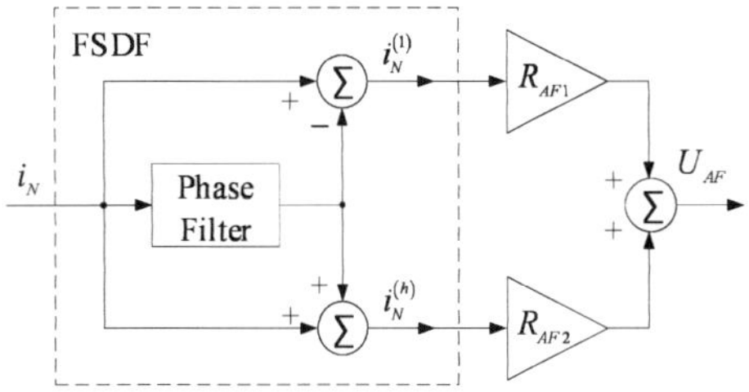

Fig. 4. Digital control signal generation system for 1-phase active filter.

The second-order phase filter in Fig. 4 has a transfer function whose module is equal to 1 in the entire frequency range. FSDF implements simultaneously the transfer functions of the band pass filter (BPF) and the band separating filter (BSF):

$$H_{BSF}(z)=\frac{i_N^{(h)}}{i_N}=\frac{1}{2}\left[1+A(z)\right]$$
$$H_{BPF}(z)=\frac{i_N^{(1)}}{i_N}=\frac{1}{2}\left[1-A(z)\right] \quad (5)$$

$A(z)$ is a transfer function of the phase filter.

The purpose of a 3-phase hybrid filter is to obtain fundamental phase voltages of the 3-phase system at the PCC, forming a positive sequence system. The remaining components of the phase stresses at the PCC caused by the asymmetry and non-linear nature of the loads must be reduced. To do this, in accordance with (3), the inverter voltages must be proportional to the components of the phase voltages to be compensated.

The block diagram of the 3-phase active filter control signal generation system is shown in Fig.5. Signals proportional to the line voltage of the network at the filter connection point are fed to the inputs of bandpass filters.

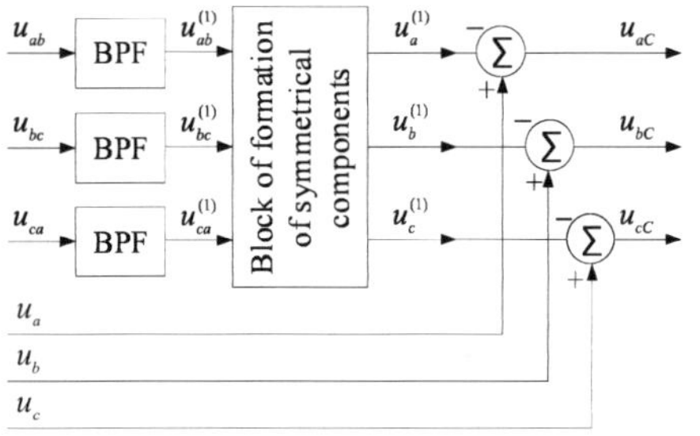

Fig. 5. Block diagram of the control signal generation system for a 3-phase active filter.

After selecting the components with the frequency of the main harmonic, the signals are fed to the input of the module for forming symmetrical components. The output signals of the module are the main frequency voltages that form a symmetrical system of direct sequence.

3-phase active filter control signals are defined by expressions:

$$i_a=\frac{P}{(u_{a1}^{(1)})^2+(u_{b1}^{(1)})^2+(u_{c1}^{(1)})^2}(u_a-(u_{a1}^{(1)}-u_0)) \quad (6)$$

$$i_b=\frac{P}{(u_{a1}^{(1)})^2+(u_{b1}^{(1)})^2+(u_{c1}^{(1)})^2}(u_b-(u_{b1}^{(1)}-u_0)) \quad (7)$$

$$i_c=\frac{P}{(u_{a1}^{(1)})^2+(u_{b1}^{(1)})^2+(u_{c1}^{(1)})^2}(u_c-(u_{c1}^{(1)}-u_0)) \quad (8)$$

where $u_0=\frac{1}{3}(u_a+u_b+u_c)$, $P=u_a i_a+u_b i_b+u_c i_c$; $u_{a1}^{(1)}$, $u_{b1}^{(1)}$, $u_{c1}^{(1)}$ are the components of phase voltages of the main frequency that form a positive sequence system.

4 HPQC modelling

To study the characteristics of the proposed HPQC, a model was developed in the MatLab (Simulink) environment (Fig. 6). The M1 module is implemented with a passive broadband filter and a 1-phase inverter. A 3-phase active filter is used as the M2 module.

The model parameters are the following: the resistance of the network R_C = 0.4 Ohms, inductance L_C = 0.318 mH.

Three 1-phase bridge rectifiers represent symmetric nonlinear load in the diagram in Fig. 6. Its parameters

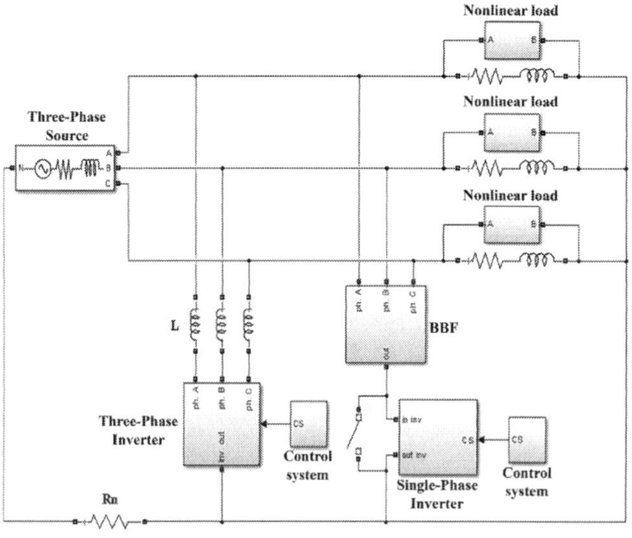

Fig. 6. MatLab Simulink model of HPQC.

Table 1. Simulation results.

Mode	I_a, A	I_b, A	I_c, A	I_n, A	K_{2U}, %	K_{UA}, %	K_{UB}, %	K_{UC}, %
Without filter-compensating devices	29,71	16,5	16,9	17,9	5,3	9,5	8,4	8,4
With a zero sequence filter	30,7	17,2	17,4	8,2	5,3	4,23	4,39	4,51
Zero sequence filter + negative sequence filter	32,66	17,65	17,83	9,1	1,1	1,91	1,87	1,85

are the following: resistance R_L = 40 Ohms, smoothing filter capacity C = 937.5 μF.

Parameters of an asymmetric linear load in phase A, the load resistance R_L = 5 Ohms, inductance L_L = 15.9 mH, in phases B and C, the load resistance R_L = 15 Ohms, inductance L_L = 47.8 mH.

The results of the simulation are presented in Table 1.

The simulation results show that after switching on the conditioner, the current of the neutral wire has reduced almost twice. The voltage asymmetry coefficient for the negative sequence K_{2U} has reduced by more than three times and is within the norm defined by GOST standard 32144-2013. Distortion of phase currents and voltages has also been significantly reduced.

5 Conclusion

The article considers a HQPC for 3-phase 4-wire systems that provides compensation for current and voltage distortions caused by the nonlinear nature and asymmetry of single-phase loads. There are proposed configurations of modules that compensate the components forming symmetric systems of negative and zero sequence.

A mathematical model of the proposed HQPC in the MatLab environment is developed. The model is used to study the compensatory characteristics of the HQPC. Analysis of the simulation results showed that the use of the proposed device can normalize the power quality in 3-phase 4-wire systems under different operating modes. With a physical model it is planned to conduct a detailed analysis of the capabilities of the conditioner.

References

[1] A.K. Shidlovsky, A.F. Zharkin, *Higher Harmonics in Low Voltage Electrical Networks* (210, 2005)

[2] T. Key, J-S Lai, Analysis of harmonic mitigation methods for building wiring systems, IEEE Transactions on Power Systems **13**, 3, 890 (1998)

[3] J. Pomilio, S. Deckmann, Characterization and compensation of harmonics and reactive power of residential and commercial loads, IEEE Transactions on Power Delivery **22**, 2, 1049 (2007)

[4] H. Hafezi, G. D'Antona, A. Dede, D.D. Giustina, R. Faranda, G. Massa, Power quality conditioning in LV distribution networks: results by field demonstration, IEEE Transactions on Smart Grid, **8**, 1, 418 (2017)

[5] H.L. Jou, J.C. Wu, K.D. Wu, W.J. Chiang, Y.H. Chen, Analysis of zig-zag transformer applying in the three-phase four-wire distribution power system, IEEE Transactions on Power Delivery **20**, 2, 1168 (2005)

[6] B. Singh, P. Jayaprakash, T. Somayajulu, D. Kothari, Reduced rating VSC with zig-zag transformer for current compensation in a three-phase four-wire distribution system, IEEE Transactions on Power Delivery **24**, 1, 249 (2009)

[7] J.C. Wu, H.L. Jou, H.H. Hsiao, S.T. Xiao, A new hybrid power conditioner for suppressing harmonics and neutral-line current in three-phase four-wire distribution power systems, IEEE Transactions on Power Delivery **29**, 4, 1525 (2014)

[8] H. Fujita, H. Akagi, The unified power quality conditioner: The integration of series- and shunt-active filters, IEEE Transactions on Power Electronics **13**, 2, 315 (1998)

[9] V. Khadkikar, Enhancing power quality using UPQC: a comprehensive overview, IEEE Transactions on Power Electronics **27**, 5, 2284 (2012)

[10] V.P. Dovgun, M.O. Chernyshov, O.E. Maloshenok, Hybrid power filters for three-phase four-wire networks, News of Higher Educational Institutions, Energy issues 1-2, 11 (2016)

[11] K. Uyyuru, M. Mishra, A. Ghosh, An optimization-based algorithm for shunt active filter under distorted supply voltages, IEEE Transactions on Power Electronics **24**, 5, 1223 (2009)

[12] E.I. Gracheva, O.V. Naumov, Evaluation criteria of contact group technical state concerning electrical appliances, International Journal of pharmacy and Technology **4**, 27084–27091 (2016)

[13] D.D. Micu, I.V. Ivshin, E.I. Gracheva, O.V. Naumov, A.N. Gorlov, Influence of resistance of contact units of switching devices on the electric power losses in shop networks of low tension, SES-2019, E3S Web of Conferences **124**, 02013, 1–4 (2019)

[14] E.I. Gracheva, V.N. Petrov, N.V. Russova, D.V. Samuilov, G.P. Svintsov, I.V. Protosovitski, On modeling on a single-winding forced valve

electromagnet in a circuit with a ballast resistor, SES-2019, E3S Web of Conferences **124**, 02012, 5–9 (2019).

[15] E. Gracheva, A. Alimova, Calculation Methods and Comparative Analysis of Losses of Active and Electric Energy in Low Voltage Devices, 2019 International Ural Conference on Electrical Power Engineering (UralCon), Chelyabinsk, Russia, 361–367 (2019)

[16] A.N. Egorov, Ya.S. Kharitonov, V.M. Khubieva, I.A. Yakushev, N.V. Golubtsov, Z.M. Shakurova, The impact of higher voltage and current harmonics on the operation of electrical systems, SES-2019, E3S Web of Conferences **124**, 05067, (2019)

IMPROVING ENERGY AND ENVIRONMENTAL EFFICIENCY OF COMBINED HEAT-AND POWER PLANT BASED ON ABSORPTION HEAT TRANSFORMERS

Yu. V. Yavorovsky[1], *A. I.* Bartenev[1], *I. A.* Sultanguzin[1,*], *A Sh* Alimgazin[2], *S. A.* Prishchepova[1], and *I. D.* Kalyakin[1]

[1]Moscow Power Engineering Institute, Moscow, Russia
[2]Eurasian National University, Nur-Sultan, Republic Kazakhstan

Abstract. In the present paper the scheme for increasing the energy efficiency of combined heat and power plants (CHPP) using absorption heat transformers (AHT) is considered. The aim of the study is to increase the energy and environmental efficiency of natural gas use in power supply systems by application steam turbine plants and absorption heat transformers. The simulation of CHPP modernization and a harmful emissions dispersion assessment were carried out using the following software tools: ISC Manager, Thermoflex. All the calculations were made for one power unit of the CHPP in Moscow. Subsequently, the obtained data on energy efficiency increase and harmful emissions reduction were multiplied on the whole power system of Moscow, the main source of heat and electricity of which are the CHPP's of PJSC "Mosenergo".

1 Relevance of the topic

Energy-efficient and environmentally friendly production of heat and electric energy is the key to the development of society and improving people's living standards.

Depreciation of the main equipment of the energy sector is the main reason for the increased costs of generating energy resources, as well as an increase in harmful emissions.

The development of CHPP schemes to reduce fuel consumption and the amount of emissions of harmful substances is a promising direction in the development of heat and electricity.

In world practice, there are examples of the use of absorption heat transformers at thermal power plants. These are coal stations in China, none of which can now be built without using AHT on them. China is currently a leader in the production of absorption heat transformers and a pioneer in the use of AHT at thermal power plants.

The main studies of the use of AHT at CHP plants were also conducted in China [1,2,3,4,5]. Unfortunately, these articles do not provide specific values for the magnitude of the energy effect from the use of AHT at CHP plants. In addition, there is no assessment of the environmental effect of reducing harmful emissions from thermal power plants.

It should be noted that articles [1-5] are based on data on coal-fired thermal power plants that are used in China, and there are practically no studies on the use of AHT on gas-fired thermal power plants specific to our region.

In this regard, modeling the modernization of a real thermal power plant to obtain reliable data on the energy and environmental effect of such modernization is a paramount task for the energy-efficient and environmentally friendly development of the fuel and energy complex (FEC) of Moscow, in particular, and Russia as a whole.

As a first step to calculate the effect of the modernization of the CHPP, an analysis of the energy system of Moscow was carried out.

2 Analysis of the energy system of Moscow

To select the object of study, the heat and power supply system of Moscow was considered [6]. It was revealed that the main energy sources are Mosenergo CHPPs. Figures 1 and 2 show the distribution of the installed electric and heat capacities of the CHPPs in Moscow.

* Corresponding author: SultanguzinIA@mpei.ru

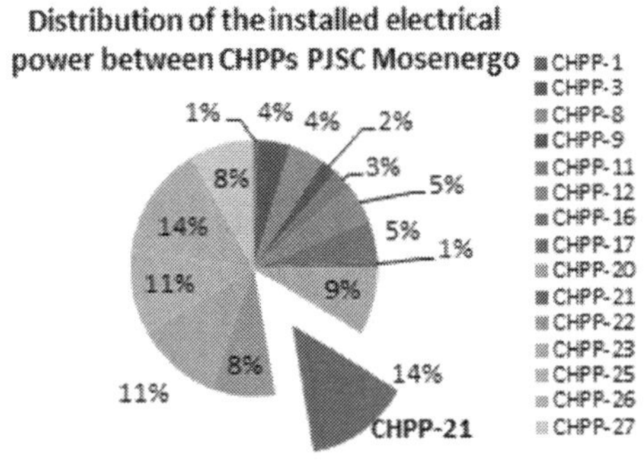

Fig. 1. Distribution of the installed electrical power between CHPPs PJSC Mosenergo.

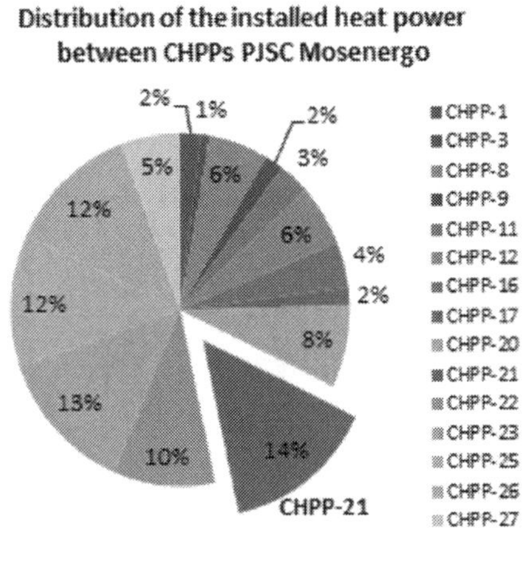

Fig. 2. Distribution of the installed heat power between CHPPs PJSC Mosenergo.

As can be seen from the diagram presented in Figure 1, CHPP-21, along with CHPP-26, is one of the largest in terms of installed electric capacity. The diagram in Figure 2 demonstrates that the installed heat capacity of CHPP-21 is the largest. This makes it a suitable object for modeling modernization, since installing an AHT on it, presumably, can lead to a greater energy and environmental effect.

Thermal Power Station №21 is a typical heat and power plant for Moscow, since its equipment includes power units with a pressure of 130 atm, more modern ones with a pressure of 240 atm, as well as a combined cycle gas turbine unit commissioned in 2008. CHPP-21 is located in the Northern Administrative District of Moscow. It supplies thermal energy to residents of the Northern, North-Eastern and Central administrative districts of Moscow, as well as Khimki, Moscow Region.

Along with the energy effect, an important factor determining the efficiency of modernization of thermal power plants using AHT is the reduction of harmful emissions as a result of modernization. In the work, the calculation of the dispersion of emissions from the CHPP before and after modernization was performed.

3 Polluting emissions

To assess the impact of emissions of harmful substances on the population, the Impact Pathways methodology is used [7, 8]. The basis of this methodology is the implementation of the following steps [9]:

- Source selection (emission determination)
- Calculation of dispersion of harmful emissions (increase in concentration)
- Calculation of exposure (dose-effect function)
- Economic damage assessment (cost).

This methodology allows estimating the emissions of harmful substances into the atmosphere, pollution of water bodies and soil.

To determine the environmental impact of harmful emissions, ISC Manager software was used [7], which calls the Industrial Source Complex (ISC) program to calculate dispersion. ISC is recommended for use by the US Environmental Protection Agency (EPA). It is based on the statistical theory of dispersion of harmful emissions in the atmosphere, based on the assumption that the distribution of the concentration of impurities in a diffusing cloud of smoke is determined in accordance with the Gauss law. To display the calculation results, ISC Manager uses the GIS "Map 2008" geographic information system.

As a function of the "dose-effect", the results of medical research are used, which are carried out in many countries of the world separately for each pollutant and for each physical effect. To determine the total physical impact in the region, the ISC Manager program calculates the population density and determines the effect of the increase in surface concentration of harmful substances on the deterioration of public health in accordance with the dose-effect function [7, 8].

To calculate the values of emissions of harmful substances from CHPPs of PJSC Mosenergo, information was collected on the number of chimneys of TPPs, their height, and also the annual emission of harmful substances into the atmosphere [10].

As a result of the simulation, it was found that the maximum concentration of emissions of nitrogen oxides (6.32 μg / m3) and carbon oxides (0.28 μg / m3) is achieved in the district of CHPP-21. The maximum permissible concentration for these pollutants is 0.06 and 3.0 mg / m3, respectively. The dispersion and impact of nitrogen oxide emissions is shown in Figures 3 and 4. The maximum impact (0.0223 units) is achieved in close proximity to CHPP-21, which makes reducing its emissions an urgent task.

 https://doi.org/10.1051/e3sconf/202017801010

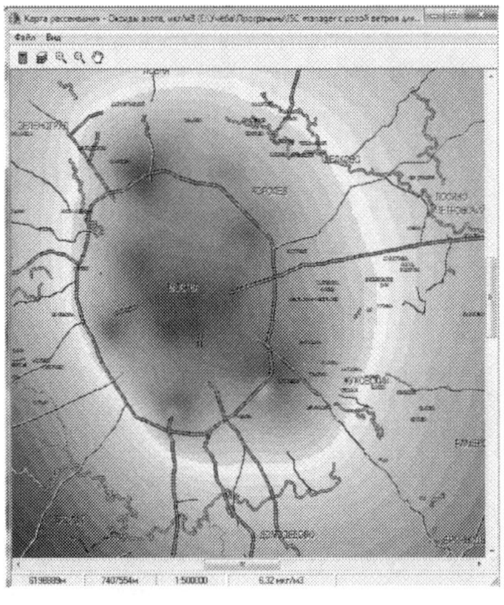

Fig. 3. Map of the dispersion of nitrogen oxides from the CHPPs of PJSC Mosenergo.

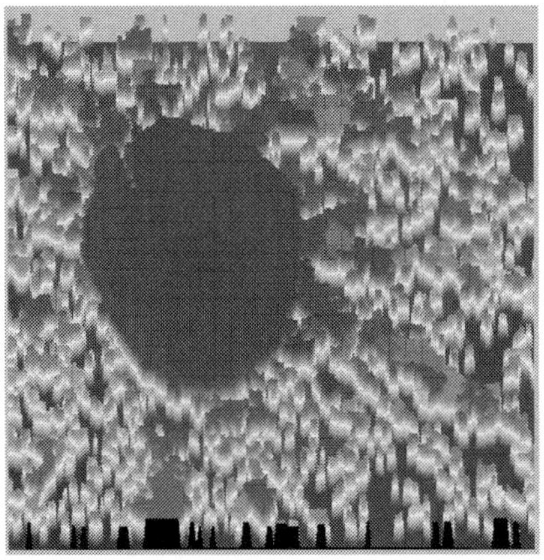

Fig. 4. Three-dimensional matrix of the effect of nitrogen oxides on the population from the CHPPs of PJSC Mosenergo.

Summary data on the effects of harmful emissions from CHPPs on the population of Moscow are presented in table 1.

Table 1. Damage from emissions of harmful substances from CHPPs of PJSC Mosenergo.

Pollutants	Mortality, ppl / year	Damage, million rubles / year(Cost of average life = 68 million rubles)
SOx	39,97	2 717,96
Solids	29,18	1 984,24
CO	0,15	10,20
NOx	160,42	10 908,56
In total	229,72	15 620,96

To assess the energy effect of the proposed modernization, it is necessary to consider in more detail the schemes of thermal power plants before and after modernization and the procedure for their calculation.

4 Structural schemes of thermal power plants before and after modernization

Figure 5 shows the initial simplified scheme of the CHPP [11]. Steam turbine, steam turbine. After triggering in the high-pressure cylinder (HPC), part of the steam enters the low-pressure cylinder (LPC). After the LPC has exhausted steam, it enters the condenser, from where, after condensation, the circulation pump is pumped back to the boiler. The condenser cooling circuit consists of a cooling tower and a pump. Part of the steam after the HPC is sent to the steam-water heat exchanger, where it condenses and fed back into the circuit. Water from the heated circuit of the steam-water heat exchanger is pumped to the water heat exchanger, which is installed at the central heating statiion (CHS), and then fed back to the steam-water heat exchanger.

The heat consumer circuit consists of water heat exchanger, a pump and the heat consumer itself.

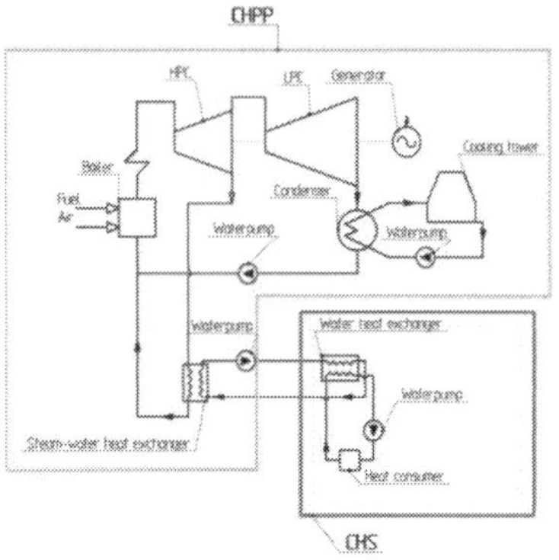

Fig. 5. Simplified CHPP scheme.

The modernized scheme of the CHPP shown in Figure 6 additionally includes two AHTs (at the CHPP and CHS). The main circuit works without changes.

Part of the steam from the selection of HPC is sent as a heating source to the AHT installed at the CHPP. Water from the condenser cooling circuit, taken in front of the cooling tower, acts as the AHT cooling medium. At the same time, part of the water taken in front of the tower from the condenser cooling circuit enters the AHT for cooling. After heating in the AHT, the cooling water is directed to the steam-water heat exchanger, where it is heated and then pumped to the AHT installed on the central heating station, where it acts as a heating coolant.

The heating water from the CHPP, passing through the AHT, transfers part of its thermal energy and then enters the water-water heat exchanger, where it also acts as a heating coolant. After the heat exchanger, water from the CHPP is sent back to the AHT, but already as a cooled coolant. After AHT, chilled water is directed to the cooling circuit of the CHPP condenser, while mixing with part of the water that was previously supplied to the AHT for cooling [12].

The use of AHT at CHPPs and CHS will allow to utilize the low-temperature waste heat of the condenser by changing the temperature schedule of the main networks from 120/70 °C to 120/20 °C, while the temperature schedule of consumers changes from 95/70 °C to 60/40 °C. Utilization of the condenser waste heat is achieved by supplying reverse network water from the line to the condenser cooling circuit, after which this water is heated in the AHT and the steam-water heat exchanger and is fed back to the heat consumers.

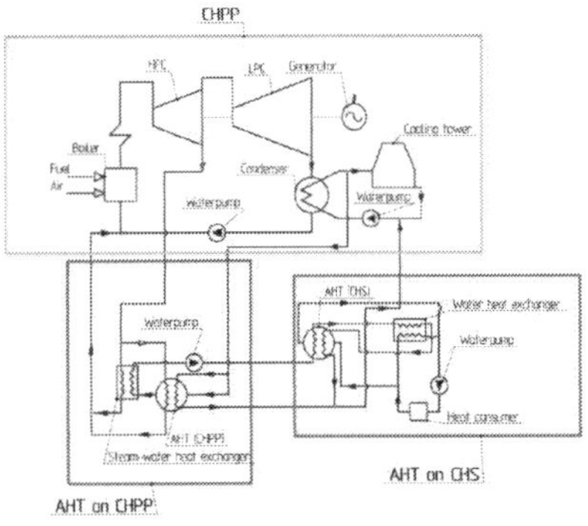

Fig. 6. Simplified CHPP scheme with AHT.

Two steam-turbine power units with a pressure of 240 atm were selected as the object for modeling the use of AHT at the CHPP, as the most modern and powerful at the station. To simplify the calculation process, the construction of CHPP models was performed using the Thermoflex software package.

Thermoflex is a universal program. The objects of modeling can be gas turbine units (GTU), GTU with recovery boilers, combined cycle plants (CCP) with recovery boilers, CHPP, CCGT with coal gasification, water desalination plants, steam power plants on solid, liquid and gas fuel, cogeneration systems , biomass power plants, solar and nuclear power plants, waste incinerator power plants, Kalina cycles, organic Rankine cycles, etc. [13].

The models of the initial and modernized circuits are shown in Figures 7, 8, 9, 10. Due to limitations in the program by the number of elements, the model of the modernized thermal power plant circuit with AHT was divided into 3 parts.

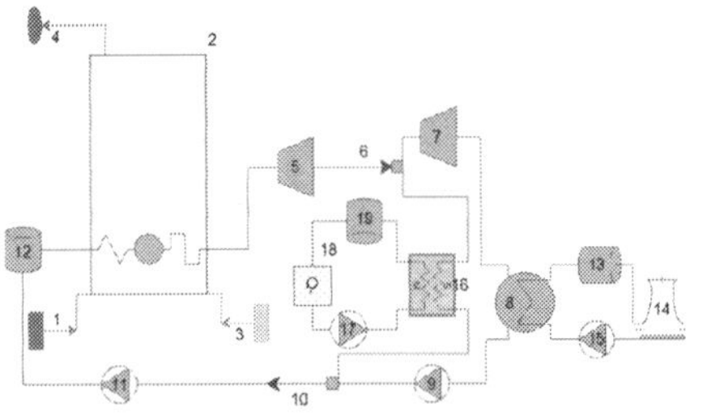

Fig. 7. Simplified CHPP model (1 - air source; 2 - boiler; 3 - fuel source; 4 - exhaust; 5 - HPC; 6 - stream splitter; 7 - LPC; 8 - condenser; 9 - water pump; 10 - stream connector; 11 - water pump; 12 - element for setting the type of coolant; 13 - element for setting the type of coolant of the condenser cooling circuit; 14 - cooling tower; 15 - circulation pump of the condenser cooling circuit).

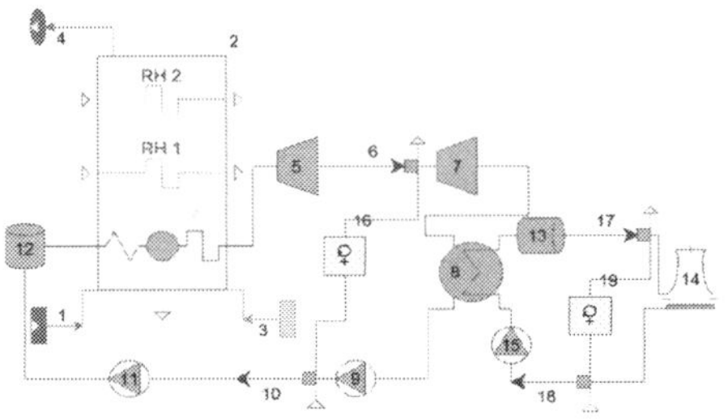

Fig. 8. Simplified CHPP model with AHT (part 1) (1 - air source; 2 - boiler; 3 - fuel source; 4 - exhaust; 5 - HPC; 6 - stream splitter; 7 - LPC; 8 - condenser; 9 - water pump; 10 - stream connector; 11 - water pump; 12 - element for setting the type of coolant; 13 - element for setting the type of coolant of the condenser cooling circuit; 14 - cooling tower; 15 - circulation pump of the condenser cooling circuit. 16 - connecting element of the steam circuit of the CHP and the ATT circuit at the CHP (Figure 9, elements 1 and 4); 17 - stream splitter; 18 - stream connector; 19 - a connecting element of the condenser cooling circuit and the ATT circuit at the TPP (Figure 9, element 12)).

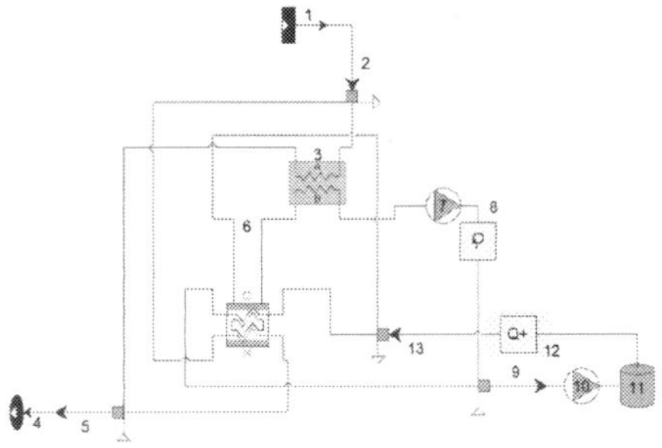

Fig. 9. Simplified CHPP model with AHT (part 2) (1 - steam from the selection of HPC (entry to element 16 of figure 8); 2 - stream splitter; 3 - steam-water heat exchanger; 4 - return water to the boiler (exit from element 16 of figure 8); 5 - stream connector; 6 - AHT; 7 - pump; 8 - connecting element of the AHT circuit at the CHPP and the heat consumer circuit (Figure 10, elements 1 and 4); 9 - stream splitter; 10 - pump; 11 - element for setting the type of coolant; 12 - a connecting element of the condenser cooling circuit and the AHT circuit at the CHPP (Figure 8, element 19); 13 - stream splitter).

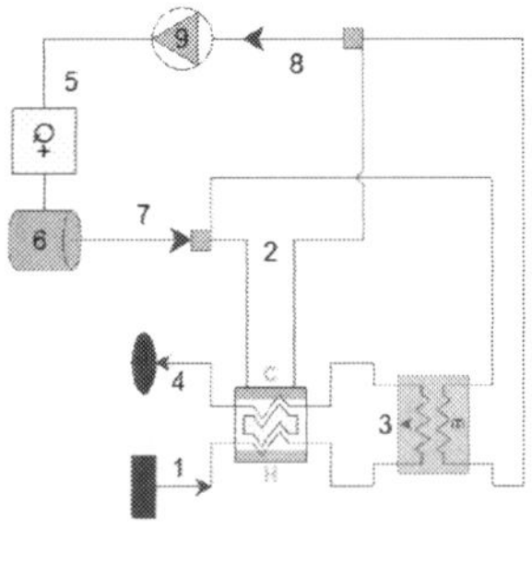

Fig. 10. 1 - hot water from the CHPP (input to element 8 of figure 9); 2 - AHT; 3 - water heat exchanger; 4 - return water to the CHPP (exit from element 8 of Figure 9); 5 - heat consumer; 6 - element for setting the type of coolant; 7 - stream splitter; 8 - stream connector.

According to the simulation results, the annual savings from using AHT at a power unit with a pressure of 240 at CHPP-21 amounted to about 7.1% or 54.9 thousand tons of equivalent fuel per year.

Annual CO2 reduction for power unit 240 at. CHP-21 amounted to 88.3 thousand tons per year, or about 7.09% of the total CO2 emissions. Based on the data obtained on reducing emissions during the modernization of CHPP plants, it is possible to assess the reduction in the impact on public health using the ISC Manager program.

5 Emissions after modernization

With the joint modernization of 130 at. and 240 at. harmful emissions will be reduced according to the data indicated in table 2.

Table 2. Reducing emissions of harmful substances from CHPP-21 after modernization of power units 240 and 130 at, t / year.

Pollutants	Emissions before modernization, t / year	Emissions after modernization, t/ year	emission reduction
Solids	0,0	0,0	0,0
SOx	9,0	8,4	0,6
NOx	3 726,0	3 461,8	264,2
CO	174,0	161,7	12,3
In total	3 909,0	3 631,9	277,1

The obtained data can be multiplied to other thermal power plants of PJSC Mosenergo and, thus, to a first approximation, obtain the energy and environmental effects from the modernization of the main sources of thermal and electric energy in Moscow

6 Multiplication of simulation results at the CHPP of Mosenergo

The annual fuel consumption for all CHPPs of PJSC Mosenergo amounted to 23,327 thousand tons of equivalent fuel. With the use of AHT at all CHPPs plants in Moscow, annual savings of 7.1% will reduce fuel consumption by 2,809 thousand tons of equivalent fuel. or, of the order, 2,442 million m3 of natural gas, which at a gas price of 4,694 rubles / thousand m3 will save 11,463 million rubles a year.

Analysis of the results allows us to draw the following conclusions:

1. The use of AHT at CHPPs and central heating plants leads to a transition to lower coolant temperatures in heat supply systems, which confirms the trend that has emerged in many countries of the world.

2. This modernization allows to reduce the load on the cooling tower, which leads to lower thermal pollution, another important environmental effect.

The use of AHT makes it possible to increase the supply of thermal energy without upgrading the power equipment of a thermal power plant.

To assess the reduction in the impact of harmful emissions from the Mosenergo CHPPs on the population, using the results of modeling of modernization of the power unit 240 at.

Table 3 presents data on economic damage from operation.

Table 3. Damage from emissions of harmful and polluting substances from the CHPPs of PJSC Mosenergo.

Pollutants	Mortality before modernization, ppl / year	Mortality after modernization, ppl / year	Difference, ppl / year	Damage reduction, million rubles / year
SOx	39,97	3,58	36,39	2 475
Solids	29,18	0,12	29,06	1 976
CO	0,15	0,14	0,01	0,7
NOx	160,42	125,38	35,04	2 383
In total	229,72	129,22	100,50	6 834

The data presented in table 3 give a clear idea of the environmental effect of modernization using AHT of all CHPPs plants owned by Mosenergo PJSC. The modernization method presented in this work is effective not only in terms of energy efficiency, but also from an environmental point of view. Damage reduction during modernization of all CHPPs amounts to 6,834 million rubles per year.

7 Conclusion

Reducing fuel consumption by a 240 at power unit is amounted to 7.1% or 54.9 thousand tons of equivalent fuel per year, which, with a gas price of 4.7 rubles per m3, will save about 224 million rubles per year. With the joint modernization of power units 130 and 240 at, the savings will be about 180 thousand tons of fuel equivalent per year, which is equivalent to 733 million rubles per year.

It is also necessary to pay attention to reducing damage from the effects of harmful emissions from CHPP-21. With the modernization of the 240 at power unit, the damage reduction will be 21 million rubles / year, and with the joint modernization of the 240 and 130 at power units, the damage reduction will be 69 million rubles per year.

Multiplication of the results obtained in the work shows that the application of the proposed technology will bring significant energy and environmental effects and will reduce fuel consumption by 2 809 thousand tons of equivalent fuel. or, of the order, 2,442 million m3 of natural gas, which at a gas price of 4700 rubles per thousand. m3 will save 11,463 million rubles a year.

Emissions from CHPPs lead to mortality of about 300 people per year, which entails damage in the amount of 15 620 million rubles per year. A decrease in these indicators is a prerequisite for the development of a heat supply system.

From an environmental point of view, such modernization will reduce emissions of harmful substances by 45,000 tons per year, which will reduce mortality by 101 people per year, and taking into account the cost of average life in monetary terms, it will save 6,868 million rubles per year.

The proposed modernization will reduce the consumption of fuel and energy resources at thermal power plants, and will also favorably affect the environmental situation in Moscow, which makes it a promising way to modernize existing CHPPs.

References

[1] S. Jian, F. Lin, Zh. Shigang, Experimental study of heat exchanger basing on absorption cycle for CHP system, Applied Thermal Engineering **102**, 1280–1286 (2016)

[2] H. Tianle, X. Xiaoyun, J. Yi, Simulation research on a variable-lift absorption cycle and its application in waste heat recovery of combined heat and power system, Energy **140**, 912–921 (2017)

[3] X. Xiaoyun, J. Yi, Absorption heat exchangers for long-distance heat transportation, Energy **141**, 2242–2250 (2017)

[4] S. Jian, F. Lin, S. Fangtian, Z. Shigang, Experimental study on a project with CHP system basing on absorption cycles, Applied Thermal Engineering **73**, 1, 732–738 (5 December 2014)

[5] H.S. Zhang, H.B. Zhao, Z.L. Li, Performance analysis of the coal-fired power plant with combined heat and power (CHP) based on absorption heat pumps, Journal of the Energy Institute **89**, 70–80 (2016)

[6] Updating the heat supply scheme of the city of Moscow for the period until 2032. Approved by order of the Ministry of Energy of the Russian Federation of 19 March 2019, No. 249 (2019)

[7] I. Sultanguzin, *Environmental Safety and Energy Efficiency of Industrial Heat and Power Systems: a Training Manual* (Publishing House MPEI, 288, 2013)

[8] A. Rabl, J. Spadaro, M. Holland, *How Much Is Clean Air Worth? Calculating the Benefits of Pollution Control* (Cambridge University Press, 671, 2014)

[9] I. Sultanguzin, *The application of the methodology of the sequence of environmental impacts of harmful emissions in industrial power engineering: a methodological manual* (Publishing House MPEI, 48, 2013)

[10] Preparation of the initial information necessary for the formation of the consolidated fuel and energy complex of Moscow, The report on the 1st stage as part of the work under state contract no. 9P4 / 01/0173200001116000223 of 27 September 2016, Moscow, 271 (2016)

[11] V. Ryzhkin, *Thermal Power Stations: A Textbook for High Schools* (Energoatomizdat, 328, 1987)

[12] A. Bartenev, S. Prishchepova, I. Sultanguzin, *Evaluation of the effectiveness of using AHT at CHPPs and central heating plants in heat supply systems,* Proceedings of the ninth international school – a seminar for young scientists and the specialist Energy Saving – Theory and Practice (Publishing House MPEI, 163–166, 2018)

[13] A. Fedyukhin, I. Sultanguzin, S. Kurzanov, R. Belov, A. Bakulin, T. Shomova, Application of applied software for solving industrial heat power problems: a training manual (Publishing House MPEI, 88, 2016)

[14] THERMOFLEX. Fully-Flexible Heat Balance Engineering Software, Thermoflow, Inc, THERMOFLEX User's Manual **3**, 1139 (February 2013)

[15] *User's Guide for the Industrial Source Complex (ISC3) Dispersion of Model Algorithms* [Electronic resource] (US Environmental Protection Agency 2, 1995). Available at: http://www.epa.gov/scram001/userg/regmod/isc3v2.pdf

Modeling an induction motor based on the equations of a generalized electric machine, taking into account the saturation of the magnetic circuit

Yuri Mikhailovich Frolov[1], *Valery Petrovich* Shelyakin[1], *Nikolay Vasilievich* Sitnikov[1], *Sergey Aleksandrovich* Goremykin[1], and *Dmitry Aleksandrovich* Tonn[1]

[1]Voronezh State Technical University

Abstract. Writing the differential equations of an asynchronous electric motor on the basis of a generalized electric machine in different coordinate axes can not only reduce the number of model equations, but also introduce various factors in the mathematical description that increase the modeling accuracy. It is proposed to write the equations of the mathematical model of an induction motor taking into account the saturation of the magnetic circuit and solve them using the Matlab application package. The proposed approach improves the accuracy of calculating the mechanical characteristics of an asynchronous electric drive.

1 Introduction

An important step in the synthesis and analysis of electric drive systems is the development of a model of an electric machine that meets the challenges of vector control [1–6].

In most cases, electromechanical energy converters are made in multiphase design. This fact complicates the mathematical apparatus that describes the transient processes in such converters. This is due to an increase in the number of equations of the mathematical model of the experimented electric machine. Therefore, as a rule, they strive to make an equivalent transition from a real machine to a generalized double-phased [7, 8]. The generalized machine in the mathematical description is characterized by the following generally accepted assumptions: "the phase windings are symmetrical, the air gap is uniform, there are no losses in steel and tooth phenomena, the magnetizing circuit is linear, the magnetizing force of each winding and the magnetic induction in the air gap are sinusoidal around the circumference of the stator (rotor), rotor parameters are reduced to the stator circuit "[7–9]. The equations of the mathematical model of a generalized machine can be written in different forms depending on the coordinate systems used. The most commonly used coordinate systems are: u,v; α,β; d,q [7, 8].

2 Methods

We write the equations of the generalized machine in the coordinate system u,v rotating synchronously with the speed of the electromagnetic field:

$$
\begin{aligned}
&U_{1v} = (R_1 + p\cdot L_1)\cdot i_{1v} + p\cdot L_{12}\cdot i_{2v} - \\
&- L_1\cdot\omega_k\cdot i_{1u} - L_{12}\cdot\omega_k\cdot i_{2u}, \\
&U_{1u} = (R_1 + p\cdot L_1)\cdot i_{1u} + p\cdot L_{12}\cdot i_{2u} - \\
&- L_1\cdot\omega_k\cdot i_{1v} - L_{12}\cdot\omega_k\cdot i_{2v}, \\
&U_{2u} = (R_2 + p\cdot L_2)\cdot i_{2u} + p\cdot L_{12}\cdot i_{1u} - \\
&- L_{12}\cdot(\omega_k - \omega_R)\cdot i_{1v} - L_2\cdot(\omega_k - \omega_R)\cdot i_{2v}, \\
&U_{2v} = (R_2 + p\cdot L_2)\cdot i_{2v} + p\cdot L_{12}\cdot i_{1v} + \\
&+ L_{12}\cdot(\omega_k - \omega_R)\cdot i_{1u} + L_2\cdot(\omega_k - \omega_R)\cdot i_{2u}, \\
&M = P_n\cdot L_{12}\cdot(i_{1v}\cdot i_{2u} - i_{1u}\cdot i_{2v}).
\end{aligned} \tag{1}
$$

where R_1, R_2 – active resistance of stator and rotor windings, respectively;

L_1, L_2 – the total inductances of the stator and rotor windings, respectively;

L_{12} – mutual inductance of stator and rotor windings;

ω_k – angular velocity of rotation of the coordinate axes;

$\omega_R = P_n\cdot\omega_{rotor}$ – angular velocity of the rotor of the machine in electrical degrees;

p – differentiation operator;

P_n – number of pole pairs.

Equations of a generalized electric machine in a coordinate system u,v – most common. Transition from coordinates v,u to a fixed coordinate system α,β occurs by direct coordinate transformation at $\omega_k = 0$.

$$U_{1\alpha} = U_{1max}\cdot cos(\omega_{0R}\cdot t), \tag{2}$$

$$U_{1\beta} = U_{1max}\cdot sin(\omega_{0R}\cdot t), \tag{3}$$

where $U_{1max} = \sqrt{3} \cdot U_{1phasevoltage}$.

Transition from coordinate system *u,v* to the coordinate system *d,q* rotating at the speed of the rotor, if we assume that $\omega_k = \omega_R$. From here, the system of phase voltages applied to the windings will be equal:

$$\begin{aligned} U_{1d} &= U_{1max} \cdot cos^2(\omega_{0R} \cdot t) + \\ &+ U_{1max} \cdot sin^2(\omega_{0R} \cdot t) = U_{1max}, \\ U_{1q} &= -U_{1max} \cdot sin(\omega_{0R} \cdot t) \cdot sin(\omega_{0R} \cdot t) + \\ &+ U_{1max} \cdot sin(\omega_{0R} \cdot t) \cdot cos(\omega_{0R} \cdot t) = 0. \end{aligned} \tag{4}$$

The system of equations (1) corresponds to the structure of the electromechanical energy converter (EMF), shown in figure 1. This structure was compiled in the Simulink application of the Matlab application package [10–12]. Here, the voltages $U_{1u}, U_{1v}, U_{2u}, U_{2v}$ are transformed control actions connecting the engine with the control system in the axes x, y.

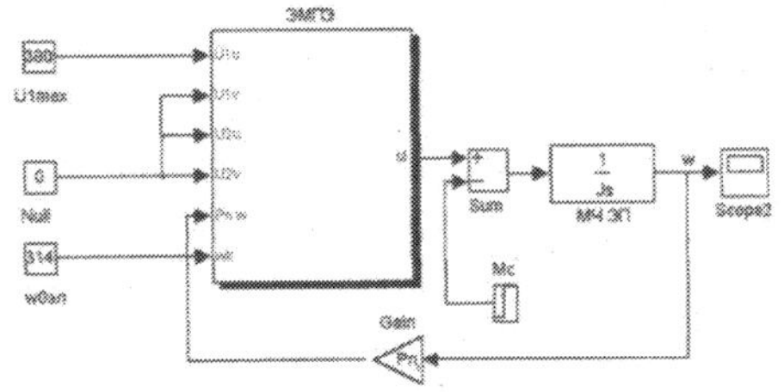

Fig. 1. Structural diagram of EMF and the mechanical part of ES.

Transients in electromechanical converters are classified as follows: during start-up, braking, reverse, restart, and also when the load changes [13–16]. To obtain information about the features of transient processes in an electric machine, reliable data on the parameters and independent variables of the model are needed.

In most cases, the nature of transient processes in EMF affects the configuration of the drive as a whole. Namely: overall dimensions, permissible electromagnetic loads, operating modes, including the features of starting and braking these systems.

3 Results

Figures 2, 3 below show the laws of change in the angular velocity of rotor 1 and electromagnetic moment 2 in time for an induction motor 4A80B6Y3 of power *1,1 kW*, $P_n = 3$, *U=220/380*. So, figure 2 illustrates the idle mode at the moment of inertia equal to the intrinsic moment of inertia of the rotating parts of the EMF ($J = J_M$).

Figure 3 characterizes the transient processes when the load on the shaft is equal to the nominal and at a double value of the moment of inertia ($J = 2 J_M$).

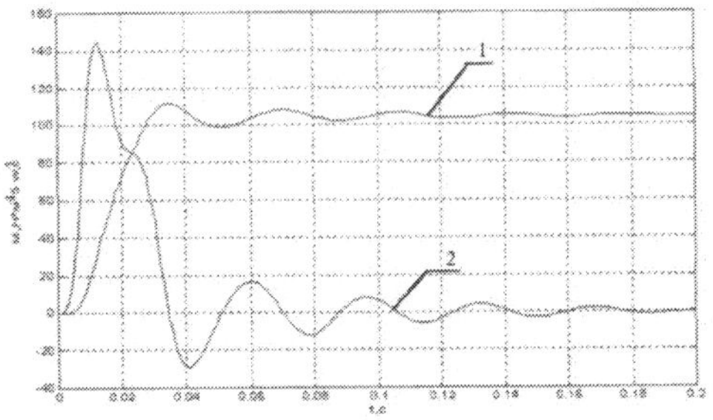

Fig. 2. Transient characteristics at a static moment equal to $M_C=0$ and moment of inertia $J = J_M$.

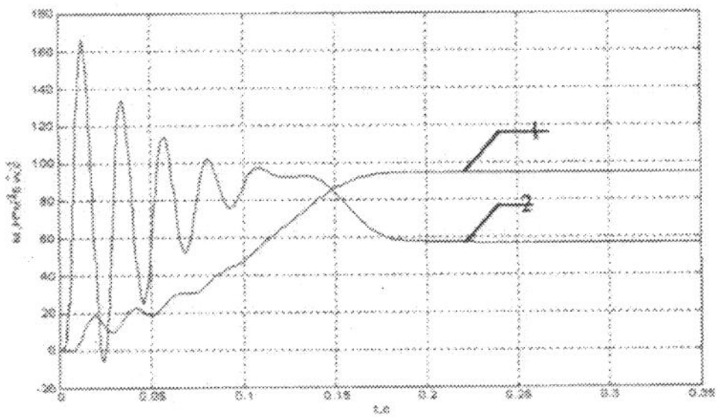

Fig. 3. Transient characteristics at a static moment equal to the rated $M_C = M_N$ and moment of inertia $J = 2J_M$.

In order to optimize the EMF by mass and size indicators, the "operating point" is selected on a non-linear section of the magnetization characteristic. When changing the parameters of the supply network, the degree of saturation of the magnetic circuit of the machine changes. This leads to the fact that the intrinsic and mutual inductances between the windings change, therefore, it is necessary to take into account changes in the inductive parameters of the electric machine over time [17, 18]. Variation of these parameters, in turn, will lead to a variety of EMF characteristics, both in steady state and in transient modes.

When starting the above-considered asynchronous electric motor due to the presence of alternating components in the amplitudes of the stator and rotor currents, and, accordingly, the magnetization current, the magnetic circuit of an electric machine can enter a deep saturation state. The inductance of the magnetization and scattering circuit of the windings of a saturated machine significantly differ from the inductances of the nominal mode, and with increasing saturation the inductance decreases. Within the limits of calculation error, it can be assumed that the magnetic flux of scattering closes through the air, therefore, we can assume that the inductive parameters of the stator and rotor will not depend on saturation.

The graphs of changes in inductance versus magnetization current will be obtained from the relationships shown in figure 4.

From curve 2, for each value of the magnetization current, we find the value of the open-circuit emf, which is equal (in relative units) to the value of flux linkage in the air gap. Knowing the magnitude of flux linkage corresponding to a certain current value I_{μ}, we find the reciprocal of the inductance (in relative units) $1/L_{\mu}$ along curve 1.

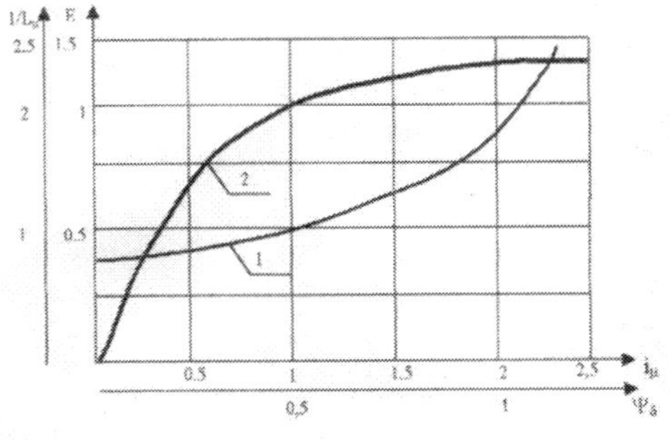

Fig. 4. Universal BP magnetization curves.

Approximate relationship $L_{\mu} = f(I\mu)$ is shown in figure 5.

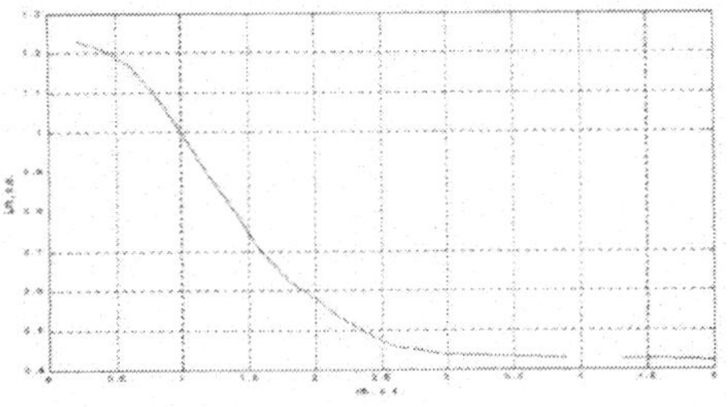

Fig. 5. Relationship $L_{\mu} = f(I\mu)$.

The magnetization current can be determined by adding the projections of the stator and rotor currents on the axis *u,v*. As $I_{\mu} = I_1 + I_2$, then from the vector diagram shown in figure 6, it can be seen that:

$$I_{\mu u} = I_{1u} + I_{2v} \tag{5}$$

$$I_{\mu v} = I_{1v} + I_{2v} \tag{6}$$

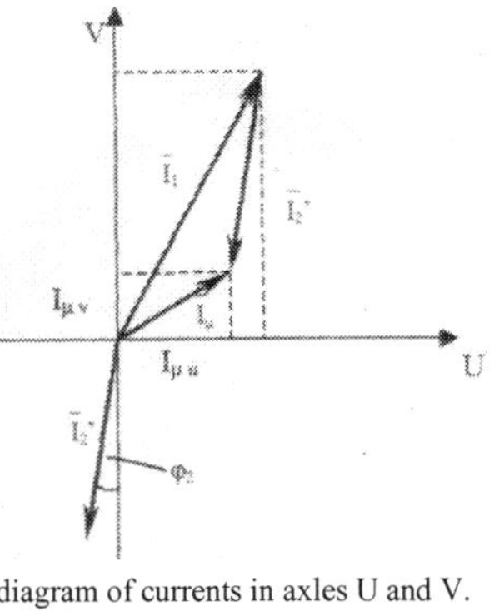

Fig. 6. Vector diagram of currents in axles U and V.

The magnitude of the magnetization current from equation (5), (6) is determined by the following formula:

$$I_{\mu} = \sqrt{(I_{1u} - I_{2u})^2 + (I_{1v} - I_{2v})^2} \tag{7}$$

The equations of an asynchronous machine, taking into account the saturation of the magnetic circuit, are written as follows

$$\begin{aligned}
&U_{1v} = (R_1 + p \cdot L_1(I\mu)) \cdot i_{1v} + p \cdot L_{12}(I\mu) \cdot i_{2v} - \\
&- L_1(I\mu) \cdot \omega_k \cdot i_{1u} - L_{12}(I\mu) \cdot \omega_k \cdot i_{2u}, \\
&U_{1u} = (R_1 + p \cdot L_1(I\mu)) \cdot i_{1u} + p \cdot L_{12}(I\mu) \cdot i_{2u} - \\
&- L_1(I\mu) \cdot \omega_k \cdot i_{1v} - L_{12}(I\mu) \cdot \omega_k \cdot i_{2v}, \\
&U_{2u} = (R_2 + p \cdot L_2(I\mu)) \cdot i_{2u} + p \cdot L_{12}(I\mu) \cdot i_{1u} - \\
&- L_{12}(I\mu) \cdot (\omega_k - \omega_R) \cdot i_{1v} - L_2(I\mu) \cdot (\omega_k - \omega_R) \cdot i_{2v}, \\
&U_{2v} = (R_2 + p \cdot L_2(I\mu)) \cdot i_{2v} + p \cdot L_{12}(I\mu) \cdot i_{1v} + \\
&+ L_{12}(I\mu) \cdot (\omega_k - \omega_R) \cdot i_{1u} + L_2(I\mu) \cdot (\omega_k - \omega_R) \cdot i_{2u}, \\
&M = P_n \cdot L_{12}(I\mu) \cdot (i_{1v} \cdot i_{2u} - i_{1u} \cdot i_{2v}).
\end{aligned} \tag{8}$$

Graphs of transients of the magnetization current (curve 1) and magnetization inductance (curve 2) when starting an induction motor 4A80B6Y3 of power *1,1 kW*, $P_n = 3$, *U=220/380*, when the load on the shaft is zero is shown in figure 7. Transition charts speed (curve 1) and electromagnetic moment (curve 2), under similar experimental conditions, are shown in figure 8.

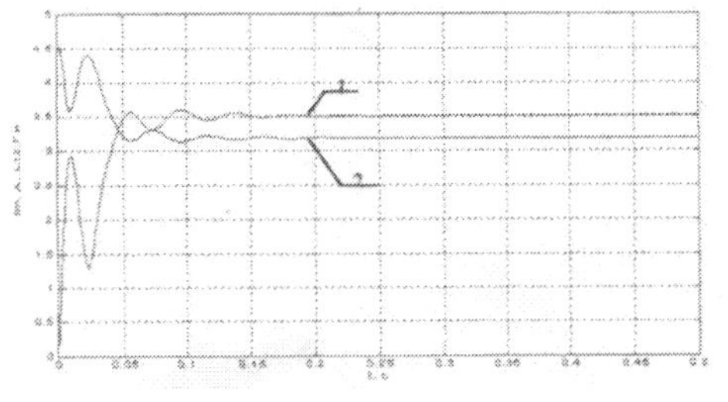

Fig. 7. Graphs of transients of the magnetization current and magnetization inductance at start-up.

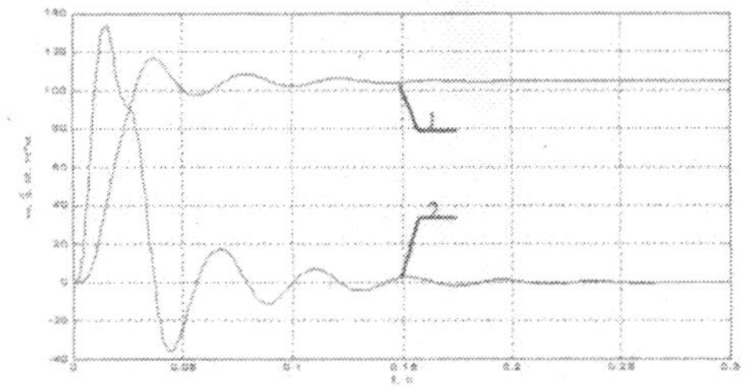

Fig. 8. Graphs of transient processes of speed and electromagnetic moment at start-up to idle.

Graphs of transients of the magnetization current (curve 1) and magnetization inductance (curve 2) when starting an induction motor with a load on the shaft equal to the nominal value is shown in figure 9. Graphs of transients of speed (curve 1) and electromagnetic moment (curve 2), under similar conditions conducting experiments are shown in figure 10.

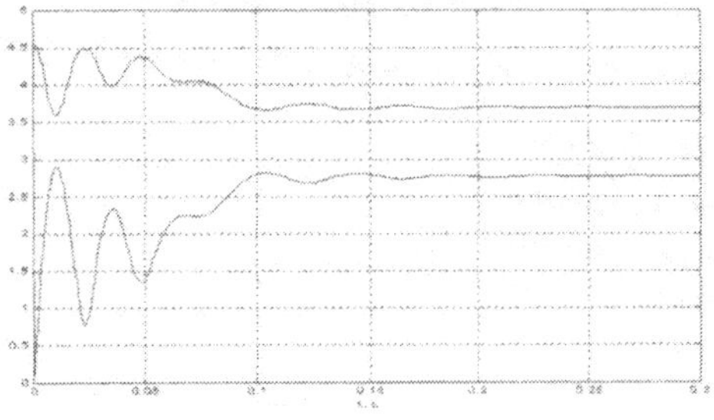

Fig. 9. Graphs of transients of the current and magnetization inductance at start, loaded with a rated torque, of the motor.

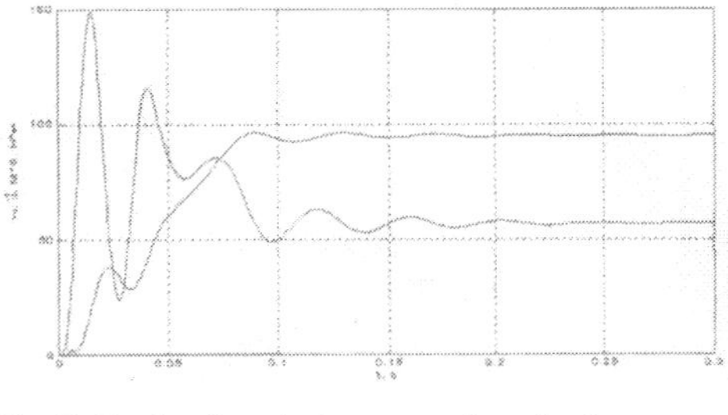

Fig. 10. Graphs of transient processes of speed and electromagnetic moment loaded with rated torque of the engine.

4 The discussion of the results

The availability of reliable information about the nature of the transition processes in modern electric drives makes it possible to increase the reliability and accuracy of the operation of critical actuators. For a more accurate mathematical description of transients, various generally accepted assumptions are removed. When trying to take into account an assumption, a differentiated approach is needed. The essence of this approach is that we take into account only those factors that, for the considered group of machines and their operating conditions, have a significant impact on the accuracy of the calculation of the mathematical model. In the framework of this work, the possibility of taking into account such an important factor as the saturation of the magnetic circuit in the calculation of transients in a specific induction motor using modern Matlab application software is shown.

References

[1] A.E. Kravchik, M.M. Schlaf, V.I. Afonin, et al. *A Series Induction Motors: Reference Book* (Moscow: Energoizdat, 504, 1982)

[2] V.N. Brodsky, E.S. Ivanov, Drives with frequency – current control (M .: Energy, 168, 1974)

[3] A.V. Ivanov-Smolensky, *Electric cars: Textbook for high schools* (M.: Energy, 928, 1980)

[4] V.I. Klyuchev, *Theory of electric drive: Textbook for universities* (M.: Energoatomizdat, 560, 1985)

[5] S.A. Kovchin, Yu.A. Sabinin, *Theory of electric drive: Textbook for high schools* (St. Petersburg: Energoatomizdat, St. Petersburg branch, 496, 2000)

[6] V.V. Rudakov, I.M. Stolyarov, V.A. Dartau, *Asynchronous electric drives with vector control* (L.: Energoatomizdat, 136, 1987)

[7] I.P. Kopylov, *Mathematical modeling of electrical machines* (M.: Higher School, 318, 1994)

[8] G.A. Sipailov, E.V. Kononenko, K.A. Ferrets, Electric machines (special course) (M.: Higher School, 287, 1987)

[9] Yu.M. Frolov, V.A. Trubetskoy, P.Yu. Tarasov, Modeling the system of frequency – current control of an asynchronous electric drive, Automation and robotization of technological processes: Materials of regional scientific and technical. conferences, Voronezh: Publishing House of VSTU, 121–129 (2000)

[10] V.P. Dyakonov, *MATLAB 6.5 SP1 / 7 + Simulink 5/6* (Basics of application, Series "Professional Library", M.: SOLON-Press, 800, 2005)

[11] V.G. Potemkin, *The system of engineering and scientific calculations in Matlab 5.x: 2 tons* (M.: Dialog-MEPhI, 1999)

[12] V.E. Kondrashov, S.B. Korolev, *MATLAB as a system for programming scientific and technical calculations* (M.: Mir, Institute for Strategic Stability, Minatom of the Russian Federation, 350, 2002)

[13] Yu.A. Sabinin, V.L. Cargo, Frequency – regulated asynchronous electric drives (L.: Energoatomizdat, 128, 1985)

[14] O.V. Sledzhanovsky, L.K. Datskovsky, I.S. Kuznetsov, et al., *Systems of subordinate regulation of AC electric drives with valve converters* (M.: Energoatomizdat, 256, 1983)

[15] D. White, G. Woodson, *Electromechanical energy conversion* (M.- L.: Energy, 528, 1964)

[16] Yu.M. Frolov, R.A. Kosmatykh, S.A. Vlasov, *AC electric drive according to the inverter – HELL system: textbook, allowance* (Voronezh: Publishing House of VSTU, 130, 1999)

[17] Yu.M. Frolov, A.V. Romanov, Generalized electric machine in the automated design of electric drives, Modern problems of informatization in engineering and technology: Tr. VI International Open Scientific Conference, Voronezh, 53–54 (2001)

[18] Yu.M. Frolov, P.Yu. Tarasov, A.N. Seleznev, The study of a generalized electric machine, taking into account the saturation of the magnetic circuit, Automation and robotization of technological processes: Materials of regional scientific and technical. conferences, Voronezh: Publishing House of VSTU, 112–119 (2001)

Experimental test of Water-Oil Emulsion Combustion

Askar Baubek[1], *Michael* Zhumagulov[2,*], *Nurlan* Kartjanov[2], *Samal* Sadykova[3], and *Muratbek* Arpabekov[1]

[1]L.N. Gumilyov Eurasian National University, Dept. of Transport logistic, 2 Satpayev str, 010003 Nur-Sultan, Kazakhstan
[2]L.N. Gumilyov Eurasian National University, Dept. of Power engineering, 2 Satpayev str, 010003 Nur-Sultan, Kazakhstan
[3]Almaty University of Power Engineering and Telecommunications, , Dept. of Power engineering, 050013 Baytursynuly, 126/1, Almaty c., Kazakstan

Abstract. This report dwells on the scientific study of the vortical device. The device is designed for efficient combustion of fossil fuels. Oil-water emulsion was taken as an example of alternative fuel. Although the functionality of the burner device is not limited to this fuel only. The innovative burner device under investigation is based on principal of vortical motion of fuel air mixture. There is a precombustion chamber specially designed, where inflammation and partial combustion of the swirling flow takes place. The article describes experiment on defining the geometric dimensioning of the burner discharge nozzle in relation to a combustion chamber. The position of the outlet nozzle inside of the burner is the main input parameter of this research. The authors analyzed the influence of this parameter on the pressure inside the combustion chamber and the electric load of the fan for the supplying combustion air. The position ranged from 0 to 120%. The optimal position is at 50% of the total length of the burner chamber. The pressure graph has an extremum in this position, and the electric energy consumption by the fan changes less significantly with further deepening of the output nozzle than earlier. The Results are presented for cold (air blowing only) and hot experiment (burning). This experiment proved the need for the design of the inner part of the discharge nozzle, which is a design feature of the burner device for which the European patent was obtained. The patent link is presented in the list of references.

1 Introduction

Today, fuel-oil residual is used as a backup and sometimes primary fuel in most of the traditional thermal power plants in the world.

Fuel-oil residual is a type of oil fuel obtained from heavy refinery residues. Fuel-oil residual is, in fact, a mixture of hydrocarbons with a molecular mass of 4001000, petroleum resins (with a molecular weight of 500-3000), carbenes, carboids, asphaltenes and other organic compounds.

Watering of fuel-oil residual occurs already at the stage of petroleum recovery -when it leaves the well together with the oilfield water and in oil desalting process.Further increase of moisture in residual fuel oil occurs when they are transported, prepared and stored [1].

In petroleum products, water can be contained as simple suspensions, then it is easily settled during storage, or in the form of a stable emulsion. The formation of stable oil emulsions leads to large financial losses. After separation of water from oil in sedimentation tanks and reservoirs, some part of oil is discharged together with water in the form of an emulsion and pollutes waste waters [1].

When fuel-oil residual of high viscosity is used in boiler houses as a liquid fuel, difficulties arise in preparing them for combustion. In most operating and projected industrial and heating boiler houses, the heating of fuel-oil residual delivered in railway tanks is provided by means of water vapor, which leads to wide range watering the fuel (from 5 to 20%). Burning of watered fuel-oil residual without preliminary preparation causes disturbance of burning process (down to extinction of the flame) and reduces the economical efficiency of fuel-burning plants [1]. Creating a water-oil emulsion with a stable degree of homogeneity allows us to burn the mixture effectively without separating the fuel from the water [2-5]. The ecological performances are also increasing [6-11].

The purpose of this study is to justify the burning of fuel-oil residual in the form of a water-fuel emulsion replacing a long and sometimes costly settling and burning of fuel-oil residual as such.

For successful implementation of the task it is necessary to fulfill 2 conditions:

1. Homogenize the mixture of fuel and moisture in a special device (for example, a dispersant). This issue is widely studied in world practice, therefore it is not worthwhile to dwell on it in detail. The dispersant used by us is cavitating.
2. Effectively mix water-fuel emulsion with an oxidant (air) in a burner.

Further information in the article concerns the second condition.

* Corresponding author: zhmg_9@mail.ru

Fuel atomizeris used for the combustion of pure fuel-oil residual. Fuel oil atomization in the combustion chamber is carried out by the induction steam or the pressure of the fuel itself. It is common knowledge that such a design is ineffective for burning water-oil emulsion. The main reason for low-efficiency combustion is the poor contact of the fuel with the oxidant.

The basis of the combustion process is the aerodynamic structure of the gas-air flow, in where the combustion process takes place [12, 13]. The role of the aerodynamic structure of the flow is to ensure a perfect mixture of fuel and oxidizer, without which neither the intensity of the combustion process nor the completeness of the heat release can be achieved.

A series of actions were taken at L.N.Gumilyev Eurasian National University (Astana, Kazakhstan), resulting in creation of a vortex combustion chamber (Figure 1) with a tangential fuel and air supply [14, 15]. Vortex principle organizing fuel combustion is used in the device. The burner device has a horizontal cylindrical shape and a nozzle. After the adjustment of the vortex combustion chamber at the pilot industrial unit, it was supposed to be installed on the boiler installation. The installation has advantages over all known analogs due to the part of the exhaust nozzle extending inside. The proposed burner device allows the stable burning of a water-oil emulsion with a moisture content of up to 30% due to intensive mixing of fuel and air. The installation has advantages over all known analogs due to the part of the discharge nozzle extending inside. Patent was received for this device for the utility model of the Russian Federation No. 128284 dated June 21, 2012 for the "Device for burning fuel" [14] and European patent EP No. 2684700 of September 24, 2016 [15].

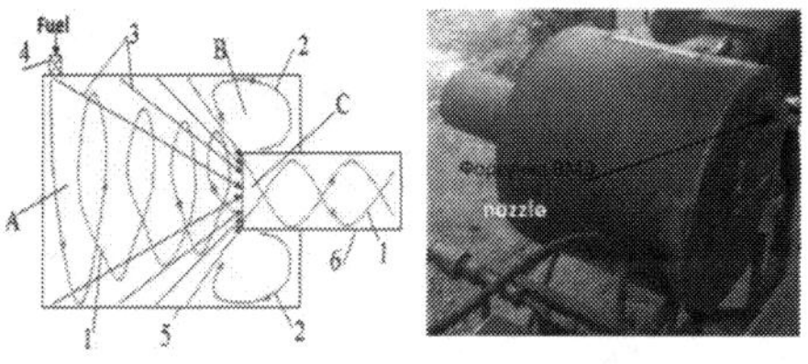

Fig. 1. Gas flow in vortex burner: 1- eddy motion of combustion products nozzle end, 2 - reverse current, 3 –flame vector in combustion chamber; 4 – supply nozzle; 5 – main chamber; 6 – discharge nozzle. Zones *A*, *B*, *C*.

2 Description of the burner

The fuel and oxidizer is fed into combustion chamber 5 (Figure 1) tangentially via supply nozzle 4. The fuel and oxidizer enter the burner separately in form of swirling jet, then mixing in the main chamber 5. As it moves, before meeting with the secondary air, pre-heating of aeromixture occurs due to mass transfer with drawn-in hot gas stream and irradiation by burning torch, leading to oxidative reactions and ignition of the fuel particles. Mixing effect peak occurs at the inlet section of discharge nozzle 6. Initiation of combustion begins in the combustion chamber 5, coming with heat treatment of the fuel, the release of volatile fractions, partial coking etc. Complete and incomplete combustion products, twisting, form a stable vortex flow (Figure 1 shows the path of the gases with lines 1). Under the influence of the vortex gas movement discharge is formed in the axial zone A, and in zone B - reverse currents of 2 combustion products. 2, Reverse currents create overpressure in zone B, whereby the pressure in zone B exceeds the zone C. As a result, flame vector 3 will be directed to the zone C. Thus, vortex flow in the combustion chamber take the form of tornado-shaped vortex with a constriction vortex radius as the gas goes to the outlet section. It should also be noted that the main vortex flow of variable section (mini tornado) practically does not touch the precombustor casing. Due to the part of the discharge nozzle extending inside the friction occurs in contact area of the main vortex of zone A with the reverse currents of zone B, as a result of which the overheating of the prechamber was almost completely eliminated.

In addition, an innovative cyclone-vortex burner is simple to manufacture and reliable in operation and modernization of the boiler does not require a significant investment.

3 Laboratory tests

The researches carried out had two stages: laboratory tests and mathematical modeling of aerodynamic flows. The present article describes the first stage. Initially, the research team set out the aim to explain experimentally the need of embedding discharge nozzle into the chamber. For this purpose, an experimental burner was assembled (Figure 2, 3) with a variable position of part of the exhaust nozzle 2 extending inside. In the combustion chamber along the flow of the fuel-air medium, nozzles are installed for measuring pressure in three (I, II, III) points, apart from the discharge nozzle. The pressure value was determined by a U-shaped manometer. Current consumption of the installation was determined at the same time, upon readings of the ammeter connected to the power supply of the air fan motor, The value of the current consumed can characterize the total load on the fan, due to the aerodynamic resistance of the main path of the burner device.

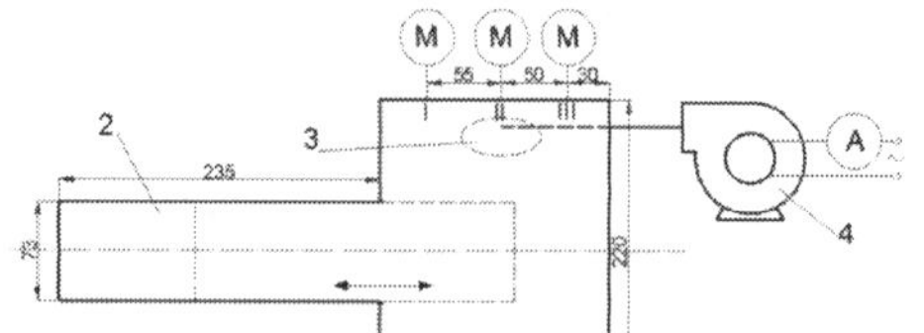

Fig. 2. Schematic layout of an experimental burner: the main chamber; 2 - discharge nozzle (mobile); 3 - inlet nozzle (for fuel and oxidizer supply); 4 - fan (for the supply of oxidizer); M-U-shaped manometers at points I, II, III; A-ammeter.

Fig. 3. Experimental burner with a mobile discharge nozzle.

The experiment was carried out in two stages:

1. «Cold» blow off of the combustion chamber-without supply and burning of the fuel. Only air supply.
2. «Warm» stage - simultaneous supply of fuel and air with subsequent combustion and achievement of a stable flame.

During the experiment, the position of the projecting part of the discharge nozzle 2 gradually changed. The step of changing the position of the projecting part was 20 mm. The first measurements were made in the absence of a nozzle part extending inside the furnace, which corresponds to the widely known analogues [13]. Then, step by step, the protruding part was buried inside.

As a result, the following experimental dependences were obtained:

Figure 4 shows the experimental dependence of the pressure inside the main chamber at points I, II and III on the position of the discharge nozzle. Measurements were made with a "Cold" blow off. The initial position (without embedding) is characterized by minimum pressure values. The first extremum is at the depth of the nozzle of 60 mm, which corresponds to the maximum pressure at all three points. Subsequent embedding of the nozzle inside the chamber positively affects the pressure and leads to its decrease. This goal was set by the research team. The increased pressure in the chamber results in increased loads (over-expenditure of energy) on the blowing mechanisms, and also is an indicator of the increase in aerodynamic resistance. The second extremum is 100 mm. When analyzing figure 4 without comparing it with other figures in this article, it is obvious that embedding the nozzle beyond the 100 mm mark will not lead to a positive result.

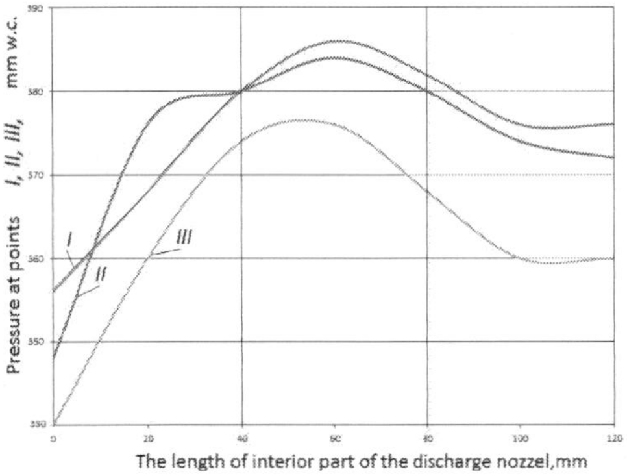

Fig. 4. The experimental dependence of the pressure inside the main chamber on the position of the discharge nozzle (cold stage).

Figure 5 shows the dependence of the fan current consumption on the position of the discharge nozzle. The level of power consumption on the fan is an indirect indicator of the aerodynamic resistance of the main path of the combustion chamber 1 (Figure 2). The curve clearly demonstrates a reduction in the current consumption with an increase in the degree of embedding of the discharge nozzle. The extremum of the curve lies at the level of the nozzle position 80 mm deep. Further embedding of the nozzle is not effective. Further, as noted above, the experiment was carried out under the conditions of the full-fledged functioning of the proposed burner device - the "warm" stage. The burner was supplied with diesel fuel, spark igniter was used for burning and similar measurements were carried out under real combustion conditions.

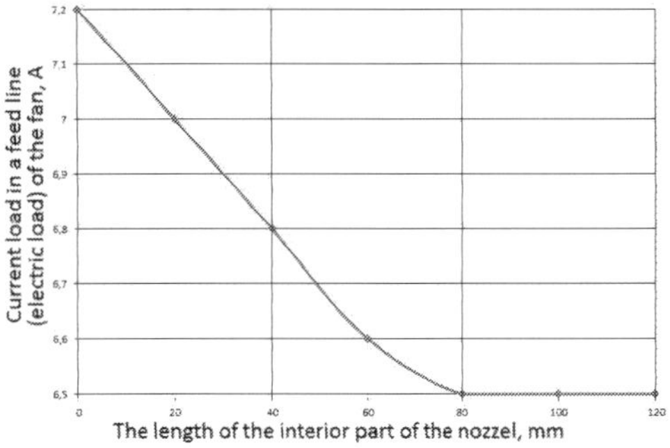

Fig. 5. The experimental dependence of the current consumption of the fan on the position of the discharge nozzle (cold stage).

Figure 6 is a graph of the pressure dependence within the main chamber on the position of the discharge nozzle in the conditions of a "warm" start of the burner device. In "warm" conditions, the initial position of the burner demonstrates a different behavior compared to cold blows. The maximum value of the pressure inside the chamber is measured in the absence of part of the discharge nozzle extending inside. With subsequent increase of extending part, the pressure drops.

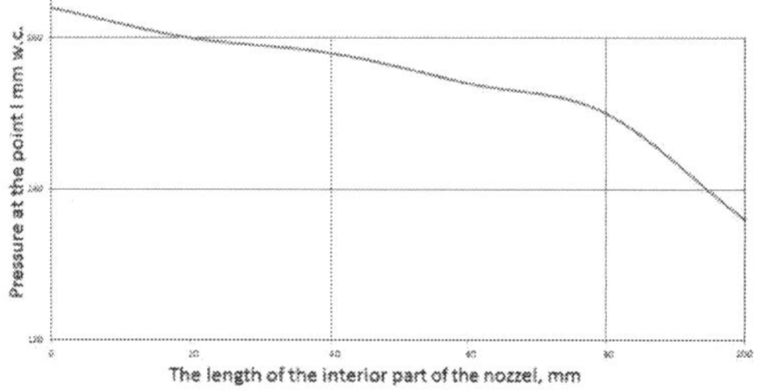

Fig. 6. The experimental dependence of the pressure inside the main chamber on the position of the discharge nozzle ("warm" stage).

Figure 7 shows the power consumption at the fan motor terminals. In general, the trend repeats the conditions of cold start (without burning fuel).

As a result, based on the results of the experiment, it is possible to draw a logical conclusion: the use of the discharge nozzle as a part extending inside of the main prechamber of combustion positively affects the aerodynamic characteristics of the fuel- air mixture. As the part extending inside increases, the pressure within the chamber decreases, alongside general reduction in energy consumption on the fan. For the experimental sample, the optimum position of the discharge nozzle inside the chamber is 80 mm, which is 47% for a total length of a chamber in 170 mm.

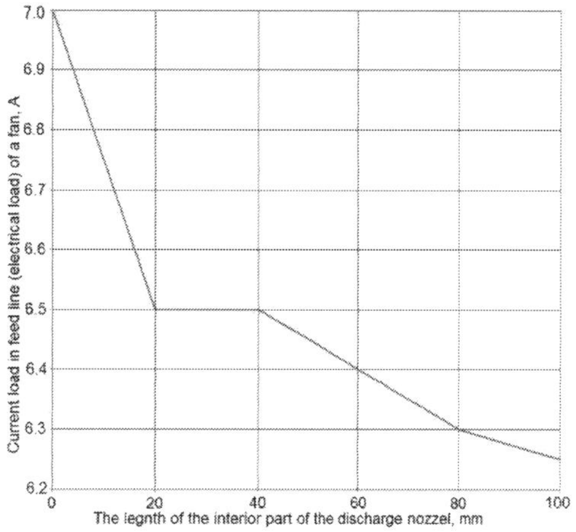

Fig. 7. The experimental dependence of the fan current consumption on the position of the discharge nozzle ("warm" stage).

At the next stage, it was decided to describe mathematically the aerodynamic flows in the proposed burner device in support of the experiment or to refute it. However, the format of this conference limits the number of pages in the report. Therefore, we are not provide simulation results here.

4 Conclusion

1. The necessity of changing the inner part of the discharge nozzle is substantiated. The recommended degree of embedding was 47-50% of the total length of the burner chamber;

2. Paten of Russian Federation and the European Patent Office was obtained for the structural characteristics of the combustion chamber (precombustor) and its elements;

3. The prototype of the burner showed perspective introduction of cyclone- vortex technology in boiler technology, as evidenced by the increase in their heating capacity, efficiency and a significant reduction in harmful emissions;

4. The minimum value of the horizontal coordinate for the outlet nozzle of small cross section was determined.

References

[1] G.F. Knorre, *Combustion processes* (Moscow: State Energy Publisher, 396, 1959)

[2] A.K. Hasannuddin, J.Y. Wira, R. Srithar, S. Sarah, M.I. Ahmad, S.A. Aizam, M.A.B. Aiman, M. Zahari, S. Watanabe, M.A. Azrin, S.S. Mohd. Effect of emulsion fuel on engine emissions-A review, Clean Technologies and Environmental Policy **18**, 1 (2015)

[3] M.Y. Khan, Z.A.A. Karim, F.Y. Hagos, A.R.A. Aziz, I.M. Tan, Current Trends in Water-in-Diesel Emulsion as a Fuel, Scientific World Journal, 1–15 (2014)

[4] M.Z. Abu-Zaid, Performance of single cylinder direct injection diesel engine using water fuel emulsions, Energy Convers. Manag. **45**, 697–705 (2004)

[5] A. Alahmer, J.Yamin, A. Sakhrieh, M.A. Hamdan, Engine performance using emulsified diesel fuel, Energy Convers Manag. **51**, 1708–1713 (2010)

[6] R.K. Gopidesi, S.R. Premkartikkumar, Performance, emission and combustion analysis of diesel engine fuelled with emulsified biodiesel, Progress in Industrial Ecology – An International Journal **12**, 1/2, 59–66 (2018)

[7] M. Kim, J. Oh, C. Lee, Study on Combustion and Emission Characteristics of Marine Diesel Oil and Water-In-Oil Emulsified Marine Diesel Oil, Energies **11**, 1830 (2018)

[8] E.M. Afify, S.S. Korah, The effect of air charge temperature on performance, ignition delay and exhaust emissions of diesel engine using w/o emulsions as fuel, SAE Paper no. 870555 (1987)

[9] T.Xavier, M. Alain, NOx and PM emissions reduction on an automotive HSDI Diesel engine with water-in-diesel emulsion and EGR an experimental study, Fuel **31**, 79–92 (2011)

https://doi.org/10.1051/e3sconf/202017801012

[10] Y.B. Zeldovich, Oxidation of nitrogen in combustion and explosion, Dokl Akad Nauk SSSR Seriya A51, 217–220 (1946)

[11] M. Tadashi, M. Yasushi, T. Minoru, M. Noboru, Experimental reduction of NOx, Smoke and BSFC in a Diesel engine using uniquely produced water (0-80 %) to fuel emulsion, SAE Paper no. 780224 (1978)

[12] K.A. Shtym, T.A. Solov'eva, Conversion of KVGM-100-150 boilers to cyclone-swirl burning of gas, Thermal Engineering **62**, 3, 202–207 (2015)

[13] A.N. Shtym, *Aerodynamics of cyclone-vortex chambers* (Vladivostok: Far Eastern University Publisher, 200, (1984)

[14] A. Baubek, N. Baubek, Fuel combustion device, Patent of Russian Federation no. 128284 (2012)

[15] A. Baubek, N. Baubek, Device for burning fuel [Electronic resource], European Patent no. 2864700 (2016). Available at: https://data.epo.org/gpi/EP2864700B1-DEVICE-FOR-BURNING-FUEL

Experience of designing a low-pressing turbocharger compressor using the modern version of a Universal modelling method

Yuri Galerkin[1], *Aleksandr* Drozdov[1,*], and *Sergey* Sibiriakov[2]

[1]Peter the Great St.Petersburg Polytechnic University St.Petersburg, Polytechnicheskaya, 29, Russia
[2]NPO "Turbotekhnika" Protvino, Zavodskoy passage, 4., Russia

Abstract. The paper presents the joint experience of the NPO "Turbotekhnika" and the R&D Laboratory "Gas Dynamics of Turbomachines" of SPbPU for designing a centrifugal compressor operating at pressure ratio 1.61 and a mass flow rate of 0.62 kg/s. The design was executed using the Universal modelling method and the inviscid quasi-three-dimensional calculation program 3DM.023. At the first step, by the preliminary design program, the stage dimensions were determined. The expected gas-dynamic characteristics are calculated. At the final design step, stator elements were offered by NPO "Turbotekhnika" and the configuration of the impeller blades was optimized based on the non-viscous quasi-three-dimensional calculations. NPO "Turbotekhnika" designed, manufactured and tested the 140E compressor at blade velocities 150, 200, 250 and 300 m/s. A comparison and analysis of experimental and calculated characteristics is presented. The design pressure ratio was calculated almost exactly for the design flow rate at a blade velocity of 300 m/s. The expected efficiency was confirmed. The mathematical model slightly overestimates the efficiency and the pressure ratio for the off-design flow rates.

1 Introduction

For optimal gas-dynamic design of centrifugal compressors, various approaches are used: CFD calculations, appropriate engineering mathematical models, etc. [1-4]. The CFD calculations of the turbocharger gas-dynamic characteristics are presented in paper [5]. The measured and calculated efficiency characteristics of a single-stage turbocharger TK23N-06 with a vaneless diffuser and a scroll are shown in Figure 1.

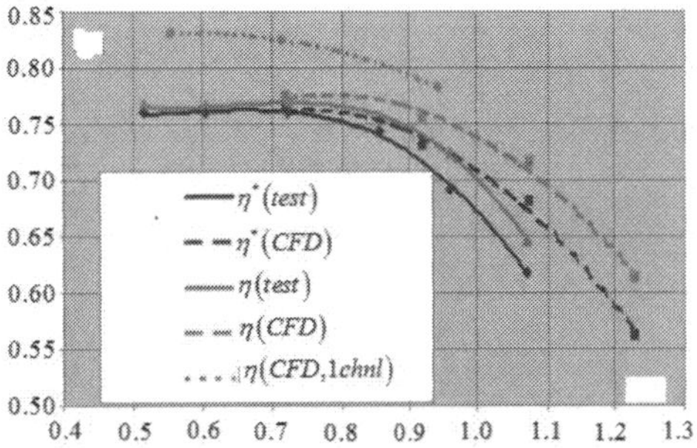

Fig. 1. The efficiency versus mass flow rate (kg/s) of the compressor TK23N-06 at n = 21000 rpm [5].

In the Figure 1: η^* is efficiency depending on total parameters, η is efficiency depending on static parameters, "chnl" has been calculated for one blade sector.

The maximal stage efficiencies calculated by the CFD more or less agree with the experimental data. But the calculated characteristic is biased towards high flow rates. The simulation is unsatisfactory.

The application of engineering mathematical models for calculating the turbocharger compressor is presented in paper [6]. The calculation is performed in a one-dimensional setting. The flow parameters are determined in five control sections: at the impeller inlet, at the vaneless section inlet, at the vane diffuser inlet, at the scroll inlet and the exit from the stage. Figure 2 presents a comparison of experimental and TuCMS calculated performances. Calculated performances are significantly different from experimental. This indicates a lack of model elaboration and the need for its further development.

There are many other publications that demonstrate inefficiency of commercial CFD programs for centrifugal compressor characteristics simulation. The sample in Figure 2 demonstrates that correct engineering simulation is not easy, too.

2 Aims and problems

NPO "Turbotekhnika" works on the turbocharger for the 8CN26/26 gas engine. Preliminary matching analysis of the turbocharger and the engine has demonstrated that

* Corresponding author: A_drozdi@mail.ru

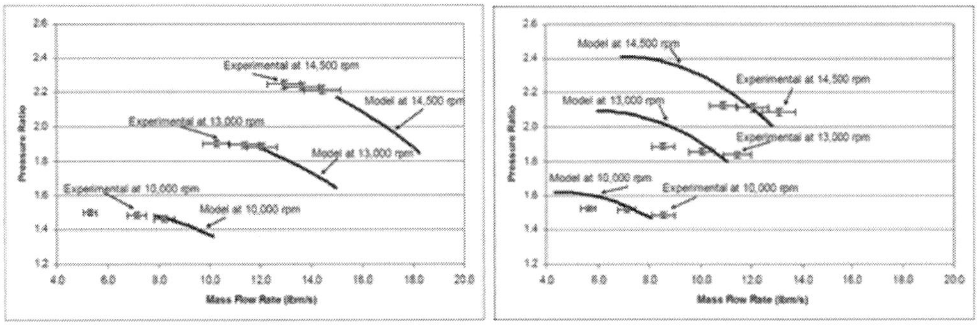

Fig. 2. Comparison of experimental characteristics and calculations performed using the program TuCMS [6].

the compressor pressure ratio at the design flow rate must be about 1.6. A large range of flow rate is needed.

In gas-dynamic design of the compressor the R&D laboratory "Gas Dynamics of Turbomachines" (LGDTM) of SPbPU has applied its Universal modelling method [0-0]. It allows to perform end-to-end optimal design using the following steps:

- search for the optimal compressor version using a simplified mathematical model without the flow path dimensions;
- primary design, i.e., determining the basic dimensions of the flow path and the blade configuration, corresponding to a design flow rate and pressure ratio of the compressor;
- optimization and calculation of characteristics using the last (8-th) version of mathematical models of efficiency and work coefficient;
- digital description of the flow path and the formation of compressor flow path solid models.

Aim of the research is to design a low pressure ratio turbocharger compressor, to manufacture and test an experimental turbocharger. Evaluate the applicability of the Universal modelling method to effective design of turbocharger compressors.

3 140E compressor parameters and candidates comparison

Figure 3 (left) shows the menu of the candidates' comparison and preliminary design PC software with compressor parameters ($\bar{m}$ =0.62 kg/s, $\pi^* = p^*_{ex} / p^*_{inl}$ = 1.61, where $\bar{m}$ is mass flow; π^* is pressure ratio; p^*_{inl} is inlet total pressure; p^*_{ex} is exit total pressure). The turbine with 41 000 rpm was chosen by NPO "Turbotechnica" for the turbocharger. Figure 3 (right) shows the compressor candidates' efficiency, calculated using a simplified mathematical model [10-12].

The compared candidates have different loading factors $\psi_{T\,des} = c_{u2} / u_2$ (where c_{u2} is the circumferential velocity component; u_2 is blade velocity; $\psi_{T\,des}$ is the design loading factor).

The bigger $\psi_{T\,des}$ is, the smaller is an impeller diameter and bigger is flow rate coefficient $\Phi_{des} = \dfrac{\bar{m}_{des}}{\rho^*_{inl} \cdot 0{,}785 D_2^2 u_2}$ ($\bar{m}_{des}$ is design mass flow rate, ρ^*_{inl} is gas density depending on total parameters at an

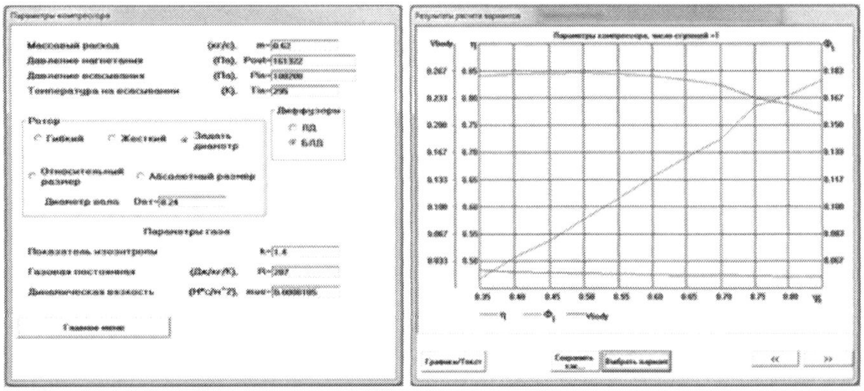

Fig. 3. The program of single-shaft centrifugal compressor candidates' comparison and preliminary design (SSCCPD). On the left – the compressor parameters. On the right – candidates' efficiency, flow rate and the approximate case volume versus the loading factor.

Table 1. SSCCPD program. Part of the output file of the candidates' calculation. 140E compressor variant selected for the design.

**** Selected variant ****

#	Stage	Φ_{des}	M_u	D2	$\bar{D}_h$	$\psi_{T des}$	Re_u	η^*
1	3D+VLD	0.1141	0.8708	0.1397	0.2400	0.5720	2.680E+06	0.8421

```
Compressor efficiency, ETc=0.8421
RPM                  ,   n=41000.00 1/min
Power consumption    ,   N=32.35 kW
Tip speed            ,   U2=299.80 m/s
Body volume          ,   Vb=1.558E-02 m^3
```

inlet).

From Figure 3 it follows that a simplified mathematical model to obtain maximum efficiency recommends choosing the candidate with $\psi_{T des}$ = 0.50 with the blade velocity 325 m/s and the impeller diameter being 151 mm. The expected efficiency is 0.847.

For the final design, a candidate with a slightly larger loading factor $\psi_{T des}$ = 0.572 is taken. Its parameters are presented in Table 1. The selected compressor variant has a blade velocity of 300 m/s, impeller diameter 140 mm, but the efficiency is lower by 0.49%. Despite this, for a medium-speed engine with a very long service life, an impeller with a lower blade velocity is preferable. The designed compressor is named 140E. The turbocharger with this compressor is named TKR 140E.

In the Table 1: $M_u = u_2 / \sqrt{kRT^*_{inl}}$ is the blade's Mach number; $\bar{D}_h = D_h / D_2$ is hub ratio; k is isentipic coefficient; R is gas constant; T^*_{inl} is total temperature at the inlet.

4 Preliminary design

After the preferred compressor candidate is selected, a preliminary design program calculates all the dimensions necessary for characteristics calculating. The dimensions are automatically transferred to the program for characteristics calculating of a single-stage compressor (OPTIM 2 program).

Compressor TKR 130 stator elements were offered by NPO "Turbotechnica" in accordance with their design method. According to the primary design, the stator element's dimensions are close to the dimensions offered by NPO "Turbotechnica".

In the preliminary design, the dimensions of the impeller blades are also calculated. This allows to check the compliance of the impeller preliminary design with the technical specification. Figure 4 shows velocity diagrams of an inviscid flow on the preliminary designed impeller blades. Calculations of an inviscid quasi-three-dimensional flow are carried out according to the program 3DM.023 [13].

The flow character corresponds to the recommendations in the monograph [13], but at the shroud blade-to-blade surface, the velocity peak at the leading edge indicates a negative incidence angle. Not optimally large deceleration of the relative velocity is noticeable. Flow deceleration on the blades suction side on the shroud blade-to-blade surface should also be reduced.

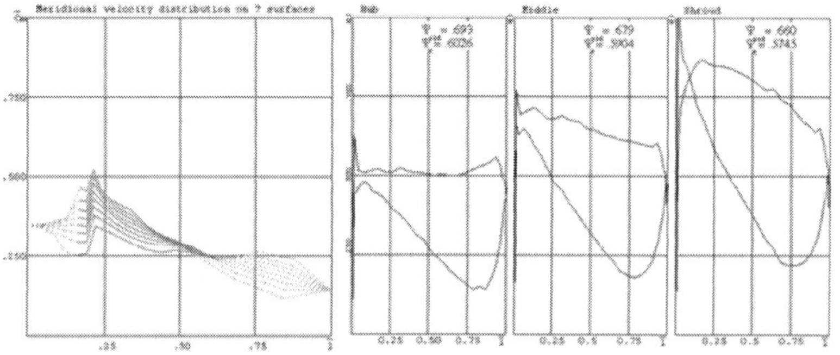

Fig. 4. 3DM.023 program. The primary design of the compressor 140E impeller. On the left are the meridional velocities on eight blade-to-blade surfaces, on the right are the velocity diagrams on the three blade-to-blade surface.

 https://doi.org/10.1051/e3sconf/202017801013

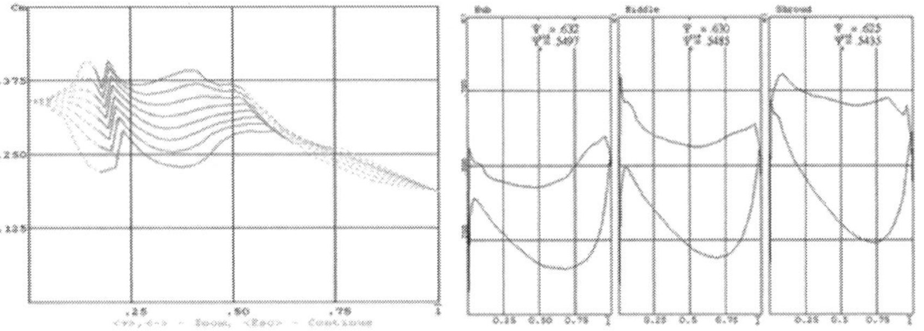

Fig. 5. 3DM.023 program. Final design of the compressor 140E impeller. On the left are the meridional velocities on eight blade-to-blade surfaces, on the right the velocity diagrams on the three blade-to-blade surface.

5 Final design TKR 140E

The quasi-three dimensional inviscid flow calculation using the 3DM.023 PC software presents the value of the loading factor $\psi_{T\,id}$. Viscosity reduces the loading factor, which is taken into account by the empirical coefficient $K_\mu = \psi_{T\,des} / \psi_{T\,id}$. The value adopted at the preliminary design K_μ = 0.92 is based on experiments with stages with return channels. Turbocharger TKR stages include a scroll. In this stage, the flow is characterized by a circumferentially uneven pressure field, which creates a non-stationarity of the flow in the impeller. The test data of the NPO "Turbotekhnika" were reduced. The value of empirical coefficient is K_μ = 0.98 for the impellers similar to 140E.

The value of the loading factor of the impeller 140E $\psi_{T\,des}$=0.585 is 2% more than that of the adopted candidate (Table 1). This is a common design practice. The well-known European standard [0] requires unconditional provision of a design pressure ratio, allowing an excess of power consumption up to 4%. In order to obtain the desired value $\psi_{T\,id} = \psi_{T\,des} / K_\mu$ = 0.585/0.98 = 0.597 with non-incidence flow on shroud and to optimize the velocity diagrams the variants of the blade row were considered.

The form of the function $\beta_{bl\,s} = f(\bar{l}_m)$ (where $\beta_{bl\,s}$ is the impeller blade angle on shroud; $\bar{l}_m$ is the impeller blade relative length in meridional plane) with double curvature is important. This form is explained by the desire to eliminate the "saddle" in the velocity diagram on the shroud current line. The velocity diagrams are shown in Figure 5 on the right.

A complete elimination of the "saddle", i.e. to obtain a linear decrease in velocity along the blade suction side $\bar{w}_s$ (where $\bar{w}_s$ is relative velocity at the flow separation point on the blade profile) was not possible. In accordance with the recommendations [0-0], an increase in the exit blade angle from the hub to the shroud is taken. The diagrams of meridional velocities are shown in Figure 5 on the left.

Figure 6 shows the 140E impeller blade row according to the data of the 3DM.023 software and the view of the solid model calculated by the "3D-Compressor" program included in the system of the end-to-end design programs. The software reads the dimensions obtained as a result of the design using the Universal modelling method and transfers the information to the commercial package SolidWorks,

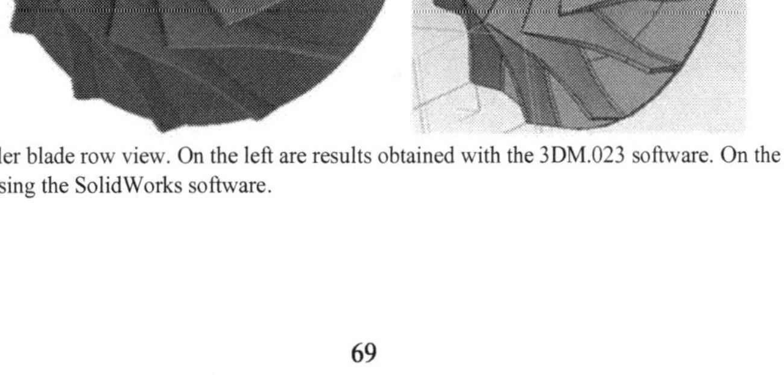

Fig. 6. TKR 140E impeller blade row view. On the left are results obtained with the 3DM.023 software. On the right is a view of a solid model obtained using the SolidWorks software.

which automatically builds a 3D model.

A family of 140E compressor characteristics is calculated $\eta^*, \pi^*, N = f(\bar{m})$ (where N is shaft power; η^* is polytrophic efficiency) at blade velocities 150, 200, 250 and 300 m/s using the direct gas dynamic solution software. The characteristics are presented below in comparison with the test results.

6 Test results. Comparison with the project

Figure 7 shows a section of the turbocharger TKR 140E of NPO "Turbotekhnika". The measured and calculated characteristics of the 140E compressor are compared on Figure 8.

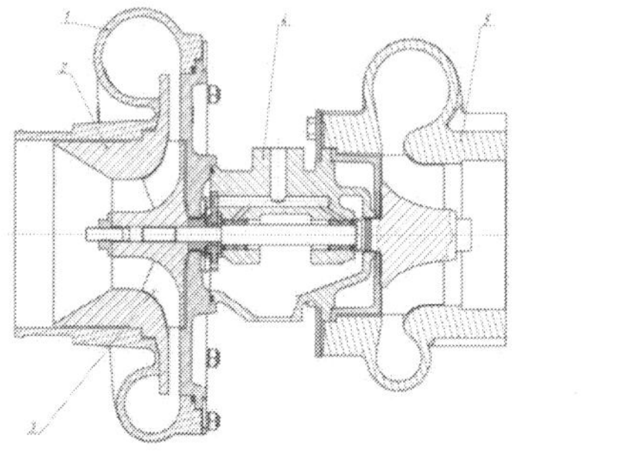

Fig. 7. Sectional drawing of turbocharger TKR 140E. 1 – compressor housing; 2 – insert; 3 – impeller (R&D laboratory "Gas dynamic of turbomachines" project); 4 – bearing housing; 5 – turbine.

At the design flow rate $\bar{m}_{des}$ = 0.62 kg/s at u_2 = 300 m/s the specified pressure ratio is ensured almost

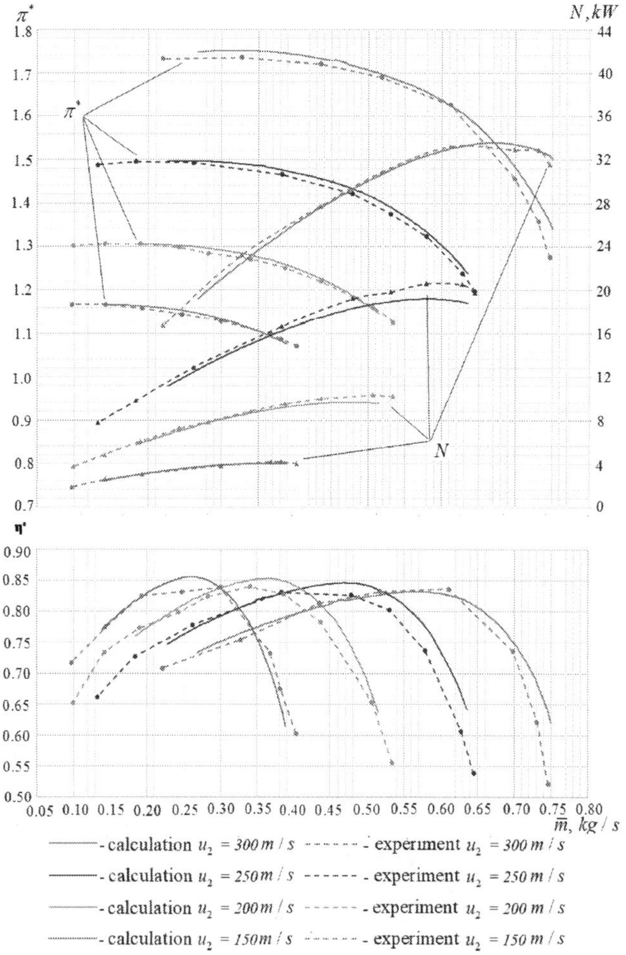

Fig. 8 Comparison of design and experimental characteristics of model 140E compressor at blade velocities u_2 = 150, 200, 250 and 300 m/s.

exactly. The expected efficiency is confirmed. The mathematical model overestimates the efficiency and pressure ratio in off-design modes. At the same time, the calculated maximum pressure ratio flow rate is calculated closer to the design flow rate in comparison with the experiment. As experimental data accumulate, the empirical coefficients of the mathematical model will be refined.

7 Conclusion

The new version of the Universal modeling method has confirmed the effectiveness with regard to the design of relatively small-sized turbochargers compressors.

The compressor 140E has good gas-dynamic characteristics. This design method can be applied for other turbocharger deigns.

Acknowledgements

The research was performed by a Grant of the President of the Russian Federation for young PhD MK-1893.2020.8.

References

[1] V.E. Evdokimov, Turbines and compressors **3**, 4 (1997)

[2] A.T. Lunev, Compressors and pneumatics **10**, 4–7 (2001)

[3] K. Kabalyk, W. Kryłłowicz, Transactions of the institute of fluid-flow machinery **131**, 41–53 (2016)

[4] W. Kryłłowicz, P. Świder, Z. Kozanecki, K. Kabalyk, Jr.Z. Kozanecki, 12th European Conference on Turbomachinery Fluid Dynamics and Thermodynamics, 127934 (2017)

[5] V. Melnikov, A. Proskurov, CAD and graphics **4**, 92–96 (2005)

[6] K. Lusardi, Thesis for the Degree of Master of Science, 90 (2012)

[7] Y. Galerkin, *Proceedings of the science school of compressor engineering SPbGPU* (M.: pub. SPbGPU, 2010)

[8] A. Drozdov, Y. Galerkin, AIP Conf. Proc. 030052-1–030052-10 (2018)

[9] Y. Galerkin, K. Soldatova, A. Drozdov, A. Rekstin, ASME 2015 Gas Turbine India Conference, 119665 (2015)

[10] A. Rekstin, Y. Popova, A. Ucehovscy, AIP Conf. Proc., 030036-1–030036-11 (2018)

[11] A. Rekstin, K. Soldatova, Y. Galerkin, University news 9, 702 (2018)

[12] A. Rekstin, B.V. Bakaev, Scientific and technical statements SPbPU **24**, 4, 24–38 (2018)

[13] Y. Galerkin, *Turbocompressors* (Information and publishing center KHT, 596, 2010)

[14] ISO 5389 Turbocompressors Performance test code (1992)

[15] Y. Galerkin, A. Drozdov, ASME 2015 Gas Turbine India Conference, 119665 (2015)

[16] Y. Galerkin, A. Drozdov, Scientific and technical statements SPbPU **4**, 231, 179–188 (2015)

Mathematical model of centrifugal compressor vaneless diffuser based on CFD calculations

Olga Solovyeva[1,*], and *Aleksandr* Drozdov[1]

[1]Peter the Great St.Petersburg Polytechnic University St.Petersburg, Polytechnicheskaya, 29, Russia

Abstract. The approximate engineering techniques based on mathematical modelling are used in centrifugal compressor design. One of such methods is the well-proven Universal Modelling Method, developed in the scientific and research laboratory "Gas dynamics of turbo machines", SPbPU. In the modern version of the compressor model, vaneless diffusers mathematical model was applied based on a generalization of the CFD calculations. The mathematical model can be used for vaneless diffusers with a relative width in the range of 1.4 – 10.0%, with a radial length up to 2.0, in the range of inlet flow angles 10 to 90 degrees, the inlet velocity coefficient in the range of 0.39 – 0.82, Reynolds number varying from 87 500 to 1 030 000. The model was also used for calculating low-flow-rate model stages with narrow diffusers with diffusers' relative width in the range of 0.5 – 2.0%. The mathematical model showed lesser accuracy. To widen the model applicability, new series of CFD-calculations were executed. A series of vaneless diffusers was designed with relative width in the range of 0.6 – 1.2%, The gas-dynamic characteristics of loss coefficients and outlet flow angle versus inlet flow angle of diffuser were calculated. Regression analysis was used to process the calculated data. System of algebraic equations linking geometric, gas-dynamic parameters and similarity criteria was developed. The obtained equations are included in a new mathematical model of the Universal Modelling Method.

1 Introduction

CFD calculations are widely used for the calculation and design of centrifugal compressors and their elements [1-4], but the problem of verification of the calculated data is important. The calculation results largely depend on the object of research, applied software packages, selected boundary conditions, etc. Some foreign authors claim good agreement between the calculated and experimental data [5-8]. In many cases, the authors either do not indicate the calculation methodology, or use programs of their own design. Other authors point to a number of unsolved problems with the verification of CFD calculations. The main problems are associated with the calculation of the centrifugal compressor impellers. Calculations for various CFD programs overestimate the loading factor by 7 – 12%, depending on the type of impeller, and shift the calculated gas-dynamic characteristics of efficiency to high flow rates.

The authors' experience confirms the presence of such problems [10-13]. In our opinion, solving the problem of overstating the loading factor will also lead to the matching of the ranges of the calculated and measured efficiency characteristics. The experience of the R&D Laboratory "Gas Dynamics of Turbomachines" in the application of CFD calculations to the modeling of stator elements of centrifugal stages shows good results [14-15]. CFD calculations have been applied in the implementation of gas-dynamic projects of centrifugal compressors and replaceable flow paths in the interests of industrial partners. In particular, the shape of the VLD and the crossover was optimized. Based on this experience, O.A. Solovyeva as part of a PhD thesis research [16] developed a new mathematical model of VLD. Its features are presented below.

2 Materials and methods

2.1 The mathematical model of VLD, based on a generalization of the computational experiment for VLD with b_3 / D_2 = 0.014 – 0.100 (model No. 1)

At the first stage, researches were conducted on the method of setting, conducting and processing the results of CFD calculations of VLD [15]. Recommendations are made on choosing the density of the computational grid and the turbulence model. Before conducting virtual aerodynamic tests of the VLD series, it is necessary to make sure that the flow pattern calculated using the CFD program corresponds to the physical meaning, and the integral characteristics match the experimental results with the required accuracy. For this, a comparison of the results of a computational experiment with the known experimental data was performed. Figure 1 shows a comparison of VLD characteristics calculated by the ANSYS CFX program and experimental ones tested together with impellers [16]. For comparison, the

* Corresponding author: Solovyeva.oa@yandex.ru

calculated diffusers closest in parameters were chosen $b_3 / D_2 = 0.014$, 0.016, 0.033, 0.038, with radial length $D_4 / D_2 = 1.6$, Mach number $M_{c2} = 0.75$ (where b_3 is VLD width, D_2 is impeller diameter, D_4 is VLD outlet diameter, M_{c2} is Mach number). In experimental and computational researches, the pressure coefficient increases with increasing ratio c_{r2} / c_{u2} (where c_{r2} is the radial component of absolute velocity, c_{u2} is the circumferential component of velocity, α_2 is the flow angle) ($c_{r2} / c_{u2} = tg\alpha_2$), i.e., the diffuser works less and less efficiently as the flow direction approaches the tangential one. Some of the results, which correspond to a small value c_{r2} / c_{u2}, were obtained with a rotating stall; accordingly, these results are less reliable [16]. As a result, for a small value c_{r2} / c_{u2}, the difference between the experimental (b_3 / D_2=0.0390) and calculated (b_3 / D_2=0.0380) characteristics is large. Characteristics calculated by the ANSYS CFX program are higher than the experimental characteristics. The difference is explained by the fact that the experimental data presented in [16] relate to stages where mixing losses occur in the VLD, caused by the appearance of a "wake" in the impeller. The uneven flow pattern at the impeller outlet also increases VLD losses. In the calculation of an isolated VLD, there is no negative influence of the impeller.

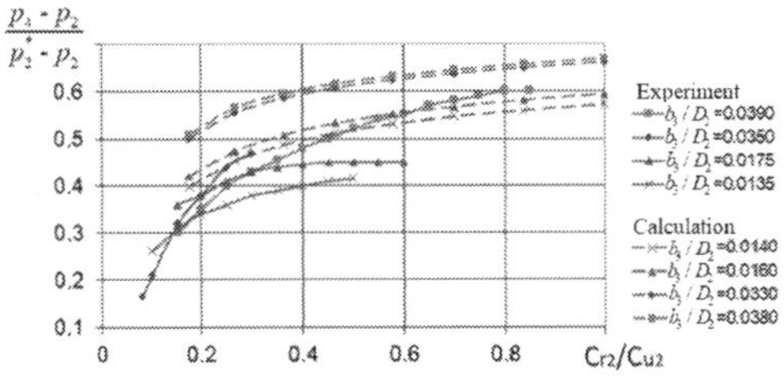

Fig. 1. Comparison of experimental and calculated pressure recovery coefficients in a vaneless diffuser depending on the parameter c_{r2} / c_{u2} for different values of the relative width of the diffuser. Experiment (solid line): b_3 / D_2 =0.0135, 0.0175, 0.0350, 0.0390; calculation (dashed line): b_3 / D_2 = 0.0140, 0.0160, 0.0330, 0.0380.

Comparison with the provisions of the theory showed the regularity of the flow of gas-dynamic characteristics, and with well-known experiments – a good agreement on the flow structure. This allowed us to draw conclusions about the reliability of the data obtained as a result of the calculation, and the validity of replacing the physical experiment in the research of VLD by a computational experiment. The flow and gas-dynamic characteristics of the VLD with a relative width b_3 / D_2 = 0.014 – 0.100, with a radial length D_4 / D_2 up to 2.0, and in the range of inlet flow angles α_2 = 10 – 90° [15, 16]. Similarity criteria varied between λ_{c2}=0.39 – 0.82, Re_{b2}=87 500 – 1 030 000 (where λ_{c2} is velocity coefficient, Re_{b2} is Reynolds number). The approximation of the computational experiment was performed using the EXCEL program by the method of regression analysis. To calculate the flow parameters at the VLD outlet, we need to know any two of five coefficients – efficiency, loss coefficient, velocity ratio, recovery coefficient, and outlet flow angle. The loss factor and the outlet flow angle from the VLD were selected as modeling objects:

$$\zeta, \alpha_4 = f\left(\bar{b}_3,\ \bar{D}_4,\ \alpha_2,\ \lambda_{c2}, \mathrm{Re}_{b2}, \bar{k}_{roug}\right) \qquad (1)$$

where k_{rough} is relative surface roughness, ζ is the loss coefficient.

The loss coefficient was simulated in three stages: first, the dependence $\zeta = f\left(\bar{b}_3, \alpha_2, \lambda_{c2}\right)$ was approximated, then correction factors were introduced that take into account the influence of the VLD length, Reynolds criterion, and relative roughness. The correction for the influence of the Reynolds criterion takes into account the surface roughness if the surface is not hydraulically smooth. The general structure of the formula is:

$$\zeta = \frac{A \cdot \alpha^B \cdot K_{D4}}{K_{\mathrm{Re}, k_{roug}}}, \qquad (2)$$

where: $A = c \cdot (b_3 / D_2)^2 + d \cdot (b_2 / D_2) + e;$

$c = 2312.7 \cdot \lambda_{c2}^2 - 2234.9 \cdot \lambda_{c2} + 830.63;$

$d = -207.36 \cdot \lambda_{c2}^2 + 266.98 \cdot \lambda_{c2} - 88.333;$

$e = 6.4922 \cdot \lambda_{c2}^2 - 6.9342 \cdot \lambda_{c2} + 3.5363;$

$B = f \cdot (b_3 / D_2)^2 + g \cdot (b_3 / D_2) + h;$

$f = -124.32 \cdot \lambda_{c2}^2 + 186.87 \cdot \lambda_{c2} - 17.242;$

$g = 19.241 \cdot \lambda_{c2}^2 - 29.915 \cdot \lambda_{c2} - 3.3289;$

$h = -0.6906 \cdot \lambda_{c2}^2 + 0.6577 \cdot \lambda_{c2} - 0.527.$

To account for the VLD length, a rather complicated system of approximating formulas was also obtained:

$$K_{D4} = j \cdot (D_4 / D_2)^2 + k \cdot (D_4 / D_2) + l, \qquad (3)$$

where: $j = m \cdot (b_3 / D_2)^2 + n \cdot (b_3 / D_2) + p$;

$m = 3455.2 \cdot \lambda_{c2}^2 - 3513.1 \cdot \lambda_{c2} + 669.16$;

$n = -443.08 \cdot \lambda_{c2}^2 + 427.91 \cdot \lambda_{c2} - 74.406$;

$p = 8.756 \cdot \lambda_{c2}^2 - 8.6285 \cdot \lambda_{c2} + 0.5291$;

$k = r \cdot (b_3 / D_2)^2 + s \cdot (b_3 / D_2) + t$;

$r = -12860 \cdot \lambda_{c2}^2 + 13344 \cdot \lambda_{c2} - 2726.1$;

$s = 1633.8 \cdot \lambda_{c2}^2 - 1613.6 \cdot \lambda_{c2} + 306.13$;

$t = -31.076 \cdot \lambda_{c2}^2 + 31.02 \cdot \lambda_{c2} - 1.7055$

$l = u \cdot (b_3 / D_2)^2 + v \cdot (b_3 / D_2) + w$;

$u = 11401 \cdot \lambda_{c2}^2 - 11984 \cdot \lambda_{c2} + 2574.4$

$v = -1439.3 \cdot \lambda_{c2}^2 + 1440.6 \cdot \lambda_{c2} - 290.69$;

$w = 26.521 \cdot \lambda_{c2}^2 - 26.682 \cdot \lambda_{c2} + 2.2384$

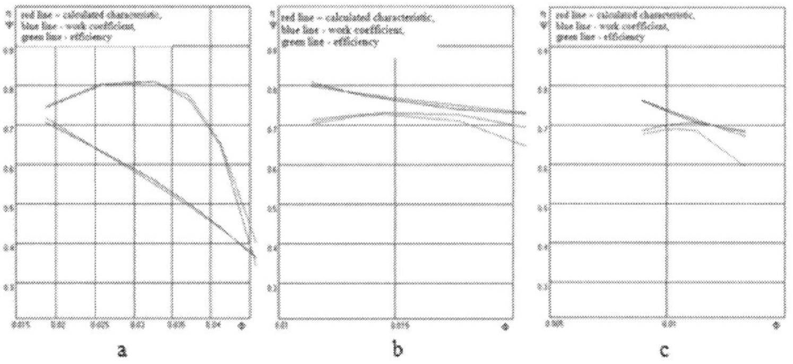

Fig. 2. Comparison of calculated and experimental data for the stage a) 0028-056-0373 $\Phi_{des} = 0.028$, $\psi_{T\,des} = 0.56$, $\bar{D}_{hub} = 0.373$, $\bar{b}_3 = 0.033$; $\bar{D}_4 = 1.45$, $M_u = 0.60$, $\mathrm{Re}_u = 4.8\cdot10^6$; b) XXX3-Q with parameters: $\Phi_{des} = 0.0203$, $\psi_{T\,des} = 0.69$, $\bar{D}_{hub} = 0.337$, $\bar{b}_3 = 0.01$; $\bar{D}_4 = 1.40$, $M_u = 0.785$, $\mathrm{Re}_u = 9.8\cdot10^6$; c) XXX3-T with parameters: $\Phi_{des} = 0.013$, $\psi_{T\,des} = 0.62$, $\bar{D}_{hub} = 0.337$, $\bar{b}_3 = 0.008$; $\bar{D}_4 = 1.40$, $M_u = 0.803$, $\mathrm{Re}_u = 10\cdot10^6$.

Correction for the influence of the Reynolds criterion – relative roughness – was made by analogy with the resistance coefficient for the plate, but taking into account the specifics of the formulation of the computational experiment. For a hydraulically smooth surface:

$$K_{\mathrm{Re}} = \frac{\lambda_{calc}}{\lambda_{MM}} = \frac{0.0032 + \dfrac{0.221}{\left(8.73\cdot10^6\cdot\bar{b}_3\right)^{0.237}}}{0.0032 + \dfrac{0.221}{\mathrm{Re}_b^{0.237}}}; \qquad (4)$$

- for rough surface:

$$K_u = \frac{\lambda_{calc}}{\lambda_{MM}} = \left(0.0032 + \frac{0.221}{\left(8.73\cdot10^6\cdot\bar{b}_3\right)^{0.237}}\right)\left(2\lg\frac{2}{\bar{k}_{roug}} + 1.74\right)^2. \qquad (5)$$

In the calculations, we choose the formula, in which the loss coefficient is greater. The system of equations is characterized by a good accuracy of 98.6%. The approximation for calculating the outlet flow angle from the VLD is performed using the same method as for the loss coefficient – the method of regression analysis. The calculated data are approximated by a power polynomial of the following form:

$$\alpha_4 = \alpha_2 + a\cdot\alpha_2^2 + b\cdot\alpha_2 + c\,, \qquad (6)$$

where: $a = d\cdot\lambda_{c2} + e$;

$d = 0.0013\ln(b_3 / D_2) + 0.01$;

$e = -0.0021$;

$b = f\cdot\lambda_{c2} + g$; $f = -0.13\ln(b_3 / D_2) - 0.9844$;

$g = 0.0563\ln(b_3 / D_2) + 0.3345$;

$c = h\cdot\lambda_{c2} + i$; $h = 2.208\ln(b_3 / D_2) + 9.7999$;

$i = -5.137\ln(b_3 / D_2) - 13.136$.

In the mathematical model there are no empirical coefficients that need to be identified. In addition, due to the fact that the entire diffuser is calculated, and not 20 of its parts, as in previous versions of the model, the calculation speed increases. The proposed formulas are implemented using the gas-dynamic function of the current density [16]. The result is an equation containing one unknown – the velocity coefficient λ_{c4} at the VLD outlet:

$$\lambda_{c4} = \lambda_{c2}\frac{\left(\dfrac{1-\dfrac{k-1}{k+1}\lambda_{c4}^2}{1-\dfrac{k-1}{k+1}\lambda_{c2}^2}\right)^{\frac{k-1}{k}}}{\bar{D}_4\dfrac{\sin\alpha_4}{\sin\alpha_2}\left(\dfrac{1-\dfrac{k-1}{k+1}\lambda_{c2}^2}{1-\dfrac{k-1}{k+1}\lambda_{c4}^2}\right)^{\frac{k-1}{k}\frac{\zeta}{1-\left(\frac{\lambda_{c4}}{\lambda_{c2}}\right)^2}}}, \qquad (7)$$

where k is isentropic coefficient. The equation is solved by bruteforcing. Further calculation is based on the well-known equations of thermogasdynamics. The developed mathematical model was applied in the 8th version of the Universal Modeling Method programs. A comparison of the calculated by the 8th version of the mathematical model and the measured gas-dynamic characteristics was carried out for model stages of the 20CE series with VLD. The range of basic design parameters for model stages: $\Phi_{des} = 0.028 - 0.080$, $\psi_{T\,des} = 0.40 - 0.67$, $\bar{D}_{hub} = 0.25 - 0.373$, $b_2 / D_2 = 0.03 - 0.061$; $\bar{D}_4 = 1.42 - 1.60$, $M_u = 0.60 - 0.80$, $\mathrm{Re}_u = 5.9\cdot10^6 - 8.0\cdot10^6$ (where $\bar{D}_{hub}$ is relative hub diameter, Φ_{des} is design relative flow rate coefficient, $\psi_{T\,des}$ is design loading factor). Comparison showed the average error in modeling the design (maximum) efficiency of 0.977%, the average error at five points (except for the point of highest flow rate) being 1.27%. An example of a comparison is shown in the figure 2a.

2.2 Application of the model for Clarke model stages

The second object of assessing the applicability and effectiveness of the developed VLD model was the

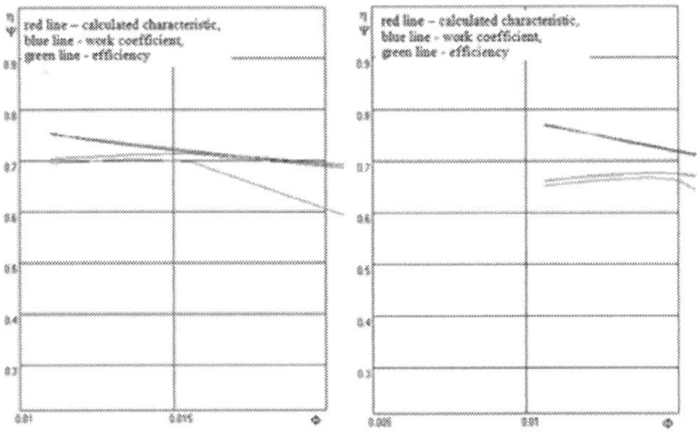

Fig. 3. Comparison of calculated and experimental data for stage XXX3-S with parameters: Φ_{des} = 0.015, $\psi_{T\,des}$ = 0.63, $\bar{D}_{hub}$ = 0.337, $\bar{b}_3$ = 0.008; $\bar{D}_4$ = 1.40, M_u = 0.803, Re_u = $10\cdot10^6$ (left) and stage XXX3-U with parameters: Φ_{des} = 0.011, $\psi_{T\,des}$ = 0.62, $\bar{D}_{hub}$ = 0.337, $\bar{b}_3$ = 0.0057; $\bar{D}_4$ =1.40, M_u =0.803, Re_u = $10\cdot10^6$.

Clarke series of low-consumption model stages, designed in the 1950s. The range of basic design parameters for model stages of the Clarke company is: Φ_{des} = 0.007 – 0.024, $\psi_{T\,des}$ = 0.60 – 0.69, $\bar{D}_{hub}$ =0.3214 – 0.3375, b_2 / D_2 = 0.0094 – 0.0476; $\bar{D}_4$ = 1.33 – 1.4, M_u = 0.366 – 0.843, Re_u = $4.8\cdot10^6 - 1.1\cdot10^7$. Comparison of calculated and experimental data showed an average error of modeling the design (maximum) efficiency of 1.15%, an average error of five points (except for the point of highest flow rate) being 2.15%. An example of comparison of calculated and measured gas-dynamic characteristics is presented in the figures 2b),c) and 3. We can see that for low-flow-rate stages, there is a tendency to overestimate the calculated efficiency compared to the measured one. One reason for this is the underestimation of the VLD loss coefficient. The stages considered have VLD with a width of $\bar{b}_3$ = 0.005 – 0.019, which for most stages extends beyond the width of the VLD involved in virtual wind tunnel tests and the creation of a VLD mathematical model.

This shows that it is incorrect to extrapolate previously calculated data to narrower VLD. In this regard, it became necessary to refine the already existing mathematical model by including narrow VLD in the sample for approximation. This assumption required verification. The reason for the difference in the calculated and experimental characteristics may also be underestimation of the impact of losses in other elements of the flow part. Thus, in low-flow stages, friction losses (due to the fact that the channels are narrow and long) have a greater impact on the reduction of efficiency than mixing losses. Since there are no element-by-element characteristics for Clarke model stages, it is possible to verify the conclusions made only with the help of CFD calculations. For this purpose, four diffusers were calculated with a relative width corresponding to that used in the Clarke stages b_3 / D_2 =0.006, 0.008, 0.010, 0.012.

Calculation setup corresponds to the above recommendations. For these objects, the characteristics of the loss coefficient and the change of the flow angle in the VLD were calculated. They were compared with the calculation using formulas (2) and (7), see figure 4.

In the calculated range of narrow diffusers, the width decreases by 2 times, while the loss coefficient increases

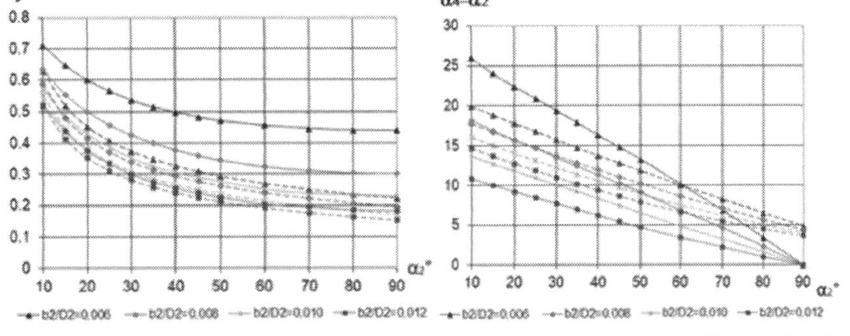

Fig. 4. Comparison of approximating (dashed line) and calculated (solid line) values of VLD loss coefficient (left) and $\alpha_4 - \alpha_2$ (right) with b_3 / D_2 =0.006, 0.008, 0.010, 0.012, $\bar{D}_4$ =1.60, λ_{c2} =0.64

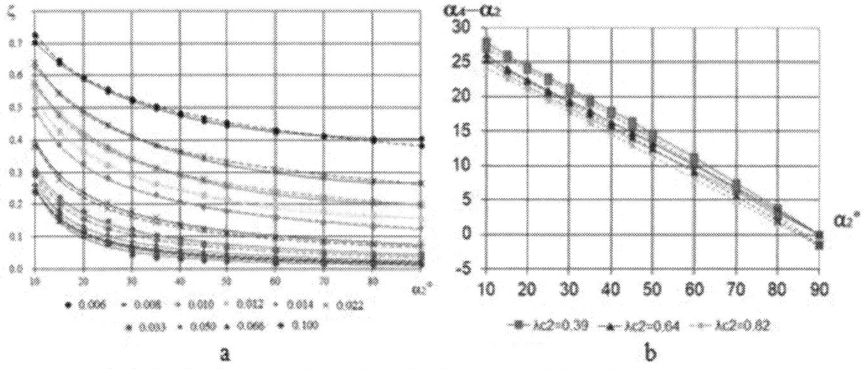

Fig. 5. Comparison of calculated and approximating values of VLD a) loss coefficients depending on the angle α_2. $D_4/D_2 = 1.6$, $\lambda_{c2} = 0.39$; b) $\alpha_4 - \alpha_2$ depending on the inlet flow angle at λ_{c2}=0.39, 0.64, 0.82. b_3/D_2 =0.006; D_4/D_2 =1.6. Solid lines – calculation, dashed lines – approximation.

by 1.5 times, the efficiency decreases by 1.6 times. For comparison, in the previously calculated series of diffusers, the width decreased by 7.15 times, and the loss factor increased by 2.6 times. This shows that for narrow diffusers, width has a greater influence on the value of the loss coefficient. If we apply the previously proposed mathematical model for narrow vaneless diffusers, then for a diffuser with a relative width b_3/D_2=0.006, the model understates the loss coefficient by 1.13 – 2.0 times, and the outlet flow angle by 1.24 – 0.95 times for the inlet flow angles of 10 and 90 degrees, respectively.

2.3 Objects

To increase the accuracy of modeling low-flow-rate stages, additional calculations of narrow VLD were performed in the ANSYS CFX program. The characteristics of diffusers with a relative width b_3/D_2=0.006, 0.008, 0.010, 0.012 and with a radial length D_4/D_2 = 1.6 – 2.0 in the range of velocity coefficients λ_{c2}= 0.39 – 0.82 and the corresponding Reynolds numbers Re_{b2} =36 800 – 73 600. Calculations were made at inlet flow angles α_2 =10°– 90°.

2.4 The mathematical model of VLD, based on a generalization of the computational experiment for VLD with b_3/D_2 = 0.006 – 0.100 (model No. 2)

In general, the approximating dependence for the loss coefficient $\zeta = f(\bar{b}_2, \bar{D}_4, \alpha_2, \lambda_{c2}, \mathrm{Re}, \bar{k}_{roug})$ is presented in the same form as for model No. 1, however, the system of approximating formulas has a completely different structure and coefficients. At the first step, the influence of $b_3/D_2, \alpha_2, \lambda_{c2}$ was approximated at a fixed value of the relative length of the diffuser D_4/D_2 =1.6:

$$\zeta = \frac{A \cdot \alpha^B \cdot K_{D4}}{K_{\mathrm{Re},k_{roug}}}, \quad (8)$$

where:

$A = c \cdot (b_3/D_2)^4 + d \cdot (b_3/D_2)^3 + e \cdot (b_3/D_2)^2 + h \cdot (b_3/D_2) + i;$

$c = 111633 \cdot \lambda_{c2}^2 - 208134 \cdot \lambda_{c2} - 184432;$

$d = 64704 \cdot \lambda_{c2}^2 - 61330 \cdot \lambda_{c2} + 71806;$

$e = -10988 \cdot \lambda_{c2}^2 + 12468 \cdot \lambda_{c2} - 7411.4;$

$h = 465.59 \cdot \lambda_{c2}^2 - 499.03 \cdot \lambda_{c2} + 245.24;$

$i = -1.7437 \cdot \lambda_{c2}^2 + 2.468 \cdot \lambda_{c2} - 0.17;$ $B = f \cdot \ln(b_3/D_2) + g;$

$f = 0.0796 \cdot \lambda_{c2}^2 - 0.194 \cdot \lambda_{c2} - 0.2775;$

$g = -0.7042 \cdot \lambda_{c2} - 1.6742.$

Corrections for the influence of the Reynolds criterion – relative roughness and relative length of the diffuser are made by analogy with model No. 1, using the formulas (3), (4), (5).

The approximation for calculating the outlet flow angle from the VLD is performed differently for the VLD range with a relative width $0.006 \le b_3/D_2 \le 0.012$ and $0.012 < b_3/D_2 \le 0.100$.

For VLD with relative width $0.006 \le b_3/D_2 \le 0.012$:

$$\alpha_4 = \alpha_2 + A \cdot \alpha_2^2 + B \cdot \alpha_2 + C;, \quad (9)$$

where: $A = d \cdot \lambda_{c2} + e;$

$d = 0.0036 \cdot \ln(b_3/D_2) + 0.0199;$

$e = -0.15 \cdot (b_3/D_2) - 0.0004;$ $B = f \cdot \lambda_{c2} + g;$

$f = -0.431 \cdot \ln(b_3/D_2) - 2.2826;$

$g = 0.4487 \cdot \ln(b_3/D_2) + 2.0442;$ $C = h \cdot \lambda_{c2} + i;$

$h = 6.8893 \cdot \ln(b_3/D_2) + 29.521;$

$i = -27.87 \cdot \ln(b_3/D_2) - 110.55.$

For diffusers with a relative width $0.012 < b_3/D_2 \le 0.100$, the model for calculating the outlet flow angle remains the same, the formula (6).

The resulting system of 42 algebraic equations is characterized by a good accuracy of 98.9%, figure 5.

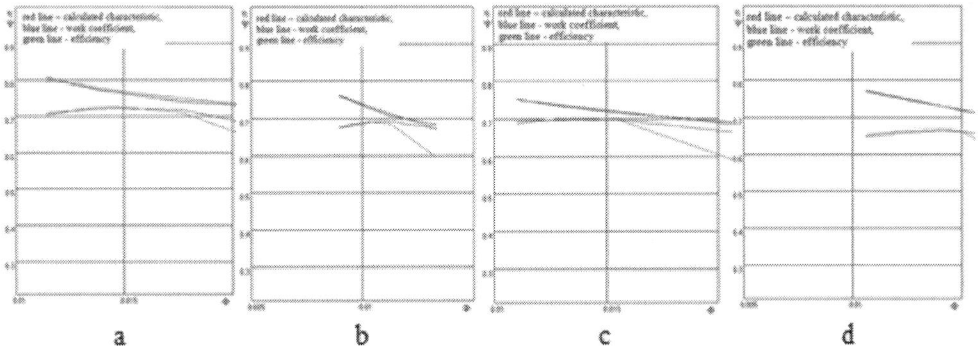

Fig. 6. Comparison of calculated and experimental data for stage a) XXX3-Q with parameters: $\Phi_{des} = 0.0203$, $\psi_{T\,des} = 0.69$, $\bar{D}_{hub} = 0.337$, $\bar{b}_3 = 0.01$; $\bar{D}_4 = 1.40$, $M_u = 0.785$, $\mathrm{Re}_u = 9.8 \cdot 10^6$; b) stage XXX3-T with parameters: $\Phi_{des} = 0.013$, $\psi_{T\,des} = 0.62$, $\bar{D}_{hub} = 0.337$, $\bar{b}_3 = 0.008$; $\bar{D}_4 = 1.40$, $M_u = 0.803$, $\mathrm{Re}_u = 10 \cdot 10^6$; c) XXX3-S with parameters: $\Phi_{des} = 0.015$, $\psi_{T\,des} = 0.63$, $\bar{D}_{hub} = 0.337$, $\bar{b}_3 = 0.008$; $\bar{D}_4 = 1.40$, $M_u = 0.803$, $\mathrm{Re}_u = 10 \cdot 10^6$; stage XXX3-U with parameters: $\Phi_{des} = 0.011$, $\psi_{T\,des} = 0.62$, $\bar{D}_{hub} = 0.337$, $\bar{b}_3 = 0.0057$; $\bar{D}_4 = 1.40$, $M_u = 0.803$, $\mathrm{Re}_u = 10 \cdot 10^6$.

3 Results and discussion

3.1 Application of model No. 2 in the programs of the Universal Modelling Method

Formulas (8) – (9) were introduced into the mathematical model of the Universal Modeling Method. Since the proposed VLD model both by formulas (3) – (6) and by formulas (8) – (9) does not imply the presence of empirical coefficients, identification was not required. A comparison was made between the calculated and experimental gas-dynamic characteristics for model stages 20CE and model stages of the Clarke series. For model stages 20CE, the comparison showed an average error of modeling the design (maximum) efficiency of 1.01%, an average error of five points (except for the point of highest flow rate) of 1.31%.

For Clarke model stages, a comparison of the calculated and experimental data showed an average error of modeling the design (maximum) efficiency of 0.99%, an average error of five points (except for the point of highest flow rate) of 1.97%. Figure 6 shows a comparison of the calculated and measured gas-dynamic characteristics of the same Clarke model stages, which were presented above in figures 2 - 3. They clearly show the improvement of the modeling accuracy of the gas-dynamic characteristics of low-flow-rate model stages.

The results showed a slight increase in the error in calculating the characteristics of model stages 20CE (about 0.05%) and a decrease in the error in calculating Clarke model stages (about 0.2%). This confirms the accuracy of the analysis and the effectiveness of the developed approximating dependencies.

4 Conclusion

The mathematical model of flow parameters calculation in VLD with relative width $b_3 / D_2 = 0.014 - 0.100$ created in [16] was applied for calculation of medium-flow model stages of 20CE series (relative width of VLD $b_3 / D_2 = 0.030 - 0.061$) and low-flow model stages by Clarke Tooling (relative width of VLD $b_3 / D_2 = 0.005 - 0.019$). Application experience has shown good match between calculated and experimental gas dynamic characteristics of medium-flow model stages and overestimation of efficiency for low-flow stages. Comparison of the results of CFD calculations of narrow VLD with the calculation using approximating formulas confirmed the assumption that the reason for the overestimation of the efficiency of low-flow stages is the insufficient accuracy of modeling the loss coefficient and the outlet angle of the flow in narrow VLD. A series of CFD calculations of narrow VLD in the range $b_3 / D_2 = 0.006 - 0.012$ was performed and the obtained characteristics of the loss coefficient and the change of the flow angle were approximated. The mathematical model of VLD created on the basis of new approximating formulas allowed to increase accuracy of efficiency modeling of low-flow stages by 16% for the mode with the maximum efficiency and by 9% for the characteristic as a whole, without losing in accuracy of modeling of medium-flow stages at the same time.

Acknowledgemetns

The research was performed by a Grant of the President of the Russian Federation for young PhD MK-1893.2020.8.

The calculations were carried out using the supercomputer center "Polytechnic" SPbPU.

References

[1] A. Startsev, Yu. Fokin, Eu. Steshakov, CFD design and analysis of a compact single-spool compressor for a heavy transport helicopter's powerplant, 29th congress of the international council of the aeronautical sciences (St. Petersburg, Russia, 2014)

[2] C. Xu, W.J. Chen, Computational analysis on a compressor blade, Int. Conf. On Jets, Wakes and Separated Flow, Proceedings of Int. Conf. On Jets, Wakes and Separated Flow (Toba-shi, Mie, Japan, 2005)

[3] Jindra Kosprdova, Jiři Oldřich, The Development of Centrifugal Turbo Compressor Stage Using CFD, The 20th International Conference on Hydraulics and Pneumatics (Prague, 2008)

[4] R. Puzyrewski, Y. Galerkin, P. Flaszynski, Direct and inverse numerical calculation for the tested centrifugal impeller, XI Internationale tagung forschung praxis und didaktik im modernen maschinenbau (Germany, Stralsund 2001)

[5] M. Mosdzien, M. Enneking, A. Hehn, D. Grates, P. Jeschke, Journal of the Global Power and Propulsion Society **2**, 1 (2018)

[6] J.M. Sorokes, B.R. Hutchinson, The Practical Application of CFD in the Design of Industrial Centrifugal Compressors, ASME International Mechanical Engineering Congress and Exposition, Challenges and Goals in Pipeline Compressors, edited by A. Engeda (USA, Florida, 2000)

[7] J.M. Sorokes, Sidestream optimization through the use of computational fluid dynamics and model testing turbomachinery, sympos. Proceed. Texas: A&M (2000)

[8] G.E. Guidotti, *Towards Centrifugal Compressor Stages Virtual Testing: Ph.D. thesis* (Università degli Studi di Bologna, 2014)

[9] R. Matas, T. Syka, L. Hurda, EPJ Web of Conferences **180**, 02060 (2018)

[10] Y. Galerkin, I. Voinov, A. Drozdov, Comparison of CFD-calculations of centrifugal compressor stages by NUMECA Fine/Turbo and ANSYS CFX programs, International Conference on Compressors and their Systems 2019, IOP Conf. Series: Materials Science and Engineering (London, 2017)

[11] A.I. Borovkov, I.B. Voinov, M.A. Nikitin, Yu.B. Galerkin, A.F. Rekstin, A.A. Drozdov, Experience of performance modeling the single-stage pipeline centrifugal compressor, Oil and Gas Engineering (OGE-2019), AIP Conf. Proc. 2007 (Omsk, 2019)

[12] A.I. Borovkov, I.B. Voinov, Yu.B. Galerkin, A.A. Drozdov, K.V. Soldatova. Experimental characteristic simulation for two-stage pipeline centrifugal compressor, International Conference on Compressors and their Systems 2019, IOP Conf. Series: Materials Science and Engineering (London, 2019)

[13] A. Borovkov, I. Voinov, Y. Galerkin, A. Nikiforov, M. Nikitin, O. Solovyeva, K. Kabalyk, Issues of gas dynamic characteristics modeling: a study on a centrifugal compressor model stage, *E3S Web of Conferences* (2019)

[14] L. Marenina, Y. Galerkin, K. Soldatova, International Journal of Mechanical Engineering and Robotics Research 7, 6, 656–661 (2018)

[15] Y. Galerkin, O. Solovieva, Flow behavior and performances of centrifugal compressor stage vaneless diffusers, International Conference on Numerical Methods in Industrial Processes, International Journal of Mechanical, Aerospace, Industrial, Mechatronic and Manufacturing Engineering **9**, 1 (France, Paris, 2015)

[16] O.A. Solovyova, *A mathematical model for calculating the gas-dynamic characteristics and optimizing the vaneless diffusers of centrifugal compressor stages: PhD thesis* (SPbPU. 2018)

E3S Web of Conferences **178**, 01015 (2020) https://doi.org/10.1051/e3sconf/202017801015
HSTED-2020

Modeling of asynchronous motor operation modes for the correct selection of voltage regulation devices

Azat Akhmetshin[1,*], *George* Marin[1,2], and *Dmitrii* Mendeleev[1,2]

[1]Kazan State Power Engineering University, 420066 Kazan st. Krasnoselskaya 51, Russia
[2]JSC "Tatenergo" branch "Kazan CHP-2", 420036 Kazan st. Tatsevskaya 11, Russia

Abstract. Many electrical appliances are used in production and in everyday life, which include elements that are extremely sensitive to voltage deviations from acceptable values. Failure in their work can cause equipment failure or a breakdown in technological processes. There are a number of technical solutions to solve this problem, one of which is the using of voltage control devices such as boost transformers. The principle of operation of booster transformers is the introduction of a longitudinal EMF into the electric circuit, which provides booster. The choice of voltage regulation devices consists in determining its power and the required transformation ratio. The latter needs some justification, because it cannot be formally accepted: if it is necessary to increase the voltage, for example by 5%, it is enough to introduce a longitudinal EMF of 5% of the nominal voltage into the electric circuit. This is due to the fact that with increasing voltage from the load side, the power consumption also increases, which causes an increase in voltage drop compared to modes in the absence of voltage regulation devices. Thus, the load will receive a slightly lower voltage level in comparison with the desired one.

1 Introduction

It is extremely difficult to maintain the voltage at the consumer unchanged and equal to the nominal. The characteristics of individual electrical receivers set the limits for voltage deviations.

Slow changes in power supply voltage (typically longer than 1 minute) are usually due to changes in electrical load.

Electric power quality indicators related to slow changes in the supply voltage are negative $\delta U_{(-)}$ and positive $\delta U_{(+)}$ deviations of the supply voltage at the point of transmission of electrical energy from the nominal value, %:

$$\delta U_{(-)} = \left[\frac{(U_0 - U_{m(-)})}{U_0}\right] * 100,\% ,$$

$$\delta U_{(+)} = \left[\frac{(U_{m(+)} - U_0)}{U_0}\right] * 100,\% ,$$

where $U_{m(-)}$, $U_{m(+)}$ are the values of the power supply voltage, lower than U_0 and larger U_0, respectively, averaged over a time interval of 10 minutes in accordance with the requirements of State Standard 33073-2014 "Electric energy. Electromagnetic compatibility of technical equipment. Control and monitoring of electric power quality in the public power supply systems"[8].

U_0 - voltage equal to standard rated voltage U_n or matched voltage U_c.

The standard rated voltage U_n is 220V in low voltage electrical networks (between phase and neutral conductors for single-phase and four-wire three-phase systems) and 380V (between phase conductors for three- and four-wire three-phase systems). The following standards are established for the above-mentioned indicators of electric power quality: positive and negative voltage deviations at the electric energy transmission point must not exceed 10% of the nominal or agreed voltage value during 100% of the time interval of one week.

Permissible values of positive and negative voltage deviations at the points of common connection should be established by the network organization, taking into account the need to comply with the norms of this standard at the points of transmission of electrical energy.

Conditions must be provided in the consumer's electrical network under which the deviations of the supply voltage at the terminals of the electrical receivers do not exceed the permissible values established for them when meeting the requirements of this standard for the quality of electricity at the point of transmission of electrical energy.

Induction regulators, adjustable autotransformers, voltage boosting transformers and linear regulators can be classified as booster devices [20]. Such devices act on the resulting voltage at the consumer or receiver by creating additional EMF, the vector of which is superimposed on the main voltage according to, counter or at some angle, as is the case in some schemes of

* Corresponding author: ahmetshin.ar@mail.ru

booster transformers that perform "transverse" regulation.

A booster transformer is a device that serves to regulate voltage and consists of two independent units, a "series" transformer, the primary winding of which is connected in series to the line cut, and a special control transformer or autotransformer. Booster transformer is mainly used for local voltage regulation. The voltage in the regulated line differs from the voltage of the mains by the amount of additional EMF serial transformer.

Typically, voltage regulation is carried out in steps; its smoothness depends on the number of control branches of the supply transformer. Booster transformer allows you to get additional EMF phase-shifted relative to the main voltage. The shear angle depends on the switching circuit of the control transformer and for the simplest schemes is 0°, 30°, 60° and 90°.

Transformers creating an additional voltage that coincides in phase with the main voltage are called boost voltage transformers with "longitudinal" regulation, and creating additional voltage, shifted by a certain angle, with "transverse" regulation.

The booster transformer is usually characterized by "throughput power", i.e. power transmitted over the line into which the serial winding is included, and "own power" - the make-up device itself.

The operating mode of the power source, circuit resistance. Voltage deviations are not always in the range of acceptable values.

Voltage regulation is the process of changing the voltage level in a characteristic accurate electrical system using special technical means. The historical development of methods and methods for regulating voltage and reactive power occurs from the lowest hierarchical levels of energy system control to the highest. In particular, voltage regulation in the central power supply networks of distribution networks - at regional substations [1, 4, 5], where the transformation coefficients (k_T) change - is supported by the voltage of consumers when changing their operating mode.

A large number of induction motors are used in agriculture and industrial enterprises. The change in the consumed active and reactive power when choosing voltage control devices has not been considered by anyone [7].

A mechanism with a "ventilatory" mechanical transmission was investigate in the work. The mechanism in which the coefficient of proportionality of speed and generalization of the results were used.

Consider the circuit diagram, figure 1. As a load, we consider three asynchronous motors with a capacity of 15 kW, 37 kW, 110 kW with a rated voltage U_n = 380V.

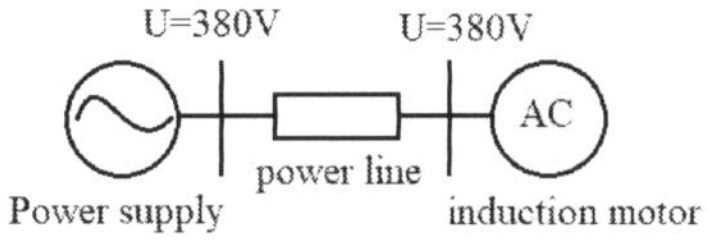

Fig. 1. Schematic diagram of connecting an induction motor.

The effect of voltage boosting depends on the ratio between the resistances of the power supply line and the load and generally acts to reduce the resulting voltage [2, 3].

2 Modeling induction motor

We model the circuit diagram in the MATLAB/ Simulink environment to plot the dependences of the power consumption on the voltage level, taking into account the different operating conditions of the induction motor [4-6].

The moment of resistance of an induction motor is represented by the formula:

$$M_c = M_{TP} + \left(k_{el} * M_{nom} - M_{TP}\right) * (\omega / \omega_{synch})^{\gamma} , \quad (1)$$

where M_{TP} is the initial moment at s = 1 or ω = 0, usually determined by friction forces; k_{el} is the engine load factor at synchronous angular velocity (the moment of resistance of the mechanism at ω = 1, expressed in fractions of the nominal moment of the engine); γ is an exponent characterizing this mechanism [9, 10].

The engine operation model for γ = 2 is shown in figure 2, for γ = 5 it is shown in figure 3. To determine the dependence of power consumption on engine load factor k_{el} which varies from 30% to 70%, and the initial moment of M_{TP} which varies from 10% to 50%.

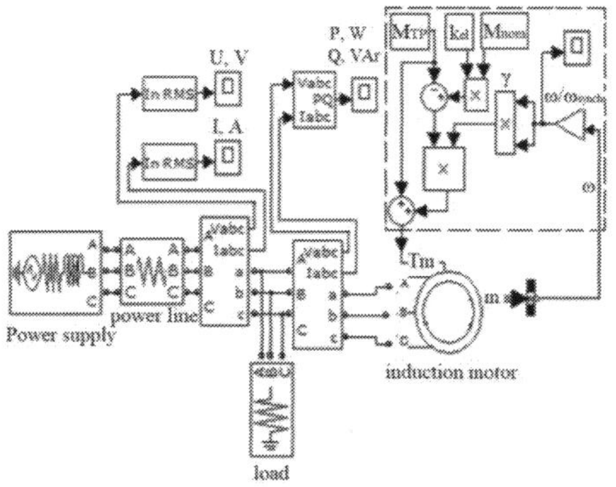

Fig. 2. Engine operation diagram for γ = 2, k_{el} = 30-70%, M_{TP} = 10-50%.

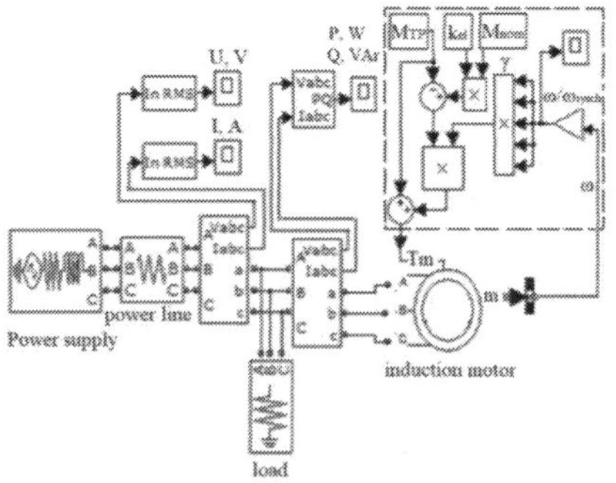

Fig. 3. Engine operation diagram for γ = 5, k_{el} = 30-70%, M_{TP} = 10-50%.

A 3-Phase Source block (designed to simulate a three-phase voltage source) is used as a power center, which includes three voltage sources connected to a star with a neutral wire [13-16]. The Three-Phase Series RLC Branch block was used as the power line because the inductive resistance of the line is neglected in the 0.4 kW network due to the small length [17-19]. The block in our case models a three-phase circuit consisting of three active resistances. An asynchronous motor is represented by the Asynchronous Machine block simulating an asynchronous electric machine in a motor mode [11,12]. The operation mode is determined by the law of the electromagnetic moment of the machine. The block contains several preset models of the induction motor. As an asynchronous motor we use an asynchronous motor with a power of 15 kW, 37 kW, 110 kW and a rated voltage of 380V and a frequency of 50 Hz specified by the MATLAB / Simulink program. The Tm port is designed to provide a moment of resistance to movement. The moment of resistance to movement is given by formula (1).

Define the synchronous angular velocity:

$$\omega_{synch} = \eta * \frac{2*\pi}{60},$$

where η is rotational speed of an induction motor.

Define the rated torque at a synchronous speed for the considered induction motors:

$$M_{nom.s} = \frac{P_m}{\omega_{synch}}$$

where P_m is the rated engine power indicated in the engine model of the MATLAB/Simulink program.

For an induction motor with a rated power of P_M = 15kW: η = 1460 rpm, $M_{nom.s}$ = 95.49 N*m, for an induction motor of P_M = 37kW: η = 1480 rpm, $M_{nom.s}$ = 235.56 N*m, for an induction motor P_m = 110kW: η = 1487 rpm, $M_{nom.s}$= 700.28 N*m.

The load factor for active and reactive power is determined by:

$$k_l = \frac{P_{mw}}{P_n},$$

where P_{mw} is the active power consumed by an induction motor at various levels of operating voltage, U_w, which is calculated relative to the nominal (0.7-1.1*U_n) with the condition that k_{el} = 30-70% and M_{TP} = 10-50%; P_n is the rated active power of the engine.

The dependence of k_l on the supply voltage for two variants of mechanisms is presented in figures 4 and 5.

Figures 4 and 5 show that engines of different power, initially equally loaded, have approximate dependencies that slightly depend on the voltage level, this suggests that for them it is possible to take the load factor the same within a few percent of the error.

Also from figures 4 and 5 it can be seen that the M_{TP} value has little effect on the load factor, which makes it possible to group the motors in one equivalent exactly by the value of this coefficient.

Determine the resistance at rated voltage by the formula:

$$r_{nb} = \frac{U_n^2}{P'_{mk}}$$

where U_n is the rated voltage, P'_{mk} is the consumed active power at the rated voltage U_n and under various conditions of change M_{TP} = 10-50%, k_{el} = 30-70%.

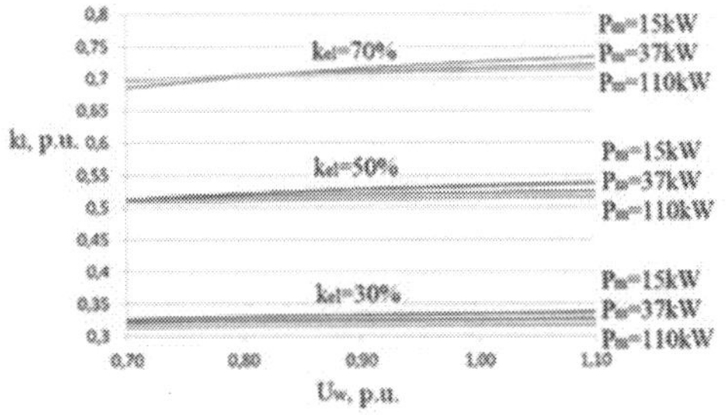

a - M_{TP}=10%

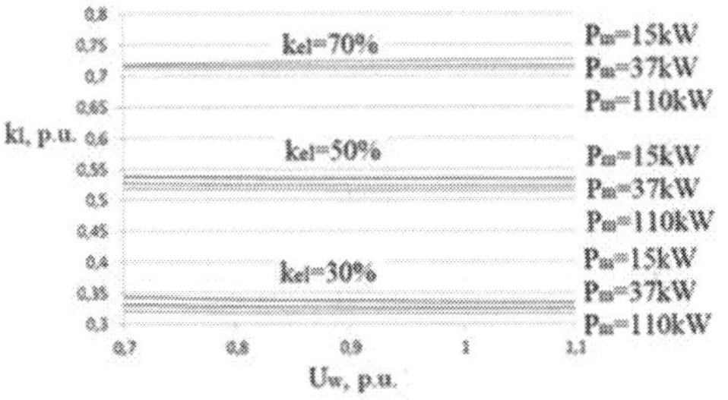

b - M_{TP}=50%

Fig. 4. The dependence of the load factor of the active power on the mains voltage for an induction motor γ=2.

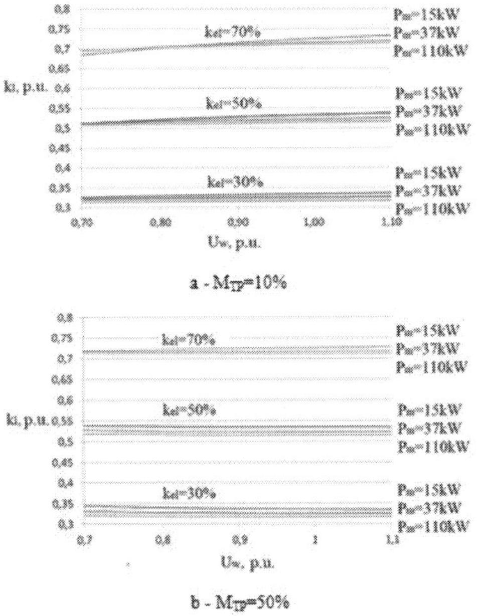

Fig. 5. The dependence of the load factor of the active power on the mains voltage for an induction motor $\gamma = 5$.

We determine the active resistance under the condition U_w= 0.7-1.1*U_n by the formula:

$$r_w = \frac{U_w^2}{P_{mw}}$$

Figure 6 shows a graph of resistance changes versus voltage level. From which (figure 6) it is seen that the active resistance at various values of M_{TP} = 10-50%, k_{el} = 30-70% and engine power varies according to one law and is within 5% of the error, and varies significantly from the voltage level. The actual resistance value can be determined by the formula:

$$r_w = \frac{r_{U_w}}{r'_{U_w}} * r'_w , \quad (2)$$

$$x_w = \frac{x_{U_w}}{x'_{U_w}} * x'_w , \quad (3)$$

where r'_{U_w}, x'_{U_w} are the load resistance at U_p, expressed in p.u. .; r_{U_w}, x_{U_w} - load resistance at U_P, expressed in named units; r'_p, x'_p - load resistance at U_n, expressed in p.u.

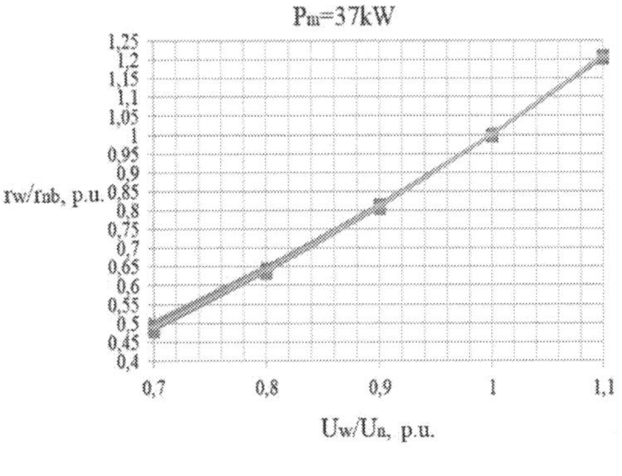

Fig. 6. Graph of changes in active resistance against voltage at M_{TP} = 10-50% and k_{el} = 30-70%.

It is necessary to determine tgϕ_n to determine the correction coefficient k_P. The tgϕ_n change diagram is shown in figures 7 and 8.

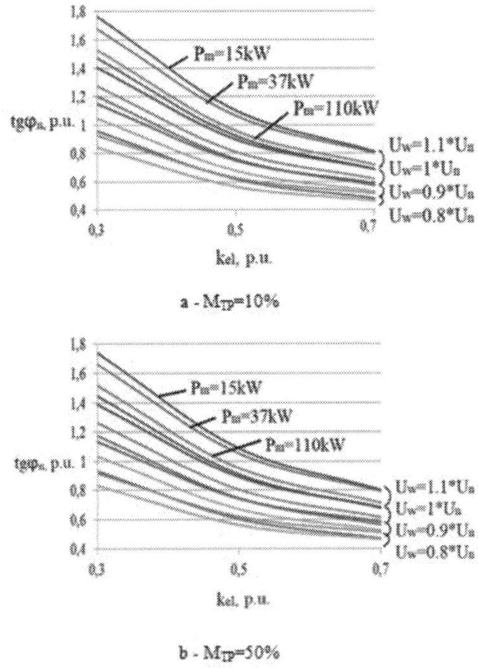

Fig. 7. . Change in tgϕn at different levels of U_P relative to k_{el} for different values of M_{TP} for an induction motor $\gamma = 2$.

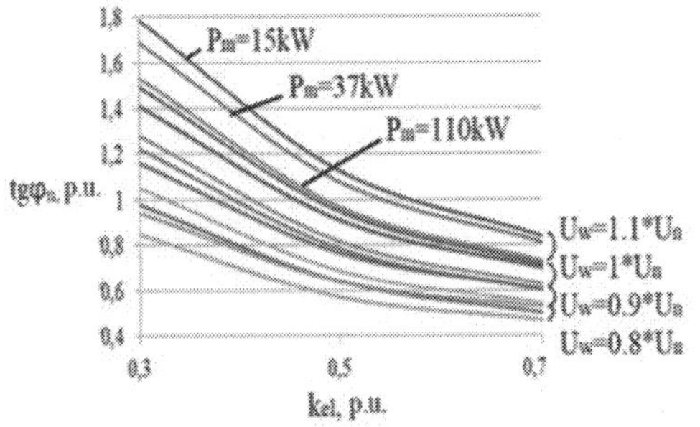

a - M_{TP}=10%

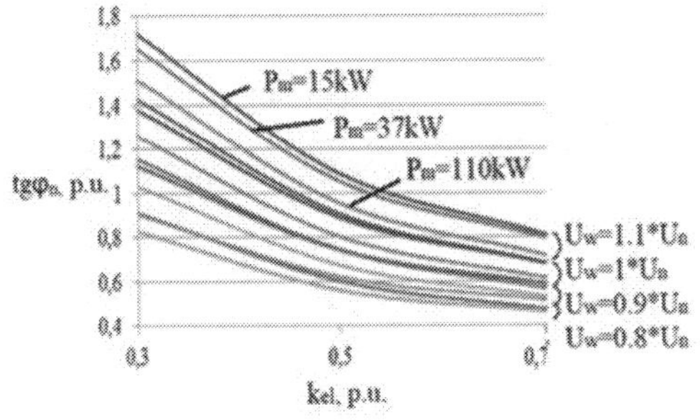

b - M_{TP}=50%

Fig. 8. Change in tgϕ_n at different levels of U_P relative to k_{el} for different values of M_{TP} for an induction motor $\gamma = 5$.

It can be seen from figures 7 and 8 that with increasing voltage, the value of tgϕ_n also increases.

The change in tgϕ_n obeys one law with the variation of k_{el}, M_{TP}. In order to generalize the obtained tgϕ_n values for various values of M_{TP}, we use the least squares method.

As a result, generalized characteristics are obtained that make it possible to determine tgϕ_n for engines of different powers, neglecting M_{TP}.

To determine k_w, all parameters are known, and k_w is determined by expression (4).

$$k_w = \left| \frac{r_{nb}*(1+jtg\phi_n)}{(r_{l0}+jx_{l0})*k_l*l+r_{nb}*(1+jtg\phi_n)} \right| = \left| \frac{(1+jtg\phi_n)}{(\alpha+j\beta)*k_{kl}+(1+jtg\phi_n)} \right|, \quad (4)$$

where $\alpha = {r_{l0}}/{r_{nb}}$; $\beta = {x_{l0}}/{r_{nb}}$, $k_{kl} = k_l * l$.

From formula (4) it can be seen that the ratio between the network and load resistances affects the magnitude of the voltage boost.

In real conditions, to determine k_{el}, tgϕ_n, r_{nb}, you can use clamp meters with the function of measuring power.

3 The choice of voltage regulation device

The problem of choosing the transformation coefficient (k_T) of voltage regulation devices is to correctly take into account the value of the consumer load power, which changes with voltage [9,10,19].

Let's consider two options:

1. When the voltage of the consumer changes, they remain constant (r_n, x_n = const);
2. When the voltage changes, the consumer resistance changes.

Figure 9 shows a diagram based on the calculation results for tgϕ = 0.4

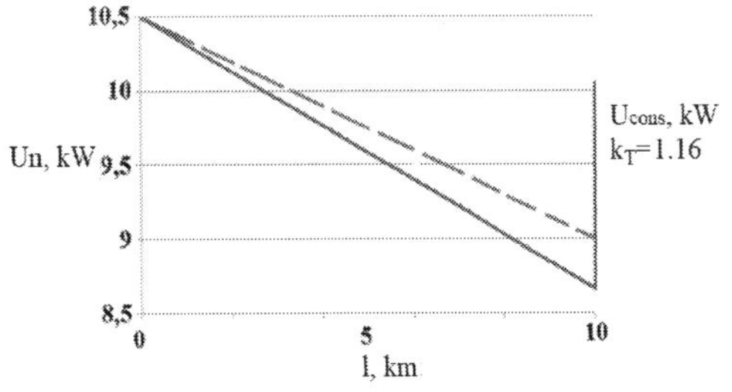

Fig. 9. The change in voltage (r_n, x_n = const).

In figure 9, the voltage drop for the initial values of the load power, provided that the consumer load is constant, k_T, = 1.1, is shown by a dashed line, and the voltage drop with a constant resistance (r_n, x_n = const) is shown by a solid line.

Figure 9 shows that when the voltage on the consumer increases with a constant load resistance of the consumer, the voltage drop increases: this indicates the need to increase k_T from 1.1 to 1.16. In this scheme, it will be necessary to install two nodes, two boost transformers.

Subject to voltage changes the consumer resistance changes. To determine the resistance under the condition U_{cons} = 10kV, we use the graph of the dependence of resistance on voltage in p.u., figure 6, for fan mechanisms with $\gamma = 5$ constructed using the MATLAB / Simulink simulation software.

As an example, consider the load represented by three asynchronous motors P = 110kW with k_l = 0.5, P = 37kW with k_l = 0.7, P = 37kW with k_l = 0.4, the length of the power line varies 10 km and is made by AS -16, figure 10.

We neglect the power loss and voltage drop in the power transformer.

 https://doi.org/10.1051/e3sconf/202017801015

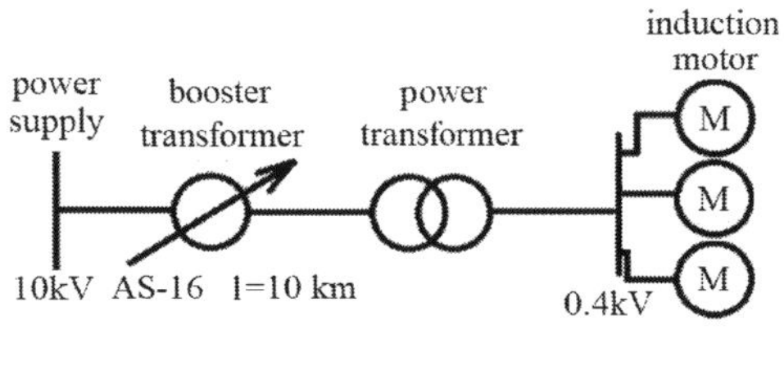

Fig. 10. Single line diagram.

To determine the resistance values at U_{cons} = 10kV, we use the formulas (2 and 3):

Having determined the resistance values using formulas (2-3), we calculate the power consumption at the beginning of the line at $U_{cons} = U_n$ = 10kV.

Based on the calculation results for tgϕ = 0.4, a diagram is constructed, figure 11.

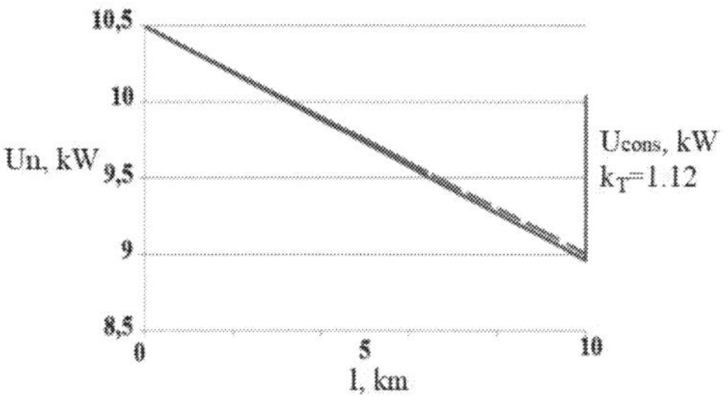

Fig. 11 Voltage change (r_n, x_n depends on U_{cons})..

From figure 11 it can be seen that to increase the voltage level to the nominal value, k_T should be equal to 1.12.

4 Conclusion

As can be seen from the results obtained, when choosing the coefficient of transformation of voltage regulation devices, the nature of the dependence of the load power on its voltage should be taken into account. So the transformation coefficient, calculated under the condition of constant power when the load voltage changes, will turn out to be somewhat underestimated, which does not provide the required voltage level in reality (since the magnitude of the load power changes with voltage change).

On the other hand, the calculation of the transformation coefficient of voltage regulating devices (resistance constant), when the supply voltage changes, will lead to the choice of an oversized transformation coefficient, this fact affects the number of boost transformers installed in one node (two or three), and the number of installed knots in a line.

The variant with which the load resistance depends on the voltage allows you to take into account the actual changes, which ensures the correct selection of the transformation coefficient.

References

[1] Y.I. Soluyanov, A.I. Fedotov, A.R. Ahmetshin, Calculation of electrical loads of residential and public buildings based on actual data, IOP Conference Series: Materials Science and Engineering **643**, article no. 012051. DOI 10.1088/1757-899X/643/1/012051 (2019)

[2] K. Suslov, V. Piskunova, D. Gerasimov, E. Ukolova, A. Akhmetshin, P. Lombardi, P. Komarnicki, Development of the methodological basis of the simulation modelling of the multi-energy systems, International Scientific and Technical Conference Smart Energy Systems 2019 (SES-2019), E3S Web of Conferences **124**, article no. 01049. DOI: 10.1051/e3sconf/201912401049 (2019)

[3] D.I. Mendeleev, G.E. Marin, A.R. Akhmetshin, The implementation and use of gas turbines with absorption refrigerating machine in the technological schemes of thermal power plants, 2019 International Multi-Conference on Industrial Engineering and Modern Technologies Publisher IEEE INSPEC, accession no. 19229407. DOI 10.1109/FarEastCon.2019.8934431 (2019)

[4] K. Bakhteev, A. Fedotov, R. Misbakhov, The improving efficiency of electric receivers on the industrial enterprises in case of short-term power outages, Proceedings of the 2019 20th International Scientific Conference on Electric Power Engineering Ostrava Czech Republic, 347–352. DOI 10.1109/EPE.2019.8778181 (2019)

[5] E.I. Gracheva, A.N. Alimova, Calculation methods and comparative analysis of losses of active and electric energy in low voltage devices, 2019 International Ural Conference on Electrical Power Engineering INSPEC, accession no. 19080057. DOI 10.1109/URALCON.2019.8877627 (2019)

[6] E.I. Gracheva, A.N. Alimova, Calculating probability of faultless work of shop nets with the help of coefficients of ratio, 2019 International Russian Automation Conference INSPEC, accession no. 19090885. DOI 10.1109/RUSAUTOCON.2019.8867727 (2019)

[7] A.B. Chubais, E.V. Ametistova, Yu.A. Udaltsova, Y.M. Urinson, *Economics and management in the modern electric power industry of Russia* (Moscow NP KONTs UES, 616, 2009)

[8] GOST 33073-2014, Electrical energy. Electromagnetic compatibility. Control and monitoring of the quality of electric energy in general-purpose power supply systems (2014)

[9] Yu.S. Zhelezko, *Electricity losses. Reactive power. Electricity quality: A guide for practical calculations* (Moscow ENAS, 456, 2009)

[10] I.A. Danilov, K.V. Lototsky, *Electric motors* (Moscow Ed., 527, 1972)

[11] I.V. Chernykh, *Modeling of electrical devices in MATLAB, SimPowerSystems and Simulink* (Moscow DMK Press, 288, 2007)

[12] I.P. Kryuchkov, V.A. Starshinov, Yu.P. Gusev, M.V. Pirates, *Transients in electric power systems* (Moscow Publishing House MPEI, 416, 2008)

[13] D.I. Mendeleev, Yu.Ya. Galitskii1, G.E. Marin, A.R. Akhmetshin, Study of the work and efficiency improvement of combined-cycle gas turbine plants, International Scientific and Technical Conference Smart Energy Systems 2019 (SES-2019) E3S Web of Conferences **124**, article no. 05016. DOI: 10.1051/e3sconf/201912405061 (2019)

[14] E.I. Gracheva, O.V. Naumov, Evaluation criteria of contact group technical state concerning electrical appliances, International Journal of pharmacy and Technology **4**, 27084–27091 (2016)

[15] I. Monedero, C. Leon, J. Ropero, Classification of electrical disturbances in real time using nueral networks, IEEE Trans. Power Deliv, 1–9 (2007)

[16] J. José, G. De, A. Agüera-Pérez, J.C. Palomares-Salas, J.M. Sierra-Fernández, A. Moreno-Muñoz, A novel virtual instrument for power quality surveillance based in higher-order statistics and case-based reasoning, Mesurement **45**, 1824–1835 (2012)

[17] International Conference on Renewable Energies and Power Quality (ICREPQ'17) Malaga (Spain), 4th to 6th April 2017, no. 15, Optimization of Residential Load Consumption during Energy Peaks using Smart Metering Rovan Refaat El-Razky, Ahmed Ali Daoud, Kamel El-serafi 227–232 (2017)

[18] I. Junquera, J. García-Villalobos, I. Zamora, P. Eguía, J.I. San Martín, International conference on renewable energies and power quality (ICREPQ'17) Malaga (Spain), 4 th to 6 th April 2017, Renewable Energy and Power Quality Journal (RE&PQJ), no. 15, Home energy management systems and electric vehicles: challenges and opportunities, 334–339 (2017). ISSN: 2172-038 X

[19] A.A. Gerasimenko, *Transmission and distribution of electric energy: a manual* (D: Phoenix Krasnoyarsk Publishing projects, 720, 2006)

[20] A.F. Zobaa, Shady H.E. Abdel Aleem, Power Quality in Future Electrical Power Systems, The Institution of Engineering and Technology, 440 (2017)

https://doi.org/10.1051/e3sconf/202017801016

Modification of the synchronous motor model for topological optimization

Timur Petrov[1,*], and *Alfred* Safin[1]

[1]Kazan State Power Engineering University, Kazan, Russia

Abstract. Synchronous motors are increasingly used in industry, and they need to be made more energy efficient. To do this, it is possible to optimize the rotor of the synchronous motor by changing the design, obtaining the best option due to topological optimization. But the standard design of a synchronous motor is quite complicated for calculations by the finite element method, so it is necessary to simplify it as much as possible. To make all elements the simplest from the geometric point of view (rectangles, straight lines as much as possible, etc.), and to make for calculation not the complete model, but the smallest sector for which topological optimization can be adequately calculated. To do this, it is necessary to calculate the magnetic moment for a standard design, for each step of structural change, and compare these values. If there is a slight discrepancy in the data, it will be necessary to choose the sector correctly and confirm the adequacy of the model by comparison. All these calculations should be performed for the predicted model, which should be obtained after topological optimization.

1 Introduction

It has long been known that permanent magnet synchronous motors (PMSM) have better technical characteristics than conventional asynchronous motors, but from an economic point of view, they are much more expensive [1,2]. One reason is the high cost of rare earth magnets, which are used in permanent magnet synchronous motors [3]. Thus, if you use fewer magnets, then we can reduce the payback period, and make the use of synchronous motors more attractive.

According to the design of the rotor, synchronous motors can be classified into 2 types:

- surface permanent magnet synchronous motor (SPMSM);
- interior permanent magnet synchronous motor (IPMSM).

Built-in magnets can provide a higher torque value, but they are more difficult to manufacture, and the exact same number of magnets is used as for surface magnets [4,5]. However, it is possible to make the rotor in such a way that the magnets were located in it, not just in the form of trapezoid or rectangles, but of arbitrary shape, in order to reduce the number of magnets. And in order to determine the optimal design of the rotor, it is necessary to use topological optimization, and iterate over the options not just by enumeration, but by a genetic algorithm [6]. However, if you try to optimize the standard design of the synchronous motor, then there will be difficulties with the number of finite elements, and it is necessary to simplify the design element as much as possible, which will be shown in this article.

2 Initial data

For optimization, the engine was selected 11 DVM (valve engine), hereinafter referred to as the "engine".

The engine refers to general-purpose products, both continuous and cyclic, non-repairable. The engine is designed to work with a special semiconductor converter (electric drive).

This machine has the following characteristics:

- shaft rotation frequency of 750 rpm;
- voltage in the DC link of the electric drive 520 V;
- without rotor position sensor, climatic modification B2;
- rated power 15 kW;
- long-term permissible motor current at a moment of load on the shaft - 40 A;
-lonely permissible moment of load on the shaft at a speed of 100 Nm;
- operating ambient temperature from minus 60 to plus 50;
- relative humidity of the environment of 80% at a temperature of plus 27;
- degree of protection of the IP55 engine.

According to the number of permissible operable states, engines belong to products of type 1, which during operation can be either in an operable state (state with nominal efficiency) or in an inoperable state (state of failure) [7].

Depending on the application mode, the engines belong to equipment operating in continuous use mode and the main part of the operating time is in operation mode [8,9].

* Corresponding author: tobac15@mail.ru

We have chosen this engine, since it is a rather simple design for topological optimization, and already based on the algorithm for our PMSM we will develop a universal optimization program.

To check the optimization, we will use the ELCUT program. ELCUT is a modern program for engineering modelling of electromagnetic, thermal and mechanical problems by the finite element method, which is suitable for calculating torque. ELCUT contains a magnetostatics module that can be used to calculate and analyse devices such as a solenoid, electric machines, magnetic screens, permanent magnets, magnetic disks, and the like.

3 Model modification

An example of calculating the magnetic induction for an engine is presented in Figure 1, these data agree with the values that we obtained during the tests.

The grooves were set as follows A +A + B-B-C + C + A-A-B + B + C-C-, the current density was calculated based on the current and the area of the groove, and amounted to 7 A/mm2.

The magnets are made of neodymium, and the direction of the coercive force is selected based on the location of the magnet, the stator, concentrator and air are set by standard values for magnetostatics tasks.

For the same example, we can get the value of the torque, however, for the optimization to be correct, we need to get the moment for the engine to rotate 90 degrees, and for these parameters calculate the root mean square moment. The Labelmover subroutine allows you to rotate the engine, in our case it is 18 steps of 5 degrees.

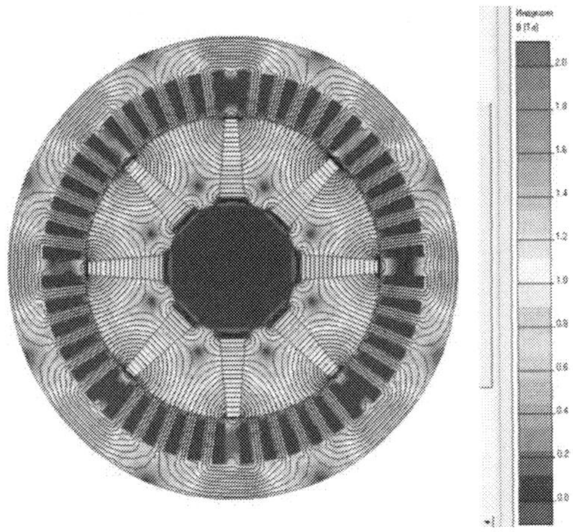

Fig. 1. Magnetic induction motor.

Modification of the synchronous motor model will be as follows:

1. Change the design of the grooves. Let's move from the complex design of the grooves to the simplest in terms of geometry (Figure 2). The difference between the moments of the first and last option is 9%, and the form of the graph of the change in the moment in time is preserved, so that it can be argued that this simplification can be used.

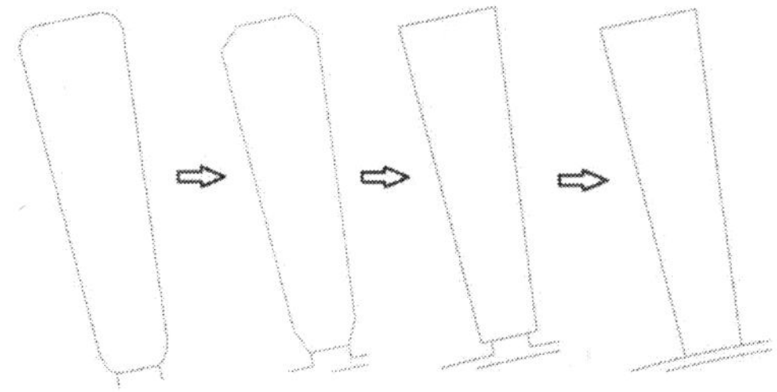

Fig. 2. Changing the design of the groove.

2. In the ELCUT program, the most accurate torque is calculated using the automatic grid sampling step, and thus for the engine the number of grid elements is about 100,000, which is impossible to calculate topological optimization. Therefore, it is considered possible to make a grid with a manual sampling step of 2.5 mm, the shape of the graph of the change in time is also saved.

It is important to clarify that during optimization we will compare a specific design and design after optimization; therefore, the accuracy of the obtained torque is not important to us, but an increase in its value is important.

3. Given the symmetry of the engine, you can take half the engine, and check whether the torque decreases by about half. In the calculation, we got a decrease of 2.2 times, and maintaining the shape of the graph (Figure 3), and based on the shape of the magnetic field, we came to the conclusion that you can use the sector at 45 degrees.

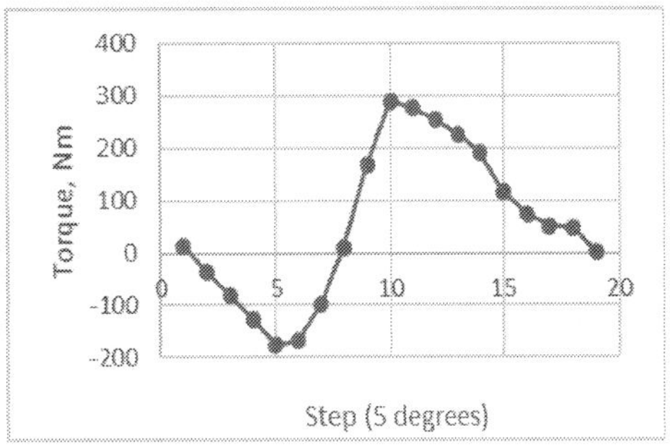

Fig. 3. Engine sector torque.

4 Actions to Improve Optimization Efficiency

First you need to take cells of 5 mm each, select local maxima from them (you cannot compare them at this stage, just collect the set). We can say that we have a rms moment of more than 10 Nm for example - these are all local maxima. And then, based on these local maxima, carry out the calculation, reducing the cell size. But in general, anyway, one local maximum will be obtained with one version and the geometry of the windings [10,11].

Accordingly, the problem can be rephrased: not searching for the best option, but searching for local maxima, because at this stage we will not find the best option. Cells should be taken at 3 mm for optimal construction. Anything more than 50 Nm is acceptable - we assemble into a set.

Then, at the next stage, these local maxima can be clustered by neural networks. And then on clustered to conduct in-depth study. Or manually cluster, but it will be too time-consuming task.

And for the most effective solution, it is necessary to divide each square diagonally in both directions and sort through both options for dividing each square with a diagonal. So, an even shape can be obtained, but not square faces, because of which the moment can be considered 2 times more or less [12].

And the turn can be removed and collect everything that at the 10th step more than 50 Nm gives the moment, so reduce the search time [13]. And plus, we use sectors of 22.5 degrees (easier calculation). If we perform all these operations, we can also count for 3 mm cells.

And it will also be necessary to carry out cleaning, it looks as follows:

We check the cell, if the cell above and below and left and right are not equal to the value of this cell, then the cell takes such a value, which are larger around. If all the cells around are different, then any value presented is accepted [14].

It is also important to consider that this optimization will not contain cells with air material for ease of calculation, however, it is obvious that this is impractical, therefore this point will be taken into account in the subsequent optimization of thermal characteristics, though also topological optimization. And in the end, a strength calculation will be performed to verify the resulting design [15].

5 Conclusion

Topological optimization provides the concept of a magnetic system, i.e. a rough picture of the arrangement of magnets. And then you can optimize the design by varying the width and height of the magnets in order to achieve the practical implementation of the rotor design.

This article demonstrates how you can go from the complex standard design of a permanent magnet synchronous motor to the simplest one (sector - 45 degrees, shown in Figure 4).

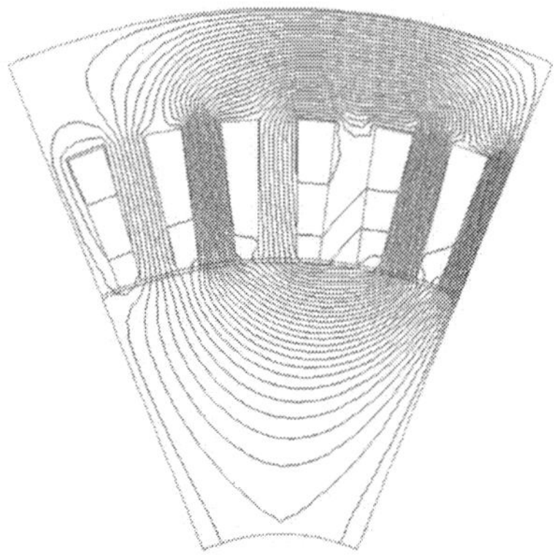

Fig. 4. The final version of the sector for calculation.

While the adequacy of the model remains at a high level, and with this model it is possible to carry out topological optimization through genetic an algorithm. The speed of calculations will increase with the help of computer software more than 10 times.

Acknowledgments

The reported study was funded by RFBR, project number 19-37-90134.

References

[1] B.E. Vasiliev, L.A. Magerramova, Analysis of the possibility of using topological optimization in the design of uncooled turbine blades, Vestnik SGAU **3-1** (2015)

[2] C. Li, I.Y. Kim, Multi-material topology optimization for automotive design problems, Proceedings of the Institution of Mechanical Engineers, Part D: Journal of Automobile Engineering **232**, 14, 1950–1969 (2018)

[3] E.I. Gracheva, O.V. Fedorov, Forecasting Reliability Electrotechnical Complexes of In-Plant Electric Power Supply Taking into Account Low-Voltage Electrical Apparatuses, 2019 International Conference on Industrial Engineering, Applications and Manufacturing (ICIEAM), Sochi, Russia, 1–5 (2019)

[4] K.A. Bashin, R.A. Torsunov, S.V. Semenov, Methods of topological optimization of structures used in the aerospace industry, Bulletin of the PNRPU, Aerospace engineering **4**, 51 (2017)

[5] K.L. Shenoy, M.S. Kumar, Design topology and electromagnetic field analysis of Permanent Magnet Brushless DC motor for electric scooter application, International Conference on Electrical, Electronics, and Optimization Techniques, ICEEOT 2016, 1541–1545 (2016)

[6] T.I. Petrov, A.R. Safin, I.V. Ivshin, A.N. Tsvetkov, V.Y. Kornilov, The Prospects of Using a Synchronous Machine with Permanent

Magnets in the Oil Industry, 14th International Scientific-Technical Conference on Actual Problems of Electronic Instrument Engineering, APEIE 2018 – Proceedings, article no. 8546157, 336–338 (2018)

[7] H. Shim, S. Wang, K. Hameyer, Topology optimization of magneto thermal systems considering eddy current as Joule heat, IEEE Trans. Magn. **43**, 4, 1617–1620 (2007)

[8] D.N. Dyck, D.A. Lowther, Automated design of magnetic devices by optimizing material distribution, IEEE Trans. Magn. **32**, 3, 1188–1193 (1996)

[9] S. Wang, D. Youn, H. Moon, J. Kang, Topology optimization of electromagnetic systems considering magnetization direction, IEEE Trans. Magn. **41**, 5, 1808–1811 (2005)

[10] A. Safin, A. Kopylov, R. Gibadullin, T. Petrov, R. Khusnutdinov, O. Fedorov, Thermal Model of a Linear Electric Machine, 1st International Conference on Control Systems, Mathematical Modelling, Automation and Energy Efficiency (SUMMA), Lipetsk, Russia, 426–428 (2019)

[11] D.H. Kim, J.K. Sykulski, D.A. Lowther, The implications of the use of composite material in electromagnetic device topology and shape optimization, IEEE Trans. Magn. **45**, 3, 1154–1157 (2009)

[12] J. Lee, N. Kikuchi, Structural topology optimization of electrical machinery to maximize stiffness with body force distribution, IEEE Trans. Magn. **46**, 10, 3790–3794 (2010)

[13] E.I. Gracheva, O.V. Naumov, A.N. Gorlov, Modelling Characteristics of Reliability Low-Voltage Switching Devices on the Basis of Random Checks on the Example of Contactors, Proceedings – 2019 1st International Conference on Control Systems, Mathematical Modelling, Automation and Energy Efficiency, SUMMA 2019, article no. 8947595, 641–643 (2019)

[14] J.S. Choi, J. Yoo, Structural topology optimization of magnetic actuators using Genetic algorithms and ON/OFF sensitivity, IEEE Trans. Magn. **45**, 5, 2276–2279 (2009)

[15] L.L. Myagkov, S.M. Sivachev, E.E. Strizhov, S.P. Chirsky, Topological optimization of a high-performance diesel piston, Dvigatelestroyeniye, 2, 3–10 (2018)

The use of heat transformers for the low-temperature secondary energy resources recovery in non-ferrous metallurgy enterprises

A Sh Alimgazin[2], *S A* Prishchepova[1], *I A* Sultanguzin[1,*], *A V* Fedyukhin[1], *Yu V* Yavorovsky[1], and *A I* Bartenev[1]

[1]National Research University "Moscow Power Engineering Institute" (NRU "MPEI"), 17 Krasnokazarmennaya Street, Moscow, 111250

[2]L.N. Gumilyov Eurasian National University, Satpaev st., 2, 010008, Nur-Sultan, Republic Kazakhstan

Abstract. The paper discusses the concept of developing schemes for the utilization of low potential secondary energy resources waste heat in heat transformers (HT). There ara several examples of using HT in non-ferrous metallurgy plants in Russia and Kazakhstan: the production of nickel, zinc and ferroalloys. In the case of nickel production, two schemes are proposed for consideration: for summer and winter modes. The main object of study is a nickel production scheme using a lithium bromide absorption chiller (AbCh) in the summer. The nickel production process needs cold. In the proposed scheme, the main source of utilized low-grade waste heat for AbCh is recycled water. The basic parameters were calculated under various extreme conditions, the energy effect of the circuit modernization was estimated, and equipment was selected. Another example of the HT use is the use of a heat pump unit (HPU) in the production of zinc and ferroalloys. HPU is designed to heat feed water before chemical water treatment and at the same time to "cool down" the circulating water, which is then used for electrolysis, replacing purchased artesian water.

1 Introduction

Due to a decrease in the quantity of fossil fuels, its rise in price on the world market and aggravation of environmental problems, energy efficiency increasing through the use of secondary energy resources (SER) has become more relevant and is gaining world significance The scarce of SER use leads to thermal pollution of the environment [1]. Non-ferrous metallurgy should be given special attention, since there is a large amount of excess heat in it in the form of industrial effluents or reverse cycles. This excess heat can be used in HT to generate cold or heat with a higher potential [2, 3]. This heat transforms into the heat of hot water, suitable for the needs of hot water supply (HWS), heating or ventilation of an enterprise [4-8]. The conversion of waste heat into cold became possible with the advent of absorption chillers. The resulting cold can be used both in the technological processes of the enterprise, and for the needs of air conditioning in buildings.

The article discusses the modernized nickel production technology in Russia, which requires constant maintenance of the circulating water temperature level at the inlet to the production of not more than 20 °C in winter and summer. It also provides an example of the successful introduction of HPU in zinc production and discusses the prospects for their application at a ferroalloy plant in the Republic of Kazakhstan.

2 HT use in the nickel production unit

Below is a calculation of tubular furnaces reaction excess heat utilization of a nickel production unit (NPU). In accordance with the technology, cooling of 400 m3 / h from 85 to 20 ° C is required.

The most responsible is not only the use of SER, but also the provision of a stable technological process with cooling of the coolant to predetermined temperatures, both in winter and summer periods.

It is proposed to consider two ways of heat utilization (Figure 1):

- in the winter period, it is necessary to use heat pumps to heat the return water of the nickel production workshop heat network and cold feed water of the thermal power plant boiler. In winter the reverse water circuit cooling of the NPU in the heat exchangers 9 and 11 is quite sufficient and additional cold from the AbCh is not required.

- in summer period after NPU reaction heat recovers for heating the return water of the heating network and cold feed water of the thermal power plant boiler it is cooled to 20°C in AbCh [9].

* Corresponding author: SultanguzinIA@mpei.ru

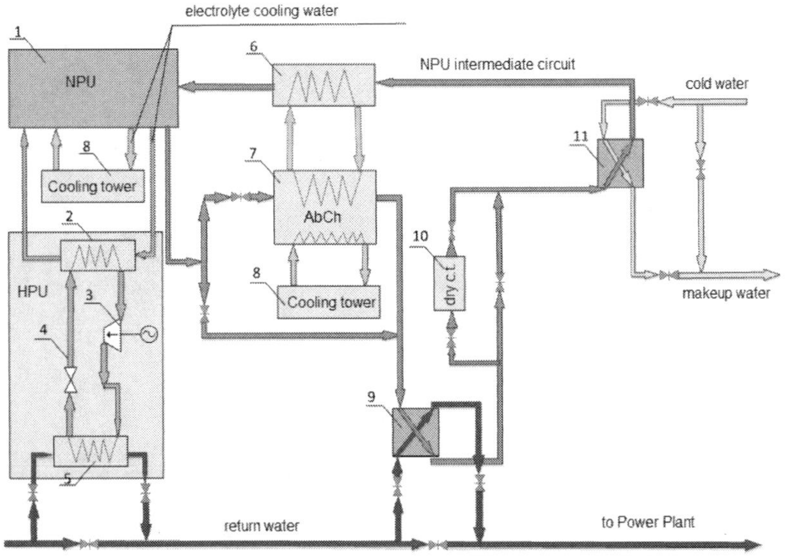

Fig. 1. The general scheme of heat recovery. 1 –nickel production unit (NPU); 2 – heat pump evaporator; 3 – heat pump compressor; 4 – throttle; 5 – heat pump capacitor; 6 – water-water heat exchanger of AbCh; 7 – AbCh; 8 – cooling towers; 9 – water-to-water heat exchanger for return water heating; 10 – a complex of dry cooling towers; 11 – water-to-water heat exchanger for thermal power station make-up water heating.

Figure 1 shows the general scheme of heat recovery in winter and summer modes.

3 NPU SER recovery in winter period

Figure 2 shows a diagram of low-grade heat utilization using a heat pump to heat the return water of the heating network in the NPU in winter at an outdoor temperature of -15 °C [8, 10, 11].

The use of a heat pump reduces the water temperature from 50 °C to 40 °C at a flow rate of 550 t/h from an electrolyte cooling heat exchanger.

The compressor capacity of the vapor compression HPU is 1017 kW with a compression ratio of 2.2. The coefficient of performance (COP) is quite high: COP = Q_T/N_K = 7226 kW/1017 kW = 7.11

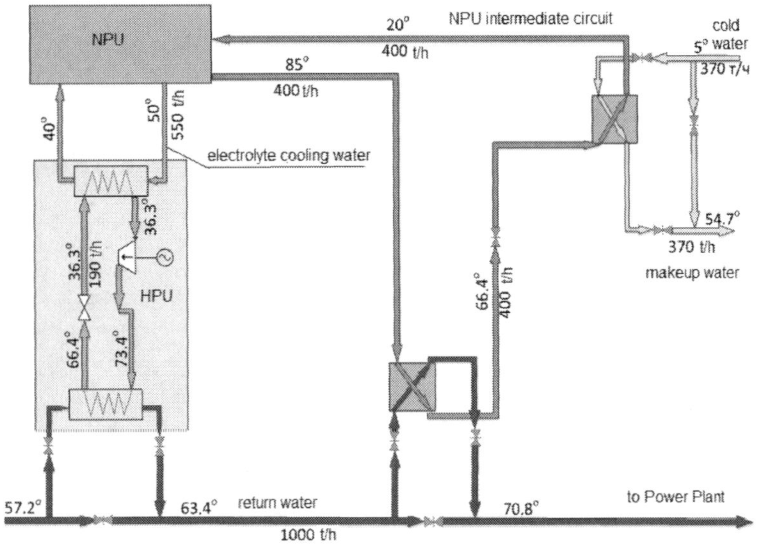

Fig. 2. Scheme of low-grade heat utilization using a heat pump in winter at an outdoor temperature of -15°C.

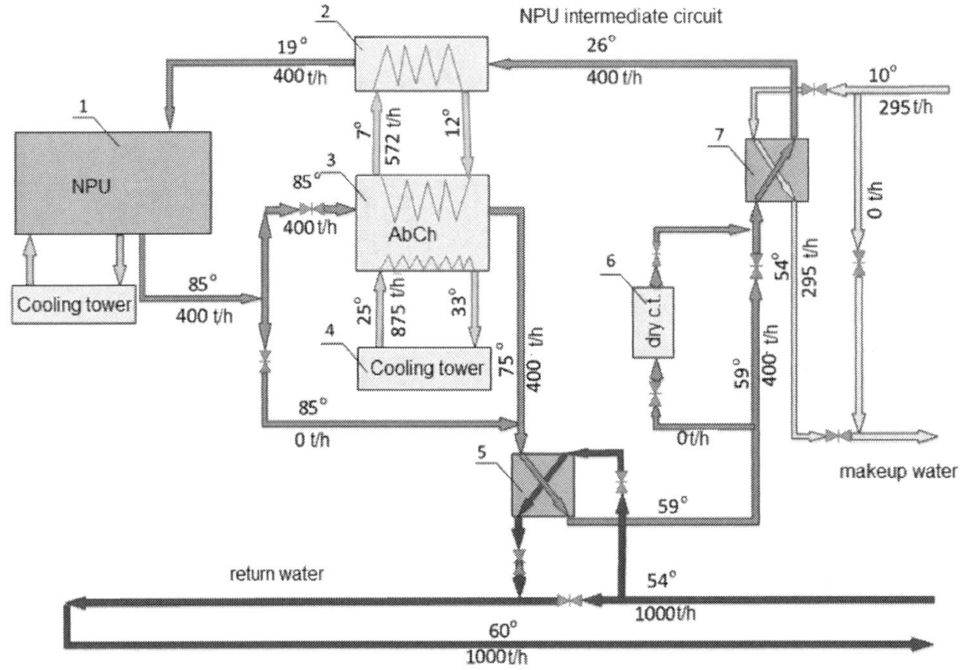

Fig. 3. Schematic diagram of the nickel production installation water circuit with make-up water parameters Gmw=295 t/h, tmw=10°C. 1 – nickel production installation (NPI); 2 – water-water heat exchanger of AbCh; 3 – AbCh; 4 – cooling towers; 5 – water-to-water heat exchanger for return water heating; 6 – a complex of dry cooling towers; 7 – water-to-water heat exchanger for thermal power station make-up water heating.

The heat pump allows to produce thermal energy to heat the water of the NPU heating network in amount Qt = 7226 kW.

The amount of obtained useful heat energy after NPU reactionary heat recovery for heating water of the heating network and feed water of the thermal power plant boiler is 37.2 MW.

4 NPU SER recovery in summer period

The modernized nickel production technology requires constant maintenance of the circulating water temperature level at the inlet to the production of not more than 20 °C. In accordance with this, as well as for the outlet circulating water heat utilization, it is proposed to use the AbCh.

The following advantages compared to compression refrigeration machines dictate the choice of AbCh:

- Full automation.
- Long service life (at least 20 years).
- Minimum power consumption. Electricity is required for the operation of pumps and automation.
- The minimum noise level.
- Environmentally friendly. The refrigerant is ordinary water. [12]
- Recovers the thermal energy of the discharged hot water, flue gases or production processes.
- Fire and explosion safety.

The principle of diagram operation is as follows: a stream of water heated during production (1) to a temperature t_1 = 84-85°C is used as heating water in AbCh circuit (3), where it is cooled to Δt_2 = 10-11°C, then in the return water heat exchanger (5), after which it is additionally cooled in the make-up water heat exchanger (7), leaves it with a temperature t_3 = 25-27°C and is additionally cooled in the heat exchanger (2) to a temperature t_4 = 19.5-20 °C, after which it returns to NPU. (Figure 3)

To ensure the required temperature level of the NPU circulating water under extreme conditions, a complex of dry cooling towers is provided (6). In these cases, part of the circulating water stream is taken from the main collector and cooled in dry cooling towers, after which it is mixed with the main stream.

As part of the development and modeling of circuit solutions, calculations were carried out at three critical points with different indicators of temperatures (t) and flow rates (G) of power plant makeup cold water, only one will be considered in the work:

- G = 250 t/h, t = 10°C;
- G = 295 t/h, t = 10°C, the flow rate is taken as the hourly average for the warmest day of summer, the water temperature is also taken with a reserve of Δt = 1°C (Figure 3);
- G = 295 t/h, t = 15°C.

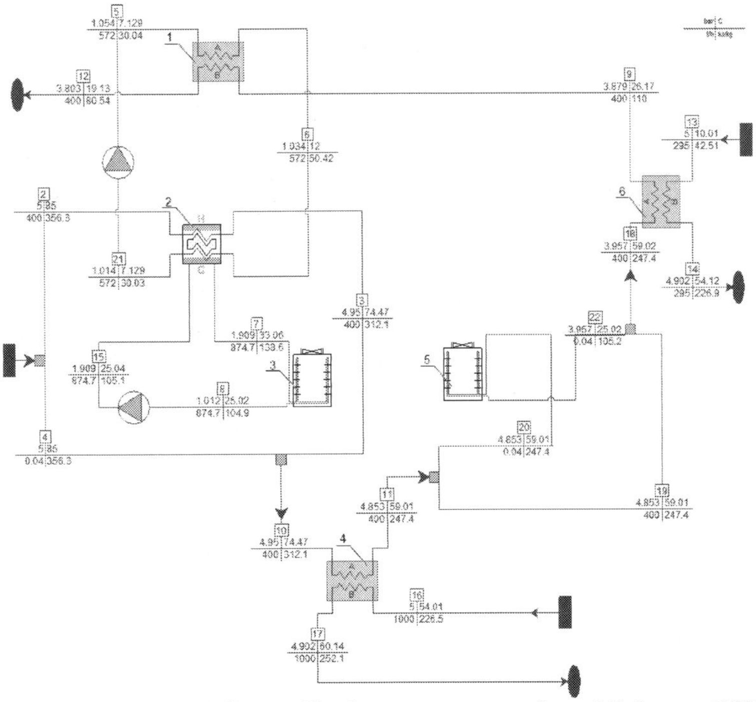

Fig. 4. Calculation of the NPI water circuit diagram with make-up water parameters Gmw = 295 t/h, tmw = 10 °C in THERMOFLEX. 1 – water-water heat exchanger of AbCh; 2 – AbCh; 3 – cooling towers; 4 – water-to-water heat exchanger for return water heating; 5 – a complex of dry cooling towers; 6 – water-to-water heat exchanger for thermal power station make-up water heating.

Computer simulation of the new nickel production scheme was carried out in the Thermoflex software environment (Figure 4). Thermoflex is a modular program with a graphical interface. The user can create a model of any station using more than one hundred and seventy-five symbolic images contained in its database. It is also possible to simulate almost any type of modern installations and heat sources in conjunction with heating networks [13, 14].

Computer simulation has shown the sufficiency of this design for maintaining technological needs and the optimal use of the production water recycling potential.

The main characteristics of the simulated chiller are presented in Table 1.

Table 1. Characteristics of the simulated AbCh.

Num.	Indicator	Value
1	Heating water flow, m3/h	400
2	Cooled water flow, m3/h	572
3	Cooling water flow, m3/h	874
4	Cooled water temperature °C	12/7
5	Cooling water temperature °C	25/33
6	Heating water temperature °C	85/74

Other indicators of temperatures and flow rates of the circuit solution, see Figure 3.

Waste heat recovery in AbCh not only reduces primary fuel consumption, but also thermal pollution of the environment. Water of the intermediate circuit for NPU reaction heat utilization with a temperature of 85 °C and a flow rate of up to 400 m3/h is used as a source of thermal energy for the AbCh.

Heating water is cooled from 85°C to 74.5°C in AbCh. The heat energy consumption with heating water in AbCh is 4.9 MW. Then it enters the heat exchanger for heating the reverse water of the NPU heat network from 54°C to 60°C with a flow rate of 1000 m3/ h. The heating water is cooled from 74.5 °C to 59 °C in the heat exchanger of the NPU heating system water. The heat load of this heat exchanger is 7.2 MW.

Then it enters the heat exchanger of power plant boiler feed water, where it is cooled from 59°C to 26.17°C. Makeup water is heated from 10 °C to 54.1°C. The heat load of this heat exchanger is 15.3 MW.

In AbCh, water with a temperature of 26.2°C is cooled to 19.1°C due to chilled water, which is heated from 7°C to 12°C. After that, water with a flow rate of 400 m3/h and a temperature of 19.1°C is returned to the intermediate circuit for the utilization of NPU reaction heat.

Thus, we obtain 27.4 MW as useful heat energy during the utilization of NPU reaction heat, and 3.3 MW as a cold in AbCh. Two AbCh-2000Vn produced by Teplosibmash OJSC with a total cooling capacity of 3600 kW can be considered for installation based on the calculation results.

a) b)

Fig. 5. The first industrial NT-3000 heat pump at Kazzinc JSC in the Republic of Kazakhstan.

A cooling water circuit with cooling parameters from 33 °C to 25 °C is also required for AbCh,. As a cooling circuit, the plant's water recycling system or its own local small-sized cooling tower can be used.

5 HT use in the zink production

Currently, the use of HT in Russia and the CIS countries is just beginning to become popular. So in the Republic of Kazakhstan, an example of the use of HPU is the NT-3000 heat pump with a capacity of 3 MW at Kazzinc JSC (Ust-Kamenogorsk). HPU is intended for heating feed water before chemical water purification from + 8°C to + 40°C and at the same time for "cooling" circulating water, which is then used during electrolysis. The circulating water cooling allows to replace the purchased artesian water, thereby reducing material production costs. Table 2 shows the results of the heat pump NT - 3000 adoption [8].

Figures 5a and 5b show how NT-3000 looks like.

Table 2. The results of the heat pump NT – 3000 adoption.

Num.	Indicator	Value
1	Coefficient of performance	6.8
2	Recycled water cooling	7.7 °C
3	Cost of thermal energy at HPU	198 tenge / Gcal
4	Cost of cold	440 tenge / Gcal
5	Capital Costs, including:	27.63 million tenge
6	Cost of HPU	24.9 million tenge
7	Total heat and cold production for 1999-2004 in cash	57.2 million tenge
8	Cost of electricity to drive the compressor	4.55 million tenge
9	Estimated simple payback period	About 2 years

Later, in order to utilize the circulating water low-grade heat of the whole enterprise of Kazzinc JSC, a design of a 30 MW heat pump station was agreed. With the help of this project, it becomes possible to completely utilize the heat of circulating water (29.8 MW) and use this heat in the winter period for the needs of heating, ventilation and hot water supply, which will make it possible to refuse the services of a third-party power plant [10].

6 Prospects for the use of TT at the Aksu Ferroalloy Plant - a branch of Kazchrome JSC

In the Republic of Kazakhstan in 2018-2020, scientists of the Research Institute “Energy Saving and Energy Efficient Technologies” (“ES&EET”) of ENU named after L.N. Gumilyov and the Start Company BMTU LLC (Astana) conduct joint research to evaluate the possibilities of using HPU at the Aksu Ferroalloy Plant (AFP) - a branch of Kazchrome TNC JSC using alternative energy sources at various enterprise facilities [15].

The prospects for the use of HPU in the smelting shop No. 4 thermal circuit using the water recycling system (WRS) low-grade heat with temperatures up to +35°C directed to the HPU evaporators were studied.

HPU recovers the WRS waste heat of the AFP (from 4,000 m3/h to 6,000 m3/h), which is used for cooling in the enterprise cooling towers. Thus, all utilized heat can be sent to heat the various facilities of the workshop. During the implementation of this project, it is planned to cover the total heat load (2.01 MW = heating + hot water supply) of enterprise workshop No. 4.

In accordance with the initial data, presented by AFP specialists, an enlarged calculation of the HPU efficiency use for heat supply of objects in the workshop No. 4 (Table 3) is carried out. From which it follows that the annual economic effect from the use of HPU in the workshop amounts to 42.812 million tenge, the payback period of the proposed project is 3.53 years [15].

a) b)

Fig. 6. Production site of the melting shop No. 4 of the AFP, where heat pumps installation with 2.34 MW total capacity of is planned.

Table 3. Comparison of annual operating thermal energy generation costs of the AFP smelting shop №4 from various heat sources.

Heat source	Annual operating costs, million tenge*	Cost 1 Gcal/hour Tenge/Gcal (including VAT)
Heat supply of workshop №4 from the district heating plant (ROK-2) of the AFP	50.928	2 500
Heat pump boiler room	8.116	490
The data are for the period summer 2019, the exchange rate is - 1 $ = 381 tenge		

7 Conclusion

The article provides examples of the HT use in industrial enterprises.

Utilization of NPU low-grade heat in HPU with a heat capacity of 7.2 MW allows to get up to 30 MW of usable heat energy for heating the water of the heating network and makeup water in winter.

The presented scheme for the low-grade heat utilization using AbCh in the summer period allows to obtain 27.4 MW as useful heat energy. The source of thermal energy for the AbCh is the water of the NPU intermediate circuit with a temperature of 85°C and a flow rate of up to 400 m3/h. Calculations show that the AbCh is able to produce cold in an amount of 3.3 MW with cooling parameters for chilled water from 12°C to 7°C.

The HPU at Kazzinc JSC allows not only to meet the needs of the enterprise in thermal energy, but also allows to reduce material costs for the purchase of artesian water. A larger project of the 30 MW station allows to completely decline the services of a third-party power plant.

The utilization of waste heat from the recycling water supply system of the AFP smelting shop No. 4 in HPU makes it possible to:

- decline the traditional source of heat - the coal boiler ROK-2 with significant operating costs;
- significant decrease greenhouse gas emissions;
- ensure year-round reliable heat supply;
- reduce the capacity of circulation pumps and reduce water consumption for production needs.

Similar events can be carried out at almost any large metallurgical enterprises in the industrial regions of Russia and Kazakhstan.

References

[1] I. Sultanguzin, *Environmental safety and energy efficiency of industrial heat power systems: a training manual* (Moscow: Publishing House MPEI, 288, 2013)

[2] L. Zhiqiang, X. Nan, Y. Sheng, Thermodynamic and parametric analysis of a coupled LiBr / H2O absorption chiller / Kalina cycle for cascade utilization of low-grade waste heat, Energy Conversion and Management **205,** 112370 (2020)

[3] M.L. Chougui, S. Zid, Energy analysis of single effect absorption chiller (LiBr / H2O) in an industrial manufacturing of detergent, Energy Procedia **50**, 105–112 (2014)

[4] S. Prishchepova, I. Sultanguzin, *The use of absorption chillers for house cooling in the summer using the solar collectors excess heat, Proceedings of the eighth International School-Seminar for Young Scientists and Specialists "Energy saving – theory and practice"* (Moscow: Publishing House MPEI, 411–414, 2016)

[5] C. Peizhe, Y. Mengxiao, L. Zhiqiang, Zh. Zhaoyou, Y. Sheng, Energy, exergy, and economic (3E) analyses and multi-objective optimization of a cascade absorption refrigeration system for low-grade waste heat recovery, Energy Conversion and Management **184,** 249–261 (2019)

[6] Y. Sheng, Q. Yu, W. Yifan, Y. Siyu, A novel cascade absorption heat transformer process using

low grade waste heat and its application to coal to synthetic natural gas, Applied Energy **202,** 42–52 (2017)

[7] E. Khosrow, F.J. Gerard, S.F. Amy, Thermo-economic analysis of steady state waste heat recovery in data centers using absorption refrigeration, Applied Energy **139**, 384–397 (2015)

[8] A. Alimgazin, *Development of schemes and technologies for the use of non-traditional and renewable energy sources for heat supply of isolated objects* (author. doc tech. sciences. Almaty: AUPET, 44, 2010)

[9] S. Prishchepova, I. Sultanguzin, Example of energy efficiency improving of the enterprise using AbCh in mining and metallurgical industry, Radioelectronics, electrical and power engineering, Moscow: Publishing House MPEI **2**, 356 (2016)

[10] A. Alimgazin, S. Alimgazina, A. Bakhtiyarov, Analysis of the prospects for the use of heat pump technologies at metallurgical enterprises of the Republic of Kazakhstan, Proceedings of the VIII International Scientific and Practical Conference Energy-efficient and resource-saving technologies in industry. 100 years of national metallurgical furnaces design, 238–250 (2016)

[11] I. Sultanguzin, T. Shomova, P. Shomov, V. Dostovalov, The use of heat pumps in gas processing plants. Abstracts and reports of the XV Minsk International Forum on Heat and Mass Transfer **3**, 420–424 (2016)

[12] D. Keppler, Absorption chillers as a contribution to a climate-friendly refrigeration supply regime: Factors of influence on their further diffusion, Journal of Cleaner Production **172**, 1535–1544 (2018)

[13] A. Fedyukhin, I. Sultanguzin, S. Kurzanov, R. Belov, A. Bakulin, T. Shomova, *The use of applied software to solve the problems of industrial heat power: a training manual* (Moscow: MEI Publishing House, 88, 2016)

[14] THERMOFLEX. Fully Flexible Heat Balance Engineering Software, THERMOFLEX User's Manual, Thermoflow, Inc. **3,** 1139 (2013)

[15] A. Alimgazin, S. Alimgazina, Prospects for the use of energy-saving heat pump technologies using alternative energy sources at the Aksu Ferroalloy Plant, a branch of Kazchrome TNK JSC, Bulletin of PSU named S. Toraigyrova, Series Power engineering **4**, 35–38 (2019)

Study of processes of artificial freezing and thawing of soils when developing a model of energy-efficient radiation-convection setup

Anatoly Velsovskij[1], *Liliya* Mukhametova[2], *Anton* Sinitsyn[1] and *Denis* Karpov[1]

[1]Vologda State University, LeninaStr.. 15, 160000, Russia
[2]Kazan State Power Engineering University, Kazan, Russia

Abstract. This paper considers one stage of development of a new power plant for thawing frozen soil during emergency repairs. It includes experimental study of temperature conditions for freezing and artificial thawing of various types of soils in field and laboratory conditions. The main results of laboratory research are presented. The operation process of an industrial prototype of power plant is modelled on a test bench using analogues of radiation-convective heat generators. The obtained calculated and experimental characteristics of artificial thawing of frozen soil by a surface radiation-convective method will allow one to calculate the thermal power of a power plant, and to carry out its design and thermal calculation.

1 Introduction

The search for new technological solutions and soil preparation systems during repair, construction and emergency operations is an important task in our country due to its harsh climatic conditions. The increased mechanical strength of frozen soils prevents their excavation by conventional technical means, which necessitates the preliminary preparation of such soils. The following basic methods of soil preparation in winter conditions are used: thawing with subsequent excavation of thawed soil, preliminary mechanical loosening and excavation of frozen soil. Mechanized methods are currently the most common, but the use of special means of loosening or cutting into separate pieces in complex urban underground economy is difficult, and sometimes impossible for technical requirements and safety reasons. In urban conditions, the problem of excavation of frozen soils is dealt by, first of all, emergency, communal and energy supplying services and organizations. A large number of emergency situations in winter are associated with ruptures of pipelines for various purposes. The recovery time of site directly depends on the speed of opening the damaged pipeline by the repair team. Another important aspect of the problem raised above is soil preparation before construction - a task for builders conducting work in winter. The use of special means of soil loosening or cutting does not enable to use the mixture without a sufficient amount of thawed soil for construction of embankments, backfilling of trenches and sinuses of foundation pits due to the impossibility of high-quality compaction, since it leads to soil subsidence after thawing. In such conditions, the use of alternative methods of soil preparation during repair, construction and emergency works expands the technical capabilities of the organization and allows ensuring the work efficiency. These tasks can be solved using soil thawing.

The performed literature and patent review showed a variety of methods for thawing frozen soils. They are classified according to the used heat source (electricity, steam, gas, fuel oil, coke, hot slag, etc.) and operation principle (thermal, chemical, etc.)

General scientific principles for the design of engineering systems and devices in frozen soils are formulated in the works of N.A. Tsytovich [1], Pavlov A.R. [2], Sagala T.A. and Kologrivov M.M. [3]. Nakano, Y., Brown, J., [4], Harlan, R. L. [5], Guymon, G. L. and J. N. Luthin [6] dealt with the problems of heat and mass transfer in frozen soils.

Despite the fact that significant scientific forces are being attracted to solve this problem in the world, in practice there still remain a number of issues related to the design, calculation, execution and safe operation of devices and mechanisms for the development of frozen soils.

The existing methods are of limited application due to high energy costs and long time required for heating (from five hours or more to several days), or they are unacceptable for safety reasons for certain types of work (excavation of gas, electric networks, etc.).

The development of an effective and uninterrupted power supply system for low outdoor temperatures was previously considered in [7]. This article is focused on the solution to the problem of eliminating accidents in case of failure of a particular section of engineering communications.

The need to create a small-sized setup for thawing frozen soils appeared as a result of professional cooperation of the university and the city emergency-

https://doi.org/10.1051/e3sconf/202017801018

Table 1. Characteristics of the developed setup.

No.	Characteristics	Value
1	Operation principle	Radiation-convective heat transfer
2	Fuel type	Natural gas, propane-butane
3	Thawing time, h	Not more than 4
4	Maximum thawing depth, m	Not less than 1.5
5	Installation size, m	1 x 1.5 x 0.4
6	Allowable weight, kg	Not more than 90
7	Operating personnel, man	2

maintenance and public utilities, which needed such a device. The importance of task for creating a new setup was determined by the unsolved problem of rapid development of frozen soil in cramped conditions of complex urban underground utilities. In such conditions it is impossible to use conventional technical means of loosening or cutting frozen soils due to technical requirements and safety reasons [8].

Currently, there are a large number of methods for calculation heating of frozen soils, but the model proposed in [9] is the most acceptable. In accordance with [10], the thermal performance of the installation for thawing sand building material was calculated according to empirical formulas used in the calculations for construction of facilities in permafrost areas.

2 Problem statement

To determine the requirements and parameters of setup being developed, the exploratory studies were carried out aimed at identifying its main technical and operational characteristics and their analysis (table 1). The radiation-convective thawing principle was chosen, and gas fuel was the main energy source.

The time required to thaw the soil to a given depth, shown in table 1, was determined on the basis of requirements for emergency repairs by various repair services in one work shift. In order to ensure soil thawing within the allotted time interval (not more than 4 hours), it was necessary to determine the power of setup being created.

A number of methods were applied to identify the required power and experimentally confirm the calculated parameters by conducting a series of tests on freezing and thawing various types of soils. Sand of medium size and light loam were used as the main soils. For them, physical properties and classification indicators were determined in accordance with the requirements of regulatory documents. To create conditions for experiment conducting in close to field

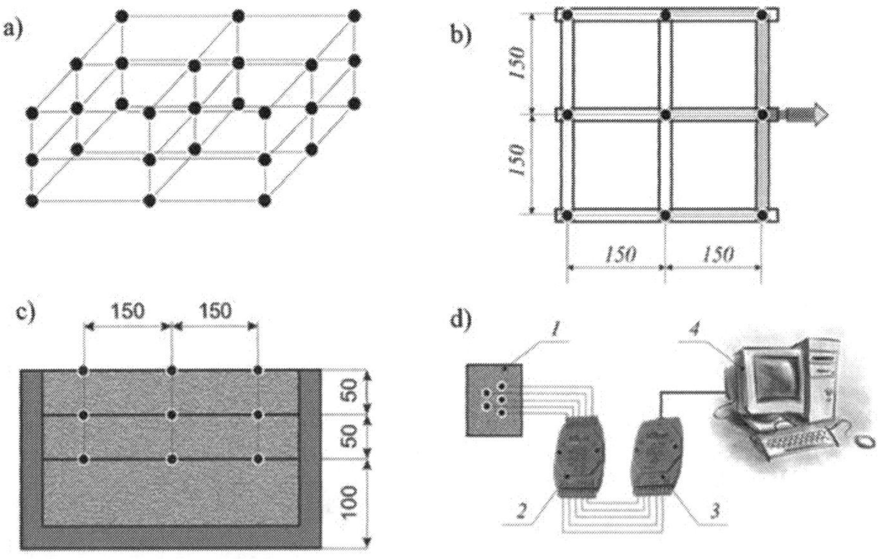

Fig. 1. Preparation for an experiment on soil freezing: a) Matrix of thermocouple location points; b) Frame for accurate placement of sensors in the plan; c) Mold with compacted soil and location of thermocouples in the profile; d) Temperature measurement scheme (1 - mold with soil and installed temperature sensors, 2, 3 - analogue block-digital converter, 4 - computer), dimensions are shown in millimeters.

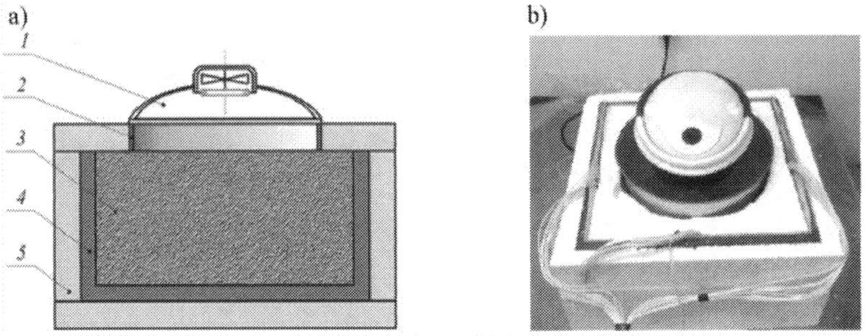

Fig. 2. Scheme of experimental setup: a) Scheme of setup, b) Photograph of the laboratory setup, 1 - Radiation-convection device, 2 - Metal ring, 3 - Frozen soil, 4 - Mold for soil, 5 - Thermal insulation.

conditions, the maximum density, optimal humidity and density in natural occurrence at a depth of up to 1.5 m were determined.

3 Methods

The methodology of experiment on freezing molds with soil included:

1) Preparation of the necessary elements (molds for soil, grids with thermocouples for measuring temperature fields, block of converter equipment and computer for recording readings from thermocouples);

2) Soil preparation (drying to constant weight, mixing to obtain a homogeneous environment, moistening, laying and layer-by-layer compaction in the prepared mold);

3) Preparation of a stationary cold store (type KHN-1) for testing, setup of freezable molds with soil and setting the temperature regime (from –4 to –6 °C) for uniform freezing.

The temperature change in soil during freezing and subsequent thawing was monitored similar to [11-12] using temperature sensors, location matrix of which is shown in Figure 1, a. The measurements were carried out in three horizons; nine temperature sensors were installed in each of them. To achieve accurate placement of temperature sensors, they were fixed on a specially designed frame, the view of which and dimensions in plan and profile are shown in Figures 1, b, c.

As in [7], chromel-alumel junctions of thermocouples connected to an analogue-to-digital converter were used as temperature sensors during the experiment. The RS 232 cable was used to connect the analogue-to-digital converter with PC, which was used to display the freezing-thawing processes in real time (Figure 1, d).

During experiments, the mold with frozen soil was covered with heat insulation from the sides and from the bottom to prevent their heating and to create conditions for thawing from above (Figure 2).

4 Results

The experiment consisted of preliminary freezing the soil, and its subsequent artificial thawing. Every two seconds the readings of temperature sensors were registered, and the depth of soil freezing (thawing) was measured using the metering probe. The measured freezing depths for sandy soil are presented in Figure 3.

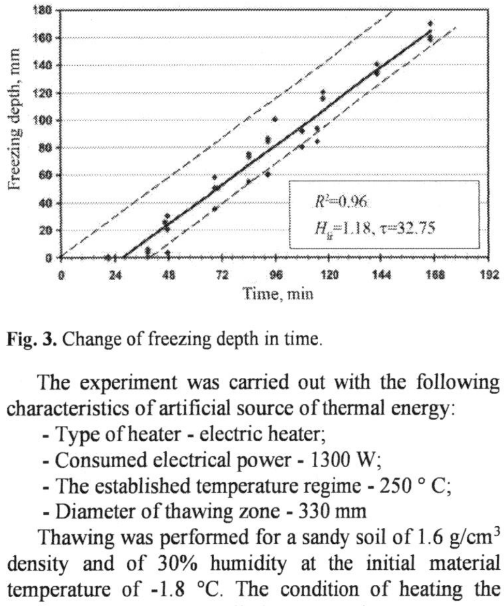

Fig. 3. Change of freezing depth in time.

The experiment was carried out with the following characteristics of artificial source of thermal energy:

- Type of heater - electric heater;
- Consumed electrical power - 1300 W;
- The established temperature regime - 250 ° C;
- Diameter of thawing zone - 330 mm

Thawing was performed for a sandy soil of 1.6 g/cm^3 density and of 30% humidity at the initial material temperature of -1.8 °C. The condition of heating the frozen soil by the radiation-convective setup was experimentally studied in field and laboratory conditions. Comparison of the obtained results is shown in Figure 4.

The relationship between the thawing depth H_{th} and the time of thermal energy supply τ can be expressed by the following empirical expression:

$$H_{th} = K \cdot \tau^{0.5},$$

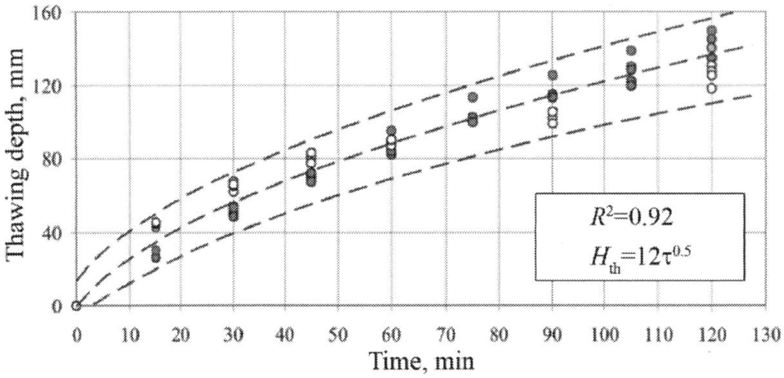

Fig. 4. Relationship between the thawing depth (dark dots are experimental data obtained in field conditions, light dots are experimental data obtained in the laboratory).

where $K = f(K_1, K_2,..K_N)$ is the coefficient depending on the soil type, the power of heat source, the condition of thawing and removal of thawed soil, environmental conditions, etc .; τ is time of heat energy supply,.

Calculations were performed according to the indicated empirical dependence and according to the engineering method for calculating the thickness of melt zone of dispersed material as a function of its thawing time. Comparison of the calculation results showed sufficient accuracy of the proposed dependence.

The conditions for removing the thawed part of frozen soil have a strong influence on thawing rate of frozen soil. The generalizing experimental dependences of depth of artificial thawing of frozen soil change over time were obtained for energy expenditures of 1300 W and thawing area of $85 \cdot 10^{-3}$ m^2, and shown in Figure 5.

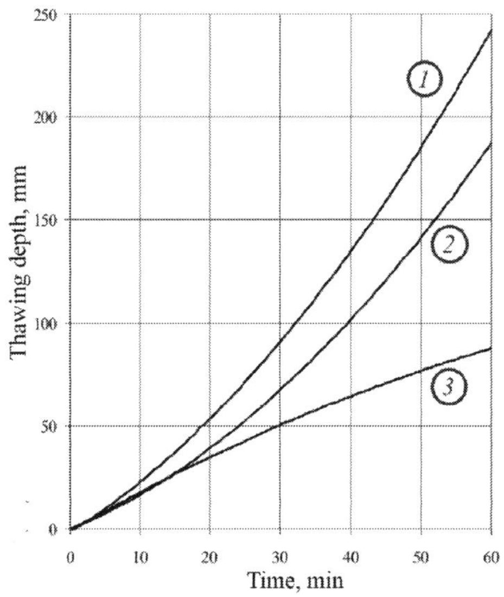

Fig. 5. Generalizing experimental dependence of depth of artificial thawing of frozen soil change over time: 1 - permanent removal of thawed part of frozen soil; 2 - periodic removal of thawed part of frozen soil; 3 - without removing the thawed part of frozen soil.

From Figure 5 it can be seen that the rate and, accordingly, the depth of thawing depends on the conditions for removal of thawed soil. With permanent removal of soil (every 5 min), the depth of thawing is 3 times greater than when thawing is performed without removing it. With periodic soil removal (every 15 min), the thawing depth is 2 times greater than thawing is performed without removing it. The ratio of thawing depth at constant and periodic soil removal was 1.5 times. This must be taken into account when developing technology for thawing frozen soil.

The installation is an integrated mobile device (Figure 6), divided into three separate but interconnected

Table 2. Comparative characteristics of technical means.

No.	Characteristics	The developed device	Thermoelectric mat TERMOMAT	The device for heating soil and concrete E700M (WACKER NEUSON)
1	Device power, W/m^2	**7500**	340	180
2	Thawing time to a depth of 0.5 m, hour	**15**	30	24
3	Costs for thawing, rub./h·m^2	**4,9**	23	147
4	Heating temperature, °C	**250**	70	82
5	Dimensions, mm x mm x mm	1200x1600x700	**550x2740x15**	3390x1500x1970
6	Mass, kg	110	**20**	3346

sections: working, electric and technical. It includes a burner, a smoke pipe, a smoke exhaust, a reflective screen, a fan, a heat exchanger and an air heater. The fan is operated by an electric motor connected to an electric battery. The electric motor and the battery are located in the electric motor section. The burner, smoke tube, smoke exhaust, air heater and heat exchanger are located in the technical section, the reflective screen is in the working section.

Thawing of frozen soils by a surface artificial method is achieved through the use of a device for radiation-convective heating of working bodies. The invention is consisted of the following main equipment: a burner device, which is used to mix fuel and air with formation of a fuel-air mixture and to maintain its stable combustion; a smoke tube in which the flame is developed and flue gases are moved, as well as a pipe is heated to create radiation -convective heat transfer from the pipe to the environment; a smoke exhauster that stimulates motion of flue gases from the burner along the entire length of the smoke pipe to its outlet to the air heater. In the air heater, heat is exchanged between the exhaust flue gases and cold air, as a result of which the source air is heated to the required temperature to increase the efficiency of burning the mixture. A reflective screen is located in the upper part of the working space of the apparatus above the smoke tubes and fan. To create radiation-convective heat transfer, a fan is placed in the apparatus, which performs motion of gaseous heated medium in the working space area, limited by a reflective screen and heated material. The fan operation is organized by an electric motor connected to an electric battery.

Efficiency is ensured by optimizing the location of the main elements of the installation and by combination of convective and radiation heat transfer methods.

To evaluate the efficiency of the obtained results, we compared the two closest analogues to the developed technical means for soil thawing: thermoelectric mat TERMOMAT and the WACKER NEUSON E700M device - a unit for thawing soil to a depth of 1.5 meters and for service of concrete, running on liquid fuel. Table 2 shows the comparison of the listed technical means. The highlighted values show the advantages of the considered technical tool over others.

5 Conclusions

The obtained calculated and experimental dependences of artificial thawing of frozen soil by a surface radiation-convective method will allow us to calculate the thermal power of the setup, perform its design calculation and develop a working test sample.

The proposed installation is more efficient in terms of specific heating power per 1 m2 of soil; thawing time to a depth of 0.5 m; costs for thawing expressed in monetary terms – rub./h•m2, as well as heating temperature. Thermomat is more attractive from the point of view of movement and transportation as it has low dimensions and weight, however, taking into

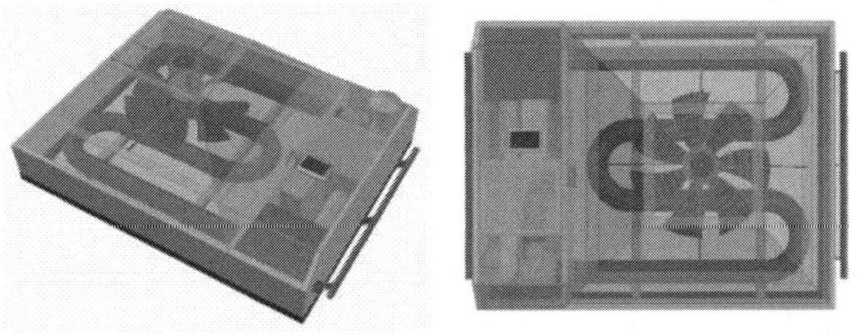

Fig. 6. Prototype of a mobile radiation-convection device. General view.

account its low efficiency during thawing (about 30 hours), these advantages are not so important. In view of the fact that the second analogue - the device for heating the soil and concrete E700M - is designed for large heating areas (up to 400 m2), it is inefficient and unreasonable for our conditions.

Thus, it is proved that the developed installation is efficient and competitive as compared with the existing technical means.

In addition, the following main features of the proposed installation are highlighted:

- Construction and design have no analogues in the modern market of technical equipment for repair and construction works;
- It has the minimum time for warming up frozen soil in comparison with the existing installations;
- Low specific energy consumption of the installation and the combined heat transfer method provide high installation efficiency;
- The mobility and compactness of the installation enables it transportation in a truck and carrying out work by a team in the amount of two people under any conditions of tightness of the urban landscape;
- The flexibility of regulating the heating capacity of the installation ensures its use by thawing building materials and heating building structures for any purpose;
- An energy-efficient combination of installation elements and a thermo-insulated housing allow repair and construction work in any climatic conditions.

References

[1] N.A. Tsytovich, Mechanics of frozen soils (Moscow: Higher School, 446, 1973).

[2] A.R. Pavlov, Mathematical modeling of heat and mass transfer processes during phase transitions: textbook (Yakutsk, 55, 2001).

[3] T.A. Sagala, M.M. Kologrivov, The use of mathematical models of non-stationary thermal conductivity with a phase transition of component in the calculations of defrosting bulk cargo, Refrigeration equipment and technology 3, 46-51 (2008).

[4] Y. Nakano, J. Brown, Effects of a freezing zone of finite width on the thermal regime of soils, Water Resour. Res. 7, 1226-1233 (1971).

[5] R.L. Harlan, Analysis of coupled heat-fluid transport in partially frozen soil, Water Resour. Res. 9, 1314-1323 (1973).

[6] G.L. Guymon, J.N. Luthin, A coupled heat and moisture transport model for arctic soils, Water Resour. Res. 10, 995-1001 (1974).

[7] A. Sinitsyn, A. Mihin, D. Zaripova, Development of a mobile power plant for thawing frozen soils, IOP Conference Series: Earth and Environmental Science 337, 1, 012064 (2019).

[8] A. Sinitsyn, Device for Thawing of Frozen Soils, Life Science Journal 10, 4, 1651-1656 (2013).

[9] A.A. Sinitsyn, D.A. Belyanskii, The Development and Experience in Using a Series of Celsius Sensors for Monitoring the Temperature of the Working Body in a System for Controlling the Drying of Ceramic Articles, Measurement Techniques 56, 11, 1269-1273 (2014).

[10] A.A. Sinitsyn, D.A. Belyanskiy, I.A. Sukhanov, Development of New Sensors for the Temperature Control of a Working Body, Life Science Journal 10, 4, 1057-1061 (2013).

[11] M.V. Pavlov, D.F. Karpov, A.A. Sinitsyn, Experimental Research of the Thermal Balance of Loose Disperse Materials Exemplified by Milled Peat Exposed to Infrared Heating. Vestnik MGSU, Proceedings of Moscow State University of Civil Engineering 5, 114-119 (2012).

[12] M.V. Pavlov, D.F. Karpov, A.A. Sinitsyn, Solution to the Boundary Problem of Heat and Water Transmission in a Layer of a Loose Disperse Material Exemplified by Milled Peat Exposed to Infrared Heating. Vestnik MGSU, Proceedings of Moscow State University of Civil Engineering 6, 92-98 (2012).

Analysis of power quality in presence of frequency distortions

Violetta Zatsepina[1], *Evgeniy* Zatsepin[1], *Oleg* Shachnev[1], *Alexander* Shachnev[2], and *Timur* Petrov[3,*]

1Department of electrical equipment, Lipetsk state technical University, Lipetsk, 30 Moskovskaya street, 398055, Russia
[2]Department of housing and communal services of Lipetsk city administration, Lipetsk, 8 Kommunalnaya square, 398000, Russia
[3]Kazan State Power Engineering University, Kazan, Russia

Abstract. Today, semiconductor elements are becoming more popular in power supply networks. Their presence in any objects of electrical equipment provides high performance with high operating efficiency. However, modern research shows the semiconductor element base negatively affects quality of electricity when they are used in parallel. In low- and medium-voltage networks, this is fraught with the appearance of frequency «noise», which negatively affects parameters equipment with presence of latter, as well their control system. To detect them, a special approach is necessary, since such harmonics can appear both from directly semiconductor elements and from operating mode of equipment where element base is applied. The purpose of this article identifies higher harmonic distortions generated by objects containing semiconductor elements.

1 Theoretical substantiation

Modern electric power industry is closely connected with the use of semiconductor element base. It is used in various converters, automation, sensors and other elements where performance, reliability and high technical and economic indicators are required. It is known that the efficiency of their operation depends on the quality of electricity, since the vast majority of semiconductors must comply with the voltage level, and even a deviation from the normal voltage level within the permissible 5% can reduce their efficiency and reliability, which is not rational for conveyor production and utilities [1].

Reducing the probability of failure semiconductor elements lies in comprehensive analysis of operating voltage level and identification of higher harmonic components with their subsequent elimination. The study is useful in low- and medium-voltage objects that are widely used and where their compensation unlikely. It should also be taken into account that in control rooms for monitoring voltage level in cells, a graph of active values is often used, which does not give staff an understanding level of harmonics, since averaging also affects «white noise», so for greater objectivity, the article will use graphs instantaneous values of electrical quantities.

2 Practical experiment and explanations

We will conduct an experiment in Matlab R2014a for an electric network with voltage of 0.4 kV with a diode-type lighting system, where the primary voltage converters (drivers) often fail. The arguments of standard Matlab function of signal itself, dimension of result vector (FftL), and measurement. The last argument defines which dimension signal is located along if input a multidimensional array. Since signal under study is vector, it can be omitted. Consider first the electrical load without consumers. As result, we get a vector of complex numbers. This is exponential representation of our signal in frequency domain [2, 3].

The modules of these complex numbers represent amplitudes of corresponding frequencies, and arguments represent their initial phases. If resulting phase is uniquely calculated in radians or degrees, then additional tools must be used with amplitude and frequencies, which we will discuss later [4, 5].

For a clear comparison, let's give a graph with disconnected consumers (fig. 1, upper graph) and connected ones (fig. 1, lower graph). Below, we present the decomposition according to fast Fourier transform in amplitude-frequency spectrum in fig. 2 (lower and upper graph, respectively). Note that this decomposition is useful in case of continuous analysis of current charts and their dynamic change will be presented as quickly as possible, since it affects the minimum amount of resources.

The current analysis used a sample rate of F_d = 1024 Hz, and the number of Fourier lines in construction of spectrum – 5120. To ensure higher accuracy, the F_d indicator can be taken higher, but this is fraught with obtaining imaginary frequency perturbations, since calculation in Matlab R2014a is performed in a discrete region [6].

After receiving results in figures 1 and 2, it is not difficult to notice that frequency disturbances are visible in spectral region, but signal from them is quite weak, and in case of analyzing graphs active values, it is almost not noticeable due to data averaging.

* Corresponding author: tobac15@mail.ru

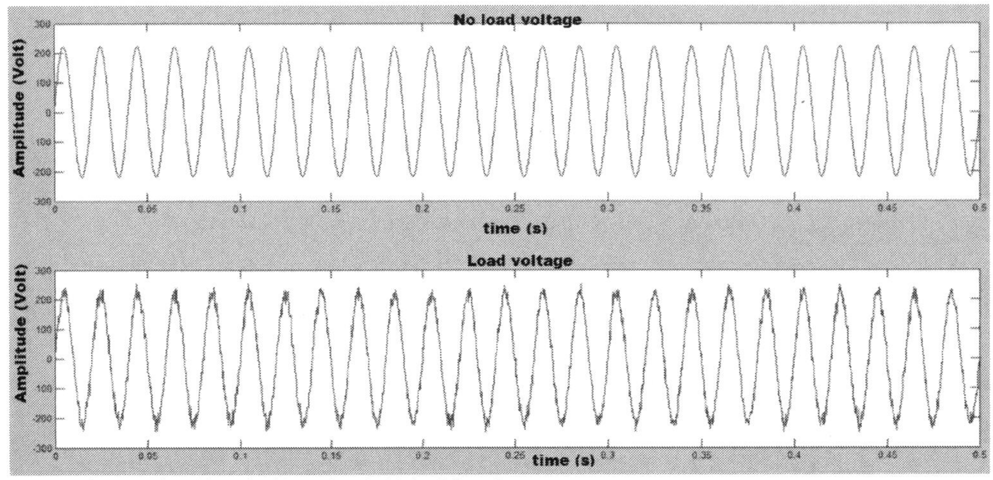

Fig. 1. Voltage graphs without load (upper) and with load (lower).

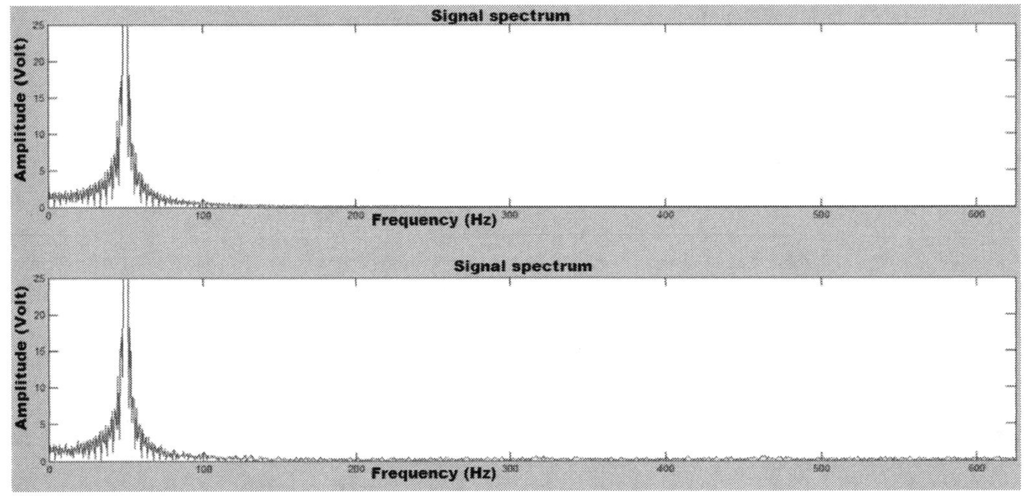

Fig. 2. The amplitude-frequency spectrum without load (top) and with load (bottom).

So, if there is a main graph with a frequency of 50 Hz, you can see fluctuations in 3, 5, 7, 9 (especially clearly), 11 (faintly noticeable) order, which corresponds to frequencies in 150 Hz, 250 Hz, 350 Hz, 450 Hz, 550 Hz. It is possible to determine the frequency level from spectral signal, but it is quite difficult to determine amplitude of each order. To get out of this situation, it is advisable to use either other Fourier transforms, which is fraught with large resources and will affect calculation time, or use scaling coefficients, which are used by default in wavelet decomposition [7, 8].

We will conduct an experiment using a wavelet transform of the current signal. Note that 2025 points obtained over time t = [0...0.5] were taken for analysis. Since analysis uses a one-dimensional graph, we will use section Wavelet 1-D in paragraph of calling wavemenu Matlab R2014a (Wavelet Toolbox). Given that it is planned to use real-time wavelet analysis, the expansion to third level by Daubechies wavelets (db10) will be sufficient. Figure 3 shows the complete decomposition of the studied graph of the lighting network voltage [9].

Since graph is one-dimensional, the study does not have a time domain and takes the numbering of points. You can link to time by writing additional scripts, but this is not included in this study. Figure 3 shows that current graph S is decomposed into final approximated graph a_3 and detailed graphs d_1, d_2, d_3. In discrete domain, addition of all graphs will give the initial one, therefore, the probability of extra frequencies appearing during conversion is very small, which is very important when analyzing this kind of load graphs [10].

In order to determine frequency indicators, it is necessary to take inversely proportional values of scale the distribution histogram according to figure 4 on original graphs.

Here S is histogram of initial graph under study, a_3 – is the histogram of final approximated graph with an enlarged scale, as evidenced by exit of abscises axis

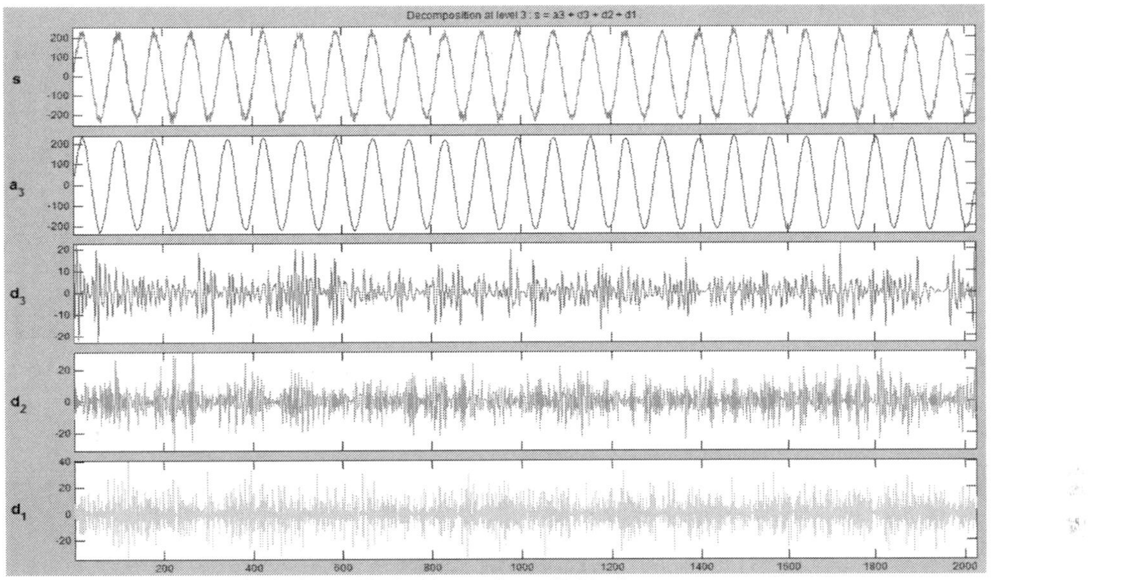

Fig. 3. The decomposition graph of voltage on Daubechies wavelets to third level.

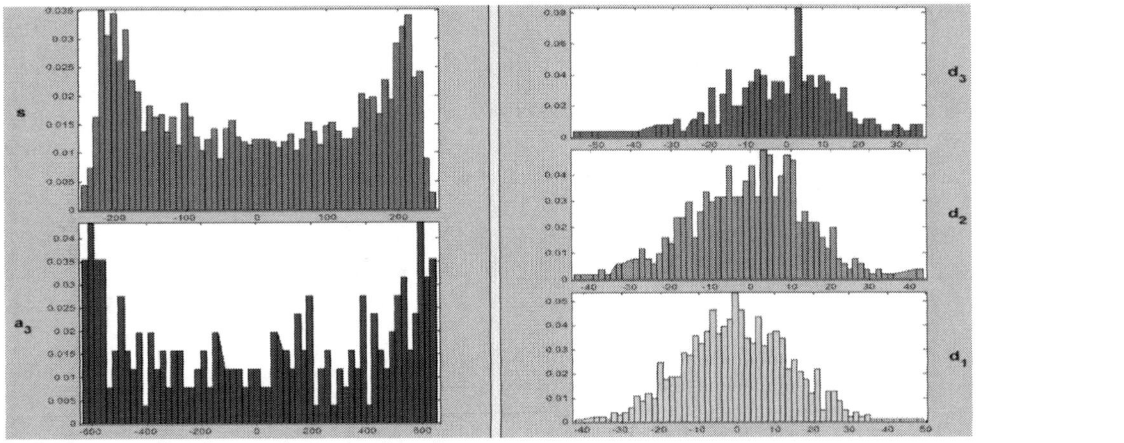

Fig. 4. Histograms of wavelet expansion graph under study.

beyond boundaries of values of 220 amplitude components; d_1, d_2, d_3 – corresponding level of decomposition histogram of detailing graphs, as evidenced by low scale values. For example, most of high-frequency bursts were identified at first decomposition corresponding d_1 graph, which corresponds to scale 0.0015...0.01 according to histogram of amplitude 20-40. Taking inversely proportional values, we get frequency characteristics, that is, the corresponding 666.67...100 Hz. You can get more accurate frequency values by calculating each column of histogram [11-14].

3 Processing of results and conclusion

Comparing the results obtained with fast Fourier transform and wavelet analysis, we can note the following:

- wavelet analysis provides more accurate indicators for frequency perturbations;
- wavelet analysis is more in demand when loads change or when transient modes occur;
- data of frequency characteristics in wavelet analysis can be used in automated control systems due to high performance;
- wavelet analysis is difficult to understand, so it is possible to use wavelets Daubechies or Haar (the most simple when translated into the usual frequency characteristics);

- Fourier analysis is useful for surface analysis of load graphs in the absence of weak frequency perturbations.

Semiconductor elements that generate mostly «white noise» are poorly suited for the usual Fourier analysis. Its low accuracy may affect the failure of semiconductor voltage sensors that are used in systems of APM, smart home, smart city, lean manufacturing, and others [15-18]. Given that above things are being developed and placed everywhere, elimination of frequency noise and its diagnosis will be a priority in near future. Using method described in article will allow you to diagnose and further take measures to eliminate frequency disturbances, which will have a positive impact on their technical and economic indicators [19, 20].

References

[1] V.I. Zatsepina, E.P. Zatsepin, O.Ya. Shachnev, Improving Eefficiency of High-Power Plants through Modernization STATCOM Devices, Proceedings – 2019 1st Int. Conf. on Control Systems, Mathematical Modelling, Automation and Energy Efficiency, SUMMA 2019, 1–5 (2019)

[2] Z. Dianmin, D. Shijie, G. Xiaohong, E-Commercial-e is applied on market modem Power Syst. Machine, 11–14 (2001)

[3] V.I. Zatsepina, E.P. Zatsepin, O.Ya. Shachnev, Analysis of load indicators power grid complex when feeding differentiated consumer, Int. Conf. on Industrial Engineering Application and Manufacturing (ICIEAM), 8742998 (2019)

[4] L. Fernández-Robles, G. Azzopardi, E. Alegre, N. Petkov, Machine-vision-based identification of broken inserts in edge profile milling heads, Robotics and Computer-Integrated Manufacturing, 276–283 (2017)

[5] A.N. Shpiganovich, A.A. Shpiganovich, O.V. Fedorov, Concerning the opportunity of fail-safe systems development, Int. Conf. on Industrial Engineering Application and Manufacturing (ICIEAM), 8076255 (2017)

[6] E.I. Gracheva, O.V. Fedorov, Forecasting reliability electrotechnical complexes of in-plant electric power supply taking into account low-voltage electrical apparatuses, Int. Conf. on Industrial Engineering Application and Manufacturing (ICIEAM), 8743057 (2019)

[7] T.I. Petrov, A.R. Safin, I.V. Ivshin, A.N. Tsvetkov and V.Y. Kornilov, The Prospects of Using a Synchronous Machine with Permanent Magnets in the Oil Industry, 14th International Scientific-Technical Conference on Actual Problems of Electronic Instrument Engineering, APEIE 2018 – Proceedings, article no. 8546157, 336–338 (2018)

[8] E. Gracheva, A. Alimova, Calculation methods and comparative analysis of losses of active and electric energy in low voltage devices, Proceedings – 2019 Int. Ural Conf. on Electrical Power Engineering, UralCon 2019, 361–367 (2019)

[9] V.I. Zatsepina, E.P. Zatsepin, Reliability of power supply systems under influence of negative factors, Int. Conf. on Industrial Engineering Application and Manufacturing (ICIEAM), 1–4 (2017)

[10] E.I. Gracheva, A.N. Alimova, Relationship between the characteristics of power transformers and their load capacity, News of higher educational institutions in Chernozem region **1**, 48–55 (2019)

[11] O.V. Fedorov, Some features of structure internal systems power supply, Reliability and security of energy **3**, 30–33 (2015)

[12] B.N. Parsunkin, O.V. Petrova, E.I. Polukhina, Research influence of electric mode DSP on cost produced steel, Theory and technology of metallurgical production, 44–46 (2014)

[13] A. Kiyoumarsi, M. Ataei, R. Hooshmand, Electric Arc Furnace Voltage Flicker Mitigation by Applying a Predictive Method with Closed Loop Control of TCR/FC Compensator, Journal of Electrical Engineering & Technology **1**, 116–128 (2010)

[14] E.I. Gracheva, O.V. Naumov, Evaluation criteria of contact group technical state concerning electrical appliances, International Journal of pharmacy and Technology **4**, 27084–27091 (2016)

[15] D. Maihöfner, I. Talavera, J. Hanson, C. Bott, Vertical reactive power flexibility through distributed energy resources for a reactive energy management, 11th IEEE Int. Conf. on Compatibility Power Electronics and Power Engineering, 100–105 (2017)

[16] L.S. Czarnecki, P.D. Bhattarai, A Method of Calculating LC Parameters of Balancing Compensators for AC Arc Furnaces, *IEEE Transactions on Power Delivery* **32**, 688–695 (2017)

[17] W. Bialy, V.E. Sovin, V.I. Zatsepina, E.P. Zatsepin, O.Ya. Shachnev, Ensuring efficient operation of electromechanical systems with frequency regulation with periodic voltage sags, E3S Web of Conferences, 05059 (2019)

[18] V.V. Moroshkin, V.A. Skopintsev, Analysis and forecast of emergency situations in electric power systems based on statistical methods, *Electric power plants*, 12, 2–6 (2012)

[19] M.T. Esfahani, B. Vahidi, A Predictive Reactive Power Measuring Based on Time Series and DLSL Algorithm for Compensating Applications, IEEE Transactions on Instrumentation and Measurement **64**, 2646–2656 (2015)

[20] J. Wang, A wavelet denoising method based on the improved threshold function, Int. Conf. on wavelet Analysis and Pattern Recognition, 70–74 (2014)

Stator elements optimization of centrifugal compressor intermediate type stage by CFD methods

Lyubov Marenina[1,*], *Yuri* Galerkin[1], and *Alexandr* Drozdov[1]

[1]Peter the Great St.Petersburg Polytechnic University, St.Petersburg, Polytechnicheskaya, 29, Russia

Abstract. Optimal gas-dynamic design is a complex and time-consuming process. Modern CFD methods help in solving optimization problems and reliably calculating characteristics of stator elements of centrifugal compressor stages. To carry out such calculations, it is necessary to create a parametrized model, which facilitates automation of the process of changing the flow path geometry, rebuilding its dimensions and the computational grid. Using the Direct Optimization program of the ANSYS software package, we have optimized the flow path of the stator elements of a centrifugal compressor intermediate type stage consisting of a vaneless diffuser and a return channel. In this paper, the MOGA (Multi-Objective Genetic Algorithm) optimization method was used. The object of the study was stator elements of one of the model stages designed by the Problem Laboratory of Compressor Engineering, SPbPU. The goal was to achieve the minimum value of the loss coefficient of stator elements when changing 5 geometric parameters: the number of vanes, the inlet vane angle, the height of the vane at the inlet to the return channel vane cascade, the radius of curvature of the leading edge and the thickness of the vane profile. For the best variants based on the results of optimization, the characteristics of the loss coefficient depending on the flow rate coefficient were calculated, their characteristics were compared with the initial variant of the stator elements. The best variant in the design mode has a loss coefficient 4.4% lower than the reference model. With a flow rate coefficient of 1.63 times greater than the calculated one, the optimized variant's loss coefficient is 33% less.

1 Introduction

The process of selection of return channel sizes and calculation of their gas-dynamic characteristics is described in the corresponding sections of the monographs [1-4] and in separate research papers [5-12]. Calculations of the stator elements of the flow path of the centrifugal stages by the methods of computational gas dynamics provide information, on the basis of which it is possible to formulate design recommendations. The stator elements of the model stage 0028-058-0373 (design flow coefficient 0.028, design loading factor 0.58, hub to impeller ratio 0.373) by the Problem Laboratory of Compressor Engineering, SPbPU [1] were selected as the object under study. Fig. 1 shows sketches of the stage in the meridional and radial planes with the indication of the control sections and its main geometric parameters.

The goal of our work is to optimize the return channel in order to reduce the loss coefficient in the entire work area for flow rate coefficient. The set of geometric parameters that have the greatest impact on

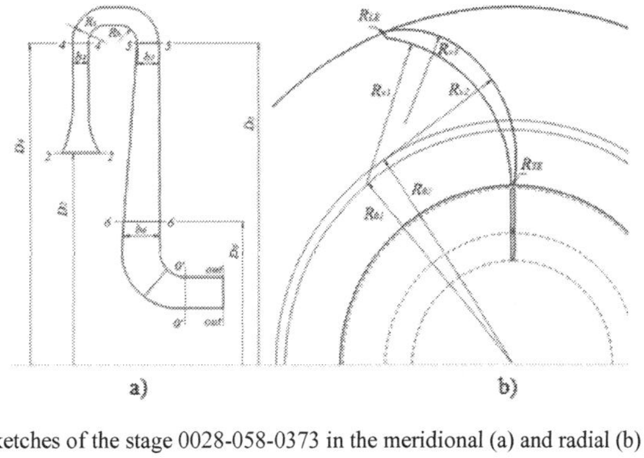

Fig. 1. The stator elements sketches of the stage 0028-058-0373 in the meridional (a) and radial (b) planes.

* Corresponding author: marenina_ln@mail.ru

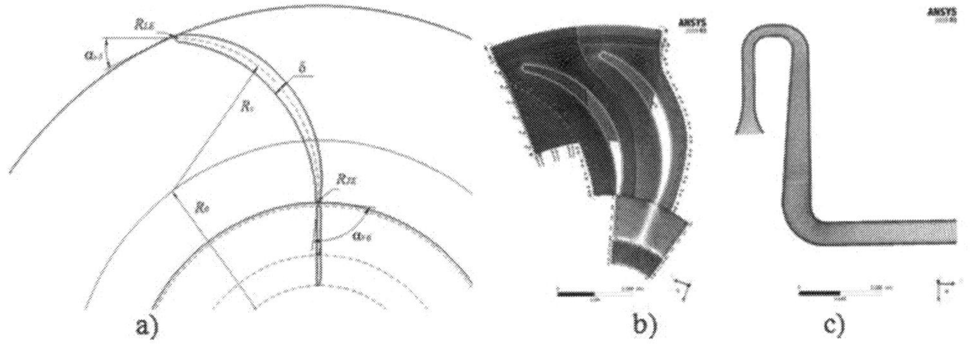

Fig. 2. Vane sketch, built along the vane midline in the form of a circular arc (a), general view of the computational domain in the radial (b) and meridional (c) plane.

the shape of the flow path is optimized.

To carry out the calculations, it is necessary to create a parametrized model, which allows us to automate the processes of changing the geometry of the flow path according to its new dimensions and of rebuilding the computational grid. When constructing a sketch of the vane in the DesignModeler program, the initial vane profile was replaced with a variant made with the traditional shape of the midline of the vane in the form of a circular arc [1] (Fig. 2 a). Vanes of this shape ensured high efficiency of compressors designed by a scientific team of prof. Galerkin at SPbPU [13]. We will demonstrate the advantage of two-arc vanes compared with the profile of the model stage vanes with the calculations below. In the DesignModeler CAD program, the midline of the vanes is defined by the radii R_0 and R_v, which are calculated according to the formulas from [14]:

$$R_v = \frac{D_6^2 - D_5^2}{4(D_6 \cos\alpha_{v6} - D_5 \cos\alpha_{v5})} \quad (1)$$

$$R_0 = \sqrt{R_v^2 + 0,25D_6^2 - R_v D_6 \cos\alpha_{v6}} \quad (2)$$

where D is a diameter, α_v is a vane angle, «5» and «6» are indices of control sections according to the Fig. 1. The radii of the surface of the vane are automatically selected taking into account the specified radii of curvature of the leading and trailing edges R_{LE}, R_{TE} and the thickness of the vane δ.

2 CFD calculation methodology

For the correct formulation of the optimization problem, it is necessary to create a parametrized model of the flow path and determine the dimensions of the computational grid common to all calculations, select the appropriate turbulence model, and set the boundary conditions. The methodology for setting up CFD calculations was studied by the authors in [15, 16]. At all stages of model creation and calculations, the Workbench environment of the ANSYS CFX 19 software package was used. The construction of a parameterized model of the flow path was described in detail in [17]. In DesignModeler, the calculation area is built on the basis of sketches, and all their sizes can be set in Workbench as parameters, so that when their values change, the sketch is automatically rebuilt. Also, instead of fixed parameter values, it is possible to specify expressions that describe relationships between sizes. In the TurboGrid mesh generator, design meshes for the main and straightening vanes were built. The values $y^+ < 20$ were set in accordance with the requirements of the correct modeling of the boundary layer using the SST turbulence model (Shear-Stress-Transport). In the CFX-Pre program, computational grids were combined into one area consisting of two main return channel vanes and one straightening vane (Fig. 2 b, c). The total number of elements in the resulting area is 893 000. The following conditions were set at the boundaries:

- at the inlet: total pressure P_2^*, total temperature T_2^* and flow direction α_2;
- at the outlet: mass flow $\bar{m}$;
- on the walls from the «0'» section (Fig. 1) to the outlet, the «free slip» condition is accepted;
- between the grids of the main vanes, the condition «general connection»;
- between the grids of the main and straightening vanes, the condition «frozen rotor»;
- on the «periodic» surfaces, the condition of the radial periodicity «rotational periodicity» was set.

The CFD-Post program defines a formula for calculating the loss coefficient (3) as a function of the values averaged by the program in sections «2» and «0'».

$$\zeta_{2-0'} = \frac{p_2^* - p_{0'}^*}{0,5(\rho_2 + \rho_{0'})} \cdot \frac{2}{c_2^2} \quad (3)$$

where ρ is density, c is velocity, and «2» and «0'» are the indices of control sections according to the Fig. 1. For total and static pressure, averaging over the area is used, for other values – over the mass flow rate. The value of this function is displayed as the output parameter of the Workbench and is used in the Direct Optimization program, where we can set the output function as the efficiency parameter, according to which the program will change the values of the input parameters to achieve the best result, in this case, the lowest value of the loss coefficient. The Direct Optimization program allows us to choose an optimization method that depends on the number of objectives, the set limits and the desired number of design points. For the research conducted in the work, the MOGA method was used. The initial conditions for the search for optimal sizes were set in accordance with the initial operating conditions of the stator elements of the stage 0028-058-0373 in the design mode with flow rate coefficient Φ_{des} = 0.028. For the initial return channel and optimized variants, the characteristic $\zeta_{2-0'} = f(\Phi)$ was calculated at six points corresponding to the operation of the stage when it was tested with a conditional Mach number M_u = 0.6. Fig. 4 shows the result of modeling the characteristics of the stage 0028-058-0373 according to the IDENT program [18].

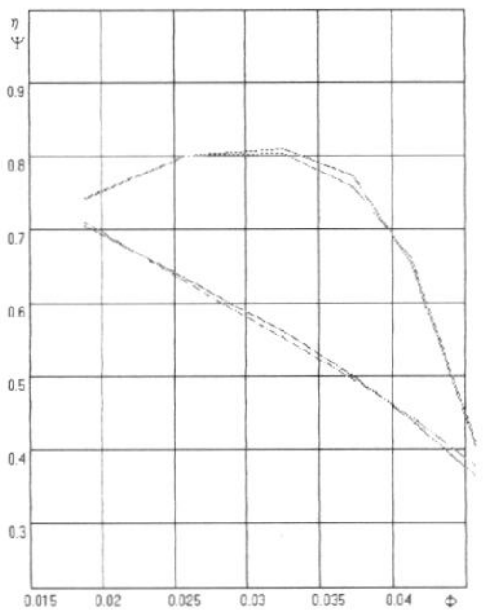

Fig. 3. The IDENT program. Stage characteristics 0028-058-0373 according to the test results (green) and calculated using the Universal Modeling Method (red).

The Universal modeling method is widely used in design practice [19-23] and accurately calculates the characteristics of stages and compressors and the flow parameters in the control sections of the flow path. It is sensible to carry out our CFD study at 6 values of the flow coefficients at which the stage was tested. Table 1 shows the data from the calculation performed in the IDENT program. Since the calculation corresponding to the conditional flow rate coefficient Φ_{des} = 0.028 is of interest, we assume that the parameters vary linearly between their values at Φ = 0.0326 and 0.0258. The data thus obtained for the design point are also listed in Table 1.

The test results of model stage 0028-058-0373 in Fig. 3 are presented in a dimensionless form. For calculations in the CFX-Pre program, the initial parameters must be specified in dimensional form. The stage was tested under the following conditions: p_0^* = 101 000 Pa, T_0^* = 288.15 K, M_u = 0.6. The transition to dimensional parameters was carried out according to known ratios. The parameters are presented in Table 2. The values of φ_2 are calculated according to the mathematical model when processing the results of testing the model stage that were presented in the Table 1 (above).

3 Results and discussion

3.1 Optimization of the initial return channel with a two-arc profile by 5 parameters

To carry out optimization in the design mode with Φ_{des} = 0.028, the following geometric parameters were chosen as variable values and the limits of their change were set:

- for the number of vanes even values were set: z = (18 ... 28);
- for the inlet vane angle: α_{v5} = (16° ... 30°);
- for the height of the vane at the inlet to the stator elements: b_5 = (15 ... 32) mm;
- for the radius of curvature of the leading edge R_{LE} = (1.4 ... 4) mm;

Table 1. Data from the calculation of the stage 0028-058-0373 executed in the IDENT program.

No	1	2	3	4	5	6	7
Φ	0.0457	0.0413	0.0373	0.0326	0.028	0.0258	0.0187
φ_2	0.248	0.215	0.189	0.163	0.139	0.127	0.0925
ψ_T	0.3599	0.4242	0.4727	0.5238	0.570	0.5925	0.6601
ψ_i	0.3761	0.4435	0.4952	0.5508	0.603	0.6283	0.7114

Table 2. The calculated gas-dynamic parameters necessary for setting boundary conditions.

Φ	0.0457	0.0413	0.0373	0.0326	0.028	0.0258	0.0187
T_2^*, K	303.75	306.55	308.69	311.00	313.18	314.22	317.66
P_2^*, Pa	114182	120813	125663	129170	132266	133981	136455
α_2^0	34.6	26.9	21.8	17.3	13.8	12.1	8.0
$\bar{m}$, (kg/s)	1.4990	1.3570	1.2260	1.0720	0.9204	0.8480	0.6147

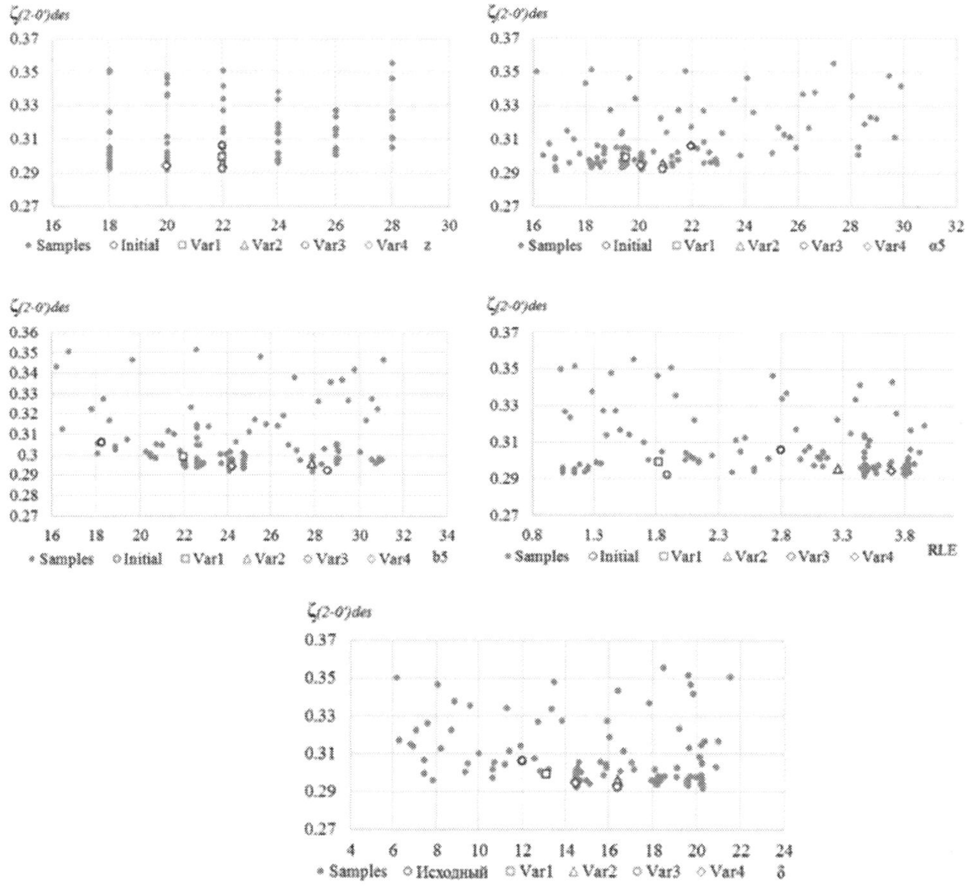

Fig. 4. ANSYS CFX 19. Results of calculation of 120 variants of stator elements obtained during optimization in the Direct Optimization program.

- for vane thickness δ = (6 ... 22) mm.

During the optimization process, 120 variants of stator elements with loss coefficient $\zeta_{2-0'des}$ ranging from 0.356 to 0.292 were calculated in the Direct Optimization program. Fig. 4 presents the values of the loss coefficient for all the calculated variants for each variable parameter separately. The parameters of the four return channels with the lowest value of the loss coefficient are presented in Table 3.

The loss coefficients of optimized variants are 2.3 – 4.5% lower than of the initial one. For the three variants, the optimal number of vanes was found to be 22. The optimized variants have optimal vane angles ranging from 19.5 to 21°, which is 1 – 2.5° less of initial, while the height of the vanes at the inlet is much greater. The biggest difference in variant 4 is that it has the maximum vane height at the inlet, almost equal to the vane height at the outlet (b_6 = 30.5 mm). In addition to a smaller angle, the variant 4 profile is 36% thicker. The variant 3 has a smaller number of vanes z = 20 and a smaller value of b_5 compared to the variant 4. Fig. 5 presents the characteristics of the loss coefficient of the compared

Table 3. The geometric parameters of the stator elements variants considered best by optimization results in Direct Optimization.

№	z	α_{v5}	b_5, mm (b_5 / b_4)	R_{LE}, mm	δ, mm	$\zeta_{2-0'}$
Initial	22	22.000	18.300 (1.34)	2.800	12.000	0.3057
Variant 1	22	19.538	20.727 (1.57)	1.819	13.137	0.2989
Variant 2	22	20.862	27.909 (2.04)	3.263	16.454	0.2956
Variant 3	20	20.083	24.173 (1.76)	3.694	14.453	0.2943
Variant 4	22	20.900	28.600 (2.09)	1.819	16.400	0.2921

https://doi.org/10.1051/e3sconf/202017801020

variants.

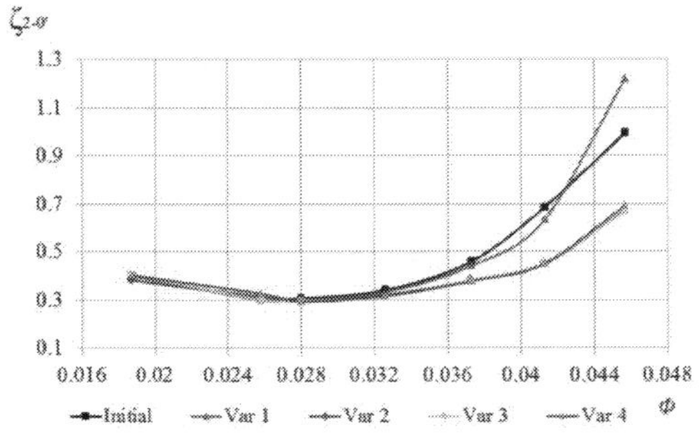

Fig. 5. ANSYS CFX 19. Loss coefficient versus flow rate coefficient.

Variant 1 does not exceed the initial variant. Variants 2, 3 and 4 are almost equivalent and noticeably more effective than the initial variant at high flow rates. With a maximum flow rate Φ = 0.0457, their loss coefficient is 33% less. Fig. 6 shows the flow structure in the regime with the maximum flow coefficient Φ = 0.0457 for the initial variant and the variant 3. The flow in the meridional plane is more organized for the initial variant with lower diffusivity of the crossover (b_5 / b_4 = 1.34). However, in the radial plane, a more favorable flow for variant 3 is with b_5 / b_4 = 1.76. At the maximum flow rate, the initial variant has a larger incidence angle, which leads to a significant separation of the flow. For the optimized variant, flow separation is virtually nonexistent.

3.2 Verification of optimization results by CFD calculation of whole stage

Despite considerable efforts, CFD calculations of the characteristics of centrifugal stages still do not provide reliable results [24, 25]. The results of modeling stage 0028-058-0373 characteristics were not entirely satisfactory either [26]. The authors nevertheless believe that a comparison of the calculated characteristics of the stage with the initial and optimized return channel can confirm or refute the result of optimization. According to the authors of [26], the problem of the mismatch between the characteristics of the efficiency and the effective pressure is to a large extent connected with the fact that CFD calculations significantly overestimate the head coefficient. In order to approximate the conditions of gas movement in return channel during CFD calculation, the blade height at the outlet of the impeller was reduced. This reduced the head coefficient. Fig. 7 (left) shows the head coefficient of the impeller of the stage 0028-058-0373 in comparison with the calculation of the head coefficient of this impeller and its modification with a reduced blade height at the outlet.
For a modified impeller near the design mode Φ_{des} = 0.028, the head coefficient coincides with its actual value. In Fig. 7 (right), the polytrophic head coefficients are compared, and Fig. 8 compares the characteristics of

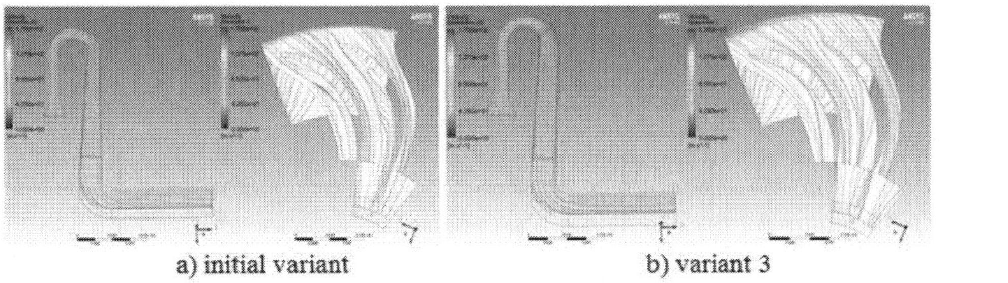

Fig. 6. ANSYS CFX 19. Streamlines in the meridional (above) and radial (below) planes, Φ = 0.0457.

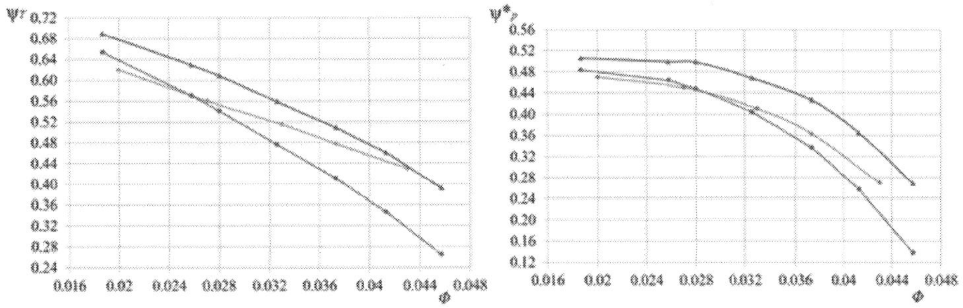

Fig. 7. Characteristic of the head coefficient (left) and the polytrophic head coefficient characteristic (right). Impeller 0028-058-0373 (experiment) – green, impeller 0028-058-0373 (calculation) – red, impeller 0028-058-0373 with reduced b_2 (calculation) – blue.

efficiency.

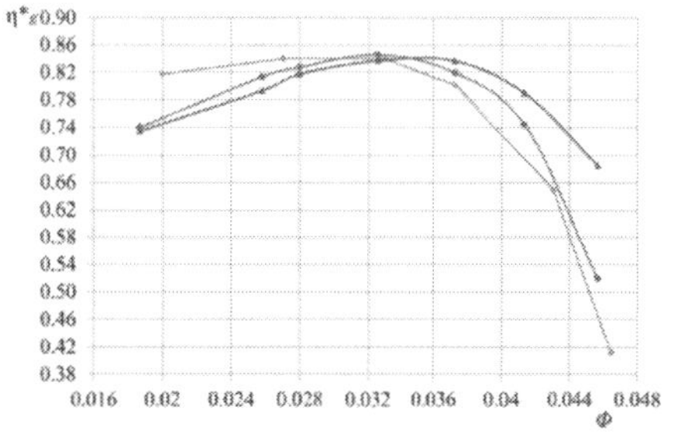

Fig. 8. Efficiency characteristic of the stage 0028-058-0373 (experiment) – green, impeller 0028-058-0373 (calculation) – red, impeller 0028-058-0373 with reduced (calculation) – blue.

The calculation of the stage with the initial variant of the stator elements and with the optimized variant 3 qualitatively confirmed the optimization result presented in the Fig. 5. Optimization of return channel slightly increased the efficiency and polytropic head coefficient at flow rates higher than the calculated one.

4 Conclusion

In this paper we have investigated the stator elements of the model stage 0028-058-0373 (design flow coefficient 0.028, design loading factor 0.58, hub to impeller ratio 0.373). Comparison of the initial and two-arc vane profiles showed a less favorable flow around the suction surface of the vane near the leading edge of the initial variant of the vane profile. The loss coefficient in the design mode of the two-arc profile is 0.2% higher. Optimization of return channel by five parameters allowed to reduce the values of the loss coefficient by 4.4% in the calculated mode and by up to 33% in the modes higher than the calculated one. The main influence on the efficiency was exerted by an increase in the vane height at the vane cascade inlet b_5 with a simultaneous decrease of the inlet vane angle α_{v5}. The best optimized variants have values b_5 / b_4 = 1.76 ... 2.08. The values of vane thickness δ / D_2 = 0.035 ... 0.04, which is also higher than that of the initial variant. As a result of optimizing the crossover shape, the stator elements loss coefficient decreased by a small amount of about 1% in the design mode. The optimized variant has a larger sleeve radius R_h / b_4= 1.77 compared with the initial R_h / b_4 = 1.09, which leads to an undesirable increase in the overall dimensions of the compressor. The principle of choosing the vane height b_6 based on the conditions of the necessary acceleration at the entrance to the next stage [1], at the same time ensures good efficiency of this section of return channel. The calculation of the characteristics of the stage with the initial and optimized return channel confirmed the somewhat greater efficiency of the optimized variant at flow rates greater than the designed one. The positive effect of optimization is small, which indicates a good level of design techniques presented in [1]. The authors believe that with the planned studies of stages with medium and large flow coefficients, optimization will give a more noticeable increase in efficiency.

Acknowledgements

The calculations were carried out using the supercomputer center Polytechnic, SPbPU. The research was performed by a Grant of the President of the Russian Federation for young PhD MK-1893.2020.8.

References

[1] K. Seleznev, Y. Galerkin, *Centrifugal compressors* (L.: mechanical engineering, 271, 1982)
[2] Yu. Galerkin, *Turbocompressors* (LTD information and publishing center KHT, Moscow, 596, 2010)
[3] R.H. Aungier, Centrifugal compressors: a strategy for aerodynamic design and analysis, The American Society of Mechanical Engeneers (2000)
[4] N.A. Cumpsty, *Compressor Aerodynamics* (Longman Scientific and Technical, England, 1989)
[5] C. Rube, T. Rossbach, M. Wedeking, D.R. Grates, P. Jeschke, Experimental and numerical investigation of the flow inside the return channel of a centrifugal process compressor, Turbine Technical Conference and Exposition, Montreal, Canada (2015)
[6] F.D. Bellis, E. Guidotti, D.T. Rubino, Centrifugal compressors return channel optimization by means of advanced 3D CFD, Turbine Technical Conference and Exposition, Montreal, Canada (2015)
[7] C. Aalburg, A. Simpson, J. Carretero, T. Nguyen, V. Michelassi, Extension of the stator vane upstream across the 180deg bend for a multistage radial compressor stage, Proceedings of ASME Turbo Expo 2009: Power for Land, Sea and Air, GT2009-59522 (8-12 June 2009)
[8] M. Yagi, T. Nishioka, H. Kobayashi, H. Nishida, S. Yamamoto, Effects of return channel with splitter vanes on performance of multistage centrifugal compressor, Turbine Technical Conference and Exposition, Montreal, Canada (2015)
[9] A. Veress, R. Van den Braembussche, Inverse design and optimization of a return channel for a multistage centrifugal compressor, Journal of Fluids Engineering **126**, 799806 (2004)
[10] Y. Nishida, H. Kobayashi, H. Nishida, K. Sugimura, Performance improvement of a return channel in a multistage centrifugal compressor using multiobjective optimization, Journal of Turbomachinery **135**, 031026, 1–8 (2013)

[11] A. Hildebrant, Aerodynamic optimization of a centrifugal compressor return channel and U-turn with Genetic Algorithms, ASME Turbo Expo 2011, ASME Paper GT201145067 (2011)
[12] A. Hildebrandt, Numerical analysis of overall performance and flow phenomena of an automatically optimized treedimensional return channel system for multistage centrifugal compression systems, ASME Turbo Expo 2012, ASME Paper GT2012-68559 (2012)
[13] Yu.B. Galerkin, A.F. Rekstin, A.A. Drozdov, K.V. Soldatova, O.A. Solovyeva, E.Yu Popova, Through a system of optimal gas-dynamic design of industrial centrifugal compressors. Scientific basis, application practice, XVIII International Scientific and Technical Conference, Kazan (2019)
[14] V. Riess, *Centrifugal compressors* (1981)
[15] Yu.B. Galerkin, L.N. Marenina, Compressors and pneumatics **1**, 30–36 (2014)
[16] L. Marenina, Y. Galerkin, K. Soldatova, Computational fluid dynamics application for analysis of centrifugal compressor stage stator part, International Journal of Mechanical Engineering and Robotics Research **7**, 6, 656–661. DOI: 10.18178/ijmerr.7.6.656-661 (2018)
[17] Yu.B. Galerkin, A.F. Rekstin, L.N. Marenina, K.V. Soldatova, Creation of parametrizes model of return channel flow path for CFD-researches, Oil and Gas Engineering-2020, Materials of the 10th International Scientific and Technical Conference, Omsk, 127–128 (2020)
[18] Y. Galerkin, K. Soldatova, Gas-dynamic process in industrial centrifugal compressors modeling, Scientific basis, stapes of development, current state, Publishing house of Polytechnic University, 327 (2011)
[19] Y. Galerkin, A. Rekstin, K. Soldatova, Gas dynamic design of the pipe line compressor with 90% efficiency [Electronic resource], Model test approval, International Conference on Compressors and their Systems, London, City University, UK. Available at: http://iopscience.iop.org/article/10.1088/1757-899X/90/1/012044/meta. DOI: 10.1088/1757-899X/90/1/012044 (2015)
[20] Y. Galerkin, A. Rekstin, K. Soldatova, A. Drozdov, Gas dynamic designs of centrifugal compressors for gas industry, Specific features, ASME Gas Turbine India Conference, GTINDIA2015-1215. Available at: www.asmeconferences.org/GTIndia2015/PaperAccepted.cfm?noToolbar=yes. DOI: 10.1115/GTINDIA2015-1215 (2015)
[21] Y. Galerkin, K. Soldatova, A. Drozdov, The application of mathematical models for industrial centrifugal compressor optimal design, ICCMS 2018: 10th International Con-ference on Computer Modeling and Simulation, Australia, Sydney, 008, 187–191 (2018)
[22] A.F. Rekstin, V.B. Semenovskiy, Y.B. Galerkin, A.A. Sokolov, The analysis of design and measured gas-dynamic characteristics of the centrifugal compressor within turboexpander aggregate, Oil and Gas Engineering (OGE-2018), AIP Conf. Proc. 2007, Omsk, 030028-1–030028-11. DOI: 10.1063/1.5051889 (2018)
[23] A. Rekstin, V. Semenovskiy, K. Soldatova, Y. Galerkin, K. Sokolov, The simulation of gas-dynamic characteristics of centrifugal compressors in turbo-expander units, International Scientific and Technical Conference Smart Energy Systems, SES 2019, Kazan, E3S Web of Conferences **124**, 01005. DOI: 10.1051/e3sconf/201912401008 (2019)
[24] P. Le Sausse, P. Fabrie, D. Arnou, F. Clunet, CFD comparison with centrifugal compressor measurements on a wide operating range, EPJ Web of Conferences **45**, 01059. DOI: 10.1051/epjconf/20134501059 (2013)
[25] A. Borovkov, I. Voinov, Y. Galerkin, A. Nikiforov, M. Nikitin, O. Solovyeva, K. Kabalyk, Issues of gas dynamic characteristics modeling: a study on a centrifugal compressor model stage, International Scientific Conference on Energy, Environmental and Construction Engineering (EECE-2019), E3S Web Conf. **140**. DOI: 10.1051/e3sconf/201914006003 (2019)
[26] N. Kortikov, A. Borovkov, I. Voynov, A. Kirillov, A. Drozdov, Modeling the gas-dynamic characteristics of the low-flow and mid-flow model stages for an industrial centrifugal compressor, International Scientific Conference on Energy, Environmental and Construction Engineering (EECE-2018), MATEC Web Conf. **245**. DOI: 10.1051/matecconf/201824504019 (2018)

Mathematical Simulation of the Synchronized Asynchronous Electric Drive

Victor Meshcheryakov[1,*], *Dmitry* Sibirtsev[1], and *Ekaterina* Mikhailova[2]

[1]Lipetsk State Technical University, Department of Electric Drives, Lipetsk, Russia
[2]Kazan State Power Engineering University, Department of Power Stations, Kazan, Russia

Abstract. At the end of the last century asynchronous motors with a phase rotor were most widely used among adjustable-speed AC drives. They were used for conveyors, transporters, cranes. That was due to the relative simplicity to adjust the motor speed by acting on the rotor chain. The introduction of a frequency control method in such drives is now complicated by the fact that most frequency converters are designed to be used in drives with a cage asynchronous motor. Shorting of a phase winding leads to highly increased losses during acceleration and motor speed control. If the stator winding of a wound-rotor asynchronous motor is connected to a frequency converter and the rotor winding is connected to a DC link of this converter, the motor will have properties of a synchronous one. The electric drive is able to work in a wide range, while motor characteristics are absolutely rigid. The implementation of such control method is presented in this article. The control system operation has been tested with simulation in the Matlab Simulink software pack. The obtained results are defined as follows.

1 Introduction

There is a number of production mechanisms that require control systems to obtain absolutely rigid performance characteristics and a wide range of speed control. In such mechanisms it is advisable to use a synchronized electric drive based on a wound-rotor asynchronous motor [1,2]. A wound-rotor asynchronous motor is put into synchronous mode by supplying a direct current to the rotor winding [3]. Since DC flows through the rotor windings of the motor and the rotor speed of the motor does not depend on the load on the shaft, which is typical for a synchronous machine, the wound-rotor asynchronous motor operating in such mode is called synchronized [3]. The synchronous mode of the motor has among others one advantage over the asynchronous mode which is the ability of the motor to change the magnitude and mark of the reactive power when adjusting the magnitude of the excitation current [4]. With direct connection of the stator winding to the supply mains providing motor overexcitation, reactive power is supplied to the supply mains network and, if underexcited, is consumed from the network, there is therefore a fundamental possibility of influencing the energy exchange process between the motor and the power supply system by controlling the excitation current of powerful synchronized asynchronous motors [5].

2 Approaches

The connection diagram of the synchronized motor is shown in Fig. 1. The rotor winding is connected to a constant-voltage source in accordance with the diagram shown in the figure. The diodes in the diagram are used to exclude free currents in a closed circuit formed by two paralleled rotor windings. On completing the electromechanical transient process, connected with the start-up of the electric drive and retraction of the motor in synchronism, the motor continues to work in a static mode, possessing all the characteristics of a synchronous nonsalient pole motor of the classical type.

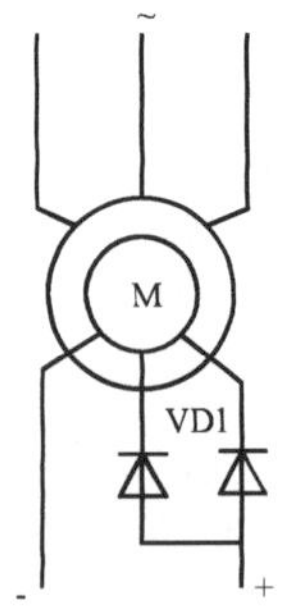

Fig. 1. The connection diagram of the synchronized asynchronous motor.

* Corresponding author: mesherek@stu.lipetsk.ru, 467816@mail.ru

 https://doi.org/10.1051/e3sconf/202017801021

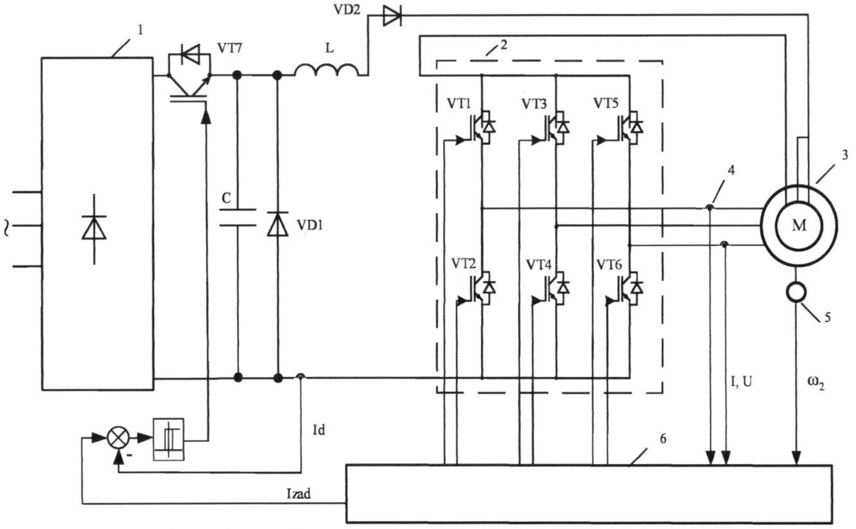

Fig. 2. Connection diagram of the synchronized motor.

A wound-rotor asynchronous motor operating in synchronous mode requires an accurate setting of the acceleration rate with its correction when the moment of inertia of the mechanism changes to ensure stable operation in the start-up mode carried out with a power-up sensor. Therefore, an additional internal stator current control circuit subject to the motor speed control circuit is introduced into the automatic control system of a synchronized electric drive built on a current inverter to increase the stability. The internal circuits for regulating phase currents of the stator are based on a relay current controller. The block diagram of the frequency synchronized asynchronous electric drive is shown in Fig. 2. The system of the frequency-controlled synchronized asynchronous electric drive uses a standard converter that feeds the stator winding. It contains an uncontrolled rectifier, the output of which is connected to the input of a controlled autonomous inverter. The electromagnetic field produced by the stator winding is determined by the output parameters of the frequency converter: the current I_1 and the frequency f_1. The device contains a rectifier 1 and an inverter 2, an electric motor 3. The inverter is made on fully controlled valves - IGBT-transistors. The speed sensor 5 is mounted on the motor shaft, the sensors 4 supply the voltage and current signals to the control system 6. The block diagram of the control unit implemented in the Matlab Simulink software pack is shown in Fig. 3.

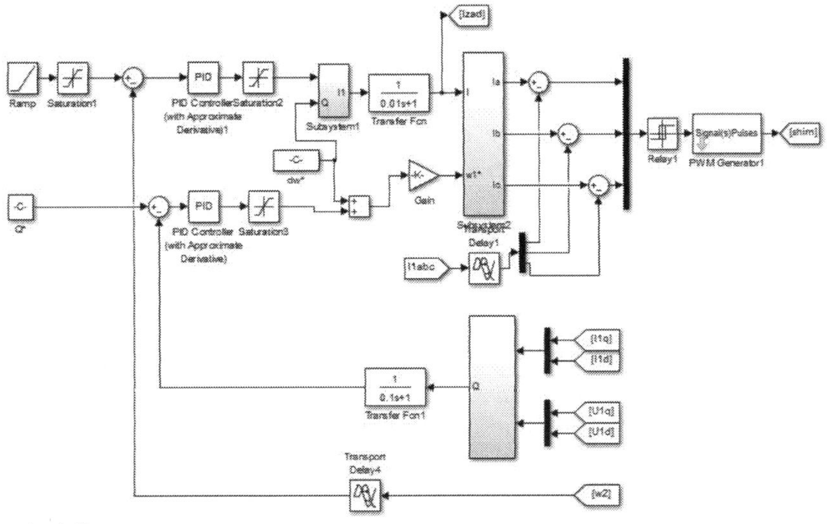

Fig. 3. Control unit diagram.

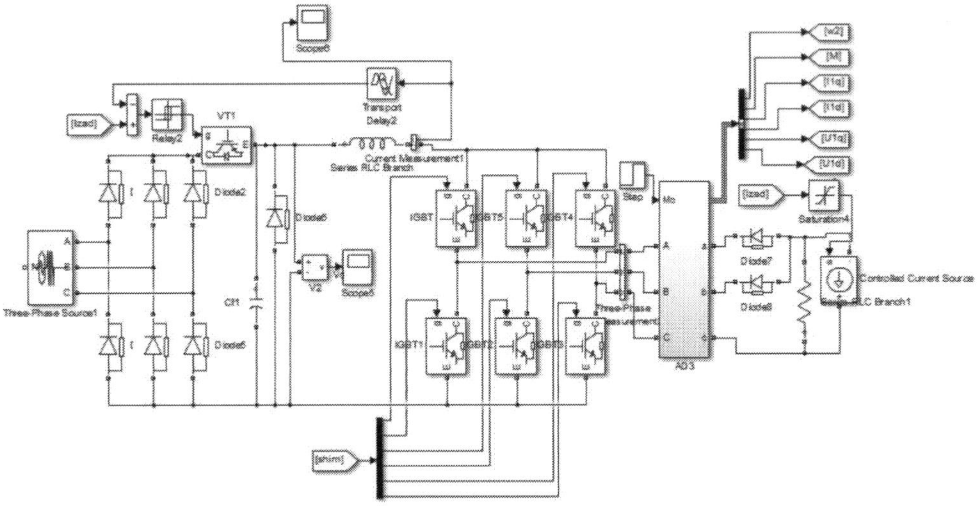

Fig. 4. Model of the synchronized asynchronous electric drive.

The stator current frequency is generated using the power-up sensor in the start-up mode. When reaching the given stator current frequency, the control system switches to the operating mode at the required speed. The current regulator operates according to the relay control. As a result of the comparison, we get the difference between the given and the actual values of the current in phases at the adders outputs [6]. The received signals are sent to the hysteresis inputs operating according to the following algorithm:

- when reaching the difference between the phase current set value and the phase current measured value of the threshold upper limit, the upper-arm valve is switched on and the lower-arm valve of the corresponding phase is switched off;
- when reaching the difference between the phase current set value and the phase current measured value of the threshold lower limit, the upper-arm valve is switched off and the lower-arm valve of the corresponding phase is switched on.

Then the signal from the hysteresis output is fed to the output of the transistors control of the corresponding phase. The switching frequency of power transistors can be 10 kHz.

The torque angle θ is an important parameter for synchronous mode of an asynchronous motor [3]:

$$q = \int \omega_1 dt - \int \omega_2 p_n dt \tag{1}$$

where ω_1 – the circular frequency of the stator voltage; ω_2– the rotor speed; p_n – the number of pairs of motor poles. The torque angle control circuit operates in the control system, which corrects the stator current frequency setting signal to reduce the rotor oscillations arising during the motor start-up. Adjustment of the torque angle control circuit is made considering the following features:

- the oscillation amplitude of the torque angle depends both on the electrical and mechanical parameters of the motor;
- static load on the drive shaft is determined by the average torque angle for the period of free oscillations [7];
- to quench the oscillations arising in the drive, according to classical mechanics, it is necessary to apply the control action in one phase with the oscillations of the system. The adjusting factor is chosen out of necessity to adjust the torque angle rapidly [8].

The factor is set to less than unity to have a stable system, but increasing it, the operating speed of the torque angle control circuit increases. Hence, it is practical to determine it independently for each separate electric drive [9]. The time constant T is determined by the free oscillations frequency of the electric drive system, it should not be less than this frequency. Excessive increase leads to a degradation in the system dynamic characteristics when the load on the shaft changes. Thus, the time constant is recommended to be chosen in the limit:

$$(1 \div 2)T_0 \tag{2}$$

where T_0 is the period of free oscillations of the electric drive open system.

The corrected current signals at the output of the phase currents unit are determined by the formulas:

$$\begin{aligned} I_a^* &= I_m sin((\omega + \omega_0)t) \\ I_b^* &= I_m sin((\omega + \omega_0)t + 120^o) \\ I_c^* &= I_m sin((\omega + \omega_0)t + 240^o) \end{aligned} \tag{3}$$

where ω_0 is the correction signal.

In the start-up mode, the rate of change of the frequency signal is set by the power-up sensor tuned in accordance with the value of the total inertia moment of the drive and the required acceleration [10].

If the selected acceleration is slightly overrated and the motor rotor does not have time to change its speed with a given tempo, the signal corrects the acceleration rate and stator current frequency.

3 Result

The proposed control system was modeled in the Matlab Simulink software environment (Fig. 4) [11]. The results of the simulation are shown in Fig. 5.

4 Conclusions

The synchronous asynchronous electric drive control system is designed and simulated in the work. Since the stability of the considered electric drive is greatly influenced by the value of the torque angle, the prompt speed setting was corrected in the control system in order to maintain this angle constant in the start-up mode. The vector correction of variables introduced into the scalar control system allows to orient the stator current vector relative to the rotor flux-linkage vector at any moment. Thus, the starting torque is stabilized in the synchronized frequency drive control system. This allows to use this ED on mechanisms with a cyclic operation mode.

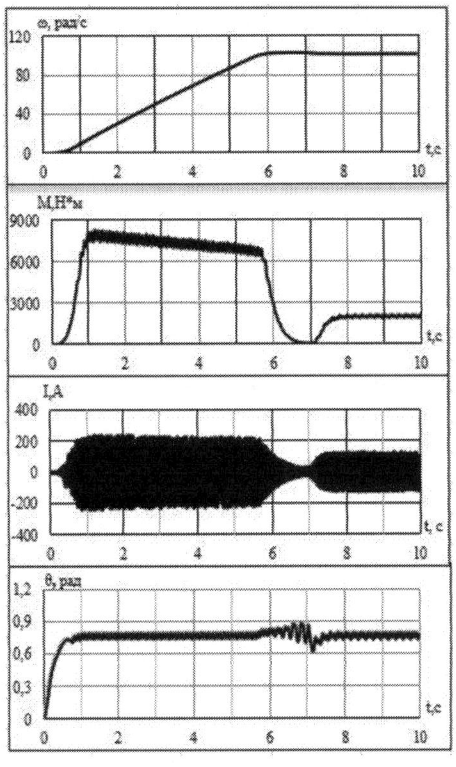

Fig. 5. Simulation results of a synchronized asynchronous electric drive.

Acknowledgements

The paper was powered by research grant RFBR 19-48-480001 "Development, investigation and optimization of energy-saving electrical and electrically driven automated systems for plasma, electrometal slag and induction technologies and units".

References

[1] V.N. Meshcheryakov, *Systems of an adjustable asynchronous electric drive for hoisting and transport mechanisms: monograph* (Lipetsk, LSTU, 112, 2005)

[2] G.G. Sokolovsky, *AC electric drives with frequency control* (Moscow, Academy, 272, 2006)

[3] V.N. Meshcheryakov, P.N. Solomatin, *Synchronized asynchronous electric drive with frequency control*, News of HEIs, Electromechanics **2**, 51–57 (2009)

[4] Y.I. Gracheva, N.V Chernova, A.I Fedotov, E.A. Fedotov, Local Fourier transformation application for mathematic modeling of synchronous machine valve actuator, Journal of engineering and applied sciences **11**, 1, 2939–2945 (2016)

[5] V.N. Meshcheryakov, A.M. Bashlykov, O.V. Meshcheryakova, System of scalar frequency control of synchronized asynchronous machine with power supply of stator and rotor windings from autonomous current inverter, HTCS'2012: materials of International scientific and technical conference, Stary Oskol TNT, 50 (2012)

[6] V.N. Meshcheryakov, D.S. Sibirtsev, Frequency asynchronous electric drive with correction of phase shift between moment-forming vectors, Control systems and information technologies **2**, **68**, 48–57 (2017)

[7] Y.I. Gracheva, O.V. Naumov, Operating Mode Influence on Probability Characteristics of Electric Devices, Journal of engineering and applied sciences **11**, 1, 2934–2938 (2016)

[8] H. Rehman, L. Xu, Alternative energy vehicles drive system: Control, flux and torque estimation, and efficiency optimization, IEEE Trans. Veh. Technol **60**, 3625–3634 (2011)

[9] J. Rodriguez, R.M. Kennel, J.R. Espinoza, M. Trincado, C.A. Silva, C.A. Rojas, High performance control strategies for electrical drives: an experimental assessment, IEEE Trans. Ind. Electron. **59**, 812–882 (2012)

[10] V.N. Meshcheryakov, D.V. Lastochkin, Z.M. Shakurova, S. Valtchev, Energy saving system of cascade variable frequency induction electric drive, SES 2019, Kazan, E3S Web of Conferences, 01037 (2019)

[11] I.V. Chernykh, *Modeling of electrical devices in MATLAB, SimPowerSystems and Simulink* (St. Petersburg, St. Pete, 288, 2008)

Optimization of energy spending with accelerated ventilation

Pustovaya Olessia[1,*], and *Pustovoy* Evgeniy[1]

[1]Federal State Budget Educational Institution of Higher Education "Far Eastern State Agrarian University", Russia

Abstract. Currently the cost of electric power to maintain indoor microclimate of a production facility is one of the highest. This issue is especially acute for livestock farming, where chambers and placements are not sealed from outside temperature and other factors. One of the most difficult issues of energy systems efficiency improvement is the development of adaptive climate controls for production areas and placements. A large number of factors, some of which are caused by the influence of random variables, makes the development of an algorithm for controlling ventilation systems a multi-faceted task. Classical methods of calculating microclimate parameters and their management do not allow to effectively use the adaptive control capabilities of energy efficiency of the processes under consideration. In this case the ventilation systems operation requires an excessive consumption of electric power since their design is calculated according to the maximum temperature indicators of the facility under study. One way to optimize the operation could be the introduction of an automated control system using strict algorithm or fuzzy logic algorithm. In this case, the economization of power consumption can reach up to 50% per a ventilation unit output. The authors propose using a concept of wind-cold index as a basis for constructing adaptive control of an accelerated ventilation system. Its assessment was made for the fixed indoor air movement speed. The introduction of a wind-cold index concept allows using a coefficient of correction due to the combined influence of an air movement speed in the room and air temperature, which, in turn allows to increase the minimum adjustment temperature from 16 degrees Centigrade to 18 degrees Centigrade (0C) while maintaining comfortable conditions for animals. The introduction of correction coefficients allows reducing energy expenditure by 32% in accelerated ventilation system use, that is, by the 5th month. The calculations made by authors show that to improve the adaptive control process, it is necessary to include air humidity into the algorithm as another parameter. To determine the impact of this parameter on the adaptive control process will be the goal of further research.

1 Introduction

Maintaining the microclimate at the production premises is one of the most difficult academic areas to analyze and research. A large number of interacting factors, such as humidity, temperature, environmental conditions, and others, significantly complicate the analysis. The most difficult in this regard are premises for keeping livestock, the productivity of which depends on the quality of maintaining the microclimate. This issue is particularly relevant to cattle-keeping facilities.

Recently, the tethered method of keeping livestock with periodic walking of farm animals on specialized sites has become more widespread. In summer, high indoor temperatures cause overheating which leads to livestock death. One solution to the problem is the use of high-speed ventilation which cools the surface of the animal's body. In this case, the air flow withdraws excess heat and discharge it to the external environment. According to calculations, the heat efflux when the temperature drops by 3-400C can reach up to 128 kJ/h. [1-15]

The most promising for this purpose are automated high-speed ventilation systems, the operation of which is based on the use of software with integrated microclimate parameter. In addition to optimizing these parameters, automated controls significantly reduce the cost of power required for ventilation of the premises.

1.1 Research objective

The purpose of this study is to consider the possibility of using a high-speed ventilation system with a hard algorithm to control the microclimate parameters in the cowshed and to assess the energy efficiency of such a system.

1.2 Experimental base, research progress

The research had been conducted from May to October at OOO "Pogranichnoe" in the Konstantinovskiy District of the Amur Region. The assessment was carried out in a cowshed for 400 heads of dairy cattle with tethered housing. The cowshed is made of reinforced concrete structures, measuring 77.2 x 21m. the height of the

* Corresponding author: pus14@rambler.ru

cowshed is 6.5 m along the ridge (includes two identical rooms for 200 heads). In the premises of the first team, a hydro-wash system is installed for cleaning manure. The second team is located in a completely identical room of the cowshed with a scraper installation for manure cleaning.

The room has forced ventilation. During the period of study, high-speed ventilation based on VOK-6,3 fans (Fig. 1. a) was used. The number of installed fans in the room is 8 on each side. The air is supplied through windows, the doors at the ends of the room, and the ventilation shafts on the ridge of the roof. For comparison, the design and the structure of the second team room was identical to that of the working room of the first team. The room is equipped with 1.5 kW Multifan fans, the temperature control is not enabled, and the fan operation mode is round-the-clock.

a.

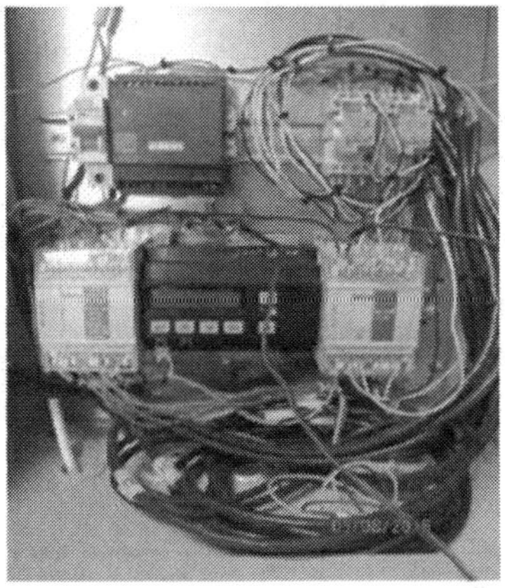

b.

Fig. 1. Cowshed premises (a. interior of a 400-head cowshed; b. automated control system based on PR200 manufactured by "Owen" company).

The fans were controlled by a PR200 microcontroller (Fig. 1. b), which has software that allows you to operate the fans in stages depending on the room temperature, starting from the temperature of 16^{0}C in increments of 2^{0}C. The fans are started in pairs with output at full power at a temperature of 22^{0}C. The temperature registration for the automated system was carried out by 6 temperature sensors installed at key points of the room.

The study of the operation of the high-speed ventilation system involved the use of LOGGER100 loggers manufactured by the "Owen" company. Temperature and humidity were recorded in the cowshed. Such registration was performed every 30 minutes. The obtained data could be presented in a graphic form, or in the MS Excel format

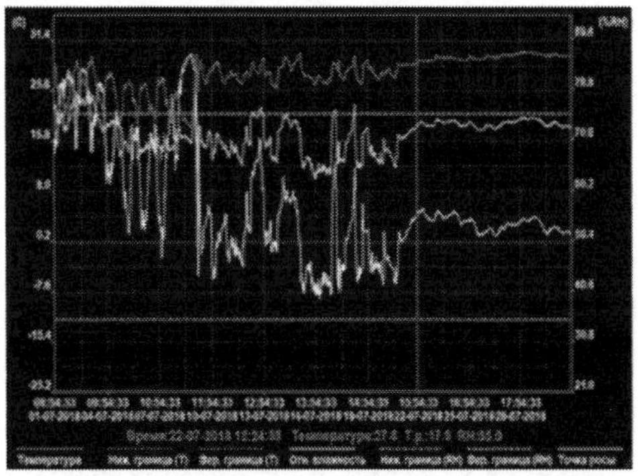

a

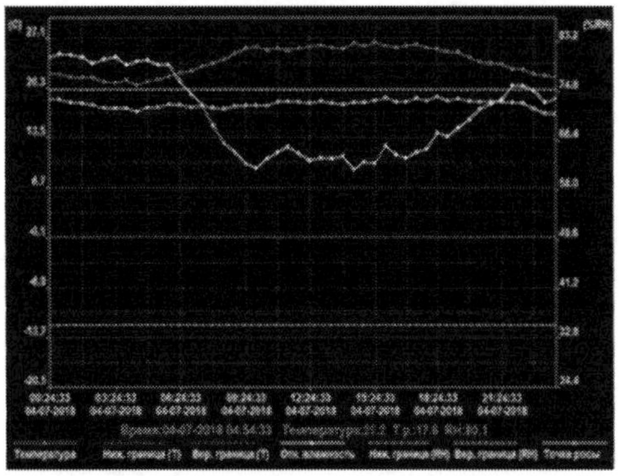

b.

Fig. 2. LOGGER 100 data (a. graph reflecting changes in temperature and humidity conditions within a month; b. graph reflecting changes in temperature and humidity conditions within a day).

2 Research results

Monitoring of microclimate parameters was carried out in the period from May to October in 2018 and 2019. An example of registered characteristics is given in Figure 2a. (data from the logger for July 2018). This figure shows the temperature, humidity, and dew point in the room of the first team for the period from July 1 to July 31, 2018. Figure 2b reflects the diurnal variation of temperature and humidity for July 4, 2018.

The obtained graphs (Fig.3) allow us to estimate the power consumption of the ventilation unit during the day in different periods. If we consider the data for May 24,

2018 (Fig. 3a.), the room temperature ranged from 14.5^{0}C to 19^{0}C, humidity — from 63.5% to 67.8%. Similar parameters for May 18, 2018 were as follows: humidity varied from 58.3% to 65.1%, temperature — from 18.1^{0}C to 24^{0}C (Fig. 3b).

All considered parameters can be assessed for the entire observation period (Fig. 3). The registered sawtooth characteristic in the spring and autumn period is due to a decrease in the ambient temperature at night and, as a result, a delayed decrease in the temperature in the cowshed.

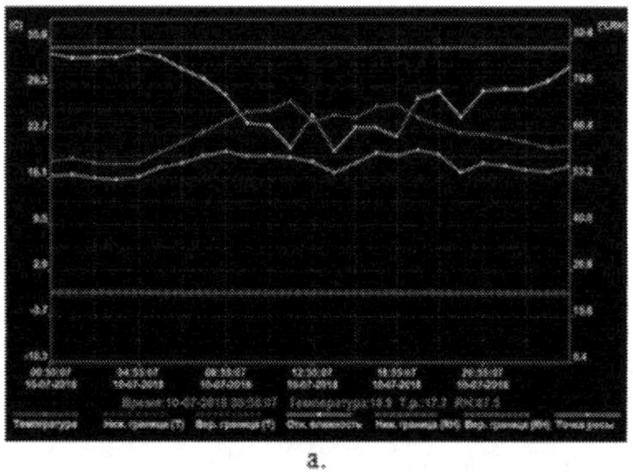

a.

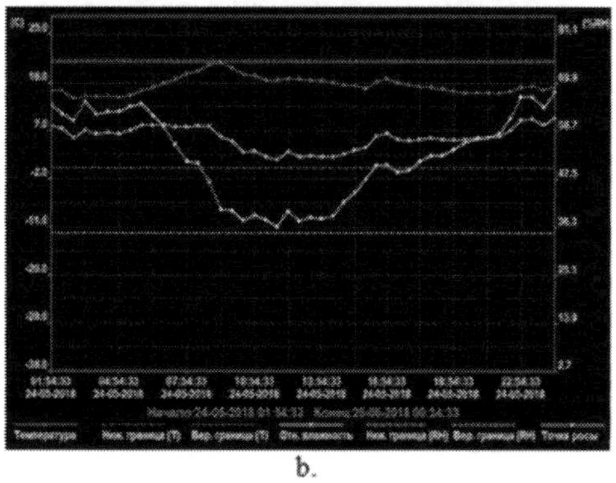

b.

Fig. 3. LOGGER 100 data (a. graph of changes in temperature and humidity for May 24, 2018; b. graph of changes in temperature and humidity changes for May 18, 2018).

The most interesting mode of operation was noted in the spring and autumn period, where there are significant fluctuations in the studied parameters, which affected the amount of power consumed by the ventilation unit greatly.

The analysis of the obtained experimental data conducted for one day (May 18, 2018) showed that the power consumption (Table 1) significantly depends on climate conditions.

For this reason, in the spring and autumn period, the difference between daytime and night temperatures allows farms to reduce daily power consumption as shown by the analysis of the graph in Figure 3b to 149.3 kW, while without automation, power consumption would be about 288 kW, which would reduce daily consumption by 138.7 kW. At the same time, in the summer the difference between day and night temperatures is not so significant, so the power consumption for July 4, 2018 is 264 kW

As shown by the results obtained (Fig. 4), the use of an automated control system for high-speed ventilation allows you to smooth out diurnal temperature fluctuations and reduce humidity in the room. In our case, the operation of the ventilation system with automatic temperature control (Fig.4b) allowed to stabilize humidity (from 52.2% to 57.2%) and temperature (from 15.5^{0}C to 27.9^{0}C) in the room, as well as to improve air exchange, which resulted in the increase of the feed intake and milk yield. In the room of the second team, humidity fluctuations ranged from 60.6% to 88.1% and the temperature varied from 16.5^{0}C to 28.2^{0}C (Fig. 4a.), the air in the room was more humid and warm, which creates discomfort.

Table 1. Calculation of the ventilation unit power consumption for May 18, 2018.

Control temperature, ^{0}C	The start time of the fan operation	The end time of the fan operation	Working time, min	Number of fans put into operation, pcs.	Power consumption, kW
16			24 y	4	36
18	6 h 24 m	24 h 00 m	17 h 36 m	4	52.4
20	7 h 54 m	19 h 24 m	11 h 36 m	4	34.4
22	8 h 24 m	13 h 24 m	8 h 54 m	4	26.5
	15 h 54 m	17 h 54 m			
Total					149.3

The obtained experimental data characterize the energy consumption for the system with strict algorithm, programmed by software. However, modern automated systems have adaptive software, which makes quite relevant the study of the influence of small variables, such as changes in heat generated by animals while the external conditions change.

The heat released by animals or humans is most often considered as a design value that does not require clarification when planning the microclimate of production facilities. However, practical, and theoretical studies show that the amount of heat generated by animals is closely related to inside temperature, humidity, and air velocity. Currently, a comprehensive assessment of these values for industrial facilities is not performed, however, to improve the accuracy of automated systems will require taking into account possible changes in the amount of heat generated by animals under the complex influence of factors mentioned above.

To assess the feasibility of using these parameters as a tool to increase the energy efficiency of automated control of accelerated ventilation, the authors evaluated air movement speed in a cow house. Based on the performed studies, the air flow rate ranges from 12.2 to 3.6 meters per second while the ventilation is on. To calculate possible influence of the considered complex of factors, the authors evaluated the wind-cold index in the facility under consideration. Since the cowsheds are identical, the first team's facility was taken under study. The temperature values for the calculation were selected in accordance with the fixed values in the facility. The calculation of the wind-cold index is given in table 2. The index's calculation is done according to the formula provided by U.S. National Weather Service

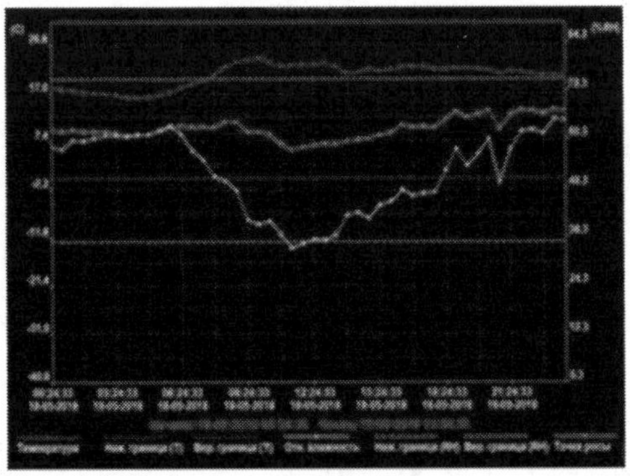

a.

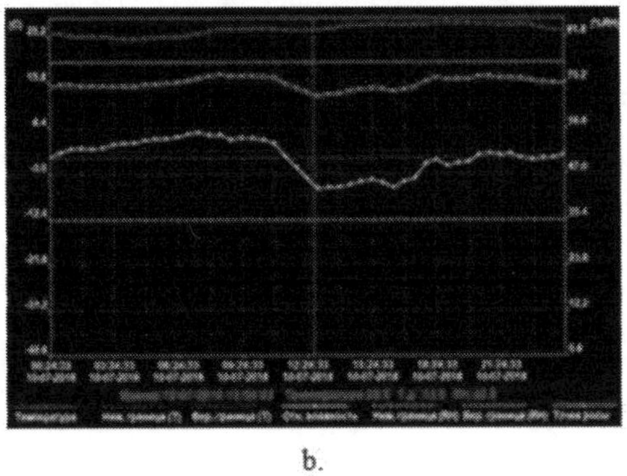

b.

Fig. 4. LOGGER 100 data (a. graph of changes in temperature and humidity conditions for July 10, 2018 (second team); b. graph of changes in temperature and humidity for July 10, 2018 (first team)).

Table 2. Calculated values of the wind-cold index for the facility under study.

#	Inside temperature, degrees Centigrade (^{0}C)	Air Velocity, meters per second (m/s)				
		12.2	11.9	10.2	6.8	3.6
1	16.0	13.9	13.9	14.1	14.7	15.5
2	16.5	14.5	14.6	14.8	15.3	16.1
3	17,0	15,2	15,2	15,4	16,0	16,7
4	17,5	15,9	15,9	16,1	16,6	17,3
5	18,3	17,0	17,0	17,2	17,6	18,3
6	18,7	17,5	17,5	17,7	18,1	
7	18,9	17,8	17,8	18,0	18,4	
8	19,1	18	18,1	18,2	18,7	
9	19,5	18,6	18,6	18,8	19,2	
10	20,0	19,2	19,3	19,4	19,8	
11	20,5	19,9	19,9	20,1	20,5	
12	21,0	20,6	20,6	20,9		
13	21,2	20,9	20,9	21,0		

As it is shown by the calculations (table #2) the lower is the ambient temperature, the greater is its cooling capacity. Thus, at a temperature value of 16^0 C, the felt temperature along with the air flow speed of 12.2 m/s will amount to 13.9^0 C, likewise with the air flow speed of 6.8 m/s, the felt temperature will amount to 14.7^0 C. With increasing temperature, this difference decreases significantly and when a certain value is reached, a heat inflow to animal's body is observed instead of cooling. With options considered, the air flow rate for 6.8 m / s is 21^0 C, and at a speed of 3.6 m / s it is 18.7^0 C.

The table above allows to make adaptive control corrections in the accelerated ventilation system management, based on the provisions that the air flow takes heat from body's surface, cooling it to a certain temperature. Basing on this notion, the reference point for the temperature setting in accelerated ventilation systems should be the temperature value to which the surface of animal's body is cooled rather than the inside air temperature in the facility. Therefore, taking this fact into account, the automatic control settings can be adjusted accordingly. Currently the system is configured to turn on / off the fans pairwise when the minimum value reaches 16^{0}C and start upwards with the intervals of 2^{0}C with the temperature reaching 22^{0}C, all fans are deployed.

The introduction of wind-cold index allows to shift the temperature control border for 16^{0}C with an increment of 2^{0}C; for 18^{0}C it's 1.5^0 C; for 20^{0}C - 1.0^{0}C; for 22^{0}C - 0^{0}C, accordingly, which will reduce the ventilation system operating time to achieve the desired effect.

The analysis of power consumption was carried out for a ventilation system with strict control algorithm without considering the wind-cold index but taking into account its influence. The calculation of operating time for an accelerated ventilation system was done for the period from May to September according to the diagrams obtained with the help of loggers installed in the facility (Fig. 2, 3, 4). The total operating time of the ventilation system is shown in Figure 5.

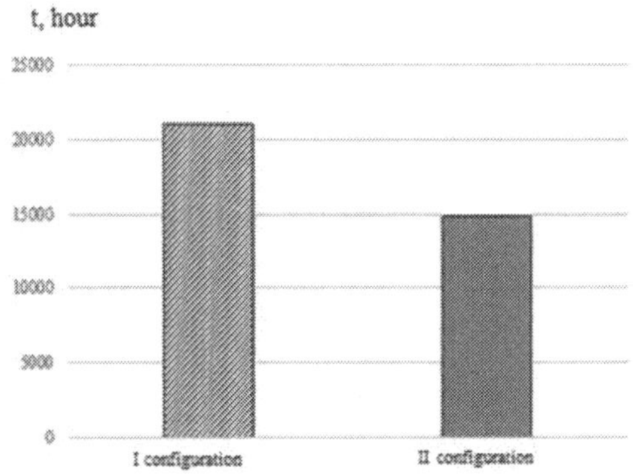

Fig. 5. The total operating time of the accelerated ventilation system (I configuration – without the wind-cold index; II configuration - with the wind-cold index).

The presented diagram illustrates that the number of operating hours of the accelerated ventilation system using the wind-cold index drops significantly with the difference of about 30%, which leads to substantial savings of electric energy (Fig. 6).

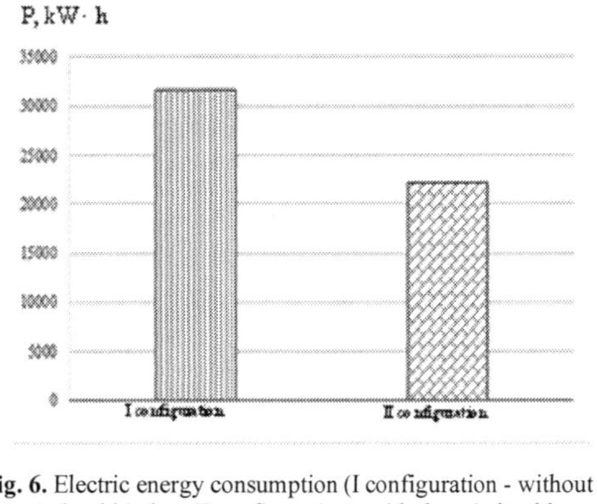

Fig. 6. Electric energy consumption (I configuration - without the wind-cold index; II configuration - with the wind-cold index).

The difference between power consumption in the first and second configurations made up 30%, which is significant for agricultural enterprises.

The efficiency of wind-cold index use in a livestock facility will vary essentially depending on external temperatures. Considering system efficiency during the season (Fig. 7), the total energy savings will amount 14,246 kW · h, which is about 32%, with the total energy consumption of 44064 kW · h by the ventilation system during the season.

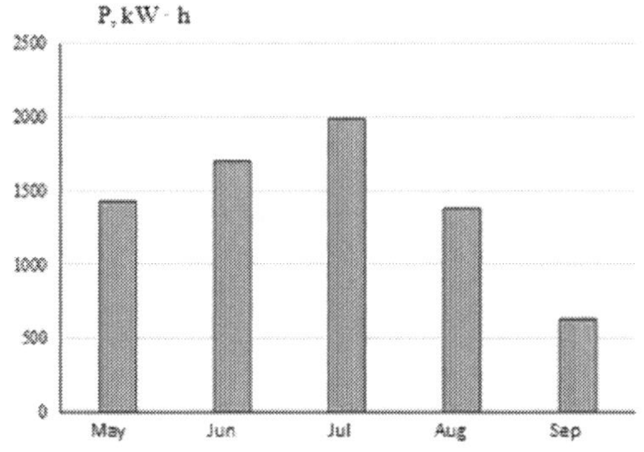

Fig. 7. The difference in power consumption between I and II configurations during the operation of the accelerated ventilation system (I configuration - without the wind-cold index; II configuration - with the wind-cold index).

Consequently, the use of regulation with the wind-cold index in adaptive control systems will lead to significant savings in electric energy consumption.

It should be noted however, that the considered factors require further investigation of the relationship between the wind-cold index and the effect of humidity on power consumption by the accelerated ventilation system.

3 Conclusions

The optimization of temperature and humidity modes in a cow house is one of the key indicators for improving the conditions of cattle maintenance. The results obtained during the research indicate that the use of automated control systems based on microcontrollers, PR200, manufactured by OWEN company in particular, allows to smooth out fluctuations of indoor temperature and humidity, as well as to reduce power consumption for maintaining the desired microclimate inside the facility. This way electric power savings can reach up to 50-52% during spring-autumn period, and up to 10-15% in summer.

The introduction of wind-cold index as one of microclimate characteristics will allow to move to the adaptive control of the microclimate parameters in industrial facilities. The analysis of inside temperature changes with the use of wind-cold index together with air flow speed showed that the automated control system can be reconfigured to higher temperatures without losing comfortable conditions for animals. For the facilities of the first and second teams, the introduction of correcting temperature coefficients allowed to reduce electric power consumption by 32%.

The study assessment shows that it is necessary to introduce air humidity as an additional parameter into the algorithm to improve the adaptive control management. To determine the impact of this parameter on the adaptive management process will be the goal of further research.

The use of automated systems is based on a strict algorithm with exact parameters, however, the use of intelligent systems will allow to remove strict restrictions from process control and make the system work variable, that is, provide a further reduction in electric power consumption.

References

[1] V.F. Vtoriy, V.V. Gordeev, S.V. Vtoriy, E.O. Lantsova, Weather influence on temperature and humidity conditions in the cowshed, Bulletin of the All-Russian Research and Development, Design and Technology Institute of Livestock Breeding Mechanization **3**, 23, 67–72 (2016)

[2] V.F. Vtoriy, V.V. Gordeev, S.V. Vtoriy, E.O. Lantsova, Assessment of the temperature and humidity regime in the cowshed using graphical information modeling, Bulletin of the All-Russian Research and Development, Design and Technology Institute of Livestock Breeding Mechanization **4**, 24, 67–72 (2016)

[3] S.S. Trunov, Analysis of energy modes of the cowshed, Bulletin of the All-Russian Research and Development Institute of Electrification of Agriculture **4**, 21, 68–72 (2015)

[4] S.A. Rastimeshin, S.S. Trunov, An energy-efficient method of combating air overheating in the cowshed, Don Agricultural Science Bulletin **2**, 34, 24–27 (2016)

[5] O.A. Pustovaya, E.A. Pustovoy, Calculations for the high-speed ventilation system, Rural Mechanization Expert, 1, 24–25 (2018)

[6] H.H.R. Zanetoni, I.F.F. Tinôco, M. Barbari, L. Conti, G. Rossi, F.C. Baêtal, M.O. Vilela, C.G.S. Teles Jr., R.R. Andrade, Alternative form to obtain the black globe temperature from environmental variables, Agronomy Research **17**, 3, 900–906 (2019)

[7] Li. Rong, Mechanisms of natural ventilation in livestock buildings: Perspectives on past achievements and future challenges. Biosystems engineering **151** (2016). ISSN: 1537-5110. Online ISSN: 1537-5129

[8] L. Rong, B. Bjerg, T. Bartzanas, G. Zhang, Review on mechanism of natural ventilation in livestock buildings past studies future challenges, Aarhus, Denmark (26.06.2016–29.06.2016)

[9] D. Tikhomirov, A.N. Vasilyev, D. Budnikov, A.A. Vasilyev, Energy-saving automated system for microclimate in agricultural premises with utilization of ventilation air, J. Mobile Communication, Computation and Information, 1–8 (2019)

[10] M.A.A. El-Rahman, M.S.El Shal, M.K.A. El-Wahab, M.M.A. El-Aziz, Study on ventilation in enclosed poultry houses [Electronic rcsourcc], Zagazig J. Agric. Res. **43**, 6A (2016). Available at: http:/www.journals.zu.edu.eg/journalDisplay.aspx?Journalld=1&queryType=Master

[11] F. Rojano, P.-E. Bournet, M. Hassouna, P. Robin, M. Kacira, C.Y. Choi, Modelling the impact of air discharges caused by natural ventilation in a poultry house. Biosystems Engin. **180**, 168–181 (April 2019)

[12] Y. Wang, W. Zheng, B. Li, Li Xuanyang, A new ventilation system to reduce temperature fluctuations in laying hen housing in continental climate Biosystems Engineering **181**, 52–62 (May 2019)

[13] T. Nicoletta, V. Francesca, C. Giovanni, P. Simona, Development of a CFD Model to Simulate Natural Ventilation in a Semi-Open Free-Stall Barn for Dairy Cows, M. C., Buildings (2075-5309), DB Applied Science & Technology Source **9**, 8, 183–183, 1. DOI: 10.3390/buildings9080183 (August 2019)

[14] V. Ronja, W. Lutz, Z. Werner, J.H. Stefan, B. Johannes, N. Knut, P. Martin, A. Ivonne, A.Konrad, H.-P. Isabel, S. Martin, S. Günther, Biosystems Engineering **164**, 85–97. DOI: 10.1016/j.biosystemseng.2017.09.013 (December 2017)

[15] N. Tomás, G. Jim, F. Richard, S. Da-Wen, Optimising the ventilation configuration of naturally ventilated livestock buildings for improved indoor environmental homogeneity Building and Environment **45**, 4, 983–995. DOI: 10.1016/j.buildenv.2009.10.005 (2010)

[16] C. Mikovits, W. Zollitsch, S.J. Hörtenhuber, J. Baumgartner, K. Niebuhr, M. Piringer, I. Anders, K. Andre, I. Hennig-Pauka, M. Schönhart, G. Schauberger, Impacts of global warming on confined livestock systems for growing-fattening pigs: simulation of heat stress for 1981 to 2017 in Central Europe, Int J Biometeorol 63, 2, 221–230 (February 2019). PMID: 30671619. ISSN: 1432-1254

Cordline multi-motor electric drive with energy recovery

Rishat Ganiev[1,*]

[1]Kazan National Research Technological University, Institute of chemical technology, 423570 Avenue of Builders, 47, Nizhnekamsk, Republic of Tatarstan, Russian Federation

Abstract. The article offers a variant of developing a cordline electric drive in the production of automobile tires to increase the energy-efficient due operation of production lines based on interconnected electric drives with energy recovery. The modes of operation of the technological line of cord production with energy recovery is possible in the network are shown, requirements for the drive system are drawn up. Simulations in Matlab 10, Simulink and SimPowerSystems carried out the research. The developed version of the frequency-controlled electric drive allows increasing the energy efficiency of production lines in the production technology of automobile tires by using generator modes of asynchronous motors as a part of frequency-controlled electric drives.

1 Introduction

Production of cord consists of impregnation of synthetic fabric with latex followed by rubberization of the fabric on multi-roll calender rolls. Calender rolls consist of mechanisms driven by reversible multi-quadrant electric drives with varying signs of speed and torque [8]. Before cutting, a roll of fabric is loaded into the cutting line at a speed of 3 m / min with a tension of about 50 kgf using a rolling device and a pulling station. After loading the fabric and beginning to rubberize, the speed of movement of the fabric increases to 45-50 m/min. Hence, the speed control range of the drive must be at least (1: 15).

Braking modes when the line is stopped are accompanied by generator modes in the electric drive of the seaming device. The braking energy released when the line is stopped for loading can be used to power the drives of the pulling device. To research on the possibility of recovery, let's consider the general requirements for a multi-motor drive of a cord line.

2 Requirements to drive

Since the volume of the supplied rubber mixture changes during the calendering process, which causes a change in the load moment, it is necessary to provide control of the drive moment. According to the technology on the line for impregnating cord fabric, the electric drive must provide a number of additional requirements:
- speed matching of individual line mechanisms;
- setting and stabilization of tension on various sections of the cord line, as well as ensuring stable operation of the multi-motor electric drive;
- maintain a steady speed in steady mode (with an accuracy of ±5 % when changing the static torque from 0.25 to 1.00 of the rated motor torque);
- compensation of static and dynamic moments;
- maintaining the degree of filling of the compensator and the continuity of the technological process when replacing rolls in the electric drives of rolling and rolling devices.
- reversing and limiting the time of emergency braking of the rolls (no more than ¼ of the circumference of the roll). Precision maintain set parameters of the electric drive should provide a steady-state deviation of maximum operating speed of the line in the static operation – no more than ± 1 % of nominal operating speed at a static voltage mains three-phase AC +10%, -15%, frequency ± 1 %, the ambient temperature ±10 degrees C [1].

3 The existing scheme of drives

At now each cordline motor is powered individually from frequency converters. For this purpose, a single three-winding power transformer is used, the output voltage of the secondary winding of which is connected to the input of frequency converters. Figure 1 shows the General connection diagram for frequency converters. The scheme also includes: for further research, two electric drives of the "KAMA-3 "equipment based on the "Repique" line are selected: M20, which is an electric drive of the tension roller ,and M70, which is an electric drive of the rolling device, one of which is powered by a frequency Converter Mitsubishi FR-A741-22k, and the second is powered by a frequency Converter Mitsubishi FR-A740-00620 EC.

For researching the operating modes of the cordline electric drives, an increase in the voltage in the DC link

* Corresponding author: n7007@mail.ru

 https://doi.org/10.1051/e3sconf/202017801023

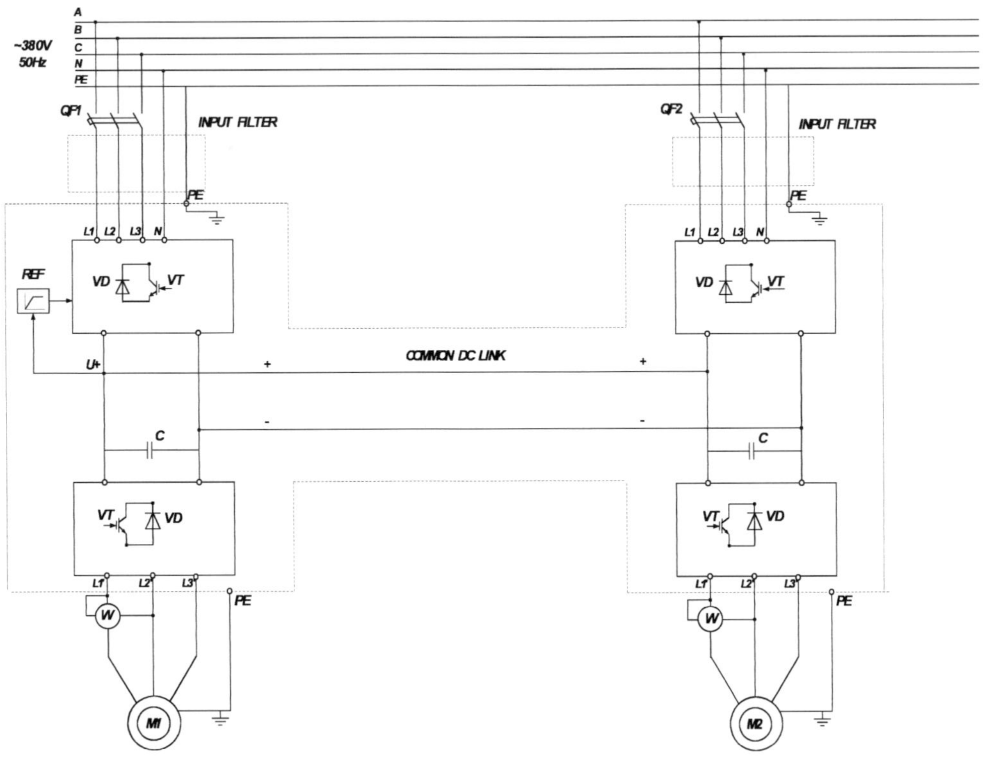

Fig. 1. Experimental scheme for research.

of the tension roller electric drive at the time of acceleration of the process line was registered. It is determined that the line speed is ahead of the specified rotor speed of the tensioner motor, and the rotor speed begins to exceed the speed of rotation of the stator electromagnetic field, and the motor goes into generator mode.

A Mitsubishi FR-A741 frequency Converter with recovery capability is used to power the electric motor . The recovered energy of the tension roller electric drive can be used to power another electric drive of position M70. Combining converters with a common DC bus will solve the problem of using energy recovery.

Frequency Converter FR-A741 for low and medium power ranges is an inverter with a built-in function of returning (recovering) stored energy from the engine to the network. The idea to combine a conventional inverter (for example, a series of frequency converters Mitsubishi FR-A740) with a power return device will eliminate the brake resistors and significantly reduce the time of stopping the drive,as well as increase the efficiency of the line as a whole.

4 Experimental scheme for research

Two frequency converters with the possibility of energy recovery were adopted for the study. The connection diagram with a common DC link of these converters is shown in figure 1.

The scheme works as follows: at the moment when M1 (M2) is in braking mode, and M2 (M1) is in acceleration mode, the energy stored in M1 (M2) will flow to M2 (M1), respectively, the energy from the network will be consumed in a smaller amount, which can be observed with the help of wattmeters installed at the output of the if. In the case when both engines are running in braking modes, the energy will be recovered to the network. The intensity setter (ZI) serves as a regulator for switching the rectifier to the inverter mode when the voltage on the bus bridge of the DC link increases in the case of simultaneous operation of both electric motors in generator mode

5 Computer Simulation of recovery modes

In the Matlab Simulink program, a model of an electric drive was assembled to study the generator mode according to thc schcmc "Nctwork – sccondary power source (SPS) – power semiconductor Converter – electric machine". The model is shown in figure 2. [2]. In this scheme, the motor and its Converter are represented by a DC link, respectively, a passive active-inductive circuit RL and a regulated voltage source (EMF). This substitution scheme is valid for both a DC

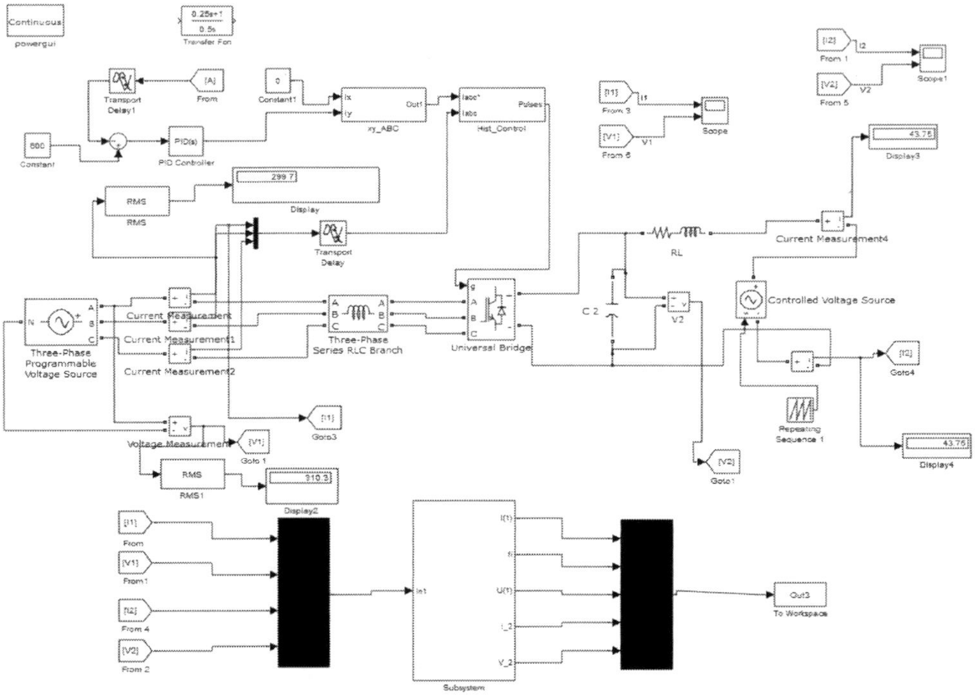

Fig. 2. Model of a secondary power source with energy recovery in the network.

motor with a pulse-width Converter and an asynchronous motor with a self-contained voltage inverter. The filter capacitor is always connected to the secondary power supply terminals. The coupling equation for the load current has the form:

$$L\left(\frac{di_H}{dt}\right)+Ri_H=U_d-E(t) \qquad (1)$$

If the motor is motoring mode of operation and motor connected to the output of the Converter, the voltage Ud on the capacitive element is greater than the EMF value of the machine, the current sign is positive, the energy flow is directed from the power source to the motor. In the case of generator mode of the motor, the EMF exceeds the value of the capacitor voltage, the load current is negative, and the energy flow is directed from the motor to the power source.

To implement the secondary power supply model in the current control loop, the Simulink function package is used, as well as a package with a set of special functions of SimPowerSystems. To configure the main block variables in the specified packages, there is a parameters window. Each the Simulink and SimPowerSystems package block has configuration window for basic parameters. In the block x, y-ABC, the coordinate system (x, y) is converted from rotating to stationary A, B, C according to the following equations:

$$u_\alpha = u_x cos(\omega_1 t) - u_y sin(\omega_1 t) \qquad (2)$$

$$u_\beta = u_y cos(\omega_1 t) + u_x sin(\omega_1 t) \qquad (3)$$

$$u_A = u_\alpha \qquad (4)$$

$$u_B = -0,5u_\alpha - \sqrt{0,865}u_\beta \qquad (5)$$

$$u_C = -0,5u_\alpha + \sqrt{0,865}u_\beta \qquad (6)$$

In this computer simulation, the load current is changed due to the change of the electromotive force of the machine. The xy_ ABC subsystem converts the XY coordinates (coordinates of DC signals) into a three-phase ABC current system, where they are set for motor control. The Hist_Control subsystem includes relay controllers. The Subsystem model is a measurement model. In this scheme, input and output connectors and two harmonic analysis blocks from the SimPowerSystems Extras Measurement library are connected. [1].

6 The results of computer simulation

Figure 3 shows quasi-steady processes in the system at the moment when the current at the output of the secondary power supply is measured according to the trapezoidal law. At a time interval where the load current is constant, the system operates in a steady state (motor at a positive current, generator – at a negative). During the time interval, where the current changes according to the linear law, a transient mode of operation of the system will be observed.

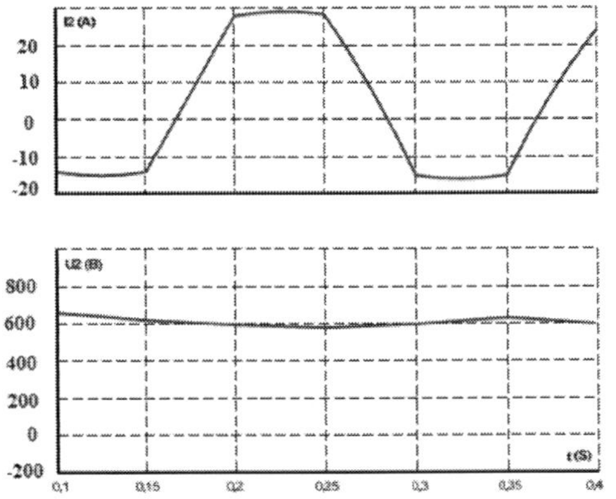

Fig. 3. Electromagnetic processes in SPS with electric drive energy recovery in the network.

Figure 4 shows the electromagnetic processes in the steady-state operation mode. To obtain the characteristics of electromagnetic processes in the entire range of changes in the load current of the secondary power supply, a signal is formed that changes according to a linear law in the Repeating Sequence block. The current takes a negative sign in the generator mode and a positive sign in the motor mode.

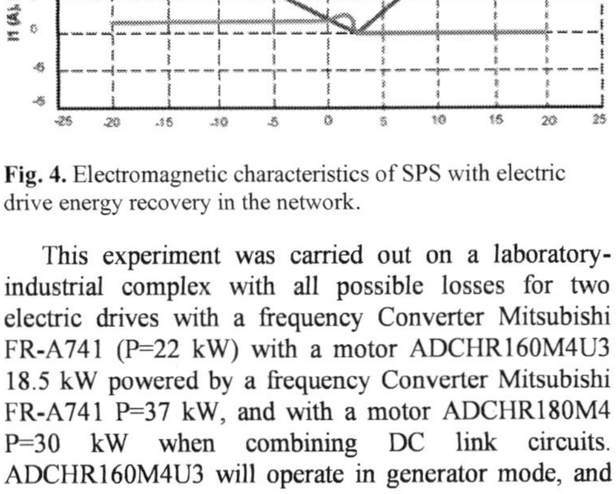

Fig. 4. Electromagnetic characteristics of SPS with electric drive energy recovery in the network.

This experiment was carried out on a laboratory-industrial complex with all possible losses for two electric drives with a frequency Converter Mitsubishi FR-A741 (P=22 kW) with a motor ADCHR160M4U3 18.5 kW powered by a frequency Converter Mitsubishi FR-A741 P=37 kW, and with a motor ADCHR180M4 P=30 kW when combining DC link circuits. ADCHR160M4U3 will operate in generator mode, and ADCHR180M4U3 in motoring mode. The simulation results shown in figure 5 confirm the possibility of recovery during acceleration and deceleration of the line (current I2 changes its direction).

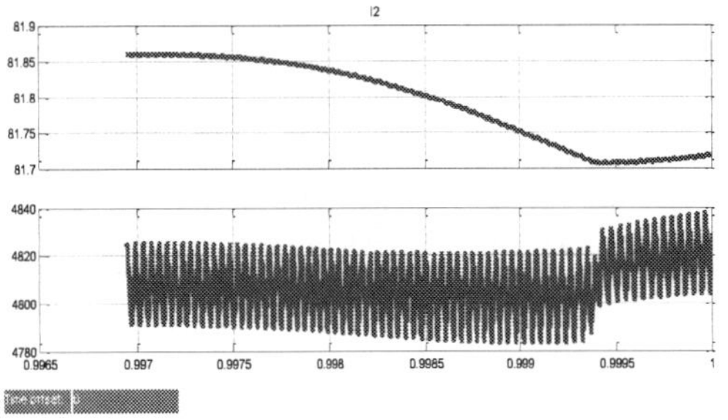

Fig. 5. Characteristics of the system output (upper graph – current, lower-voltage).

7 Conclusion

The graphs obtained from the simulation (Fig. 4) showed that the current I1 in the network increases when the load current of the secondary power source increases in both the motor and generator modes of the electric drive. The current phase relative to the voltage determines the power consumption from the network or recovery. When energy is recovered to the network, the current phase relative to the voltage is equal to 180°, and when energy is consumed in the motor mode-0 ° The research results confirm the possibility of using a common DC bus for frequency drives as part of a multi-motor electric drive of production lines in order to use recovery energy to reduce energy consumption. Economic calculations have shown that if this scheme is implemented, the energy consumed from the network is reduced by 5652.14 W and will be 26386.54 W. The approximate energy savings will be 25,787,9 kWh per year.

References

[1] R.N. Ganiev, S.N. Shatunov, Frequency-controlled electric drive with recovery as part of the cordline production of truck tires, Bulletin of the Chuvash University, 3, 44–52 (2018)

[2] R. N. Ganiev, Reversible frequency converters in the composition of multimotor electric drives, IOP Conference Series: Materials Science and Engineering, International Scientific Electric Power Conference 23–24 May 2019, Saint Petersburg, Russian Federation, **643** 012068 (2019)

[3] J.E. Valdez-Resendiz, J.C. Mayo-Maldonado, A. Llamas-Terres, J.C. Rosas-Caro, Quadratic boost converter based on stackable switching stages IET Power Electronics **11**, 8 (2018)

[4] A. Rehaoulia, H. Rehaoulia, F. Fnaiech, Output voltage quality analysis of three-phase multilevel inverters Electrical Motorering, Archiv für Elektrotechnik **100**, 2, 733-739 (2018)

[5] S. Zambou, R. Nuessl, F.M. Azeutsap, S.S. Zekeng, J.M. Ndjaka, B. Magunje, S.D. Walton, E.O Jonah, M. Harting, D.T. Britton, Switching alternating current (ac) using a fully screen-printed current-driven transistor, IET Power Electronics **11**, 1, 168–174 (2018)
[6] P. Záskalický, Mathematical model of a five-phase voltage-source pwm-controlled inverter, Electrical Motorering, Archiv für Elektrotechnik **99**, 4, 1179-1184 (2017)
[7] K. Bernacki, Z. Rymarski, Electromagnetic compatibility of impedance source inverters, Elektronika ir Elektrotechnika **23**, 3, 55-63 (2017)
[8] N. Prabaharan, K. Palanisamy, Analysis of cascaded h-bridge multilevel inverter configuration with double level circuit, IET Power Electronics **10**, 9, 1023-1033 (2017)
[9] A.M. Airabella, G.G. Oggier, L.E. Piris-Botalla, G.O. García, C.A. Falco, C. Ramos, A. Martins, A. Carvalho, Semi-conductors faults analysis in dual active bridge dc-dc converter, Active filtering of dfig stator and rotor current harmonics caused by distorted stator voltages, European Power Electronics and Drives Journal **21**, 1, 43-54 (2011)

The influence of main parameters of regenerative heat exchanger on its energy efficiency

Nikolay Monarkin[1,*], *Anton* Sinitsyn[1], *Mikhail* Pavlov[1] and *Timur* Akhmetov[2]

[1]Vologda State University, Institute of civil engineering, 160000 Vologda Lenina str. 15, Russia
[2]Kazan State Power Engineering University, Kazan, Russia

Abstract. The influence of various parameters of stationary switching regenerative heat exchangers used for ventilation on its thermal efficiency was studied. Considered are the geometric (length, diameter and wall thickness of a single equivalent nozzle channel), thermophysical (density and heat capacity of the nozzle material) and operation (air flow through the regenerator and the time of one stage of accumulation/regeneration of thermal energy) parameters.

1 Introduction

In recent years, compact energy-saving devices have been used for ventilation of civil buildings as an alternative to central systems. The most energy-efficient among them are stationary switching regenerative heat exchangers (SSRHE). Such devices can decrease the high heat consumption for ventilation and provide a fairly simple solution to the problem of insufficient air exchange in some rooms [1,2].

Given the relevance of SSRHE usage for premises ventilation, an important issue is the design of such devices. In this case, first of all, it is necessary to consider a heat exchanger called a nozzle, which is the SSRHE basis. A nozzle, like any heat exchanger, must effectively transfer thermal energy between flows, and this characteristic depends on its design and material. In addition, the important SSRHE parameters are the volume of air flows through the device and the duration of their passage.

Most of the works that give recommendations on the design of ventilation heat exchangers consider devices based on recuperative heat exchangers, including rotary heat exchangers [3-5]. The existing works devoted to SSRHE do not give specific information on the most challenging design [6,7]. Therefore, the purpose of this paper is to determine the nature and degree of influence of changes of the main SSRHE parameters on its energy efficiency, which can be used in design of such devices.

2 Materials and methods

As part of SSRHE construction, the regenerative nozzle is a cylindrical bunch of longitudinal channels with a small cross-sectional size. The SSRHE operation principle consists in alternate changing of modes of external air inflow and internal air exhaust within a certain time period. In the exhaust mode, the internal warm air of room passes through the nozzle and heats it in order to transfer part of thermal energy to the outside air after switching the mode to the inflow. Thus, the device changes the stages of accumulation and regeneration of thermal energy. To induce air motion in the heat exchanger, a reversible fan is used. For effective ventilation of one or several adjacent rooms, SSRHE are installed in pairs in antiphase mode [8].

Parameters of heat exchanger (regenerator) are determined from numerical implementation of the constructed mathematical model, which describes the processes of heat transfer in a single nozzle channel [9]:

$$c_n \cdot \rho_n \cdot S_n \cdot \frac{\partial T_n(z,\tau)}{\partial \tau} = \Pi \cdot \alpha \cdot \left(T_a(z,\tau) - T_n(z,\tau)\right), \quad (1)$$
$$0 < \tau < n \cdot \tau_0, \quad 0 < z < L$$

$$G \cdot c_a \cdot \frac{\partial T_a(z,\tau)}{\partial z} + \Pi \cdot \alpha \cdot \left(T_a(z,\tau) - T_n(z,\tau)\right) + \quad (2)$$
$$+ c_a \cdot \rho_a \cdot S_a \cdot \frac{\partial T_a(z,\tau)}{\partial \tau} = 0, \; 0 < \tau < n \cdot \tau_0, \; 0 < z < L$$

$$T_a(0,\tau), \quad T_a(z,0), \quad T_n(0,\tau), \quad T_n(z,0) \quad (3)$$

Equations (1) and (2) describe the temperature fields of nozzle (T_n (z, τ)) and air (T_a (z, τ)), respectively. The parameters (3) are set as the boundary conditions of the problem.

In the model (1) - (3): G is the mass air flow in channel, kg/s; c_n, c_a is the mass heat capacity of nozzle and air, respectively, J/(kg·K); ρ_n, ρ_a are the densities of nozzle and air, respectively, kg/m^3; P is the channel perimeter, m; S_n, S_a are the cross-sectional area of the nozzle channel and air, respectively, m^2; α is the heat transfer coefficient, W/(m^2·K); $\tau_0=\tau_{acc}+\tau_{reg}$ is the regenerator cycle time, s; τ_{acc} is the duration of accumulation stage; τ_{reg} is the duration of regeneration stage.

* Corresponding author:nikolay-monarkin@yandex.ru

The evaluation criterion is the dimensionless energy efficiency coefficient, which is the ratio of the actual energy used by the regenerator to the maximum possible energy for use.

$$E_{acc} = \frac{\delta T_1}{\Delta T_{max\ reg} \dfrac{\delta T_2}{\Delta T_{max}}} \qquad (4)$$

where $\Delta T_{max}=(T_{in} - T_{out})$ is the maximum possible cooling or heating of air in nozzle from the internal medium temperature T_{in} to the outside temperature T_{out}, °C;

δT_1 is the cooling of internal nozzle air at the end of the accumulation phase, °C;

δT_2 is the heating of outside nozzle air at the end of the regeneration phase, °C.

The average stage performance coefficients are equal $E_{acc}= E_{reg}=E$.

During operation, the following nozzle parameters are varied: the single channel diameter (d), length (l), thickness of single channel wall (δ), volumetric air flow rate through the nozzle (L), density (ρ_n), specific isobar heat capacity (c_n). Also the time of one stage of accumulation/regeneration of thermal energy varies ($\tau=\tau_{acc}=\tau_{reg}$).

3 Results and discussion

The heat transfer efficiency in regenerator is primarily affected by the contact area (heat transfer area) between air and nozzle. Consequently, the parameters of regenerator nozzle, which affect the change in heat transfer area, also affect the efficiency of regenerator. The smaller is the heat transfer area, the lower the efficiency is and vice versa.

The heat transfer area decreases when nozzle length decreases and the total number of nozzle channels is reduced. It occurs with an increase in transverse dimensions and with an increase in wall thickness of a single nozzle channel.

If the transverse dimensions of a single nozzle channel increase due to the limited overall transverse size (diameter) of nozzle, the total heat transfer area decreases, which is associated with a decrease in the total number of channels in nozzle. The overall diameter of nozzle is limited as the possible amount of through holes in outer walls is limited. In this case, the nozzle diameter is limited to 0.2 m (area of 0.03 m^2).

The relationship between the energy efficiency coefficient and the equivalent diameter of a single channel is shown in Figure 1. It is seen that when diameter increases from 1.6 to 5.8 mm (by 3.63 times), the efficiency decreases: for a flow rate of 15 m^3/h by 3%, for a flow rate of 20 m^3/h by 4%, for a flow rate of 31 m^3/h by 7%, for a flow rate of 50 m^3/h by 10%. That is, an increase in the channel diameter with a constant air flow through the nozzle reduces its efficiency, and the sharpest decrease is observed at the maximum air flow rate.

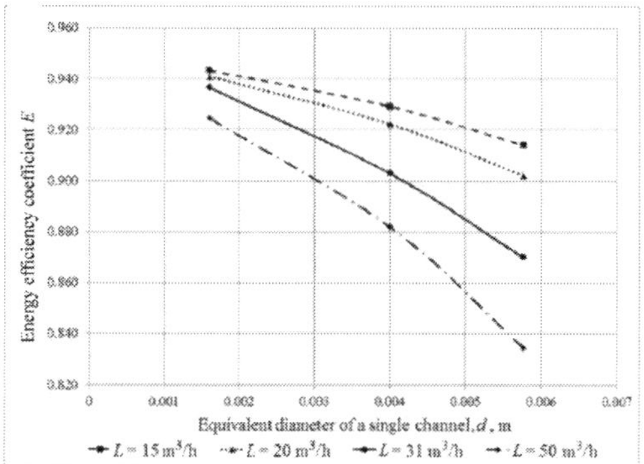

Fig. 1. The relationship between the energy efficiency coefficient E and the equivalent diameter of a single channel (l=0.64 m, δ=0.5 mm, τ=40 s).

According to Figure 2, an increase in the length of channel (nozzle) contributes to an increase in coefficient E. If the channel length increases from 0.2 to 0.64 m (by 3.2 times), then the efficiency coefficient increases by 5% for d=1.6 mm, by 10% for d=4.0 mm; by 9% for d=5.8 mm.

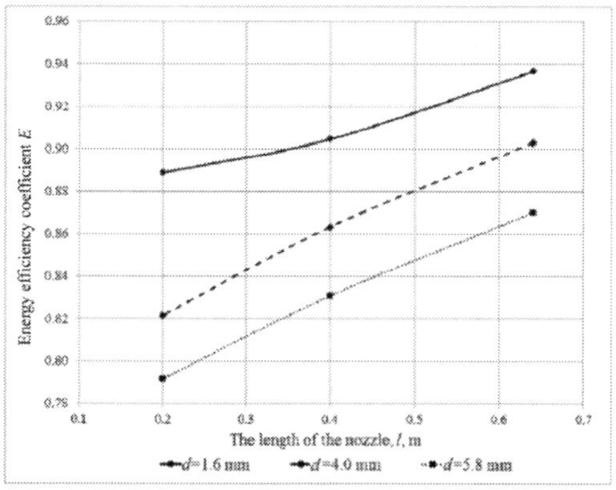

Fig. 2. The relationship between the energy efficiency coefficient E and the channel length (L=31 m^3/h, δ=0.5 mm, τ=40 s).

If the wall thickness of a single channel increases, the total number of nozzle channels decreases due to the larger filling of nozzle section with solid material. Thus, the smaller channel wall thickness results in a more efficient regenerator (Fig. 3). In this case wall thickness is limited only by the condition of structural strength.

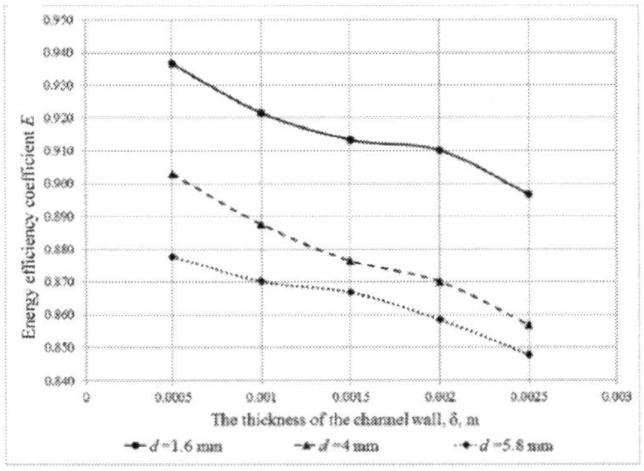

Fig. 3. The relationship between the energy efficiency coefficient E and the wall thickness of a single channel for three channel diameters (l=0.64 m, τ=40 s, L=31m^3/h).

In general, in Figure 3, a decrease in efficiency is observed with an increase in the diameter and wall thickness of channel, and for a curve for one diameter the decrease in efficiency is weak. For diameter of 1.6 mm, the efficiency coefficient decreases by 4%, for diameter of 4.0 mm it decreases by 5%, for diameter of 5.8 mm it decreases by 3%.

The greater is the amount of air passing through the SSRHE per unit time, the more its efficiency decreases (Fig. 4). So, for an increase in air flow from 15 to 50 m^3/h (by 3.33 times), the efficiency decreases as follows: by 2% for diameter of 1.6 mm, by 5% for diameter of 4.0 mm, by 9% for diameter of 5.8 mm.

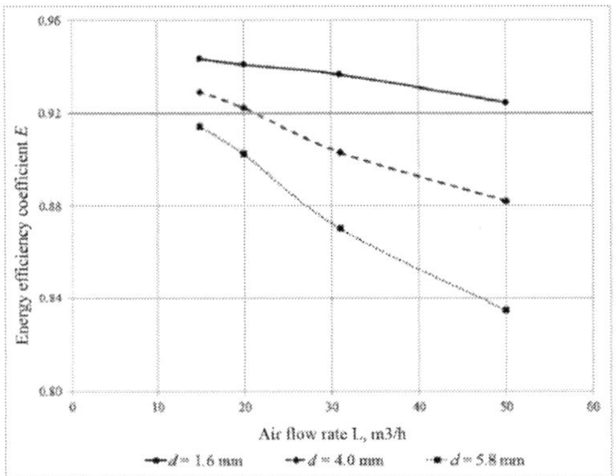

Fig. 4. The relationship between the energy efficiency coefficient E and air flow rate for three channel diameters (l=0.64 m, τ=40 s, δ=0.5 mm).

An increase in time of stage of accumulation/regeneration of thermal energy in regenerator leads to a decrease in its energy efficiency (Fig. 5). Figure 5 shows that when the stage time changes from 5 to 300 s (by 60 times), the efficiency decreases by 3% for a length of 0.64 m, by 6% for a length of 0.4 m, by 9% for a length of 0.2 m. That is, in the worst conditions there exists an option with a minimum length and maximum stage time. And it is worth noting that for a decrease in length with a constant diameter, the regenerator efficiency decreases.

Figure 6 shows the relationship between the efficiency coefficients and the density of the nozzle material for three different air flow rates at constant heat capacity (channel length l=0.64 m, diameter d=4.0 mm, wall thickness δ=0.5 mm). The figure shows that behavior of efficiency coefficients is the same for all three flow rates: there is a noticeable increase in the beginning (from 500 to 2000 kg/m^3), then the process is practically stabilized. When the density of nozzle increases from 500 to 4000 kg/m^3, the efficiency coefficient increases by 1-2% for all these flow rates.

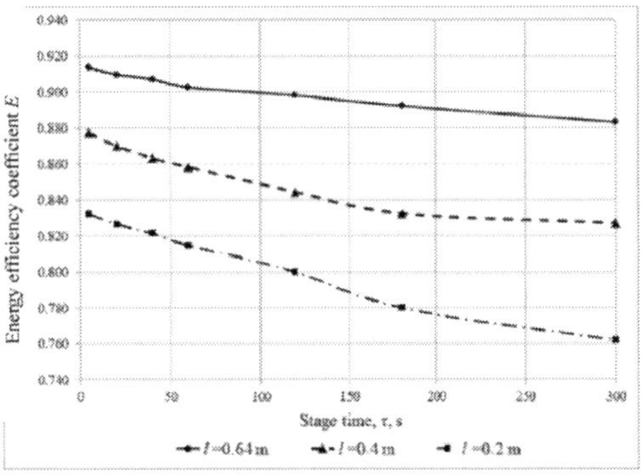

Fig. 5. The relationship between the energy efficiency coefficient E and the time of one stage of accumulation/regeneration for various channel lengths (d=4.0 mm, δ=0.5 mm, L=31m^3/h).

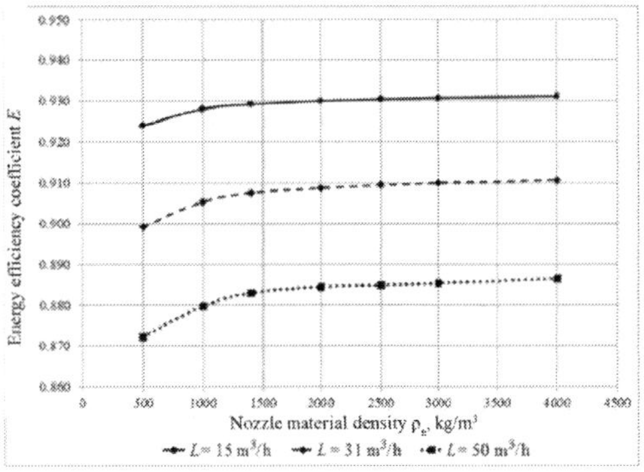

Fig. 6. The relationship between the energy efficiency coefficient E and the nozzle material density (d=4.0 mm, l=0.64 m, δ=0.5 mm, τ=40 s).

Variation in heat capacity at a constant density of nozzle material of 1400 kg/m^3 gives similar results (Fig. 7). With an increase in the heat capacity of nozzle from 1000 to 4000 J/(kg·K), the efficiency coefficient increases for all these flow rates by 1-2%.

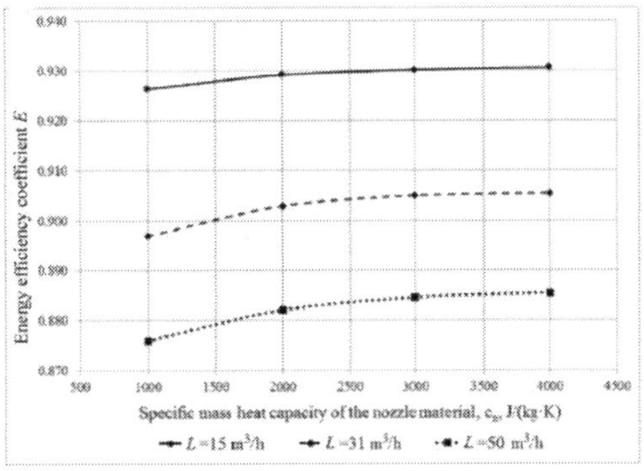

Fig. 7. The relationship between the energy efficiency coefficient E and the heat capacity of nozzle material (d=4.0 mm, l=0.64 m, δ=0.5 mm, τ=40 s).

4 Conclusions

The performed studies show that the diameter and length of a single nozzle channel, as well as the air flow rate through nozzle, have a decisive influence on the energy efficiency of regenerator. When diameter increases from 1.6 to 5.8 mm (by 3.63 times), the efficiency decreases by 3-10% for the range of air flow rates of 15-50 m3/h. With an increase in the channel length from 0.2 to 0.64 m (by 3.2 times), the efficiency increases by 5-10% in the same range of air flow rates. The increase in air flow rate from 15 to 50 m3/h (by 3.33 times) reduces E by a maximum of 9%. The wall thickness of a single channel weakly affects the efficiency.

To obtain significant changes in efficiency when varying the time of one stage of accumulation/regeneration of thermal energy, its increase or decrease should be too strong, which in this case is unacceptable. Therefore, we can assume that the stage time slightly affects the regenerator efficiency.

An increase in density by 8 times and heat capacity by 4 times increase E by only 2%, which allows us to consider the change in these parameters as insignificant.

Thus, the presented study allows one to determine the degree of influence of the listed parameters on the energy efficiency of SSRHE. It also determines the ranges of optimal parameters that can be used in designing.

References

[1] A.L. Naumov, S.F. Serov, A.O. Budza, HVAC: Ventilation, Heating, Air Conditioning, Heat Supply and Building Thermal Physics, 1 (2012).

[2] N. Monarkin, S. Lukin, Yu.M. Anurov, B.A. Tihomirov, G.A. Agasiants, S.M. Galileev, T.R. Akhmetov, IOP Conf. Series: Earth and Environmental Science, 337 (2019).

[3] V.V. Nemirovskaya, A.V. Kuzovlev, Scientific journal of NRU ITMO. Series: Refrigeration and air conditioning 2 (2015).

[4] M.N. Stepanenko, A.Ya. Shelginsky, Reliability and safety of energy 2, 25 (2014).

[5] M.R. Nasr, M. Fauchoux, R.W. Besant, C.J. Simonson, Renewable and Sustainable Energy Reviews 30 (2014).

[6] V.A. Vasiliev, A.I. Gavrilov, K.K. Kamenetsky, E.V. Sobol', Bulletin of the International Academy of Refrigeration 1 (2010).

[7] A.E. Aleshin, A.V. Tsygankov, Bulletin of the Dagestan State Technical University. Technical science 2 (2016).

[8] N.N. Monarkin, S.V. Lukin, A.A. Kochkin, BST: Bulleten' stroitel'noy tehniki 6, 1018 (2019).

[9] N.N. Monarkin, S.V. Lukin, A.A. Kochkin, Volga Scientific Journal 3, 47 (2018).

Provision of High-efficient Milling Methods Using a PMSM Electric Drive with a Controlled Information-Measuring System

Sergey Ambrosimov[1], *Victor* Meshcheryakov[2,*], and *Igor* Ivshyn [3]

[1]Institute of Mechanical Engineering, Lipetsk State Technical University, Lipetsk, Russia
[2]Electric Drive Department, Lipetsk State Technical University, Lipetsk, Russia
[3]Department of Power supply of industrial enterprises, Kazan State Power Engineering University, Kazan, Russia

Abstract. The article presents new highly effective methods of milling shaped surfaces, which are implemented on four-and five-coordinate CNC machines. For their implementation, three simultaneously coordinated movements of the machine's organs are used. To implement the proposed milling methods on CNC machines, an electric drive is proposed, built on the basis of a Permanent Magnet Synchronous Motor (PMSM) with a vector control system with a proportional P-speed controller. For this purpose, an ecstatic control information and measurement system with zero static error is proposed. Comparative results of mathematical modeling of transient processes in electric drives with the standard and proposed control systems are presented.

1 Introduction

In mechanical engineering, controlled electric drive systems are widely used on CNC machines, while the requirements for drive and control systems are becoming more complex in terms of ensuring high speed and accuracy of working out control actions. In these conditions, AC electric drive systems based on a Permanent Magnet Synchronous Motor have proved to be the best.

An urgent problem in mechanical engineering is the search for ways to improve the efficiency of machining processes, increase labor productivity, tool life and cutting quality. Milling is one of the most productive processing methods due to the high concentration of cutting edges simultaneously involved in the cutting process, and at the same time high cost due to the high cost of equipment and tools. In addition, milling is associated with large volumes of cut material, especially when processing complex shaped surfaces, which affects the cost of the tool: disk shaped or end radius cutters.

One of the most effective ways to increase the durability of a cutting tool, as shown by recent research, is the constant displacement of the cutting edge tip during processing relative to the cutting surface, which leads to an increase in the active length of the cutting edge and a decrease in thermal stresses. For example, milling of a complex profile surface with convex sections of profiles such as Cams of couplings and gears is carried out with a tool with two conical producing surfaces that roll over the nominal surface of the part with nonlinear coordination between the three elementary movements of the workpiece (Fig. 1) [1]. One of them, the rotational wx1, is carried out in such a way that the straight forming tool is consistently tangent to each point of the processed profile, and the other two Sz and Sy agree with the rotational ωx1, so when processing cross sections of a straight profile, the forming rolls are applied to the workpiece.

The screw milling method is used for processing complex shallow narrow and wide grooves (Fig. 2) [2, 3]. When milling wide grooves, the workpiece rotates around the Y axis and simultaneously moves in the direction of the Y axis until the nominal surface is reached and a lenticular surface area is selected. Then three movements are made simultaneously in the direction of rotation of the movement in the direction of the formation of the groove, i.e. X and Z.

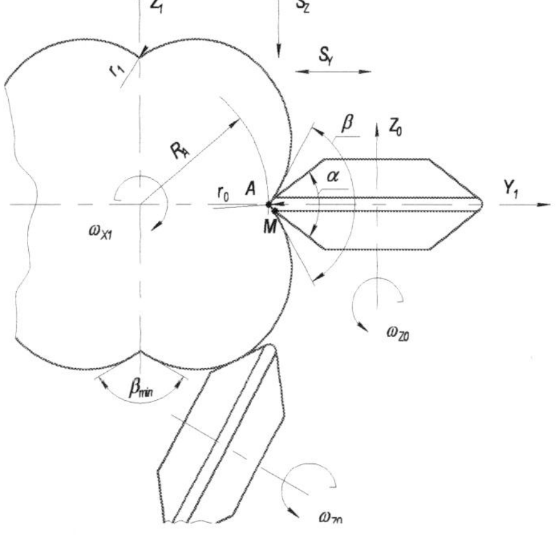

Fig. 1. Scheme of nonlinear rolling of a complex profile of a part with a complex mill.

* Corresponding author: mesherek@yandex.ru

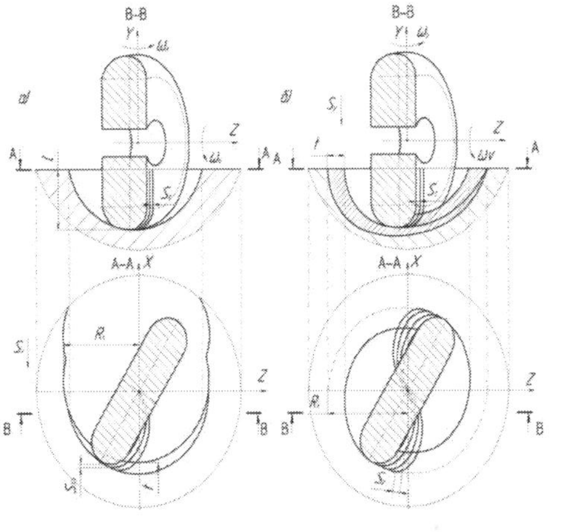

Fig. 2. The scheme of formation of the spiral groove for moving the cutter.

When milling narrow grooves, a lenticular area is formed after milling, which is bounded by the edges of the groove. The mill performs a relative reciprocating motion until the processing surface touches the sides of the groove, continuing to sink to its bottom. The advantage of this method is the displacement of the upper part of the cutting edge (CE) along the entire length of the RC, starting from the right side of the cutter and ending with the left in one section. The cutting zone moves quite intensively relative to the CE, providing a good heat sink and low thermal load on the cutting edges.

Theoretical studies have confirmed a significant increase in processing productivity and tool life due to an increase in the number of teeth simultaneously involved in the work, increasing the length of the cutting edges due to their constant displacement, i.e., the working contact of the vertices of the cutting teeth relative to the cutting surface.

CNC machines are currently widely used in engineering production for processing complex shaped surfaces, but their capabilities are already used by 10%... 20 %, for example, when processing pockets of complex shapes, the end and disk radii of the cutters move along an equidistant surface, which significantly reduces the processing performance and tool life.

The developed milling methods require a fairly complex kinematics of forming-three or more complex coordinated movements of the machine's organs, and the movements are usually non-linearly coordinated, i.e. the quantitative ratio between individual movements changes, constantly obeying nonlinear laws. The provision of such kinematics is possible only with the use of CNC machine tools with four or five simultaneously controlled coordinates.

2 Relay-vector control

To ensure high performance of the machine drive, it is advisable to use a PMSM motor that has a minimum moment of inertia of the Rotor. An electric drive based on a PMSM motor with a vector control system with a proportional P-speed controller (Fig.1), provides typical transients when starting and adjusting the output coordinates of the drive.

At the same time, to implement the proposed high-efficiency milling methods on CNC machines, it is advisable to perform an astatic system with zero static error and a reduction in over-regulation at the time of transition from the start-up mode to the steady-state mode. As a result of the research, a control system with a nonlinear speed controller (NLRS) and flexible speed feedback was developed, shown in Fig.3 with a dotted line.

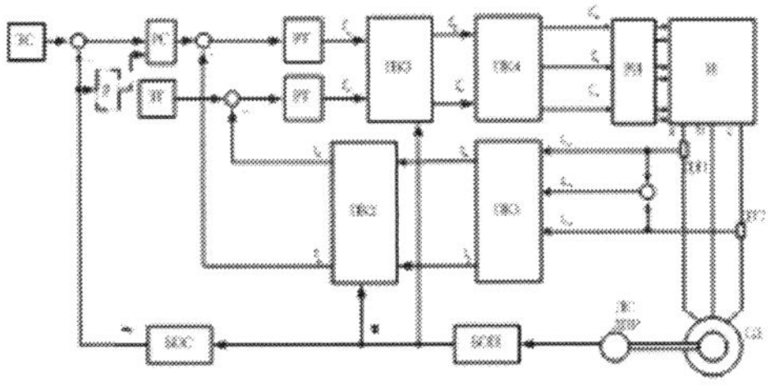

Fig. 3. Functional diagram of a PMSM electric drive.

3 Vector control with sinusoidal PWM

The results of mathematical modeling of transient processes in electric drives with standard and proposed control systems are shown in Fig. 4. the graphs show that the electric drive system with a nonlinear speed controller and flexible speed feedback is more suitable for use on CNC machines that implement the proposed innovative milling methods.

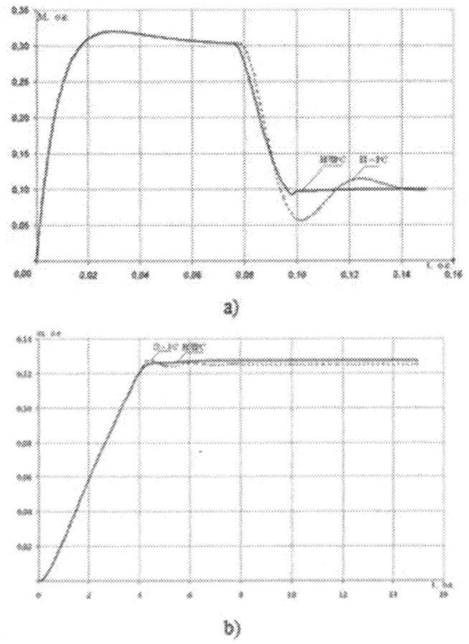

Fig. 4. Comparative results of transient modeling: a – M(t), b - ω(t) electric drive start mode.

Acknowledgements

The paper was powered by research grant RFBR 19-48-480001 «Development, investigation and optimization of energy-saving electrical and electrically driven automated systems for plasma, electrometal slag and induction technologies and units».

References

[1] S.K. Ambrosimov, A.A. Petrukhin, A method for processing curved surfaces, pat. no. 2167746 (RF), Byul., no. 15 (2001)

[2] S.K. Abrosimov, M.G. Sterkin, Method of processing shaped concave surfaces, pat. no. 22008502 (RF), Byul., no. 20 (2003)

[3] S.K. Ambrosimov, K.S. Ambrosimov, Method of screw black-and-white milling of shaped surfaces, pat. no. 2344023 (RF), Byul., no. 2 (2009)

[4] Y.I. Gracheva, O.V. Naumov, Evaluation criteria of contact group technical state concerning electrical appliances, International Journal of pharmacy and Technology **4**, 27084–27091 (2016)

[5] Y.I. Gracheva, N.V. Chernova, A.I. Fedotov, E.A. Fedotov, Local Fourier transformation application for mathematic modeling of synchronous machine valve actuator, Journal of engineering and applied sciences **11**, 1, 2939–2945 (2016)

[6] Y.I. Gracheva, O.V. Naumov, Operating Mode Influence on Probability Characteristics of Electric Devices, Journal of engineering and applied sciences **11**, 1, 2934–2938 (2016)

[7] Y.I. Gracheva, O.V. Naumov, Evaluation criteria of contact group technical state concerning electrical appliances, Journal of engineering and applied sciences **11**, 1, 26763–26770 (2016)

[8] Y.I. Gracheva, I.V. Ivshin, O.V. Naumov, A.N. Gorlov, D.D. Mitsu, Influence of resistance of contact units of switching devices on the electric power losses in shop networks of low tension, E3S Web of Conferences **124**, 02013, SES-2019, 1–4 (2019)

[9] Y.I. Gracheva, V.N. Petrov, N.V. Russova, D.V. Samuilov, G.P. Svintsov, I.V. Protosovitsky, On modeling on a single-winding forced valve electromagnet in a circuit with a ballast resistor, E3S Web of Conferences **124**, 02012, SES-2019, 5–9 (2019)

[10] Y.I. Gracheva, A.N. Alimova, Calculating Methods and Comparative Analysis of Losses of Active and Electric Energy in Low Voltage Devices, International Ural Conference on Electrical Power Engineering (UralCon), 361–367 (2019)

[11] Y.I. Gracheva, A.N. Alimova, Calculating Probability of Faultless work of Shop Nets with the Help of Coefficients of Ratio, International Russian Automation Conference (RusAutoCon), 3–6 (2019)

[12] V.N. Voyekov, V.N. Meshcheryakov, A vector control system of the valve electric drive on the basis of the independent inverter of tension with relay regulation of entrance current of the inverter and phase currents of the stator, Messenger YUURGU, Power series **17**, 2, 48–57 (2017)

[13] V.N. Meshcheryakov, V.V. Danilov, Sh.R. Khasanov, S. Valtchev, Minimization of the stator current in induction motor with defined load on the shaft by maintaining optimum absolute slip, Kazan, SES 2019, E3S Web of Conferences, 01036 (2019)

[14] V.N. Meshcheryakov, D.V. Lastochkin, Z.M. Shakurova, S. Valtchev, Kazan, SES 2019, E3S Web of Conferences, 01037 (2019)

The use of porous metals in the design of heat exchangers to increase the intensity of heat exchange

Natalia Rydalina[1,*], *Oleg* Stepanov[1], and *Elena* Antonova[1]

[1]Industrial University of Tyumen, Department of Industrial heat engineering, 625000, Tyumen, 38 Volodarskogo St., Russia

Abstract. Heat exchangers are widely used in heat supply systems. To increase the efficiency of heat supply systems, heat exchangers with porous metals are proposed to design. There was a test facility set up to study new types of heat exchangers. The countercurrent flow of heat carriers was activated in those heat exchangers. Freon moved through the heat exchanger pores, and water moved through the inner tubes. It should be noted that the porous materials in the heat exchangers differed in the coefficient of porosity. To be compared, one of the heat exchangers did not contain any porous material. The first test cycle proved the feasibility of using porous metals in heat exchange equipment. Afterwards, a simplified mathematical model of the heat exchanger was compiled. Such an analytical form makes a solution convenient for engineering calculations. Numerical calculations based on this model were compared with the experimental data. Heat transfer intensity of materials with different porosity was compared.

1 Introduction

Heat exchangers make one of the main elements in heat supply systems. Basically, heat supply systems use sectional counterflow water heaters or plate heat exchangers. Their heat transfer process is well-known, and it is difficult to obtain significant increase in the heat transfer coefficient. Recently, the problem of designing new heat exchangers (for example with porous metals) has become urgent [1-8].

Porous heat-conducting materials made of powdered aluminum, copper and other materials make it possible to create new efficient and compact heat exchangers [9-12].

There are special heat exchangers of the “liquid-gas” type with a phase transition of a heat carrier and without a phase transition. There are also various designs of such devices. For instance, they use a triple layer of porous metal adjacent to a heat transfer surface with a different degree of porosity. They also apply the method of transmitting a hot coolant with a phase transition through a porous metal in order to increase the heat transfer surface and its efficiency. Heat exchangers in the form of porous fins with different angles of inclination are also used. However, it should be noted that sometimes the units are expensive or it is hard to clear them out. [12-18]

Thus, nowadays heat exchangers with porous metals are being designed.

The Department of Industrial Heat Power Engineering in Tyumen Industrial University became a platform for setting up a test facility to study the effectiveness of using such materials.

The test facility is a stand where three heat exchangers with porous inserts and one heat exchanger without them are fixed. Heat exchangers are countercurrent, i.e. water flows through the central copper tubes, and freon moves through the pores of the inserts in the opposite direction. Water is pumped. The water temperature can be changed since the pump is followed by a boiler. Freon moves through a refrigeration circuit consisting of a compressor, an evaporator and a condenser. Thus, two circuits were set up in the test facility: one for water circulating, and the other for freon circulating.

The object of the research is the heat transfer process in heat exchangers of the "liquid - gas" type with a phase transition of the coolant and without a phase transition. The purpose of the work is an experimental and theoretical study of the heat transfer intensity in heat exchangers of the "water-gas" type with a phase transition of a coolant containing porous inserts of different porosity.

2 Methods

Heat exchangers are comprised of 19 copper tubes for water flow. Four cylindrical inserts of aluminum with different porosity are on them. The porosity of the inserts in the first heat exchanger is 0.4901, in the second heat exchanger the porosity of the inserts is 0.6169, in the third one the porosity of the inserts is 0.4739, in the fourth there is no porous insert. The inside of the heat exchanger with porous inserts is shown in Figure 1.

* Corresponding author: rydalinanv@tyuiu.ru

 https://doi.org/10.1051/e3sconf/202017801026

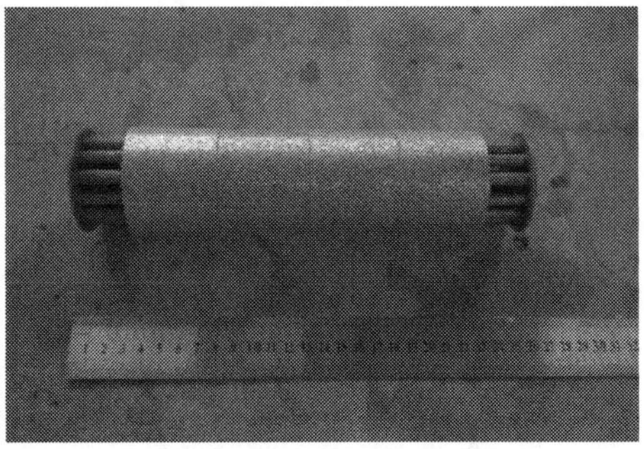

Fig. 1. Porous aluminum inserts.

Porous inserts are cylinders of porous aluminum with a height of 50 mm and a diameter of 49 mm. There are 19 holes with a diameter of 6 mm in each of the inserts. Water moves through the tubes installed into the holes. Freon moves through the pores of the insert.

A series of experiments was carried out at the set up. The first part of the experiment was as follows: water consumption was regulated by a ball valve. The temperature changes were measured at the inlet to and from the heat exchanger. The flow rate was also recorded. The average value of a series of experiments was found. The obtained values determined the freon mass flow rate and the amount of heat Q transferred from water to freon [19].

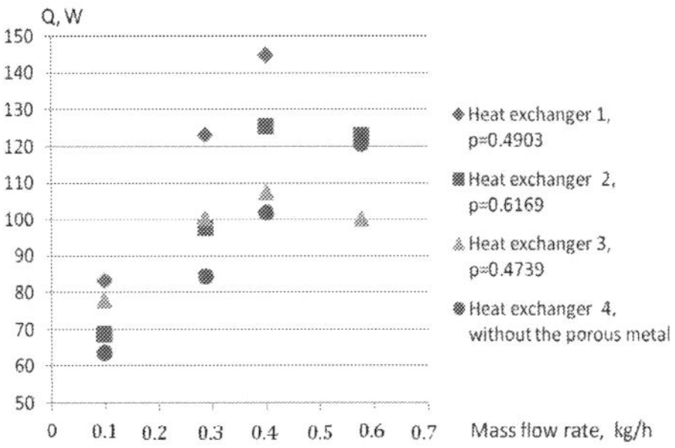

Fig. 2. Results of measurements of heat transfer intensity in heat exchangers with and without porous inserts.

The heat transfer intensity is higher for heat exchangers with porous metal inserts. Moreover, heat exchanger 4 without porous inserts has lower heat transfer intensity than other heat exchangers. The results are shown in fig.2. It can be concluded that heat exchangers with porous metals are more efficient than a conventional counterflow heat exchanger.

The next part of the experiment was carried out with an unvarying flow rate of the cooling heat carrier - water. Water of room temperature of about 20 - 22 ° C was cooled to a temperature of about 3 - 50 ° C in the circuit with freon running. The water temperature and other measurements were recorded every 2 minutes. As a result, the following data are presented in Fig. 3.

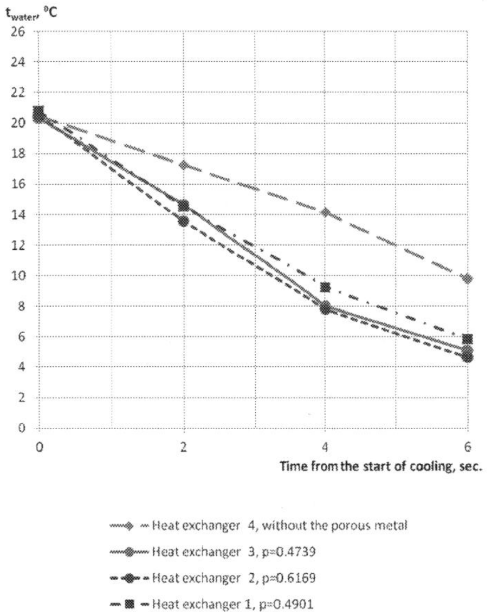

Fig. 3. The graph of temperature changes in water cooled by freon.

Summarizing the results of measurements throughout the experiment, we can register more efficient heat transfer in heat exchangers with porous aluminum. Besides, the most effective is the heat exchanger with the largest porosity among the tested heat exchangers.

Then we considered the heat transfer process inside a porous cylinder (porous insert) made of aluminum with a constant coefficient of thermal conductivity λ_c. The cylinder was well-insulated; therefore, we assumed that there was no heat exchange with the external environment. There are 19 copper tubes inside the porous cylinder. Chilled water flowes through them. An inlet temperature of water is tin (Fig. 4).

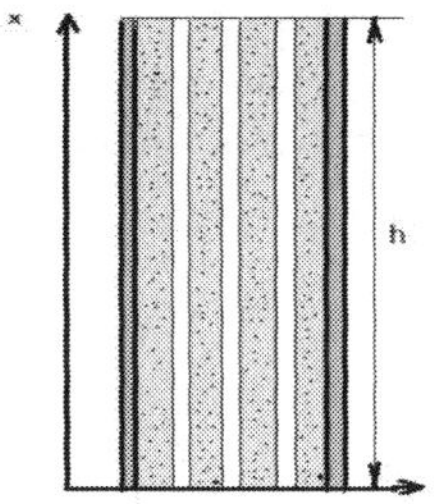

Fig. 4. Porous tube cylinder.

Freon is continuously supplied from the bottom up with a constant specific mass flow rate G_c.

Water is supplied to the tubes from top to bottom (counterflow circuit) with a constant specific mass flow rate G_B. The input water temperature was specified tin. While passing through the heat exchanger, the water is cooled and at the outlet has a temperature tout (found by calculation). The cooled water passes through the water circuit and gets into the heat exchanger with a temperature of t_{in1} = t_{out}. At the outlet of the heat exchanger the temperature will be tout1, so that the value of tin will be different in the next cycle. It is also found by calculation. Further on, the cycle is repeated manifold. So the temperature of the water circulating through the heat exchanger decreases continuously when the refrigeration circuit is operating. In order to justify such a change in temperature, it is necessary to find a dependence making it possible to determine the temperature of water at the outlet of the heat exchanger, and to compare the calculated data with the experimental ones.

The inner surface area of the tubes S is known. The volume of the porous insert V is known. The temperature of the freon at the inlet and outlet is t (0) = t_{c1} and t (h) = t_{c2}. The specific mass flow rates of freon and water (G_c and G_B) are also known. They are measured with the flow rate meter during the experiment.

Also, the specific heat of freon and water is set - c_{pc} and c_{pB}.

The porosity of the inserts p is considered as the ratio of the pore volume to the entire volume of the material. The porosity is considered uniform. Therefore, the cross section for the gas passage on a surface unit normal to the direction of the gas flow is f_w=p, and the cross section of the solid skeleton involved in heat transfer is f_c=1-f_w=1-p.

Here is the equation for the one-dimensional problem of cooling a porous body [20]:

$$\frac{d^2t}{dx^2} - \xi_c \frac{dt}{dx} = 0 \qquad (1)$$

where $\dfrac{G_c \cdot c_{pc}}{\lambda_c (1-p)} = \xi_c$.

The temperature field becomes two-dimensional if there are tubes with water, and Eq. (1) does not work.

This equation is proposed to supplement with a function of distributed heat sources (sinks). The function describes the process of heat transfer from a porous material through the walls of copper pipes to water, with some error though.

$$q = \frac{\alpha \cdot (t_в - t) \cdot S}{\lambda_в \cdot V_в} \qquad (2)$$

Consequently, the following equation succeeds:

$$\frac{d^2t}{dx^2} - \xi_c \frac{dt}{dx} + \frac{\alpha \cdot (t_в - t) \cdot S}{\lambda_в \cdot V_в} = 0 \qquad (3)$$

where V_B is the volume of porous inserts, α is the coefficient of heat transfer from the copper wall to water, t_B is the water temperature at a given value of x.

Single-valuedness conditions:

$$0 \le x \le h,\ t(0) = t_{c1},\ t(h) = t_{c2} \qquad (4)$$

We denote $\dfrac{\alpha \cdot (t_в - t) \cdot S}{\lambda_в \cdot V_в} = A$

Eq. (3) takes the form:

$$\frac{d^2t}{dx^2} - \xi_c \frac{dt}{dx} + A = 0 \qquad (5)$$

Boundary value problem Eq.(3) and (4) was solved by standard methods. The function of temperature changes of the porous metal in the heat exchanger along the x axis was obtained:

$$t = t_{c1} + \frac{A}{\xi_c} x + \left(e^{\xi_c x} - 1\right) \cdot \frac{t_{c2} - t_{c1} - \frac{A}{\xi_c} h}{e^{\xi_c h} - 1} \qquad (6)$$

We differentiate the obtained temperature function on variable x:

$$\frac{dt}{dx} = \frac{A}{\xi_c} x + \frac{(t_{c2} - t_{c1}) \cdot \xi_c - A \cdot h}{e^{\xi_c h} - 1} \cdot e^{\xi_c x} \qquad (7)$$

With function Eq. (7) known, we derive the formula for the heat flow rate:

$$q = -\lambda_c \cdot (1-p) \cdot \left(\frac{A}{\xi_c} x + \frac{(t_{c2} - t_{c1}) \cdot \xi_c - A \cdot h}{e^{\xi_c h} - 1} \cdot e^{\xi_c x} \right) \qquad (8)$$

The quantity ΔQ = Δq · S = (q (h) - q (0)) · S, where Q is the amount of heat. It is equal to the heat transferred from water to freon without heat loss. The volume of water in the tubes is V_T. This is the capacity of the tubes. The heat capacity and density of water are c_{pv} and ρ_v respectively. The initial water temperature is known and is equal to t_{out}. The outlet water temperature will be found at a known ΔQ by the formula:

$$t_{in} = t_{out} - \Delta t_T \qquad (9)$$

where

$$\Delta t_T = \frac{\Delta Q}{c_{pв} \cdot V_T \cdot \rho_в} \qquad (10)$$

3 Results and Discussion

The developed model was verified through a series of experiments evaluating its accuracy and correctness. The experiments were carried out with the following values of the main parameters: the diameter of the porous sample - d = 0.049m; the total length of the porous insert is h = 0.2 m; number of tubes - n = 19; the inner diameter of the tubes is d = 0.004 m; heat capacity of water - cf = 4187J / kg • K; thermal conductivity coefficient of aluminum - λs = 209.3 W / m · K. The same values of the main parameters were specified for theoretical calculations.

Comparative graphs below present theoretical calculations and the results of experimental measurements for each of the heat exchangers (Fig. 5-7).

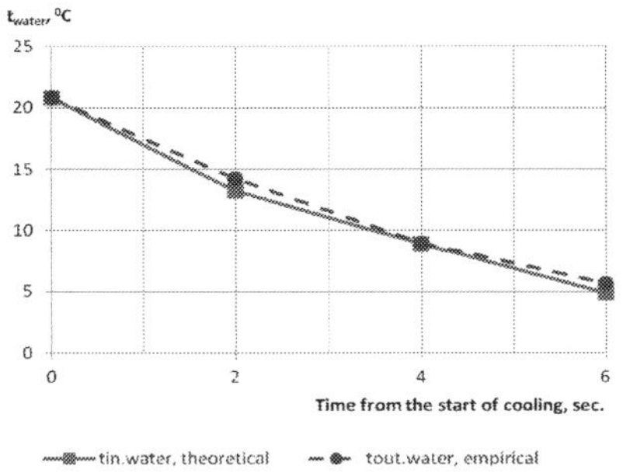

Fig. 5. Theoretical and empirical lines of temperature changes in water cooled by freon heat exchanger 1, p = 0.4901.

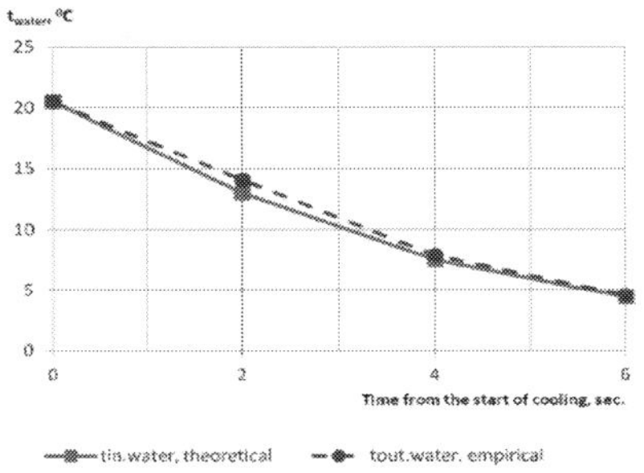

Fig. 6. Theoretical and empirical lines of temperature changes in water cooled by freon heat exchanger 2, p = 0.6169.

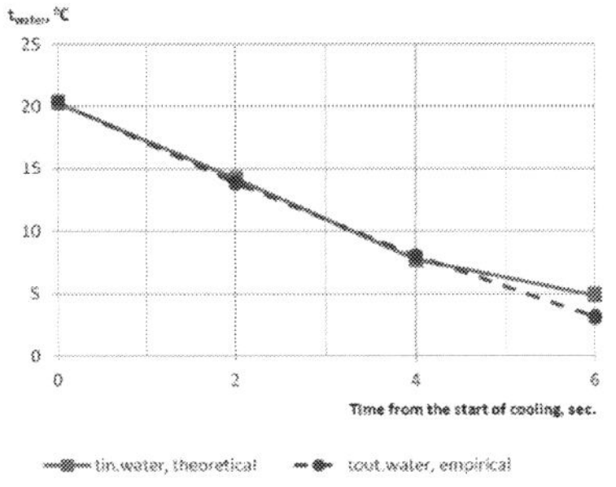

Fig. 7. Theoretical and empirical lines of temperature changes in water cooled by freon: heat exchanger 3, p = 0.4739.

The graphs show that the empirical and theoretical curves almost coincide. This verifies the model having small errors within acceptable values.

4 Conclusions

Based on the research and calculations, the following conclusions can be drawn:

1. The heat transfer rate is higher in heat exchangers with porous metal inserts compared to heat exchangers without porous metal inserts.
2. The heat transfer intensity is higher in the heat exchanger with the highest porosity p = 0.6169 compared to other tested heat exchangers with porous inserts.
3. A mathematical model has been developed to describe the heat transfer process in porous inserts of a counterflow heat exchanger.
4. A calculation equation has been derived to determine cooling degree of the hot heat carrier at the outlet of a heat exchanger, as well as determine temperature in the porous structure in any section of a heat exchanger.
5. Laboratory research proves the possibility of designing new heat exchangers with porous metals inside, they can be used in the power industry, namely in heat supply systems.

References

[1] S. Rashidi, M. Kashefi, K. Kim, O. Samimi-Abianeh, Appl. Enl **243**, 206–232 (2019)

[2] W. Tan, L. Saw, H. Thiam, J. Xuan, Z. Cai, M. Yew, Renew. and Sustain. Ener. Rev. **96**, 181–197 (2018)

[3] M. Sarkar, K. Velusamy, P. Munshi, O.P. Singh, Progr. in Nucl. Ener. **118**, 103142 (2020)

[4] K. Anirudh, S. Dhinakaran, Renew. Ener. **145**, 428–441 (2020)

[5] A. Robone, S. Kuruneru, M. Islam, S. Saha, Appl. Ther. Engin. **162**, 114232 (2019)

[6] C. Soto, C. García-Rosales, J. Echeberria, E. Platacis, A. Shisko, F. Muktepavela, M. Malo, Fus. Engin. and Des. **146**, 1983–1987 (2019)

[7] C. Soto, C. García-Rosales, J. Echeberria, E. Platacis, A. Shisko, F. Muktepavela, M. Malo, T. Hernández, Appl. Ther. Engin. **159**, 113980 (2019)

[8] A. Vikulin, N. Yaroslavtsev, V. Zemlyanaya, Therm. Engin. **66**, 397–401 (2019)

[9] X. Yang, J. Yu, Z. Guo, L. Jin, Y.-L. He, Appl. Ener. **239**, 142–156 (2019)

[10] A. Korsun, I. Merinov, V. Kharitonov, M. Bayaskhalanov, V. Chudanov, A. Aksenova, V. Pervichko, Therm. Engin. **66**, 225–234 (2019)

[11] M. Sarkar, K. Velusamy, P. Munshi, O. Singh, Progr. in Nucl. Ener. **118**, 103142 (2020)

[12] X. Lu, Y. Zhao, Internat. Jour. of Heat and Fluid Flow **80**, 108482 (2019)

[13] O. Stepanov, N. Rydalina, E. Antonova, I. Akhmetova, P. Zunino. Internat. Jour. of Civil Engin. and Technol. **10**, 2, 2130–2137 (2019)

[14] M. Liu, Z. Cui, Y. Li, Proc. Metallur. and Mater. Proces. Scien. **50**, 1908–1920 (2019)

[15] B. Buonomo, H. Celik, D. Ercole, O. Manca, M. Mobedi, Appl. Ther. Engin. **159**, 113980 (2019)

[16] K. Venkateshwar, S. Ebadi, H. Simha, S. Mahmud, Jour. of Therm. Scien. and Engin. Applicat. **11**, 041004 (2019)

[17] J. Qu, Q. Sun, H. Wang, D. Zhang, J. Yuan, Internat. Jour. of Heat and Mass Transfer **137**, 20–30 (2019)

[18] A. Akinshilo, Appl. Ther. Engin. **156**, 506–513 (2019)

[19] V. Davletbaev, N. Rydalina, E. Antonova, MATEC Web of Conf., **245**, 07002 (2018).

[20] V. Isachenko, V. Osipova, A. Sukomel, 3nd ed. Energy, Moscow (1975)

Natural Gas-Oxygen Combustion in a Super-Critical Carbon Dioxide Gas Turbine Combustor

Ivan Komarov[1,*], *Daria* Kharlamova[1], *Bulat* Makhmutov[1], *Sofia* Shabalova[1], and *Ilya* Kaplanovich[1]

[1]National Research University "Moscow Power Engineering Institute", Moscow, Russia

Abstract. The paper presents results for chemical kinetics of combustion process in the combustor of oxy-fuel cycle super-critical carbon dioxide gas turbine based on the Allam thermodynamic cycle. The work shows deviation of the normal flame propagation velocity for the case of transition from the traditional natural gas combustion in the N2 diluent environment to the combustion at super-high pressure up to 300 bar in CO2 diluent. The chemical kinetics parametric study involved the Chemkin code with the GRI-Mesh 3.0 kinetic mechanism. This mechanism provides good correspondence between calculation results and test data. The CO2 and N2 diluents temperature, pressure and contents influence the flame propagation velocity and the chemical kinetics parameters in the two gas turbine types. It is demonstrated that the CO2 diluent slows down chemical reactions stronger than the N2 one. The flame propagation velocity in carbon dioxide is four time smaller than in the N2 one. In the oxy-fuel cycle combustor a pressure increase reduces the flame propagation velocity. Increase of the CO2 content from 60 to 79% reduces the flame propagation velocity for 65% at atmospheric pressure and for 94% at super-critical pressure. An increase of the combustor inlet mixture temperature from 300 to 1100 K at super-critical pressure causes the flame propagation velocity increase for 94%. The flame propagation velocities compatible with the traditional gas turbines may be reached at the CO2 diluent content of the O2 + CO2 mixture in the active combustion zone must be below 50%.

1 Introduction

Transition to the environmentally friendly, low toxic and greenhouse emissions cycles is one of the most prospective directions for the hydro-carbon power production industry. The closed gas turbine cycles on super-critical carbon dioxide with organic fuel combustion in pure oxygen seem very attractive. In these terms, the Allam cycle (Figure 1) promises maximal production efficiency and nearly zero emissions [1-5].

Introduction of the Allam cycle requires solutions of a range of research problems that includes development of the power equipment analysis to operate with new heat carriers at non-traditional thermodynamic parameters.

The Allam cycle parameters of oxy-fuel combustion differ from the traditional gas turbines by the following parameters:

- The oxy-fuel cycle combustor pressure of 300 bar instead of the traditional gas turbines values below 30 bar;
- Methane is burned in the carbon dioxide environment instead of the nitrogen diluent in traditional gas turbines.

The common combustor analysis is based on the detailed description of methane combustion in air environment at pressures up to 30 bar. The analysis determines the complete fuel burnout conditions, oxygen-fuel mixture velocity at the combustor entry and the components residence time for the high temperature reaction zone. The oxy-fuel cycle combustor development involves methane oxygen combustion in the carbon dioxide environment at super-critical pressure up to 300 bar and investigation of the flow aerodynamic structure at the oxygen-fuel mixture formation and CH4-O2 combustion [6-9]. This paper discloses the chemical kinetics investigation that makes a base for further 3-D simulation of the CH4-O2-CO2 combustion and the results that are adequate to the actual processes. Especially important is the transition from N2 to CO2 combined with the combustor inlet parameters increase that is related to the normal flame propagation velocity. This parameter depends upon temperature, pressure and the components contents but is not related to the combustor burner design, or the flame tube structures. The analysis results provide a conclusion on the necessity to change the combustor inlet oxygen-fuel mixture velocity and the diluent content (γ – CO2 / (CO2 + O2)) that provide the flame stability and complete burn-out. The combustor envelope dimensions are similar to the traditional gas turbines [10-12].

* Corresponding author: komarov_vanya@bk.ru

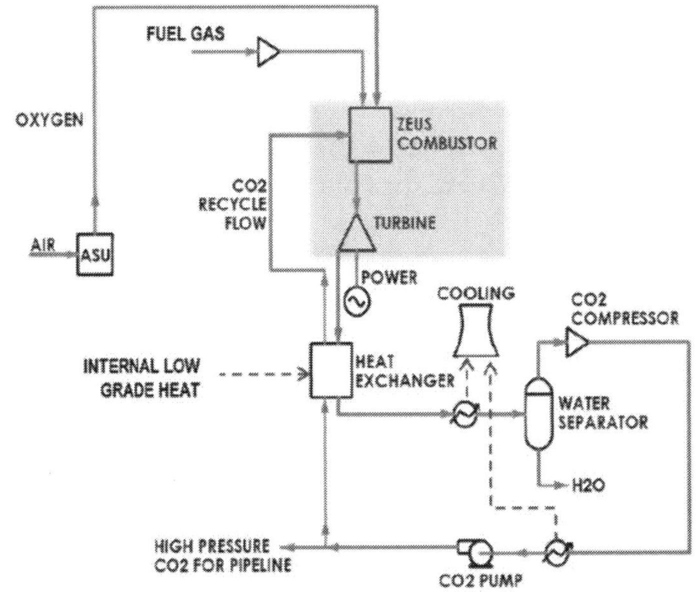

Fig. 1. Flow chart of Allam cycle on natural gas fuel.

2 Materials and Methods

The parametric study on influence of the diluent type and concentration upon the CH4-O2- CO2 and CH4-O2-N2 mixture combustion kinetic parameters was carried out with the Chemkin computer simulation code.

In the study were selected the following detailed kinetic mechanisms that allow calculations with numerous mixture contents and initial parameters. The following two kinetic mechanisms were considered [13-17]:

- Mechanism #1: GRI-Mesh 3.0 developed for the natural gas combustion including the chemical incomplete combustion. The mechanism basic kinetics involves 53 intermediate reaction products and 325 convertible chemical reactions of the methane transformation into complete combustion products;
- Mechanism #2: USC II involves 325 components and 725 convertible reactions applied to a broad spectrum of combustion scripts. The USC II model was developed on the combustion mechanism GRI-Mesh 3.0, the hydrogen and carbon monoxide combustion, the fuel combustion model C3 and the complex model of ethylene and acetylene combustion. The both models include the reaction velocity data from the recent actual investigations.

The results were verified by comparison with the test data of the papers [18-20].

The results show that at atmospheric pressure the

Table 1. Summarizes input data for the chemical kinetics parametric studies on the N2 and CO2 diluents at different pressures, diluent contents and reactor inlet temperatures.

Analysis series	Pressure, bar	Diluent type and mass content, γ	Gas mixture temperature upstream combustion reactor, K	Oxidizer excess coefficient
Diluent CO2				
1	1	0.60	300	0.7-1.4
2	1	0.64	300	0.7-1.4
3	1	0.68	300	0.7-1.4
4	1	0.79	300	0.7-1.4
5	20-300	0.60	300	1.0
6	20-300	0.64	300	1.0
7	20-300	0.71	300	1.0
8	20-300	0.74	300	1.0
9	20-300	0.79	300-1200	1.0
Diluent N2				
10	1	0.79	300	0.7-1.4
11	20-300	0.79	300-1200	1.0

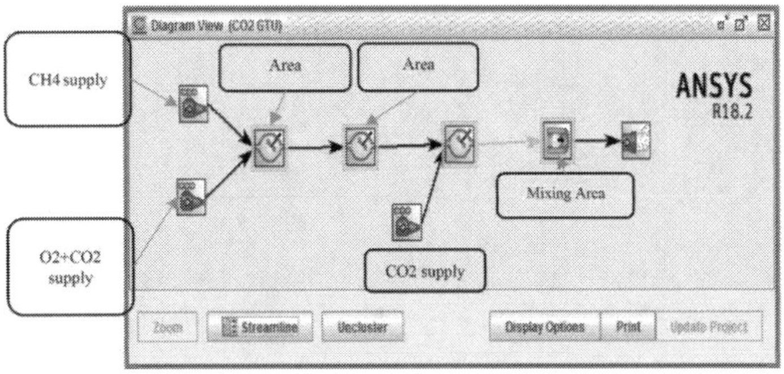

Fig. 2. Combustor structure comprised of the Chemkin reactors.

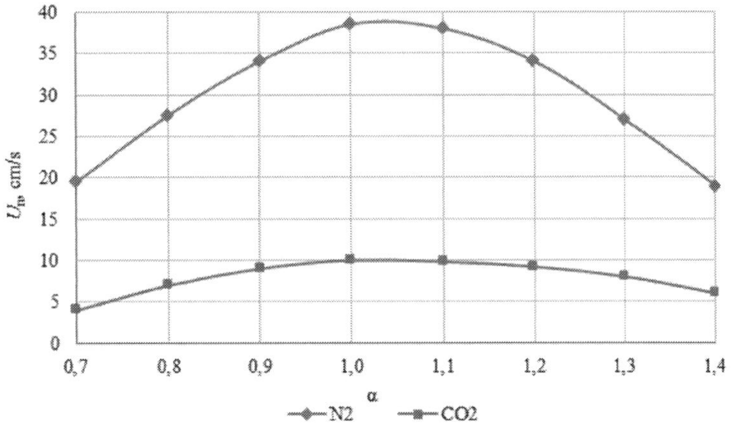

Fig. 3. Normal flame propagation velocity in the N2 and CO2 diluents (γ = 0.79) at atmospheric pressure and gas temperature 300 K.

both detailed mechanisms demonstrate moderate deviations from the test data. At high pressure the USC II shows overrated estimations of the flame propagation velocity and the GRI-Mesh 3.0 underrates it. The GRI-Mesh 3.0 calculation error is below 3.5% and the USC II One is above 6%. The further calculations were carried out with the GRI-Mesh 3.0.

Figure 2 shows the combustor structure model comprised of a set of elementary reactors. The model includes the staged supply of CO2 diluent into the combustor. The mixture is ignited at the CO2 volumetric content equal to the value γ. The second CO2 flow cools the combustor walls, the third flow supplied to the end of active combustion zone mixes with the main flow as to reach the target temperature.

3 Results

Figures 3-6 show the normal flame propagation velocities in the N2 and CO2 environments at different initial values of oxygen-fuel mixture temperature and pressure, oxidizer excess and the diluent contents in diluent-oxygen mixture.

The calculation results show that the carbon dioxide is a stronger chemical reactions inhibitor than nitrogen. The maximal flame surface propagation velocity is 10 cm/s, and with the N2 diluent at other conditions being equal it is almost 40 cm/s. In average at the oxidizer excess coefficient from 0.7 to 1.4 the normal flame propagation velocity in the carbon dioxide environment is four times smaller.

Reduction of the CO2 content in its mixture with oxidizer at γ reduction for 0.1 increases the flame propagation velocity for 2 cm/s in average. The maximal increase occurs at the CO2 content reduction from 0.74 to 0.6. In the whole range of oxidizer excess coefficients the maximal flow propagation velocity occurs at α = 1.

Pressure increase reduces the normal flame propagation velocity. The maximal reduction occurs at moderate pressure values below 70 bar. The further pressure increase also reduces U_n, but in a remarkably smaller degree.

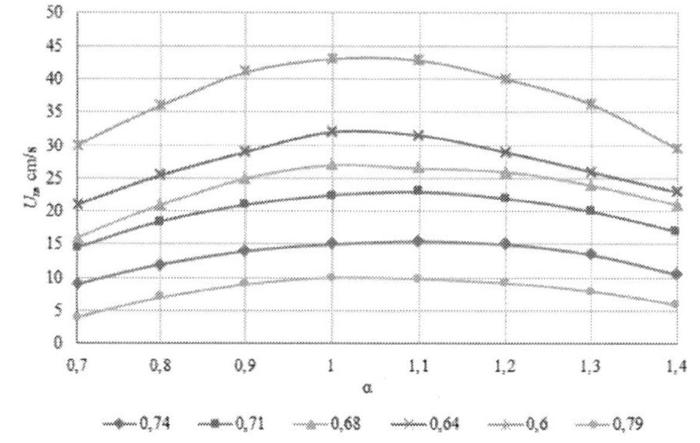

Fig. 4. Normal flame propagation velocity at different γ (CO2) and oxidizer excess values at atmospheric pressure and gas temperature 300 K.

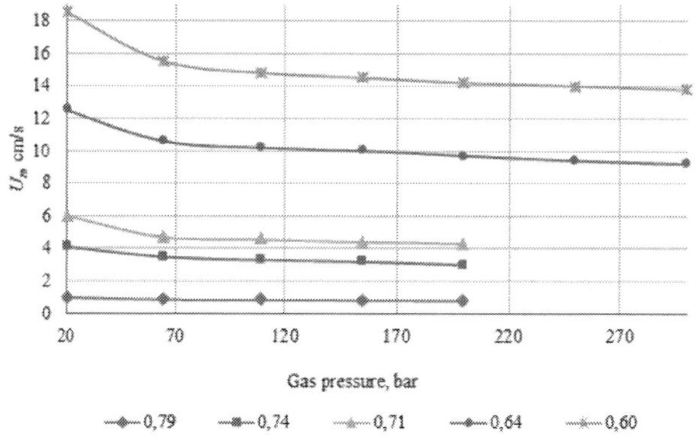

Fig. 5. Normal flame propagation velocity at different pressure and γ (CO2) values at gas temperature 300 K.

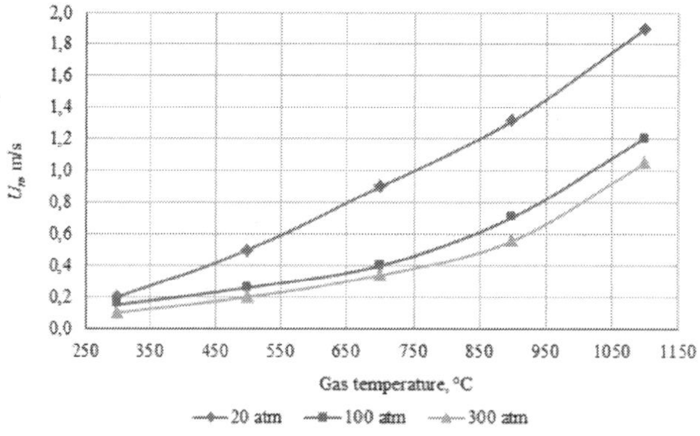

Fig. 6. Normal flame propagation velocity at in the CO2 diluent at different combustor inlet gas pressure and temperature (γ = 0.71).

According to the **Ошибка! Источник ссылки не найден.** results increase of the gas mixture temperature is followed by the normal flame propagation velocity growth at all pressure values, and the growth rate is non-linear. The temperature increase from 900 to 1100°C increases the velocity for average 22 cm/s. At lower mixture temperatures the similar 200°C temperature increase causes only a 8-10 cm/s increase. At the initial mixture temperature T_0 convection and radiation heat transfer heats the mixture up to the ignition temperature T_i that begins exothermal reactions and the mixture temperature grows. It is obvious that when the mixture

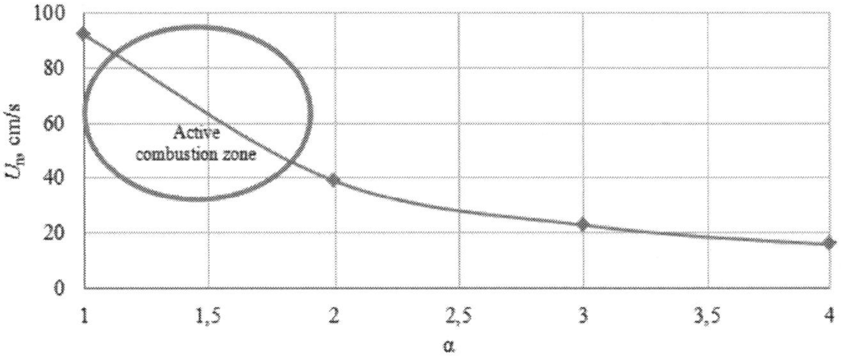

a) Gas pressure 30 bar, gas temperature 866 K, diluent – N2 (γ = 0.79)

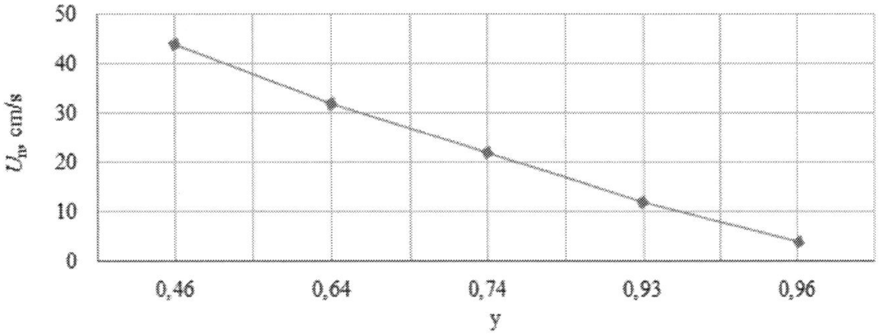

b) Gas pressure 300 bar, gas temperature 933 K, diluent – CO2

Fig. 7. Normal flame distribution velocity in the traditional and oxy-fuel cycle gas turbines.

temperature T_0 reaches the ignition border it causes a faster mixture ignition.

Ошибка! Источник ссылки не найден.a shows the normal flame propagation velocity for a traditional gas turbines combustor, pressure 30 atm and T_{in} = 866 K that corresponds to the compressor exit at compression up to 30 bar. **Ошибка! Источник ссылки не найден.**b demonstrates a similar velocity relation with the CO2 diluent rate γ in the oxy-fuel cycle combustor.

The **Ошибка! Источник ссылки не найден.**b results show that the flame propagation velocity in oxy-fuel cycle combustor is near to U_n of the traditional gas turbines if the CO2 diluent content in the mixture is below 0.5.

4 Discussion

The main research results are the following:

- The CO2 diluent is a stronger chemical reactions inhibitor than N2. At other conditions equal the normal flame propagation velocity in CO2 is four times smaller than the N2 diluent one, the 75% velocity reduction from 39 to 10 cm/s at the diluent content in oxidizer 79%;
- The combustor pressure increase reduces the flame propagation velocity, at the pressure increase from 20 atm up to super-critical pressure of 200 atm the reduction is 36%;
- A change in the CO2 concentration of the diluent in the mixture with the oxidizing agent in the range of the volume content from 60 to 79% leads to a decrease in the flame propagation rate by 65% (from 42.5 to 10 cm/s) at atmospheric pressure and by 94% at supercritical pressure (from 14.2 to 0.8 cm/s).

In relation with the ballasting gas content supplied together with oxidizer at CO2-O2 diluent mass contents from 0.945 to 0.460 the following parameters are determined:

- The normal flame propagation velocity verifies from 7 to 45 cm/s. A 1% reduction of the diluent content in CO2-O2 mixture increases the U_n velocity for 1.4%;
- The U_n amount for the natural gas combustion in the O2 and CO2 mixture is compatible with the traditional gas turbine combustors at the γ values from 0.47 to 0.65. This γ range corresponds to the CO2 content in the O2 + CO2 mixture in the active combustion zone below 15% of the total mass.

5 Conclusion

The investigation results show the problems concerned with transition to the new diluent and higher combustor pressure. In the oxy-fuel cycle conditions CO2 remarkably slows the combustion process. Stable efficient methane combustion in the oxygen and carbon dioxide mixture may be reached under the following conditions:

- Carbon dioxide supply into the active combustion zone must be below 20% of its total mass flow. This flow must be split into two parts, one as a mixture with oxidizer with the γ of CO2-O2 mixture below 0.5, and second along the combustor walls for its cooling. This CO2 flow split allows application of the traditional gas turbine combustor recommendations;
- The gas-fuel mixture velocity in the oxy-fuel cycle combustor must be below 20-25 m/s;

Further 3-D combustion simulation may employ the GRI-Mesh 3.0 kinetic mechanism, its calculation results fit test results with errors below 3.5%.

Acknowledgemetns

This study conducted by Moscow Power Engineering Institute was financially supported by the Ministry of Science and Higher Education of the Russian Federation (project No. FSWF-2020-0020).

References

[1] F. Weizhong, The research on design and technology of new high efficiency supercritical unit – a kind of cross-compound steam turbine generator unit in a manner of elevated and conventional layout, Proceedings of the 2-nd IEA CCC Workshop on Advanced Ultras-Supercritical Coal-Fired Power Plants, Rome, Italy (2014)

[2] S. Martin, B. Forrest, N. Rafati, X. Lu, J. Fetvedt, M. McGroddy, D. Freed, Progress update on the Allam cycle: commercialization of NET Power and the NET Power Demonstration Facility, Proceeedings of the 14-th Greenhouse Gas Control Technologies Conference, Melbourne, Australia (2018)

[3] A.B. Amato, P.D. Hudak, D. Carlo, D. Noble, J. Scarborough, T. Seitzman, T. Lieuwen, Methane oxy-combustion for low CO2 cycles: Blowoff measurements and analysis, Journal of Engineering for Gas Turbines and Power **133**, 6 (2011)

[4] S. Suzuki, Y. Iwai, M. Itoh, Y. Morisawa, P. Jain, Y. Kobayashi, High pressure combustion test of gas turbine combustor for 50MWth supercritical CO2 demonstration power plant on Allam cycle, Proceedings of the International Gas Turbine Congress, Tokyo, Japan (2019)

[5] A. Alenezi, J.S. Kapat, Thermodynamic analysis of CO2 Allam cycle for concentrated solar power complemented with oxy-combustion, Proceedings of AIAA Propulsion and Energy Forum, Indianapolis, Indiana, USA (2019)

[6] I.L. Polezhaev, V. Yu, Normal flame propagation speed and analysis of the influence of system parameters on it, Thermophysics of high temperatures **43**, 6, 933–942 (2005)

[7] E.L. Petersen, M. Röhrig, D.F. Davidson, R.K. Hanson, C.T. Bowman, High-pressure methane oxidation behind reflected shock waves, Proceedings of the Symposium (International) on Combustion **26**, 1, 799–806 (1996)

[8] A.H. Lefebvre, *Gas turbine combustion: Alternative fuels and emissions* (Boca Raton: CRC Press, 557, 2010)

[9] G. Kim, Y. Kim, Y.J. Joo, Conditional moment closure for modeling combustion processes and structure of oxy-natural gas flame, *Energy & Fuels* **23**, 9, 4370–4377 (2009)

[10] B.T. Chorpening, K.H. Casleton, G.A. Richards, M. Woike, B. Willis, Stoichiometric oxy-fuel combustion for power cycles with CO2 sequestration, Proceedings of the 3-rd Joint Meeting of the US Sections of the Combustion Institute, Chicago, Illinois, USA (2003)

[11] S. Coogan, X. Gao, A. McClung, W. Sun, Evaluation of kinetic mechanisms for direct fired supercritical oxy-combustion of natural gas, Proceedings of the Turbomachinery Technical Conference and Exposition, Seoul, South Korea (2016)

[12] V.P. Zhukov, V.A. Sechenov, A.Y. Starikovskii, Spontaneous ignition of methane-air mixtures in a wide range of pressures, Combustion, Explosion and Shock Waves **39**, 5, 487–495 (2003)

[13] M.D. Smooke, *Reduced kinetic mechanisms and asymptotic approximations for methane-air flames: A topical volume* (Berlin: Springer-Verlag, 245, 1991)

[14] N. Slavinskaya, M. Braun-Unkhoff, P. Frank, Reduced reaction mechanisms for methane and syngas combustion in gas turbines, Journal of Engineering for Gas Turbines and Power **130**, 2, 021504 (2008)

[15] H. Wang, X. You, A.V. Joshi, S.G. Davis, A. Laskin, F. Egolfopoulos, C.K. Law, USC Mech Version II, High-temperature combustion reaction model of H2/CO/C1-C4 compounds (2007) [Electronic resource]. Available at: http://ignis.usc.edu/Mechanisms/USC-Mech%20II/ USC_Mech%20II.htm

[16] D. Healy, D.M. Kalitan, C.J. Aul, E.L. Petersen, G. Bourque, H.J. Curran, Oxidation of C1-C5 alkane quinternary natural gas mixtures at high pressures, Energy & Fuels **24**, 3, 1521–1528 (2010)

[17] X. Gao, S. Yang, W. Sun, A global pathway selection algorithm for the reduction of detailed chemical kinetic mechanisms, Combustion and Flame, 167, 238–247 (2016)

[18] A. Frassoldati, T. Faravelli, E. Ranzi, The ignition, combustion and flame structure of carbon monoxide/hydrogen mixtures, Note 1:

Detailed kinetic modeling of syngas combustion also in presence of nitrogen compounds, International Journal of Hydrogen Energy **32**, 15, 3471–3485 (2007)

[19] W.P. Fan, Y. Gao, Y.M. Zhang, C.L. Chow, W.K. Chow, Experimental studies and modeling on flame velocity in turbulent deflagration in an open tube, Process Safety and Environmental Protection **129**, 291–307 (2019)

[20] G. Gai, S. Kudriakov, B. Rogg, A. Hadjadj, E. Studer, O. Thomine, Numerical study on laminar flame velocity of hydrogen-air combustion under water spray effects, International Journal of Hydrogen Energy **44**, 31, 17015–17029 (2019)

The optimal gas dynamic design system of industrial centrifugal compressors based on Universal modeling method

Yuri Galerkin[1], *Aleksey* Rekstin[1], *Aleksandr* Drozdov[1,*], *Kristina* Soldatova[1], *Olga* Solovyeva[1], and *Elena* Popova[1]

[1]Peter the Great St.Petersburg Polytechnic University St.Petersburg, Polytechnicheskaya, 29, Russia

Abstract. We present the modern stage of development of Universal Modeling Method, a complex of mathematical models and software for optimal design of centrifugal compressors - a new version of simplified mathematical model of efficiency and new software for variation calculations of multistage compressors. Based on this numerical calculation complex we have created a method for preliminary design of flow paths of stages - 2D and 3D impellers, vane and vaneless diffusers and return channels. The new, 9th version of its mathematical model features a quasi-3D calculation method of 2D and 3D impellers design, a new principle of pressure characteristic calculation, a new model of vaneless diffusers and much more. "Digital twin of a centrifugal compressor stage" and "3D compressor" software create digital descriptions of the flow part and its solid model ("digital twin").

1 Introduction

Despite the successes of gas dynamic theory and computational gas dynamics, the issue of creating a reliable design of the centrifugal compressor flow path corresponding to the requirements of technical specifications for mass flow rate and pressure ratio at the highest possible efficiency has not been fully resolved.

Since an analytical description of the workflow is impossible, the search for the optimal solution can be done only by comparing the flow path options. For each of these options, it is necessary to calculate the efficiency and pressure ratio for a design flow. Russian and foreign methods of mathematical modeling [1 - 6] more or less correspond to the task.

The Polytechnic Scientific School of Compressor Engineering works with the Universal Modeling Method [7 - 9]. Using the 4th version of the Method, new-generation gas pipe line compressors were created as part of the Ural-Gazprom conversion program [9] and dozens of other successful projects were implemented [10]. The mathematical models for calculating the efficiency are significantly improved in the models of the 5th version [11] and 6th version [12], and they are used in the design of new compressors [13, 14].

The disadvantage of the Method was the relative complexity of the design process. Design was carried out by consistent application of several indirectly related programs. During the process of optimization, the user needed to make a number of decisions requiring a lot of experience. Additionally, a solid model of a flow path was not created.

The aim of the research is to minimize the number of user interactions in the process of optimal design, improve the quality of mathematical models and create a closed system of computer programs. At the end of the design process, the developed end-to-end system produces a series of compressor gas-dynamic characteristics, and a solid model of the flow path. In order to improve the quality and reliability of the design, two main steps of design have been significantly improved - preliminary design and final design with the search for the optimal solution. The developed new design techniques and the improvement of mathematical models are based on computational and experimental researches.

2 Variant calculation of a multistage compressor. New simplified mathematical model of efficiency

Compliance with compressor parameters established by specification can be ensured with a virtually uncountable number of compressor variants. In general, designer can change: number of shafts on which the impellers are installed; number of stages on the shafts; shaft RPM; number of intermediate cooling; impellers diameters; types of impellers (2D or 3D); types of diffusers (vane or vaneless).

The same mass flow passes through all stages of the compressor, the volume flow is determined by the change in gas density ρ_0^*. One of the two main parameters of the design stage [0] - the design flow rate coefficient Φ_{des} depends on the impeller diameter D_2 and the RPM n. The second main design parameter is the design loading factor $\psi_{T\,des}$. The compressor pressure ratio π^* is equal to the product of the pressure

* Corresponding author: A_drozdi@mail.ru

ratios of the stages (taking into account the pressure loss in the following gas cooler).

The idea of a mathematical model for evaluating the stages efficiency depending on the design parameters Φ_{des}, ψ_{Tdes}, structural constraint $\bar{D}_{h\,min}$ and relative similarity criteria was formulated and implemented by E. Popova [0]. The authors of [0] proposed and verified a modern version of the equations system:

$$\eta^*_{des} = 1 - X_1 K_\Phi K_{\psi_T} K_{D_h} K_{M_u} + \Delta\eta_{VD} - \Delta\eta_{inl} \qquad (1)$$

where η^* is efficiency, the coefficients K take into account the influence of the above parameters.

In expanded form, the equation system appears as follows:

$$\begin{aligned}\eta^*_{des} = 1 - X_1\left[1 + X_5\left(\Phi_{des} - 0.085\right)^{X_6}\left(1 + X_7\bar{D}_h^{X_8}\right)\right]\times \\ \times\left[1 + X_9\left(\psi_{Tdes} - 0.5\right)^{X_{10}}\right]\times\left[1 + X_{11}\bar{D}_h^{X_{12}}\left(1 + \Phi_{des}\right)^{X_{13}}\right]\times \\ \times\left[1 + X_{14}(M_u - 0.5)^{X_{15}}(\Phi - 0.01)^{X_{16}}\right] + \\ + X_{19}\left(\psi_{Tdes} - 0.50\right)^{X_{20}} - X_{17}\frac{\Phi_{des}^{X_{18}}}{\psi_{Tdes}},\end{aligned} \qquad (2)$$

where $\bar{D}_h$ is relative hub ratio.

The availability of a method for assessing the efficiency of a stage opens up the possibility of calculating the efficiency of compressors taking into account losses in the inlet nozzle and intermediate coolers.

The free choice of RPM does not change the essence of the presented variant calculation, it only increases the number of compared variants.

For a compressor with a more complex scheme was created the PPMVTSK-G9R program, that provides an opportunity to compare variants with a different number of shafts, stages, loading factors and flow rate coefficients, the number and location of gas coolers. Examples of programs application for variant calculation are presented in [0].

After selecting the desired compressor variant, software handles the preliminary design - calculation of the compressor flow path sizes for calculating the gas-dynamic characteristics by the Universal Modeling method.

3 Preliminary impeller design

In previous versions of the Method, the preliminary design of the stage flow path was carried out based on the principles of a Polytechnic scientific school [0] taking into account the recommendations of leading Russian scientists [0 - 0]. The experiments of the second half of the last century, on which design recommendations are based, do not reflect the complexity of modern compressor technology.

Since the Universal Modeling Method calculates the characteristics of any flow paths, automatically or interactively, the optimal design program is formally able to improve any preliminary project. However, the ability of a mathematical model to find the right solution should not be overestimated.

The most important part of the impeller design is an accurate estimate of the flow coefficient, at which the impeller blades, the diffuser vanes, and return channel (RCh) vanes have non-incidence inlet flow. It is also necessary to determine the loading factor with maximum accuracy.

The authors set the aim to solve these problems with the maximum approximation to reality already at the step of preliminary design, at which user participation is not necessary. This reduces the difference between the final project and the preliminary project.

To this end, 124 2D and 3D impellers with design parameters in the range $\Phi_{des} = 0.015 - 0.15$, $\psi_{T\,des} = 0.40 - 0.70$, $\bar{D}_h = 0.25 - 0.40$ were optimally designed by the Universal Modeling Method, and their blade rows were optimized based on the analysis of inviscid flow velocity diagrams. Both the main dimensions and the shape of the blades for each of the impellers are compared in many ways, so the proposed new preliminary design technique is based on the calculations of many hundreds of impeller variants.

The dimensions of the impeller entrance according to tradition, confirmed by successful practice, are selected on the basis of minimizing the relative velocity.

A well-tested recommendation - non-incidence flow around the blades at a set design mode - ended up ineffective in the case of low-flow impellers [0].

Comparison of dozens of variants of low-flow rate 2D impellers allowed to propose the following methodology for selecting sizes: inlet blade angle should be at least 23°; the inlet blade height corresponds to such a positive incidence angle, at which the peak of velocity at the leading edge is dosed - the outlet blade angle should be at least 16°, etc.

The preliminary design of medium-flow rate 2D impellers and high-flow rate 3D impellers basically corresponds to that presented in the monograph by Yu. B. Galerkin [0]. The results of researches by CFD methods from [0, 0] were used. The main novelty of the preliminary design is the exact recommendations on choosing the sizes of the flow path needed to calculate the characteristics according to the efficiency model of the Universal Modeling Method.

For sizes $\bar{D}_{1s}$ (blade inlet shroud diameter), $\bar{D}_0$, $\bar{D}_h$, $\bar{L}_m$ (relative axial length of impeller), $\bar{D}_{1h}$ (blade inlet hub diameter), $\bar{R}_h$ (relative hub radius of curvature), $\bar{R}_s$ (relative shroud radius of curvature), l_{mbl} (length of a blade in meridional plane), $\bar{b}_2$ (relative outlet blade height), formulas that approximate the results of a computational research are recommended for calculating the inlet blade angles at the shroud and at the hub $\beta_{bl1s}{}^0$, $\beta_{bl1h}{}^0$, the outlet blade angle $\beta_{bl2}{}^0$, the number of blades z_{imp} and the shape factors of the blades in the 3DM.023 program.

4 Preliminary design of stator stage elements

Stator elements of centrifugal compressor stages consist of vane and vaneless diffusers, return channels.

Modern vaneless diffusers have a narrowed initial section to increase the flow angle to the main part with a constant width. If the flow angle at the exit of the impeller is sufficient for the VLD to work well, a narrowed initial section is not needed. Preliminary design subjects are relative width $\bar{b}_3 = b_3 / D_2$ (where b_3 is VLD width) and radial extent $\bar{D}_4 = D_4 / D_2$ (where D_4 is VLD outlet diameter). In the proposed method of preliminary design, the choice of the relative width is linked for the first time not with the parameters of the design mode, but with ensuring a non separate flow at the pre-surge mode.

4.1 Assessment of the surge limit

In the practice of calculations, it is generally accepted to understand the surge line as a regime with a maximum pressure ratio [0].

The surge line corresponds to a flow coefficient at which the head polytropic coefficient $h_p = h_T - h_w$ (where h_p is polytropic head; h_T is Euler head; h_w is loss of head) reaches its maximum:

$$\frac{d(\psi_p)}{d\Phi} = 0 \tag{3}$$

The linear dependence of the loading factor on the relative flow coefficient was analyzed in publications [0, 0]. Within the practically used part of the compressors dimensional characteristics, the dependence of the work coefficient on the flow rate is also linear [0]. For preliminary design purposes, it is acceptable to consider the function $\psi_T = f(\Phi)$ to be linear as well, although the dependence between Φ and φ_2 is not quite linear (where φ_2 is flow rate coefficient at an impeller exit). After transformations and a search for Φ_{cr} corresponding to the maximum value of $\psi_{p\max}$:

$$\frac{\Phi_{cr}}{\Phi_{des}} = 1 - \left[\frac{\frac{\psi_{T0}}{\psi_{T\,des}} - 1}{17\left(1 - \eta_{hyd\,des}\right)}\right]^{0,5} \tag{4}$$

4.2 Choosing the width of the VLD main part

After calculating the ratio of Φ_{cr} / Φ_{des}, it is necessary to determine how much the exit flow angle from the impeller α_{2des} (where α_{2des} is the flow angle at design flow rate) decreases at the surge line. The obtained value should be compared with the angle α_{3sb} (where α_{3sb} is the flow angle at VLD inlet) at which separation occurs in the main part of the VLD ("separation boundary"), and choose $\bar{b}_3$ at which $\alpha_{3cr} = \alpha_{3sb}$.

The strong dependence of the critical flow angle at which separation begins on the relative width of the VLD is obvious. Processing the results of the calculation experiment established the following dependence of the tangent of the flow angle at the separation boundary in the VLD, i.e. on the surge line:

$$tg\alpha_{3sb} = 0.0875 + 3.5\bar{b}_3 \tag{5}$$

The relative width of the VLD at which $\alpha_{2cr} = \alpha_{2sb}$:

$$\bar{b}_3 = \frac{-0.0875 + \sqrt{0.00766 + 14 tg\alpha_{2sb}\bar{b}_2}}{7} \tag{6}$$

4.3 VLD radial extent

Even with an inlet flow angle 10°, in a narrow and long diffuser static pressure increases [0, 0]. Obviously, a rational ratio D_4 / D_2 should be sought in conjunction with the subsequent element of the flow path. In [0] the problem is discussed on the basis of the expression for the loss of efficiency in the stator elements of the stage. Below this expression is transformed like this:

$$\zeta_{SE} = \zeta_{VLD} + \frac{\zeta_{RCh}}{\bar{D}_4^2}\left[ctg^2\alpha_2 + \left(\frac{tg\alpha_2}{b_3 / b_2}\right)^2\right] = f\left(\bar{D}_4\right) \to \min \tag{7}$$

where ζ is the loss coefficient.

With an increase in the radial extent of VLD, the loss coefficient increases, but the kinetic energy at the RCh inlet or scroll becomes smaller. In a longer RCh friction losses are higher. Parameters analysis of the series of model stages 21CV developed by K.V. Soldatova [0] and the calculation research by the Universal Modeling Method made it possible to propose the following equation for the preliminary design of VLD in stages with RCh:

$$\bar{D}_4 = 1.45 + 1.4\left(\Phi_{des} - 0.015\right)^{0.5} \tag{8}$$

For stages with scrolls, the radial extent of the VLD is limited by design considerations.

4.4 Preliminary RCh design

In the formation of the design method, the results, generalized in the monograph [0] and a number of additional calculations are used.

Characteristic features of RCh vanes by the Universal Modeling Method: the shape of the vane midline is a circle arc; the shape of the vane profiles is a two-arc profile with a maximum thickness in the middle of the profile equal to $\delta_{\max} / D_2 = 0.040$; outlet vane angle $\alpha_{v6} = 95°$; the number of vanes is even; the number of straightening vanes is half the number of main vanes.

In [0] the characteristics of the efficiency and work coefficient of one of the 20CE stages series compares, obtained as a result of repeated tests of different stage versions. On average, the "24 vanes with double-arc profile" vane row increases the stage efficiency by 1.2%.

According to the experience of model stages series designing [0, 0] and CFD calculations [0], RCh vanes with an inlet vane angle of less than 25° are ineffective. Empirical formulas are recommended for the range of stages $\Phi_{des} = 0.015 - 0.15$:

$$\alpha^{o}_{v5} = 25^{o} + 21.6^{o}\left(\Phi_{des} - 0.015\right)^{0.35} \quad (9)$$

$$\frac{\bar{b}_5}{\bar{b}_4} = 1.45 - 0.5\left(\Phi_{des} - 0.015\right)^{0.4} \quad (10)$$

$$z_{RCh} = 20 + 12\left(\Phi_{des} - 0.015\right)^{0.35} \quad (11)$$

where $\bar{b}_5$ is vane height at RCh inlet; z_{RCh} is number of RCh vanes; α^{o}_{v5} is vane angle (to tangential direction) at RCh inlet.

The vanes height at the outlet is determined from the condition of mandatory acceleration of the flow in the outlet annular confuser (OAC) of at least 15% and the need for a smooth rotation of the flow in the OAC [0].

Given the accepted ratios $R_{s6} / b_6 = 0.6$ and $F_6 / F_{0'} = 1.15$ (where $\bar{b}_6$ is blade height at RCh outlet):

$$\bar{b}_6 = 0.417\left[\left(\bar{D}^2_{0'} + 1.38\left(\bar{D}^2_{0'} - \bar{D}^2_{h'}\right)\right)^{0.5} - \bar{D}_{0'}\right] \quad (12)$$

The RCh vane outlet diameter is equal to:

$$\bar{D}_6 = \bar{D}_{0'} + 2\bar{R}_{s6} = \bar{D}_{0'} + 2\frac{R_{s6}}{b_6}\bar{b}_6 = \bar{D}_{0'} + 1.2\bar{b}_6 \quad (13)$$

where $\bar{D}_6$ is RCh outlet diameter.

4.5 Capabilities of the preliminary design method

An example of a quick analysis of the stages efficiency with different design parameters is shown in the data below for calculating stages with Φ_{des} = 0.015, 0.025, 0.045, 0.0625 and 0.080. All stages have identical remaining design parameters $\psi_{T\,des}$ = 0.050, $\bar{D}_h$ = 0.350, relative Mach number M_u = 0.70, impeller diameter Reynolds number Re_u = 6000000, isentropic coefficient k = 1.40. Figure 1 (left) shows the characteristics of efficiency, polytropic work coefficient, work coefficient and pressure ratios of a stages series.

After optimizing the flow path sizes during the final design, the efficiency can be increased. But for a comparative assessment, the data presented are sufficient.

The effect of Φ_{des} on the stage characteristics is logical. Efficiency reaches its maximum at such Φ_{des} that the decrease in friction loss with increasing channels hydraulic diameter ceases to compensate for the growth of the lost pressure from an increase in the growth of flow velocities in the impeller and in stator elements (SE).

5 The main features of the Universal Modeling Method 9th version

In the end-to-end system, after the preliminary design, optimization of the flow path main dimensions and calculation of the gas-dynamic characteristics of the compressor are carried out by the 9th version of the model. To calculate the compressor characteristics, mathematical models of efficiency and work coefficient are used. In the 9th version, both models are different from the previous ones. An example of a formula for calculating the static efficiency of the impeller demonstrates the general features of the models of the Universal Modeling Method and the features of the 9th efficiency model:

$$\eta_{1-2} = \frac{1 - \lambda^2_{u0}\left[\begin{array}{l}\sum_{i=1}^{8}\left(\zeta_{frbl\,i} + \zeta_{mix\,i}\right)\bar{w}''^2_1 + \zeta_{mix\,b3/b2}\bar{w}''^2_1 + \zeta_{hss\,i=1}\bar{w}''^2_{1i=1} + \\ + \zeta_{hss\,i=8}\bar{w}''^2_{1i=8} + \sum_{i=1}^{8}\left(\zeta_{in\,i}\bar{w}''^2_{1i}\right) + 2\psi_T\left(\beta_{lk} + \beta_{fr}\right)\end{array}\right]}{\frac{2k}{k+1}\lambda^2_{u0}\left(\varphi'_1 - \psi_i\right)} \quad (14)$$

where λ is velocity coefficient; $\bar{w}''_1$ is relative velocity in impeller throat; φ'_1 is flow rate coefficient at an impeller inlet.

The structure of the formula follows from the well-known connection of efficiency with the indicators of polytropic and isentropic. As in all models of the Method, friction, mixing, and incidence losses are calculated separately. Losses are calculated on the blades (separately on both sides) and on the hub and shroud surfaces. Loss coefficients are algebraic equations with empirical coefficients. A feature of the 9th model for the impeller is that profile losses are calculated on eight axisymmetric current surfaces. This brings the impeller geometric model to the real flow path.

In papers [0, 0] it is shown the linear characteristic $\psi_T = f\left(\varphi_2\right)$, on which the loading factor model in 9th version is based. Two parameters determine the nature of the loading factor characteristic:

$$\psi_T = \psi_{T0} - \varphi_2\cos\beta_T \quad (15)$$

where:

$$\beta^0_T = 10.9655\left(\beta^0_{bl2}\right)^2 + 17.4802\left(\frac{l}{t}\right)^{0.4} + \\ + 18.22\bar{b}^{0.5}_2 - 5.702\left(\frac{b_2}{b_1}\right)^{0.8} \quad (16)$$

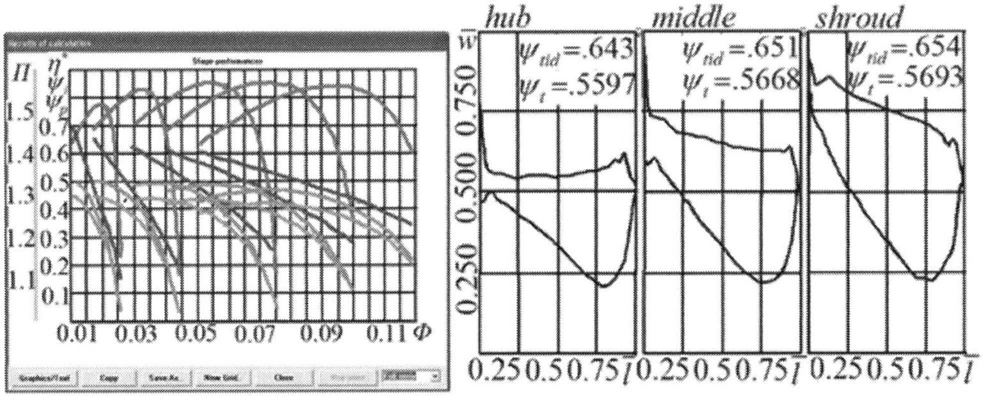

Fig. 1. "OPTIM 2" program, preliminary design. A series of stages with Φ_{des} = 0.015, 0.025, 0.045, 0.0625 and 0.080. Red - efficiency, blue – work coefficient, red – polytropic work coefficient, green – pressure ratio (left); 3DM.023 program. The final project of the 3D impeller. Velocity diagrams on three streamlines (right).

$$\psi_{T0} = 1 - 0.0479\left(\frac{l}{t}\right)^{0.9} - 0.0025\left(\frac{b_2}{b_1}\right)^2 + \\ +0.0255\left(\frac{\beta_{bl2}}{40}\right)^{0.7} - 3.7462\bar{b}_2^{1.5} \qquad (17)$$

Unlike the well-proven but complex mathematical model of VLD of previous versions, the introduced model is based on a generalization of the calculated CFD experiment [0, 0, 0]. To calculate the parameters at the VLD outlet, it is sufficient to know the loss coefficient and the flow angle at the VLD outlet. In [**Ошибка! Источник ссылки не найден.**] approximating equations of the form $\zeta, \alpha_4 = f\left(\bar{b}_2, \bar{D}_4, \bar{\kappa}_{rg}, \alpha_2, \lambda_{c2}, \mathrm{Re}_{b2}\right)$ were proposed, where $\bar{\kappa}_{rg}$ is relative surface roughness.

6 Optimization of the flow path and calculation of characteristics

After the preliminary design, the files necessary for calculating the velocity diagrams of the inviscid quasi-three-dimensional flow on the impeller blades are formed.

Advantages of the preliminary design of the blade row:
- the loading factor corresponds to its design value;
- the general nature of the flow around blades satisfies the requirements formulated in [0].

The disadvantage of the preliminary design is that the cross-section at the row inlet is slightly smaller than necessary, therefore, at the shroud streamline there is a peak of velocities corresponding to a negative incidence angle.

In the final project, to eliminate the negative incidence angle, it turned out to be sufficient to slightly increase the impeller entrance diameter $\bar{D}_0$. According to [0] it is advisable to equalize the circulation of velocity on different axisymmetric current surfaces.

This is due to a variable outlet angle β_{bl2} with an increase from the hub to shroud. To reduce the unevenness of the meridional flow, the blades are slightly inclined in the direction opposite to the rotation direction. Figure 1 (right) shows velocity diagrams on three axisymmetric streamline surfaces of the final impeller version.

Characteristics calculated by the program "OPTIM 2" showed that the correction of the stator elements basic dimensions is not required. 3D impeller size changing increased the efficiency at the design point by only 0.2%. More importantly, the adjustment of the impeller dimensions provides non-incidence flow around the 3D impeller blades at the shroud streamline.

7 Example of optimal compressor design

The design of a single-stage compressor for a turbo-expander unit was made for one of industrial partners. Figure 2 shows a compressor cross-section and results of preliminary design, efficiency and other parameters of candidates with different loading factors.

The hub ratio is big that influences efficiency. The compressor candidate with $\psi_{T\,des} = 0.435$ was chosen. The candidate has slightly lesser efficiency than candidates with larger loading factors. But the chosen candidate has a wider operating area towards lower flow rates, and the maximum power coincides practically with the design flow rate. This is important for a turbine drive.

For the selected candidate, the PC program automatically performs the preliminary design. By preliminary design the radial impellers have blades with controlled velocity diagrams. For a cast impeller of a turbo-expander compressor, a simpler midline is adopted - a circular arc. At the final design, the vanes of the

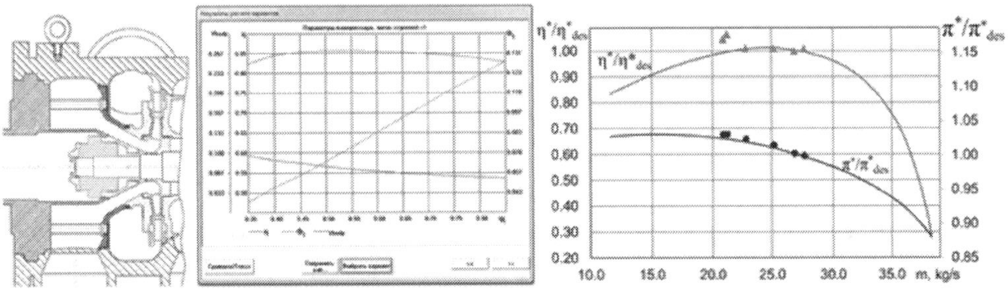

Fig. 2. Turbo-expander compressor, preliminary design program menu and compressor candidates with different loading factor $\psi_{T\,des}$ = var (above), design characteristics (solid line) and the results of plant test (dots).

diffuser were adjusted for a set design flow rate of 27.26 kg/s. The design characteristics of the compressor are shown in Figure 2.

The factory "air" compressor tests are recalculated to the design conditions shown by dots in Figure 2. The tests confirmed the design level of efficiency and showed a margin in pressure ratio. Cast impellers have irregular shapes. The margin guarantees necessary pressure ratio for impellers those shape deviates of design one.

8 Conclusion

The authors constantly use the engineering methodology in design work and are engaged in the improvement of this tool. In the near future, it is planned to carry out an extensive analysis of vane diffusers and return channels by CFD calculations, approximate the results and use simpler and more reliable math models of these elements in the compressor stage model. Another task is the accumulation of experimental data on the characteristics of high flow rate stages with 3D impellers.

Acknowledgements

The research was performed by a Grant of the President of the Russian Federation for young PhD MK-1893.2020.8.

References

[1] A.T. Lunev, Compressors and pneumatics **10**, 4–7 (2001)

[2] A. Lunev, O.A. Viachkilev, Y.V. Drozdov, Design and research of compressor machines **3** (1997)

[3] A.T. Lunev, Development of high-efficiency replacement flow parts of centrifugal compressors of gas-pumping units: Dis ... cand. tech. sciences (2005)

[4] D. Japikse, Design system development for turbomachinery (turbopump) designs and a decade beyond, JANNAF Conference (1998)

[5] D. Japikse, Agile design system in the age of concurrent engineering, JANNAF Conference (1996).

[6] D. Japikse, J. Bitter, Effective two-zone modeling of diffusers and return channel systems for radial and mixed-flow pumps and compressors, 11th International symposium on transport phenomena and dynamics of rotating machinery (2006)

[7] Y. Galerkin. *Turbocompressors* (Information and publishing center KHT, 596, 2010)

[8] Y. Galerkin, *Proceedings of the science school of compressor engineering SPbGPU* (M.: pub. SPbGPU, 2000)

[9] Y. Galerkin, *Proceedings of the science school of compressor engineering SPbGPU* (M.: pub. SPbGPU, 2010)

[10] Yu. Galerkin, A. Rekstin, K. Soldatova, A. Drozdov, Yu. Popov, The development of the scientific school of turbocompressor engineering LPI-SPbPU Peter the Great, the results of cooperation with compressor builders, Proceedings of 17th International Scientific and Technical Conference, Kazan (2017)

[11] K.V. Soldatova, Creation of a new mathematical model of the flowing part of centrifugal compressors and a database of model stages: Dis ... doc. tech. sciences (2015)

[12] A. Drozdov, Design method of centrifugal compressors with 3D impellers: Cand. of tech. Sci. (2017)

[13] A.F. Rekstin, V.B. Semenovskiy, K.V. Soldatova, Y.B. Galerkin, K.K. Sokolov, Compressors and pneumatics **1**, 13–20 (2018)

[14] A.F. Rekstin, B.V. Bakaev, St. Petersburg Polytechnic University Journal of Engineering Science and Technology **4**, 24–38. DOI: 10.18721/JEST.24403 (2018)

[15] K. Seleznev, Y. Galerkin, *Centrifugal compressors* (1982)

[16] E. Popova, Optimization of the main parameters of turbomachine stages on the basis of mathematical modeling: Dis ... cand. tech. sciences (1990)

[17] A.F. Rekstin, E.Y. Popova, A.A. Uchekhovsky, Compressors and pneumatics **1**, 33–40 (2018)

[18] V.F. Ris, *Centrifugal compressors* (1964)

[19] G.N. Den, *The mechanics of the flow in centrifugal compressors* (1973)

[20] V. Schnepp, *Design and calculation of centrifugal compressor machines* (1995)

[21] C.P. Livshits, *Aerodynamics of centrifugal compressor machines* (1966)
[22] A.F. Rekstin, Y.B. Galerkin, Bulletin of the Perm national research Polytechnic University **2**, 43–54. DOI: 10.15593/2224-9877/2018.2.06 (2018)
[23] Y.B. Galerkin, A.F. Rekstin, K.V. Soldatova, A.A. Drozdov, Compressors and pneumatics **4**, 21–31 (2015)
[24] J.M. Sorokes, *Sidestream optimization through the use of computational fluid dynamics and model testing turbomachinery sympos*, 2000.
[25] Yu.B. Galerkin, K.V. Soldatova, Compressors and pneumatics **1**, 24–34 (2016)
[26] Yu. Galerkin, A. Rekstin, K. Soldatova, A. Drozdov, Compressors and pneumatics **6**, 11–19 (2016)
[27] Yu. Galerkin, O. Solovieva, Compressor technology and pneumatics **3**, 35–41 (2014)
[28] Yu. Galerkin, O. Solovieva, Compressor technology and pneumatics **4**, 15–21 (2014)
[29] A.F. Rekstin, K.V. Soldatova, Y.B. Galerkin, Proceedings of higher educational institutions **9**, 702/2018. DOI: 10.18698/0536-1044-2018-9-44-56 (2018)
[30] L. Marenina, Compressor technology and pneumatics **3**, 27–35 (2016)
[31] A.F. Rekstin, A.A. Drozdov, O.A. Solovyeva, Y.B. Galerkin, AIP Conf. Proc. 2007, 030035-1–030035-12. DOI: 10.1063/1.5051896 (2018)
[32] O.A. Solovyova, *A mathematical model for calculating the gas-dynamic characteristics and optimizing the vaneless diffusers of centrifugal compressor stages: PhD thesis* (SPbPU, 2018)

The new ways to increase the efficiency of thermal hydraulic processes in systems of separation and reheating of NPP wet steam turbines

Mikle Egorov[1,2,*], *Ivan* Kasatkin[1], *Ivan* Kovalenko[2], *Irina* Krectunova[1], *Nataliya* Lavrovskaya[2], and *Nadezhda* Litvinova[2]

[1]Peter the Great St. Petersburg Polytechnic University, Russia, 195251 Saint-Petersburg, Polytechnicheskaya 29
[2]St. Petersburg State University of Aerospace Instrumentation, Russia, 190000 Saint Petersburg, Bolshaya Morskaya st. 67

Abstract. Systems of moisture separators - reheaters (SMSR) were created simultaneously with the powerful turbines development of nuclear power plants operating on a wet steam from the early 60s of the 20th century. This lengthy experience made the opportunity to identify the moisture separators - reheaters (MSRs) drawbacks, to determine the objectives for their design and computation improvement in order to increase thermodynamic efficiency, MSR devices and MSR systems reliability in general. Based on maintenance experience and results of our and other researchers' model and field studies this article discusses potential ways of further development and improvement, taking into consideration both SMSR components (separation and reheating) under conditions of their flow in one apparatus.

1 Introduction

The moisture separation and reheating organisation of low-pressure steam in turbine units of nuclear power plants with water-cooled reactors is the problem, appeared with the steam turbines creation operating on saturated steam, i.e. simultaneously with the large-scale nuclear power advent in the USSR and in other countries (in early 1960s) [1]. The steam parameters are significantly lower than the standard reheated steam parameters ~~of~~ used in organic fuels power engineering. This does not apply to nuclear power installations with supercritical parameters that are promising, but have not been implemented yet. Despite measures, taken to dry saturated steam in separating devices of boiling reactors and steam generators to a moisture level of $y <(\sim 0.5)\%$, with the steam expansion in the flowing turbine part, it is humidified, and, if it is not dehydrated after the high pressure (HPP) part, then the humidity reaches values unacceptable for a turbine blade. Therefore, when steam reaches the so-called separation pressure, the steam is sent to special devices – moisture separators-reheaters (MSRs) [2–5]. They separate moisture from the steam and overheat the dried steam to a temperature that ensures reliable operation of the LP part. Overheating is carried out at the expense of the heat of the condensing high-pressure steam. Two-stage overheating is the most common at MSR of Soviet (Russian) design: in the first stage, due to the heat of the condensing steam withdrawn from the turbine's turbocharger, and in the second stage, due to the heat of the hot steam in front of the turbine.

MSR constructive schemes of Russian NPPs with water-cooled reactors that have been in operation since early 1970s (SPP-220-1, SPP-500, SPP-500-1, SPP-1000, SPP-1000-1), are presented in table 1 [6]. Their main design parameters (NPP turbine is working at rated power) are shown in the table.

2 Materials and methods

2.1 Some thermal-hydraulic processes features in MSR, revealed during their operation

At the initial period of devices design, naturally, there was no their long-term operation experience. Moreover, many features of the medium flow processes, heat and mass transfer in them had not been sufficiently studied [7–10]. They were investigated only on small-scale models and did not take into account the scale factor, the geometry specifics of the systems and moisture separation and reheating processes [11–17]. Natural designers' tendency to maximize the devices' compactness (minimum metal capacity) made the problem even more complex. This, in particular, led to a constructive solution, consisting in the placement of both parts – separation and reheating – in one housing, which means that the device has:

– steam and water phases with moisture and vapor content varying from zero to one (dried reheated steam, condensate),

– zones with high velocities (up to ~ 60 m/s in the inlet steam part) and with low velocities (~ 0.4 m/s at the condensate outlet of the heating steam),

– sections with high heat transfer ($\alpha \sim 15000$ W/(m^{2}K) in the condensation zone of steam) and with relatively low ($\alpha \sim 300$ W/(m^2 K) for reheated steam.

All of the aforementioned processes create operating

* Corresponding author: mikhail.yu.egorov@gmail.com

Table 1. . Design characteristics of Russian systems of moisture separators- reheaters (SMSR) operating at rated power.

Power unit	**WWER-440**		**RBMK-1000**		**RBMK-1000**		**WWER-1000**		**WWER-1000**
Notation	**SMSR-220 M**		**SMSR-500**		**SMSR-500-1**		**SMSR-1000**		**SMSR-1000-1**
Reheater tube types	finned		smooth-tube		smooth-tube		finned		finned
Stages' number; reheater type	2; straight-tube		2; spiral coils		2; straight-tube		2; straight-tube		1; straight-tube
Thermal power, MW	28.2		37.5		33.1		46.2		59.2
SMSR number for one turbine, pcs	2		4		4		2 or 4		4
Heating steam (condensing):	in tubes		in tubes		in the intertubular space		in tubes		in tubes
inlet temperature, ° C	206	253	210	278	210	278	228	272	272
pressure, MPa	1.8	4.2	1.9	6.3	1.9	6.3	2.6	5.6	5.7
consumption, kg / s	8.92	7.50	9.08	11.6	8.75	11.7	13.7	17.7	36.1
Heated steam:	1 stage	2 stage	1 stage	2 stage	1 stage	2 stage	1 stage	2 stage	
inlet / outlet temperature, ° C	136/189	189/241	137/197	197/265	137/192	192/263	185/208	208/250	157/250
pressure, MPa	0.304		0.334		0.334		1.15		0.567
steam consumption at the inlet, kg / s	136		128		144		328		313
humidity at the inlet of SMSR, %	13.0		15.5		15.0		11.6		14.2
humidity at the outlet of the separator, %	~0.5		~1		<(~0.5)		~0.5		~0.5
Total SMSR height , m	13.98		6.71		9.05		13.35		14.60
Case diameter (internal), m	3.44		4.15		4.13		3.44		4.00
Steel grade of housing material / tubes	16GS[2]/20[3]		08H18N10T[4]		08H18N10T[4]		22K[3]/20[3]		22K[3]/20[3]
Heating tube bundle:									
diameter, mm	16 x 2		18 x 1.4		14 x 1.2		16 x 2		16 x 2
length, m	4.6		90	102	2.0		4.6		3.9
number, pcs.	3478	3441	432	216	14760	17640	2960	3959	8288
overheating cassettes (modules) number, pcs.	94	93	spiral coils		60	70	80	103	224
Tubes number in the cassette (module), pcs.	37		–		254		37		37
Weight of dry apparatus / case, t	110/22		105/23		119/38		128/40		152/79
Weight of tubes, t	22		60		50		22		27
Hydraulic resistance, kPa	18.1		17.7		22.6		33.4		16.3
Mass ratio to thermal power, t / MW	3.90		2.80		3.15		2.48		2.58
The ratio of the overall volume to the thermal output, m^3 / MW	2.70		2.74		3.28		2.46		2.50

modes that decrease efficiency and the device reliability, leading to:

1) erosion and vibration wear of the separation devices, see Figure 1, showing the wear of the separation unit of Kalininskaya NPP [18];

2) increased moisture of vapor at the separation devices' outlet, see Figure 2, where the moisture after the separation units at Smolensk and Kursk NPPs [19] is compared with its design value;

3) lower temperature of the working vapor of low pressure in comparison with the design values, see Figure 3, where the output temperature of superheated steam at Leningrad NPP [20] is compared with its design value;

4) unstable flow of heating steam condensate, see Figure 4, which shows the development of fluctuations in condensate flow [21]. The condensate supercooling below the saturation temperature causes temperature pulsations of the heat exchange surfaces, thermal cyclic stresses and heat fatigue damages in welded joints and other reheater elements;

5) separator accumulation at the apparatus bottom and / or casting of the separated moisture portion into the supply or the reheater discharge tubes, having much higher temperature (see Table 1).

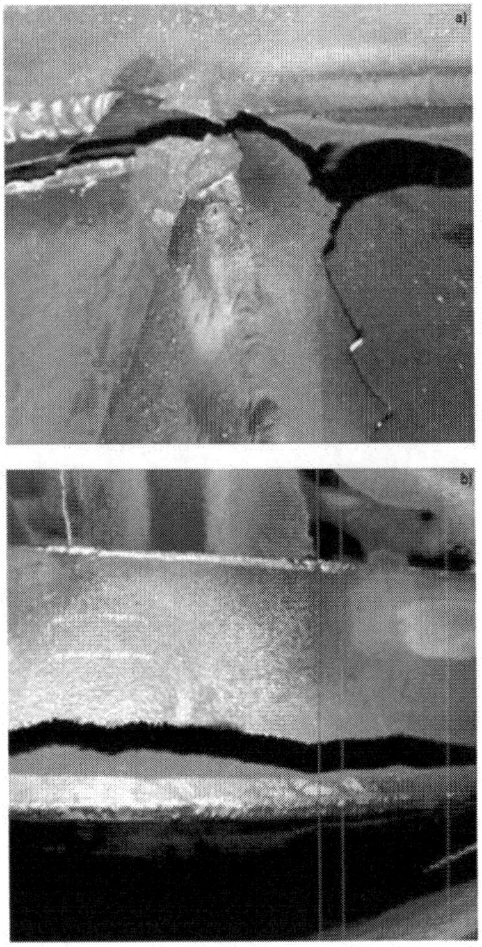

Fig. 1. Erosion wear of the separation unit, SMSR, Kalinin NPP [18]: a) – block adjacent to the hull, b) – bottom view.

The impact of these negative consequences usually appears only after a number of years of SMSR operation. As a result, it turned out that the actual overhaul SMSR life is significantly lower than the planned one.

All the mentioned factors and operating conditions should be taken into account when designing devices at the stage of their thermal and hydraulic calculations. We present below some factors analysis, emphasizing factors detected during long-term (during 30-40 years) operation at Russian NPPs with boiling reactors such as RBMK-1000 with SMSR-500-1.

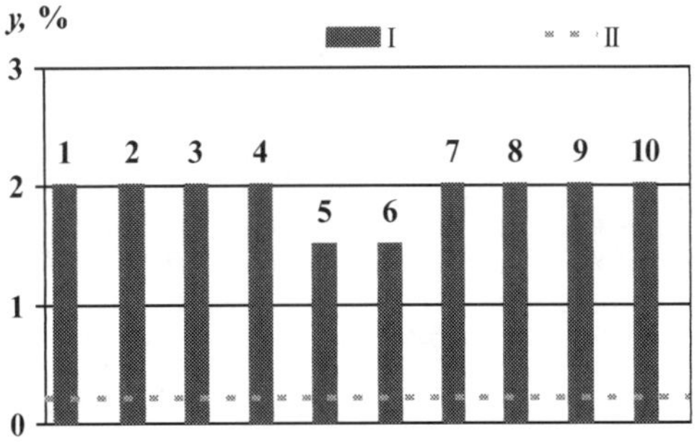

Fig. 2. Moisture (y) behind the separation units at the turbine units of Smolensk NPP (1-6), 6 turbines and Kursk NPP (7-10), 4 turbines: I – values according to the working data [19]; II – project value.

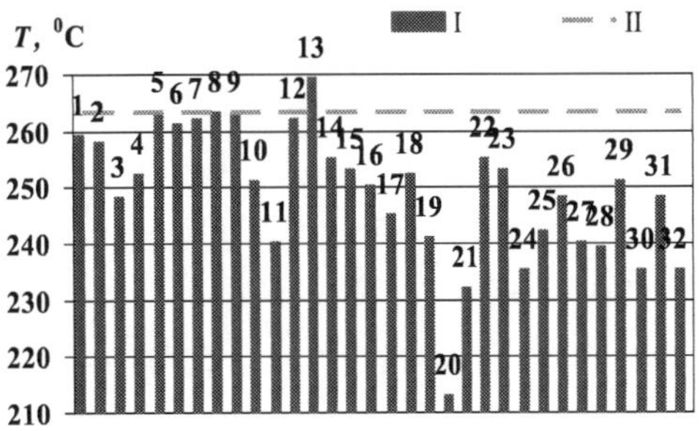

Fig. 3. The temperature of reheated steam of low pressure at the output of SMSR: I – actual [20]; II – design, Leningrad NPP.

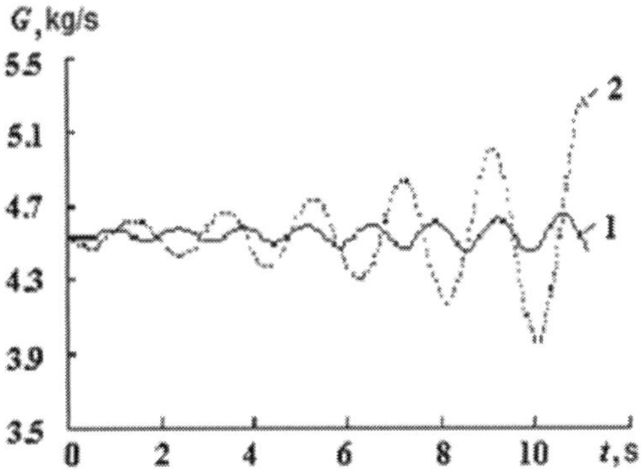

Fig. 4. Estimated evaluation of the development of self-oscillations of the condensate flow at the outlet of the reheater tube: the height of the condensate column in the tube is 1 – l_k = 0.3 m; 2 – l_k = 0.9 m [21].

3 Results and Discussion

3.1 Analysis of the ways to increase the efficiency of thermal hydraulic processes

The common problem of designing and calculating the

separation part of all SMSR, and SMSR-500-1 in particular, is the organization of moisture uniform distribution and vapor flows between separation devices.

Another problem is to ensure stable discharge of the separator from the SMSR. The path of the separator evacuation is one of the links in the general system for separating the moisture from the steam, draining it down into the separating cavities and unimpeded removal into a special container - separator collector. Problems in operation of this system, i.e. if there are fluctuations in consumption, steam inclusions capture, and water effervescence with a drop in pressure, etc., decreases effectiveness of the apparatus separation part as a whole. Poor quality of separator removal leads to the above-mentioned possibility of contact with the reheater tubes, which is especially characteristic for devices with an upper separation part (SMSR-220-1, SMSR-500, SMSR-500-1, SMSR-1000). Placing separation bags at the bottom of the device, as applied at SMSR-1000-1, significantly reduced the damage to the system.

Motion of the steam-moisture flow in the inlet chamber is rather complicated and occurs under conditions of significant influence of the mass forces. In spite of the small characteristic size of the water droplets, their trajectories can differ substantially from the steam current lines. Up to date none of the popular software packages have tools for solving this kind of tasks.

In Russia V.A. Barilovich, L.I. Zaichik, I.V. Derevich and others scientists were developing approaches to solving motion of two-phase disperse flows problems [22–23]. L.I. Zaychik and I.V. Derevich offered successful approach for aerosol flows, where deviations from homogeneity are not very high.

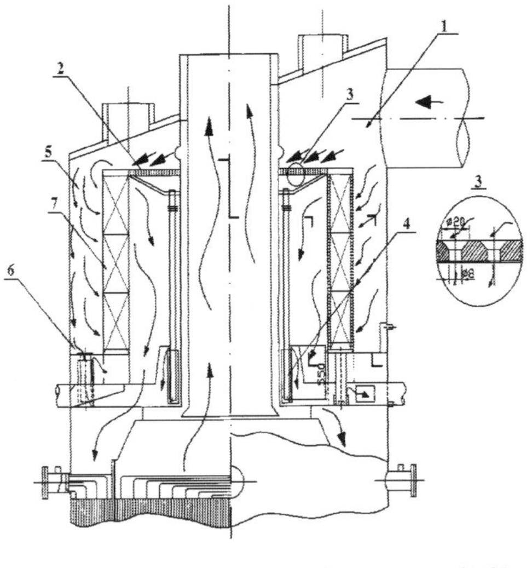

Fig. 5. General view of the modernized separation part SMSR-500-1: 1 – entrance chamber, wet steam inlet; 2 – perforated plate; 3 – holes with openings; 4 – openings for taking the separator from the perforated plate, 6 pcs.; 5 – drainage of moisture from the walls of the inlet chamber; 6 – openings for taking the separator into an external annular collector, 6 pcs.; 7 – separation blocks.

For the streams formed in the input chamber SMSR-500-1, Figure 5, the formation of liquid films and their subsequent destruction significantly complicates the solution of two-phase disperse flows problem.

Unfortunately, present day designers use solutions obtained in homogeneous approximation. Due to the complexity and originality of the two-phase flow regimes in the separation part of the SMSR it is advisable to carry out additional experimental research [24], as, for example, it was done when developing the upgraded version of SMSR-500-1 [25–26].

4 Conclusion

1. Analysis of Russian material constructions accumulated over the years of long-term operation of SMSR has revealed the features of the workflow that affect the efficiency and reliability of both the SMSR itself and the SMSR system as a whole. It is established that opportunities to intensify and improve the thermal and hydraulic characteristics of SMSR systems at Russian NPPs are not fully used.

2. The paper analyzes features of thermal-hydraulic processes in SMSR systems and the adverse consequences that have been manifested in the course of long-term operation, caused by insufficient consideration of these features.

3. In this article we also looked at the characteristic aspects of design improvement, leading to the enhancement of thermal and hydraulic processes, and the concept of the whole SMSR apparatus and its elements.

4. Regulatory materials that are newly created or updated need to include requirements that preclude the transfer of the separator to the reheater and the pulsation of the supercooled condensate flow.

5. It is important to eliminate the unevenness in both calculations and in SMSR tests. It is necessary to ensure equalization of the velocity field at the perimeter of the inlet chamber, the separating devices and the separating pipeline for heated steam. It is required to ensure the equalization of the distribution of the heating steam flow at reheater modules.

6. The upper location of the separator in the SMSR is undesirable for reliability reasons, since this eliminates the contact of the separator with the supply and removal tubes of the heating steam.

References

[1] A. Ulasen, A. Kalyutik, A. Blagoveshchenskii, MATEC Web Conf **245** (2018)

[2] L. Liu, B. Bai, Nuclear Engineering and Design **298**, 229 (2016)

[3] P. Peceny, M. Kolovratnik, International Journal of Engineering **XI**, 4 (2013)

[4] J. Manabe, 22nd Intern. Conference on Nuclear Engineering ICONE22-30297 **V02AT09A038** (2014)

[5] V.V. Legkostupova, A.V. Sudakov, Matec Web of Conferences **23**, 01058 (2015)

[6] A.V. Sudakov, B.S. Fokin, *Steam generators and heat exchanging equipment of NPPs* (SPb, NPO CKTI, 2005)

[7] N.D. Agafonova, I.L. Paramonova, Thermal Engineering **60**, 3 (2013)

[8] S. Alekseev, Y. Ilyukhin, V. Kukhtevich, I. Paramonova, S. Svetlov, V. Sidorov, High Temperature **37**, 4 (1999)

[9] N.D. Agafonova, I.L. Paramonova, Thermal Engineering **61**, 8 (2014)

[10] N.D. Agafonova, I.L. Paramonova, Journal of Engineering Physics and Thermophysics **89**, 4 (2016)

[11] G. Li, Ch. Yan, J. Wang, Annals of Nuclear Energy **73** (2014)

[12] M. Le, J. Kim, J. Kim, H. Do, J. Lee, Nondestructive Testing and Evaluation **33**, 1 (2018)

[13] R. Rafee, H. Rahimzadeh, Iranian Journal of Chemistry and Chemical Engineering **29**, 3 (2010)

[14] E. Narimani, S. Shahhoseini, Applied Thermal Engineering **31**, 2 (2011)

[15] H. Zhang, Q. Liu, B. Qin, H. Bo, Annals of Nuclear Energy **75** (2015)

[16] X. Wang, P. Cheng, J. Xie, Z. Zhou, Journal of Mechanical Science and Technology **29**, 8 (2015)

[17] F. Kavousi, Ya. Behjat, Sh. Shahhosseini, Chemical Engineering Research and Design **91** (2013)

[18] V. Solomeev, Rosenergoatom **7** (2006)

[19] N. Davidenko, V. Solomeev, P. Kruglikov, Thermal Engineering **1** (2008)

[20] V.V. Legkostupova, A.V. Sudakov, Thermal Engineering **62**, 3 (2015)

[21] A. Kozlov, Proc. of the Odessa Polytechnic university **15** (1998)

[22] L. Zaichik, Proc. of the 2nd Russian Nat. Conf. on Heat Transfer **1** (1998)

[23] V.A. Barilovich, *Bases of thermogasdynamics of two-phase flows and their numerical solution* (SPb, Polytechnic University Press, 2009)

[24] M. Le, J. Kim, J. Kim, H. S. Do, J. Lee, Nondestructive Testing and Evaluation **33**, 35 (2018)

[25] V.V. Legkostupova, A.V. Sudakov, St. Petersburg Polytechnic University Journal of Engineering Science and Technology **4**, 254 (2016)

[26] V.V. Legkostupova, A.V. Sudakov, St. Petersburg Polytechnic University Journal of Engineering Science and Technology **23**, 3 (2017)

E3S Web of Conferences **178**, 01030 (2020) https://doi.org/10.1051/e3sconf/202017801030
HSTED-2020

Method of accounting the dynamics of technological connections when forming a list of objects for the reconstruction and construction of electrical networks

A.V. Vinogradov[1], *V.E.* Bolshev[1,*], *A.V.* Vinogradova[1], *A.A.* Lansberg[2], *M.O.* Ward[3], and *N.K.* Miftakhova[4]

[1]Federal Scientific Agroengineering Center VIM, 5 1st Institutsky Proezd, Moscow, 109428, Russia
[2]Orel State Agrarian University, 69 General Rodina St., Orel, 302019, Russia
[3]Damascus University, Baramkah, Damascus, 30621, Syria
[4]Kazan State Power Engineering University, 51 Krasnoselskaya St., Kazan, 420066, Russia

Abstract. The paper gives a method of accounting the dynamics of technological connections when forming a list of objects for the reconstruction and construction of electrical networks. The specified method allows maintaining a database of technological connections taking into account the information received from the monitoring systems for electric network operating modes. It makes it possible to assess the dynamics of technological connections by settlements and regions. Based on this, it allows determining the availability of a power reserve for new technological connections and pre-compile a list of objects for reconstruction and construction of electrical network taking into account the dynamics of technological connections.

1 Introduction

All year round, applications are sent to a power grid company for technological connection of electric energy consumers as well as electric energy production facilities. The service of technological connection is provided in the case of connecting both the first time commissioned and reconstructed energy receiving devices in case of increasing connected capacity and changing the category of power supply reliability, connection points, types of production activities [1].

Technological connection is carried out on the basis of an agreement concluded between a power grid company and a legal or natural person in a timely manner. According to the decree of the Russian Federation [1], the power grid company does not have the right to refuse a consumer a technological connection. In case of unreasonable refusal, a consumer may appeal to the court with a claim for concluding a contract. The investment component is included in the amount of payment for technological connection in order to develop power supply centers and power lines, build electric grid facilities necessary for technological connection. [2]

Lack of accounting dynamics of technological connections when planning modernization, construction or reconstruction of electric networks leads to the fact that the transmission capability of power transmission lines (PTL) and the reserve capacity of substations and transformer substations (TS) become insufficient to connect new consumers. It results in damage to both power grid companies and consumers [3-6]. Power grid companies in this case receive less payment for electricity while consumers cannot expand the technological process and therefore do not receive profit from production.

Losses of the power grid company can be determined as follows [7, 8]. To do this, the volume of unapproved electric energy V is calculated associated with the violation of the terms of technological connection. It arises from the lack of reserve transmission capability of power lines or the capacity of substations and TS:

$$V = D \times \tau, \qquad (1)$$

where D is deficit of capacity or capability, kW; τ is time of exceeding the deadlines established for technological connection implementation.

The cost of unsupplied electricity $C_{unsup.\ el.}$ is determined by equation (2):

$$C_{unsup,\,el.} = V \times T, \qquad (2)$$

where T is tariff for electric power transmission service, dollars/kWh.

The damage to consumers $D_{consumer}$ is determined by equation (3):

$$D_{consumer} = V \times d_0, \qquad (3)$$

where d_0 is the specific damage from the lack of electricity supply to consumers, dollars/kWh.

d_0 = 1.5-2 dollars/kWh if there is prevailed production load [9] or different categories of agricultural

* Corresponding author: vadimbolshev@gmail.com

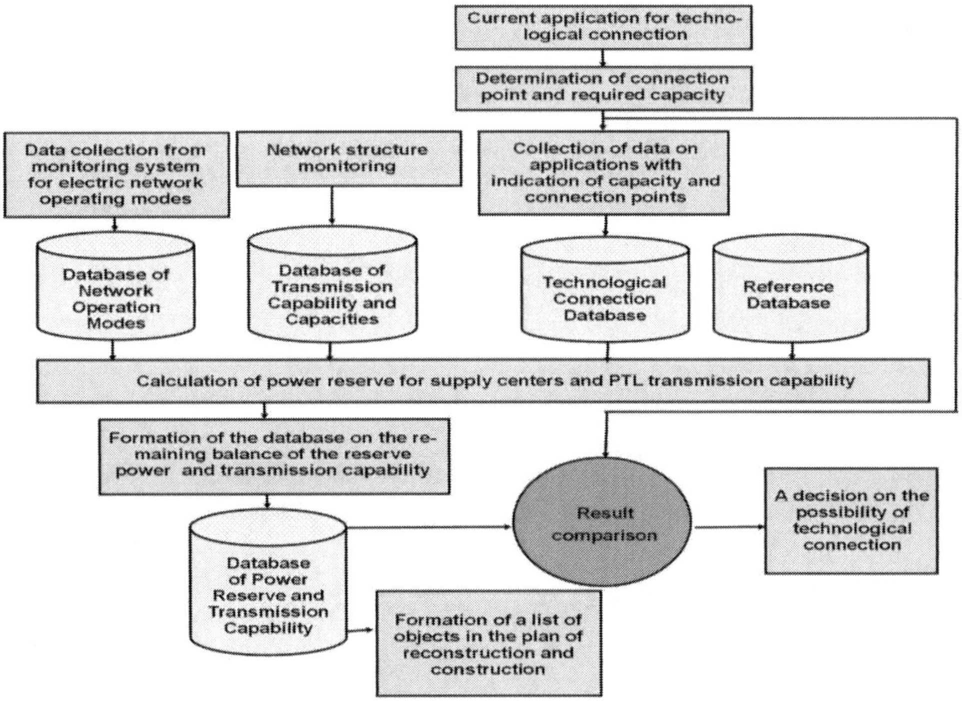

Fig. 1. Method of accounting the dynamics of technological connections when forming a list of objects for the reconstruction and construction of electrical networks.

consumers [10].

The analysis carried out in [11] using the example of Orelenergo (a branch of IDGC of Center, PJSC) showed that there were 83.397% of applications for technological connection in violation of deadlines in 2013, 66.499% in 2014, 55.484% in 2015, 52.606% in 2016, 53.476% in 2017. The percentage of fulfillment of applications for technological connection is increasing every year, but remains not high enough. One of the reasons for it is that the dynamics of technological connections is not taken into account when planning investment programs aimed at the reconstruction and development of electric networks. At the same time, the full implementation of this task is possible only with creating intelligent electric networks equipped with monitoring, control and accounting systems [12-16]. It is necessary due to constantly monitor the load of all network objects, monitor the change in this load, determine reserve of power for all substations and TS and power transmission line capability included in the controlled power supply system. So, keeping track of all applications for technological connection and the facts of their implementation is an important task along with controlling other factors affecting the possibility of technological connection on time.

Monitoring systems allow taking into account the dynamics of the growth of loads at the electrical network sections. As a consequence, it gives information about the possibiliffigty of technological connection to a particular section of an electric network, about the need for timely reconstruction of this section to increase its transmission line capability and about the possibility of increasing the power reserve for power sources.

Figure 1 shows a method for taking into account the dynamics of technological connections when forming a list of objects for the reconstruction and construction of electrical networks. It uses data from a system for monitoring electric network operating modes and a system for monitoring consumer connections. The method is as follows. A database of network operation modes is generated on the base of collecting data from monitoring systems. A database is also being formed on the capacities of substations, transformer substations, transmission capability of power transmission lines based on their passport data and acceptable operating modes. To do this, the network structure is monitored and all changes are made to the specified database. The reference database is used in the calculations and contains information on calculation methods and the necessary reference materials. Upon receipt of a new application for technological connection to electrical network, the necessary capacity and connection point are determined. This data is entered into the technological connection database, which takes into account completed and submitted applications for connection. At the same time, data on applications are sorted taking into account the supply centers and power transmission lines. Based on the information available in the databases, the

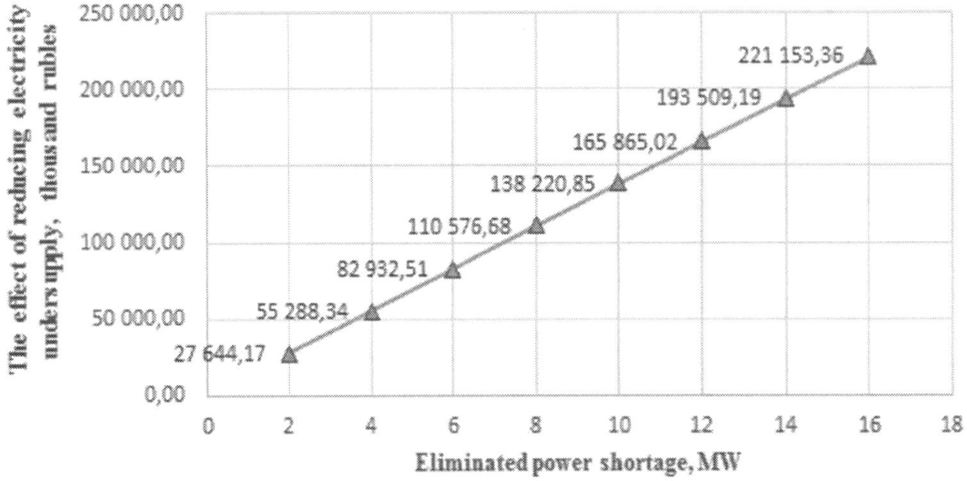

Fig. 2. Dependence of the effect on reducing electricity undersupply due to the elimination of the power shortage for Orelenergo substations.

calculation of the power reserve for the supply centers (substations and TS) and the PTL transmission capability is performed. Since the declared capacity practically does not match the consumed and usually above it, the calculation is performed using the actual values of the consumed power obtained from monitoring systems. Based on the calculated data, a database of power reserve and transmission line capability is formed.

If reserve for a substation, a transformer substation or transmission capability for a power transmission line reach the permissible minimum values, these objects are included in the list of objects for the reconstruction plan, network construction and further into the investment program of a power grid company. In this case, the minimum allowable power reserve is determined depending on the dynamics of the connections to a specific supply center or a power transmission line. That is, with high dynamics of technological connections, the minimum allowable reserve is set larger than with low dynamics of the connections. As for the current connection, the required capacity is compared with the power reserve for the supply center and the transmission capability for PTL the object is planned to connected to. Based on this comparison, a decision is made about the possibility of connecting at this point without reconstructing a substation, a transformer substation or a power transmission line.

The practical result of the implementation of the method is to decrease losses power grid companies due to reducing the time for technological connections.

Figure 2 shows a graph of the effect of reducing electricity undersupply due to the elimination of the power shortage calculated for scarce energy facilities of Orelenergo. There are the substations: 110/10 kV Volodarskaya, 110/10 kV Yuzhnaya, 110/35/10 kV Kulikovskaya, 35/10 kV Zvyaginki, 35/10 kV Streletskaya. The total power undersupply with the existing dynamics of technological connections in for five years it can be up to 16 MW.

Timely reconstruction of substations will allow to obtain the indicated effect for power grid companies. For consumers, this effect will be significantly higher since timely technological connection to electric networks will allow them to start the technological process in a timely manner.

2 Conclusion

Accounting the dynamics of technological connections to electrical networks requires equipping all power grid facilities with monitoring means that control the current load level of these facilities. The transmission line capability, the power reserve of substations and transformer substations with reference to the point of technological connection are evaluated when each new application for technological connection. If reserve for a substation, a transformer substation or transmission capability for a power transmission line reach the permissible minimum values, these objects are included in the list of objects for the reconstruction plan, network construction and further into the investment program of a power grid company.

References

[1] On the approval of the Rules for non-discriminatory access to services for the transmission of electric energy and the provision of these services, the Rules for non-discriminatory access to services for operational dispatch control in the electric power industry and the provision of these services, the Rules of non-discriminatory access to the services of the administrator of the trading system of the

wholesale market and the provision of these services, and the Rules for the technological connection of power receivers of electric energy consumers, facilities for the production of electric energy as well as electric grid facilities owned by grid organizations and other persons, to electric grids, Decree of the Government of the Russian Federation from 27 December 2004 No. 861 (2004)

[2] A.V. Vinogradov, A.V. Vinogradova, A.A. Kuchinov, Vremya osushchestvleniya tekhnologicheskogo prisoedineniya k elektricheskim setyam kak faktor effektivnosti sistem elektrosnabzheniya [The time of technological connection to electric networks as a factor in the efficiency of power supply systems], Vestnik NGIEI [Bulletin of the NIEGI] **6**, 73, 54–60 (2017)

[3] H.L. Willis, *Power distribution planning reference book* (CRC press, 2004)

[4] J.J. Grainger, W.D. Stevenson, W.D. Stevenson, *Power system analysis* (McGraw-Hill Inc, Singapore, 2003)

[5] E. Gracheva, A. Alimova, Calculation Methods and Comparative Analysis of Losses of Active and Electric Energy in Low Voltage Devices, International Ural Conference on Electrical Power Engineering (UralCon), 361–367. DOI: 10.1109/URALCON.2019.8877627 (2019)

[6] A. Vinogradov, A. Vinogradova, V. Bolshev, A.I. Psarev, Sectionalizing and Redundancy of the 0.38 kV Ring Electrical Network: Mathematical Modeling Schematic Solutions, International Journal of Energy Optimization and Engineering (IJEOE) **8**, 4, 15–38. DOI: 10.4018/IJEOE.2019100102 (2019)

[7] A. Vinogradov, A. Vasiliev, V. Bolshev, A. Semenov, M. Borodin, *Time factor for determination of power supply system efficiency of rural consumers, Handbook of Research on Renewable Energy and Electric Resources for Sustainable Rural Development* (Hershey, PA: IGI Global, 394–420, 2018). DOI: 10.4018/978-1-5225-3867-7.ch017

[8] A. Vinogradov, V. Bolshev, A. Vinogradova, M. Borodin, A. Bukreev, I. Golikov, *Mobile Measuring Complex for Conducting an Electric Network Survey, Handbook of Research on Energy-Saving Technologies for Environmentally-Friendly Agricultural Development* (Hershey, PA: IGI Global, 243–267, 2020)

[9] V.Ya. Khorolsky, M.A. Taranov, D.V. Petrov, *Tekhniko-ekonomicheskie raschety raspredelitel'nyh elektricheskih setej* [Technical and economic calculations of distribution electric networks] (Terra Print, Rostov-on-Don, Russia, 2009)

[10] N.M. Zul, A.M. Agalarov, N.F. Molosnov, *Metodicheskie rekomendacii po opredeleniyu ushcherba sel'skohozyajstvennomu proizvodstvu ot pereryvov v podache elektroenergii* [Guidelines for determining damage to agricultural production from interruptions in the supply of electricity] (VIESH, Moscow, Russia, 1974)

[11] A.V. Vinogradov, M.V. Borodin, Yu.A. Volchenkov, ZH.V. Peshehonova, *Sovershenstvovanie deyatel'nosti po energosberezheniyu i po osushchestvleniyu tekhnologicheskih prisoedinenij filiala OAO "MRSK CENTRA" – "ORYOLENERGO"* [Improvement of energy saving and technological connections for the branch of ORELENERGO, a branch of IDGC of Center] (Publishing House of Orel State Agrarian University, Oryol, Russia, 2015)

[12] J.A. Momoh, Smart grid design for efficient and flexible power networks operation and control, Power Systems Conference and Exposition, PSCE '09, 1–8. DOI: 10.1109/PSCE.2009.4840074 (2009)

[13] A. Ipakchi, F. Albuyeh, Grid of the future, IEEE Power and Energy Magazine **7**, 52–62. DOI: 10.1109/MPE.2008.931384 (2009)

[14] A. Vinogradov, V. Bolshev, A. Vinogradova, T. Kudinova, M. Borodin, A. Selesneva, N. Sorokin, A System for Monitoring the Number and Duration of Power Outages and Power Quality in 0.38 kV Electrical Networks, In Vasant P., Zelinka I., Weber GW. (eds) Intelligent Computing & Optimization. ICO 2018. Advances in Intelligent Systems and Computing **866,** 1–10. DOI: 10.1007/978-3-030-00979-3_1 (2019)

[15] A. Vinogradov, A. Vinogradova, I.O. Golikov, V. Bolshev, Adaptive Automatic Voltage Regulation in Rural 0.38 kV Electrical Networks, International Journal of Emerging Electric Power Systems **20**, 3. DOI: 10.1515/ijeeps-2018-0269 (2019)

VIBRATION DIAGNOSTICS OF POWER EQUIPMENT BEFORE COMMISSIONING

Mikolay Hruntovich[1,*], *Deniz* Moroz[2], *Alexey* Panfilov[1], *Yegor* Zhuk[1], and *Ekaterina* Mikhailova[3]

[1]Sukhoi State Technical University of Gomel, Prospect Octiabria, 48, 246746, Gomel, Republic of Belarus
[2]State Enterprise "Research Institute Belgiprotopgaz", Domashevskij lane,11a, Minsk, Republic of Belarus
[3]Kazan State Power Engineering University, str. Krasnoselskaya, 51, 420066, Kazan, Russia

Abstract. The article describes how to use vibration diagnostics to detect defects in the installation of new equipment and to determine the quality of repairs based on vibroacoustic characteristics. Especially relevant is the vibration diagnostics of rolling bearings before installation on the mechanism. If there are errors in the calculations of the magnetic system of synchronous generators, vibration occurs, which is not eliminated during operation. Design errors, poor assembly of pumping units, asynchronous motors, power transformers cause increased vibration, which leads to damage to the specified equipment. The main reasons causing the vibration of the transformer tank are considered and the frequency range from 50 to 6000 Hz is determined on which these defects can occur. Full-scale experiments of the vibration of the transformer tank have shown that certain defects can occur in a narrow frequency band. The practical application vibration diagnostics of the pump unit describes in this article. Technical diagnosis of pumping units of the sewer station was carried out for various included composition. The results obtained made it possible to exclude reverse hydraulic impacts as a possible cause of the failure of the pump unit.

1 Introduction

Quality control of purchased equipment is controlled by various regulatory documents [1, 2]. The necessary conditions for performing maintenance and repair of equipment is the presence of a modern system of maintenance and repair at the enterprise, which includes: products – objects of maintenance and repair; diagnostic tools [2]; performers (organizations, specialists); documentation (design, including operational and repair, regulatory, organizational, technological, etc.) establishing requirements for components of the maintenance and repair system and the relationship between them [3].

Efficiency of product maintenance and repair can be improved due to: improvement of product design as object of maintenance and repair; Improvement of the product maintenance and repair strategy in accordance with operational data on its reliability; testing of operational and maintenance documents; providing diagnostic control of both the unit as a whole and its individual units during repairs; development of normative documents based on modern scientific and technological developments; qualification of maintenance and repair contractors; application of modern methods of technical diagnostics of various equipment.

2 Research results

Consider vibration diagnostics of electric machines and rolling bearings.

2.1 Vibration diagnostics of rolling bearings before installation on the mechanism

Current studies have shown that about 60% of bearings delivered to enterprises are low-quality due to increased vibration. A vibration speed of 2.8 mm/s was adopted to reject new bearings. Measurement of vibration acoustic characteristics of bearings on vibration test bench showed that 40% of tested group of bearings in frequency range from 5 to 600 Hz and 60% of bearings in the range of 600 Hz - 5000 Hz do not pass by vibration level. Typical defects of new rolling bearings are: different dimensions of rolling bodies; not roundness of bodies of swing; ovality of an internal ring; Defects of inner and outer rings. Vibration frequencies of bearings in the range of up to 600 Hz for the listed defects can be calculated by formulas [3, 4].

2.2 Technical diagnostics of magnetic vibration of synchronous generators

During technical diagnostics of magnetic vibration, acceleration vibration sensor is placed on synchronous

* Corresponding author: gruntovich@tut.by

generator housing. The magnetic vibration of the housing of a synchronous generator with a capacity of 60 MW of thermal power plant at the frequency of 100 Hz was more than 4.5 mm/s. At this vibration level, long-term operation is not allowed. After the conducted regime studies, it was suggested that errors were made in the calculations of the magnetic system. This type of synchronous generator was subsequently discontinued.

2.3 Technical diagnostics of magnetic vibration of asynchronous motors

Under operating conditions, the magnetic vibration of asynchronous motors is practically not diagnosed.

2.3.1 Technical diagnostics of uneven gap between rotor and stator of asynchronous electric motors

The vibration value of the motor housing depends on the induction value in the air gap:

$$B_\varepsilon(\alpha,\, t) \;= F_{cp}(\alpha,\, t)\lambda_\varepsilon(\alpha,\, t),$$
$$F_{cp}\cos(\omega_1 t \;-\; p\alpha)\times \qquad (1)$$
$$\times\left(\lambda_0 + \lambda_{1\varepsilon}\cos\left(\omega_1(1-s)\frac{p_\varepsilon}{p} - p_\varepsilon\alpha\right)\right),$$

or:

$$B_\varepsilon(\alpha,\, t) = F_{cp}\lambda_0\cos(\omega_1 t \;-\; p\alpha) +$$
$$+\cos\left(\left(\frac{p_\varepsilon}{p}(1-s)\;\pm\;1\right)\omega_1 t \;-\; (p_\varepsilon \pm p)\alpha\right), \qquad (2)$$

where F_{cp} – the resultant magnetic moving force of stator and rotor in the air gap; ω_1 – the angular frequency; p – number of pole pairs; α- angular coordinate in geometric radians; λ_ε – the magnetic conductivity in the gap; λ_0 – constant component of magnetic conductivity; $\lambda_{1\varepsilon}$ – amplitude of the first harmonic of conductivity; s – engine slip; p_ε – magnetic conductivity poles.

The above expression shows that induction waves are formed in the air gap, having different speeds, polarities, and amplitudes. The speed of the induction wave determines the frequency of the magnetic gravity forces, and the order or polarity – moment arm causing deformation of stator ring. Forces with a large moment arm, i.e. polarity, are most dangerous. Table 1 shows numerical values of frequencies and order of magnetic forces arising due to eccentricity in asynchronous engine with different numbers of pole pairs p for medium sliding modes s = 0.05 and f_1 = 50 Hz.

Table 1. Values of frequencies and order of magnetic forces arising due to eccentricity in asynchronous engine with different numbers of pole pairs p for medium sliding modes s = 0.05 and f1 = 50 Hz.

Group	$p_\varepsilon = 1$ (shaft deflection)						$p_\varepsilon = 2$ (rotor ellipsis)					
	$p = 1$		$p = 2$		$p = 3$		$p = 1$		$p = 2$		$P = 3$	
	f_r	r	f_r	r	f_r	r	f_r	r	f_r	r	f_r	r
I	100	2	100	4	100	6	100	2	100	4	100	6
II	148	3	124	5	116	7	195	4	148	6	132	8
	48	1	24	1	16	1	95	2	48	2	32	2
	48	1	24	1	16	1	95	2	48	2	32	2
	52	1	76.5	3	84.5	5	95	0	52.5	2	68.5	4
III	95	2	48	2	32	2	192	4	95	4	63.5	4
IV	195	4	148	6	132	8	280	6	195	8	163	10
	4÷6	0	52	2	68.5	4	90	2	4÷6	0	36.5	2

Note: $f_r = 2f_1$; r - order of forces $r = 2p\text{-}p_\varepsilon$

In case of rotating eccentricity, more dangerous forces of the order r = 1 are generated at frequencies equal to or close to unbalanced ("reverse"). Magnetic vibrations at these frequencies are superimposed on the unbalance vibrations, increasing the vibration level at this frequency.

Only eccentricity explains presence of significant magnetic vibrations at frequencies close to frequency of **stator winding splicing** network 50 Hz.

When the numbers of poles of the main magnetic field and conductivity $p = p_\varepsilon$ are matched, magnetic forces are generated at low frequencies f_r = 4÷6 Hz. These forces have zero order r = 0 and can cause pulsation in the air gap. As a result, amplitude modulation of high-frequency magnetic (e.g. tooth) vibrations occurs. The oscillations at this frequency are close to the own oscillations of the machine mounted on the shock absorbers.

2.3.2 Technical diagnostics of stator winding splicing

During long-term operation of asynchronous electric motors, stator winding insulation dries and wedges in slots perform their function poorly. During operation of electric motor winding vibrates. Vibration frequency is calculated by formulas:

$$f_{st1} = Z_p f_p\,;\; f_{st2} = 2Z_p f_p\,, \qquad (3)$$

where Z_p – number of slots in the stator; f_p – speed of rotation of the rotor.

Vibration sensor is installed along the stator barrel circumference. The larger the measurement points, the more accurate the sector of the stator barrel with the weakened split winding is determined.

2.4 Vibrodiagnosing of power transformers

Vibration of the transformer tank housing can be caused by the following reasons: magnetostrictive vibration appearing up to 700 Hz; attenuation of winding and magnetic core pressing (loss of dynamic resistance), which manifests in the whole frequency range; damage to the branch lines in 110 kV transformers and formation of bubbles in oil, which appears in the frequency range from 700 to 1500 Hz; resonance of frequencies of individual transformer elements measured at frequency above 1500 Hz [1, 2].

From the established practice, losses of dynamic resistance of windings and magnetic circuit of transformers are determined by the total vibration level in the frequency range of 50-1000 Hz [3, 4, 5]. Vibration diagnostics of two new autotransformers of the railway traction substation showed that there are defects that cause vibration in the range of 1000-5000 Hz. Vibration spectra are shown in Figure 1: in autotransformer № 1 the limit vibration level at frequency 2800 Hz, and in autotransformer № 2 the limit vibration at frequencies 3500-3600 Hz. These vibrations are due to a defect in the installation of the autotransformers. It should be noted that in both transformers in the zone of increased vibration there are registered partial discharges, which in turn cause release of hydrogen (H_2) in transformer oil.

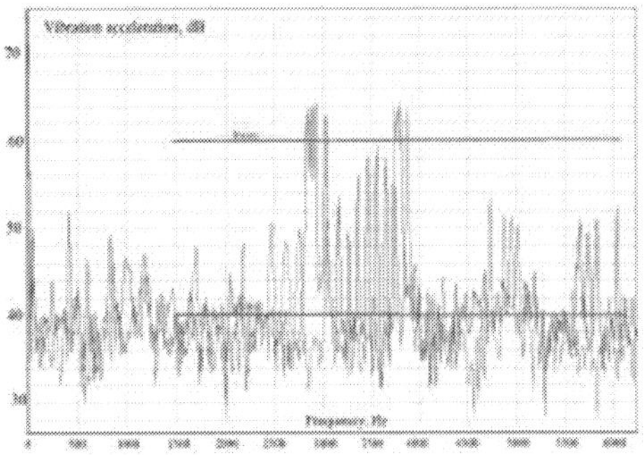

Fig. 1. Comparison of vibration levels of two autotransformers in the tank bottom of two autotransformers: green – autotransformer №. 1, blue – autotransformer № 2.

2.5 Vibrodiagnosing of pump units

Let's consider an example of practical application of the results of vibration diagnostics to identify the causes of failure of a pump unit with a capacity P_{nom} = 95 kW of pump units at a sewage station [6]. The pump unit that failed suffered damage to the impeller due to the self-unscrewing of the impeller mounting bolt. After replacement of shaft and impeller of pump unit the situation was repeated and repeated damage to impeller and shaft occurred (Figure 2). The supplier concluded that the pump failure was caused by a hydraulic shock due to an incorrectly selected pump unit and operating conditions that did not correspond to the pump. It was decided to measure vibroacoustic characteristics of Hidrostal snail pump with capacity of 75 kW installed instead of failed pump unit for three possible modes of pump station operation: 1st mode - parallel operation of two pump units; 2nd mode - parallel operation of three pump units; 3rd mode - serial disconnection of parallel operating pump units. In this case, removal of vibration acoustic characteristics of three possible modes of operation made it possible to confirm or refute the occurrence of reverse hydraulic impacts. Vibration acoustic characteristics were removed using a multichannel computer vibration acoustic diagnostic measuring complex.

Fig. 2. Electric motor shaft damages by pump impeller.

According to GOST 32106-2013 "Monitoring and diagnostics of machines. Monitoring of the state of hazardous production equipment. Vibration of centrifugal pump and compressor units" [7, 8, 9] limit value of vibration speed of new pump equipment is 4.1 mm/s, which corresponds to vibration acceleration of 78.5 dB. As a result of the operation, it was found that for the first mode the maximum vibration acceleration was 72 dB, or vibration speed - 1.9 mm/s (Figure 3).

3 Conclusions

For the second mode, the maximum vibration acceleration was 74 dB, or vibration speed was 2.4 mm/s. For the third mode, the vibration of the pump did not change, which indicates that there is no effect of the change in the composition of the station equipment on the vibration of each of them [10, 11].

Thus, thanks to the technical diagnosis of the pump units, the cause of failure of the pump unit due to hydraulic shocks in the system was eliminated.

The use of vibrodiagnostic control for power equipment and its individual units and elements can significantly increase the efficiency of the existing system of maintenance and repair:

- studies have shown that vibration diagnostics of rolling bearings before installation on the mechanism allows you to identify factory defects and avoid accidents with significant material damage. The magnetic vibration of induction motors causes not only accelerated wear of the insulation, but also damage the bearings;
- the measurements of vibration of the transformer tanks made it possible not only to determine the defects causing them, but also the frequencies at which these defects can occur. In this case, partial discharges with the release of combustible gases may occur. The conducted vibration diagnostics of the tank of two autotransformers (without turning them off) of the railway traction substation made it possible to establish installation defects;
- vibrodiagnostics of pumping units of the sewer station made it possible to evaluate all possible operating

modes of the pumping station in order to exclude the reason for the failure of the pumping unit due to water hammer in the system.

Deep knowledge of the physical bases of vibration of different objects allows to solve practical problems of vibration diagnostics of these objects.

References

[1] GOST 16504-81 The state system of testing products, Product test and quality inspection. General terms and definitions (in Russian)

[2] GOST 24297-2013 Verification of purchased products, Organization and methods of control (in Russian)

[3] Nic.Vas. Hruntovich, Improving the maintenance and repair system, Energy strategy, 21–26 (in Russian) (2013)

[4] Nic.Vas. Hruntovich, I.V. Petrov, D.V. Kirdishev, Hypocycloid vibration frequency of rolling bearings, Modern issues of energy saving and energy efficiency in technical systems, Tambov, 25-27 April 2016, 288–289 (2016)

[5] N.V. Hruntovich, et al., Vibration diagnostic of electric motor roller bearings, E3S Web of Conferences **124**, EDP Sciences (2019)

[6] Nic.Vas. Hruntovich, *Installation, commissioning and use of electrical equipment* (Minsk, 271, 2013) (in Russian)

[7] Y.I. Gracheva, V.N. Petrov, N.V. Russova, D.V. Samuilov, G.P. Svintsov, I.V. Protosovitsky, On modeling on a single-winding forced valve electromagnet in a circuit with a ballast resistor. E3S Web of Conferences **124**, 02012, SES-2019, 5–9 (2019)

[8] Nic.Vas. Hruntovich, O.V. Chaus, On some methods of obtaining information in conditions of uncertainty: deterministic and stochastic aspects, GSTU Bulletin **3**, 99–107 (2012)

[9] Nic.Vas. Hruntovich, O.V. Fedorov, D.R. Moroz, B.B. Tretyakov, P.M. Kolesnikov, Analysis of problematic issues in the operation of oil-filled transformers, Energy and Management **3**, 96, 2–6 (2017)

[10] N.V. Hruntovich, A.A. Kapanski, D. Baczynski, G.V. Vagapov, O.V. Fedorov, Optimization of a variable frequency drive pump working on a water tower, E3S Web of Conferences **124**, 05060, EDP Sciences (2019)

[11] N.V. Hruntovich, E.A. Zhuk, Improving the reliability of technical diagnostics of power oil-filled transformers is the basis of their durability and reliability, GSTU Bulletin **4**, 60–68 (2019)

[12] B.T. Holm-hansen, R.X. Gao, Structural design and analysis for a sensor-integrated ball bearing / Finite elements in analysis and design, 257–270 (2000)

[13] A.V. Barkov, A.V. Barkova, A.Y. Azovtsev, Monitoring and diagnosis of rotary machines by vibration (159, 2000)

[14] Y.P. Aslamov, I.G. Davidof, Sparse wavelet vibrating signal decomposition for pattern recognition task, PSU Bulletin **12**, 12–23 (2018)

[15] A.V. Doronichev, K.V. Konstantinov, Expert system for vibrodiagnosting rotating bearings status and repair planning in connection with current technical conditions on the basis of wavelet-technologies introduction in the field railway transportation objects, Modern technologies, System analysis, Modeling **1**, 29, 92–95 (2011)

Computational experiment on determination of load on the rollers of continuous casting machine

Navak Bashirov[1], Daniya Zaripova[2] and *Irina* Bashirova[1]

[1]Vologda State University, Institute of civil engineering, 160000 Vologda Lenina st. 15, Russia
[2]Kazan State Power Engineering University, Kazan, Russia

Abstract. The article is devoted to calculations of load on rollers of a curvilinear continuous casting machine during bending of the ingot in the levelling section. The calculation takes into account the elastic and plastic characteristics of metal and the flexibility of the roller apparatus. Determination of load on rollers together with meeting other requirements such as reducing the cooling water consumption, minimizing the energy consumption enables optimizing the technology of slab cooling in the secondary cooling zone (SCZ) and obtaining metal of the required quality.

1 Introduction

Currently, advanced continuous casting machines (CCM) are used for metal casting. In curvilinear-radial CCM, the most loaded section of the technological axis of machine is the levelling section, where the slab is bent. Here an ingot is straightened from a radial shape to a straight line and it is moved to a horizontal plane. Calculation of loads due to levelling is a complex and challenging procedure both from mathematical and technological points of view as many parameters of the machine and cast metal must be taken into account. These are the plastic and elastic properties of the slab metal itself, which also change both along the slab cross section and along the technological axis of the machine; compliance (displacement) of the roller apparatus itself due to the enormous forces and moments of forces acting during bending.

The mathematical model for calculating the load on rollers at a constant speed of slab withdrawal, i.e. when the load on slab during its straightening in each roller, the rigidity of the slab metal, and the thickness of the formed slab do not change in time, was considered in [1]. In this paper the differential equations with 4n boundary conditions were solved using the iterative method. At the first stage, the position of the bent slab axis is found from deflection of rollers. At the second stage, the deflection of rollers is found from the position of the bent axis. Since these differential equations are not solved directly, a whole series of techniques of both mathematical and computational nature was used to solve the problem.

In [1], the load on rollers was found by a modification of mathematical model, where the curvature of the elastic component of the technological path $y''(x)$ was adopted instead of the true curvature $y''_y(x)/(1+(y'_y(x)^2)^{0.5}$.

It is impossible or difficult to determine the load on rollers from field experiments and by construction a physical model based on them, since there are several tens of pairs of rollers on the levelling section. Therefore, it is more affordable to determine the load on CCM rollers from computational experiment. Firstly, it is based on creation of a mathematical model of object under study, and secondly, when conducting it, the initial mathematical model for calculating the load on rollers must undergo a series of transformations. These transformations, ultimately, should lead to such a computational algorithm that can be implemented on a computer. The ways of transforming mathematical models are directly related to mathematical methods that are used for the quantitative analysis of these models. Methods of transforming mathematical models of different levels (meta-level, micro-level) are given in [2]. Next, computer calculations, processing and analysis of the obtained results are carried out.

The purpose of this study is to determine the load on rollers at the levelling section based on the methods of computational experiment by creating a mathematical model of object using iteration method.

2 Materials and methods

Consider sequentially the main stages of computational experiment for determining the load on rollers at the levelling section of CCM with constant stationary casting of metal [3].

1. Construction of mathematical model for determining the load on rollers. Figure 1 shows a model of displacements (trajectories) along the technological

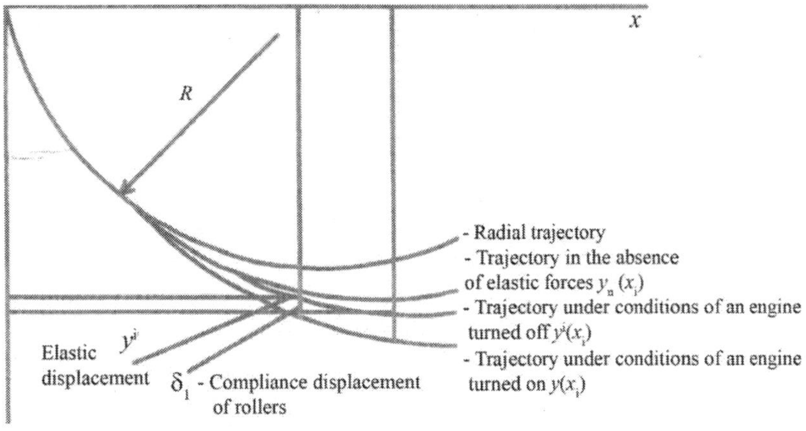

Fig. 1. Trajectories of technological axis during straightening of the ingot at CCM.

axis in slab during straightening of the ingot at CCM in the presence (absence) of various casting parameters.

The trajectories are described as follows:

$$\begin{cases} \dfrac{y_y''(x)}{(1+(y_y'(x))^2)^{\frac{3}{2}}} - z(x) = 0, \\ z''(x) = 0, \quad x_{i-1} < x < x_i, \\ y_n'''(x) = \dfrac{1}{\nu}(\dfrac{p(x)}{g(x)} \cdot y_y''(x))^5, \quad x_{i-1} < x < x_i, \quad i = 1,2,\dots,n, \end{cases} \tag{1}$$

and meet the levelling conditions at points x_i:

$$y_y(x_i) + y_y(x_i) + k_i\left(\frac{z'(x_i+0)}{g(x_i+0)} - \frac{z'(x_i-0)}{g(x_i-0)}\right) = y_i \tag{2}$$

$i = 0,1,\dots n$

$$\begin{aligned} &\frac{z(x_i+0)}{g(x_i+0)} - \frac{z(x_i-0)}{g(x_i-0)} = 0, \\ &i = 1,2,\dots,n-1, \\ &z(x_0+0) = 0, \\ &z(x_n+0) = 0, \end{aligned} \tag{3}$$

where (x_i, y_i) are the coordinates of rollers from 0-th to n-th; k_i is the roller compliance coefficient, mm/MN; $g(x)$, $p(x)$ are the elastic and plastic functions at each interval (x_{i-1}, x_i) of the slab section characteristic; ν is the slab casting rate. The following notations are also used: $n+1$ is the amount of rollers; $z(x)$ is the curvature corresponding to elastic deformation caused by external forces. The figure shows that the relations between the displacements are written by the following equations:

$$y_n(x_i) + y_y(x_i) = y(x_i)$$
$$y^i(x_i) + \delta^i = y(x_i)$$
$$y_y(x_i) = y^i(x_i) + \delta^i - y_n(x_i)$$
$$y_y(x_i) = y_y^i + \delta^i,$$

where

δ^i is the compliance displacement of the i-th roller;

$y_n(x_i)$ is the coordinate of the technological axis of the slab as a function of plastic displacement of the i-th roller;

$y_y(x_i)$ is the sum of displacements from the plastic and compliance displacements of the slab of the i-th roller;

$y(x_i)$ is the coordinate of the technological axis of the machine in working condition of the i-th roller;

y_y^i is the displacement of the technological axis due to elastic forces;

$y^i(x_i)$ is the coordinate of the technological axis of the machine at rest of the i-th roller.

The conditions (2) assume that the values $z_e'(x_0-0)$ and $z'(x_n+0)$ are equal to zero: $z'(x_0-0)=0$, $z'(x_n-0)=0$.

In addition, we assume that functions $y_n(x)$ and $y_y(x)$ are continuous and have continuous derivatives $y_n'(x)$, $y_y''(x)$ and $y_y'(x)$ when crossing the rollers:

$$\begin{aligned} &y_n^{(j)}(x_i-0) = y_n^{(j)}(x_i+0) \\ &i = 1,2,\dots,n-1; \; j = 0,1,2 \end{aligned} \tag{4}$$

$$y_y^{(j)}(x_i-0)=y_y^{(j)}(x_i+0) \quad (5)$$
$$i=1,2,...,n-1;\ j=0,1.$$

Besides this, the function $y_n(x)$ at $x=x_0$ meets the following initial conditions:

$$y_n(x_0)=y_{n0},\ y_n'(x_0)=y_{n1},\ y_n''(x_0)=\frac{1}{R}, \quad (6)$$

where $y_{n0}=y_0$ is the coordinate of the zero roller, y_{n1} is the angle of inclination of the tangent to slab at the point of contact with the zero roller, R is the radius of curvature of a CCM section.

2. The algorithm for calculating the load on straightening section, taking into account the full curvature formula, will be as follows.

To obtain the algorithm, the initial mathematical model must undergo a series of transformations. These transformations, ultimately, should lead to such an algorithm that can be implemented on computer. The ways of transforming mathematical models are directly related to mathematical methods that are used for the quantitative analysis of these models. Methods of transforming mathematical models of different levels (meta-level, micro-level) are given in [2].

In this paper, we consider two ways of converting a mathematical model for calculating the load on rollers at the CCM straightening section for the quasi-stationary mode of metal casting. The mathematical model of loading the CCM roller apparatus, which takes into account the elastic and plastic properties of the slab material, varying both in the cross section of the slab and along the technological axis of machine, as well as flexibility of the roller apparatus itself, which has the ability to redistribute the loads on these rollers, is given in [2].

3 Results and discussion

The procedure for solving differential equations with 4n boundary conditions in [2] was performed using the iterative method. The calculation algorithm was as follows: at the first stage, the position of the bent axis of slab $y_y(x)$ is found from deflection of rollers $y_y^i(x)$, $i=0,1,..n$. Formally, this procedure is designated as applying the operator F to vector $\{y_y^i(x)\}$: $y_y(x)=F[\{y_y^i\}]$. At the second stage, the deflection of rollers $y_y^i(x)$ are found from position of the bent axis. Formally, this procedure is designated as applying the operator H to function $y_y(x)$: $\{y_y^i\}=H[\{y_y(x)\}]$.

To find the deflection of rollers from the curvilinear path $y_y^i(x)$, the first step is to find the trajectory of the points of the neutral axis by the ratio of elastic and plastic properties of slab metal, provided there are no external forces

$$y_n(x)=y_n(x_0)+y_n'(x_0)(x-x_0)+$$
$$+y_n''(x_0)\frac{(x-x_0)^2}{2}+\frac{1}{2\cdot\upsilon}\int_{x_0}^{x}(x-s)\cdot\left(\frac{p(s)}{g(s)}\cdot z(s)\right)^5 ds$$

The second step is to determine from the condition $y_y^i=y^i(x_i)-y_n^i$ the deflection of rollers from the known coordinates of rollers $y^i(x_i)$ of the idle machine by introducing the deflection parameter of rollers in the presence of external elastic forces.

To solve the problem, the authors used a whole series of mathematical and computational techniques and $4n$ boundary conditions.

In [1], the load on rollers was found by a modification of mathematical model, where the curvature of the elastic component of the technological path $y''(x)$ was adopted instead of the true curvature $y_y''(x)/(1+(y_y'(x)^2)^{0.5}$. To calculate the load on rollers and find the bending moment in this work, only $2n+2$ boundary conditions are needed. One of the options for algorithmization of mathematical model for calculating the load on rollers in the straightening section can be as follows.

1. The matrix of parameters of all rollers and characteristics of metal are formed on the straightening section: (x_i, y_i) are the coordinates of rollers from 0-th to n-th; k_i is the roller compliance coefficient, mm/MN; $g(x)$, $p(x)$ are the elastic and plastic properties of metal, continuously limited functions on each interval (x_{i-1}, x_i); v is the slab rate; $n+1$ is the amount of rollers.

2. The coefficients of linear function $z(x)=A_i\cdot(x-x_i)+B_i$ of the general solution of the differential equation $z''(x_i)=0$, $x_{i-1}<x<x_i$, $i=1,2,...,n$, are set at each interval of straightening: $A, A_1, A_2,..., A_n, A_{n+1}, B_1, B_2,..., B_n$.

3. The equation

$$y_y(x)=A_0+A_{n+1}\cdot(x-x_0)+\int_{x_0}^{x}(x-s)\ \cdot z(s)ds, \quad x_0<x<x_n,$$

and the set coefficients are used to determine the elastic deflection of rollers $y_y(x)$ at each interval of straightening.

4. The relation

$$\eta_i=y_y(x_i)-y_y(x_i)-k_i\left(\frac{y_e'''(x_i+0)}{g(x_i+0)}-\frac{y_e'''(x_i-0)}{g(x_i-0)}\right),$$

$i=0,1,..n$ is calculated using the preset coefficients at each interval.

5. The system of equations is solved

$$\begin{cases} y_n(x_i)=\eta_i, & x_{i-1}<x<x_i, \\ B=C\cdot A, & x_{i-1}<x<x_i, \quad i=1,2,...,n, \end{cases}$$

and the coefficients for each interval of the technological axis are determined $A,\ B,\ A_1,\ A_2,...,A_n$.

6. Item 3 is repeated and $y_y(x) = A_0 + A_{n+1}\cdot(x-x_0) + \int_{x_0}^{x}(x-s)\cdot z(s)ds,\quad x_0 < x < x_n$ are determined using newly calculated coefficients $A,\ B,\ A_1,\ A_2,...,A_n$.

7. The relation $\eta_i = y_y(x_i) - y_y(x_i) - k_i\left(\frac{y_y'''(x_i+0)}{g(x_i+0)} - \frac{y_y'''(x_i-0)}{g(x_i-0)}\right)$, $i = 0,1,...n$ is calculated using the newly obtained coefficients.

8. Thus, the iteration continues until sufficient computational accuracy is achieved.

By substituting the following data: slab width $b = 1450$ mm, slab thickness $a = 250$ mm, surface temperature of upper edge $T_{us} = 1100$ °C, surface temperature of lateral edge $T_{ls} = 1000$ °C and performing the calculation, we get

$g_{42} = 1.071\cdot10^{-14}\frac{1}{\text{N}\cdot\text{mm}^2}$, $g_{43} = 1.062\cdot10^{-14}\frac{1}{\text{N}\cdot\text{mm}^2}$,

$g_{44} = 1.054\cdot10^{-14}\frac{1}{\text{N}\cdot\text{mm}^2}$, $g_{45} = 1.046\cdot10^{-14}\frac{1}{\text{N}\cdot\text{mm}^2}$,

$g_{46} = 1.038\cdot10^{-14}\frac{1}{\text{N}\cdot\text{mm}^2}$, $g_{47} = 1.021\cdot10^{-14}\frac{1}{\text{N}\cdot\text{mm}^2}$.

And we obtain:

$$p_{42} = 3.894\cdot10^{-10}\left(\text{N}\cdot\text{mm}^{0.5}\cdot\text{s}^{0.2}\right)^{-1},$$

$$p_{42} = 3.855\cdot10^{-10}\left(\text{N}\cdot\text{mm}^{0.5}\cdot\text{s}^{0.2}\right)^{-1},$$

$$p_{42} = 3.818\cdot10^{-10}\left(\text{N}\cdot\text{mm}^{0.5}\cdot\text{s}^{0.2}\right)^{-1},$$

$$p_{42} = 3.784\cdot10^{-10}\left(\text{N}\cdot\text{mm}^{0.5}\cdot\text{s}^{0.2}\right)^{-1},$$

$$p_{42} = 3.751\cdot10^{-10}\left(\text{N}\cdot\text{mm}^{0.5}\cdot\text{s}^{0.2}\right)^{-1},$$

$$p_{42} = 3.722\cdot10^{-10}\left(\text{N}\cdot\text{mm}^{0.5}\cdot\text{s}^{0.2}\right)^{-1}.$$

The roller compliance coefficients *ki* taking into account the slab width $b = 1450$ mm, the roller elastic modulus $E = 0.5\frac{MN}{mm^2}$, its length $l_p = 2500$ mm, outer $D_p = 380$ mm, and inner diameters $d_p = 100$, $k_{42} = k_{43} = k_{44} = k_{45} = k_{46} = k_{47} = 0.225\,{}^{mm}\!/_{MN}$.

Knowing these parameters, the force on rollers and bending moments are determined. Further some calculated values are presented. For calculation, the levelling section of CCM with a base radius of R = 10000 mm was considered. Metal was casted into slabs with a thickness of 250 mm, a width of 1700 mm, and a casting speed of 1.2 m/min with a flexibility of a roller apparatus of 0.3 mm/MN and levelling step of 312 mm. The calculated loads on rollers are the following: roller No.34 F = 0.21 MN, roller No.39 F = 0.76 MN; roller No.43 F = 1.33MN.

The initial results are obtained after the *k*-th iteration, depending on the accuracy of the task. The initial approximations should be chosen as close as possible to the true solution, since if the initial approximations are poorly defined, it is impossible to solve the system of equations.

4 Conclusions

As a result of the studies, an algorithm was developed for determining the load on the rollers in the secondary cooling zone of CCM during ingot straightening based on a computational experiment using the iteration method. Using the created mathematical model of levelling and a computational experiment, the loads on the rollers are determined when the slab is straightened in a continuous casting machine.

References

[1] N.G. Bashirov, E.M. Mukhamadiev, A.N. Naimov, A mathematical model for calculating the load on rollers at the levelling section of a curvilinear CCM, University science – to the region: Materials of the Tenth All-Russian Scientific and Technical Conference, in 2 vols, vol. 1 (Vologda: VSTU, 3-5, 2012).

[2] L.V. Bulanov, L.G. Korzunin, E.P. Parfyonov, N.A. Yurovsky, Continuous casting machines (Kazan, 349, 2003).

[3] V.S. Zarubin, Mathematical modelling in technology: textbook for universities, 3-ed. (Moscow: Publishing House of MSTU, 495, 2010).

Study of the effect of fuel temperature on gas turbine performance

George Marin[1,2], *Dmitrii* Mendeleev[1,2,*], *Boris* Osipov[1] and *Azat* Akhmetshin[1]

[1]Kazan State Power Engineering University, 420034, Krasnoselskaya St., 51, Kazan, Russian Federation;
[2]JSC "Tatenergo" branch "Kazan CHP-2", 420036, Tatsevskaya St., 11, Kazan, Russian Federation

Abstract. The development of gas turbine technologies requires the search for new solutions to improve the efficiency of gas turbine plants. The energy characteristics of a gas turbine depend on many external factors the temperature and pressure of the surrounding air including the composition of the fuel gas. This paper considers the effect of fuel gas temperature. Natural gas, synthesis gas, and aviation kerosene are considered as fuel for a gas turbine. For research, a gas turbine of the GE 6FA model was selected and its mathematical model was created. The results of calculations of the effect of fuel temperature on the energy and economic characteristics of a gas turbine are presented. The dependences of the main characteristics of the turbine were obtained. At the same time, the turbine power is assumed constant since during possible operation as part of a combined cycle gas turbine unit it is necessary to maintain a constant temperature of the exhaust gases after the gas turbine. The assumption was made - the costs of heating the fuel were not considered.

1 Introduction

Currently, widespread modernization of thermal power plants with gas turbine units. Gas turbines are used when upgrading or replacing used equipment. Nevertheless, at the same time, a lot of research is devoted to studying the work and increasing the efficiency of gas turbine plants. There are many ways to increase the efficiency of the gas turbine cycle and reduce fuel consumption for electricity generation. [1-3]

For example, such methods include:

1. Cascade air-cooling in the compressor.
2. The increase in gas temperature in the combustion chamber.
3. The use of supplying an additional working fluid to the flow part.

In this paper, as an alternative to the above methods, fuel heating is considered before it is fed into the combustion chamber of a gas turbine. Fuel heating can be carried out using the heat of the exhaust gases of the turbine itself. If a gas turbine operates as part of a combined cycle gas turbine unit it can be used to heat steam from steam turbine offsets. [4-6].

2 Materials and methods

The article describes an example of a gas-turbine unit (GTU) of PG6111FA type (from "GE Energy"). The main characteristics of the gas turbine installation are presented in table 1.

During GTU operation, filtered atmospheric air passes through the inlet mixing air chamber assembly and then is compressed in an 18-speed axial compressor. Compressed air from the compressor enters the annular space surrounding the six combustion chambers then the air enters the space between the outer casing and the inner lining of the combustion chambers and enters the combustion zone through the metering holes in the lining of each combustion chamber. Fuel injectors supply fuel to each of the six combustion chambers in which the fuel mixes with air and inflames. The hot gases from the combustion chambers expanding enter six separate transition compartments attached to the discharge side of the inner lining of the combustion chambers and then to the section of the three-stage turbine. Each stage consists of a series of fixed nozzle devices behind which a number of turbine blades is located. In each row of nozzle devices the kinetic energy of the jet increases that is accompanied by a corresponding pressure drop and in each subsequent row of rotating blades, part of the kinetic energy of the jet is spent on the useful work of the turbine rotor. After the passage of the blades of the third stage, the exhaust gases are sent to the exhaust housing and the diffuser. Then the gases are sent to the waste-heat recovery boiler. The resulting rotation of the shaft leads through the reduction gear to the rotation of the generator rotor which converts it into electrical energy. [6-10]

* Corresponding author: dylankn@ya.ru

 https://doi.org/10.1051/e3sconf/202017801033

Table 1. Technical characteristics of a GTU (PG6111FA).

No	Characteristic	Meas. unit	Value
1	Power at the generator terminals	kW	80000
2	Atmospheric pressure	kgf/cm^2	1.013
3	Compressor inlet temperature	°C	15
4	Relative humidity at the compressor inlet	%	60
5	The pressure of the fuel before the gas module	kgf/cm^2	25.9 – 30.8
6	The number of stages in the compressor	pcs	18
7	The number of steps in the turbine	pcs	3
8	Air flow	m^3/s	166
9	Compression ratio		15.8
10	Air temperature after the compressor	°C	385
11	Flue gas temperature	°C	603
12	The temperature of the gases after the combustion chamber	°C	1325

The low NO_x 2.6 (DLN 2.6) nitrogen fuel combustion system controls the distribution of fuel supplied to the multi-nozzle pre-mixers and reduces NO_x levels in the exhaust without the need for water or water vapor to be added to the fuel gas. The distribution of fuel flow in each assembly unit of fuel injectors is calculated to maintain the load of the unit and the distribution of fuel to ensure optimal turbine emissions.

The combustion system with a counter flow includes six combustion chambers located on the periphery of the compressor discharge housing. The combustion chambers are numbered counterclockwise when viewed from the top of the machine in the discharge direction. This system also includes fuel injectors, a spark plug system, flame control sensors and cross-ignition tubes.

Hot gases formed during the combustion of fuel in the combustion chambers pass through the transition compartments with forced cooling to the turbine.

High-pressure air from the compressor discharge housing is guided around the transition compartments. Part of the air enters the openings in the forced cooling sleeve to cool the transition compartments and is sent to the flow sleeve. The remaining air enters the space between the flow sleeve and the lining of the combustion chamber through the holes in the far end of the flow sleeve in the direction of flow. This air enters the combustion zone through the metering holes to ensure proper combustion of the fuel and through the grooves for cooling the skin of the combustion chamber. Air enters the combustion chamber through six nozzles that disperse the fuel and mix it with the appropriate amount of air supplied to the combustion chamber.

Natural gas, synthesis gas, and aviation kerosene are considered as fuel in this study. For research, the conditional formula for each type of fuel is calculated (table 2).

Table 2. The composition of the conditional formula for the studied fuels.

	Natural gas	Gas synthesis	Kerosene
Enthalpy	-4483.583	-2915.0	-2815.0
Carbon (C)	6.040	7.0726833	7.0768
Hydrogen (H)	22.99232	14.932036	14.6788
Nitrogen (N)	0.241048	-	-
Oxygen (O)	0.05642	-	-
Sulfur (S)	-	-	0.013595

The functional diagram of the mathematical model is presented in figure 1. Selection window for elements for design scheme is presented in figure 2. For the study, a mathematical model of a gas turbine engine was created in the software package «Automated System of Gas-dynamic Research of Energy Turbomachines» (AS GRET).

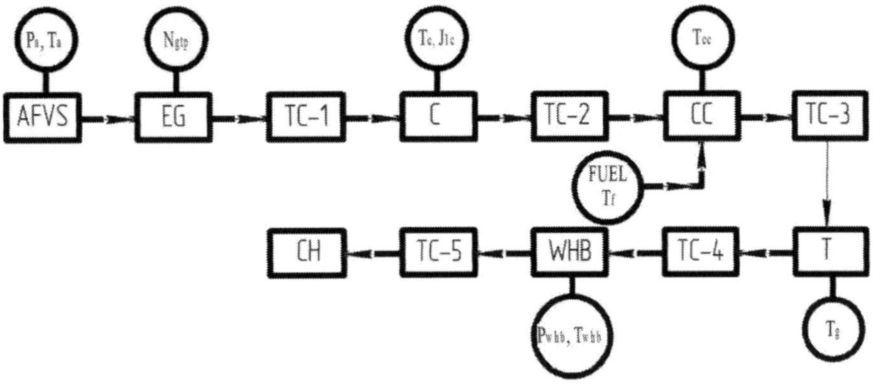

Fig. 1. The functional diagram of the mathematical model (AFVS – an air filtering and conditioning system; EG – an electric generator; TC–1 – a transition channel between AFVS and compressor; C – a compressor; TC-2 – a transition channel between the compressor and the combustion chamber; CC - a combustion chamber; TC-3 - a transition channel between the combustion chamber and the turbine; T – a turbine; TC-4 – a transition channel between the turbine and the boiler utilizer; WHRB – a waste heat recovery boiler; TC-5 - a transition channel between the waste-heat boiler and the chimney; CH – a chimney).

https://doi.org/10.1051/e3sconf/202017801033

The source data in this study:

Gas turbine power: 77 MW

Air temperature: T_a = + 15 ^{0}C,

Compressor inlet pressure: P_a = 101.3 kPa,

Humidity: 65%.

To assess the effectiveness of the influence of fuel, the following ranges of fuel heating were selected:

For fuel gas from 3 ^{0}C to 450 ^{0}C. The lower limit is defined at 3 ^{0}C since condensation of hydrocarbons and moisture in the fuel gas is possible at a lower temperature which may affect the operation of the fuel injectors. At 3 ^{0}C the absence of liquid in the feed gas is guaranteed. The maximum gas temperature of 450 ^{0}C is because decomposition of natural gas is possible at higher temperatures. [3, 11-14]

Synthesis gas is supplied in the range from 3 ^{0}C to 350 ^{0}C. (The conditions for choosing this range are the same as for fuel gas) [3]

The maximum temperature of aviation kerosene is 65 ^{0}C. The fact that the flash point of kerosene is from +28 to +72 ^{0}C (the lowest temperature of volatile condensed matter at which the vapor above the surface of the substance is able to flash in the air under the influence of an ignition source, however, is stable after removal of the ignition source does not occur). [15-17].

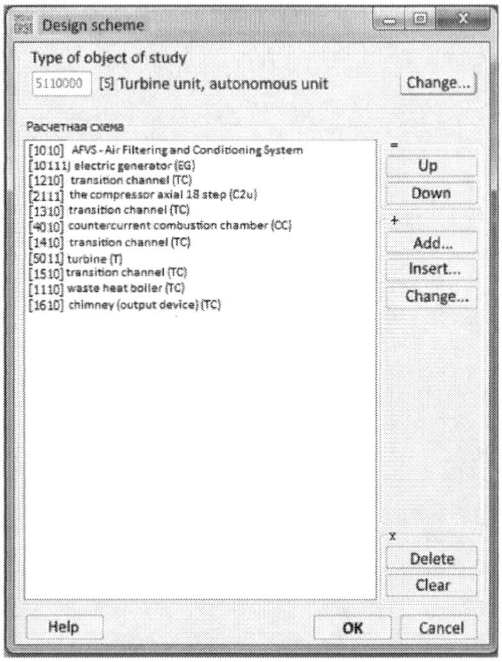

Fig. 2. Selection window for elements for design scheme.

3 Results and Discussion

When modeling such complex energy systems as a gas turbine engine, problems arise in setting parameters in

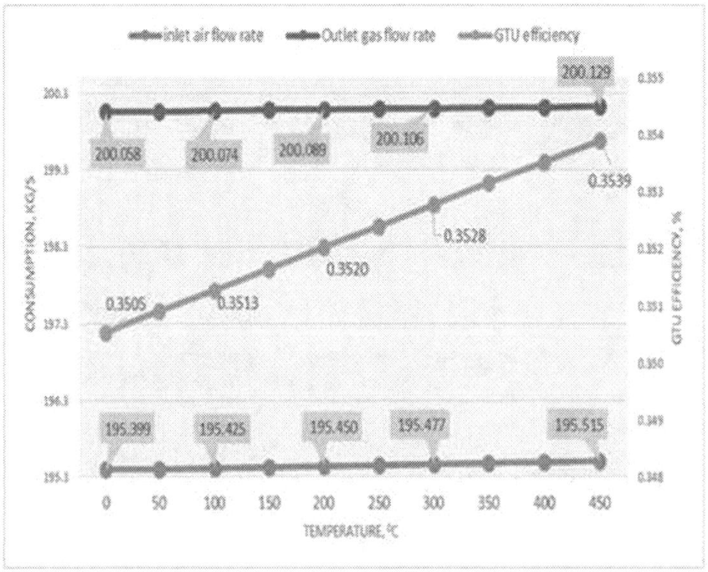

Fig. 3. The dependence of the airflow rate at the inlet and gas flow rate at the outlet of the gas turbine and the efficiency of the gas turbine from the temperature of natural gas

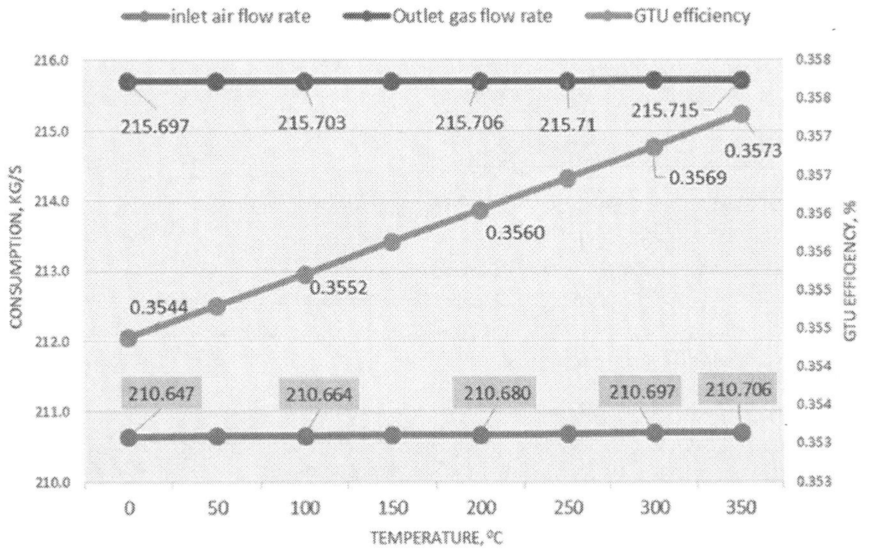

Fig. 4. The dependence of the airflow rate at the inlet and gas flow rate at the outlet of the gas turbine and the efficiency of the gas turbine from the temperature of the synthetic gas.

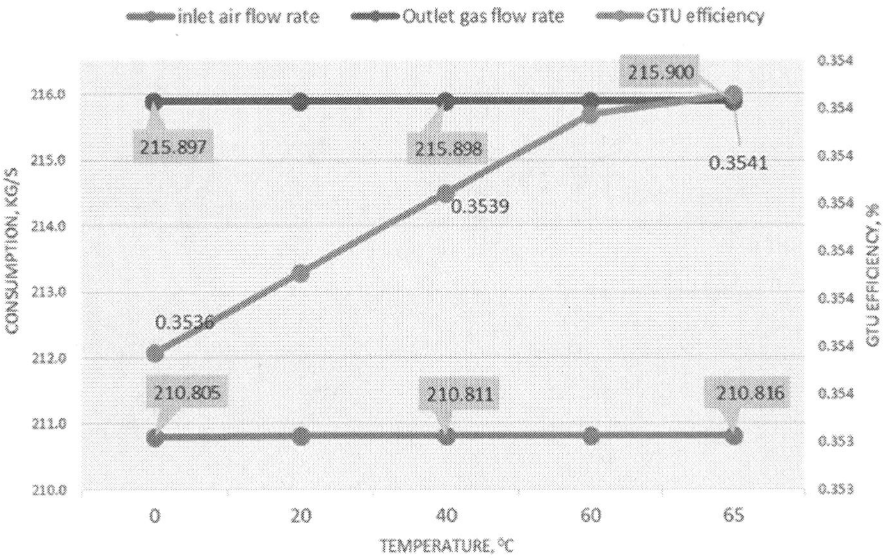

Fig. 5. The dependence of the airflow rate at the inlet and gas flow rate at the outlet of the gas turbine and the efficiency of the gas turbine from the temperature of kerosene.

the zeroth approximation, since the exact value will be determined only after the calculations are completed. To solve this problem, in this study, the method of residual systems is used. [18-20]

The simulation results are presented as graphical dependencies in Figures 3-6.

4 Conclusion

In this paper, we consider how gas temperature affects the parameters of a gas turbine plant. In particular, the following elements were noted:

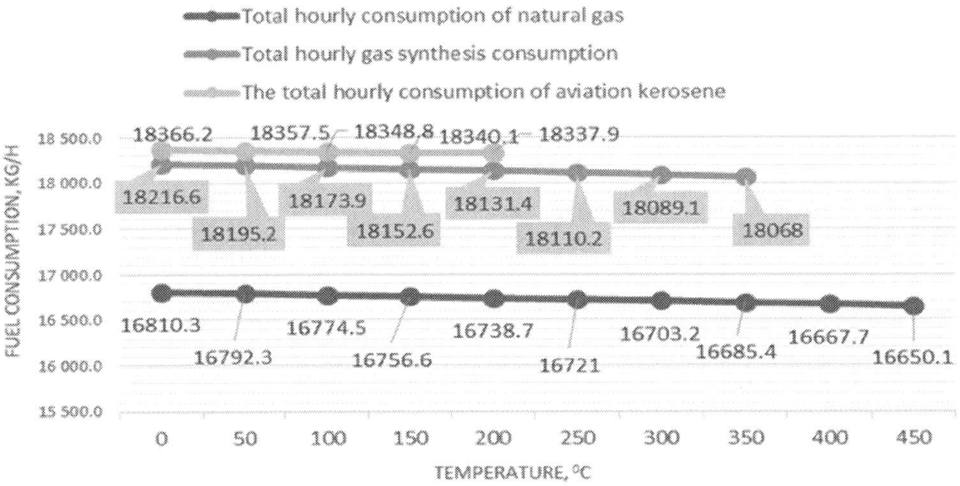

Fig. 6. The dependence of the total hourly consumption of the studied fuels on their temperature.

1. Fuel heating affects the efficiency of a gas turbine installation. When using natural gas, the increase in efficiency was from 35.05 to 35.39%. When working on syngas, the efficiency increase was from 35.44 to 35.73%. When working on aviation kerosene, the efficiency increase from 35.36 to 35.41%.
2. There is a decrease in consumption for each fuel when heating fuel; however, fuel gas is significantly less than for other types of fuel. Given that, gas turbines operate about 8000 hours per year fuel savings when using natural gas (when comparing fuels) can be about 11000 tons. The natural gas savings from heating will be 1280 tons. However, it is worth considering technical issues - how to carry out fuel heating. The costs of this modernization can be comparable with the savings from fuel costs, so it is necessary to conduct research in this direction and look for compromises in various ways.

References

[1] O.O. Badran, Appl. Energy (1999)

[2] M. Basha, S.M. Shaahid, L. Al-Hadhrami, Energy Procedia (2012)

[3] G.E. Marin, D.I. Mendeleev, A.R. Akhmetshin, 2019 Int. Multi-Conference Ind. Eng. Mod. Technol. FarEastCon 2019 (2019)

[4] E. Bocci, M. Sisinni, M. Moneti, L. Vecchione, A. Di Carlo, M. Villarini, Energy Procedia (2014)

[5] T.A. Clifford, S. Cruz-Manzo, Y. Zhang, V. Panov, A. Latimer, Proc. IECON 2017 – 43rd Annu. Conf. IEEE Ind. Electron. Soc. (2017)

[6] L. Esclapez, P.C. Ma, E. Mayhew, R. Xu, S. Stouffer, T. Lee, H. Wang, M. Ihme, Combust. Flame (2017)

[7] L. Esclapez, P.C. Ma, E. Mayhew, R. Xu, S. Stouffer, T. Lee, H. Wang, M. Ihme, Combust. Flame (2017)

[8] E. Gracheva, A. Alimova, Proc. – 2019 Int. Russ. Autom. Conf. RusAutoCon 2019 (2019)

[9] E. Gracheva, A. Alimova, Proc. – 2019 Int. Ural Conf. Electr. Power Eng. Ural. 2019 (2019)

[10] M. Henke, J. Kallo, K.A. Friedrich, W.G. Bessler, Fuel Cells (2011)

[11] J.G. Kiernan, A.D. Foster, D.T. Harden, Power Eng. (Barrington, Illinois) (1978)

[12] J. Kotowicz, M. Job, M. Brzeczek, Energy (2015)

[13] M.C. Lee, S. Bin Seo, J.H. Chung, S.M. Kim, Y.J. Joo, D.H. Ahn, Fuel (2010)

[14] A.H. Lefebvre, J.H. Whitelaw, Int. J. Heat Fluid Flow (1984)

[15] K. Liu, S. Sadasivuni, N. Parsania, Fuel (2019)

[16] W.E. Bill Forsthoffer, *Forsthoffer's Rotating Equip: Handbooks* (2005)

[17] M.M. Rahman, T.K. Ibrahim, A.N. Abdalla, Int. J. Phys. Sci. (2011)

[18] D. Rochelle, H. Najafi, Renew. Sustain. Energy Rev. (2019)

[19] P. Weiland, Appl. Microbiol. Biotechnol. (2010)

[20] L. Zheng, J. Cronly, E. Ubogu, I. Ahmed, Y. Zhang, B. Khandelwal, Appl. Energy (2019)

Simulation of wind power plant with double-fed machine when operating in autonomous system in parallel with synchronous commensurable power generator

Rauf Mustafayev[1], *Laman* Hasanova[1,*]

[1]Azerbaijan Scientific–Research and Designed–Prospecting Institute of Energetics, pr. Zardabi 94 , Azerbaijan

Abstract. Autonomous system, containing synchronous generators in a classic design, driven by either diesel or gas turbine, which operates in parallel with a wind power plant (WPP) containing an asynchronous double-fed machine (ADFM) through external network, is considered in the paper. At the same time, the power transmission scheme and the corresponding equivalent circuit are also drawn up. The equations of the synchronous generator and the external network are written in the axes rotating at the speed of the synchronous generator rotor, and the equations of the WPP's generator in the axes rotating at the speed of its rotor. By means of "docking" equations a digital model of the entire system is compiled with the external network parameters and loads on the side of the synchronous generator and WPP's asynchronous generator, which are represented by constant-conductivity shunts. Static and dynamic characteristics of the system are calculated.

1 Introduction

Autonomous commensurable power systems of are widely used in various sectors of national economy. When using wind power plants in the autonomous power supply system, naturally in places with high wind power potential, the problem of studying the conditions of parallel operation arises, when the output power of wind power plants changes depending on the wind speed value. Such systems can be used in regions without centralized power supply – mountainous villages, oil production platforms ("Oil rocks" type), etc.

2 Materials and methods

The electricity transmission diagram is presented in Fig. 1. Here synchronous generator SG, generator voltage of which is about 10.5 kV, feeds the load 1 from generator buses. Further, the step-up transformer T_{r1} through the high-voltage power line L_1 and the transformer T_{r2} feeds the load 2, at which the asynchronous generator ADFM WPP also operates.

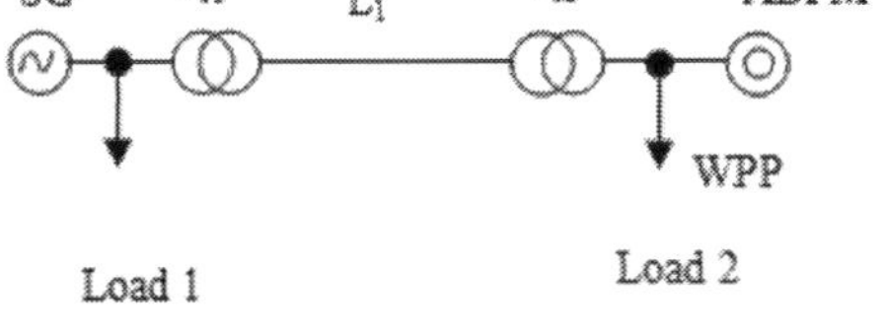

Fig. 1. Equivalent Schemes.

This scheme, of course, cannot be the only one –T_{r1} and T_{r2} may be absent, power takeoff may be in different places of the line L_1, etc. But this scheme corresponds to the semantic content of the question under study, i.e., when the output power of the WPP's asynchronous generator changes depending on the wind speed, the power distribution on load 2 will change – when the output power of the WPP decreases, the main synchronous generator will be loaded, and when it increases-vise versa [1, 2, 3, 4, 5].

The equivalent circuit of the power transmission scheme, presented in Fig.1, is shown in Fig. 2.

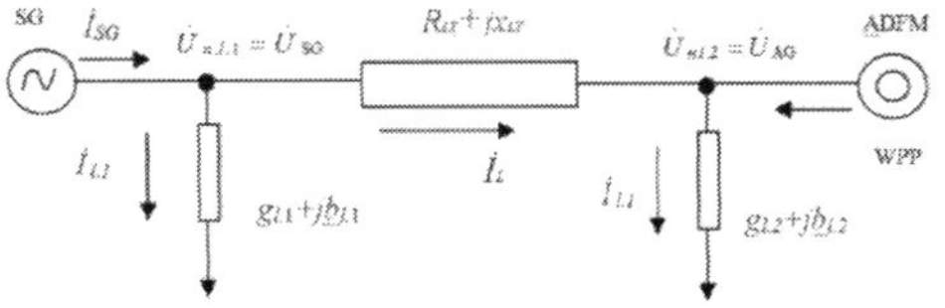

Fig. 2. Power Transmission Equivalent Schemes.

Here $\dot{I}_{l1}$, $\dot{I}_{l2}$ – currents of load 1 and load 2 in complex form, $\dot{I}_L$ – line current; $(g_{n1}+jb_{n1})$ and $(g_{l2}+jb_{l2})$ – components of the active g_{l1}, g_{l2} and reactive b_{l1}, b_{l2} conductivities of first and second loads respectively; $(r_{LT}+jx_{LT})$– concurrent resistances (active r_{LT} and reactive x_{LT}) of power transmission line L_1 and

* Corresponding author: gasanovalg@mail.ru

transformers T_{r1} and T_{r2}; $\dot{U}_{l.n.1}=\dot{U}_{SG}$, $\dot{U}_{l.n.2}=\dot{U}_{AG}$ – voltages of load nodes $\dot{U}_{l.n.1}$ and $\dot{U}_{l.n.2}$, which are equal to generator voltages of synchronous and asynchronous machines, $\dot{I}_{SG}$, $\dot{I}_{AG}$ – currents of synchronous and asynchronous generators in complex form.

Direction of 1 node is determined from the relationship:

$$\dot{U}_{l.n.1}-\dot{U}_{SG}=\dot{U}_{l.n.2}+\dot{I}_{L}\left(r_{LT}+jx_{LT}\right) \quad (1)$$

Currents in nodes 1 and 2 are respectively as follows:

$$\left.\begin{aligned}\dot{I}_{l1}&=\dot{U}_{l.n.1}(g_{l1}+jb_{l1})=\dot{U}_{SG}(g_{n1}+jb_{l1})\\ \dot{I}_{l1}&=\dot{U}_{l.n.2}(g_{n2}+jb_{l2})=\dot{U}_{AG}(g_{l2}+jb_{l2})\end{aligned}\right\} \quad (2)$$

Current balance:

$$\left.\begin{aligned}\dot{I}_{SG}&=\dot{I}_{L}+\dot{I}_{l1}\\ \dot{I}_{AG}&=\dot{I}_{l2}-\dot{I}_{L}\end{aligned}\right. \quad \text{or} \quad \left.\begin{aligned}\dot{I}_{L}&=\dot{I}_{SG}-\dot{I}_{l1}\\ \dot{I}_{L}&=\dot{I}_{l2}-\dot{I}_{AG}\end{aligned}\right\} \quad (3)$$

taking into account the last expression, the voltages of synchronous and asynchronous generators will be written as:

$$\left.\begin{aligned}\dot{U}_{SG}&=\dot{U}_{AG}+(\dot{I}_{SG}-\dot{I}_{l1})\cdot(r_{LT}+jx_{LT})\\ \dot{U}_{AG}&=\dot{U}_{SG}-(\dot{I}_{l2}-\dot{I}_{AG})\cdot(r_{LT}+jx_{LT})\end{aligned}\right\} \quad (4)$$

Then, substituting the load current values from equations (2) into the expression (3), we get:

$$\left.\begin{aligned}\dot{U}_{SG}&=\dot{U}_{AG}+\dot{I}_{SG}(r_{LT}+jx_{LT})-\dot{U}_{SG}(g_{l1}+jb_{l1})\cdot(r_{LT}+jx_{LT})\\ \dot{U}_{AG}&=\dot{U}_{SG}+\dot{I}_{AG}(r_{LT}+jx_{LT})-\dot{U}_{AG}(g_{l2}+jb_{l2})\cdot(r_{LT}+jx_{LT})\end{aligned}\right\} \quad (5)$$

Grouping members, we obtain:

$$\left.\begin{aligned}\dot{U}_{SG}\left[1+(g_{l1}+jb_{l1})\cdot(r_{LT}+jx_{LT})\right]&=\dot{U}_{AG}+\dot{I}_{SG}(r_{LT}+jx_{LT})\\ \dot{U}_{AG}\left[1+(g_{l2}+jb_{l2})\cdot(r_{LT}+jx_{LT})\right]&=\dot{U}_{SG}+\dot{I}_{AG}(r_{LT}+jx_{LT})\end{aligned}\right\} \quad (6)$$

At last, the last equations can be written in the following form:

$$\left.\begin{aligned}\dot{U}_{SG}&=\dot{K}_{s}\cdot\dot{U}_{AG}+\dot{K}_{Ls}\dot{I}_{SG}\\ \dot{U}_{AG}&=\dot{K}_{a}\cdot\dot{U}_{SG}+\dot{K}_{La}\cdot\dot{I}_{AG}\end{aligned}\right\} \quad (7)$$

where $\dot{K}_{s}=\dfrac{1}{1+(g_{l1}+jb_{l1})\cdot(r_{LT}+jx_{LT})}$;

$\dot{K}_{Ls}=\dfrac{(r_{LT}+jx_{LT})}{1+(g_{l1}+jb_{l1})\cdot(r_{LT}+jx_{LT})}$;

$\dot{K}_{a}=\dfrac{1}{1+(g_{l2}+jb_{l2})\cdot(r_{LT}+jx_{LT})}$;

$\dot{K}_{La}=\dfrac{(r_{LT}+jx_{LT})}{1+(g_{l2}+jb_{l2})\cdot(r_{LT}+jx_{LT})}$.

Solving together the equations of the system (7), we get:

$$\left.\begin{aligned}\dot{U}_{SG}&=\frac{\dot{K}_{s}\cdot\dot{K}_{La}}{1-\dot{K}_{s}\cdot\dot{K}_{a}}\dot{I}_{AG}+\frac{\dot{K}_{Ls}}{1-\dot{K}_{s}\cdot\dot{K}_{a}}\dot{I}_{SG}\\ \dot{U}_{AG}&=\frac{\dot{K}_{a}\cdot\dot{K}_{Ls}}{1-\dot{K}_{s}\cdot\dot{K}_{a}}\dot{I}_{SG}+\frac{\dot{K}_{La}}{1-\dot{K}_{s}\cdot\dot{K}_{a}}\dot{I}_{AG}\end{aligned}\right\} \quad (8)$$

For simplicity of further transformations these equations can be represented as:

$$\left.\begin{aligned}\dot{U}_{SG}&=\dot{A}\cdot\dot{I}_{AG}+\dot{B}\cdot\dot{I}_{SG}\\ \dot{U}_{AG}&=\dot{C}\cdot\dot{I}_{SG}+\dot{D}\cdot\dot{I}_{AG}\end{aligned}\right\} \quad (9)$$

where $\dot{A}=\dfrac{\dot{K}_{s}\cdot\dot{K}_{La}}{1-\dot{K}_{s}\cdot\dot{K}_{a}}$; $\dot{B}=\dfrac{\dot{K}_{Ls}}{1-\dot{K}_{s}\cdot\dot{K}_{a}}$; $\dot{C}=\dfrac{\dot{K}_{a}\cdot\dot{K}_{Ls}}{1-\dot{K}_{s}\cdot\dot{K}_{a}}$; $\dot{D}=\dfrac{\dot{K}_{La}}{1-\dot{K}_{s}\cdot\dot{K}_{a}}$.

Now we write down the equations of the external network in the axes *d*, *q*, rotating at the speed of the synchronous machine rotor [0, 0]. At the same time we superpose the q axis with the real numbers axis, and the *d* axis with the imaginary numbers axis. Then the voltages and currents of generators can be written in the following form:

$$\left.\begin{aligned}\dot{U}_{SG}&=U_{SGq}+jU_{SGd}\\ \dot{U}_{AG}&=U_{AGq}+jU_{AGd}\\ \dot{I}_{SG}&=i_{SGq}+ji_{SGd}\\ \dot{I}_{AG}&=U_{AGq}+ji_{AGd}\end{aligned}\right\} \quad (10)$$

Exactly the same representation is made for the external axis parameters:

$$\left.\begin{aligned}\dot{A}&=a_{q}+ja_{d}; & \dot{C}&=c_{q}+jc_{d}\\ \dot{B}&=b_{q}+jb_{d}; & \dot{D}&=d_{q}+jd_{d}\end{aligned}\right\} \quad (11)$$

Substituting the values of the relationships (10) and (11) in equations (9), and making the corresponding multiplication, we obtain for the synchronous generator:

$$\left.\begin{aligned}U_{SGq}&=a_{q}\cdot i_{AGq}-a_{d}\cdot i_{AGd}+b_{q}\cdot i_{SGq}-b_{d}\cdot i_{SGd}\\ U_{SGd}&=a_{q}\cdot i_{AGd}+a_{d}\cdot i_{AGq}+b_{q}\cdot i_{SGd}+b_{d}\cdot i_{SGq}\end{aligned}\right\} \quad (12)$$

and for asynchronous generator:

$$\left.\begin{aligned}U_{AGq}&=c_{q}\cdot i_{SGq}-c_{d}\cdot i_{SGd}+d_{q}\cdot i_{AGq}-d_{d}\cdot i_{AGd}\\ U_{AGd}&=c_{q}\cdot i_{SGd}+c_{d}\cdot i_{SGq}+d_{q}\cdot i_{AGd}+d_{d}\cdot i_{AGq}\end{aligned}\right\} \quad (13)$$

In equations (12) and (13): U_{SGq}, U_{SGd}, i_{SGq}, i_{SGd} – components of the synchronous generator voltages and currents along the axes *d*, *q*, rotating at the speed of the synchronous generator rotor; U_{AGq}, U_{AGd}, i_{AGq}, i_{AGd} – components of voltages and currents of the WPP's asynchronous generator, performed on the basis of double-fed machine, written in axes *d*, *q* rotating also at the speed of the synchronous generator rotor. The transition from the variables $\overset{0}{i}_{AGq}$ and $\overset{0}{i}_{AGd}$, written in the axes d_a and q_a, rotating at the speed of the asynchronous generator rotor, to the variables i_{AGq} and i_{AGd}, determined by the following relationships.

$$\left.\begin{aligned} i_{AGq} &= -\overset{0}{i}_{AGd}\sin\theta_{sa} + \overset{0}{i}_{AGq}\cos\theta_{sa} \\ i_{AGd} &= \overset{0}{i}_{AGd}\cos\theta_{sa} + \overset{0}{i}_{AGq}\sin\theta_{sa} \end{aligned}\right\} \tag{14}$$

where θ_{sa} – the transverse axis angle of the WPP's asynchronous generator relative to the transverse axis of the synchronous generator, i.e. $\theta_{sa} = \theta_s - \theta_a$.

Accordingly, the components of the asynchronous generator voltage in the axes rotating at the speed of the WPP's asynchronous generator rotor can be represented as follows:

$$\left.\begin{aligned} \overset{0}{U}_{AGq} &= U_{AGd}\sin\theta_{sa} + U_{AGq}\cos\theta_{sa} \\ \overset{0}{U}_{AGd} &= U_{AGd}\cos\theta_{sa} - U_{AGq}\sin\theta_{sa} \end{aligned}\right\} \tag{15}$$

Thus, taking into account the above, the equations of the synchronous generator in the axes d, q, rotating at the speed of its rotor [8, 9], are represented in the form:

$$\left.\begin{aligned}
p\Psi_{dss} &= U_{SGd} - \omega_{rs}\cdot\Psi_{qss} - r_{ss}\cdot i_{rsd} \\
p\Psi_{qsc} &= U_{SGq} + \omega_{rs}\cdot\Psi_{dss} - r_{ss}\cdot i_{rsq} \\
p\Psi_{drs} &= -\frac{r_{drs}}{x_{drs}}\cdot\Psi_{drs} + \frac{r_{drs}\cdot x_{ads}}{x_{drs}}\cdot i_{SGd} + \frac{r_{drs}\cdot x_{ads}}{x_{drs}}\cdot i_{dfs} \\
p\Psi_{qrs} &= -\frac{r_{qrs}}{x_{qrs}}\cdot\Psi_{qrs} + \frac{r_{qrs}\cdot x_{aqs}}{x_{qrs}}\cdot i_{SGq} \\
p\Psi_{dfs} &= \frac{r_{dfs}}{x_{ads}}\cdot U_{dfs} - r_{dfs}\cdot i_{dfs} \\
p\omega_{rs} &= \frac{1}{T_j}\cdot m - \frac{1}{T_j}\cdot m_{ems} \\
m_{ems} &= \Psi_{dss}\cdot i_{SGq} - \Psi_{qss}\cdot i_{SGd} \\
i_{SGd} &= \frac{x_{qrs}}{\Delta d_s}\cdot\Psi_{dss} - \frac{\Delta d_{1s}}{\Delta d_s}\cdot i_{dfs} - \frac{x_{ads}}{\Delta d_s}\cdot\Psi_{drs} \\
i_{SGq} &= \frac{x_{qrs}}{\Delta q_s}\cdot\Psi_{qss} - \frac{x_{aqs}}{\Delta q_s}\cdot\Psi_{qrs} \\
i_{dfs} &= \frac{x_{drs}}{\Delta d_{2s}}\cdot\Psi_{dfs} - \frac{\Delta d_{1s}}{\Delta d_{2s}}\cdot i_{SGd} - \frac{x_{ads}}{\Delta d_{2s}}\cdot\Psi_{drs} \\
p\theta &= \omega_{rs} + 1
\end{aligned}\right\} \tag{16}$$

where $\Delta d_s = x_{dss}\cdot x_{drs} - x_{ads}^2$; $\Delta q_s = x_{qss}\cdot x_{qrs} - x_{aqs}^2$; $\Delta d_{1s} = x_{drs}\cdot x_{ads} - x_{ads}^2$; $\Delta d_{2s} = x_{drs}\cdot x_{dfs} - x_{ads}^2$.

The voltages U_{SGd}, U_{SGq} are determined based on the equations (13), in which the currents i_{AGd}, i_{AGq} are determined based on the expression (14).

The asynchronous generator equations – double-fed machines in axes d_a, q_a rotating at the speed of the asynchronous generator rotor [0, 0] can be represented as:

$$\left.\begin{aligned}
p\Psi_{dsa} &= \overset{0}{U}_{AGd} + \Psi_{qsa}\cdot(1-s_a) - r_{sa}\cdot\overset{0}{i}_{AGd} \\
p\Psi_{qsa} &= \overset{0}{U}_{arq} - \Psi_{dsa}\cdot(1-s_a) - r_{sa}\cdot\overset{0}{i}_{AGq} \\
p\Psi_{dra} &= -\overset{0}{U}_{AG}\cdot k_{ur}\cdot\sin(k_{fr}\cdot\tau) - r_{ra}\cdot i_{dra} \\
p\Psi_{qra} &= \overset{0}{U}_{AG}\cdot k_{ur}\cdot\cos(k_{fr}\cdot\tau) - r_{ra}\cdot i_{qra} \\
ps_a &= -\frac{1}{T_j}\cdot m - \frac{1}{T_j}\cdot m_{ema} \\
m_{ema} &= \Psi_{dsa}\cdot\overset{0}{i}_{AGq} - \Psi_{qsa}\cdot\overset{0}{i}_{AGd} \\
\overset{0}{i}_{AGd} &= k_{sa}\cdot\Psi_{dsa} - k_m\cdot\Psi_{dra} \\
\overset{0}{i}_{AGq} &= k_{sa}\cdot\Psi_{qsa} - k_m\cdot\Psi_{qra} \\
i_{dra} &= k_{ra}\cdot\Psi_{dra} - k_m\cdot\Psi_{dsa} \\
i_{qra} &= k_{ra}\cdot\Psi_{qra} - k_m\cdot\Psi_{qsa} \\
p\theta_a &= s_a \quad \text{and} \quad \theta_{sa} = \theta_s - \theta_a
\end{aligned}\right\} \tag{17}$$

where $k_{sa} = \dfrac{x_{ra}}{x_{ra}\cdot x_{sa} - x_{ma}^2}$; $k_{ra} = \dfrac{x_{sa}}{x_{ra}\cdot x_{sa} - x_{ma}^2}$; $k_m = \dfrac{x_{ma}}{x_{ra}\cdot x_{sa} - x_{ma}^2}$; $\overset{0}{U}_{AG} = \sqrt{\left(\overset{0}{U}_{AGd}\right)^2 + \left(\overset{0}{U}_{AGq}\right)^2}$.

In the system of equations (17), the stator voltage components U_{AGd} and U_{AGq} are determined from the equations (15).

3 Results and discussion

To study the adequacy of the presented equations (16) and (17), the parameters of the elements and the coefficients of the equations were calculated.

a) As a synchronous generator, a generator of the FWTG type (forced water turbo-generator) -100-2 [12] with parameters: P_l=100 MW; U_{Ln}=10.5 kV; cosφ=0.85 (sinφ=0.527); η=0.983.

As a result of the calculations, the generator parameters are determined in relative units: $x_d = 1{,}1$; $x_q = 0{,}73$; $x_\sigma = 0{,}16$; $x_{ad} = 0{,}94$; $x_{aq} = 0{,}57$; $x_f = 1{,}26$; $x_{rd} = 1{,}16$; $x_{rq} = 0{,}735$; $r_s = 0{,}008$; $r_{rd} = 0{,}0094$; $r_{rq} = 0{,}00765$; $T_j = 7s.$; $\frac{1}{T_j} = 0{,}00045$ [rad.].

b) 125 MVA 110 kV type transformer with power S_n=125 MVA is accepted as I step-up transformer, voltage 110/10.5 kV. No-load losses $\Delta P_{n.l.}$=105 kW; short-circuit losses $\Delta P_{s.c.}$=400 kW; short-circuit voltage in % $U_{n.l.}$=11 %; no-load current in % $I_{n.l.}$=0.55 %; cosφ_1 =0.8 [0].

c)AS-120 grade conductor with parameters $r_L^* = 0.27$ Ohm/km, $x_L^* = 0.379$ Ohm/km is accepted for 110 kV OHL, connecting node 1 with node 2 [0].

Accepting the total length of the line 30 km, we get: $r_L = 0.27\cdot 30 = 8.1$ Ohm; $x_L = 0.379\cdot 30 = 11.37$ Ohm.

d) 25 MVA 110 kV type transformer with power S_t=25MVA is accepted as step-down transformer Tr2, $U_{h.v.}$=100 kV, no-load losses $\Delta P_{n.l.}$=25 kW, short-circuit losses $\Delta P_{s.c.}$=120 kW, short-circuit voltage in % $U_{s.c.}$=10.5 %, no-load current in % $I_{n.l.}$=0.65 %.

Thus, total resistance of line and transformers

$$r_{LT} = 0.074 + 0.0135 + 0.0018 = 0.089 \text{ Ohm}$$

$$x_{LT} = 0.1 + 0.369 + 0.077 = 0.546 \text{ Ohm}$$

These resistances in relative units given in Z_{baz} of the synchronous generator

$$r_{LT} = \frac{0.089}{0.92} \approx 0.097$$

$$x_{LT} = \frac{0.546}{0.92} \approx 0.59.$$

e) Load parameters for steady-state mode.

The total load of the synchronous generator as indicated $P_{l.s.}$ =100 MW; $Q_{l.s.}$=60 MVAr.

Let's assume the active and reactive loads of the 1st node equal to: P_{L1} =60 MW; Q_{L1} =30 MVAr; and of the 2nd node, P_{L2}=40 MW, Q_{L2}=20 MVAr respectively.

f) Generator made on the basis of double-fed machine, which, of course, is an asynchronous machine with a phase rotor, is accepted as WPP's asynchronous generator [0, 0]. The power of one generator is assumed to be equal to P_l =20 MW, which is equivalent to 10 WPPs, each of which has the power of 2 MW.

Thus, parameters of equivalent generator of WPS in relative units at Z_{baz}=4.5 Ohm are equal to:

$$r_1^* = \frac{0.05}{4.5} = 0.011;\ r_2^{'*} = \frac{0.058}{4.5} = 0.0128;\ x_{\sigma 1}^* = \frac{0.45}{4.5} = 0.1;$$

$$x_{\sigma 2}^* = \frac{0.52}{4.5} = 0.115;\ x_m^* = \frac{13.66}{4.5} = 3.035;$$

$$x_1^* = x_{\sigma 1}^* + x_m^* = 3.035 + 0.1 = 3.135;\ x_2^* = 3.035 + 0.115 = 3.15.$$

g) Calculation of coefficients of external network equations:

$$a_q = -0{,}124;\quad b_q = 0{,}304;\quad c_q = -0{,}124;\quad d_q = 0{,}159;$$

$$a_d = -0{,}043;\ b_d = -0{,}172;\ c_d = 1{,}69;\ d_d = -0{,}13.$$

4 Conclusions

Dynamic characteristics have represented the discrete change process of the voltage and part of the current of the rotor, at which the speed of the WPP shaft changes within ± 22 %.

Model of operation of WPP with a double-fed machine in parallel with a synchronous generator is proposed. At the same time, the WPP power is assumed to be about 20% of the synchronous generator power.

References

[1] S.P. Gawande, K.B. Porate, Review of Parallel Operation of Synchronous Generator and Induction Generator for Stability, Conference: Proceedings of the Second International Conference on Emerging Trends in Engineering & Technology, ICETET 2009, Nagpur, Maharashtra, India, 16-18 December (2009). DOI: 10.1109/ICETET.2009.193 (2009)

[2] J. Liu, M. Yushi, T. Ise, J. Yoshizawa and K. Watanabe, Parallel operation of a synchronous generator and a virtual synchronous generator under unbalanced loading condition in microgrids, 2016 IEEE 8th International Power Electronics and Motion Control Conference (IPEMC-ECCE Asia), Hefei, 3741-3748. DOI: 10.1109/IPEMC.2016.7512894 (2016)

[3] V.Z. Silva, Â.J. Junqueira Rezek, R. Di, L.Corrêa, Analysis of synchronous and induction generators in parallel operation mode in an isolated electric system, IEEE 8th International Symposium on Power Electronics for Distributed Generation Systems (PEDG). DOI: 10.1109/PEDG40096.2017 (2017)

[4] D. Harikrishna, N.V. Srikanth, Dynamic Stability Enhancement of Power Systems Using Neural-Network Controlled Static-Compensator, TELKOM-NIKA **10**, 1, 9–16 (2012)

[5] I. Tamrakar, L.B. Shilpakar, B.G. Fernandes, R. Nilsen, Voltage and frequency control of parallel operated synchronous generator and induction generator with STATCOM in micro hydro scheme **1**, 5, 743–750. DOI: 10.1049/iet-gtd:20060385 (2007)

[6] L. Congwei, W. Haiqing, S. Xudong, L. Fahai, Research of stability of double fed induction motor vector control system, Shenyang, China, Proc. of the Fifth International Conference on Electrical Machines and Systems **2**, 1203–1206, 18–20 Aug. (2001)

[7] A.P. Sokolov, O.V. Gazizova, Yu.N. Kondrashova, Study of the transients with the loss of field of the synchronous generator in the industrial electric power station, Quality Management and Reliability of Technical Systems, IOP Conf. Series: Materials Science and Engineering **666**, 1–7 (2019)

[8] R.I. Mustafaev, L.G. Hasanova, Modeling and study of the operating modes of synchronous generators of wind power plants with frequency control, Electricity, 7, 34–40 (2010)

[9] R.I. Mustafaev, L.G. Gasanova, An investigation of the dynamics of a wind power unit equipped with synchronous generators with permanent magnets, Russian Electrical Engineering **86**, 6, 258–263 (2015)

[10] M. Heller, W. Schumacher, Stability analysis of doubly-fed induction machines in stator flux reference frame, Brussels, Belgium, Proc. of 7th European Conference on Power Electronics and Applications **2**, 707–710, 8–10 Sept. (1997)

[11] H. Akagi, H Sato, Control and performance of a doubly-fed induction machine intended for a flywheel energy storage system, IEEE Transactions on Power Electronics **17**, 1, 109–116. DOI: https://doi.org/10.1109/63.988676 (2002)

[12] D.P. Kothari, I.J. Nagrath, *Electric Machines* (Tata McGraw-Hill Education, 834, 2004)

[13] I.A. Baumstein, S.A. Bazhenova, *High Voltage Electrical Reference* (3rd ed., Revised. and add., Moscow: Energoatomizdat, 768, 1989)

[14] R.I. Mustafayev, L.H. Hasanov, M.M. Musayev, E.M. Mamedov, X.İ. Nabiyev, Modeling and analysis of operating regimes of hydraulic units of small hydropower plant with double fed machine, Electromechanics, 6, 59–65 (2015)

[15] A. Dittrich, A. Stoev, Grid voltage fault proof doubly-fed induction generator system, Proc. Power Electronics and Applications (EPE), Toulouse, France (2003)

[16] J.W. Park, K.W. Lee, D.W. Kim, et al., Control Method of a Doubly-Fed Induction Generator with a Grid Synchronization against Parameter Variation and Encoder Position, Industry Applications Conference, 42nd IAS Annual Meeting, Conference Record of 2007 IEEE, New Orleans, LA, 931–935 (2007)

E3S Web of Conferences **178**, 01035 (2020) https://doi.org/10.1051/e3sconf/202017801035
HSTED-2020

Identification of corner separation modelling in axial compressor stage

Sergey A. Serkov[1], *Viacheslav A.* Sedunin[1], *Vitalii L.* Blinov[1,*], *Oleg V.* Komarov[1], and *Yurii M.* Brodov[1]

[1]Ural Federal University, Turbines and Engines Department, 620002 Yekaterinburg, Russia

Abstract. The paper presents a study of corner separations in hub to blade region at various operation conditions towards compressor stall. It is known that for compressor flows with low or none separations computation fluid dynamics with RANS methods work quite well, however, for highly separated flows they are no longer entirely valid. Therefore, several criteria were applied for prediction and quantification of possible corner separation, and the main interest of this work is in predicting the separation just before it will actually happen by certain flow metrics, so these metrics can be further used as a 'pre-stall' criteria whilst the RANS CFD operating point still behave within its appropriate limits. Also the effect of shear lean is discussed in the presented context.

1 Introduction

The main requirements for all gas turbine engines, regardless of their purpose, are high efficiency, operational reliability and long life span [1,2]. One of the key factors determining the efficiency of a gas turbine engine is the performance of an axial compressor and turbine. Since the flow in axial turbomachines is generally three-dimensional, it is important to consider a appropriate flow operating point in the blade channels [3].

A common phenomenon that becomes more significant as the pressure ratio of the compressor increases is the corner separation of the flow [4-7]. It occurs in the area of transition of the end-wall into the airfoil, leading to blockage of the flow path and decrease in the stall margin. In order to reduce the size of the corner separation, it is possible to perform aerodynamic optimization of the blade profile using a shear lean [6-12]. This method allows to increase the stall margin by reducing the pressure gradients and increasing the magnitude of the surface shear on the blades near the endwall.

In the frontal stages of the compressors, the effect of corner separation is comparatively less than in the rear stages, while the thickness of the endwall boundary layer is 1-2% of the height of the blade channel, which does not introduce significant changes in the flow structure, and the deviation of the actual velocity triangles from the calculated ones is insignificant [8]. As the flow moves to the intermediate and rear stages, a substantial increase in the endwall boundary layer occurs, which is associated with an increase in the incidence angle on the blades near the meridional contours. Moreover, the blockage due to this boundary layer can reach 20% of the span height [8].

2 Materials and methods

The flow in the blade channels was studied on a two-stage low-speed axial compressor [13,14], for which the CFD operating point was verified [15,16]. Understanding of the corner separation and its effect on the flow structure is important for optimization of the blade row, including an operating point where the optimization to be performed. For this purpose, certain criteria for identifying corner separation both near the endwall and in the blade channel have been proposed.

To select an operating point for optimization, it is necessary to establish how the corner separation of the flow develops in the blade channel and by what criteria it can be identified. At the same time, it is necessary to establish a threshold value of the criteria for the identification of the vortex, at which the corner separation will affect not only the region at the endwall, but also propagate towards the midspan, and also at what size of the corner separation it is possible to optimize the blade for reducing the size of this corner separation.

The determination of the critical size of the corner separation of the flow allows us to identify the operating point at which the optimization should take place. With an acceptable level of flow separation, the blade row works with high efficiency and productivity to ensure the required characteristics. With increasing aerodynamic load (1), the corner separation increases, which leads to a local decrease in the performance of blade sections and, as a result, their approach to their stall boundary. That is, optimization will be held at the conditions with sufficient incidence to cause the corner separation, but with no separation over the entire span of the blade.

* Corresponding author: v.l.blinov@urfu.ru

$$\psi = \frac{H_{ad}}{U^2} \qquad (1)$$

where H_{ad} - adiabatic work; U - circumferential speed.

The separation point of the boundary layer on the blade surface is characterized by zero values of shear stresses on the wall – τ (Figure 1) [4,17]. At the pressure side of the blade, separation of the flow is unlikely and therefore, along the entire length of the flat section at any height of the blade, the shear stresses are above zero. Taking into account the condition for the onset of flow separation and the condition for the distribution of shear stresses on the pressure side, it is possible to obtain criterion (2), which characterizes the beginning of separation on the 2D section of the blade from its stall.

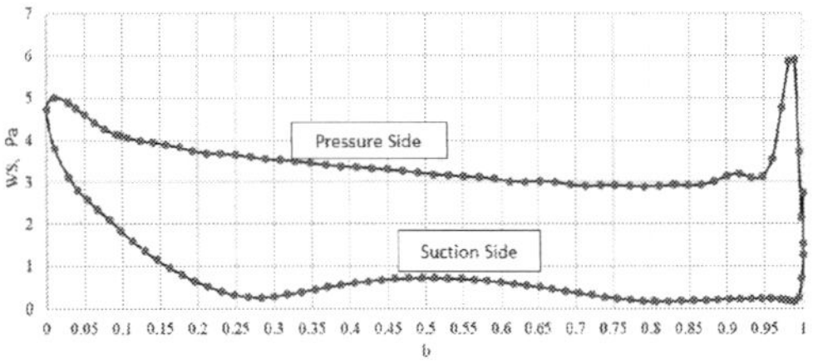

Fig. 1. Shear stress distribution (τ) along the profile chord.

The ratio of the shear stresses between the suction and the pressure sides will be a dimensionless parameter characterizing the separation of the flow regardless of the thermodynamic and kinematic quantities distributed along the chord of the blade.

$$WS_a = \frac{\int_0^1 \tau_{ps} \frac{db}{b}}{\int_0^1 \tau_{ss} \frac{db}{b}}, \qquad (2)$$

where b is the blade chord; τ_{ss}- shear stresses on the suction surface; τ_{ps} - shear stresses on the pressure surface.

Since the blade is a set of flat sections, the analysis of the operation of only a separate 2D section does not make sense due to the three-dimensional nature of the flow. Therefore, to control the parameter WS_a along the height of the channel, the blade is divided into 4 regions (Figure 2). This is done for the convenience of determining the response of the blade row performance to its geometric change. Regions 1 and 4 relate to areas at meridional contours from 0 to 20% and from 80 to 100%, respectively. It is in these parts of the blade that the corner separation and tip vortex are formed. Flat sections located in these regions are the first to approach the boundary of their stability. Regions № 2 and 3 show the row performance at the middle part of the channel from 20 to 80% of the blade height. These regions are less loaded during compressor normal operation with minimum flow separation. Each area is responsible for monitoring the aerodynamic parameters of the flow and signals an increase or decrease in the loading, as well as change in the level of flow separation from the suction side [18].

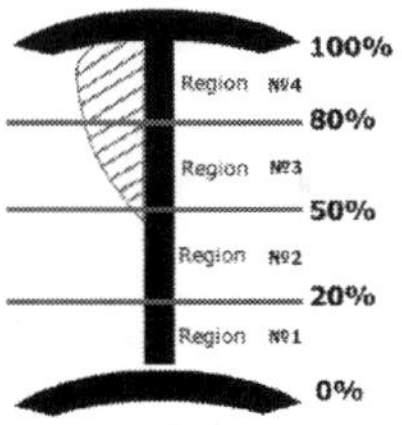

Fig. 2. Blade regions for WSa calculation.

3 Results

The study of the proposed criterion WS_a was performed on the stator blades of the first stage of a low-speed two-stage axial compressor at the entire operating range. The dependence of the parameter WS_a on the diffusion factor (3) for each selected area was drawn (Figure 3).

$$F_d = \left(1 - \frac{C_2}{C_1}\right) + \left(\frac{C_{u1} - C_{u2}}{2C_1}\right)\frac{t}{b}, \qquad (3)$$

where C_1 is the flow inlet velocity; C_2 - flow outlet velocity; C_{u1} - circumferential projection of the flow velocity at the inlet; C_{u2} - circumferential projection of the flow velocity at the exit; t/b- reversed solidity.

When loading rises, incidence, and, therefore, diffusion go up as well. That lead WS_a to grow with distinctive parabolic lines.

Recommended values of the diffusion factor for stable operation vary in a range Fd = 0.4 ... 0.6 [19,20]. In this case, in the guide vane of the first stage, the parameter Fd at the operating point is in the range from 0.1 to 0.2, and when the recommended value Fd = 0.4 is reached, a corner separation is formed near the shroud, which extends to region 3 and 4 (Figure 3).

It is worth noting that with a small size of the corner separation, which extends only along the endwall and does not occupy a large area of the channel, the mass-average value in each particular section may be lower than the real value at the endwall.

At the operating point WSa is less than one, which can be explained by positive incidence. At a given incidence angle, shear stresses on the pressure side become lower than the ones at the suction side, but they are non-zero.

As the load increases, the ratio of shear stresses goes up. The highest sensitivity to the flow separation happens, when WS_a reaches 1.6..1.8. Further compressor operation will be accompanied with a developed corner separation next to endwalls, which extends beyond the 20 percent zone. Optimization of the blade row below the critical values of 1.6 ... 1.8 will be impractical, since the flow is less affected by separation, or the separation does not create sufficient blockage of the channel.

An analysis of streamlines showed that when the row works at the value of WS_a = 1.6 ... 1.8, corner separation propagates along the blade only in region 4, and there are no saddle points on the suction surface, the presence of which would indicate the presence of vortex motion

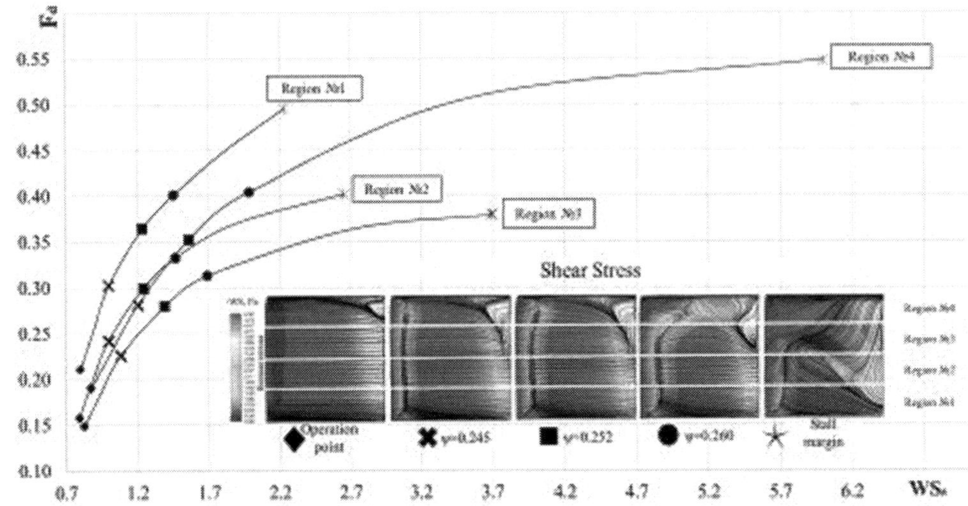

Fig. 3. WS_a parameter vs diffusion factor.

(Figure 3). With a further increase in aerodynamic loading, the corner separation extends to region 3, which increases the value of WS_a to values 1.6 ... 1.8 in zone 4, and a saddle point is formed on the surface of the blade along with focal points and nodal lines, which indicates the formation of a vortex along the blade. Therefore, when the corner separation is propagated along the blade, its transition from the region near the meridional contour to the region of the middle of the blade determines the threshold value for the beginning of the corner separation formation as WS_a = 1.6 ... 1.8 regardless of the flow in the channel, since the parameter WS_a is a dimensionless quantity and depends only on local parameters on the blade surface.

It is not possible to predict the separation of the flow using only direct physical quantities, such as velocity and pressure, since the separation of the flow depends on the derivatives of these quantities. Whereas, using the WS_a criterion, flow separation is determined by reducing the strength of interaction between the flow and the surface of the blade, which, in turn, depends on a decrease in the gradient of the flow velocity normal to the surface (du/dy) and dynamic flow viscosity (μ) (4).

$$\tau = \mu \frac{dU}{dy}, \quad (4)$$

The WS_a criterion shows only the presence of the flow separation from the wall at which the action propagates only at the surface of the blade. At the same time, calculations of channel blockage using direct physical quantities only allows us to calculate the region where the flow velocity is less than the freestream one, and separation has already happened. However, adjacent to the separation zone there is a bigger region with significantly higher turbulence intensity. This larger zone can be taken into account when predicting the operation of the row and determining the operating point for row optimization.

To determine the dimensions of the flow separation and the blockage area of the channel, a second criterion is proposed, which includes the ratio of the area that is determined in the region of the turbulence intensity above TI = 30% (5) to the cross-sectional area of the channel (F) (Figure 4, left) (6).

The turbulent flow rate is defined as:

$$TI = \frac{\sqrt{\frac{2}{3}E_k}}{V} 100\%, \quad (5)$$

where E_k is the kinetic energy of the flow; V is flow velocity.

$$TI_a = \frac{(F - TI_{area})}{F} 100\%, \quad (6)$$

The largest area of the corner separation region is expected at the trailing edge, that explains the choice of identification plane (Figure 4, left). To set a threshold value for turbulence intensity, the criterion for the area of flow turbulence intensity was limited to 30%. The streamlines that are detached from the surface of the blade at the interface between the vortex and the free steady flow zone have a high turbulence intensity, which is about 30% (Figure 4, right).

The vortex forming the corner separation has high turbulence intensity and includes not only the region of flow separation itself, but also the vortex formation region limited by the limiting current line, and the area corresponding to TI_a = 30% will be larger than the area with low Mach numbers of the corresponding average values within the separation region.

Comparison of the blockage area with isosurfaces of low Mach number (Ma) and the TI_a criterion (Figure 5) shows that near design operating points, the isosurface blockage area based on TI_a criterion is higher than blockage area based on Mach number. Further raise in

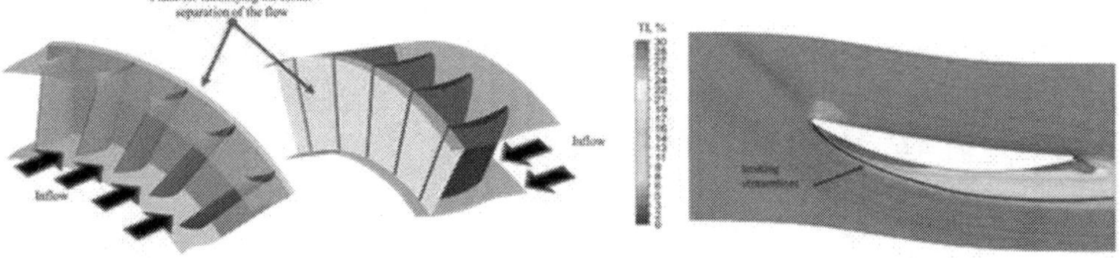

Fig. 4. Surface location for corner separation identification (left) and Turbulence intensity for limiting streamlines (right).

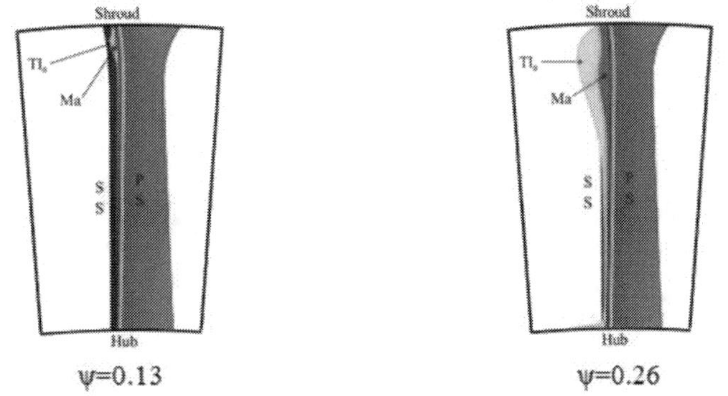

Fig. 5. Blockage area with Mach number and TIa parameter.

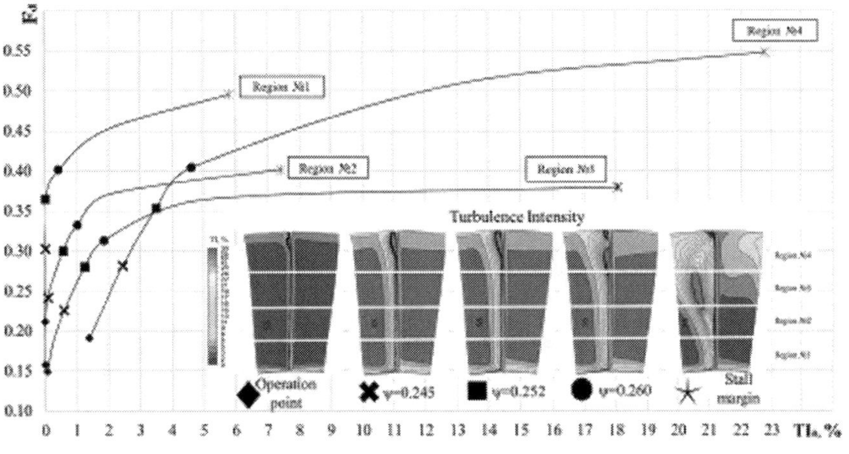

Fig. 6. Parameter TI_a vs diffusion factor for stator 1.

loading causes the TI_a parameter to reach 7%, which is 3.5 times greater than the Mach number based blockage.

Thus, this parameter allows you to identify the corner flow separation at the point when the flow separation based on the Mach number does not pose a danger to the stable operation of the row. This makes it possible to predict the behaviour of the row at higher loading, which, in turn, allows you to estimate the stall margin.

As well as the WS_a criterion, the TI_a criterion is related to the diffusion factor (Figure 6). In this case, at the 4th region of Stator 1, flow separation exists at the operating point and the blockage is 1.5% based the TIa criterion. Further loading increase leads to greater channel blockage and when the value TI_a = 4 ... 7% is reached at the shroud, the graph becomes rather horizontal.

Therefore, when the TI_a value reaches critical values of 4 ... 7%, it shows that the corner separation passes from the near-endwall region towards the midspan.

4 Discussion

Based on the distribution of the parameters TI_a and WS_a in stator blade row, a diagram was obtained for determining the row operating points depending on the

https://doi.org/10.1051/e3sconf/202017801035

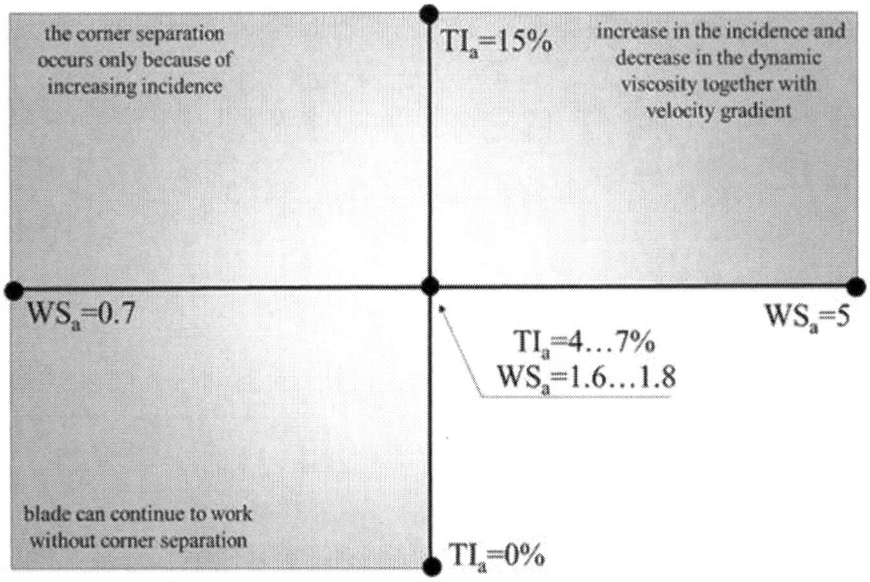

Fig. 7. The degree of corner separation development the blade channel, depending on the values of TI_a and WS_a.

degree of development of the corner separation (Figure 7). Based on this diagram, it is possible to determine how critical the analysed operating point for the compressor performance.

If the values of the parameter TI_a at the exit from approach 4..7%, but the values of the parameter WS_a are closer to 1, then the corner separation occurs only because of increasing incidence. In this case, the flow in the suction side boundary layer has the necessary velocity gradient normal to the surface, and the flow viscosity is sufficient so as not to form a swirl region along the profile surface.

Increase in both TI_a and WS_a means simultaneous increase in the incidence and decrease in the dynamic viscosity together with velocity gradient. Thus, the separation of the flow begins at the leading edge and continues along the entire suction surface of the selected area. In this case, saddle and focal points are formed on the surface, which signal a vortex motion near the wall.

5 Conclusions

It is established that the criteria WS_a is used to identify the corner separation at the blade surface, and the criterion TIa is used to determine the spanwise extension of the corner separation. At the same time, they may not necessarily be optimised at the same compressor operating point due to the different nature of the flow in the rows and the different shape of the blade.

First of all, it is necessary to evaluate the effect of corner separation on the flow in the channel using the TI_a, and additionally check the WS_a. This is because first the blockage of the channel will be determined, which affects the performance of the compressor, and then the degree of resilience to separation of each zone using the WS_a will be determined. If the threshold value is reached for any isolated distinguished zone for the TI_a (the threshold value of blockage is reached), but the ratio of the shear stresses of the suction and pressure side is below the limit WS_a = 1.6 ... 1.8, (streamlines that separate from the blade do not form vortex between the limiting streamlines and the blade surface) it is possible to optimize the selected blade row, but this region of the blade can continue to work with corner separation.

The results obtained can be used in 3D optimization of compressor blade rows by means of lean, sweep and dihedral

References

[1] N.A. Cumpsty, *Compressor aerodynamics* (Longman Scientific & Technical, 1989)

[2] V.A. Sedunin, O.V. Komarov, Blinov V.L., A.V. Skorokhodov, S.A. Serkov, Yu.M. Brodov, Redesign of an axial compressor with mass flow reduction of 30%, Third International Conference on Energy Production and Management in the 21st Century **222**, 93–101 (2018)

[3] V.A. Sedunin, O.V. Komarov, V.L. Blinov, A.V. Skorokhodov, A.O. Procopets, The application of modern computational fluid dynamics techniques for increasing the efficiency and stability of an axial compressor in an industrial gas turbine, Energy Production and Management in the 21st Century **190**, 707–719 (2014)

[4] U. Dallmann, *Topological Structures of Three-dimensional Flow Separation* (DFVLR, Gottingen, 26–45, 1983)

[5] H. To, R. J. Miller, The effect of aspect ratio on compressor performance, Proceedings of ASME Turbo Expo, GT2015-43016 (2015)

[6] J.D. Denton, L. Xu, The effects of lean and sweep on transonic fan performance, ASME Turbo Expo, GT-2002-30327 (2002)

[7] S.J. Gallimore, J.J. Bolger, N.A. Cumpsty, M.J. Taylor, P.I. Wright, and J.M. Place, The use of sweep and dihedral in multistage axial flow compressor blading – part 1: university research and methods development, Journal of Turbomachinery **124**, 521–532 (2002)

[8] R.F. Behlke, The Development of a Second Generation of Controlled Diffusion Airfoils for Multistage Compressors, Journal of Turbomachinery **108** (1986)

[9] M.Yu. SHelkovskij, Aerodinamicheskoe sovershenstvovanie osevogo desyatistupenchatogo kompressora stacionarnogo GTD, Aviacionno-kosmicheskaya tekhnika i tekhnologiya **9**, 27–32 (2014)

[10] Jr.R.W. Moore, D.L. Richardson, A study of the end wall boundary layer in an axial compressor blade row, Massachusetts Institute of Technology Report (1955)

[11] O.V. Komarov, V.A. Sedunin, V.L. Blinov, Application of optimisation techniques for new high-turning axial compressor profile topology design, Proceedings of ASME Turbo Expo, GT2014-25379 (2014)

[12] V.L. Blinov, V.A. Sedunin, O.V. Komarov, Multi-criteria optimisation of subsonic axial compressor blading, IOP Conf. Series: Materials Science and Engineering, 643 (2019)

[13] M.D. Hathaway, Axial compressoriishi Aerodynamic design and performance of a two-stage, axial-flow compressor (baseline), Asmes-84178, TCRL-24 (1983)

[14] D.L. Tweedt, T.H. Axial compressoriishi Stator blade row geometry modification influence on two-stage, axial-flow compressor aerodynamic performance, Asmes-84179 TCRL-25 (1983)

[15] O.V. Komarov, V.A. Sedunin, V.L. Blinov, S.A. Serkov, Verifikaciya zadachi chislennogo modelirovaniya techeniya vozduha v osevoj kompressornoj stupeni, Vestnik Moskovskogo gosudarstvennogo tekhnicheskogo universiteta im. N.E. Baumana. Seriya «Mashinostroenie **4**, 54–68 (2016)

[16] Yu.M. Brodov, O.V. Komarov, V.A. Sedunin, V.L. Blinov, S.A. Serkov, Osobennosti verifikacii CFD-modeli osevogo kompressora. CHast' 2: analiz integral'nyh parametrov, Kompressornaya tekhnika i pnevmatika, 4 (2018)

[17] D.J. Peake, M. Tobak, Three-Dimensional Interactions and Vortical Flows with Emphasis on High Speeds, Agardograph, 252 (1980)

[18] S.A. Serkov, *Identifikaciya i ustranenie uglovogo otryva potoka v lopatochnyh vencah pri reshenii zadach chislennogo modelirovaniya techeniya v osevyh kompressorah GTD* (Ekaterinburg, UrFU, 2019)

[19] A.R. Howell, Design of axial compressors, Proceedings of the Institute Mechanical Engineers (1945)

[20] Yu.B. Galerkin, L.I. Kozachenko, *Turbokompressory: ucheb. posobie* (SPb.: Izd-vo Politekhn. un-ta, 2008)

Automatic voltage regulation system construction for synchronous generator of a small hydro power plant using thyristor pathogens

Ishembek Kadyrov[1], *Nurzat* Karaeva[1], *Zheenbek* Andarbekov[1], *Bakytuulu* Azamat[1], *Oleg* Fedorov[2], and *Oleg* Vladimirov[3]

[1]Kyrgyz national Agrarian University named after K.I. Skryabin. 720005, Bishkek, Kyrgyzstan
[2]Nizhny Novgorod State Technical University n.a. R.E. Alekseev, 603950, Nizhny Novgorod, Russia
[3]Kazan State Power Engineering University, Kazan, Russia

Abstract. The article presents indicators of the hydropower potential of Kyrgyzstan, shows the share of small hydropower plants in the production of electric energy in the energy sector of the republic, reveals the problematic issues that need to be revealed in this work. A brief description of the existing automatic voltage regulation system and automatic field blanking system is given. Based on the analysis of existing systems, their shortcomings were identified and the main functions formulated for automatic voltage regulation systems in both static and dynamic modes associated with starting and stopping the generator, loading and dumping, as well as protective measures taken to ensure reliability operation of the hydraulic unit. The result of the research is the development of the main functional blocks in the system of automatic voltage regulation with a thyristor pathogen and a brief explanation of their work. The synthesis technique of the excitation current regulator of a synchronous generator and the main indicators obtained in the modernization process are shown. In conclusion, all the functions inherent in the modernized system for automatically controlling the excitation of a synchronous generator, aimed at the safe operation of the hydraulic unit, are noted.

1 Introduction

At present, the hydropower potential of Kyrgyzstan is concentrated in 252 large and medium-sized rivers and is estimated at 18.5 million kW of power and more than 140-160 billion kWh of electricity. Moreover, the average slope of mountain rivers varies from 5 to 20 m per 1 km of length, and the average specific power is from 2227 to 5322 kW / km. therefore, the possibilities for building small hydropower stations in the country are unlimited. So far, according to general statistics, the gross share of small hydropower plants in the production of electric energy in the energy sector of the republic is only 1 %.

The above figures are useful for investors who invest in priority sectors of the economy, such as the construction of small hydropower plants. Naturally, these small hydropower plants will be equipped with modern equipment with a perfect control system for the economical and safe operation of the hydraulic unit. In this article, we consider the issue of constructing automatic voltage regulation systems operated by small hydropower plants taking into account the realities of today.

Electricity production by several small hydropower plants, concentrated mainly in the Chui oblast of Kyrgyzstan, is carried out by the generating company OJSC «Chakan HPP». The main consumers of these products are companies and business entities of the Chui region and the southern regions of Kazakhstan.

The first several small hydropower plants were put into operation in 1945, and the rest over the next 10 years.

Consequently, the age of the main equipment in these several small hydropower plants is over 60 years old; therefore, they are practically obsolete and require modernization. This primarily concerns the system for automatically controlling the excitation of generators, since in this system the generators themselves become unusable and require continuous inspection and maintenance, most often major repairs. At the same time, frequent blackouts occurring due to the failure of direct current generators and downtime caused by the repair of a machine require the replacement of machine exciters with alternative semiconductor exciters [1-3].

The problem of constructing a system for automatic voltage regulation (AVR) considered in this article, taking into account the replacement of the DC generator with a static thyristor exciter, as a way to extend the life of hydrogenerators.

The experience gained in the development of control systems for frequency-controlled electric drives of excavators proved to be useful in the construction of AVR for synchronous generators, since the control objects in both cases are high-power alternating current machines [4-6].

At the same time, for the correct choice of the control structure of the system "Thyristor Pathogens - Synchronous Generator" (TP-SG) for a small hydroelectric power station, it is first necessary to identify and list the requirements for the AVR system of synchronous generators based on information obtained in the study of modernized systems, including including automatic field suppresion (AFS) systems.

Fig. 1 shows the most common scheme for including the excitation winding of a synchronous generator in the anchor circuit of the pathogen [7-9]. For SG with significant power (over 100 kW), the time constant of the excitation winding T_g is about 1 s or more. Such inertia delays transients in the processes of switching on and off the excitation winding in the anchor circuit of the pathogen, which may be an undesirable fact. To speed up (speed up) transients, methods are used that reduce the equivalent value of T_g or increase the rate of change of the electromotive force of the synchronous generator by controlling the excitation voltage.The decrease in T_g is possible due to the introduction of an additional resistor R_a in the field winding circuit (see Fig. 1).

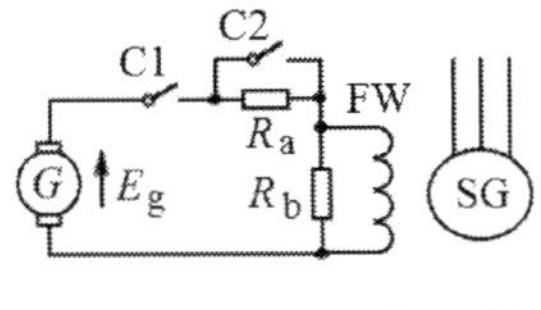

Fig. 1. The circuit of the excitation winding of the synchronous generator to the anchor circuit of the generator to a controlled pathogen.

It is obvious that in order to maintain the nominal value of the electromotive force of the generator, the excitation voltage must be increased by the magnitude of the voltage drop across this resistor in the nominal mode. Another way of boosting is to apply an increased voltage to the field winding during the transient process. At the end of the process, the voltage should be instantly reduced to the required steady state value. This is achieved by introducing a resistor R_a using contact C2.

The degree of forcing of transients in the diagram of Fig.1 is characterized by the rate of forcing:

$$k_f = \frac{R_b + R_f}{R_b R_f}\left(R_a + \frac{R_b \cdot R_f}{R_b + R_f}\right). \quad (1)$$

Where

$$R_a = k_b \frac{k_f - 1}{k_f + 1} R_f. \quad (2)$$

For a given k_f, the necessary circuit voltage can be calculated, and using (2) the resistance of the additional resistor. In fig. 2. shows the change in the electromotive force of the excitation and the excitation current during the start-up process for two forcing methods with constantly entered resistance (curves 1) and with a cutoff (curves 2). It is obvious that forcing with a cut-off is more effective, since in this case the increased excitation voltage remains unchanged over the entire interval of the transition process.

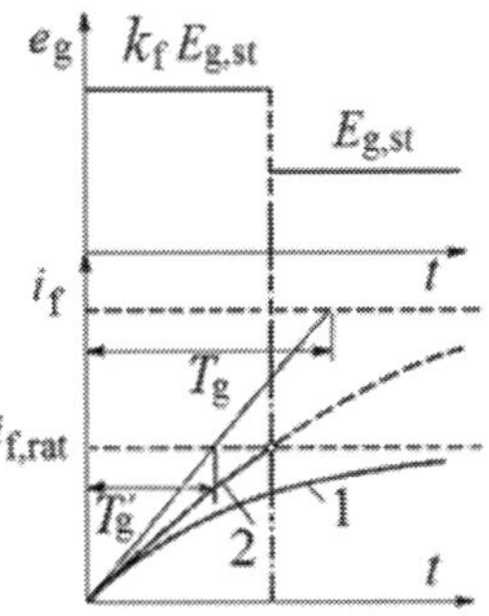

Fig. 2. Graphs of varience in the excitation voltage and the electromotive force of the pathogen during startup with forced.

Consider an automatic field blanking system of the AHL-1 type at the Electrosila plant, which is most often found at modernized small hudropower plants (Fig. 3) [10]. In the field blanking circuit of the Electrosila plant, an automatic machine is used that is equipped with three contacts: main contacts 1, arcing contacts 2 and arcing lattice 3, representing a system of parallel-connected contacts. Using this machine, the supply voltage from the exciter terminals is supplied to the excitation winding of the synchronous generator. In emergency mode, the circuit breaker trips its main contact 1 and an arc arises on it, then after a short period of time the arcing contact 2 opens. It also arises, which, under the action of an external transverse magnetic field excited by permanent magnets, is drawn into the region of the arcing lattice 3 The arc is divided into a series of series-connected short arcs that burn until the current in the field winding drops to zero.

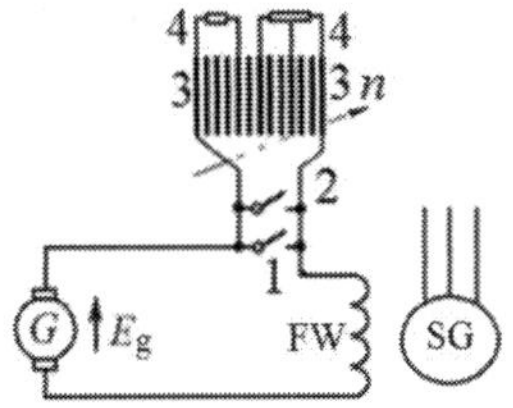

Fig. 3. Scheme automatic field suppression type AHL-1 plant "Electrosila".

The originality of this scheme is that the voltage on the arc U_a, located in the lattice, remains constant at all stages of the appearance of arcs and is equal to:

$$U_a = U_{sh} n = const,$$

whereU_{sh} – voltage on a short arc located between adjacent plates of the grating; for copper plates U_{sh} = 25 - 30 V; n is the number of gaps between the plates.

Thus, the generator protection devices together with the AHL system operate in emergency conditions so that the high-speed protection system timely disconnects the synchronous generator from the network, and the AHL system must quickly remove the excitation, i.e. pay off the field. If the AHL system does not work, then the destructive action of the emergency mode will continue due to the current generated by the generator itself [10, 11].

2 Research result

The object of modernization is synchronous generators (SG) of small power plants of the Alamedin cascade, which for 60 years of continuous operation are in a state where modernization, in the first place, of its excitation system became inevitable. In the normal power supply mode, automatic control of the SG excitation provides automatic voltage regulation (AVR) at the terminals of the current collectors, that is, due to the AVR system in the cascade, the static stability of the operation of hydroelectric power plants (HPP) is ensured.In emergency conditions or in the transient mode associated with load shedding and surge, due to the action of automatic voltage regulation, dynamic stability of each hydroelectric power station in the power supply system is connected, interconnected by a single power system. In addition, the automatic voltage control system provides a lightweight option for self-starting of hydrogenerators, creates normal conditions for the alternating voltage generators to be pulled into synchronism [16].

Before modernization, automatic control of excitation in synchronous generators of the Alamedin cascade was carried out from machine exciters, which are driven by a drive from a turbine shaft through a gearbox. These systems are built according to the type of power current compounding with a proportional device for automatic control of excitation, when continuous replenishment is made along the circuit of the main excitation winding of the pathogen. Moreover, to ensure a stable and definite distribution of reactive power between parallel-running cascade generators, the voltage control characteristic at the connection point of the generators should have positive statism in the generator's own current.

The main disadvantage of automatic voltage regulation with an electric machine exciter is that the maximum voltage can only reach twice the rated voltage. Although this voltage is provided to double boost the rotor current, it cannot satisfy the requirements, since the known inertia of the pathogen delays the increase in the excitation current. For this reason, the permissible duration of the stimulation of the excitation may be insufficient in comparison with the possible duration of emergency and post-emergency transients in the power system.

Automatic voltage regulation can also be done by regulating the excitation of a synchronous machine using controlled exciters based on modern semiconductor converters. The advantage of thyristor exciters in the automatic voltage control system is the ability to select the maximum excitation voltage to a 4-fold value, with which an accelerated increase in the rotor current to a twofold value is provided, after which the force limitation comes into effect.

Fig. 4 shows an exemplary view of the U-shaped characteristics of the generator for various values of power P and power factor $\cos\varphi$ explaining the principle of current compounding, which combines the process of inertialess control of the stator current.

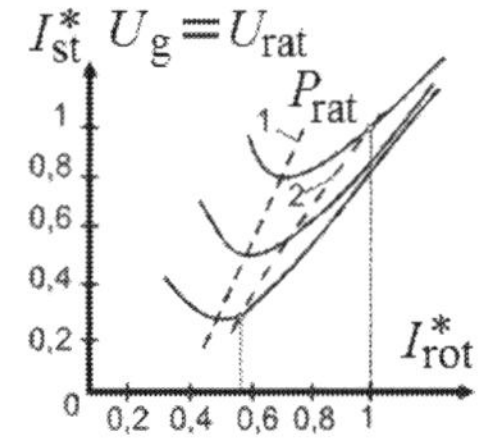

Fig. 4. Compounding characteristics of a synchronous generator.

When constructing a system for automatically controlling the excitation of a synchronous generator with a thyristor exciter, the principle of current compounding will be implemented as a separate unit, called the current compounding unit (CCU). In addition, protective functions such as protection against overloads and short circuits in the external network must be incorporated in the control system of the pathogen.

In the SG stator circuit, the controlled parameters are phase currents in the stator windings, since during synchronous generator overloads the temperature of the windings of the machine can exceed the permissible values. In this case, the protection should work if this excess is observed for a long time, since in the operating mode of the generator there may be short-term excess currents. Therefore, in the control system of the automatic excitation regulator, you must enter the block for calculating the emergency current of the stator of the synchronous generator.

The current in the rotor circuit must also be controlled, since short-term surges in the rotor current can lead to unstable operation of the synchronous generator itself. The control of the SG rotor current should not be carried out along the direct current circuit of the exciter, but along the alternating current circuit from the side of connecting the power unit to the alternating current network. In this case, the unit for calculating the emergency current of the rotor of the synchronous generator will be implemented in the same way as the unit for calculating the emergency current of the synchronous generator.

Finally, as part of the ARV control system, a block of protection against emergency values of the stator currents and the rotor of the synchronous generator must be provided, which will generate a U_{log} signal, due to which the control signals will be blocked by the power valve of the pulse-phase control system.

Blocks for calculating the emergency current of the stator (BCECS) and rotor (BCECR) of the synchronous generator. To build blocks unit for calculating the emergency current of the rotor of the synchronous

 https://doi.org/10.1051/e3sconf/202017801036

generator and unit for calculating the emergency current of the synchronous generator having the same structure, you must select the input signals.

The input signals for them can be currents at the controlled points of the synchronous generator and a signal in the control system, taken as a reference.

In fig. 5. at the input of the comparator of the Operational Amplifier 1, the rectified voltages are proportional to the stator currents (U_a, U_b, U_c) taken from the secondary winding of the CT1 current transformers with the reference voltage. As can be seen from the diagrams (Fig. 5), the output voltage of the comparator can take two values: negative in the case if $|U_{ref}| \geq |U_a, U_b, U_c|$, and positive if $|U_{ref}| < |U_a, U_b, U_c|$.

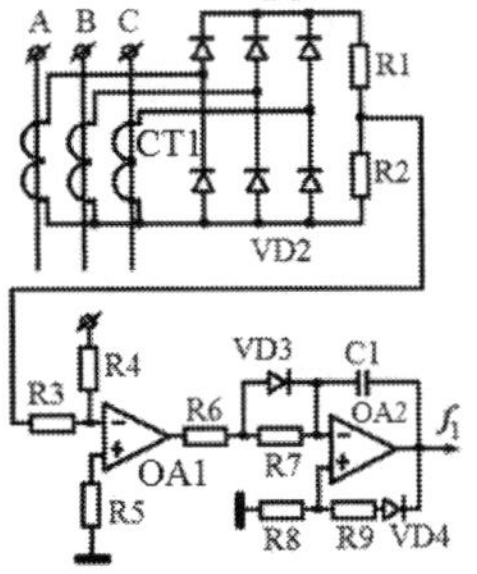

Fig. 5. Schematic diagram of the block calculating the emergency current of the stator of a synchronous generator.

The voltage from the output of the comparator is supplied to the input of the integrator of the Operational Amplifier 2, which has a variable structure, in which the integration process is performed with both a large and a small time constant. A large integrator time constant is set for cases when emergency currents occur at controlled points (the diode at the integrator input is closed). Moreover, in order to avoid false triggering of protection circuits against individual surges (Fig. 6, a), the emergency currents of the stator and rotor are monitored for several periods (see Fig. 6, b). A small time constant is set for normal currents when part of its input resistance is shunted due to an open diode at the input of the integrator.

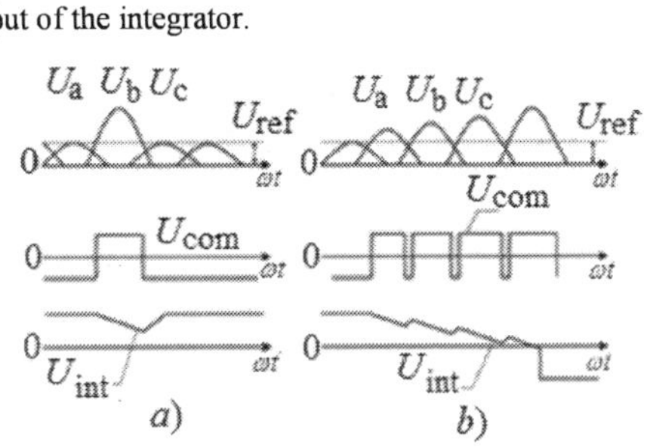

Fig. 6. Voltages at various links of the unit for calculating the emergency current of the rotor of the synchronous generator and unit for calculating the emergency current of the synchronous generator circuits.

With the appearance of a negative voltage at the output of the integrator, positive feedback comes into effect and its output voltage jumps to the bias voltage and this information f1 about the emergency mode will be stored for an arbitrarily long time. For visual observation, the red LED on the board lights up, which is included in the circuit of the external emergency alarm unit.

The block of protection against emergency values of the stator currents and the rotor of the synchronous generator. The signals generated at the output of the unit for calculating the emergency current of the rotor of the synchronous generator and unit for calculating the emergency current of the synchronous generatorare fed to the input of the comparator of the protection unit (Fig. 7) as the emergency value of the stator currents f_1 or rotor f_2 of the synchronous generator.In this case, if or, then a voltage close to zero is determined at the output of the comparator, which is determined by the leakage current through the diode VD10. The low voltage of the comparator determines the logic zero level and is applied to the corresponding input of the pulse-phase control system unit, as a result of which the supply of pulses to the control electrodes of the power valves will be blocked and the exciter of the synchronous generator will be disconnected from the network.

2.1 Pulse-phase control system unit

Currently, the most widely used synchronous phase shifting system, built as a system with a reference linear voltage. The multichannel principle of forming control pulses is laid in the block of the pulse-phase control system (PPCSU) and has a typical structure. The algorithm for constructing linear phase shift is described in detail in [15, 17], therefore, circuit solutions are not given in this paper.

2.2 Current compounding block

The system of automatic control of stator current in a synchronous generator is called a compounding device with voltage corrector in the special literature. As a specialist in the field of automated electric drives, it's easier to explain the current compounding unit as a unit of regulators with a proportional stator current regulator and proportionally an integral rotor current regulator according to the circuit shown in Fig. 8. The synthesis of the regulators was carried out using the method of sequential correction[12, 18].

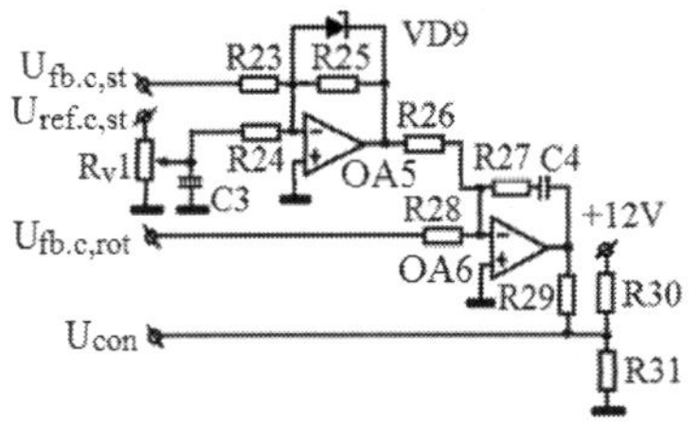

Fig. 7. Schematic diagram of the current compounding unit.

The proportional stator current regulator is implemented using the Operational Amplifier 1, and proportionally, the integral rotor current regulator is implemented using the Operational Amplifier2. The current compounding unit with a proportional device for automatic control of excitation carries out the process of regulating the stator current of a synchronous generator according to characteristic 2 shown in Fig. 4.

A controlled thyristor exciter operating on the excitation winding of a synchronous generator can be represented by a first-order linear link, as shown in the block diagram (Fig. 8). The absence of intermittent currents contributes to this idea, since the pathogen load is a large inductance [14]. It should be noted that if the thyristor exciter is represented as an inertial unit with a small uncompensated time constant $T_{ta} = 0.01$ s, and the excitation coil of the SG by an inertial unit with a large time constant $T_f = 0.7$ s, then as a result of the synthesis of the control system of the automatic excitation control device by the method of sequential correction, the excitation current controller adopts the structure of the P– link.

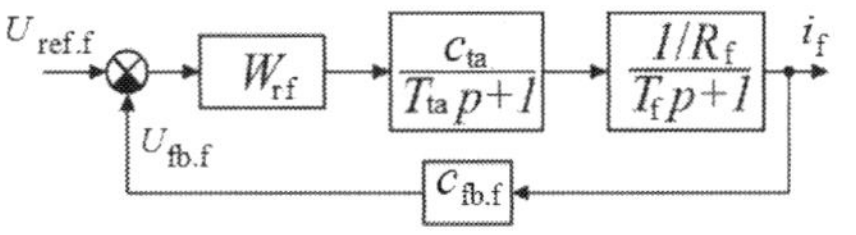

Fig. 8. Block diagram of a field circuit of a synchronous generator.

The oscillogram of the transient process of the increase in the excitation current to the nominal at a single push of the input action is shown in Fig. 9. The proportional part of the regulator ensures the excitation of the pathogen by voltage, reaching up to four times the nominal. With such forcing, an increase in current is achieved for $t_p = 0.07$ s, which indicates the advantages achieved as a result of the modernization of the automatic excitation control device. [19-25].

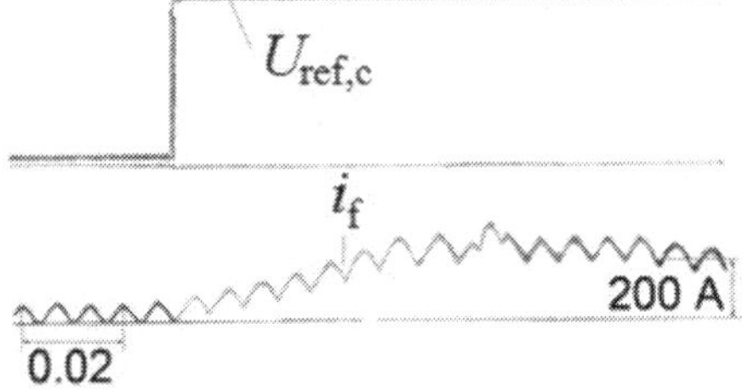

Fig. 9. Oscillogram of the increase in current in the excitation circuit during an abrupt action.

Fig. 10 a shows a schematic diagram of a device for automatically controlling the excitation of a synchronous generator, which shows that the pathogen is assembled according to a bridge circuit and connected to a three-phase network through a matching transformer (not shown in the diagram). In this scheme, the task for the stator current, which must be regulated, is carried out using the potentiometer R_{v1}. The feedback signal on the stator current is taken from the output of the diode bridge VD2, and the signal proportional to the rotor current is taken at the output of the diode bridge VD2. The emergency currents calculated by the the unit for calculating the emergency current of the rotor of the synchronous generator and unit for calculating the emergency current of the synchronous generator are stored and signaled with the help of LEDs, respectively: VD5 - excess of stator current; VD8 - excess current of the rotor.

Fig. 10, *b* shows a prototype of a device for automatically controlling the excitation of a synchronous generator, in which the control board is shown at the top, the power block in the middle, and the switching and protection equipment below. As can be seen from fig. 11, *b*, the thyristor exciter is compactly placed in a small cabinet.

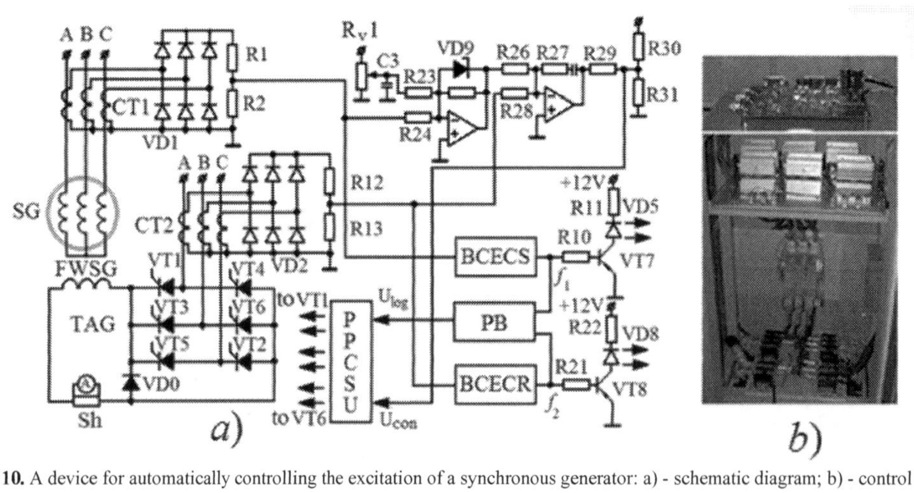

Fig. 10. A device for automatically controlling the excitation of a synchronous generator: a) - schematic diagram; b) - control board and power unit.

3 Conclusion

The current compounding unit, together with the stator and rotor emergency current calculation units, is a system for automatically controlling the excitation of a synchronous generator of small hydroelectric stations, provides the required operating modes of the synchronous generator and performs the necessary protection functions from emergency situations. The forcing of the pathogen by voltage, reaching up to four times the nominal one, ensures high speed of the pathogen, i.e. the current rises and falls during s., which is necessary both in the processes of retraction into synchronism at the time of start-up, and for the quick damping of the generator field in emergency conditions. In its dimensions and adjusting properties, an automatic excitation control device has a noticeable advantage compared to an electric machine exciter.

References

[1] National Electric Network of Kyrgyzstan [Electronic resource]. Available at: www.nesk.kg/ru/ (accessed: 15.06.2019)

[2] Small hydropower plants as an alternative for Kyrgyzstan [Electronic resource]. Available at: https://knew.kg (accessed: 15.06.2019)

[3] Construction and reconstruction of generating capacities Electronic resource]. Available at: www.invest.kg/ru/ (accessed: 15.06.2019)

[4] I.V. Bochkarev, I.Sh. Kadyrov, Microprocessor control system for the system "Direct frequency converter – AM" of the electric drive of the excavator, Electromechanics **5**, 25–30 (2007)

[5] I.Sh. Kadyrov,The study of dynamic processes in the gears of the rotation mechanism of the dragline during the sudden disconnection of the supply voltage, Mining equipment and electromechanics **6**, 29–31 (2008)

[6] I.V. Bochkarev, I.Sh. Kadyrov, Optimization of parameters of the control device of the electromechanical system of excavators, Electrical Engineering **2**, 2–7 (2009)

[7] V.M. Terekhov, *Elements of an automated electric drive* (Energoatomizdat, 224, 1987)

[8] A.I. Waldeck, *Electric Machines* (Energy, 832, 1978)

[9] I.P. Kopylov, *Electromechanical energy converters* (Energy, 400, 1973)

[10] G.L. Vulman, *Operation of generators in power plants* (Gosenergoizdat, 344, 1963)

[11] E.I. Gracheva, O.V. Naumov, Operating Mode Influence on Probability Characteristics of Electric Devices, Journal of engineering and applied sciences **11**, 1, 2934–2938 (2016)

[12] V.I. Klyuchev, *Theory of electric drive* (Energoatomizdat, 560, 1985)

[13] M.G. Chilikin, M.M. Sokolov, V.M. Terekhov, *Fundamentals of an automated electric drive* (Energy, 586, 1974)

[14] V.I. Klyuchev, V.M. Terekhov, *Electric drive and automation of general industrial mechanisms* (Energy, 360, 2000)

[15] I.Sh. Kadyrov, *Design of Electromechanical Systems for Automatic Machines* (Publishing Center Teknik, 211, 2006)

[16] P.G. Grudinsky, *Electrotechnical Handbook* (T. 2, Energy, 752, 1975)

[17] V.S. Rudenko, V.I. Senko, I.M. Chizhenko, *Fundamentals of transformative technology* (Higher school, 424, 1980)

[18] I.Sh. Kadyrov, *Principles, methods and algorithms for constructing microprocessor control systems for electromechanical machine assemblies* (Teknik Publishing Center, 200, 2007)

[19] V.A. Bessekersky, E.P. Popov, *Theory of automatic control systems* (M, Science, 767, 1972)

[20] I.Sh. Kadyrov, T.S. Borukeev, G.D. Matekova, Microprocessor-based system for pulse-phase control of a thyristor converter, News KSTU named after Razzakova **1**, 41, I, 36–43 (2017)

[21] O.V. Fedorov, Expeditious forecasting of power consumption, International Conference on Industrial Engineering, Applications and Manufacturing (ICIEAM), 8076221, 1–4 (2017)

[22] Y.I. Gracheva, O.V. Fedorov, Forecasting Reliability Electrotechnical Complexes of In-Plant Electric Power Supply Taking into Account Low-Voltage Electrical Apparatuses, International Conference on Industrial Engineering, Applications and Manufacturing (ICIEAM), 5 (2019)

[23] Y.I. Gracheva, O.V. Naumov, Estimation of Power Losses in Electric Devices of the Electrotechnical Complex, International Conference on Industrial Engineering, Applications and Manufacturing (ICIEAM), 6 (2019)

[24] Y.I. Gracheva, O.V. Naumov, A.N. Gorlov, Influence of value resistance contact units of switching devices on losses of the electric power in shop networks of low tension, IOP Conference Series: Materials Science and Engineering, 6 (2019)

[25] Y.I. Gracheva, A.N.Alimova, Calculating Probability of Faultless work of Shop Nets with the Help of Coefficients of Ratio, International Russian Automation Conference (RusAutoCon), 3–6 (2019)

Method for designing energy-efficient ventilation systems based on regenerative heat exchangers

Nikolay Monarkin[1,*], *Anton* Sinitsyn[1], *Denis* Karpov[1] and *Timur* Akhmetov[2]

[1]Vologda State University, Institute of civil engineering, 160000 Vologda Lenina str. 15, Russia
[2]Kazan State Power Engineering University, Kazan, Russia

Abstract. The method of designing ventilation systems based on stationary switching regenerative heat exchangers is presented. Recommendations for selection of rational device parameters are given and substantiated. The diameter and thickness of a single channel wall of regenerative nozzle; the total diameter and length of nozzle; the time of heat accumulation or regeneration stages; the heat capacity and density of nozzle material; air flow rate are considered.

1 Introduction

Nowadays residential multi-apartment buildings in Russia and other countries have problems of ensuring the required air exchange of rooms due to the use of hermetically sealed plastic windows and wear of existing ventilation systems [1]. Moreover, the heat consumption for ventilation can reach 50% of the total energy consumption of a building [2, 3]. One of the solutions to these problems is the use of compact ventilation devices called stationary switchable regenerative heat exchangers (SSRHE). These devices enable air exchange due to their paired installation in premises, and are also able to return (utilize) part of thermal energy of the indoor air to the room.

Given the relevance of SSRHE usage, a generalized approach to their design and application is needed. The works [4-7] present some recommendations regarding the installation of SSRHE and their use in certain rooms, but such recommendations are not exhaustive. Therefore, the purpose of this paper is to create a method for designing and selecting SSRHE for ventilation systems in building. The method is based on the previously published studies [8-10].

2 Materials and methods

SSRHE consists of a regenerative nozzle, which performs the function of a heat exchanger, and a reversible fan, enabling changing the stages of air supply and exhaust (Figure 1). In figure 1: 1 - fan; 2 - nozzle; 3 - housing; 4 - thermal insulation; 5 - outdoor enclosure; T_{in} and T_{out} are temperatures of internal and external air, respectively, °C; G_{acc} and G_{reg} are air consumptions at the stage of heat accumulation and regeneration, respectively, kg/h; D is the total nozzle diameter, m. The nozzle is a solid-state cylinder penetrated by many channels of small diameter for air passage. The SSRHE operation consists of alternately switching the stages of accumulation of the internal air heat by the nozzle (air extraction mode) and heat recovery by heat transfer from the nozzle to the external passing air (inflow mode). These devices are installed in external walls. [1].

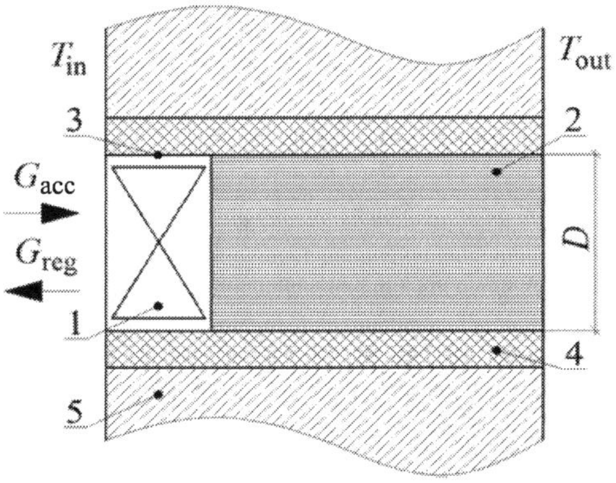

Fig. 1. Schematic diagram of SSRHE.

The purpose of designing ventilation systems based on SSRHE is to determine the nozzle parameters, the number of devices to be installed, the air flow through a separate SSRHE, and the choice of fan size.

Nozzle parameters should be selected based on rationality. The rationality of parameters was estimated by the energy efficiency coefficient of SSRHE, which is the ratio of the actual energy used by the regenerator to the maximum possible energy for use:

$$E = \frac{\delta T}{\Delta T_{max}}, \quad (1)$$

where $\Delta T_{max}=(T_{in} - T_{out})$ is the maximum possible cooling or heating of air in the nozzle from T_{in} to T_{out},

* Corresponding author: nikolay-monarkin@yandex.ru

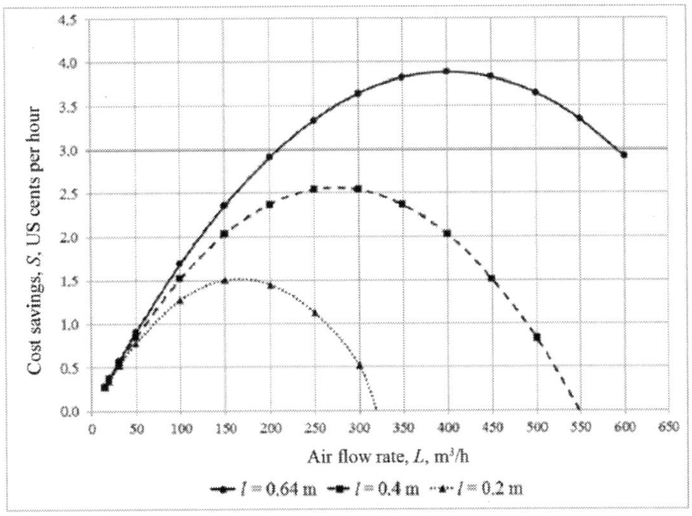

Fig. 2. Relationship between cost savings and air flow rate through SSRHE for various nozzle lengths.

°C; δT is cooling (heating) of internal air in the nozzle averaged for the accumulation (regeneration) stage, °C, it was obtained from mathematical modelling of heat transfer between air and nozzle in a single SSRHE channel [9].

The nozzle design is considered energy efficient for $E \geq 0.9$.

The air flow rate through the SSRHE is also assessed in terms of rationality. In case of increase in air flow rate, the regenerator thermal efficiency E decreases, and the aerodynamic losses in nozzle increase, but at the same time, the regenerator productivity L (m^3/h) increases. Therefore, it is possible to determine the optimal value of air flow rate as a compromise between performance and resistance. For this, the function of saving money costs S (United States cents per hour) when using SSRHE, was tested on optimum [10]:

$$S = c_a\left(T_{\text{in}} - T_{\text{out}}\right)C_H\left(-a_1L^2 + E_0L\right) - 2a_2L^2C_E, \quad (2)$$

where c_a is the mass heat capacity of air, J/(kg·K);

C_H is the cost of heat energy, US cents per kWh;

C_E is the cost of electric energy, US cents per kWh;

a_1 and a_2 are the coefficients depending on nozzle parameters;

E_0 is the initial value of coefficient E.

Figure 2 shows, as an example, the relationship between S and air flow for various regenerative nozzle lengths. The curves were calculated for the following data: C_H=28.6 US dollars per kWh; C_E=6.43 US cents per kWh; T_{in} =21°C; T_{out}=-4°C; d=4 mm; δ=0.5 mm; τ_{acc}=τ_{reg} =40 s. The figure shows that the longer the SSRHE is, the higher are the savings in cash costs. Under these conditions, the optimal air flow rates are: L=150 m^3/h for l=0.2 m and S=1.50 US cents per hour; L=250 m^3/h for l=0.4 m and S=2.54 US cents per hour; L=400 m^3/h for l=0.64 m and S=3.89 US cents per hour.

The upper curve of Fig.2 is limited by air flow rate of 600 m^3/h, since if this value is exceeded at d=4 mm, the flow mode will change from laminar to turbulent, and the used calculation method will be incorrect. The tendency of the lower curves to zero along the ordinate axis indicates that there will be no savings when exceeding the shown air flow rates.

3 Results and discussion

Further the rational parameters of the SSRHE nozzle will be discussed.

Diameter of a unit nozzle channel (d). This parameter should be selected judging from a compromise between pressure loss and energy efficiency E. It was determined that the smaller diameter of channel provides larger cost savings. However, if the size of channels is too small, they can be covered and clogged with dust and ice during moisture condensation. Therefore, it is recommended to take the channel diameter in the following range: d=0.003-0.005 m.

Wall thickness of a single channel δ. Wall thickness between the nozzle channels should not be too small. With a decrease in wall thickness, the number of channels increases and the thermal efficiency increases. However, a decrease in wall thickness leads to a decrease in the nozzle mass, which will adversely affect its heat-storage ability and require less cycle time. Also this can result in difficulties during nozzle fabrication and in a decrease in its strength. It is recommended to take the wall thickness of the nozzle channel in the following range: δ=0.4-0.6 mm.

Stage time (τ_{acc}, τ_{reg}). It was determined that a decrease in time of heat accumulation/regeneration stage leads to an increase in SSRHE efficiency. However, the time cannot be too small, since during the regenerator operation the high-quality assimilation of the supply air

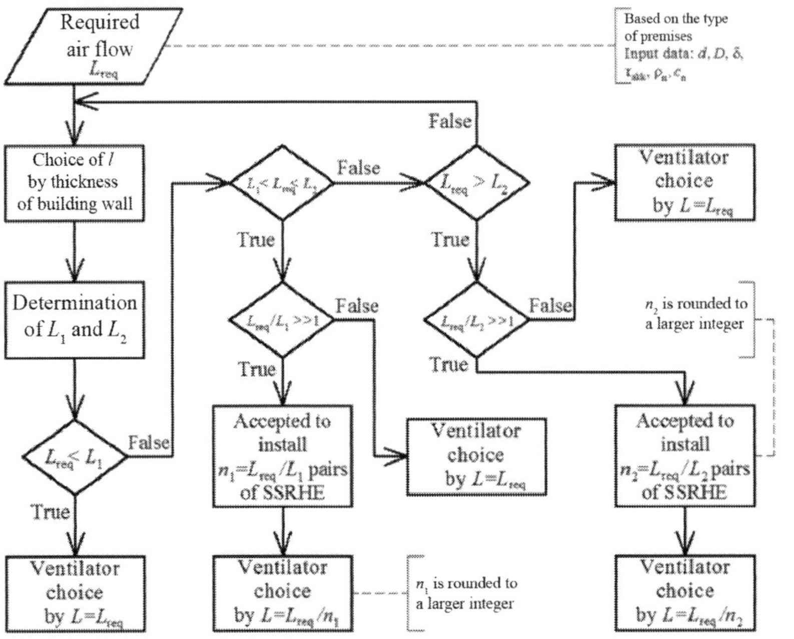

Fig. 3. Flowchart of the algorithm of the engineering method for designing a ventilation system based on SSRHE.

must be ensured, which requires sufficient time to form a stable supply jet. That is, during one stage of SSRHE operation, a sufficient amount of external air must be supplied to allow air exchange in the entire volume of room, and not just around the device itself. Therefore, it is recommended to take $\tau_{acc}=\tau_{reg}$=30-50 s.

Density (ρ_n) and heat capacity of the nozzle material (c_n). These thermophysical parameters of nozzle do not significantly affect the change in SSRHE efficiency. The following density ranges are recommended: ρ_n=1400-2000 kg/m^3. The recommended heat capacity of nozzle material is in the range c_n=2000-3000 J/kg·K. Among the existing materials, polymers, including various plastics, are the most suitable for the indicated ranges. So, structurally complex products can be constructed from them. Ceramic nozzles are also used in the existing SSRHEs, but their heat capacity hardly exceeds 1000 J/kg·K.

Total nozzle diameter (D). The total nozzle diameter affects the total heat transfer area, via the number of channels for air passage. The diameter of SSRHE housing as well as nozzle diameter is determined by the permissible size of holes in the outer walls of building, based on the strength conditions and the building appearance. For certain hole sizes, a special project for their implementation, as well as measures to strengthen the wall structure, may be required. In some regions of Russia, the maximum dimensions of holes in the external walls are stipulated, for which it is not necessary to develop a project for the construction of an architectural opening. Based on review of regulatory documents, the recommend allowable hole size is up to 0.2 m.

Based on the foregoing, the following algorithm of method for designing energy-efficient ventilation systems based on SSRHE is proposed (Figure 3):

1. The required air flow rate L_{req}, (m^3/h) for ventilation is determined according to the norms based on the type of premises, or according to the calculation based on the released hazards.

It is known that the simplest way to determine the required air exchange is to use standard multiplicity. In this case, the air exchange rate can be given either in the form of the number of room volumes, or in the form of a norm per unit area, per unit volume, per unit of equipment, per person. When determining the required air exchange by calculation, it is necessary to use the methods described in the standards and academic literature.

2. The length of regenerator nozzle l (m) is selected according to the thickness of the building outer wall.

The length of the regenerative nozzle directly depends on the entire SSRHE length, which takes into account the fan length. The total SSRHE length depends on the thickness of the building outer wall where the device is mounted. The wall thickness depends on the material used, as well as on the climatic parameters of the construction area. According to the existing wall thicknesses of buildings made of various materials, the nozzle length can vary between 0.2-0.64 m. However, the existing SSRHE models include that installed outside the wall structure, and the device is connected with the

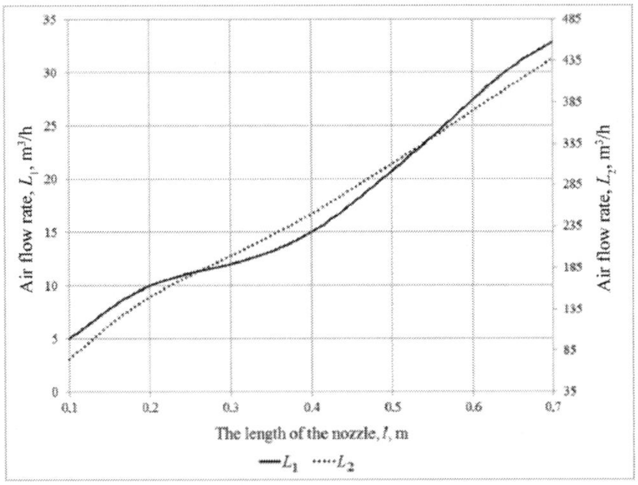

Fig. 4. Relationship between the flow rates L_1 and L_2 and the length of the nozzle.

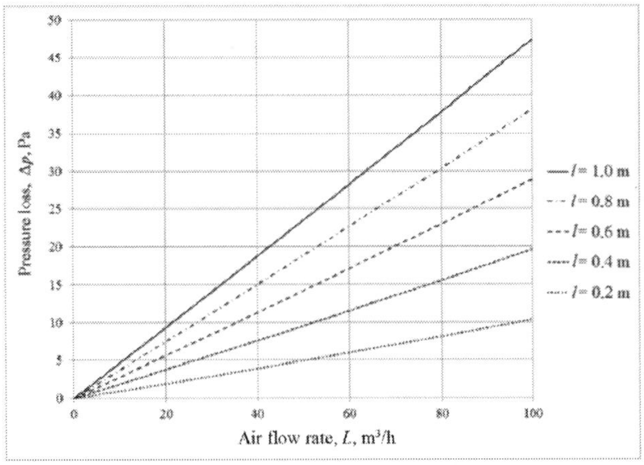

Fig. 5. Relationship between pressure loss and air flow rate and the length of the nozzle.

external environment through the air duct. In such cases, the diameter of duct passing through the outer enclosure can be reduced, and the wall thickness does not affect the nozzle length.

3. Using the calculated nozzle length, the following air flow rates (m³/h) are determined: L_1 is the flow rate, for which the thermal efficiency of the regenerator is E=0.9; L_2 is the flow rate, for which the cost savings S for heat and electric energy are maximized. Obviously, $L_2 > L_1$.

4. L_{req} is compared with L_1 and L_2 with three possible results: 1) $L_{req} < L_1$; 2) $L_1 < L_{req} < L_2$; 3) $L_{req} > L_2$.

5. If $L_{req} < L_1$, it is advisable to install a pair of regenerators to work in antiphase. From flow $L=L_{req}$ pressure losses in regenerator Δp are determined. Fan is selected according to L and Δp.

6. If $L_1 < L_{req} < L_2$, then the ratio $n_1 = \frac{L_{req}}{L_1}$ is calculated and rounded to a larger integer. Two solutions are possible here:

a) when $n_1 >> 1$ it is recommended to install n_1 pairs of SSRHE with air flow through each $L = \frac{L_{req}}{n_1}$, by which pressure losses are determined, and a fan is selected;

b) when n_1 slightly exceeds 1 it is recommended to install one pair of SSRHE with air flow through each $L=L_{req}$, by which pressure losses are determined, and a fan is selected.

7. If $L_{req} > L_2$, the ratio $n_2 = \frac{L_{req}}{L_2}$ is calculated and rounded to a larger integer. Two solutions are also possible here:

a) when $n_2 >> 1$ it is recommended to install n_2 pairs of SSRHE with air flow through each $L = \frac{L_{req}}{n_2}$, by which pressure losses are determined, and a fan is selected;

b) when n_1 slightly exceeds 1 it is recommended to install one pair of SSRHE with air flow through each $L=L_{req}$, by which pressure losses are determined, and a fan is selected.

To determine the flow rates L_1 and L_2 as a function of nozzle length, the nomogram was built, shown in Figure 4. To determine the pressure loss in nozzle Δp from the air flow rate and the nozzle length, a nomogram was built, shown in Figure 5. Nomograms are built for the source data: d=4 mm, τ_{acc}=40 s, D=200 mm, δ=0.5 mm.

It should be noted that this method does not take into account the influence of SSRHE on the moisture conditions of room. Moreover, if there is no significant release of moisture in the room, then upon steady-state air exchange, the moisture content of the internal air will be equal to that of external air. In this case, in the cold season, there may be a need for humidification of air with special devices, which is a separate issue for study.

Also, the method does not address the issue of compliance with the permissible noise level during the SSRHE operation. But the use of low-power (low-performance) fans in SSRHE design resolves this question.

4 Conclusions

Based on the performed studies a method was developed to select the SSRHE parameters, the number of installed devices, the air flow through a separate SSRHE and size fans through the required room air exchange. An appropriate algorithm has been compiled, which significantly reduces the time, which design-engineer spends for the parameters selection. This algorithm has a simplified scheme, and doesn't require solving a complex system of interconnected heat transfer equations between the air and the regenerative nozzle.

References

[1] A.A. Sinitsyn, N.N. Monarkin, T.V. Rogulin, Automation and energy saving of machine-building and metallurgical industries, technology and reliability of machines, instruments and equipment: materials of the 9th International Scientific and Technical Conference, 202-205 (2014)

[2] R.K. Calay, W.C. Wang, The 12th International Conference on Air Distribution in Rooms (2011)

[3] I.A. Sultanguzin, E.V. Zhigulina, Y.V. Yavorovsky, I.D. Kalyakin, A.V. Govorin, A.V. Fedyukhin, A.A. Krolin, S.V. Guzhov, O. Derevianko, L. Mukhametova, Development strategy of the energy supply system for MPEI campus blocks based on green building, IOP Conference Series: Earth and Environmental Science 288, 1, 012071 (2019).

[4] V.A. Vasiliev, A.I. Gavrilov, K.K. Kamenetsky, E.V. Sobol, Journal of the International Academy of Refrigeration, 1 (2010).

[5] A.A. Zakharov, M.I. Nizovtsev, Scientific Bulletin of NSTU, 1, 54 (2014).

[6] M.R. Nasr, M. Fauchoux, R.W. Besant, C.J. Simonson, Renewable and Sustainable Energy Reviews 30 (2014).

[7] D.G. Moghaddam, A. Oghabi, G. Ge, R.W. Besant, C.J. Simonson, Applied Thermal Engineering 61 (2013).

[8] N. Monarkin, S. Lukin, Yu.M. Anurov, B.A. Tihomirov, G.A. Agasiants, S.M. Galileev, T.R. Akhmetov, IOP Conf. Series: Earth and Environmental Science 337 (2019).

[9] N.N. Monarkin, S.V. Lukin, A.A. Kochkin, Volga Scientific Journal 3, 47 (2018).

[10] N.N. Monarkin, S.V. Lukin, A.A. Kochkin, BST: Bulleten' stroitel'noy tehniki 6, 1018 (2019).

Supply of additional working fluid to the flow part of the NK-8 gas turbine engine

George Marin[1,2], *Dmitrii* Mendeleev[1,2,*], *Boris* Osipov[1], and *Azat* Akhmetshin[1]

[1]Kazan State Power Engineering University, 420034, Krasnoselskaya St., 51, Kazan, Russian Federation;
[2]JSC "Tatenergo" branch "Kazan CHP-2", 420036, Tatsevskaya St., 11, Kazan, Russian Federation

Abstract. Modern energy development strategies of advanced countries are based on the construction of gas turbine units which is associated with sufficiently high values of thermal efficiency and a relatively short term for putting them into operation. In this paper, the NK-8 engine is considered. It is modernized with a mixing chamber and a power turbine for the purpose of its ground application. A study was conducted of the injection of an additional working fluid into the flow part of a dual-circuit gas turbine engine. Steam is used as an injectable substance. For research a mathematical model was created in the AS «GRET» software package. The studies were carried out under constant load, the maximum load during injection was determined. An additional worker can be supplied with summer power limitations when it is necessary to increase the power of a gas turbine installation. Studies have shown that the maximum power that can be obtained by supplying steam to the flow part is 32.2 MW.

1 Introduction

In any country, energy is a basic sector of the economy. Energy provides the prerequisites for the application of new technologies and along with other factors provides a modern standard of living for the population. Currently, the development of energy is based on gas turbine technologies. [1-2]

In order to ensure maximum thermal efficiency the initial parameters of the gas turbine (temperature and the degree of pressure increase in the compressor) are continuously increasing. At the same time, other opportunities are being used to increase the efficiency and specific power of the plants (intermediate cooling of the air in the compressor, reheating of the working fluid of the gas turbine in front of the power turbine)

2 Gas turbine description and research model

In a gas turbine engine of a simple design, an increase in the useful work of the engine can be achieved by reducing the cost of compressor drive. Reducing the work on air compression can be achieved by cooling the air in the compressor path or by supplying an additional working fluid. The working fluid may be water or steam. There is always excess steam of different temperatures and pressures at the thermal power station, therefore, steam was chosen for research as the injected substance. [3-5]

The engine NK-8 was selected as the engine under study. The main characteristics of the gas turbine installation are presented in table 1. A longitudinal section of the engine is shown in Figure 1.

The engine includes:

1. Two-stage eleven-stage compressor;
2. Multi-nozzle combustion chamber;
3. Two-stage three-stage axial gas turbine.

Table 1. Technical characteristics of a GTU (NK-8).

No	Characteristic	Meas. unit	Value
1	Air flow	kg/s	289
2	Compression ratio		12.93
3	Temperature in the combustion chamber	K	1179
4	Expansion ratio in high pressure turbine		2.278
5	Expansion ratio in low pressure turbine		2.64

The engine consists of the following main components - a low-pressure compressor, a high-pressure compressor, an annular combustion chamber, a high-pressure turbine and a low-pressure turbine. All of these units can be used in completing installations for various power values output to the drive shaft. [6-9]

The low-pressure compressor (LPC) is five-stage. It consists of a rotor and a stator. The rotating part of the LPC - the rotor has five working rings (according to the number of steps). The first two stages fan operated on both circuits. The remaining three steps compress the air in the internal circuit. The fixed part of the LPC - stator has guide vanes. There is an input guide vane (IGV) with guide vanes in front of the stator, which serves to direct airflow to the vanes of the first stage of the compressor.

* Corresponding author: dylankn@ya.ru

 https://doi.org/10.1051/e3sconf/202017801038

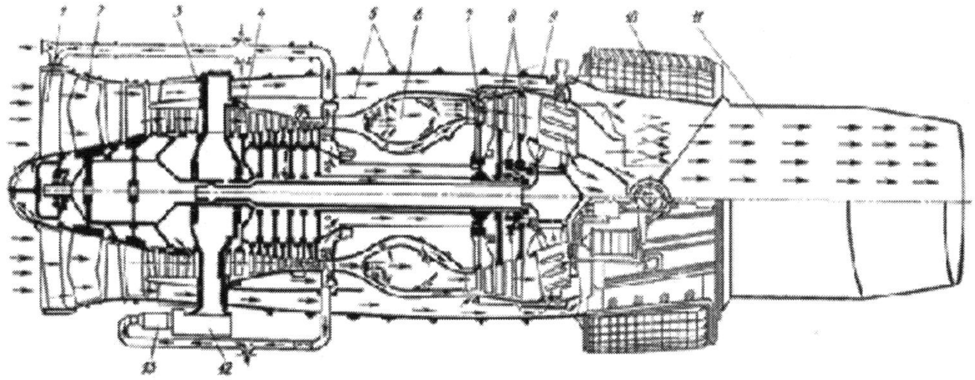

Fig. 1. A longitudinal section of the engine (1 – an input guide vane; 2 – a low pressure compressor; 3 – a middle support; 4 – a high pressure compressor; 5 – a shell; 6 – a combustion chamber; 7 – a high pressure pipe; 8 – a low pressure turbine; 9 – a spacer; 10 – a reversible device; 11 – a jet nozzle; 12 – a box of engine units; 13 – a box of aircraft units).

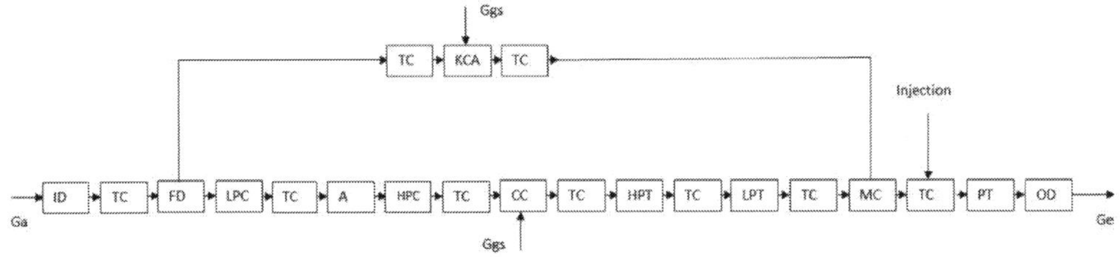

Fig. 2. The functional diagram of the mathematical model (ID – an input device; TC – a transition channel; FD – a flow distributor; LPC – a low pressure compressor; CC – a combustion chamber; A – an aggregates; HPC - a high pressure compressor; HPT – a high pressure turbine; LPT – a low pressure turbine; MC – a mixing chamber; PT – a power turbine; OD – an output device; CCWAB – a combustion chamber with afterburner device; Ga – an air consumption; Ggs – a gas supply; Ge – an exhaust gas flow).

The high-pressure compressor (HPC) is six-speed. It consists of a rotor and a stator. The rotor has six impellers as well as a front shaft. The stator has guide vanes. There is a guide vane with adjustable blades at the front of the stator.

The high-pressure turbine (HPT) is lightly loaded and operates at low temperatures, therefore, there are two options for the blades of nozzle devices - without cooling and with cooling. (The cooling option is used for higher temperatures).

The low-pressure turbine (LPT) is an axial single-stage turbine designed for high efficiency parameters.

Related design features of the engine include the presence of a second circuit, which can replace the external thermal insulation of the engine. [10-11]

For ground-based applications, this gas turbine engine has been upgraded with a free power turbine, a afterburner in the secondary circuit and a mixing chamber. The mathematical model of the engine was created in the «GRET» software package. The functional diagram is shown in Figure 2.

3 Results and Discussion

The afterburning device is necessary to prevent moisture condensation in the engine flow passage. When injecting steam into the flowing part, it is necessary to take into account the fact that the temperature of the exhaust gases after the gas turbine engine must not be lower than 460 K in order to avoid moisture condensation. For this, it is necessary to increase the temperature in the afterburner (located in the second circuit) until the temperature is reached after mixing the gas flows of the first and second circuit. [12-14]

Steam injection is performed in front of the power turbine at a pressure of 0.3 MPa and a temperature of 560K. With these parameters, no condensation occurs in the flow part.

Figure 3 shows the dependence of the change in efficiency and the total hourly fuel consumption depending on the test mode. The parameters of the modes (turbine power and injection flow rate are shown in table 2). The maximum flow rate of the injected substance is limited to 27 kg /s - this is due not only to the cross section of the flowing part of the gas engine but also to the fact that with a higher value airflow will decrease. [15-17]

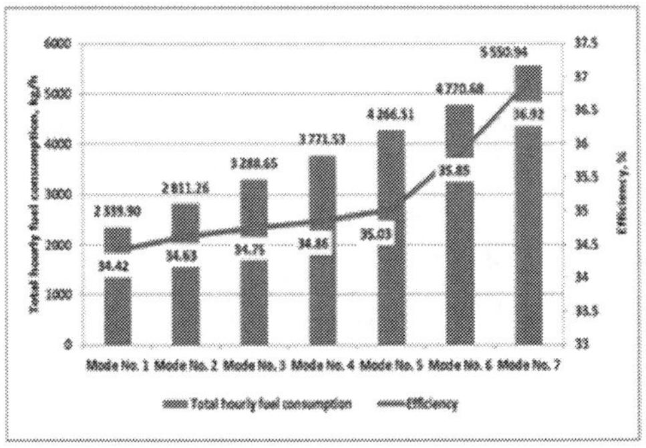

Fig. 3. Change in efficiency depending on injection flow.

Table 2. The parameters of the modes.

Mode number	1	2	3	4	5	6	7
Turbine Power, MW	12	14	16	18	20	22	25
Injection rate, kg / s	0	8	10	15	20	25	27

Figure 4 shows the change in enthalpy and useful work depending on the flow rate of the injected substance.

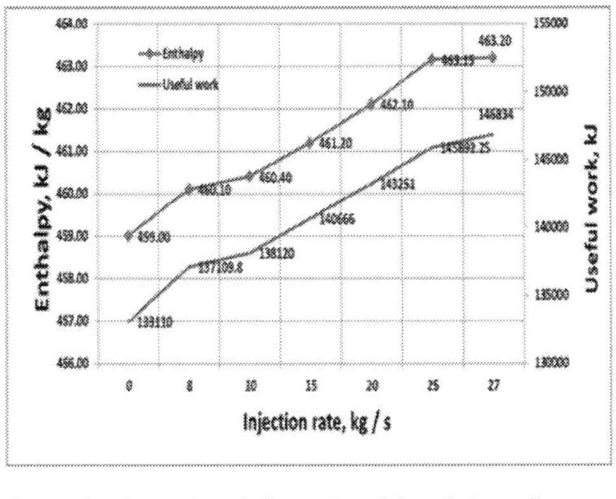

Fig. 4. The change in enthalpy and useful work depending on the flow rate of the injected substance.

During the research, it was assumed that when supplying an additional working fluid, the bypass ratio of the engine may change. Because the gas turbine should work not only in the basic mode at a load of 25 MW but also in other modes a study was conducted to find the optimal value of the bypass ratio.

Because all generating equipment operates in the wholesale electricity market, the operation of a gas turbine is possible not only at a full capacity of 25 MW but also in variable modes. [18-23]

Figures 5-7 show a graphs of changes in the main energy characteristics of the engine. Loads from 12 MW to 25 MW are considered.

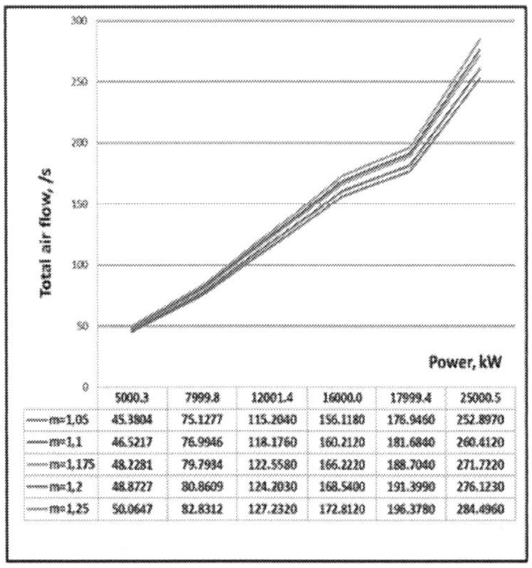

Fig. 5. Change in air flow depending on the degree of bypass and power of gas turbines.

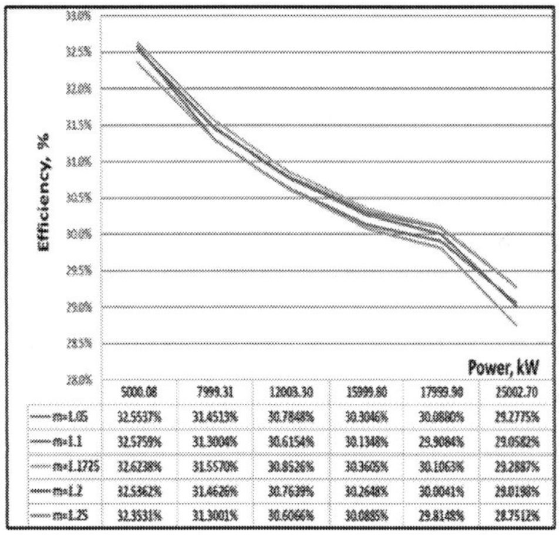

Fig. 6. Change in turbine efficiency depending on the degree of dual-circuit and power of a gas turbine.

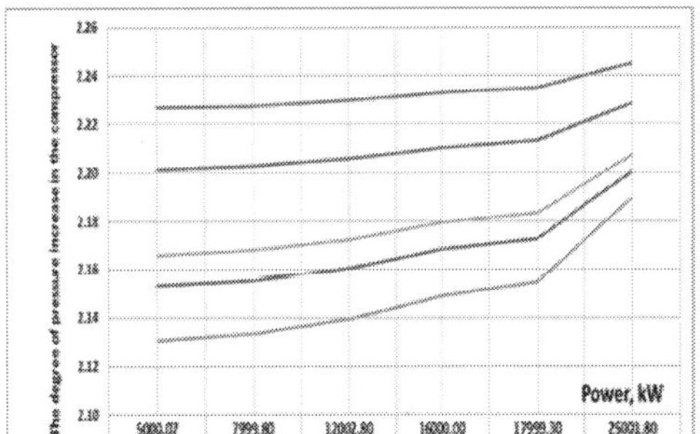

	5000.07	7999.80	12002.80	16000.00	17999.30	25001.80
m = 1.05	2.2266	2.2275	2.2298	2.2328	2.2349	2.2452
m = 1.1	2.2011	2.2025	2.2055	2.2101	2.2131	2.2285
m = 1.175	2.1660	2.1680	2.1724	2.1794	2.1832	2.2071
m = 1.2	2.1532	2.1553	2.1604	2.1683	2.1728	2.2008
m = 1.25	2.1306	2.1333	2.1395	2.1491	2.1547	2.1894

Fig. 7. Change in the degree of increase in pressure in the compressor depending on the degree of double-circuit and power of gas turbine.

4 Conclusion

This article studies the operating modes of the NK-8 gas turbine engine in double-circuit design, upgraded by a power turbine, afterburner and mixing chamber.

Steam injection in front of the power turbine leads to a steady increase in power and effective efficiency. It is necessary to exclude the possible condensation of moisture in the flow part when steam is injected.

The bypass ratio also affects the turbine performance, the optimal value is 1.175.

References

[1] O.O. Badran, Appl. Energy (1999)

[2] M. Basha, S.M. Shaahid, L. Al-Hadhrami, Energy Procedia (2012)

[3] G.E. Marin, B.M. Osipov, D.I. Mendeleev, E3S Web Conf. **124** (2019)

[4] D.I. Mendeleev, G.E. Maryin, A.R. Akhmetshin, IOP Conf. Ser. Mater. Sci. Eng. (2019)

[5] D.I. Mendeleev, Y.Y. Galitskii, G.E. Marin, A.R. Akhmetshin, E3S Web Conf. **124** (2019)

[6] E. Bocci, M. Sisinni, M. Moneti, L. Vecchione, A. Di Carlo, M. Villarini, Energy Procedia (2014)

[7] T.A. Clifford, S. Cruz-Manzo, Y. Zhang, V. Panov, A. Latimer, Proc. IECON 2017 – 43rd Annu. Conf. IEEE Ind. Electron. Soc. (2017)

[8] L. Esclapez, P.C. Ma, E. Mayhew, R. Xu, S. Stouffer, T. Lee, H. Wang, M. Ihme, Combust. Flame (2017)

[9] L. Esclapez, P.C. Ma, E. Mayhew, R. Xu, S. Stouffer, T. Lee, H. Wang, M. Ihme, Combust. Flame (2017)

[10] E. Gracheva, A. Alimova, Proc. – 2019 Int. Russ. Autom. Conf. RusAutoCon (2019)

[11] E. Gracheva, A. Alimova, Proc. – 2019 Int. Ural Conf. Electr. Power Eng. Ural (2019)

[12] M. Henke, J. Kallo, K.A. Friedrich, W.G. Bessler, Fuel Cells (2011)

[13] J. Hou, B.J. Wicks, R.A. Antoniou, Eng. Fail. Anal. (2002)

[14] J. Kotowicz, M. Job, M. Brzeczek, Energy (2015)

[15] M.C. Lee, S. Bin Seo, J.H. Chung, S.M. Kim, Y.J. Joo, D.H. Ahn, Fuel (2010)

[16] A.H. Lefebvre, J.H. Whitelaw, Int. J. Heat Fluid Flow (1984)

[17] K. Liu, S. Sadasivuni, N. Parsania, Fuel (2019)

[18] W. E. Bill Forsthoffer, *Forsthoffer's Rotating Equip. Handbooks* (2005)

[19] M.M. Rahman, T.K. Ibrahim, A.N. Abdalla, Int. J. Phys. Sci. (2011)

[20] D. Rochelle, H. Najafi, Renew. Sustain. Energy Rev. (2019)

[21] P. Weiland, Appl. Microbiol. Biotechnol. (2010)

[22] K. Weston, Energy Convers (2000)

[23] Y.G. Li, P. Nilkitsaranont, Appl. Energy (2009)

Study of gas consumption patterns for sheet glass enterprises

Yauhen Shenets[1,*], *Deniz* Moroz[2], *Nadzeya* Hruntovich[3], *Mikhail* Malashanka[4], and *Alexei* Tsvetkov[5]

[1]Public corporation "Gazprom transgaz Belarus" Nekrasova 9, Minsk, Republic of Belarus
[2]State Enterprise "Research Institute Belgiprotopgaz", Domashevskij lane,11a, Minsk, Republic of Belarus
[3]Sukhoi State Technical University of Gomel, Prospect Octiabria, 48, 246746, Gomel, Republic of Belarus
[4]Energy Efficiency Department of the State Committee for Standardization of the Republic of Belarus, Minsk, Svobody sq. 17, 220030, Republic of Belarus
[5]Kazan State Power Engineering University, str. Krasnoselskaya, 51, 420066, Kazan, Russia

Abstract. A study of the gas consumption modes of the subgroup "Enterprises of construction industry and architecture" of the regional gas supply system, as well as the production of sheet glass, which forms gas consumption in the subgroup by 94%, was carried out. The field type "daily gas consumption - outside temperature" is set for the subgroup and the plant for production of sheet glass. This is a field with a weak dependence on the growth of daily gas consumption when outside temperature decreases and without an obvious stratification into temperature regions determined by the state of the heating system ("on" or "off") and a significant vertical spread of daily gas flow at the same values of outside temperature. Additional factors affecting gas consumption were searched. The created database of 14 technological indicators over a seven-year period allowed to identify factors forming daily and annual gas consumption regimes using correlation-regression analysis. The importance of the "time worked by the line" factor and its contribution to the formation of the annual volume of gas consumption was assessed. Invention proposes a simplified method of estimating the rate of aging of furnace lining and wear of enclosing structures on the basis of a single-factor model of gas consumption from productivity by comparing the daily average annual values of specific gas consumption for sheet glass production with the same volume of output.

1 Introduction

Many years of world experience show that during the operation of furnace equipment (glass, cement enterprises) the furnace lining is burned and the enclosure structures of the furnace body are worn, as a result of which more gas is spent to compensate for losses and maintain the parameters of melting modes [1, 2, 3]. A quantitative indicator of the deterioration of the furnace lining of the sheet glass plants is the increase in the specific gas consumption (*SGC*) for the production of 1 ton of production (B_{SGC}, tons of fuel equivalent/tons – next B_{SGC}, t f.e./tons). For the world 's leading glass industry enterprises of Germany, Canada, the USA, the warranty and depreciation life of furnace equipment is 5-6 years [4, 5]. Good quality of furnace design, construction and maintenance allows to bring the life of the furnace to 10-12 years [6], and in some cases more, with acceptable reduction of energy efficiency (EE) [7]. At the end of this period, the morally obsolete furnace must be dismantled and replaced with a new one. During the operation of the furnace, its scheduled hot and cold repairs are carried out with further adjustment of melting process parameters, which allows to contain the rate of EE fall (growth of specific gas consumption per production unit output) in accordance with the approaches [8, 9 and 10]. In order to make a natural gas application, enterprises with gas furnaces in their process equipment must take into account the factor of "furnace lining burnout and wear of enclosing structures," as its disregard can lead to a significant error in the declared volume of gas even when the volume of output is maintained.

2 Methodology

According to [11] gas industry enterprises are concerned with studying the structure and regularities of regional gas supply systems functioning, as well as the regularity of natural gas consumption, their planning in different sectors of the economy. Glass industry enterprises are part of the structure of the subgroup "Enterprises of construction industry and architecture" of the enlarged balance group "Industry" of the regional gas supply system. The share of enterprises of this subgroup in the gas consumption of the of the region's industry is from 15 to 21.8% for 2012-2017. The subgroup consists of 17 balance structural elements - construction organizations, enterprises for the production of construction materials and enterprises for the production of sheet glass. At the first stage of studies of the patterns of formation of gas consumption modes, coefficients of paired correlation between the daily gas flow rate (B_{day}) and the average daily outside temperature (t_{day}) were calculated, which

* Corresponding author: shenets@mail.ru

for different years amounted to: 2012 – -0.86; 2013 – -0.18; 2014 – -0.58; 2015 – -0.72; 2016 – -0.59; 2017 – -0.53. The resulting correlation coefficient values indicate an unstable (from low to average) relationship between temperature and daily gas flow rate and require the search for additional factors affecting gas consumption. And the built structural fields over a six-year period made it possible to establish a type of field with dependence of daily gas consumption growth with reduction of outside temperature without obvious stratification into temperature areas determined by the state of the heating system ("on" or "off"), with significant vertical spread of B_{day} at the same values of outside temperature (Figure 1). In 2013, a day with low B_{day} consumption is allocated from the total daily gas consumption pattern, which may be related to the shutdown of a powerful process plant.

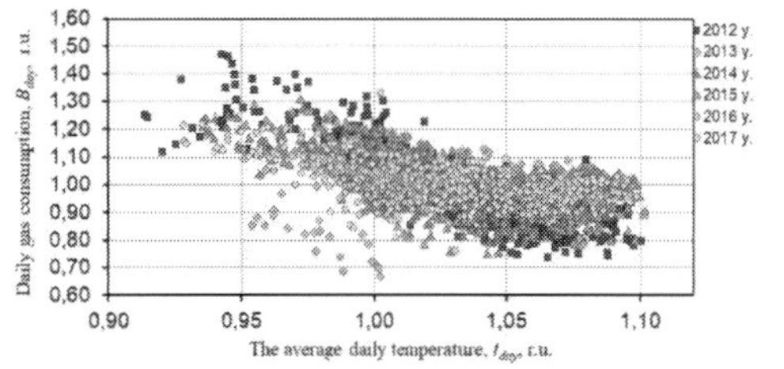

Fig. 1. Structural field "daily gas consumption - temperature" of the subgroup "Enterprises of construction industry and architecture" of the regional gas supply system.

The sheet glass production plant, which is part of the sub-group under consideration, forms 94% of its total natural gas consumption. The production of sheet glass as a consumer of energy resources is characterized by the complex nature of the relationship between energy and technology, when energy consumption is influenced by many technological factors. In order to identify and account for the factors that form the mode of gas consumption, an information database for 7 years has been formed, containing information on 14 technological indicators, such as the daily consumption of natural gas, B_{day}, thousand m^3; average daily caloric content of natural gas, kcal/m^3; daily electricity consumption; average daily productivity of the furnace, P, t/day; glass thickness, mm; density of glass, p, t/m^3; average daily outside temperature, °C; FeO content in glass, %; Fe_2O_3 content in glass, %; Fe_2O_3 content in sand, %; Fe_2O_3 content in dolomite, %; ratio of charge to broken glass, %; humidity of charge, %; number of days of furnace operation, N, days. It was found that there was no strong relationship (multicollinearity) between the factors included in the model.

For the first time it is proposed to introduce the factor "number of days of furnace operation" to build a multi-factor model of gas consumption. Further studies have shown that this has made it possible to estimate the aging rate of technological equipment and the associated increase in the volume of gas consumed.

Table 1 shows regression analysis results for significant parameters over a six-year period of operation and regression coefficient values for each parameter.

Table 1. Results of mathematical modeling after excluding insignificant factors.

Technological parameter (model factor)	Regression coefficients	Standard error	t-statistics
Productivity, P, tons/day	163.5	1.67	97.7
The outside temperature,°C	157.7	9.69	16.2
The content of FeO in the glass, %	805488.16	42592.93	18.9
Number of days of furnace operation, day	14.58	0.203	71.6

Table 1 shows that after reducing the number of factors (elimination of statistically insignificant model factors) included in the model, the values of the Student's t-distribution (t-statistics) for each parameter exceed the critical value, which indicates the statistical significance of certain parameters for the task of determining the EE of the enterprise.

Thus, the resulting multi-factor model of daily gas consumption in the production of sheet glass is described by the dependence:

$$B_{t\,f.e} = F(P,\ t_o,\ F, N) = 163.5\cdot P + \\ +157.7\cdot t_o + 805488.2\cdot F\ + 14.6\cdot N \tag{1}$$

Where:
P – productivity, t/day;
t_o – outside temperature, °C;
F – content of iron oxide in glass,%;
N – number of days of furnace operation, day.

Description of the statistically significant factors of the model:

i. The regression coefficient before the factor «performance» is 163.5. This means that every day 163.5 kilogram of fuel equivalent (next – kg. f.e.) of natural gas is spent on the technological process of glass production for the production of 1 ton of glass.

ii. The regression coefficient before the factor "outside temperature" is 157.7. This means that the 1 °C change in outside temperature for the glass manufacturing process results in an average change in the daily natural gas flow rate of 157.7 kg. f.e.. This dependence is explained by the operating conditions of the furnace in the summer (hot) period of the year, when to prevent accelerated destruction of the outer lining of the furnace, it is cooled using technological ventilation units. To compensate for this additional heat loss and maintain the process of glass formation in the furnace pool with a given temperature, an additional supply of fuel gas to the furnace burners is provided.

iii. The regression coefficient before a factor "FeO content" (iron oxide content in glass) makes 805488.3. This means that for the technological process of glass production, a change in the content of iron oxide in glass by 1 % leads to a change in the daily consumption of natural gas by 805488.3 kg. f.e.. This dependence is explained by the process of glass formation in the course of redox reaction in the glass production furnace by float method, which is based on achieving the state of equilibrium of iron ions with different degree of oxidation of Fe^{2+} and Fe^{3+} in the glass paste [12] and is based on control of temperature mode inside the furnace by adjustment in supply of fuel gas and supply of a certain portion of oxygen.

iv. Regression coefficient before factor "number of days of furnace operation" is 14.5. This means that the daily consumption of natural gas for glass production increases by 14.5 kg. f.e.. Taking into account the generally accepted type of consumption models of fuel and energy resources, the factor "number of days of furnace operation" is presented as a conditional-constant component of the model due to constant and stable loading of the furnace 700-775 tons/day during the year.

The annual reduction of EE production of sheet glass can reach up to 10% and depends on the load of production, but as a result of the repair and adjustment works carried out, the dynamics of deterioration of EE is slowed down.

The results of the processing of statistical daily data of the enterprise work showed the presence of only two significant factors for planning of specific fuel consumption for glass production, determination of EE of the enterprise for semi-annual and annual intervals.

The factors of "iron oxide content in glass" and "outside temperature" in the semi-annual interval have stable values and do not affect the determination of EE forecast values, therefore, are not taken into account in forecasting the specific consumption of natural gas of future periods. A nomogram has been developed, which determines the value of the specific consumption of natural gas for the production of sheet glass, taking into account the planned and reporting periods of future operation (6 months), when the planned average daily productivity and the time spent by the furnace were set.

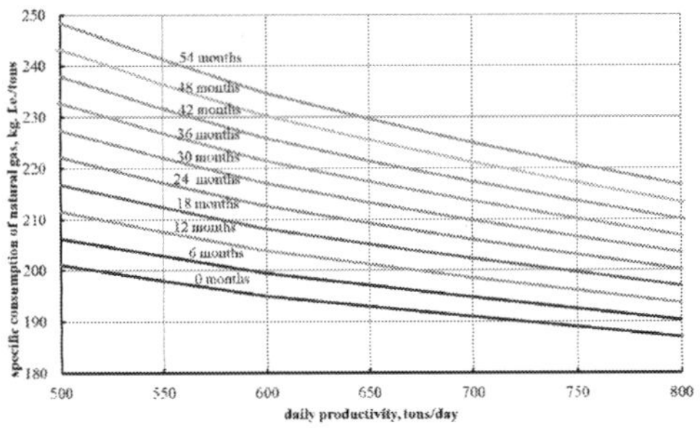

Fig. 2. Nomogram for determining the specific energy consumption of sheet glass production, taking into account the time spent by the furnace (without taking into account measures to improve energy efficiency).

A simplified method for estimating the rate of aging of the furnace lining and the wear of enclosing structures is proposed based on a one-factor model [8] of gas consumption from productivity by comparing the average daily values of specific gas consumption for the production of sheet glass at the same output volumes. In conditions of relatively stable annual values of feedstock characteristics and average annual outside temperature, it is possible to estimate the decrease in energy efficiency of a glass furnace caused by deterioration of the furnace lining and wear of the enclosing structures using one-factor models of the dependence of the total (specific) gas consumption on the volume of output:

$$B_{sgc} = B_{sgc.tech} + B_{sgc.const.} / P, t\ f.e./units\ of\ prod \quad (2)$$

Where:

$B_{sgc.tech}$ – technological specific gas consumption per product output, t f.e./units of prod;

$B_{sgc.const.}$ – conditional-constant component of gas consumption by technological installations, t f.e.

P – volume of output per day, units of prod.;

For glass production, energy resource costs associated with maintaining the working state of the furnace of the technological line form a technological conditional-constant component of fuel consumption (natural gas).

Table 2 shows the one-factor models developed over the five-year period of operation for dependence of specific gas consumption on sheet glass production [13]. The graphical interpretation of the models is shown in Figure 3.

Table 2. Specific gas consumption (sgc) models for sheet glass production.

Year	Kind of model	Bsgc.tech, t f.e./units	Bsgc.const., t f.e.
2005	*Bsgc* = 0.11+69.11/P	0.11	69.11
2006	*Bsgc* =0.07+101.8/P	0.07	101.83
2007	*Bsgc* =0.09+85.95/P	0.09	85.95
2008	Bsgc =0.04+130/P	0.04	130
2009	*Bsgc* =0,08+103.1/P	0.08	103.1

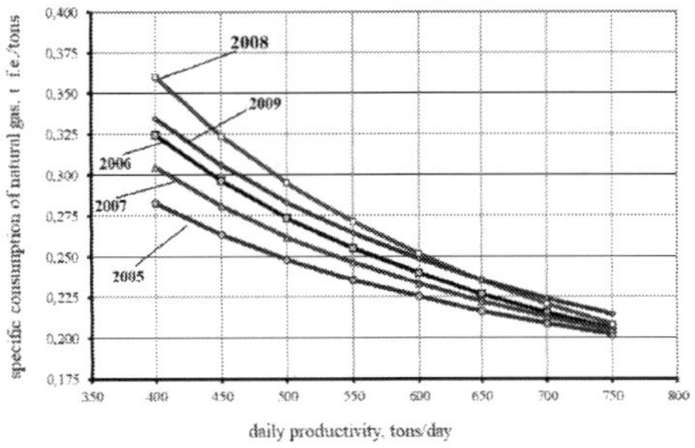

Fig. 3. Dynamics of specific gas consumption of sheet glass plant from 2005 to 2009.

As can be seen from the obtained models *Bsgc =F(P)*, in 2005, when sheet glass production was started at the design capacity, the specific gas consumption was minimal. In 2006, specific gas consumptions for sheet glass production increased. In 2007, as a result of repair works, debugging of the technological process and implementation of energy-saving measures, the values of specific gas consumption decreased relative to 2006, and in 2008 they increased again.

It has been found that at low loading of the sheet glass production line (400 tons/day), the reduction of EE due to burnout of the furnace lining and wear of the enclosing structures is more evident, since for such modes the influence of the conditional-constant component of the gas flow has more weight in the formation of the total specific flow rate. When the productivity of the process line increases, the influence of the conditional constant component of gas consumption, which is associated with maintaining the temperature regime of the furnace, decreases, as the technological component of gas consumption is growing [14]. The decrease in efficiency due to aging of the furnace lining at high productivity modes is less apparent (figure 4). For the production under study, it was found that during the five-year period the EE reduction was 18.4% with a minimum daily load of 400 tons/day on the production line.

With a capacity of 700 tons/day, the resulting change of EE (due to burning of the lining and wear of the enclosure structures of the furnace - hereinafter the coefficient of change of EE) is only 7.3% in a five-year period, or 1.8% in one year of operation, which correlates with the data of English specialists and scientific research of German scientists [6, 15]. It should also be noted that for the production under study, the main technological mode is a mode with a daily capacity of 760-775 t/day.

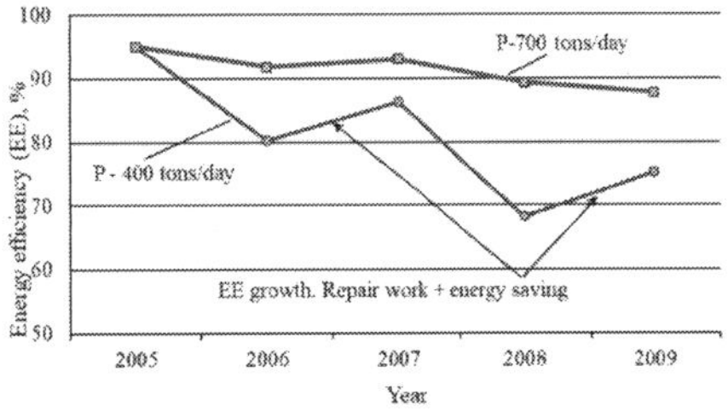

Fig. 4. Dynamics of changes of EE of sheet glass production for 2004-2009 at daily productivity of 400 and 700 tons/day.

3 Research results

Consider how ignoring the factors of "furnace lining aging and wear of enclosing structures," as well as "daily productivity" can lead to an error in the preparation of the application for natural gas. Thus, in 2009 for production the model has the following parameters: $B_{sgc} = 0,08+103,1/P$, t f.e./tons;. In the reporting year, the company operated with an average daily capacity of 550 tons/day with an actual achieved gas flow rate of 0.267 t f.e./tons. This specific gas flow rate will be recorded as a statistical report and can be used in the calculation of the planned natural gas demand for the following year.

If the company plans to operate with an average daily capacity of 700 tons/day next year, then the planned gas demand (without the model) will be calculated using its statistical and reporting value of the specific gas consumption for output and the production plan for output.

The company will make a request for technological gas in the amount of: $B_{gas} = 0,267$ t f.e./tons 700 tons/day = 186,9 t f.e./day or 68 218.5 t f.e/year, which in absolute terms is 59 320.4 thousand m^3. If we consider real dependence of specific gas consumption on production of sheet glass from the furnace productivity (table 2, figure 3), at the planned productivity of 700 t/day (growth of output volumes) the specific gas consumption will decrease to 0.227 t f.e./tons, by analogy with the example [16, 17]. Gas volume at a capacity of 700 tons/day will be 158.9 t f.e./day, or 57 998.5 t f.e/year, which in absolute terms will be 50 433.5 thousand m^3. In order to minimize the error in determining the planned demand for natural gas for the next year, the annual wear component of the furnace lining calculated according to the model should also be taken into account, that is, the predicted gas volume should be corrected taking into account the EE change factor of 1.8% according to the imputed constant component of the model. The total real value of the projected gas volume will be:

To minimize the error in determining the planned natural gas requirements for the next year should also be considered a component of the annual wear of the furnace lining, calculated by the model, in other words, the predicted volume of gas should be adjusted to the rate of change of EE +1.8% for the conditional constant component of the model. The final real value of the projected gas volume will be:

$$B_{sgc} = 0.08 + (103.1 \cdot 1.018) \; / \; 700 = 0.23 t \, f.e. / \, tons;$$
$$B_{gas} = 0.23 t \, f.e / t \cdot 700 \; tons / day = 161 t \, f.e / day$$

or 58 765 t f.e/year, that in absolute expression makes 51 100 thousand m^3.

Thus, the error in the determination of demand for natural gas is estimated at 16% towards overstatement.

4 Conclusions

The production of sheet glass as a consumer of energy resources is characterized by the complex nature of the relationship between energy and technology: energy consumption is influenced by many technological factors. In order to identify and account for the factors forming the gas consumption mode, an information database has been formed, which includes 14 indicators characterizing the gas consumption mode. The studies carried out using correlation-regression analysis allowed to justify 4 factors determining the daily mode of gas consumption:

P performance, t/day;

t_o outside temperature, °C;

F content of iron oxide in glass, %;

N number of days of furnace operation, day.

A nomogram has been developed, where, given the planned average daily productivity and time spent by the furnace, the value of the specific consumption of fuel (natural gas) for the production of sheet glass is determined taking into account the planned and reporting periods of future operation (6 months).

Invention proposes a simplified method of estimating the rate of aging of furnace lining and wear of enclosing structures on the basis of a single-factor model of gas consumption from productivity. It has been shown that ignoring the factors of "daily capacity" and "furnace lining burnout and wear of enclosing structures" can lead to an error in predicting the volume of gas consumption.

References

[1] Heat balance [Electronic resource], StaraGlass (2020). Available at: https://www.staraglass.it/wp-content/uploads/2019/11/Stara-Glass-Heat-Balance-2020-r0.pdf

[2] S. Yaitskiy, L. Bragina, N. Yaitskiy, Features of the destruction of the refractory lining of glass melting furnaces in the production of sheet glass [Electronic resource] (2012) (in Russian). Available at: http://repository.kpi.kharkov.ua/handle/KhPI-Press/10477

[3] V. Dzyuzer, V. Shvidkiy, Designing energy efficient glass furnaces (Moscow, Publishing house Tepltehnic, 340, 2009) (in Russian)

[4] L. Papko, Y. Pavlukevich, Refrigerators for glass furnaces (Minsk, BNTU, 95, 2008) (in Russian)

[5] R. Makarov, V. Tarbeev, E. Horosheva, *Quality management of flat glass (float method)* (Associations of construction universities, 152, 2004) (in Russian)

[6] GOST R 54201-2010, Resources saving, Production of domestic and container glass, Best available techniques for improving energy efficiency

[7] N.V. Hruntovich, A.A. Kapanski, D. Baczynski, G.V. Vagapov, O.V. Fedorov, Optimization of a variable frequency drive pump working on a water tower, E3S Web of Conferences **124**, 05060, EDP Sciences (2019)

[8] N. Hruntovich, Y. Shenets, S. Zinovyeva, Development of methods for a comprehensive assessment of the effectiveness of energy conservation measures using one-factor mathematical models of specific fuel and energy consumption relative to the volume of output [Electronic resource], Energy efficiency (2013) (in Russian). Available at: http://energoeffekt.gov.by/downloads/publishing/archive_journal/arhiv_journal_2013/En_11_2013.pdf

[9] N. Hruntovich, D. Moroz, S. Zhukovec, Y. Shenets, Development of methodological support for diagnosing energy efficiency, Energy and Management (2017) (in Russian). Available at: http://www.web-energo.by/files/EiM/2017/EiM-1-2017.pdf

[10] O. Fedorov, Expeditious forecasting of power consumption, 2017 International Conference on Industrial Engineering, Applications and Manufacturing (ICIEAM), IEEE (2017)

[11] N. Hruntovich, D. Moroz, V. Sholonik, On the need to study the structure and patterns of functioning of regional gas supply systems, Energy strategy, 42–46 (2017) (in Russian)

[12] F. Baucke, G. Röth. Electrochemical mechanism of the oxygen bubble formation at the interface between oxidic melts and zirconium silicate refractories, Glastechnische Berichte **61**, 5, 109–118, 9 ref. (1988)

[13] R. Petuhova, N. Hruntovich, Y. Shenets, On the problems of energy efficiency management of sheet glass production [Electronic resource], Energy efficiency (2014) (in Russian). Available at: http://energoeffekt.gov.by/downloads/publishing/archive_journal/archive_journal_2014/201403_energoeff.pdf

[14] N. Hruntovich, Y. Shenets, The influence of the structure of consumption of energy resources of the enterprise on the effectiveness of the implementation of energy saving measures [Electronic resource], Energy, Proceedings of universities and energy associations of the CIS, 58–66 (2014). Available at: https://energy.bntu.by/jour/issue/view/2/showToc

[15] R.G.C. Beerkens, Energy Efficiency Benchmarking of Glass Furnaces /62 Conference on Glass Problem, 93–103 (2002)

[16] N. Hruntovich, D. Moroz, A. Kapanskij, S. Zhukovec, Y. Shenets, Using mathematical modeling methods to solve practical problems of energy efficiency assessment [Electronic resource], Energy and Management, 21–25 (2017). Available at: http://www.web-energo.by/files/EiM/2017/eim-3-2017.pdf

[17] Y.I. Gracheva, O.V Naumov, Estimation of Power Losses in Electric Devices of the Electrotechnical Complex, International Conference on Industrial Engineering, Applications and Manufacturing (ICIEAM), 6 (2019)

Fault indicators on the Russian market

Maksim Vladimirovich Borodin[1,*], *Tatyana Anatolievna* Kudinova[2], and *Khasanov Shamil* Rashidovich[3]

[1]Department "Power Supply", Orel State Agrarian University named after N.V. Parakhin, Orel, Russia
[2]Department "Linguistics and Humanities", Orel State Agrarian University named after N.V. Parakhin, Orel, Russia
[3]"Power Supply of Industrial Enterprises", Kazan State Power Engineering University, Kazan, Russia

Abstract. The article indicates the relevance of using fault indicators. Various models and manufacturers of fault indicators are presented on the Russian market. The analysis of the characteristics of various fault indicators was produced. At present, the most suitable domestic device for electric grid companies in Russia, which allows determining the open and short circuit on power lines, as well as indicating the location of the fault, is a fault indicator CI34B of the ANTRAKS R&D&M Co, Ltd. The market analysis of fault indicators in Russia will allow specialists of design organizations, personnel of electric grid companies and other specialists to reduce the time spent on searching for information on fault indicators.

1 Introduction

The authors' works [1-11] indicate that rapid location of faults on power lines can significantly accelerate the restoration of power supply to consumers and reduce the number of attending personnel, eliminating the faults. In case of a fault on a power line, it is necessary to determine the place of wire breakage or the location of failed power equipment more accurately; in turn, it will allow the personnel to reduce the time on locating and eliminating the fault on a power line quicker.

The authors [12-20] indicate that falling of trees on power lines, breakdowns of insulators and other causes of faults greatly complicate the work of attending personnel of electric grid companies. Most often, there are no devices for monitoring power lines; so the place of insulator breakdowns and other faults must be determined visually. It takes a lot of time and leads to a long disconnection of consumers. In this regard, there is a need for the use of special equipment to detect an open and a short on power lines, as well as to indicate the location of the faults.

One of the devices that can do the above mentioned things is a fault indicator (FI). Depending on the purpose and design, the FIs are installed in the switchgear unit, on the overhead power line support, or directly on the phase line wire. The use of FIs can significantly reduce the time and cost of locating and eliminating faults in branched distribution networks.

2 Research methodology

In Russia, on the market of fault indicators, FI models of various manufacturers are widely presented. At the same time, the characteristics of FIs of various manufacturers differ from each other; it, in turn, affects their effectiveness.

The personnel, dealing with the selection of FIs, spend considerable time on it, since FIs, presented on the Russian market, have different characteristics and prices. To increase the efficiency of the selection of FIs, it is necessary to analyze their characteristics.

Analysis of the FIs, presented on the Russian market, will be made according to the following parameters:
- minimum detection current for single phase-to-ground faults;
- determination of current direction;
- presence of internal memory;
- radio communication range;
- channel used for alarm annunciation;
- fault indication;
- FI power supply;
- FI battery life;
- data transfer unit power supply;
- the largest range for line-to-line faults;
- system of fault location;
- etc.

3 Results and their discussion

Fault indicators of eight companies were presented on the Russian market. Appearance, models and manufacturers of fault indicators are presented in table 1.

* Corresponding author: maksimka-borodin@yandex.ru

Table 1. Models and manufactures of fault indicators.

Model, (manufacturer)	Appearance of the device
FI-Cl3B, (Antraks)	
JYZ-HW V2.0, (Four-Faith)	
AMKA-single phase-to-ground fault-18, (AIZ Lytkarino)	
FLA3.1VL, (INION)	
Overhead line fault indicator (OLFI), (Relematika)	
Sicam FSI, (Siemens)	
Smart Navigator (HORSTMANN)	
ASTRON-8080, (ASTRON)	

All of the above mentioned FIs are placed on the phase wire of the power line. Externally, product designs are almost identical. They are a sealed flask, inside which, depending on the manufacturer, there may be a battery, an electronic circuit, a set of LED indicators.

The installation of a FI can be carried out both on a disconnected power line using a truck crane, and on a power line under voltage using appropriate insulating rods. FIs usually operate as a group of three to six devices with information exchange with a data collection and transmission unit (DCTU).

DCTU provides communication in the appropriate protocol with the remote server. All FI models, presented in the table, are used as an alarm annunciation channel – the GSM / GPRS channel. The FI power supply is provided by built-in batteries; individual manufacturers have implemented the function of recharging the FI from power lines.

The characteristics of FIs of various manufacturers, presented on the Russian market, are shown in table 2.

The FIs, presented in Table 2, have different characteristics. So, the minimum detection current is provided by Anthrax FI-Cl3B, it is 0.5 A; determination of current direction is available in Four-Faith FI- JYZ-HW V1.0, in Relematika FI-OLFI and Anthrax FI-Cl3B; presence of internal memory is available in Anthrax FI-Cl3B; the HORSTMANN FI Smart Navigator has the longest battery life which is equal to 20 years; the smallest wire diameter for installation has the HORSTMANN FI; the largest range for line-to-line fault is available in Anthrax FI-Cl3B and ranges from 20 to 2000A; the system of fault location is available in Siemens FI- Sicam FSI, INION FI- FLA3.1VL, Relematika FI-OLFI and Anthrax FI-Cl3B.

4 Conclusions

Based on the presented analysis, we can conclude that at the moment, the most suitable domestic device for the Russian grid companies to determine the open and short circuit on power lines, as well as to indicate the location of the fault, is a fault indicator Cl34B of the ANTRAKS R&D&M Co, Ltd.

The market analysis of fault indicators in Russia will allow specialists of design organizations, personnel of electric grid companies and other specialists to reduce the time spent on searching for information on fault indicators.

Table 2. Characteristics of FIs, presented on the Russian market.

No.	Parameter	Siemens	AIZ Lytkarino	HORSTMANN		INION	ASTRON	Relematika	Antraks	Four-Faith
1	2	3	4	5	6	7	8	9	10	11
1	Model of FI	Sicam FS	AMKA-single phase-to-ground fault-18 UHL1	Smart Navigator	Smart Navigator 2.0	FLA3.1VL	ASTRON -8080 UHL1	OLFI	B 34	JYZ-FF V2.0
2	Minimum detection current for single phase-to-ground faults, A	5…1500	120	100	100	4…1500	120…630	120…600	20…2000	3…800
3	Range for line-to-line fault, A	5	3	50	5	5	20	20	0,5	3
4	Determination of current direction for single phase-to-ground fault	No	No	No	No	No	No	Yes	Yes	No
5	Presence of internal memory	No	No	No	No	No	No	No	Yes	No
6	Radio communication range, m	No data	No	30	30	50	No data	50	200	No data
7	Alarm annunciation channel	GSM/ GPRS	GSM/ GPRS	GSM/ GPRS	GSM/ GPRS	GSM/ GPRS	GSM	GSM/ GPRS	GSM/ GPRS/ SMS	GSM/ GPRS
8	Temperature range, ^{0}C	-40…+70	-40…+70	-40…+85	-40…+85	-40…+70	-40…+60	-40…+80	-55…+85	-40…+70
9	Wire Diameter for installation, mm	5-42	5-42	8-29	8-29	5-42	5-35	10-42	5-42	5-42
10	Fault indication	LEDs on the FI	Blinker and LEDs on the FI	LEDs on the FI	LEDs on the FI	Blinker and LEDs on the FI	LEDs on the FI	LEDs on the FI	LEDs on the FI, blinker on the data transfer unit	LEDs on the FI
11	FI power supply	From battery	From battery	From battery	From battery	With current in the line of more than 20 A works from power lines	From battery	From battery	From battery	From battery
12	Declared battery life, years	10	>10	20	Recharged from power lines	Recharged from power lines	7	8	10	10
13	Data transfer unit power supply	Battery powered, solar recharged	Battery powered, solar recharged	Battery powered, recharged from power lines	Battery powered, recharged from power lines	Battery powered, recharged from power lines	Battery powered, recharged from power lines	No data	Battery powered, recharged from power lines	Battery powered, solar recharged
14	System of fault location	Yes	No	No	No	Yes	No	Yes	Yes	No

References

[1] N.S. Sorokin, Improving the Operation Efficiency of Electrical Networks with Voltage of 6-35 kV by Monitoring the Parameters of the Electrical Network, National University of Life and Environmental Sciences of Ukraine, Series: Technics and Energy of the Agro-industrial Complex, 209-2, 92–95 (2015)

[2] M.V. Borodin, N.S. Uryupin, Justification of Energy-efficient Cross-section of Power Lines with Voltage Deviations of ± 5%, ± 10%, Bulletin of Nizhny Novgorod Institute of Engineering and Economics **2**, 81, 20–29 (2018)

[3] Yu.N. Glybina, R.P. Belikov, I.N. Fomin, Analysis of the Types and Amount of Faults in Electric

Networks of Voltage Class 6-10KV, Agrotechnics and Power Supply 3, 16, 43–49 (2017)

[4] A.V. Vinogradov, V.E. Bolshev, A.V. Vinogradova, M.V. Borodin, A.V. Bukreev, Technical and Economic Methods for Increasing the Efficiency of Rural Consumers' Power Supply Systems, Bulletin of Don Agricultural Science **3**, 47, 59–67 (2019)

[5] M. Borodin, A. Psarev, T. Kudinova, R. Mukhametzhanov, Improving Power Quality by Calculating Voltage Losses, E3S Web of Conferences, 1041 (2019)

[6] I. Fomin, R. Belikov, V. Zelyukin, E.V. Mikhailova, Improvement of Power Supply Reliability by Means of Remote Control of the Automatic Repeated Switching-on of Sectionalizing Circuit-Breakers, E3S Web of Conferences, 01042 (2019)

[7] A. Vinogradov, V. Bolshev, A. Vinogradova, T. Kudinova, M. Borodin, A. Selesneva, N. Sorokin, A System for Monitoring the Number and Duration of Power Outages and Power Quality in 0.38 kV Electrical Networks, P. Vasant, I. Zelinka, G.W. Weber (eds) Intelligent Computing & Optimization, ICO 2018, Advances in Intelligent Systems and Computing **866**, 1–10 (2019)

[8] S.M. Astakhov, R.P. Belikov, State and Ways to Improve the Performance of Distribution Networks in the Agro-industrial Complex, Bulletin of OrelSAU **29**, 2, 106–108 (2011)

[9] E.I. Gracheva, O.V. Fedorov, Forecasting Reliability of Electrotechnical Complexes of In-plant Electric Power Supply Taking Into Account Low-Voltage Electrical Apparatuses, 2019 International Conference on Industrial Engineering, Applications and Manufacturing, ICIEAM 2019, 8743057 (2019)

[10] E.I. Gracheva, R.R. Sadykov, R.R. Khusnutdinov, A.N. Alimova, Reliability and Layout of Low-Voltage Switchgears for Internal Shop Power Supply, Bulletin of Kazan State Power Engineering University **11**, 1 (41), 3–9 (2019)

[11] E.I. Gracheva, O.V. Naumov, Z.M. Shakurova, Information Parametric Characteristics of Reliability of Low-Voltage Switching Devices of Industrial Complexes, Bulletin of Kazan State Power Engineering University **3**, 39, 27–33 (2018)

[12] M.V. Borodin, A.A. Dorozhkina, A.E. Semenov, Methodology for the Selection and Determination of the Most Effective Measuring Transformer for a Substation, News of Higher Educational Institutions of the Black Earth Region **4**, 58, 58–68 (2019)

[13] S.M. Astakhov, A.S. Scherbakov, M.V. Borodin, Data Processing Unit of a Remote Control System for Operating Modes of a Distribution Network of 10 kV, Power supply and Safety, Collection of materials of the II International Exhibition-Internet Conference, 18–21 (2008)

[14] V.A. Chernyshov, A.E. Semenov, R.P. Belikov, V.E. Bolshev, M. Jasinski, M.S. Garifullin, The of Extending Drone Piloting Autonomy When Monitoring the Technical Condition of 6-10 kv Overhead Power Lines, E3S Web of Conferences 2019, 02010 (2019)

[15] R.P. Belikov, Remote Search and Repair of Faults in Electric Networks 6 (10) kV, Physics and Modern Technologies in the Agro-Industrial Complex, Materials of the XI All-Russian Youth Conference of Young Scientists, Students and Schoolchildren with International Participation. Ministry of Agriculture of the Russian Federation, FSBEI HE "Orel State Agrarian University named after N.V. Parakhin", 373–375 (2020)

[16] I.N. Fomin, R.P. Belikov, Development of an Algorithm for Remote Monitoring of Disconnection of a Sectionalizing Power Line Circuit Breaker, Bulletin of Agricultural Science **4**, 73, 98–105 (2018)

[17] A.V. Vinogradov, V.I. Zelyukin, A.E. Semenov, A.V. Vinogradova, The Effectiveness of the Use of Automatic Re-Inclusion in Comparison With the Causes of Damage to the OHL-10 kV, Bulletin of Don Agricultural Science **2**, 42, 74-80 (2018)

[18] A.V. Vinogradov, V.A. Chernyshov, A.E. Semenov, A.O. Moroz, A Method for Monitoring the Technical Condition of Elements of an Overhead Power Line, Agrotechnics and Power Supply **3**, 3, 64–68 (2014)

[19] A.V. Vinogradova, A.I. Psarev, Prohibition of Automatic Switching-on of a Reserve for a Short Circuit in a Reserved Section of a 0.4 kV Power Line, Bulletin of Nizhny Novgorod Institute of Engineering and Economics **4**, 107, 44–54 (2020)

[20] A.V. Vinogradov, A.N.Vasiliev, V.E. Bolshev, A.V. Vinogradova, T.A. Kudinova, N.S. Sorokin, N.V. Gruntovich, *Methods of reducing the power supply outage time of rural consumers, Renewable Energy and Power Supply Challenges for Rural Regions* (Series "Premier reference source", Hershey, PA, USA, 370–392, 2019)

Development of software for calculating the parameters of coal-water slurry fuel

Anton Sinitsyn[1,*], *Irina* Akhmetova[2] and *Ekaterina* Filippova[1]

[1]Department of Heat and Gas and Water Supply, Vologda State University, Lenina str. 15, 160000, Russia
[2]Kazan State Power Engineering University, Kazan, Russia

Abstract. The paper presents the developed software tool for calculating the combustion process. Computational studies were performed for coal-water slurry fuel. Using the developed program the calorimetric efficiency and environmental performance of CWS combustion in experimental and industrial installations can be calculated in automatic mode.

1 Introduction

Currently, there are many problems in the fuel and energy complex (FEC), one of which is the constant rise in prices for energy resources, including fuel oil and natural gas. The other problem is low efficiency of coal, as during grate firing up to 30% of the supplied volume remains unburned and goes to the dump. In results in high cost per unit of heat generated from traditional fuels.

A large number of scientists and specialists have worked and are working to increase the efficiency of the fuel and energy complex and its subsystems: A.I. Andryushchenko, V.M. Brodyansky, E.Ya. Sokolov, Yu.P. Melentiev, A.P. Merenkov, E.V. Sennova, V.A. Stennikov, M.N. Singer, K.N. Trubetskoy et al. Particularly noteworthy is the work [1], which allows us to draw conclusions about the prospects of fuel power industry and the possibility of using coal-water slurry fuels (CWS). The works of K. Sato, P. Andreussi, K. V. Thumbimuthu, etc. are the international studies related to the development of efficient combustion systems. These studies are devoted to both consideration of environmental advantages of coal liquid fuels and description of atomized combustion of coal-water slurry fuel in various devices [2-4].

In general, the issues of increasing the efficiency of heat supply sources, and, in particular, the processes of burning organic fuel, are relevant and were previously studied in [5-11].

The use of pulsating combustion of fuel in power plants, the creation of conditions for the complete combustion of the fuel-air mixture due to optimal geometric devices of the combustion space, etc. are the examples of increasing combustion efficiency. In [12], the impact of magnetic field on the completeness of combustion of gas-air mixtures was studied.

One of the solutions to these problems may be the creation and implementation of a new type of fuel, which will serve as an alternative to traditional types and will allow more efficient use of raw materials and reduce the cost of the generated heat. For these purposes coal-water slurry fuel (CWS) can be used, which is obtained by mixing finely ground coal or other carbon-containing raw materials with water or aqueous emulsion.

The tasks were set to study the efficiency of using coal-water slurry fuel as an alternative to traditional types of fuels, in particular coal, as well as to develop software for calculation the combustion process of coal-water slurry fuel. These works were performed as part of research work on the development of an effective CWS dispersion and homogenization system for the Amaltea LLC (Russia) at fuel preparation stage.

The obtained results, by analogy with [13], will make it possible to calculate the calorimetric efficiency and the environmental performance of CWS combustion in experimental and industrial installations using computer-aided procedures.

The purpose of this study is to create a mathematical model and software for the automated calculation of parameters of combustion process of water-carbon fuel based on the existing techniques.

2 Subject of study

The subject of research is coal-water slurry fuel. As already mentioned, such type of fuel is obtained by mixing crushed coal (particle sizes are <200 μm) with water or aqueous emulsion. Emulsions of fuel oil swabs or emulsions of organic origin are used to further increase the calorific value of CWS. The proportion of coal in the fuel composition is from 52 to 65%, and the proportion of liquid component can be from 35 to 48%.

The most common raw materials for CWS are:

1) Coals of most grades - brown, gas, long-flame, lean, anthracite;

2) Carbon-containing raw materials: pyrocarbon (carbon after thermal decomposition of tires and rubber

* Corresponding author: sinitsyn.science@mail.ru

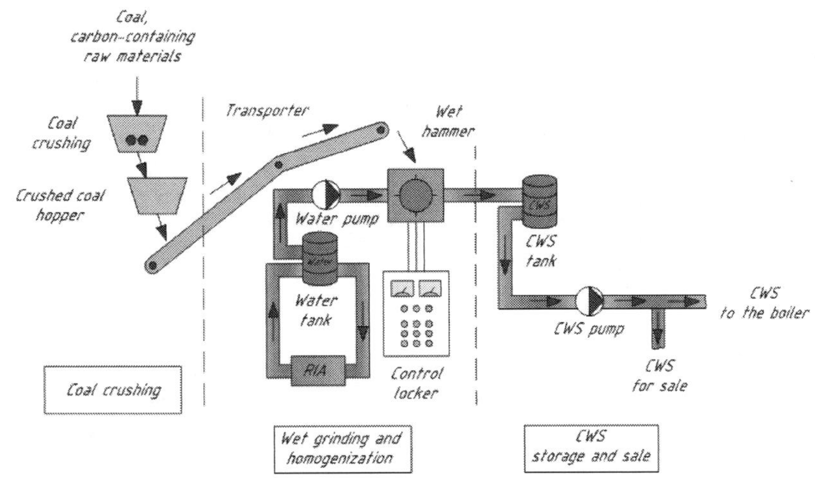

Fig. 1. The technology of preparation of coal-water slurry fuel (processes at Amaltea LLC are used as example).

under conditions of oxygen deficiency), petroleum coke (solid residue of the secondary processing of oil or oil products) and others;

3) Coal production waste (riddlings with fractions up to 1 - 6 mm, usually disposed as waste).

Preparation of coal water fuel includes three stages (Fig. 1):

1) Crushing of coal (if the initial particle size of coal exceeds 12 mm);

2) Wet grinding (up to a fraction of 100 - 150 microns);

3) Homogenization (which makes the obtained product uniform).

The prepared fuel passes quality control according to the following parameters:

1) Solids content (mass fraction of coal), which determines the calorific value of the final product and CWS viscosity;

2) Granular composition (dispersion), which depends on combustion method planned to be used at the facility;

3) Static stability, which determines the requirements for the technological regulations of fuel storage, the need for additional reagents or other measures;

4) Viscosity, which determines the requirements for pumping equipment and the quality of fuel atomization [13].

Typical characteristics of coal-water slurry fuel are presented in table 1.

Table 1. Typical characteristics of coal-water slurry fuel

Parameter	Value
1	2
Mass fraction of solid phase (coal)	58 - 70%
Grading	100% fractions less than 200 microns
Density	about 1200 kg/m^3
Ash content of solid phase	5 - 50% (depends on grade of source coal)
Net calorific value	2300 - 4300 kcal/kg (depends on grade of source coal)
Viscosity at a shear rate of 81 s^{-1}	no more than 1000 MPa·s
Flash point	450 - 650 °C
Combustion temperature	950 - 1100 °C
Static stability	- up to 5-10 days - without the use of additives - from 10 days - the use of additives is possible - from 30 days - with the use of additives
Freezing temperature	0 °C (without additives)

The other parameters (calorific value, ash content, sulfur content, etc.) depend on CWS parameters and characteristics of the source coal.

The main methods of burning such fuel are the following:

1) Autonomous flame combustion;

2) Fluidized bed combustion;

3) Burning together with other fuel.

In the first case, fuel is supplied under pressure through a spray nozzle into the combustion chamber. For this method, it is possible to implement an intermediate stage of partial gasification of CWS in order to supply generator gas - a mixture of carbon monoxide and hydrogen to the boiler. Gasification is carried out by converting the organic part of fuel into combustible gases during high-temperature heating with an oxidizing agent. The obtained gas, getting into the boiler furnace, increases the stability of CWS combustion (Fig. 2).

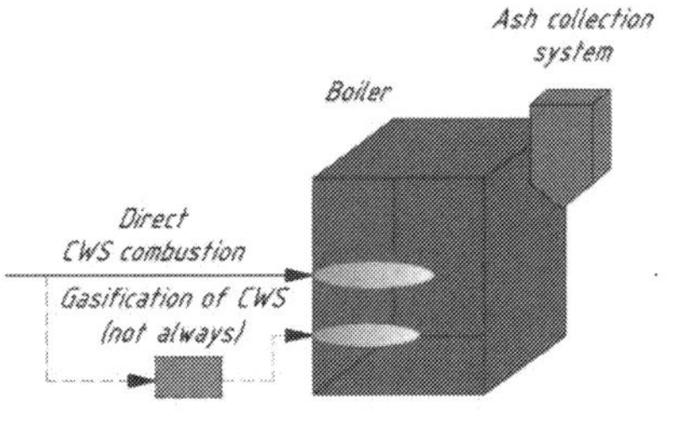

Fig. 2. Autonomous flame combustion of CWS.

The second method involves spraying fuel over a preheated layer of inert material, most often sand. Particles of CWS, falling on a fluidized bed, ignite almost instantly. Combustion using a fluidized bed can be carried out both directly in the boiler and in a special preheater (Fig. 3).

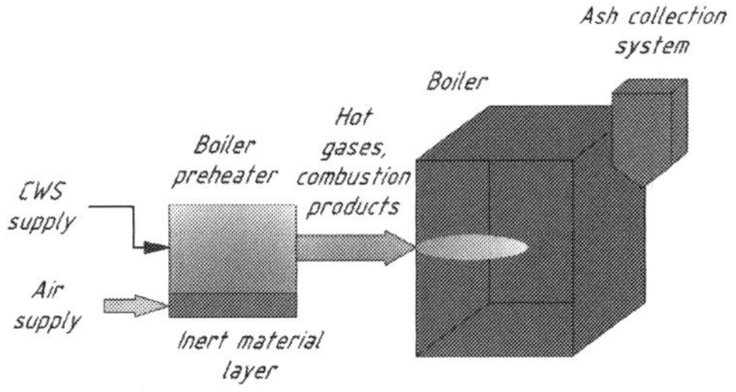

Fig. 3. Fluidized-bed combustion of CWS.

The third method involves the co-combustion of CWS with fuel already used at the heat source - coal, fuel oil or gas. Partial gasification of fuel can be also carried out here (Fig. 4).

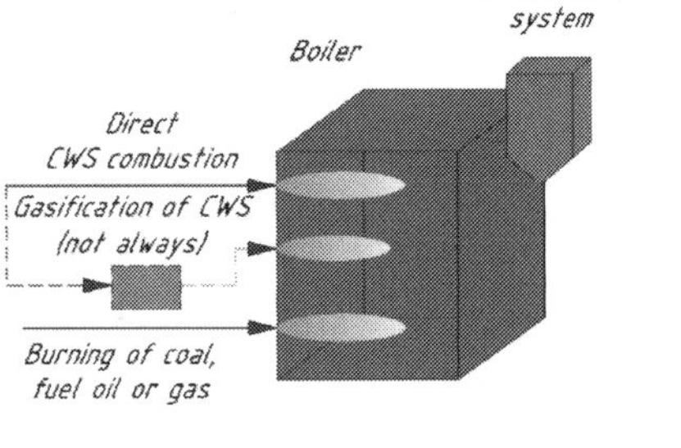

Fig. 4. Combustion of CWS together with other fuel.

3 Mathematical model for calculation the combustion process of coal-water slurry fuel

The study of efficiency of coal-water slurry fuel as an alternative to traditional types of fuels required to calculate the percentage composition of CWS and the amount of pollutants generated during the CWS combustion. To solve these problems, a program was created that allows calculating the parameters of combustion process of coal-water slurry fuel.

The physical model of process is shown in Fig. 1. Fuel of specific calorific value is supplied to the boiler at a certain temperature together with air of its own temperature. Heat released during their combustion is the available heat of the furnace space.In the work, the basic equations of the known methods will be used. [14,15].

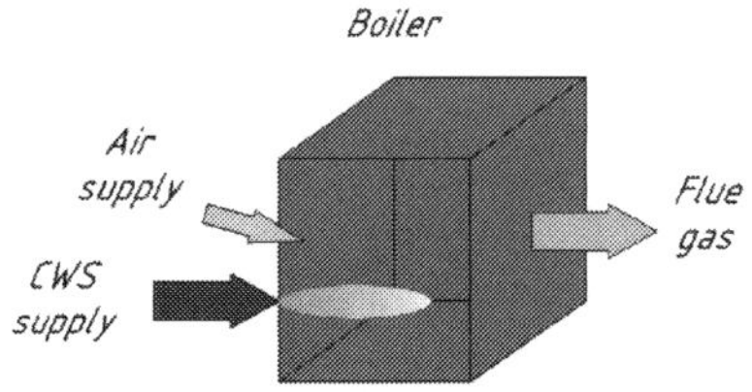

Fig. 5. Physical model of process.

To carry out the calculation, it is necessary to know the composition of burned fuel (the share of each component, its density and heat capacity), the temperature of fuel and air supplied to boiler, the calorific value of fuel, its density, the actual amount of air needed to burn the fuel, and the volumes of combustion products generated during fuel combustion.

The calorimetric combustion temperature is found from the heat balance equation of the furnace space (1), where the fuel is burned:

$$Q_{CWS} + Q_n^r + Q_a = I_{fg},\ kJ/kg, \quad (1)$$

where Q_{CWS} is the physical heat of fuel introduced into the furnace space, kJ/kg;

Q_n^r is the net calorific value, kJ/kg;

Q_a is the physical heat of air introduced into the furnace space, kJ/kg;

I_{fg} is the flue gas enthalpy at calorimetric combustion temperature, kJ/kg.

The physical heat of fuel, the heat of its combustion and the physical heat of air in total form the available heat of the furnace space (2):

$$Q_{CWS} + Q_n^r + Q_a = Q_{fuel},\ kJ/kg \quad (2)$$

The flue gas enthalpy is found by the formula (3):

$$I_{fg} = \Sigma(cV)_{fg} \cdot t,\ kJ/kg, \quad (3)$$

where $\Sigma(cV)_{fg}$ is the total volumetric heat capacity of all flue gases, kJ/(kg·°C);

t is the combustion temperature, ° C.

Thus, this computation section consists of three stages:

1) Calculation of the amount of thermal energy transferred to the furnace space;

2) Calculation of the flue gas enthalpy;

 https://doi.org/10.1051/e3sconf/202017801041

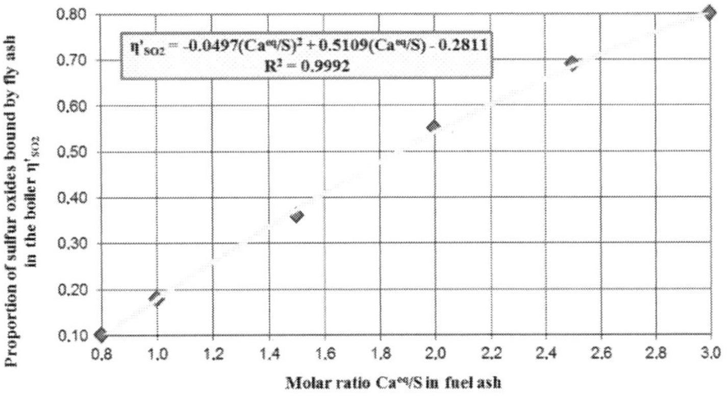

Fig. 6. Graphical representation and selection a formula for experimental data.

3) Checking the obtained value of the calorimetric temperature.

Thermal energy transferred to the combustion chamber consists of three components: heat from fuel, heat from air and heat of fuel combustion.

The thermal energy from fuel is calculated by the formula (4):

$$Q_{CWS} = \frac{T_{CWS} \cdot \Sigma\left(c_i V_i \rho_i\right)}{\rho_{WCS}}, kJ / kg, \quad (4)$$

where T_{CWS} is the fuel temperature, °C;

c_i is the heat capacity of the i-th fuel component, kJ/(kg·°C);

V_i is the share of i-th component in fuel composition;

ρ_i is the density of the i-th component, kg/m³;

ρ_{CWS} is the fuel density, kg/m³.

The thermal energy from air is calculated by the formula (5):

$$Q_a = T_a \cdot c_a \cdot V_a \cdot \rho_a, \; kJ / kg, \quad (5)$$

where T_a is the temperature of supplied air, °C;

c_a is the heat capacity of air, kJ/(kg·°C);

V_a is the actual volume of air required for burning 1 kg of fuel, m³/kg;

ρ_a is the air density, kg/m³.

The enthalpy of flue gases depends on two quantities - their total heat capacity and combustion temperature. But the heat capacity itself depends on temperature, so the enthalpy is calculated by the approximation method in the following sequence:

1) The initial temperatures t_1 and t_2 are set;

2) The volumetric heat capacity of components of flue gases at these temperatures are found;

3) The total heat capacity of all flue gases is found according to the formula (6):

$$\Sigma(cV)_{fg} = (cV)_{RO2} \cdot V_{RO2} + (cV)_{H2O} \cdot V_{H2O} + \\ +(cV)_{N2} \cdot V_{N2}, \; kJ / (kg \; °C), \quad (6)$$

where $(cV)_{RO2}$, $(cV)_{H2O}$ and $(cV)_{N2}$ are volumetric heat capacities of triatomic gases, water vapor and nitrogen vapor, respectively, kJ/(m³·°C);

V_{RO2}, V_{H2O} and V_{N2} are the volumes of triatomic gases, water vapor and nitrogen vapor, respectively, m³/kg;

4) The enthalpy of flue gases is found by the formula (3).

The calorimetric combustion temperature t_k will be calculated automatically using calculations for temperatures t_1 and t_2. Further, the volumetric heat capacities of flue gas components, their total heat capacity and enthalpy will be found for it.

The final step is to check the obtained calorimetric temperature according to the condition (7):

$$\Delta = \frac{Q_{\text{fuel}} - I_{\text{fg}}}{Q_{\text{fuel}}} \cdot 100\% \leq 3\% \quad (7)$$

4 Numerical simulation results

1) Processing of experimental data is performed in accordance with methods given in [14,15]. The developed program includes mathematical processing of data and its graphical representation. For example, Figs. 6 and 7 present relationship between emissions of sulfur oxides and molar ratio of Caeq/S, nitrogen oxides and thermal power;

2) Adaptation of method for calculation the calorimetric temperature previously used for gaseous fuels to calculate the CWS combustion temperature. The table of fuel components was completely updated, a formula showing relationship between heat capacity of mineral part of fuel and its temperature was developed (1):

$$C^A = (0.17 + T) \cdot 4.187, \; kJ / (kg \cdot °C), \quad (8)$$

where T is the fuel temperature, °C.

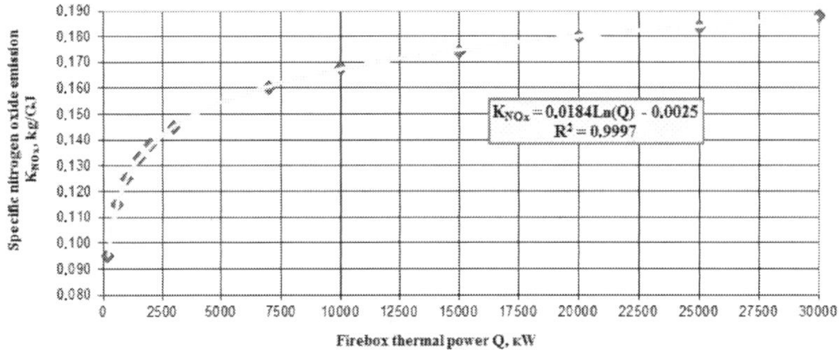

Fig. 7. Graphical representation and selection a formula for experimental data.

The program has some limitations of application. The calculations carried out using it will be correct for coal+water fuel without additives, which is burned by an autonomous flare method.

5 Conclusions

The following results were obtained during study of efficiency of coal-water slurry fuel usage as an alternative to traditional types of fuels:

1) In comparison with other types of fuels CWS usage has a number of significant technological, economic and environmental advantages. It will reduce the cost of a unit of generated heat energy, increase the efficiency of fuel resources use, make benefit from the products previously disposed as waste, and save environment. CWS is an environmentally friendly fuel in production and use and will not harm the environment during spills;

2) When developing a mathematical model, a unified methodology has been made up, which also includes experimental data presented as mathematical and graphical dependencies;

3) In the course of the study, a software tool was created that allows calculating the parameters of CWS combustion process.

Acknowledgements

The authors are grateful to Amaltea LLC for the information provided on the CWS production cycle.

References

[1] E. Sukhareva, A. Fedyukhin, O. Derevianko, M. Egorov, L. Mukhametova, I. Akhmetova, Optimization of combined heat and power plant operating mode by means of underutilized equipment mothball, IOP Conference Series: Earth and Environmental Science 288, 1, 012122 (2019).

[2] K.N. Trubetskoy, I.H. Nehoroshiy, Development of work on the use of suspensions in the energy sector of Russia, Thermal Engineering **1**, 26-29 (1994).

[3] K. Sato, K. Okira, A. Baba, et. al, Study of spray combustion of CWM, Symposium on coal slurry fuels preparation and utilization, Pitsburgh Energy Technology Centre, 178-191.

[4] P.Andrussi, M. Graziadio, L. Tognotti et. al, Atomization of coal-water slurry fuels by pneumatic internal nozzle: part 1, Optimization and spray technology 3, 187-208 (1987).

[5] P. Andrussi, M. Graziadio, L. Tognotti et. al, Atomization of coal-water slurry fuels by pneumatic internal nozzle: part 2, Since and technology 13, 34-36 (1990).

[6] N. Sovetova, A. Sinitsyn, E. Tritenko, O. Derevianko, L. Mukhametova, A. Fedukhin, S. Makoev, A. Kalyutik, *Regional management of district heating,* IOP Conference Series: Earth and Environmental Science Collection of materials International Scientific and Practical Conference, Institute of Physics and IOP Publishing Limited (2019).

[7] A.A. Sinitsyn, Study of vibrational characteristics of the operation of boiler units, News of higher educational institutions. Instrument-making **7**, 51 (2008).

[8] A.A. Sinitsyn, N.N. Boytsova, I.G. Akhmetova, S.O. Makoev, A. Asadpoori, O.A. Stepanov, N.V. Rydalina, E.O. Antonova, B.G. Aksenov, Study of operation of power-generating devices of gaseous fuels combustion, International Journal of Mechanical Engineering and Technology 10, 2, 1480-1490 (2019).

[9] A.A. Sinitsyn, The calculation of the characteristics and of the geometrical sizes of power devices, European Applied Sciences **4** (2013).

[10] E. Sukhareva, A. Fedyukhin, O. Derevianko, M. Egorov, L. Mukhametova, I. Akhmetova, Optimization of combined heat and power plant

operating mode by means of underutilized equipment mothball, IOP Conference Series: Earth and Environmental Science **288**, 1, 012122 (2019).

[11] I.A. Sultanguzin, E.V. Zhigulina, Y.V. Yavorovsky, I.D. Kalyakin, A.V. Govorin, A.V. Fedyukhin, A.A. Krolin, S.V. Guzhov, O. Derevianko, L. Mukhametova, Development strategy of the energy supply system for MPEI campus blocks based on green building, IOP Conference Series: Earth and Environmental Science **288**, 1, 012071 (2019).

[12] A.A. Sinitsyn, D. Rusakov. V.V. Izrantsev, S.A. Kravchenko, G.A. Konrashkova, Investigation of gas fuel combustion efficiency at magnetic field treatment, IOP Conference Series: Earth and Environmental Science Proceedings of the Conference the international scientific conference Efficient waste treatment – 2018, EWT-2018, 012082 (2019).

[13] E.N. Filippova, A.A. Sinitsyn, Development of a software tool for calculating the combustion process of coal-water slurry fuel, Intellectual Information Technologies and Intelligent Business, INFOS-2019 materials of the tenth international scientific and technical conference (2019).

[14] *Methodology for determining emissions of pollutants into the atmosphere when burning fuel in boilers with a capacity of less than 20 Gcal per hour* (The State Committee for Environmental Protection 09.07.1999, Intr. 01.01.2000, Moscow, 76, 1999).

[15] *Methodology for calculating emissions of pollutants during combustion of coal-water slurry fuel* (6, 1990).

Modelling a solid fuel unit for soil warming up before its exploitation

Anton Sinitsyn[1,*], *Liliya* Mukhametova[2] and *Alexander* Korobeinikov[1]

[1]Department of Heat and Gas and Water Supply, Vologda State University, Lenina str. 15, 160000, Russia
[2]Kazan State Power Engineering University, Kazan, Russia

Abstract. This work continues the previous studies on energy-efficient technologies and devices for thawing frozen soils during repair and construction works. It proceeds with the multi-year research on development of efficient devices for emergency repairs. The possibility of using a solid fuel back-draft furnace for accelerated heating of various types of soils is considered. Analysis of the existing devices is presented, the performance criteria are determined, computer modelling of the developed unit and heat and mass transfer processes for the most common types of soils are carried out.

1 Introduction

The study of properties of thawing soils is relevant due to intensive development of the permafrost zones. Research on thawing soils is important for many areas of civil and industrial construction, which are necessary to ensure the operation reliability of facilities constructed according to the second principle (foundation soils are used in thawed or thawing conditions), pipelines and roads in the permafrost zone.

Forecast of deformations is a rather complicated task since frozen soils have multicomponent compositions and differentiate in thawing conditions, and draw during thawing depends on many factors. So it requires further improvements.

A lot of research works on thermo-physical properties of frozen soils during thawing and compaction were conducted within about a hundred-year period. Thawed soil was studied by the Russian (N.A. Tsytovich, G.I. Lapkin, A.E. Fedosov, V.P. Ushkalov, S.S. Vyalov, Ya.A. Kronik, L.T. Roman, Yu.K. Zaretsky, M.N. Goldstein, I.N. Votyakov, V.F. Zhukov, G.V. Porkhaev, L.N. Khrustalev, E.P. Shusherina, G.I. Pakhomova, A.M. Pchelintsev, M.F. Kiselev, V.D. Ponomarev, E.D. Ershov, V.Z. Khilimonyuk and others) and international (Morgenstern N.R., Smith L.B., Crory F., Chamberlain E.J., Gow, Luscher A.J., McRoberts E.C., Harris C., Nixon J.F., Eigenbrod K.D., Keil L.D., Nillsen N.M., Gupta R.C., Speer T.L., Watson G.H., Ryde'n C.G. and others) scientists.

Issues of increasing the efficiency of uninterrupted heat supply sources at low temperatures, and, in particular, fossil fuel combustion processes, are relevant and were previously studied in [1–6]. The use of various fuels and intensification of fuel combustion processes in power plants, the creation of conditions for the complete combustion of fuel-air mixture due to optimal geometric arrangement of the furnace space, etc. are the examples of increasing the efficiency of thawing frozen soils.

We turn the attention to the studies [7,8], which describe in more detail the main devices, methods of thawing and methods for calculating transition phases during heating of a frozen structure. The work [9] presents an overview of the existing patents and commercial devices for elimination failures of utility systems in frozen rocks, as well as factors affecting the heat and mass transfer process. It can be concluded that the process of thermal conductivity in soils during thawing largely depends on humidity, density, soil type, cryogenic texture, mineral composition, changes in filtration coefficient, thawing rate, moisture outflow conditions, etc.

The issue of creating an efficient thawing method in the conditions of development of permafrost zones remains relevant from both practical and scientific points of view.

2 Review of the existing devices and analysis of its efficiencies

Soil exploitation associated with digging trenches in temperature conditions below zero degrees Celsius is complicated by the need for preliminary preparation and warming up of frozen soil.

In urban conditions, in the presence of a large number of active cable lines and other underground utilities, the use of pneumatic instruments (jackhammers, crowbars, wedges, etc.) is impossible due to the risk of mechanical damage to existing cable lines and other underground utilities. Therefore, the frozen soil must be pre-warmed before excavation work in the area of existing cable lines so that further work is carried out without the use of pneumatic instrument. The most

* Corresponding author: sinitsyn.science@mail.ru

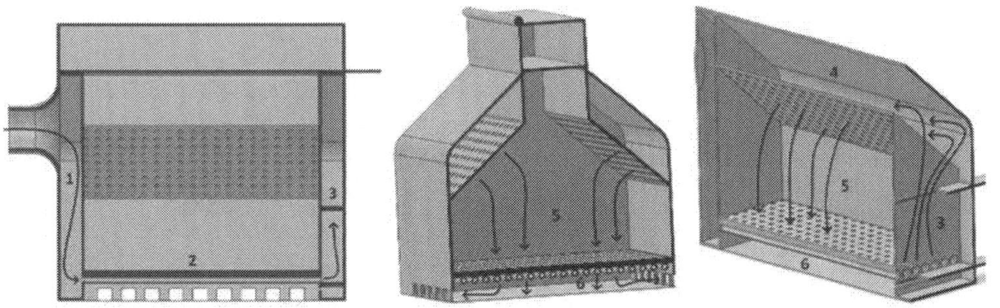

Fig. 1. Three-dimensional space model of a solid fuel furnace with back draft.

common methods of artificial thawing are thermal and electrothermal methods.

We bring into focus the following methods of soil heating during excavation in the autumn-winter period: electric reflex furnaces; electric horizontal and vertical steel electrodes; controlled fire; chemical defrosting; the use of steam and water needles; hot running coolant (sand, slag, gravel, soil, road waste - asphalt crumb); tubular electric heaters; high frequency currents; hydraulic stations, surface thermoelectric mats for soil heating, for example, TM – 800 (Russia).

The aim of development is to accelerate and reduce the cost of mining and construction work in frozen ground, associated with the need its thaw. At present, three methods for thawing of frozen soil during these operations are commonly used. The first one is to lay fire from firewood at the place where it is necessary to thaw and excavate the ground, and usually the first row of firewood does not burn and forms a kind of thermal insulation of frozen ground from the heat generated during the burning of firewood. The second way is to burn charcoal or other combustible materials on iron sheets at the place where thawing should be performed. The third one is to heat permafrost with a flame of special lamps for liquid fuel [10].

A review of patents and inventor's certificates made it possible to formulate a list of analogues of the device being developed. A back-draft portable furnace with air-cooled tubular grate is the closest analogue. In order to heat the soil, the casing of furnace is made without backstone in ash pan and has a number of flue gas vents along its edge. A pipeline for input of additional hot air into the under-the-head space is connected to the collector adjacent to the grate in addition to the pipeline for usual distribution of air above the fuel layer to improve the combustion conditions and increase the temperature of gases flowing around the soil. This device has several disadvantages, namely the low efficiency of solid fuel combustion and the probability of absence of the required draft and creation of the so-called "back draft" for sufficient temperature pressure of flue gases in the working area of furnace, as well as a small heating area. Further research is aimed at solving these drawbacks.

3 Description of the unit construction

When designing modern heat-generating devices, the following requirements should be met: it should provide a simple and economical transportation of heat to the work surface; have the most rational scheme of fuel supply and ash removal; have the best working conditions for operating personnel; be highly economical and safe to use.

The unit must be equipped with a primary air regulator. The design must ensure that during its operation, neither residues of combustion products nor unburned fuel can disrupt motion or shut-down of the air supply regulator. It should also be able to completely clean the combustion path. The minimum width of the combustion path should be 30 mm [11].

The purpose of the device being developed is to improve the quality, reliability and efficiency of devices used to heat working fluids, improve the methods of surface heating and warming of frozen soil, concrete, road pavement by radiation-convective method, with minimum energy and time requirements, taking into account the environmental safety requirements with emissions of exhaust gases into the air, the detection and study of defects in the working bodies.

The following solutions are proposed: the furnace body is formed by an iron frame with grates. The grates are covered by a double vault. The air initially enters the space between the sheets 1, through the nozzles of grate 2 passing through the regulating flow damper 3 enters the space 4, through the holes is distributed evenly over the surface of the furnace. Air is pumped by a manual fan, enters the furnace and the under-grate space for combustion of exhaust gases. The combustion products through the grates go to ash pan 6 and, meeting a soil segment, make thawing not only by hot gases, but also by the radiant heat of the burning fuel.

Fuel (firewood) is loaded through the upper door 7 (see Fig. 2), which is tightly closed during operation. Furnace can be additionally equipped by adjustable manifolds and channels to increase the path of the exhaust combustion products.

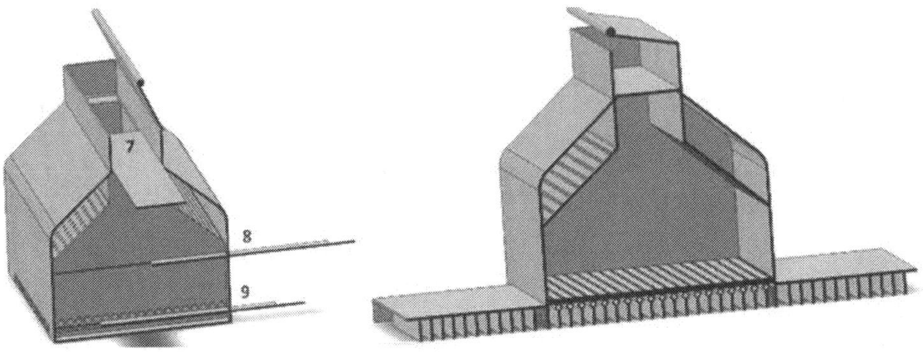

Fig. 2. Three-dimensional space model of a solid fuel furnace with back draft.

This design has the following advantages over its analogues:

- The injected air is preheated from the walls of combustion chamber, therewith cooling the grate and reducing furnace wear, while increasing the efficiency of fuel combustion;
- The area of soil heating can be increased due to adjustable manifolds;
- Solid fuel is the best option in cold conditions and does not require preliminary preparation;
- Before feeding into the combustion chamber, the wood is heated by the walls of combustion chamber (preliminary drying takes place).

4 Experimental data

As a rule the modelling process is iterative, which involves at each iteration step the refinement of solutions made at previous steps of model development. Mathematical modelling has several advantages over real experiment, including profitability (saving the resources of real system); the possibility of modelling hypothetical (not realized in nature) objects; the possibility of implementing dangerous or difficult to reproduce in nature modes, as well as the universality of hardware and software of the work performed.

In mathematical modelling, the Stefan problem was used, which is solved as a system of heat equations taking into account the phase transition.

The problem statement is as follows: the soil has the initial temperature T_{in}, and it is frozen. The phase transition temperature T_{ph} is higher than T_{in}. On the soil surface x=0 at initial time t=0, a constant temperature T_0 is established abruptly and maintained above the freezing temperature. The melt layer that appears at the surface increases with time. A front is established, the motion of which should be determined, as well as temperatures in thawed and frozen zones.

The coordinate of front freezing is denoted as x_1, temperature in the melt layer $0 \le x \le x_1$ is denoted as $T_1(x,t)$, $T_2(x,t)$ is the temperature in not warmed area $x_1 \le x \le \infty$. Consequently, the problem of warming up the soil is considered as the problem of conjugating temperature fields on a warming front that moves, so finally we get a system of equations of thermal conductivity.

$$\left.\begin{aligned} \frac{\partial T_1}{\partial t} &= a_1 \frac{\partial^2 T_1}{\partial x^2}, 0 \le x \le x_1 \\ \frac{\partial T_2}{\partial t} &= a_1 \frac{\partial^2 T_2}{\partial x^2}, x_1 \le x \le \infty \end{aligned}\right\}, \quad (1)$$

where a_1 is the coefficient of thermal diffusivity in the thawed zone and a_2 is the coefficient of thermal diffusivity in the frozen zone with initial conditions

$$T_2(x,0) = T_{in}, \quad (2)$$

$$T_1(0,t) = T_0 > T_{ph}, \quad (3)$$

$$T_1|_{x_1} = T_2|_{x_2} = T_{ph} = const, \quad (4)$$

To simplify the solution, we take $T_{ph} = 0$, thereby switching to the Celsius scale. One can change the zero point according to the requirements.

The front moves with a predetermined speed, therefore, in addition to the boundary conditions (4) for the heat equations, it is necessary to set one more parameter that determines the front speed. Assume that for some time dt the front moves a distance dx_1. At the same time, a mass of water equal to $rSdx_1$ is melt and the amount of heat $LrSdx_1$, is absorbed, where S is the front surface area, L is the specific heat of phase transition, r is the mass of water per unit volume of soil. According to the law of conservation of energy this amount of heat must be equal to the difference in heat amount passing from the frozen and thawed areas through the front:

$$\left[-\lambda_2 \left.\frac{\partial T_2}{\partial x}\right|_{x_2} + \lambda_1 \left.\frac{\partial T_1}{\partial x}\right|_{x_1} \right] Sdt = L\rho S dx_1, \quad (5)$$

or

$$\lambda_1 \left.\frac{\partial T_1}{\partial x}\right|_{x_1} - \lambda_2 \left.\frac{\partial T_2}{\partial x}\right|_{x_2} = L\rho \frac{dx_1}{dt}, \qquad (6)$$

This condition is called the Stefan condition at the front of phase transition. Solution of equations (1) should be searched in the form:

$$T_1 = A_1 + B_1 erf\left(\frac{x}{2\sqrt{a_1 t}}\right); \qquad T_2 = A_2 + B_2 erf\left(\frac{x}{2\sqrt{a_2 t}}\right), \qquad (7)$$

where A_1, A_2, B_1, B_2 are the unknown constants, which are determined from the boundary and initial conditions.

During problem solution, a formula was obtained that shows the law of defrosting front motion.

$$x_1 = \beta\sqrt{t}, \qquad (8)$$

where β is the unknown constant at this point, which is further found from the final transcendental equation.

$$\frac{\lambda_1 T_0 e^{-\frac{\beta^2}{4a_1}}}{\sqrt{a_1}\, erf\left(\frac{\beta}{2\sqrt{a_1}}\right)} + \frac{\lambda_2 T_н e^{-\frac{\beta^2}{4a_2}}}{\sqrt{a_2}\, erfc\left(\frac{\beta}{2\sqrt{a_2}}\right)} = L\rho\beta\frac{\sqrt{\pi}}{2}, \qquad (9)$$

Thus, if thermophysical parameters of substance $\lambda_1, \lambda_2, a_1, a_2, L, r$, and boundary and initial temperatures T_0 and T_{in} are given, then constant β, as well as constants A_1, A_2, B_1, B_2 in further calculation are uniquely determined.

Analyzing the motion of phase transition boundary as a function of time, we can conclude that the soil density affects the heating time, the higher the density, the longer time thawing will take. Using the formulas presented above, a graph of temperature distribution from surface in 5..10..20 minutes was constructed, which gives us a visual representation of the process taking place underground.

Further, the results of mathematical modelling are presented in plots.

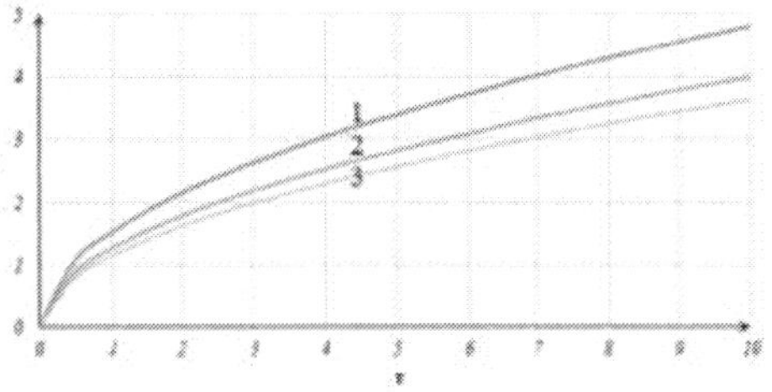

Fig. 3. Motion of phase transition boundary (H, cm) as a function of time (τ, min).

Fig. 3 shows plant layer, peat with ρ=1200 kg/m^3 (1); sands, sandy loam, loam ρ=1750 kg/m^3 (2); soil with a particle size of more than 80 mm (3), with boulders content up to 30% ρ=2100 kg/m^3.

It was determined that performance of a power unit depends on many factors: the type, structure of frozen soil and its thermo-physical properties, climatic conditions and the nature of soil surface, the volume of thawed soil and the frequency of its removal from the pit, etc. Regarding the latter factor, it is worth to note that experimental studies by the experimental unit, conducted in field conditions, showed that the character of removal of the thawed part has a strong influence on the speed of thawing of frozen soil. For 1.3 kW power expenditures and a thawing area of $85 \cdot 10^{-3}$ m^2 of bulk sandy soil, generalizing experimental dependences of the time-varying depth of artificial thawing of the soil were obtained.

The speed and, accordingly, the depth of thawing depends on the conditions for removal of thawed soil. With constant removal of soil (every 5 min), the depth of thawing is 3 times greater than for thawing without removing it. With periodic (every 15 min) removal of

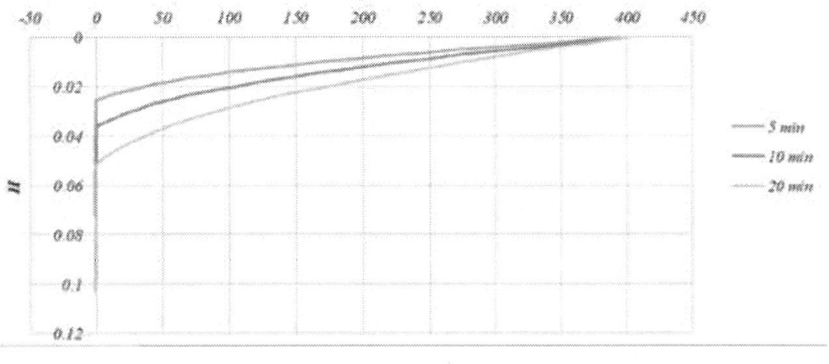

Fig. 4. Temperature distribution after 10 minutes of unit operation (*T*, ◦C), as a function of distance to surface *H*, m.

soil, the depth of thawing is 2 times greater than for thawing without soil removal, which was taken into account when developing the technology for the unit operation.

5 Conclusions

It can be said that the presented unit is suitable for use during works related to the development of frozen soil, heating of building materials and heating of complex building structures to determine defects by non-destructive testing.

In comparison with the existing technical means, the unit being developed is efficient and commercially viable.

In addition, the main features of the proposed unit are highlighted:

- Its construction and design have no analogues in the modern market of technical equipment for repair and construction works;
- It has the minimum time for warming up frozen soil compared to the existing units;
- Low specific energy consumption of the unit together with the combined heat transfer method ensure its high efficiency;
- The mobility and compactness of the unit allows its transporting in a truck and carrying out work by a team of two people under lack of space of the urban landscape;
- The flexibility of regulating the heat output of the unit ensures its use for thawing building materials and heating building structures for any purpose;
- Energy-efficient combination of unit elements and a thermo-insulated housing can provide repair and construction work in any climatic conditions.

Acknowledgements

The presented work was performed within the framework of financing under the federal target program of the Ministry of Science and Education of the Russian Federation, as well as within the framework of research funded by the Innovation Support Fund. This work was supported by a grant named after the outstanding organizers of the oil and gas industry of the International Fuel and Energy Association (IFPA). The research results received international recognition at the 2nd International Exhibition of Inventions and Inventors of Mediterranean Countries INVENTECO EXPO in Catania (gold medal), as well as at the youth competition Russia Power and HydroVision Russia 2011 (PennWell, United Kingdom).

References

[1] N. Sovetova, A. Sinitsyn, E. Tritenko, O. Derevianko, L. Mukhametova, A. Fedukhin, S. Makoev, A. Kalyutik, Regional management of district heating, IOP Conference, Series: Earth and Environmental Science Collection of materials International Scientific and Practical Conference, Institute of Physics and IOP Publishing Limited (2019).

[2] E. Sukhareva, A. Fedyukhin, O. Derevianko, M. Egorov, L. Mukhametova, I. Akhmetova, Optimization of combined heat and power plant operating mode by means of underutilized equipment mothball, IOP Conference, Series: Earth and Environmental Science **288**, 1, 012122 (2019).

[3] A.A. Sinitsyn, The calculation of the characteristics and of the geometrical sizes of power devices, European Applied Sciences **4** (2013).

[4] L.R. Mukhametova, I.G. Akhmetova, A.V. Fedukhin, S.O. Makoev, A.A. Kalyutik, About heat supply schemes for settlements and city districts, IOP Conference, Series: Earth and Environmental Science **288**, 1, 012108 (2019).

[5] A.A. Sinitsyn, Simulation of fire engineering processes in energy devices aimed at their optimization and improvement of reliability, Life Science Journal, the Acta Zhengzhou University Oversea Version, Marsland Press: New York **10** (2013).

[6] A.A. Sinitsyn, A. Mihin, D. Zaripova, Development of a mobile power plant for thawing frozen soils, IOP Conference Series: Earth and Environmental Science Proceedings of the Conference the international scientific conference Efficient waste treatment – 2018, EWT-2018 (2019).

[7] A.A. Sinitsyn, D.F. Karpov, I.A. Sukhanov, Solving the problem of thawing frozen soils during preparatory construction work in low-rise constructions, Low-rise construction within the framework of the National project Affordable and Comfortable Housing for Russian Citizens: Technologies and Materials, Problems and Development Prospects in the Volgograd Region, materials of the International Scientific and Practical Conference (2009).

[8] D.F. Karpov, A.A. Sinitsyn, A.Yu. Velsovsky, Architecture and Construction **3** (2010).

[9] D.F. Karpov, A.A. Sinitsyn, D.A. Belyansky, A.Yu. Velsovsky, Calculation and experimental studies of the distribution of temperature fields during artificial thawing of frozen soil, Bulletin of Kazan State University of Architecture and Civil Engineering **2**, 14 (2010).

[10] A.A. Sinitsyn, Device for radiation-convective heating of working bodies, Patent of the Russian Federation, 129608, 27.06.2013, application no. 2012141775/06, 01.10.2012.

[11] A.A. Sinitsyn, N.N. Tyukin, E.A. Ilyichev, A mobile device for heating the soil and road pavement during emergency repair and restoration works, Energy: Efficiency, Reliability, Safety, Proceedings of the XIX All-Russian Scientific and Technical Conference (2013).

E3S Web of Conferences **178**, 01043 (2020) https://doi.org/10.1051/e3sconf/202017801043
HSTED-2020

Asynchronous electric drive with an additional switch in the DC link of the frequency converter

D. Yu. Kukishev[1], *V. N.* Meshcheryakov[1], *V. N.* Voyekov[1], and *I. V.* Ivshin[2]

[1]LSTU, Department of Electric Drive, 398055 Lipetsk, Russia
[2]The department "power supply of industrial enterprises", Kazan State Power Engineering University, Kazan, Russia

Abstract. Regulation of power on an inverter entrance - one of key tasks in the modern electric drive. In this article the way of regulation of power on an inverter entrance by means of the additional switchboard in a link of a direct current for the electric drive on the basis of the asynchronous engine with a short-circuited rotor (ADKZ) with vector management, with relay management of transistor keys of AIT was considered. Comparison of frequency of switching of keys of the inverter is carried out, at various levels of restriction of current in a link of a direct current.

1 Introduction

Thanks to development of the microprocessor equipment and power electronics introduction of frequency electric drives for such mechanisms as fans, pumps, for the purpose of increase in efficiency of the electric drive, namely energy saving became possible. Achievement of this purpose is possible in two ways: 1. Introduction on production of highly effective electric motors.2. Introduction on production of semiconductor converters of frequency, with the improved power characteristics. Various types of the frequency converters (FC), but are known today, strong development of element base of power electronics promotes the increasing application of PCh constructed on the basis of completely operated IGBT transistors which have high switching ability. Thanks to new control algorithms of power keys of the converter of frequency, the solution of a problem of regulation of power on an inverter entrance became possible. The set of works is devoted to this problem [1][2][3][4]. Analyzing already available works, it is possible to draw a conclusion that in most cases, for control of the additional switchboard in a link of a direct current, the method of the pulse-width modulation (PWM) is offered. But also other way - relay management is possible.

The electric drive about ADKZ [5] operated by a vector control system with relay management of transistor keys of AIT is taken as a basis of a research. As the additional switchboard in a link of a direct current the IGBT transistor is taken. AIT is connected to the power supply, in our case - to the uncontrollable rectifier [6].

The function chart of system of vector management of the asynchronous electric drive is submitted in the figure 1. The power scheme of the asynchronous electric drive with the additional switchboard in a link of a direct current is submitted in the figure 2.

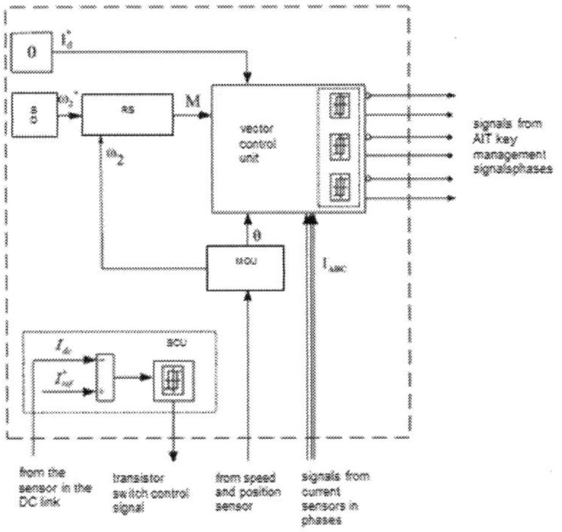

Fig. 1. The function chart of system of vector management of the asynchronous electric drive with the switchboard in a link of a direct current.

The power scheme of a link of a direct current [7] is submitted in the figure 3. In an algorithm of work of this scheme two states stand out clearly. The first state - the VT transistor is switched on, the source of constant tension is connected to loading through a throttle of L. At the same time the throttle of L and the condenser C accumulate energy from a source which is also spent on R loading. The equivalent circuit for this interval is submitted in the figure 4.

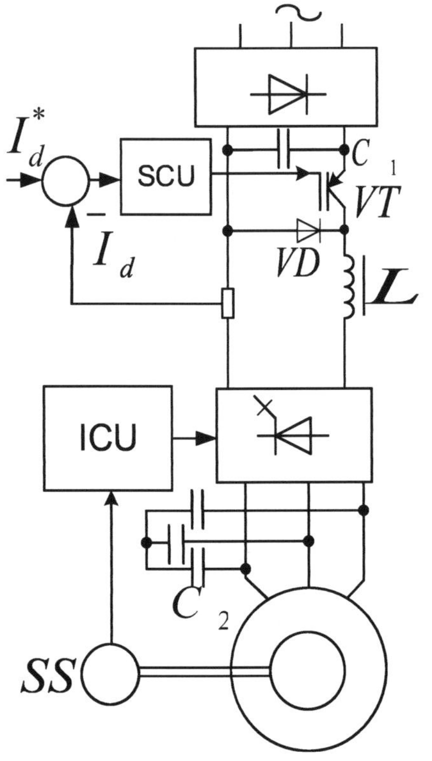

Fig. 2. The power scheme of the asynchronous electric drive with the additional switchboard in a link of a direct current.

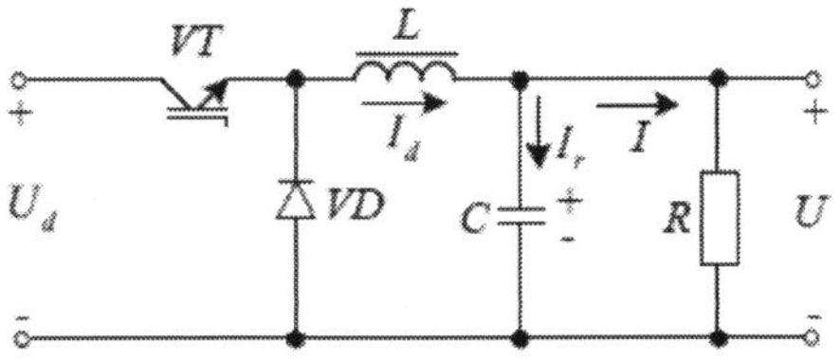

Fig. 3. Power scheme of a link of a direct current.

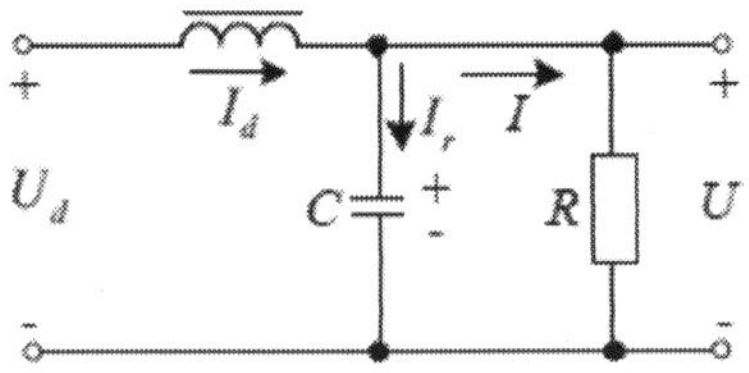

Fig. 4. The equivalent circuit in the first state.

The second state - the VT transistor is switched off and current from a throttle of L becomes isolated via the diode shown in the figure 3.18 and also current of the condenser proceed on tension converter exit in R loading [8,9]. The equivalent circuit for the second state is submitted in the figure 5.

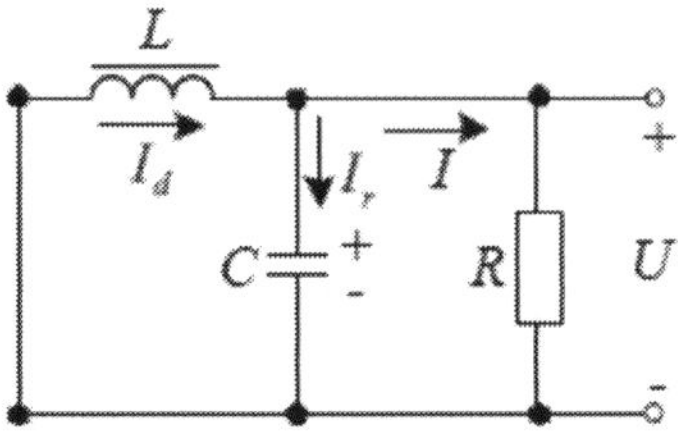

Fig. 5. The equivalent circuit in the second state.

The operated switchboard in a link of a direct current receives a task from the hysteresis regulator which is built in the control unit of the switchboard, and the frequency of switching of the operated VT switchboard depends on hysteresis module width [10,11]. The function chart of the control unit of the switchboard is submitted in the figure 6. Regulation happens by comparison of a preset value on an entrance to the measured value. In case of relay regulation of current the task for maintenance of a certain value of current of I_d* = by const in a link of a direct current is used. The key of VT becomes isolated when the difference between the set I_d* and the actual current of I_{dc} of a link of a direct current reaches the lower threshold of operation of the module of a hysteresis. The key is disconnected at the moment when the difference reaches the top threshold of turning on of the module of a hysteresis [12,13]. Thanks to it level current is stabilized about set.

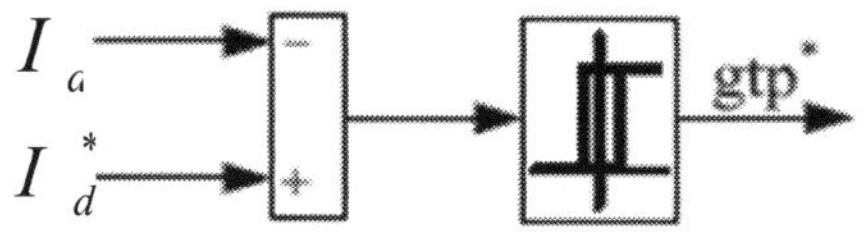

Fig. 6. Function chart of the control unit of the switchboard.

In the figure 7 the model, the asynchronous electric drive with the additional switchboard in a link of a direct current, in the environment of Matlab is shown, at management with maintenance of a constant task for I_d*=1870 A current at dispersal and 1970 A at the established speed. Feedback comes to the relay regulator from the current measuring instrument. Hysteresis module width for the relay regulator in a link of a direct current is chosen equal 1A. Results of modeling are given in figures 8-10.

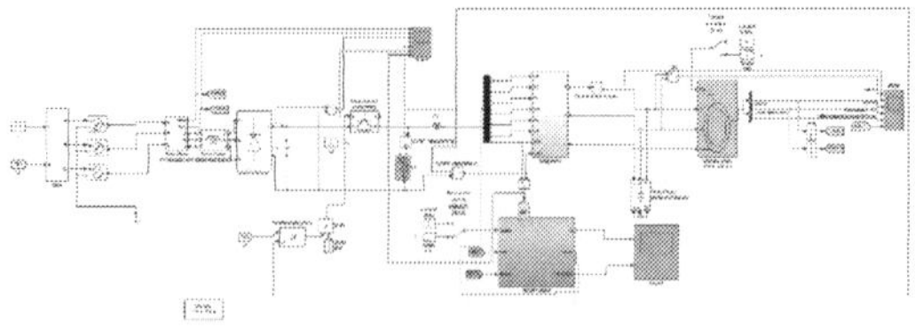

Fig. 7. Model of vector control of the asynchronous electric drive.

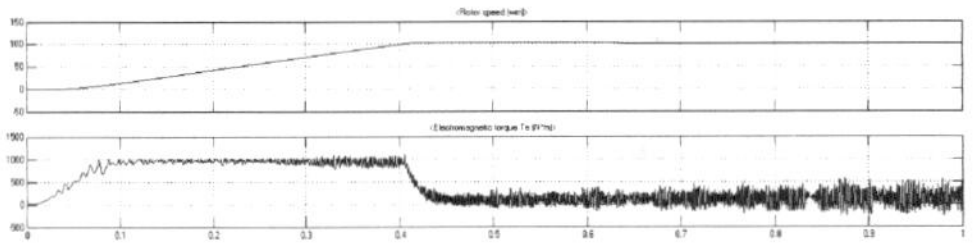

Fig. 8. Schedules of change: a) speed; b) torque.

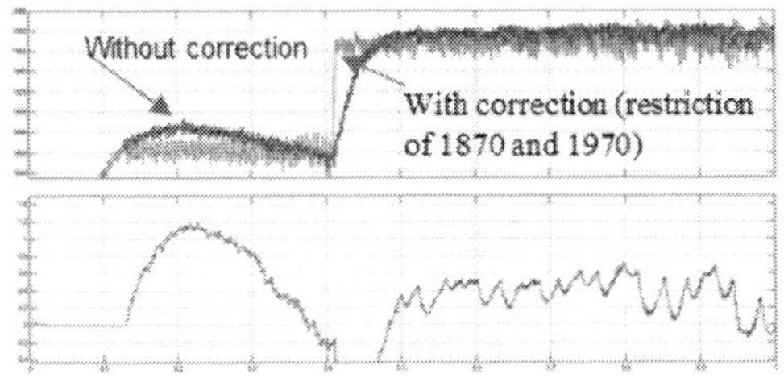

Fig. 9. Schedules of change: a) current of a link of constant tension; b) reduction of frequency of switching of keys of the inverter in %.

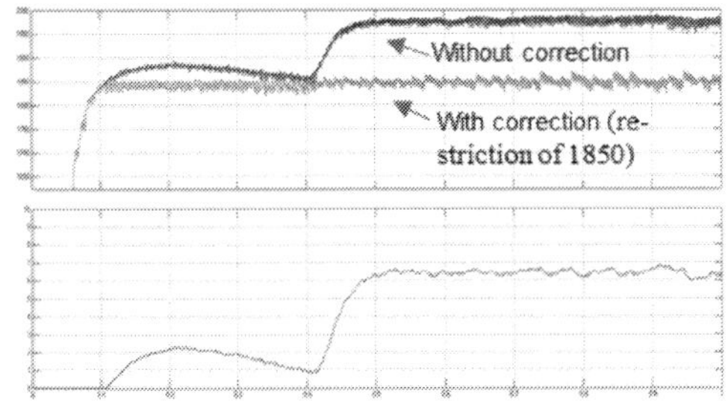

Fig. 10. Schedules of change: a) current of a link of constant tension; b) reduction of frequency of switching of keys of the inverter in %.

Analyzing the schedules submitted in figures 9, 10 it is possible to draw a conclusion that, holding current in a link of a direct current at the set level, by means of the additional switchboard, we resolve not only an issue of regulation of power on an inverter entrance, but also we regulate the frequency of switching of keys of the inverter. The less we keep current in a link of a direct current, the less frequency of switching.

Acknowledgements

Article is written with assistance of a grant of the Russian Federal Property Fund 19-48-480001 "Development, a research and optimization of the energy saving electrotechnical and electrodriving automated complexes for plasma, electrometalslag and induction technologies and units".

References

[1] V.Yu. Alekperov, R.U. Maganov, V.F. Forest warden, V.I. Grayfer, A.V. Bezzubov, N.P. Khokhlov, M.Ya. Ginzburg, V.I. Pavlenko, V.I. Sagalovsky, A.V. Sagalovsky, V.M. Volkov, G.L. Agapova, The Applicant and the patent holder Open joint-stock company The Lukoil oil company, Open joint-stock society Russian Innovation Fuel and Energy Company, Patent no. 2207700, Way of control of the valve electric motor, no. 2000108696/09, zayavl. 11.04.2000, opubl. 27.06.2003

[2] Yu.S. Usynin, *Control systems of electric drives: studies. grant* (Chelyabinsk: Publishing house YUURGU, 358, 2001)

[3] B.L. Aliyevsky, S.A. Scherbakov, S.R. Mizyurin, The valve generator for the autonomous systems of power supply of a direct current, Electricity, 1, 27–32 (2003)

[4] V.N. Meshcheryakov, V.N. Voyekov, A vector control system of the valve electric drive on the basis of the independent inverter of tension with relay regulation of entrance current of the inverter and phase currents of the stator, Messenger YUURGU, Power series **17**, 2, 48–57 (2017)

[5] V.N. Meshcheryakov, P.E. Tsvetkov, Sistema of the optimum frequency asynchronous electric drive with correction on a longitudinal component of current of the stator, News of higher education institutions, Electromecanics, 3, 36–39 (2012)

[6] Y.I. Gracheva, O.V. Naumov, Operating Mode Influence on Probability Characteristics of Electric Devices, Journal of engineering and applied sciences **11**, 1, 2934–2938 (2016)

[7] Y.I. Gracheva, N.V. Chernova, A.I. Fedotov, E.A. Fedotov, Local Fourier transformation application for mathematic modeling of synchronous machine valve actuator, Journal of engineering and applied sciences **11**, 1, 2939–2945 (2016)

[8] V.N. Voyekov, V.N. Meshcheryakov, Relay regulation in the frequency converter with the independent inverter of current, Electrical equipment: network online scientific magazine **2**, 2, 30–34 (2015)

[9] A.S. Anuchin, *Control systems of electric drives: textbook for higher education institutions* (M.: MEI publishing house, 373, 2015)

[10] V.N. Meshcheryakov, V. Voyekov, Patent holder federal state-funded educational institution of higher education of Lipetsk State Technical University, Patent no. 166655, The device for control of the electric drive of alternating current (LSTU) (RU), Bull. **34**, no. 2016113199/07, zayavl. 06.04.2016, opubl. 10.12.2016

[11] Yu.N. Kalachev, *Vector regulation (practician's note); methodical instructions* (M.: EFO, 63, 2013)

[12] S.G. Herman-Galkin, *Virtual laboratories of semiconductor systems in the environment of Matlab-Simulink: textbook* (SPb.: Lan publishing house, 448, 2013)

[13] L. Ting, Y. Tan, G.Wu, W. Shumao, Simulation of PMSM vector control system based on matlab, Simulink. Proc. IEEE Int. Conf. Measuring Technology and Mechatronics Automation (ICMTMA09) **2**, 343–346 (2009)

Estimation of the driven gas turbine unit technical performance using the standard measuring systems

Vitalii L. Blinov[1,*], *Oleg V.* Komarov[1], and *Egor A.* Zaslavskiy[1]

[1]Ural Federal University, Turbines and Engines Department, 620002 Yekaterinburg, Russia

Abstract. In large pipeline gas transport systems the operation and maintenance of gas pumping units are carried according to the current number of equivalent working hours of centrifugal gas compressors and gas turbines. Modern terms of lean production require the maintenance procedure to be done according to the current technical performance of equipment. The paper presents a designed and verified methodology of technical performance estimation of gas turbine units using the standard measuring systems. This method includes a verified high-order mathematical model based on the gas dynamic function for the precise analytical description of turbomachinery aerodynamics. The models are defined for different types of multi-shaft gas turbines. In this article the results of technical performance estimation of different gas turbine units are discussed.

1 Introduction

In large pipeline gas transport systems the operation and maintenance of gas pumping units are carried out according to the current number of equivalent working hours of centrifugal gas compressors and gas turbine units (GTU). Modern terms of lean production require the maintenance procedure to be done according to the current technical performance of equipment. The estimation of the current technical performance of GTU is essential for its proper utilization, maintenance and extension of its operation time between failures and planned overhauls [1,2].

To date, in the gas transportation systems few algorithms addressed to online control of equipment efficiency and degradation coefficient during its life time are used. The basic explanation of the degradation reasons has been extensively studied and presented in [3-5] as an example. It is worth noting that the accurate determination of the gas turbines (GT) current power is a key point.

One of the most widespread solutions for power consumption definition is the introduction of a torque meter in the shaft between a power turbine and a gas compressor [6]. However, in the current operating conditions, this method is extremely difficult since most units are not equipped with a torque meter. Another solution is the estimation of the power consumed by a gas compressor [5,7]. This method requires the proper measurement of the natural gas massflow which is not always possible. Gas path analysis methods for gas turbine performance estimation are also used [8,9]. Special attention is paid to the gas turbines with the flow part variable geometry of the power turbine (PT) [10].

In any case, a gas pumping station remains a very complicated system to have enough measuring facilities and human power for estimation of the real performance of every module, so mathematical models still has opportunities for implementation [2,9].

2 Materials and methods

Most of the GTU for the mechanical drive have two or three shaft designs [2]. Aero derivative, marine and stationary drives are used [11]. For every type of the gas turbine a similar approach is applied to adopt a system of parametrical diagnostics by thermal and aerodynamic parameters. The key feature of this approach is that only standard measured parameters are considered for a gas turbine unit. No additional power measurements, like a torque meter or measurement of the power consumed by a gas compressor, are involved.

The proposed approach can be divided into three stages (Figure 1). At the first stage, a mathematical model of gas turbines (according to nominal load mode parameters) is developed. At the second stage, it is checked by using test results. Having verified the data, the approach can be used to assess the technical condition of gas turbines only if standard measured parameters are used. This is the third stage.

2.1 Mathematical model of GTU

The following parameters are used as input data:

- Ambient pressure and temperature.
- GTUs nominal parameters taken according to the GT catalogue issued by VNIIGAZ [11]: nominal power output, efficiency of GT, fuel gas consumption, cycle

* Corresponding author: v.l.blinov@urfu.ru

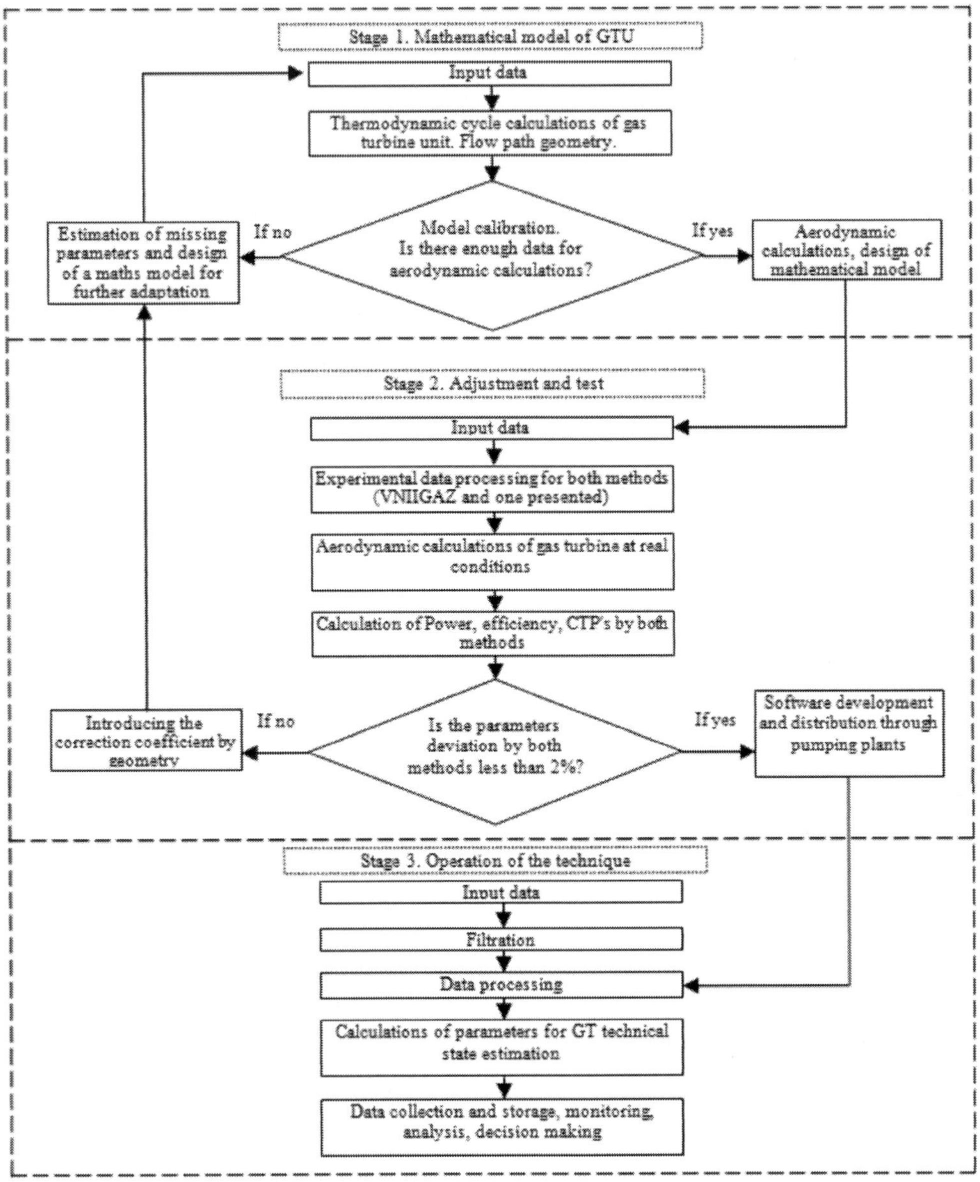

Fig. 1. Architecture of the driven gas turbine unit technical performance estimation with standard measurements.

airflow, mass flow of working gas in turbine after PT, air compressor pressure ratio, temperature after compressor, turbine inlet temperature, temperature of gases before or after PT, hydraulic pressure drop in inlet and outlet ducts, shafts speed. It should be noted that this catalogue [11] is not a producer statement declaration about any GT parameter, but really measured data after the acceptance testing of each unit on the site and further statistical averaging.

- Variable parameters: mechanical losses; efficiency of every turbine, combustor, and axial compressor; the speed coefficient in stator and rotor blades. Also the relative coefficient is used to consider the difference in the airflow through a turbine and a compressor. These parameters must be verified and adjusted within the adjustment and test stage.

- Geometry of the real flow path of turbines: diameters and height of blades.

The thermodynamic calculations were made according to [12]. At this stage the main parameters are: specific heat, temperature and pressure after compressor, specific compression work; then pressure and

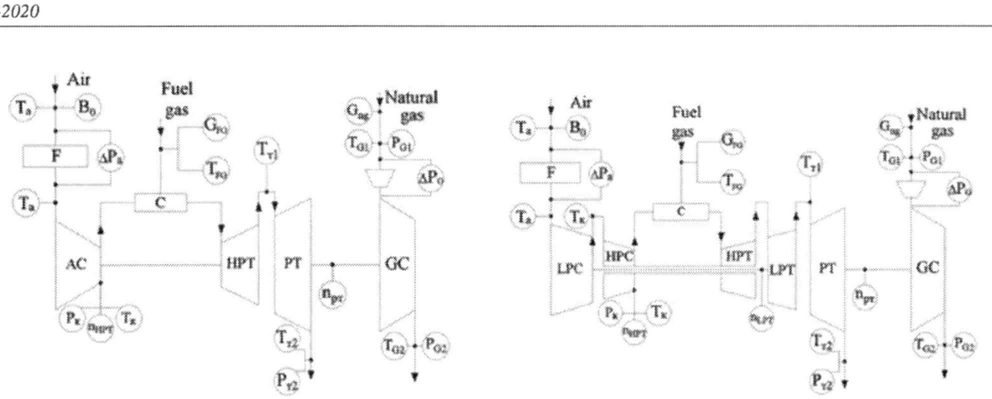

Fig. 2. Principal scheme for two (left) and three (right) shaft GTUs: F – filter; AC – axial compressor; LPC/HPC – low/high pressure compressor; C – combustor; HPT/LPT – high/low pressure turbine; PT – power turbine; GC – natural gas compressor.

temperature at turbines inlet and outlet, pressure ration distribution between turbines, airflow, specific expansion work, etc. Additionally, the basic geometric parameters of the turbines flow part are determined at this stage.

The next step is model calibration. The aim of this model calibration is to achieve good convergence between the computed integral parameters of GT at nominal mode and ones from the catalogue [11]. If necessary, some adjustable parameters can be varied, such as efficiency of modules and turbine/compressor mass flow.

The aerodynamic calculations of turbine stages are done according to the methodology described in [12] which consists of the basic algorithm for velocity diagrams correlations together with reaction, load coefficients and basic thermodynamic. As a result, pressures and temperature (static and total) at all stages together with kinematic parameters are calculated with the use of real flow path geometry.

2.2 Adjustment and test

The input data for this stage is the measured parameters of the gas turbine during testing. The principal schemes for two and three shaft GTUs are presented in Figure 2.

Also the input data is basic and adjustable parameters such as: swept areas and flow angles at each stator stage outlet, kinematic reaction at mid span, etc. (taken from nominal mode calculations at the previous stage).

Processing of the test results is carried out by two methods: VNIIGAZ [7] and one presented. In the first case, the effective power and efficiency of a gas turbine are determined by the parameters of a natural gas compressor. The gas flow was determined by an ultrasonic flow meter.

In the second case, the PT real effective work and gas mass flow are used to calculate the effective power of the GTU. The experimental data processing differs from the thermodynamic cycle calculations of GTU in the nominal mode using the measured inlet and outlet temperatures and pressures of the axial compressor and power turbine for calculating. This allows us to determine with sufficient accuracy the thermodynamic parameters of each turbine stage and the specific effective work of the PT, kJ/kg:

$$H_e = C_p \cdot T_{t1} \cdot \left(1 - \left(\frac{P_{t1}}{P_{t2}}\right)^{\frac{1-k}{k}}\right) \cdot \frac{G_t}{G_a} \cdot \eta_t \cdot \eta_{mech} \tag{1}$$

where C_p – gas heat capacity at constant pressure, kJ/kg/K; T_{t1} – PT inlet temperature, K; P_{t1}, P_{t2} – PT pressure inlet and outlet respectively, Pa; k – ratio of specific heats under total temperature; G_t – PT mass flow at the nominal operating mode of GTU, kg/s; G_a – air consumption in GTU at the nominal operating mode, kg/s; η_t – adiabatic turbine efficiency; η_{mech} – mechanical turbine efficiency.

Parameter G_t/G_a as well as mechanical and adiabatic turbine efficiency is accepted according to the results of the GTU model calculation at the previous stage. If the blades geometry of a turbine without cooling is known, then the curve of the efficiency can be estimated using the results of CFD analysis. Otherwise, the efficiency value can be taken from the known range [2,12]. For turbines with cooling the efficiency can be estimated using the known methods [13].

The next step is aerodynamic calculations of a gas turbine at real conditions. The purpose of this step is to determine the parameters for the PT mass flow calculation in the current operation mode (a test mode). The mass flow of working gases through PT for the power output calculation $N_{e\ calc}$ is based on gas dynamic functions (GDF) [14] by the following equation:

$$G = \frac{m \cdot P^* \cdot q(\lambda) \cdot F}{\sqrt{T^*}}, \tag{2}$$

where F – swept area in considered section, m^2; T^* – total temperature in considered section, K; P^* – total pressure in considered section, H/m^2; $q(\lambda)$ – GDF of reduced mass flow density; m – constant coefficient of gas:

$$m = \sqrt{\left(\frac{2}{k+1}\right)^{\frac{k+1}{k-1}} \cdot \frac{k}{R}}, \tag{3}$$

where R – universal gas constant, m^2/s^2K.

Total pressure in considered section is:

$$P^* = \frac{P \cdot C_a^2}{2R \cdot T \cdot j^*(\lambda)}, \quad (4)$$

where $j^*(\lambda)$ – GDF of dynamic pressure; P and T – static pressure and temperature in considered section that are determined in the previous step (experimental data processing); C_a – axial component of flow gas velocity in section, m/s :

$$C_a = \phi \cdot \sqrt{2 \cdot (1-\rho) \cdot \frac{H}{\eta}} \cdot \sin\alpha, \quad (5)$$

where ϕ – speed coefficient in stator or rotor blades (in considered section); ρ – kinematic reaction at mid span; H – specific expansion work of considered turbine stage; η – adiabatic efficiency of turbine stage; α – outlet flow angle in considered section. These parameters are determined at the stages «Mathematical model of GTU» and «Experimental data processing».

The GDF of dynamic pressure is:

$$j^*(\lambda) = \frac{k}{k+1} \cdot \lambda^2 \cdot \left(1 - \frac{k-1}{k+1} \cdot \lambda^2\right)^{\frac{1}{k-1}}, \quad (6)$$

$$\lambda = \frac{C_a}{a_{cr}}, \quad (7)$$

$$a_{cr} = \sqrt{2 \cdot (k/k+1) \cdot R \cdot T^*}, \quad (8)$$

$$T^* = T + \frac{C_a^2}{2 \cdot C_p}, \quad (9)$$

where λ – reduced velocity; a_{cr} – speed of gas in critical throat, m/s.

The gas dynamic function of reduced mass flow density is determined from:

$$q(\lambda) = \left(\frac{k+1}{2}\right)^{\frac{1}{k-1}} \cdot \lambda \cdot \left(1 - \frac{k-1}{k+1} \cdot \lambda^2\right)^{\frac{1}{k-1}}, \quad (10)$$

The PT specific expansion work (1) and mass flow (2) are defined according to this model at the experimental data processing stage and then the power output can be obtained.

The final iteration of calculations takes place for the following parameters: corrected power output $N_{e\ cor}$, corrected PT shaft speed $n_{PT\ cor}$, corrected massflow through PT $G_{FG\ cor}$, efficiency $\eta_{e\ cor}$ for every recorded operating mode. The turbine parameters with nominal corrected outlet temperature $T_{t2\ cor}^{nom}$ for these modes are defined.

The power at the nominal PT outlet temperature $T_{t2\ cor}^{nom}$ can be defined according to Figure 3 where it is graphically presented for one of the GTUs.

The final stage of the procedure is the calculation of GTU coefficients of the technical performance (CTP). CTP can be determined from the ratio of real power, efficiency or fuel gas consumption at the nominal operating mode to the passport (ideal) data of the gas turbine [15]. CTP calculated by power is used more often and is defined by the following equation:

$$K_{Ne} = \frac{N_{e\ cor}^{nom}}{N_e^{nom}}, \quad (11)$$

$N_{e\ cor}^{nom}$ and N_e^{nom} – power of GT, corrected to nominal outlet temperature $T_{t2\ cor}^{nom}$, and ideal (based on producer's documentation) power of GT.

At the next step the comparison takes place between the results obtained within the presented methodology and the VNIIGAZ methodology [7].

In some cases the difference between these two values exceeded 2%. This is due to the fact that a load factor and other stage parameters at particular GT type might not be chosen in the optimal range. And the behavior of this design differs from the expected one.

To adjust this universal model to the exact geometry reality corrective coefficient is introduced (called constriction coefficient) in the calculation of massflow through PT. So the mass flow is powered by this coefficient and the linear dependency is observed.

2.3 Operation of the technique

Based on the presented methodology, the in-house code can be developed and distributed among gas pumping plants for statistical information collection. Input data is standard measured parameters of gas turbines.

It is essential to filter this data before use. First of all, the start and stop modes of gas turbines should be excluded from the analysis. It is also necessary to identify transient operating modes of gas turbines, when the gas turbine operation parameters change significantly in a short period of time. For example, this happens when the load rises or falls, or when the outside temperature changes significantly. The filtration algorithm should track the change in the main parameters of the gas turbine operation (the shafts speed and the temperature of gases in different sections). If the change in parameters exceeds a predetermined value within a given time range, then this time interval is excluded from the analysis. It is also necessary to exclude the operating time of the gas turbine when there is no measurement of any parameter or the diagnostic system has identified a malfunction of the measuring equipment.

Then, calculations are carried out according to the presented approach and the current gas turbine power values in different modes are determined. After that, it is necessary to correct all the parameters to the nominal conditions. Using this data, the graphs of GTU power changes are built (Figure 3). Gas turbine power trends are linear. All "dropped points" are excluded from the further analysis and the power trend is being rebuilt. Such points are determined by the standard deviation of the current power value from the GT power trend. Then the available power, efficiency and CTP of the gas turbine (11) are determined using these trends.

Table 1. Information about the GTUs.

Designation	Type	Cycle	Shaft	Rated power	Efficiency
GTU-10	stationary	regenerative	2	10,0	29,0
GTU-16	aero derivative	simple	3	16,0	27,5
GTU-25	aero derivative	simple	3	25,0	38,7
GTU-30	stationary	simple	2	30,7	36,0

Monitoring these results makes it possible to track the performance degradation of gas turbines. The data can be used for prediction and decision-making on maintenance and repair of gas turbine units [16, 17].

3 Results

The described technique was tested on different types of gas turbine units in the gas transport system. A comparison of the results for the two approaches was carried out. Four types of gas turbines of different capacities (10, 16, 25, and 30 MW) were tested and information about the GTUs is presented in the table [11].

Measurement schemes are presented in Figure 2. After the experimental data processing, the output power was determined by two methods: VNIIGAZ ($N_{e\ exp.}$) and the presented methodology ($N_{e\ calc.}$). It should be noted that the accuracy of the test results significantly depends on the quality of the measurements and the accuracy of the measuring instruments [18], that is why the measuring equipment was calibrated to meet requirements of GOST standard [19]. The results of determining the power are presented in Figures 3-5.

4 Discussion

The experimental data for two GTUs-30 is presented in Figure 3. The trend line is developed with the use of the measured points and then extrapolated down to the

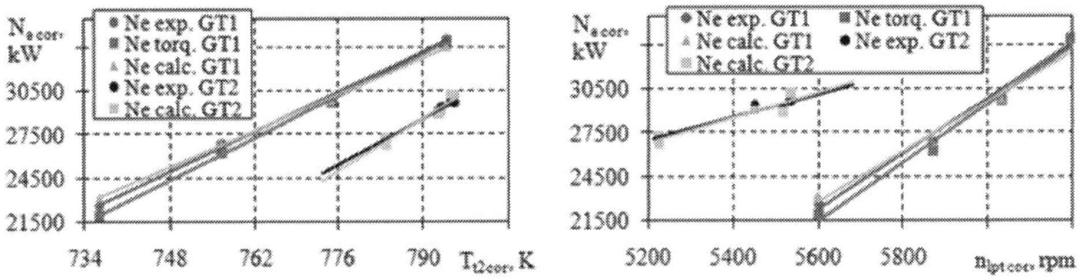

Fig. 3. Test data for GTU-30.

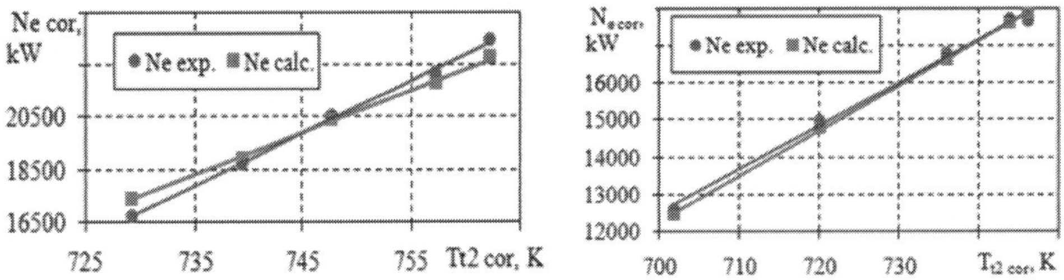

Fig. 4. Test data for GTU-25 (left) and GTU-16 (right).

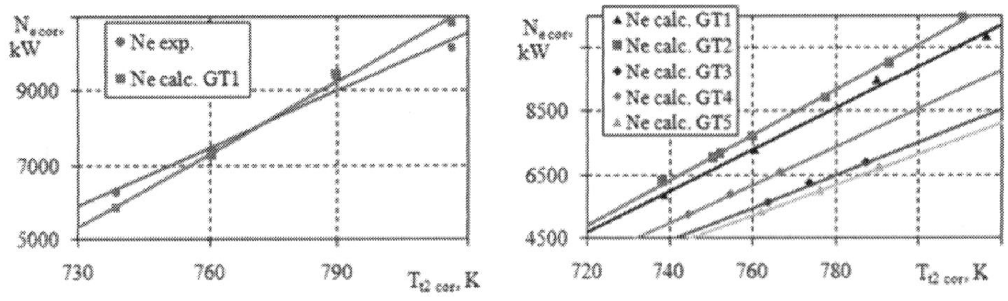

Fig. 5. Test data for GTU-10.

nominal exhaust temperature (figure 3, left) and the nominal shaft speed (figure 3, right), where the expected power output is estimated and then compared with nominal one. The GT1 was tested in winter and the GT2 – in summer. It is known that in summer gas turbine power limitation occurs due to reaching the maximum exhaust temperature of gases. In winter, one occurs when the maximum shafts speed is reached. The CTP of a gas turbine must be determined by the parameter that reaches the maximum value under the current operating conditions before others [15]. If one determines the CTP of GT1 by $T_{t2cor.}$ and CTP of GT2 by $n_{lpt\ cor.}$ then the CTP of both units will be about 1,0. However, in fact, the CTP of GT1 by $n_{lpt\ cor.}$ equals 0,82 (figure 3, right) and the CTP of GT2 by $T_{t2cor.}$ equals 0,88 (figure 3, left). It should be noticed that the line of GT2 extrapolated from a narrow set of points so there might lead to a mistake in the inclination angle of this line. Nevertheless, it can be seen that these points are significantly distant from the line of GT1.

The power values determined by both methods coincide with an accuracy of 2% for both gas turbines. For GT1, the power was also determined using a torque meter ($N_{etorq.}$). The standard deviation of $N_{etorq.}$ from the power determined by thermodynamic parameters does not exceed 2,5%. The deviation of gas turbine power determined by several approaches can be used as a criterion for the reliability of experimental data in gas turbine diagnostic systems, as well as for calibrating one approach relative to another.

Figure 4 shows the experimental data for GTU-25 (left) and GTU-16 (right). The standard deviation from the VNIIGAZ methodology was 2,1 and 1,5% for GTU-25 and GTK-16 respectively. Figure 5 represents the experimental data for several GTUs-10. The standard deviations from the VNIIGAZ methodology for different gas turbines were in the range of 1–5%.

The statistical data for five units is presented in Figure 5 (right). All points here are calculated according to the presented methodology by using only standard measurements with recalibrated measuring equipment. The lines here represent the trend of corresponding sets of points. It is clear that these units have different values of CTP. This is mainly due to the fact that the units have different operating hours after repair. Nominal exhaust temperature for GTU-10 is 780 K. All five trends crossing nominal temperature at power turbine outlet (corrected) in different locations are: 9250, 8620, 7400, 6510, and 6250 kW respectively. So GT№1 has 0,925, №2 – 0,862, №3 – 0,740, №4 – 0,651, and №5 – 0,625. It can be seen that GT № 3, 4 and 5 have an unsatisfactory technical condition. It is worth noting that the units have a total operating time of more than 130 000 hours, and in some cases the operating time reaches 200 000 hours. Today these type units are being replaced with the new ones.

The degradation coefficient of gas turbines can be estimated by the shift of the trend line. For this, the exact value of the power of a gas turbine is not necessary to determine and it is enough to track the change in power (for example, in %). Also, to start tracking is not necessary with the "ideal" state of the gas turbine. This feature may be used in diagnostic systems.

The presented method can also be used for quantitative estimation of maintenance procedures at GTU based only on standard measurements. Any unit can be tested before and after washing [20] or tip clearance recovery in an axial compressor, and, as a result, CTP gain can be defined. The same test series were carried out on marine derivative GTU before and after compressor washing [9].

5 Conclusions

The method of power output estimation is realized and showed its good convergence with the method based on the power consumption of a gas compressor. The method accuracy is within the measurement error. The use of standard measurement systems in most cases is sufficient to determine the CTP of GTU. The measurement accuracy in most cases complies with GOST [19]. The presented methodology has several adjustable parameters for gas dynamic functions that are to be defined and properly verified based on the first set of special experiments and technical documentation. After its validation, the model can be implemented for all machines of this type in real time.

References

[1] P.P. Walsh, P. Fletcher, *Gas turbine performance* (Blackwell Science Ltd a Blackwell Publishing company, 2004)

[2] B.S. Revzin, *Gas-pumping units with a gas-turbine drive (transl.)* (UGTU-UPI, Ekaterinburg, 2002)

[3] R. Kurz, K. Brun, Degradation in gas turbine systems, Journal of Engineering for Gas Turbines and Power **123**, 70–77 (2001)

[4] D. Burnes, R. Kurz, Performance Degradation Effects in Modern Industrial Gas Turbines, Proceedings of Zurich Global Power and Propulsion Forum (2018)

[5] A.P. Tarabrin, V.A. Schurovsky, A.I. Bodrov, J.P. Stalder, Influence of axial compressor fouling on gas turbine unit performance based on different schemes and with different initial parameters, ASME Paper 98-GT-416 (1998)

[6] R.D. van Millingen, J.D. van Millingenp, Phase Shift Torquemeters for Gas Turbine Development and Monitoring, International Gas Turbine and Aeroengine Congress and Exposition, Orlando (1991)

[7] V.A. Shurovskiy, Yu.N. Sinicyn, V.I. Korneev, A.V. Cheremin, G.S. Stepanov, *Guidelines for conducting thermal engineering and gas-dynamic calculations during testing of gas turbine gas pumping units (transl.)* (VNIIGAZ, 1999)

[8] Y.G. Li, Gas Turbine Performance and Health Status Estimation Using Adaptive Gas Path

Analysis, Article in Journal of Engineering for Gas Turbines and Power **132**, 4 (April 2010)

[9] O.V. Komarov, V.L. Blinov, V.A. Sedunin, A.V. Skorochodov, Parametrical Diagnostics of Gas Turbine Performance On Side At Gas Pumping Plants Based On Standard Measurements, Proceedings of ASME Turbo Expo, GT2014-25392 (2014)

[10] O.V. Komarov, V.L. Blinov, V.A. Sedunin, On Technical Performance Estimation of a Gas Turbine with Variable Power Turbine Vanes, Procedia Engineering **150**, 2nd International Conference on Industrial Engineering, 1378–1383 (2016)

[11] *Album of parameters for gas turbine gas pumping units (transl.)* (VNIIGAZ, 2006)

[12] O.V. Komarov, V.L. Blinov, A.S. Shemyakinsky, *Thermal and gas-dynamic calculations of gas turbine units: a training manual (transl.)* (Ural Federal University, Ekaterinburg, 2018)

[13] J.H. Horlock, L. Torbidoni, Calculations of Cooled Turbine Efficiency, Journal of Engineering for Gas Turbines and Power **130**, 1 (January 2008)

[14] A.A. Dimentova, F.S. Reksitin, V.A. Ryabov, *Tables of gas-dynamic functions (transl.)* (Mashinostroenie, Moscow, 1966)

[15] A.V. Semushkin, A.O. Podlozny, Ye.A. Chernikova, V.A. Shchurovskiy, Methodical grounds for parametric diagnostics of turbine gas-compressor units, News of gas science **1**, 29, 22–31 (2017)

[16] I.B. Murmanskii, K.E. Aronson, V.L. Blinov, N.V. Zhelonkin and B.E. Murmansky, Digital diagnostic complex for power turbine units equipment, IOP Conf. Series: Materials Science and Engineering **643**, 012109 (2019)

[17] S. Ahsan, T.A. Lemma,. Remaining Useful Life Prediction of Gas Turbine Engine using Autoregressive Model, MATEC Web of Conferences **131**, 04014 (2017)

[18] M. Pinelli, P.R. Spina, *Gas Turbine Field Performance Determination: Sources of Uncertainties*, Article in Journal of Engineering for Gas Turbines and Power 124, 1 (January 2002)

[19] Gas turbines, Test methods. Acceptance tests, ISO/DIS 2314 (2007)

[20] K.K. Botros, H. Golshan, D. Rogers, Effects of Engine Wash Frequency on GT Degradation in Natural Gas Compressor Stations, ASME Turbo Expo (2013)

Modeling of application the FACTS devices on the example of power system of United Power System of Center

Alexey Udaratin[1,*], *Kirill* Loginov[1], *Timur* Akhmetov[2] and *Alexandr* Nemirovskiy[1]

[1]Vologda State University, Lenina Street 15, Vologda, 160000 Russia
[2]Kazan State Power Engineering University, Kazan, Russia

Abstract. This paper considers installation of a flexible AC transmission system (FACTS) device on the 500 kV line of substation (SS) Belozerskaya - SS Vologodskaya of the United Power System of Center. Three FACTS devices in normal operation mode were modeled: a static reactive power compensator (STATCOM), a static synchronous series compensator, and a unified power flow control system. The obtained results are summarized in Table, which substantiates the choice of the best compensation for the Belozerskaya - Vologodskaya line.

1 Introduction

Development of electric power complex is one of the most important tasks of the Russian Federation nowadays. Every year, the demand for electricity is becoming higher. A growing number of new houses, shopping centers, hospitals are connected to substations that transmit energy to all corners of the country. But transmission of energy flows has some troubles. Some part of energy is lost during transmission, which includes:

1) Losses of technical nature. They arise during energy transmission via electric networks and are caused by physical processes that occur in wires and equipment.

2) Electrical energy, which is spent for operation of substations and personnel. This energy is assessed by meters installed on auxiliary transformers of energy plants.

3) Losses, which are caused by errors during its measurement [1].

The first item accounts for the highest electrical energy consumption, since reactive power arises in wires during transmission, and its loss is greater for the longer line. But this problem can be solved if one includes devices that compensate it during undervoltage or overvoltage in the line. The flexible AC transmission systems (FACTS) [2] are the devices, which over the past 20 years are acknowledged to provide the necessary steps for successful compensation of power and voltage.

FACTS devices are divided into two generations: the first and the second. The first-generation devices include shunt reactors (ShR), static capacitor banks (SCB), series compensation devices (SCD), static compensators (SC). The second generation devices include synchronous static compensators (STATCOM), static synchronous series compensators (SSSC), unified power flow control systems (UPFC), phase-shifting devices (PSD), and others.

Over the past 20 years, the second-generation devices were installed on new substations or replaced the existing first-generation devices. STATCOM, SSSC and UPFC are the most widespread devices. They differ among themselves by the installation method: the static synchronous compensator is installed in shunt, the static synchronous series compensator is installed in series with the line, the unified power flow control system is placed universally (namely, in both ways).

Series compensation of reactive power is usually carried out using a static synchronous series compensator. It is attributed to the second generation of FACTS and allows one to withdraw the shortcomings of a thyristor-controlled capacitor bank. SSSC is a static synchronous generator that operates without external energy source as a compensator, the voltage of which is squared relative to the linear one. Static synchronous compensator allows one to use the rated energy of device or absorb power to improve the dynamic component of power system by additional temporary compensation of power to increase or decrease the voltage drop in the line. Energy distribution using SSSC uses the Dynamic Voltage Restorer (DVR) principle. It is aimed to maintain the network in relation to its most vulnerable consumers under short-term disturbances. The power of commercially available DVR installations does not exceed 1 MVA. SSSC is similar to STATCOM, but the alternating voltage output is brought in series with the line. SSSCs are series-connected FACTS devices based on a converter based on voltage source and they can be considered as a perfect type of series compensation, in a similar way STATCOM is more advanced equipment than static reactive-power compensator. [3]

Reactive shunt compensation is mainly carried out by the STATCOM device. Since it belongs to a type of

* Corresponding author: alexu79@mail.ru

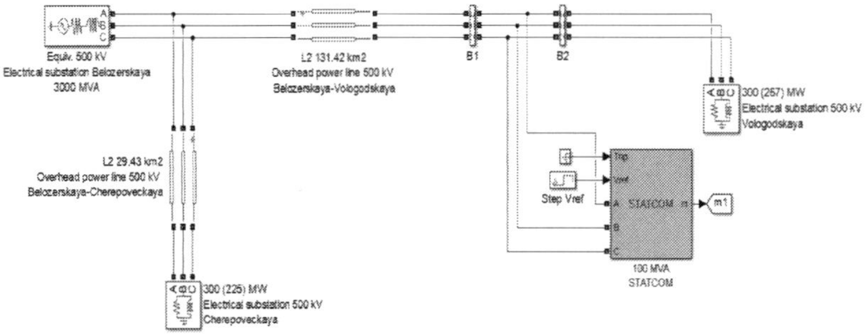

Fig. 1. Model of 500 kV SS Belozerskaya - SS Vologodskaya power system with STATCOM device.

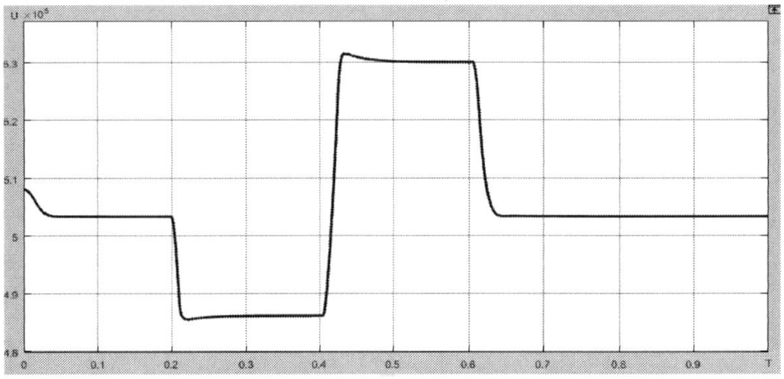

Fig. 2. Voltage on bus B1 with STATCOM device. U is voltage on line, T is device operation time.

serial compensators, this device can provide instantaneous reactive power in response to voltage of response to vibrations, STATCOM can also be used as an active harmonic filter.

Universal compensation is a combination of STATCOM and SSSC, which are interconnected by DC link. This allows power to flow from the output of one device to the output of another. Each of its parts consists of transformer and electric converter, which have a joint DC capacitor. Using angular infinite sequential introduction of voltage, UPFC can simultaneously or selectively control voltage, transmission line angle, resistance or active and reactive power [4].

The electric power industry in Russia needs the newest tools to control and transmit electric power over vast distances through existing lines with the smallest financial cost for their modernization, so the present work is of high relevance.

The aim of this work is to model various types of FACTS compensation devices on the line and to select the best one for installation on the line of the United Power System of Center. These includes the following tasks:

1) To model a power line and analyze its output active and reactive power, as well as voltage.

2) To calculate and model the line with FACTS compensation devices.

3) To make a conclusion on the use of a particular type of compensation to improve power transmission on the line and choose the best one.

Modeling was carried out using the MATLAB Simulink program, the effectiveness of which was previously tested on another power system [3]. The 500 kV line SS Belozerskaya - SS Vologodskaya of UPS Center was selected to install the devices.

2 FACTS Modeling

2.1 STATCOM Modeling

The power system SS Belozerskaya - SS Vologodskaya with STATCOM device is shown in Fig. 1. Voltage in the line with the device switched on is shown in Fig. 2. The change in flows of active (P, MW, solid line) and reactive power (Q, Mvar, dashed line) in power system with the device is shown in Fig. 3.

Consider a time interval equal to t=1 s. At t=0-0.2 s and t=0.6-1 s, STATCOM is switched off. At t=0.2-0.6 s STATCOM is switched on and connected to the network.

The modeling results are presented in Table 1.

Analysis of graphs and tables shows that STATCOM operates in two modes:

Table 1. Change in U, Q and P on bus B1 with STATCOM device.

STATCOM operation mode	t, s	U, kV	Q, Mvar	P, MW
Switched off	t=0-0.2 t=0.6-1	506	215	275
Undervoltage mode	t=0.2-0.4	487	276	254
Overvoltage mode	t=0.4-1	530	105	302

1) At undervoltage in the network, STATCOM increases the reactive power flow in it, while compensating the losses (it works as an inductive load);

2) At overvoltage, it reduces the reactive power flow and increases the generation of active power flow, while compensating its shortage [5].

2.2 SSSC Modeling

SSSC connected in series to the Belozerskaya - Vologodskaya line is shown in Fig. 4. Static synchronous series compensator, unlike STATCOM, is connected to the power line itself. SSSC also has a POD Controller, which consists of an active power measuring system, a low-pass filter, a high-pass filter and a surge suppressor [6].

At t=0-0.2 s, SSSC is switched off. At t=0.2-1 s SSSC is in operation. The modeling results are presented in Table 2.

A static synchronous series compensator can serve as a generator or consumer of reactive power, without requiring an extra power source. SSSC also allows one to simulate active and reactive power to compensate losses in the line, as well as to spend it for auxiliary power supply system, without affecting the voltage. It can be seen from Fig.5 that when active power decreases by t=0.2-0.6 s or increases by t=0.6-1 s, voltage almost does not change [7]. Its basic principle of operation is similar to that of STATCOM. When active power decreases, SSSC compensates it by reactive power and vice versa, which is seen in Fig. 6.

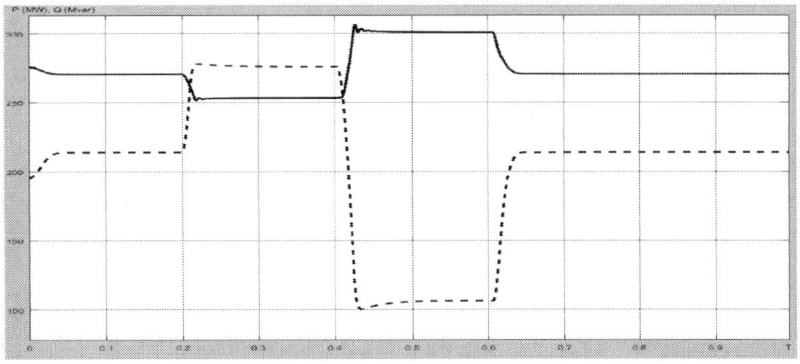

Fig. 3. Active and reactive power of line on the B1 bus in the normal STATCOM mode. P is active power, shown by solid line, Q is reactive power, shown by dashed line, T is device operation time.

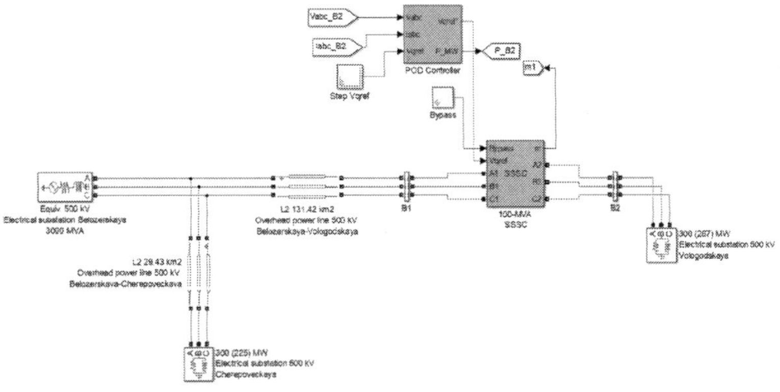

Fig. 4. Model of 500 kV SS Belozerskaya - SS Vologodskaya power system with SSSC device.

Table 2. Change in U, Q and P on bus B1 with SSSC device.

SSSC operation mode	t, s	U, kV	Q, Mvar	P, MW
Switched off	t=0-0.2	508.5	196	275
Undervoltage mode	t=0.2-0.6	508.3	204	250
Overvoltage mode	t=0.6-1	509	187	302

Table 3. Change in U, Q and P on bus B1 with UPFC device.

UPFC operation mode	t, s	U, kV	Q, Mvar	P, MW
Switched off	t=0	516	210	276
Undervoltage mode	t=0-0.5	493	215	270

2.3 UPFC Modeling

Fig. 7 shows a power network model with UPFC. The unified power flow control system consists of SSSC and STATCOM, interconnected by a DC power line.

At t=0-0.5 s, the device operates in undervoltage mode, at t=0.5-1 s, voltage in the system increases. The modeling results are presented in Table 3.

Capacitance of power storage of a DC capacitor is usually small. So, SSSC must receive (or give) active

Fig. 5. Voltage on bus B1 with SSSC device. U is voltage on line, T is device operation time.

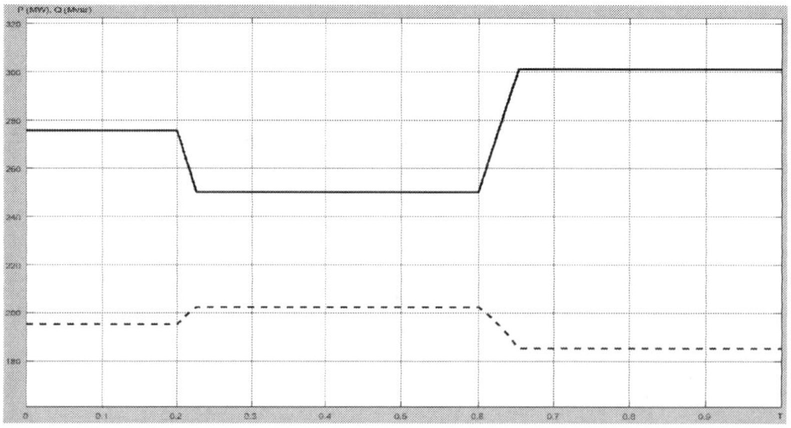

Fig. 6. Active and reactive power of line on the B1 bus in the normal SSSC mode. P is active power, shown by solid line, Q is reactive power, shown by dashed line, T is device operation time.

 https://doi.org/10.1051/e3sconf/202017801045

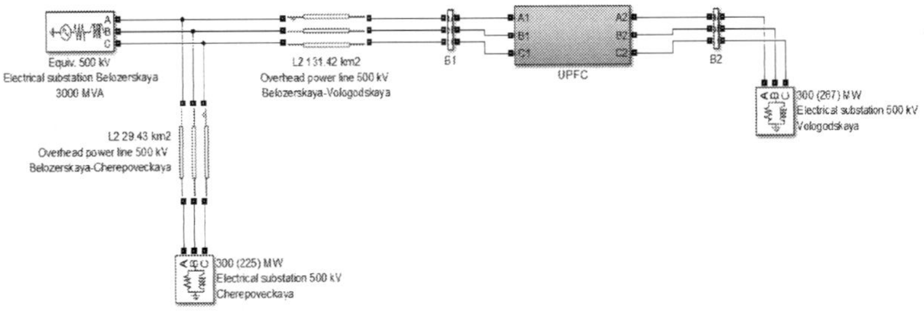

Fig. 7. Model of 500 kV SS Belozerskaya - SS Vologodskaya power system with UPFC device.

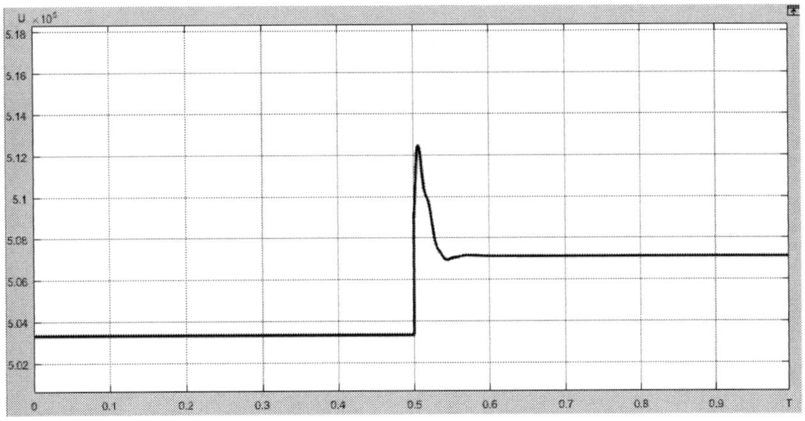

Fig. 8. Voltage on bus B1 with UPFC device. U is voltage on line, T is device operation time.

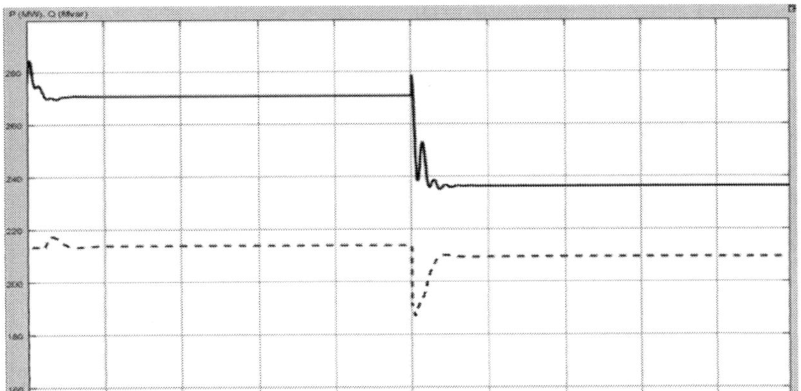

Fig. 9. Active and reactive power of line on the B1 bus in the normal UPFC mode. P is active power, shown by solid line, Q is reactive power, shown by dashed line, T is device operation time.

power from the network in exactly the same amount in which STATCOM gives (or receives) it [8]. Therefore, when active power in the system increases, the reactive power also increases. This can be seen from Figs. 8 and 9. At t=0-0.5 s, voltage decreases accompanied with an increase in reactive and active power in the network. At t=0.5-1 s, the device increases voltage, lowers the reactive power together with the active one and takes part of energy for auxiliary power supply system.

Table 4. Comparison of Q and P on bus B1 of FACTS devices.

FACTS operation mode	Q, Mvar STATCOM	P, MW STATCOM	Q, Mvar SSSC	P, MW SSSC	Q, Mvar UPFC	P, MW UPFC
Switched off	215	275	196	275	215	270
Undervoltage mode	276	254	204	250	215	270
Overvoltage mode	105	302	187	302	210	238
Average value	190.5	278	195.5	276	212.5	254

3 Conclusion

The conducted modeling shows that each FACTS device compensates reactive power in different ways. The values given in Tables allow one to choose the best compensation for the 500 kV line SS Belozerskaya - SS Vologodskaya. Since power compensation includes increasing the active power in network and decreasing the reactive power, the smallest lost power and the greatest transmitted power can be considered as criterion for selection of the most suitable option. The results of modeling were used to calculate the average values of reactive and active power, which are shown in Table 4.

This table can be used to analyze the efficiency of each device relative to the normal operating mode of system and select the best of them. STATCOM can reduce the loss of reactive power on line by 24.5 Mvar and increase the active power by 3 MW. SSSC reduces losses by 0.5 Mvar and increases active power by 1 MW. UPFC enables to reduce reactive power by 2.5 Mvar, and to reduce useful power by 16 MW. The obtained data show that STATCOM is the best choice for installation on the line, since it best reduces reactive power and increases active power.

References

[1] A. Udaratin, A. Alyunov, A. Krutikov, L.R. Mukhametova, O.O. Zaripov, I.V. Bochkarev, E3S Web of Conferences 124, 02020 (2019).

[2] A. Nemirovskiy, A. Kashin, V. Kosmach, Y. Titovec, I. Toptygin, D. Zaripova, IOP Conference Series: Earth and Environmental Science (EES) 337, 1, 012071 (2019).

[3] A. Ydaratin, T. Akhmetov, S. Makoev, IJCIET 10, 2, (2019).

[4] V. Zhukov, A. Shmelev, D. Mikheev, Russian Electrical Engineering 89, 5 (2018).

[5] M. Bashirov, A. Khismatullin, E. Sirotina, LNEE 641, 613-621 (2020).

[6] S. Tsyruk, S. Gamazin, Y. Ryzhkova, K. Charafeddine, Dynamics, 8601484 (2019).

[7] V. Kozlovsky, S. Petrovsky, V. Vakhnina, D. Skripnuk, ElConRus, 8657322, 569-573 (2019).

[8] D. Aydarov, V. Kozlovsky, V. Vakhnina, S. Kleymenov, N. Didenko, ElConRus, 8656815, 416-419 (2019).

Improving the energy efficiency of wide crossings of overhead power lines

Natalya Shevchenko[1], *Anatoly* Soshinov[1], *Olga* Elfimova[1], *Julia* Lebedeva[1] and *Olga* Akhmedova[1]

[1]Kamyshin Technological Institute (branch) of Volgograd State Technical University, Kamyshin, Russia

Abstract. The use of new generation wires in the design of wide spans of overhead power lines over water barriers and large gorges can increase their transmission capacity and increase reliability. However, when large currents flow, load losses in line also increase. Thus, it is necessary to obtain a methodology for technical and economic comparison of design options for overhead power lines over large crossings, which enables to determine the most cost-effective version of the project with high mechanical reliability. For comparative analysis, five options for wide spans of overhead power lines over the river with new-generation wires of Russian production were compiled: Aluminium Alloy Conductor Steel Reinforced, Z-type (high conductivity), Thermal-Resistant Conductor, Aluminium Compozite Core Conductor, ASk2y. Option with Aluminium Conductor Steel Reinforced wire was taken as the source. For the first option, the crossing scheme E-A-A-E was selected, for the remaining options E-I-I-E scheme was considered. For the modes of maximum loads, minimum temperature and average annual conditions, the mechanical calculation of wires was carried out using the method of permissible stresses. The wire bending deflections were determined in accordance with the theory of the catenary curve. The method of integrated indicators was used to perform a technical and economic comparison of these options. The calculation results showed that among the proposed options, the most optimal is a crossing with the ACCC wire having minimum relative investment per 1 MW of transmitted electricity and minimal power loss. The transmission capacity of the line with this wire is increased by 1.8 times, and the cost of crossing is reduced by 16%. Due to the compact design of wire, the probability of ice formation on wire is reduced, and the reduced bending deflection reduces the probability of wire break due to natural environmental influences.

1 Introduction

The urgent task of the power engineering sector is to optimize design solutions for wide spans of overhead power lines (OHL) through large gorges and water barriers in order to increase their reliability and current transmission capacity. When designing wide spans of OHL, it is important to reduce the loads on transmission towers and foundations, to reduce the wire bending deflection, to reduce the probability of wire breaks and swinging during icing and squally winds [1, 2, 3]. These tasks are solved by using innovative types of wires with high electrical conductivity and high mechanical strength [4, 5, 6].

To assess the energy efficiency of line, it is important to evaluate the level of losses of the transmitted power.

The design of wide spans of overhead power lines is carried out in accordance with the requirements of the standard "Norms of technological design of overhead power transmission lines with a voltage of 35-750 kV" [7] and the provisions of the "Rules of Electrical Installations Design" [8]. The standard recommends to design the crossing in several ways. The final choice of options is based on a feasibility study.

In Russia, construction of wide spans through water barriers includes a practical experience of using high-temperature high-strength wires of the ACS (TACSR/ACS) brand with a steel core [9], clad with Aluminium manufactured by Lumpi-Berndorf (Austria) and EM-CABLE LLC (Russia), compacted wires with reinforced high-strength steel core AACSRZ (Lamifil LLC, Russia) [10,11,12].

2 Problem statement

The goal of this work is to increase the energy efficiency of a wide span of a 220 kV double-circuit overhead power line across the river. The length of the crossing is 2080 meters.

Climatic conditions are the following: II region in terms of wind (wind speed 29 m/s), III region in terms of ice (ice wall thickness b = 20 mm) [13].

The following tasks were set:

- To develop options for crossing of overhead line over the river using wires of a new generation of Russian production in order to increase energy efficiency;
- To select crossing schemes, types of transmission towers and wire brands;

- To perform a mechanical calculation of wires taking into account climatic conditions and a feasibility study of the developed options.

3 Development of options

Based on the experience of designing and construction a wide span of 220 kV OHL through the Kama Reservoir in 2009 [14, 15], five options for the crossing were proposed for consideration.

The first option: the crossing scheme E-A-A-E using crossing towers AT 133 and terminal towers K330-2+5K. The height of the AT 133 transmission tower is 154 m, the suspension height of the lower wire is 133 m (Fig. 1.). Aluminium Conductor Steel Reinforced ACSR 500/336 is used.

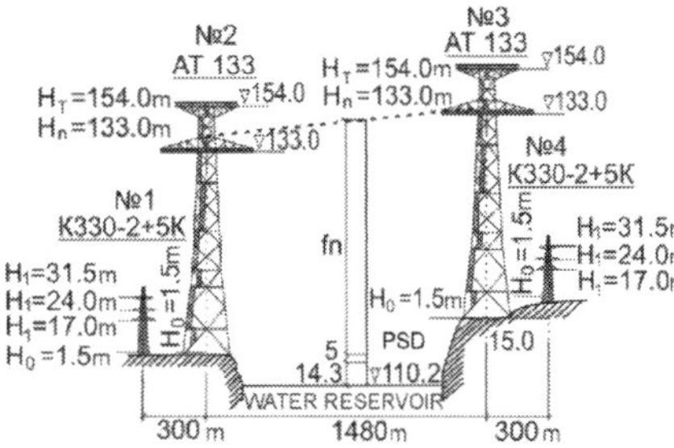

Fig. 1. The crossing scheme E-A-A-E.

The second option is using high-temperature high-strength wire TACSR/ACS 521-A20SA (Russia) with a gap between the steel wires of core and the wires of aluminium-zirconium alloys. The gap is filled with heat-resistant grease. The wire has not only increased electrical conductivity, but also increased corrosion resistance owing to the use of aluminium cladding technology.

The crossing scheme E-I-I-E (Fig. 2) was selected using intermediate towers PP330-2/79.5 and terminal towers K330-2+5K. The height of the PP330-2/79.5 tower was 104.4 m, the suspension height of the lower wire was 75.2 m.

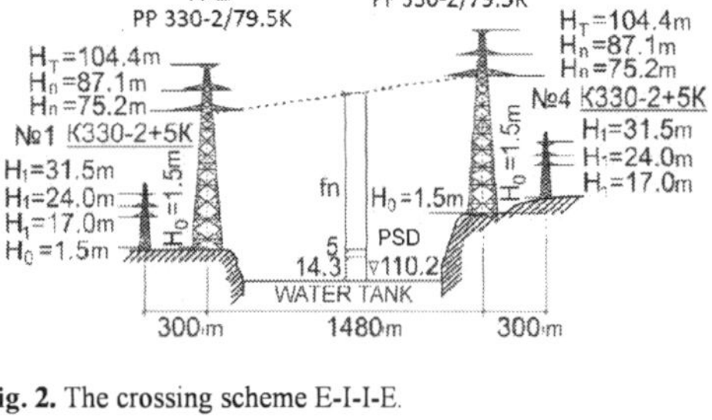

Fig. 2. The crossing scheme E-I-I-E.

Other crossing options will be performed according to the same E-I-I-E scheme with replacement of wire types, which are listed below.

Third option: AACSRZ-527 (HC) - Aluminium Alloy Conductor Steel Reinforced, Z-type (high conductivity). The conductive part has a Z-shaped profile, which reduces the wind resistance. Reinforced high strength steel core enhances mechanical strength of wire.

Fourth option: Aluminium Compozite Core Conductor ACCC™ - 527. The conductive part is made of trapezoidal wires of annealed Aluminium. Core is made from basalt and carbon fibers.

Fifth option: ASk2y-527 wire. The compacted wire is made of trapezoidal Aluminium wires and a high-strength steel core with zinc- Aluminium coating.

Data for comparative analysis of the considered crossing options are given in table 1.

Table 1. Characteristics of wires of 220 kV overhead line.

Parame-ters	Mea s.u.	Wire brand				
		AC SR	TAC SR	AAC SRZ	AC CC	ASk 2y
Options		1	2	3	4	5
Fc	mm²	826.7	521.0	526.6	574.8	562.0
dc	mm	37.5	29.7	27.5	27.72	27.8
Gc	t/km	4.005	3.483	2.081	1.54	1.853
E	kN/ mm²	114.0	154.6	85.9	116.0	79.3
RTS	kN	467	666	312.4	159.1	175.6
α (TKLP)	10^{-6}/ °K	15.5	12.6	17.39	1.61	18.1
Tower reference		AT 133	PP330-2/79.5			
Tower mass	kg	400	154.4			

4 Mechanical calculation

The mechanical calculation of wires and cables of wide spans is carried out according to the method of permissible stresses for maximum load conditions, minimum temperature and average annual conditions. The voltage at any point of wire is found from the equation [16]:

$$\sigma - \frac{\sigma}{\gamma\beta}\,\mathrm{ch}\,\frac{\gamma l}{2\sigma}\cdot\frac{1}{\frac{\sigma_m}{\gamma_m}\cdot ch\frac{\gamma_m l}{2\sigma_m}} = \sigma_m - \frac{1}{\beta} - \frac{\alpha}{\beta}(t - t_m) \quad (1)$$

where σ, σ_m are voltage and permissible voltage in the material of wire or cable in one of the considered modes, respectively;

γ, γ_m are the specific mechanical load of wires, N/m · mm²;

β, α are the coefficients of elastic elongation and thermal expansion;

t, t_m are the temperatures typical for the calculated or known mode.

The overall bending deflections when crossing a large gorge or water barriers are determined in accordance with the requirements of the "Rules of Electrical Installations Design". For high-temperature wires, it should be determined at the maximum permissible wire temperature.

The wire bending deflection is determined in accordance with the theory of the catenary curve.

At different heights of suspension points, wire bending deflections corresponding to the large and small equivalent spans are determined by formulas (2) and (3):

$$f_1 = \frac{\gamma \cdot l_{eg_1}^2}{8 \cdot \sigma_0}; \; f_2 = \frac{\gamma \cdot l_{eg_2}^2}{8 \cdot \sigma_0}; \quad (2)$$

$$f_2 = -\frac{\sigma}{\gamma} \cdot \ln \cos \frac{\gamma \cdot x_0}{\sigma}; \; f_1 = f_2 + \Delta h. \quad (3)$$

Wire deflections in the span at different heights of suspension points are shown in Fig. 3 [17].

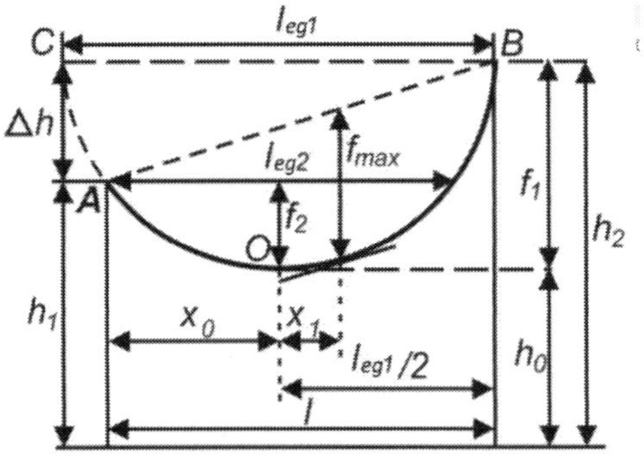

Fig. 3. Determination of wire deflections in the span for various heights of suspension points.

5 Results of mechanical and technical and economic calculations

The wire length in span L is determined taking into account the deflection f according to the formula:

$$L = l + \frac{8 \cdot f^2}{3 \cdot l}. \quad (4)$$

The integral effect Ie is taken as a criterion of profitability [18]:

$$I_e = \sum_{t=0}^{T} (R_t - Z_t) \cdot \frac{1}{(1+E)^t} - \sum_{t=0}^{T} K_t \cdot \frac{1}{(1+E)^t}, \quad (5)$$

where R_t is the income received at the t-th calculation step;

Z_t is the costs at the t-th calculation step;

T is the planning horizon, taken 13 years;

$\sum_{t=0}^{T'} K_t$ is the cost of crossing, million rubles;

E is the discount rate;

t is the number of calculation step.

The results of calculation of the considered crossing options are given in Table 2.

Table 2. Characteristics of wires of the 220 kV overhead line crossing.

Parameters	Mea s. u.	Options				
		1	2	3	4	5
Bending deflection, f	m	145. 6	86.0	66.1	49.0	62.4
Length of wire in span, L	km	12.7 4	12.0	12.6 0	12.5 1	12.6 1
Weight of wire in span	t	51.0	44.2	26.2	19.8	23.2
Cost of 1 t of wire	ths. rub.	175	525	525	700	350
Cost of wire in span	mln. rub.	8.93	23.2	13.8	13.9	8.11
The cost of crossing, K	mln. rub.	192. 7	171. 4	162. 0	162. 1	156. 3
Cost reduction	%	-	11.0	16.0	15.9	18.9

An important indicator of effective functioning of an overhead power line is the minimum loss of electricity. To assess the technical efficiency of crossing, we introduce the concept of the energy coefficient K_e [19-22].

$$K_e = \frac{W_{re}}{W_{re} + \Delta W_L + \Delta W_{CL}}, \quad (6)$$

where W_{re} is the transmitted electricity,

ΔW_L is the load losses of electricity;

ΔW_{CL} is the corona electricity losses.

Maximum transmitted power is:

$$P_{max} = 1.05 \cdot \sqrt{3} \cdot I_{max} \cdot U_{rat} \cdot \cos\varphi. \quad (7)$$

where I_{max} is the maximum line current, A;

$\cos\varphi$ is the power factor, taken 0.9;

U_{rat} is the rated voltage of the network, taken 220kV.

Transmitted electricity per year W_{re}:

$$W_{re} = P_{max} \cdot T_{max}, \quad (8)$$

where T_{max} is the number of hours of maximum load use, taken T_{max}= 6000 h.

Load losses of electricity in line are:

$$\Delta W_L = 3 \cdot I_{max}^2 \cdot L \cdot r_0 \cdot \tau, \quad (9)$$

where τ is the maximum loss time, taken 4500 h.

The calculation results are presented in Table 3.

Table 3. Parameters of the 220 kV OHL crossing wires.

Parameters	Meas. u.	Options				
		1	2	3	4	5
$t_{max.}$	°C	90	150	90	180	90
I_{all}	A	960	1344	979	1867	1360
P_{max}	MW	345.3	483.84	352.44	672.12	489.6
r_0 at 20 °C	kOhm/km	58.8	166.5	77.5	54.5	58.2
W_{re}	MW·h ·10⁵	20.7	29.0	21.2	38.4	29.5
$\Delta W\Sigma$	MW·h ·10⁵	0.14	0.75	0.20	0.46	0.29
K_e	%	99.3	97.4	99.0	98.8	99.0
I_e	bln. rub.	19,1	26.1	19.4	35.2	27.1
Relative capital investments	mln. rub./MW	0.56	0.35	0.46	0.24	0.32

The calculations show that among the proposed options, the most optimal is crossing with ACCC wire having minimum relative investment per 1 MW of transmitted power. The transmission capacity of line with this wire is increased by 1.8 times, and the cost of crossing is reduced by 16%. The smaller diameter of wire reduces the probability of ice formation on the wire, and the reduced bending deflection reduces the probability of wire break due to natural environmental influences.

6 Conclusions

The results of calculations of a wide crossing of 220 kV overhead transmission line over the river showed that the use of modern types of wires can solve the problem of both increasing the transmission capacity of overhead power lines and increasing their reliability.

The cost of span construction is reduced by 11-19% due to the possibility of using towers and foundations of lower material consumption.

References

[1] Liqiang An, Yongyu Guan, Zhijian Zhu, Jiong Wu, Ronglun Zhang, Engineering Failure Analysis 104, 911-931 (2019).

[2] Guohong Chen, Xu Wang, Jiaqing Wang, Jun-jian Liu, Wenming Tang, Engineering Failure Analysis 19, 13-21 (2012).

[3] S. Karabay, F. Kaya Önder, Electric Power Systems Research 72, 21, 179-185 (2004).

[4] S. Karabay, Materials and Design 30, 3, 816-825 (2009).

[5] R.B. Kalombo, J.M.G. Martínez, J.L.A. Ferreira, C.R.M. da Silva, J.A. Araújo, Procedia Engineering 133, 223-232 (2015).

[6] N. Fedorov, Overhead lines 1, 6, 31-34 (2012).

[7] STO 56947007, Norms of technological design of overhead power transmission lines with a voltage of 35-750 kV (2014).

[8] Rules of Electrical Installations Design, Russian Electrical Code, 7th Edition, All sections and exhibits, Saint-Petersburg, 928 (2000).

[9] R.C.R. de Mendez, J.D. Riera, L.F.F. Miguel, J. Kaminski Jr, L.F.F. Miguel, J.B.G. Ferreira da Silva, On Modeling the Dynamic Response of the 190m-high Tower for the Crossing of Trombetas River in the Amazon Region, CIGRE Session, B2-212 (2012).

[10] B. Subba Reddy, Diptendu Chatterjee, Energy Procedia 90, 179-184 (2016).

[11] G.S. Rashmi, Shivashankar, Poornima, Materials Today: Proceedings 4, 10, 11318-11324 (2017).

[12] C.R.F Azevedo, T. Cescon, Engineering Failure Analysis 9, 6, 645-664 (2002).

[13] I. Levchenko, A. Zasypkin, A. Alliluyev, E. Satsuk (Moscow: Publishing House of Mos-cow Power Engineering Institute, 447, 2007).

[14] V. Kachanovskaya, M. Yermoshina, Electricity, Transmission and distribution 1, 16, 50-54 (2013).

[15] V. Kachanovskaya, M. Yermoshina, Transmis-sion and Distribution World, USA, 52-54 (2013).

[16] A. Boshnyakovich, Calculation of substation wires and large overhead power line crossings (Leningrad: Energy, 236, 1975).

[17] M. Korotkevich, N. Prokofiev, Power engineer-ing, Energetika, Proceedings of CIS higher edu-cation institutions and power engineering associations 61, 5, 432-439 (2018) .

[18] G. Ugarov, N. Shevchenko, Yu. Lebedeva, A. Soshinov (Moscow: Pero, 186, 2013).

[19] Yu. Lebedeva, N. Shevchenko, A. Soshinov, Modern problems of science and education 6, 68-71 (2010).

[20] N. Shevchenko, Izvestiya VolgGTU, Series: Energy Conversion Processes and Power Plants 8, 3, 89-92 (2011).

[21] N. Shevchenko, Yu. Lebedeva, G. Ugarov, Vestnik SGTU 1, 3, 121-125 (2011).

[22] N. Shevchenko, Abstract of the dis. Cand. Technical Sciences 05.09.03 (Saratov, 20, 2011).

E3S Web of Conferences **178**, 01047 (2020) https://doi.org/10.1051/e3sconf/202017801047
HSTED-2020

Relationship between operational properties of peat heat-insulating materials and the content of mineral binders in them

Anton Sinitsyn[1,*], *Ludmila* Voropay[1], *Regina* Salikhova[2] and *Olga* Yukhtarova[1]

[1]Vologda State University, Lenina str. 15, 160000, Russia
[2]Kazan State Power Engineering University, Kazan, Russia

Abstract. This work presents the main results of calculation and experimental study on relation between the operational properties of peat heat-insulating materials (HIM) and the content of mineral binders in them aiming at reducing heat energy costs by energy-consuming objects. It was experimentally proved that introduction of cement increases thermal conductivity, and introduction of calcium oxide as a mineral binder, on the contrary, promotes the development of material porosity while maintaining strength characteristics and elasticity and helps to reduce the thermal conductivity and specific density. The established relationship between the content of calcium oxide and the thermal conductivity enables modelling of technology for HIM producing based on peat and inorganic binders.

1 Introduction

The President and the Government of the Russian Federation have set a strategic task to determine ways to rationally use the country's most important national treasure – natural energy resources. Currently, the policy of energy and resource conservation is one of the priority economic directions for the entire national economy. Reducing the energy consumption of residential and public buildings, various industrial facilities will create conditions for increasing the efficiency of use of fuel, energy, material resources during production of heat and electricity, a wide range of industrial, transport and construction products [1-7].

To reduce energy costs and create comfortable conditions for the use of buildings, a set of measures is used. It includes the use of high-quality environmentally friendly new types of heat-insulating materials obtained on the basis of available natural raw materials or unclaimed plant waste products (wood waste). Peat composite materials are of great practical interest in the modern market of heat-insulating materials. It is known that peat plates made from different types of peat with introduction of wood waste and mineral binders are characterized by thermal conductivity in the range 0.04-0.065 W/m·K, which meets the requirements of the Russian State Standards. The introduction of mineral components increases the prism strength together with thermal conductivity. Therefore, when choosing a chemical composition during production of heat-insulating materials (HIM), it is necessary to establish a relationship between the thermal conductivity and the content of mineral binder. It is also known that heat-insulating plates produced without addition of inorganic binders have low thermal conductivity, but at the same time low fire resistance, biostability, and temperature resistance, which are significant disadvantages.

In this regard, to increase the strength characteristics, fire resistance, biostability while obtaining thermal conductivity, almost equal to that of air, it is necessary to introduce mineral additives into peat, for example, cement, and limestone.

It is known that introduction of mineral binders in a peat mixture increases thermal conductivity and specific density of materials by reducing porosity. However, formation of a porous structure depends on the amount of mineral binders introduced into the peat mixture, for example, cement and quicklime (calcium oxide).

The aim of this work is to experimentally establish the relationship between thermal conductivity and the amount of cement or calcium oxide introduced into the peat mixture.

To achieve this aim, the following tasks were solved:

1. To perform a patent search on the specified topic; to produce samples using the known technologies in laboratory conditions, and to determine their thermal conductivity, specific density, prism strength.
2. To analyze the obtained results.
3. To experimentally determine the relationship between the amount of cement or calcium oxide introduced into the peat mixture and the operational characteristics of the obtained HIM.
4. To choose the optimal amount of inorganic components introduced into peat based on the obtained experimental results.
5. To produce model samples of HIM using new technology and examine their performance characteristics.

The object of study is the operational characteristics of composite peat heat insulators. The subject of study is

* Corresponding author: sinitsyn.science@mail.ru

Table 1. Chemical characteristics of the high-lying peat of the Vologda region.

Sample No.	Field	Peat type	Peat kind	Degree of decomposition,%	Ash-content, %	Content of organic substances, %	pH
1	Vologodskoe	High-lying	Pine-cottongrass	20.0	11.68	88.32	6.0
2				20.0	11.78	88.22	6.0
3				20.0	11.86	88.14	6.0
4				20.0	11.87	88.13	6.0

the relationship between the amount of mineral additives and the thermal conductivity.

The work was carried out in the REC "Problems of the modern technological environment". The similar works performed by the REC can be found in previous publications, for example, in [8-10].

2 Methods

During the experiment, the following methods were used: gravimetric, titrometric, combustion technique, stationary heat flow technique, compressive strength test, densimetric technique.

At the first stage of study, samples of heat-insulating materials were obtained using the known technologies. The high-lying peat taken from the peat deposit of the Vologda region was used as a heat carrier. In laboratory conditions, the chemical characteristics of peat were determined from its combustion: ash content, organic matter content in accordance with the Russian State Standards GOST 27784-88 [11] and GOST 26213-91 [12]. Samples were calcined in a muffle furnace SNOL-7-2/1100 at a temperature of 350-400 °C. The resulting solid precipitate was weighed on analytical balance and the ash content was calculated according to the following expression (1):

$$A = \left(\left(m - m_1\right)/m_2\right)\cdot 100,\ \%, \tag{1}$$

where A is the ash content, %; m is the mass of crucible with ash residue, g; m_1 is the mass of the empty crucible, g; m_2 is the mass of dry soil, g.

The content of organic substances was determined from ash-content according to formula (2):

$$X = \left(100 - A\right), \tag{2}$$

where A is the ash content, %.

The chemical characteristics of four peat samplings are presented in Table 1.

It is known that during HIM production peat acts not only as a heat insulator, but also as a framing, which forms a certain structure of the composite mixture after introduction to peat of additives (wood sawdust), inorganic cement or calcium oxide, and organic binder - hardener.

3 Results

The previous studies performed at Vologda State University experimentally proved that high-lying peat forms a more porous structure compared to that of low-lying peat. So, the resulting composites of different content always have lower thermal conductivity - 0.045-0.065 W/m·K, higher elasticity, lower specific density of 170-260 kg/m^3 while maintaining the prism strength not more than 0.3 MPa, which meets the requirements of the Russian State Standard GOST.

The performed patent search showed that cement is widely used as an additive to peat, and it is added to a mixture of high-lying peat and wood waste [13]. The following optimal composition of mixture was determined experimentally: 12% phr of high-lying peat, 38% phr of Portland cement, 25% phr of wood waste, 22% phr of water. The high-lying peat with a decomposition degree of 5–10% and a moisture content of 55–70% is used. According to the technology, it together with wood waste is soaked in solution containing methanol in the amount of 6–10 mg/l at a temperature of 80–100 °C for 2–5 min This technology enables obtaining heat-insulating plates that have compressive strength of 6.41 MPa, and thermal conductivity of 0.18 W/m·K, which does not meet the requirements of the Russian State Standard GOST and exceeds the allowable thermal conductivity by 4 to 5 times.

Thus, the introduction of Portland cement contributes to an increase in mechanical strength, resistance to water and biological damage, which is an advantage of this technology. However, the specific density also sharply increases to 920 kg/m^3 and, as mentioned above, the thermal conductivity increases above the permissible values, which limits application of this technology. The lack of technology itself should also be noted. The feasibility of using a water-methanol mixture for soaking sawdust is questioned, and this process is carried out at a temperature of 80 - 100 °C, which significantly exceeds the methanol boiling point. Therefore, at this temperature, methanol evaporates and enters the environment. To eliminate the negative impact of methanol, which belongs to the first class of toxicity, there is a need to install an additional ventilation system.

In recent years, cement cyclone dust (cement production wastes) have been used as fillers, which is added to a suspension of organic binder - polyvinyl acetate - and the suspension is introduced into cut peat

Table 2. Composition of the initial heat-insulating mixture.

Sample No.	The initial mixture composition				
	High-lying peat (wet), %	**Sawdust, %**	**Wood powder, %**	**Cement, %**	**Polyvinylacetate polymer, %**
1	72	-	-	14	14
2	72	-	-	14	14
3	72	-	-	14	14
4	62.5	12.5	-	12.5	12.5
5	62.5	12.5	-	12.5	12.5
6	62.5	12.5	-	12.5	12.5
7	62.5	-	12.5	12.5	12.5
8	62.5	-	12.5	12.5	12.5
9	62.5	-	12.5	12.5	12.5

[14]. However, the introduction of cyclone dust alone does not provide high mechanical strength of HIM, and therefore it is always recommended to add Portland cement as a supplementary material.

In laboratory conditions, HIM samples with addition of cyclone dust were obtained. The following chemical composition is maintained: 28% phr of cement, 52% phr of peat modified with a suspension of polyvinyl acetate and cyclone dust, 20% phr of water.

After thorough mixing, molding and holding in air, heat-insulating plates are obtained that have the following characteristics: specific density of 400 kg/m^3, compressive strength of 1.5 MPa, thermal conductivity of 0.1 W/m·K. Thus, with an increase in peat content and due to the impregnation of peat with polyvinyl acetate and cyclone dust, the uniformity of the structure and the durability of heat-insulating materials increase. However, the introduction of peat cement in amount of 28% contributes to an increase in thermal conductivity and specific density of material.

To increase the material porosity, the authors [13] propose to introduce sodium carbonate in addition to Portland cement and cyclone dust. The introduction of sodium carbonate as an additive helps to reduce the specific density of HIM while maintaining its strength characteristics. However, it was experimentally established that the porosity of material, which governs its heat-insulating properties, is determined by the medium pH and the processing temperature. If the medium pH is more than 7, that is, the medium is alkaline, so under these conditions it is necessary to heat the insulating mixture to the decomposition temperature of sodium carbonate, which is a significant drawback of this technology. In addition, the use of sodium carbonate to create porosity in the amount of 10% of the total mass of heat-insulating mixture increases the material costs, which reduces competitiveness of this technology in the HIM market.

Also HIM samples were obtained with the addition of cement with lower concentrations. The mixture composition is presented in table 2.

To obtain more accurate results, each experiment was repeated 3 times. The main HIM characteristics are derived from the average of three measurements. Physico-mechanical characteristics of HIM are presented in table 3.

The results show that the introduction of wood filler in the form of sawdust and powder into the samples promotes reduction of thermal conductivity and specific density. At the same time, replacing wood sawdust with wood powder results in higher values of these parameters.

The results indicate that introduction of Portland cement increases the thermal conductivity and reduces the porosity of material at minimum concentration of this binder. Therefore, it can be concluded that the use of cement in production of heat-insulating materials is not always advisable and it is better to use mineral binders

Table 3. Physico-mechanical characteristics of insulating samples.

Sample No.	Physico-mechanical characteristics		
	Density, kg/m^3	**Thermal conductivity, W/m·K**	**Mechanical strength, MPa**
1	1238	0.533	4.64
2			
3			
4	657	0.232	1.67
5			
6			
7	758	0.283	2.92
8			
9			

Table 4. Composition of thermal insulation mixture for HIM production.

Composition of thermal insulation mixture	**Sample No.**		
	1	**2**	**3**
High-lying peat (with preserved natural structure), %	40	50	60
Wood powder, %	20	15	10
Calcium oxide (CaO), %	20	15	10
Polyvinylacetate polymer, %	15	15	15
Water, %	5	5	5

that provide greater porosity of heat insulator structure, for example, quicklime (calcium oxide).

Unlike cement, when calcium oxide is introduced into wet peat, a more uniform porous structure with small inclusions from peat and sawdust is formed, which contributes to a decrease in thermal conductivity and specific density of HIM while maintaining their strength characteristics.

At the next stage of the study, the optimal composition for obtaining HIM based on peat, calcium oxide, wood powder and an organic binder - polyvinyl acetate polymer was established.

A series of samples were made according to the composition shown in table 4.

Water is added to the high-lying peat for moisturizing and, after the peat has been completely saturated with water, calcium oxide obtained by calcining the limestones is added. After thorough mixing, wood powder is added and organic hardener is introduced. After mixing, the mixture is placed in the molds of 150×150×40 mm, the mass is compacted with a press, and the mixture setting conditions (setting time and temperature) are selected. Next, the obtained samples are cooled and tested in laboratory conditions to determine their average density, compressive strength, thermal conductivity. The research results are presented in table 5.

While determining the relationship between thermal conductivity and the content of calcium oxide, it was found that addition of calcium oxide does not increase the thermal conductivity above the requirements of the Russian State Standard GOST. At the same time, the dependence of structure formation on temperature is also noted. It was found that when a heat-insulating composition is heated to 105 °C in order to remove free water, a porous structure is first formed, which undergoes cracking upon slow cooling. The sample subjected to heat treatment at a temperature of 105 °C is shown in Figure 1.

Fig. 1. The sample subjected to heat treatment at 105 °C.

Therefore, it is not recommended to heat the mixture to a temperature of 105 °C. A firm homogeneous finely porous structure forms at a temperature of 95-100 °C and at a setting time of 16 hours.

The thermal conductivity is evaluated using the device ITS-1 "150".

The measurements of thermal conductivity are based on the stationary heat flux method according to the requirements of the Russian State Standard GOST 7076-99 [15]. The obtained thermal conductivity is presented in Figure 2.

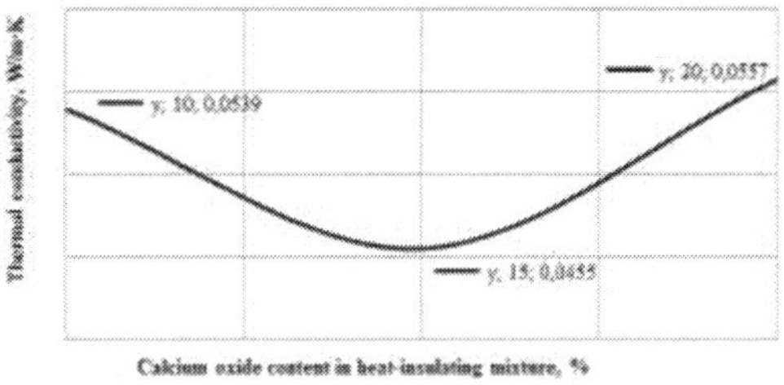

Fig. 2. Relationship between thermal conductivity and calcium oxide content in heat-insulating mixture.

Table 5. Physico-mechanical characteristics of HIM.

Physico-mechanical characteristics	**Sample No.**		
	1	**2**	**3**
Average density, kg/m³	482	457	461
Thermal conductivity, W/m·K	0.0557	0.0455	0.0539
Compressive strength, MPa	1.65	1.38	1.42

The experimental results also prove the dependence of thermal conductivity on the content of calcium oxide. With an increase in the content of calcium oxide in mixture from 10 to 15%, a decrease in thermal conductivity from 0.0539 to 0.0455 W/m·K is observed. With an increase in the content of calcium oxide in mixture from 15 to 20%, an increase in thermal conductivity from 0.0455 to 0.0557 W/m·K is observed.

The established relationship between the content of calcium oxide and the coefficient of thermal conductivity allows one to model the HIM production technology and to obtain samples with certain specified performance characteristics.

At the same time, the specific density and prism strength of the obtained samples were determined. Prism strength and specific density were determined in accordance with the requirements of the Russian State Standard GOST 17177-94 [16]. The results on average density are presented in Figure 3.

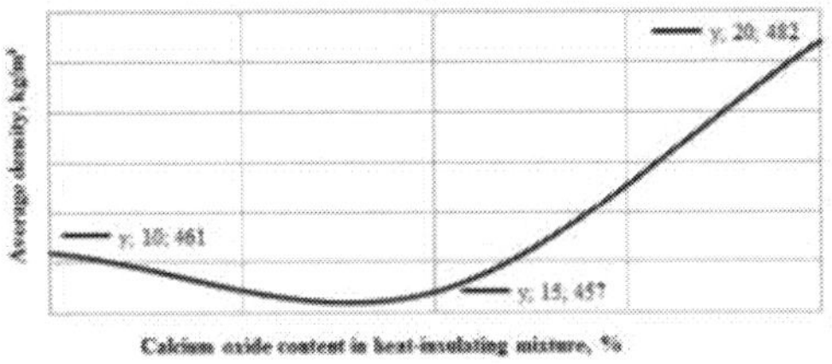

Fig. 3. Relationship between average density and calcium oxide content in heat-insulating mixture.

Analysis of results shows that with an increase in the content of calcium oxide from 10 to 15%, the average density of samples decreases from 461 to 457 kg/m^3. On the contrary, with an increase in calcium oxide from 15 to 20%, an increase in the average density of samples from 457 to 482 kg/m^3 is observed. The minimum average density is typical for samples containing 15% of calcium oxide in the initial composite mixture. The studies of mechanical strength are presented in Figure 4.

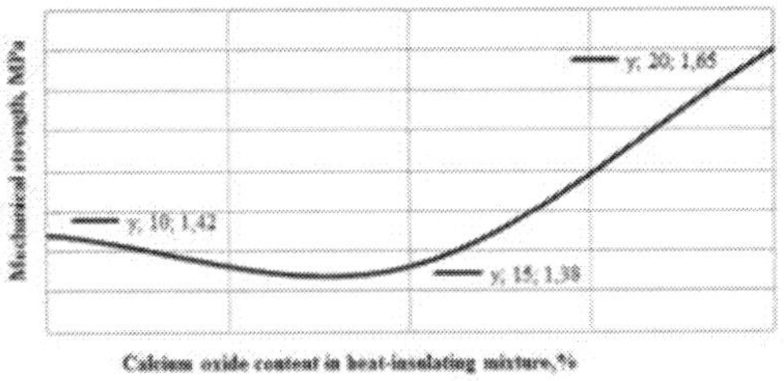

Fig. 4. Relationship between mechanical strength and calcium oxide content in heat-insulating mixture.

The obtained data indicate that with a decrease in the content of calcium oxide, the compressive strength decreases. Maximum strength is typical for samples with calcium oxide content in the composite mixture of 20%.

It was concluded that the optimal composition of HIM with introduction of calcium oxide is as follows: 50% of horse peat (with a preserved natural structure); 15% of wood powder; 15% of calcium oxide; 15% of polyvinyl acetate polymer; 5% of water.

4 Conclusions

Thus, based on the obtained experimental data, the following conclusions can be drawn:

1. It has been experimentally proved that introduction of cement contributes to an increase in thermal conductivity. The introduction of calcium oxide as a mineral binder, on the contrary, contributes to the development of material porosity while maintaining strength characteristics and elasticity and helps to reduce the thermal conductivity and specific density.
2. The established relationship between the content of calcium oxide and the thermal conductivity allows one to model the HIM production technology based on peat and inorganic binders and to obtain samples with certain specified performance characteristics.
3. The optimal composition of HIM was experimentally selected, which provides consumer properties of the composite that meet the requirements of the Russian State Standard GOST.
4. HIM model samples were fabricated using new technology and their operational characteristics were investigated.

References

[1] L.V. Larchenko, R.A. Kolesnikov, L. Mukhametova, Russian oil and gas industry as a sphere of international interests and economic cooperation, E3S Web of Conferences **161**, 01006 (2020).

[2] V. Markov, A. Fedyukhin, I. Sultanguzin, B.G. Matisov, N.V. Nikitkov, V.V. Izrantsev, V.G. Knorring, G.A. Konrashkova, L.R. Mukhametova, Improvement of energy efficiency of motor fuel production at oil refinery, IOP Conference Series: Earth and Environmental Science **337**, 1, 012080 (2019).

[3] L.R. Mukhametova, I.G. Akhmetova, A.V. Fedukhin, S.O. Makoev, A.A. Kalyutik, About heat supply schemes for settlements and city districts, IOP Conference Series: Earth and Environmental Science 288, 1, 012108 (2019).

[4] I. Akhmetova, A. Tyfetylov, A. Tamakchi, G. Khadiyllina, O. Derevianko, Z. Syed, Improving the competitiveness of automobile engineering enterprises by advancement the internal control over its indicators, International Journal of Civil Engineering and Technology **9**, 13, 1865-1876 (2018).

[5] A.A. Kalyutik, D.V. Grigorieva, A.V. Fedyukhin, O.V. Derevianko, I.G. Akhmetova, Reduction of the technological minimum by bypass method with environmental impact estimation on the example of energy blocks with T-100 and T-250/300-240 turbines, IOP Conference Series: Earth and Environmental Science **288**, 1, 012094 (2019).

[6] V.I. Sitas, A.V. Fedyukhin, I.G. Akhmetova, A. Mitrofanov, S.O. Makoev, A. Asadpoori, A.A. Sinitsyn, E.A. Kikot, Assessment of technical and economic efficiency indicators of cogeneration in modern market conditions, International Journal of Civil Engineering and Technology **10**, 2, 2106-2117 (2019).
[7] E. Sukhareva, A. Fedyukhin, O. Derevianko, M. Egorov, L. Mukhametova, I. Akhmetova, Optimization of combined heat and power plant operating mode by means of underutilized equipment mothball, IOP Conference Series: Earth and Environmental Science **288**, 1, 012122 (2019).
[8] D.F. Karpov, M.V. Pavlov, A.A. Sinitsyn, Experimental Research of the Thermal Balance of Loose Disperse Materials Exemplified by Milled Peat Exposed to Infrared Heating, Vestnik MGSU **5**, 114-119 (2012).
[9] D.F. Karpov, M.V. Pavlov, A.A. Sinitsyn, Solution to the Boundary Problem of Heat and Water Transmission in a Layer of a Loose Disperse Material Exemplified by Milled Peat Exposed to Infrared Heating, Vestnik MGSU **6**, 92-98 (2012).
[10] A.A. Sinitsyn, Experience of Small Innovative Enterprises Foundation in the Vologda Region, Middle East Journal of Scientific Research **16**, 10, 1424-1427 (2013).
[11] The Russian State Standard GOST 27784-88, *Soils, Method for determination the ash content of peat and peat soil horizons* (Standards Publishing House, 1988).
[12] The Russian State Standard GOST 26213-91, *Soil, Methods for determination of organic matter* (Standards Publishing House, 1992).
[13] A.Yu. Tarakanov, O.A. Medvedev, *Peat-polymer mixture for producing heat-insulating products*, Patent 2555180 Russian Federation, MPK S 04 V 38/08, applicant and patent holder All-Russian Research Institute for Civil Defense and Emergencies EMERCOM of Russia, 2014116521/03, declared 24.04.2014, publ. 10.07.2015, bull. no. 19.
[14] O.A. Medvedev, L.B. Khoroshavin, V.S. Rudnov, et al., *Peat-polymer mixture for producing heat-insulating products with increased fire resistance*, Patent 2513801 Russian Federation, MPK(51) S 04 V 26/00, applicant and patent holder All-Russian Research Institute for Civil Defense and Emergencies EMERCOM of Russia, 2012147661/03, declared 09.11.2012, publ. 20.04.2014, bull. no. 11.
[15] The Russian State Standard GOST 7076-99, *Method for determination of thermal conductivity and thermal resistance under stationary thermal conditions* (Standards Publishing House, 2000).
[16] The Russian State Standards GOST 17177-94, *Heat-insulating materials and products, Research methods* (Standards Publishing House, 2002).

 https://doi.org/10.1051/e3sconf/202017801048

Experimental and numerical study of non-stationary solid-liquid phase transitions of n-tetradecane

Viktoria Zakharova[1,*], *Pavel* Kuznetsov[1], *Daniyar* Kalimzhanov[1], and *Aleksandr* Baranenko[1]

[1]ITMO University, Faculty of Cryogenic Engineering, 197101, St. Petersburg, Kronverksky pr. 49, Russian Federation

Abstract. During the experiments, four spherical flasks with n-tetradecane with a total mass of 0.344 kg were placed in a test container inside which an aqueous solution of propylene glycol with a mass concentration of 50% and an initial temperature of 25 °C was circulating. As a result of studies, the values of heat flows at different time intervals during the solid-liquid phase transition of n-tetradecane were obtained. Based on the modeling of the processes, the distribution of temperature and velocity of the heat carrier in the test container as well as the values of the local heat transfer coefficients on the surface of the flasks were established. Numerical calculations of the process of phase transition being studied according to the authors' model were also performed. Numerical calculations of the process under study have been performed using the authors' model. A satisfactory convergence of the experimental and calculated values of the heat flow has been obtained. The results can be used in the development of thermal energy accumulators with n-tetradecane.

1 Introduction

Non-stationary liquid-solid-liquid phase transitions are rather common processes. They can be implemented in various technical devices. In particular, they occur in thermal energy accumulators with phase change materials (PCM) [1, 2].

Thermal energy accumulators are used in heating, hot water supply, cooling and air conditioning systems with variable load. The energy stored in accumulators is used at the maximum consumption of the corresponding resource. This allows installing equipment of lower power, equalizing daily electricity consumption, reducing electricity costs when double-tariff accounting is used and when charging thermal energy storage devices takes place at night. The accumulators can use renewable energy sources such as solar energy and night-time radiation cooling, which reduces energy consumption [3, 4].

In comparison with accumulators with liquid, the advantage of thermal energy accumulators with PCM is that they have an order of magnitude higher volume and mass density of stored energy. This is due to the fact that in them, the main contribution to the thermal energy accumulation is made by the heat of the phase transition.

Also, in such accumulators, the process of accumulating and spending energy stored is carried out at a constant temperature corresponding to the temperature of the phase transition of the PCM. In a more common design of these devices, the PCM is placed inside the container in capsules, which can be of arbitrary shape, including balls or cylinders. The heat carrier circulates in the remaining free internal volume of the accumulator [5].

A fairly large number of theoretical and experimental papers are devoted to the study of processes in such thermal energy storage devices [6-12]. In [6], a detailed experimental and theoretical study of two-dimensional temperature fields in spherical capsules was carried out (axisymmetric problem). In [7], a calculation model for a numerical solution based on Navier – Stokes equation using the enthalpy approach is presented. In [8], the calculation model was applied to study the influence of fins located inside cylindrical capsules on heat exchange. In [9], an experimental and numerical study for elliptical capsules was carried out. However, in these papers, the determination of such quantities as the total heat flow, primarily necessary for the design of cooling systems, occurs indirectly from the results of calculating temperature fields. In [11], the authors investigated the processes in dense backfills of steel capsules. Paper [12] is devoted to modeling a system with thermal energy storage devices as a whole.

For cold accumulation in air conditioning systems, PCM with a phase transition temperature of 5-10 °C are of interest. Among such substances are paraffins the advantage of which is stability during repeated multidirectional phase transitions, a relatively high specific heat of the phase transition and a slight difference in density in the liquid and solid states.

2 Research and results

The authors have performed an experimental and numerical study of the solid-liquid phase transition of n-

* Corresponding author: viktoriasju@yandex.ru

tetradecane (chemical formula $C_{14}H_{30}$). According to the passport, the mass fraction of the main substance is 99.89%, the melting point is 6.1 °C. The main properties [13, 14] are the following: specific mass heat of melting (λ_{hc}) is 215 kJ/kg, liquid phase density (ρ_{hc}) at 25 °C is 762.8 kg/m^3, dynamic viscosity at 25 °C is $2.13 \cdot 10^{-3}$ Pa·s, thermal conductivity coefficient of the liquid phase is 0.143 W/(m·K), specific heat of the liquid phase ($C_{p\,hc}$) is 2198 J/(kg·K).

Experimental studies were carried out in an isolated test container with dimensions of 240x150x115 mm (Fig. 1). The PCM was placed in round-bottom flasks of a spherical shape with an outer diameter of 66 mm, the flasks were filled to the base of the neck. In the experiments, 4 flasks were used; in them, the PCM mass was 0.344 kg. In the test container, the flasks were located in a strictly defined order in all the experiments as it is shown in Fig. 1. The heat carrier movement was carried out in a closed circuit and with a mass flow rate of 2.1 l/m. The heat carrier volume in the circulation circuit was 4.15 l. An aqueous solution of propylene glycol with a mass concentration of 50% was used as a heat carrier. In all the experiments, the initial temperature of the heat carrier was 25 °C.

The movement of the heat carrier in the test container is shown in Fig. 2, it is modeled in the Fluent package. Modeling is carried out for the steady-state operation mode of the stand. When modeling the movement of the heat carrier, the natural convective movement of the liquid is also considered, since preliminary calculations have shown its noticeable influence in this design of the stand. When modeling, the external walls of the test container are assumed to be thermally insulated, since the temperature of the heat carrier corresponds to the ambient temperature, which eliminates the heat flow between them. The temperature of the walls of the flasks is assumed to be constant and equal to the melting temperature of the SPT.

Before starting the experiments, the flasks with the PCM were cooled until the complete transition of the PCM to a solid state. In the course of these experiments, the following parameters were determined and recorded. At the beginning and end of the experiment, the temperature of the PCM as well as of the phase transition; the heat carrier temperature in regular time intervals, the time of the beginning and end of the experiment, the heat gain from the pump. The experiments were over with the complete transition of the PCM to a liquid state. To measure the temperature, electronic thermometers with an accuracy of 0.2 °C were used. In the experiments, the phase transition temperature of the PCM was 6.1 °C, which corresponds to its passport value.

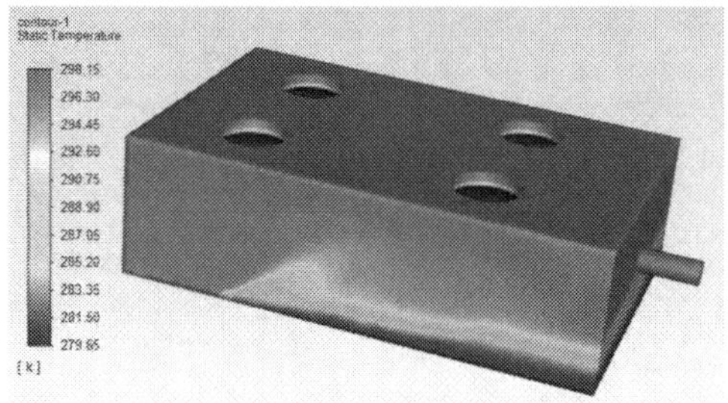

Fig. 1. Temperature distribution in the test container during the experiment.

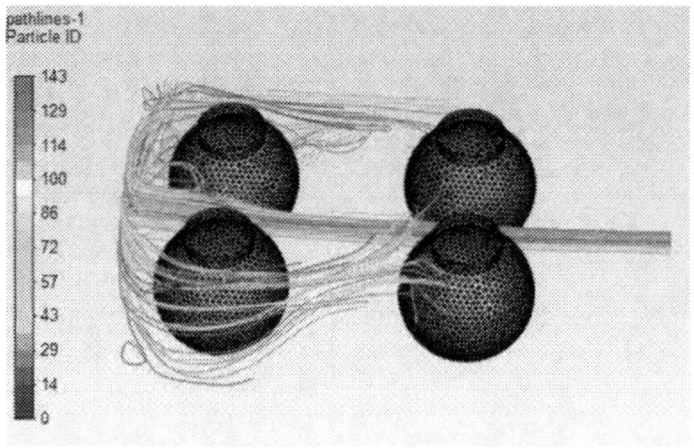

Fig. 2. Heat carrier movement in the test container.

Numerical calculations have also shown that in the test container, during the experiment, the temperature gradient of the heat carrier can reach 10°C. The lowest temperatures of the heat carrier are set in the lower part of the test container, despite the fact that the heat carrier moves in it at a certain speed. This indicates the feasibility of supplying the heat carrier from below. Local heat transfer coefficients differ by an order of magnitude (Fig. 3). Their maximum value is 3600 W/(m2·K), the average value is 380 W/(m2·K). The maximum values of the local heat transfer coefficients take place on the side of the flasks along the flow of the heat carrier of their inlet pipe in the central zone of the test container.

Table 1. Average heat transfer coefficients from flasks to theheat carrier.

Flask number	1	2	3	4
Average heat transfer coefficient, W/(m^2·K)	345	453	341	384

When processing experimental data, the convergence of thermal balances was checked. The heat absorbed by the PCM (Q_{PCM1}) should be equal to the heat released from the heat carrier (Q_{PCM2}). Q_{PCM2} includes the heat of the heat carrier (Q_{hc}), flask (Q_f) and pump (Q_p), kJ.

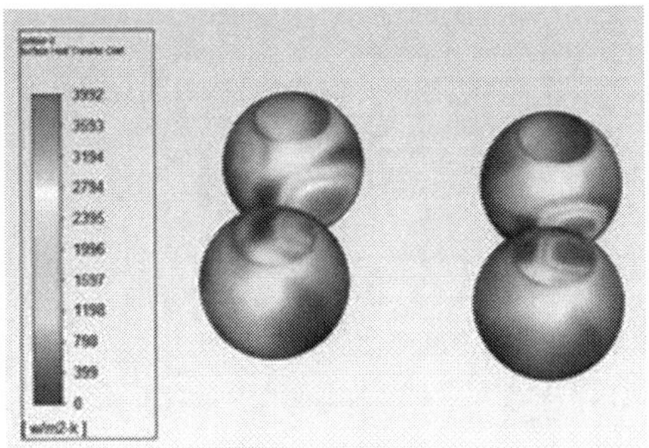

Fig. 3. Local heat transfer coefficients on the surface of the flasks.

$$Q_{\text{PCM1}} = Q_{\text{PCM2}}, \quad (1)$$

where m_{PCM} is the mass of the SPT in flasks.

In general, the heat of the PCM is determined by the formula

$$Q_{\text{PCM1}} = m_{\text{PCM}}\left(C_{\text{p PCM}} \cdot \Delta t_{\text{PCM}} + \lambda_{\text{PCM}}\right). \quad (2)$$

The heat from the heat carrier side

$$Q_{\text{PCM2}} = Q_{\text{hc}} + Q_{\text{f}} + Q_{\text{p}}. \quad (3)$$

The heat of the heat carrier

$$Q_{\text{hc}} = V_{\text{hc}}\rho_{\text{hc}}C_{\text{p hc}}\Delta t_{\text{hc}}. \quad (4)$$

V_{hc} is the heat carrier volume in the circulation circuit, m^3.

The heat of the flask

$$Q_{\text{f}} = m_{\text{f}}C_{\text{p f}}\Delta t_{\text{f}}. \quad (5)$$

where m_{f}, $C_{\text{p f}}$ are the mass and specific heat of flasks, respectively, Δt_{PCM}, Δt_{hc}, Δt_f are temperature differences of the PCM, heat carrier and flask, respectively, at the end and at the beginning of the experiment, °C.

The heat of the pump

$$Q_{\text{p}} = q_{\text{p}}\Delta\tau_{\text{e}} \quad (6)$$

q_{p} is the heat flow from the pump, kW. $\Delta\tau_{\text{e}}$ is the experiment conduction time, s.

The heat flow from the pump was determined as follows. The heat carrier circulation was carried out in a closed circuit in the absence of flasks with the SPT in the test container. The initial and final heat carrier temperature and the time of the test were recorded. Knowing the volume of the heat carrier in the circulation circuit, the increase in the temperature of the heat carrier and the duration of the measurement, the amount of heat flow from the pump to the heat carrier was determined. The measurements were made at least three times during each series of experiments. The discrepancy in q_{p} values did not exceed 5%.

The convergence of thermal balances was ± 4%.

The average heat flow from the heat carrier for a certain time interval, kW

$$q = q_{\text{hc}} + q_{\text{f}} + q_{\text{p}} \quad (7)$$

$$q_{\text{hc}} = \frac{V_{\text{hc}}\rho_{\text{hc}}C_{\text{p hc}}\Delta t_{\text{hc}}}{\Delta\tau_{\text{i}}}, \quad q_{\text{f}} = \frac{m_{\text{f}}C_{\text{p f}}\Delta t_{\text{f}}}{\Delta\tau_{\text{i}}} \quad (8)$$

Δt_{i} is the change in temperature of the heat carrier and flask for a certain time interval $\Delta\tau_{\text{i}}$.

The values of q calculated from experimental data are presented in Fig. 4. This figure also shows the calculated dependence $q_{\text{PCM}} = f(\tau)$ computed by the authors' model [15].

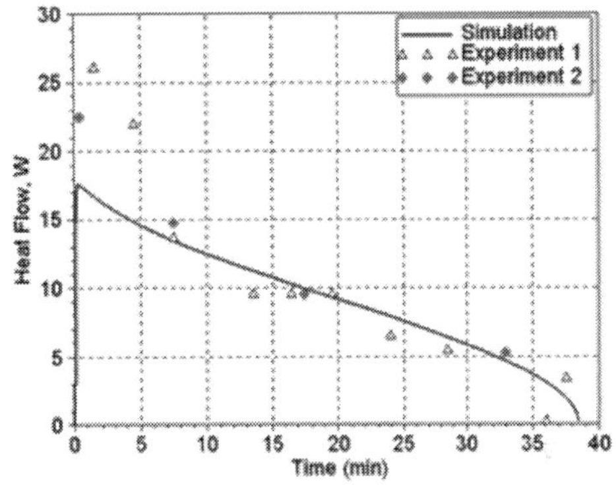

Fig. 4. The average calculated and experimental values of heat flow per flask.

At the initial moment of time, the heat flow increases sharply and after a fairly short time, it begins to decrease due to an increase in the thermal resistance of the liquid phase of the PCM (since its thickness increases) and a decrease in the contact surface of the phases of the PCM. The results obtained correspond to the regularities of the process under study as well as to the data of other authors [16]. At the same time, the calculated and experimental values of the duration of the solid-liquid phase transition process differ by only 2 %. The greatest interest present the data on the values of the heat flow in the time interval of the phase transition process of 5-30 min, when the dependence of q_{p} on time is monotonous. Based on these data, it is most advisable to model the calculations of thermal energy accumulators.

3 Conclusions

The duration of the experiment was 36-38 minutes. The heat flow reached its maximum value of 26 W in 90 seconds after the start of the experiments. In the interval of 5-30 minutes, the dependence of the heat flow on time is monotonous. The results obtained by the authors have confirmed the adequacy of the previously developed model of the phase transition process. They can be used in the development of methods for calculating thermal energy accumulators with the PCM.

References

[1] C. Veerakumar, A. Sreekumar, Int J Refrig **67**, (2016)

[2] B. Al-Aifan, R. Parameshwaran, M. Kushagra, R. Karunakaran, Int J Refrig **76** (2017)

[3] G.T. Kim, Y.U. Choi, Y. Chung, M.S. Kim, K.W. Park, M.S. Kim, Int J Refrig **88** (2018)

[4] L. Mishra, A. Sinha, R. Gupta, Green Buildings and Sustainable Engineering, Proceedings of GBSE 2018 (2019)

[5] A. Sharma, V.V. Tyagi, C.R. Chen, D. Buddhi, Renew. Sust. Energ. Rev. **13** (2009)

[6] F.L. Tan, S.F. Hosseinizadeh, J.M. Khodadadi, L. Fan, Int. J. Heat Mass Transf. **52** (2009)

[7] H. Sattari, A. Mohebbi, M.M. Afsahi, A.A. Yancheshme, Int J Refrig 7**3** (2017)

[8] M.M. Joybari, F. Haghighat, S. Seddegh, A.A. Al-Abidi, Energy Convers. Manag. **152** (2017)

[9] D.H. Shina, J. Parka, S.H. Choib, H.S. Kob, S.W. Karnga, Y. Shina, Energy Convers. Manag. **182** (2019)

[10] M.A.M. Rosli, Renew. Sust. Energ. Rev. **42**, (2015)

[11] G. Zsembinszki, G.F. Angel, L.F. Cabeza, Appl. Sci **10**, (2020)

[12] R. Gulfam, W. Zhu, L. Xu, I.I. Cheema, P. Sheng, G. Zhao, Y.D. Deng, Energy Convers. Manag. **156**, (2018)

[13] T.S. Khasanshin, A.P. Shchemelev, High Temp. **40**, (2002)

[14] D. Mackay, et al., *Handbook of physical-chemical properties and environmental fate for organic chemicals, 2nd ed* (New York, 2006)

[15] V.Y. Zakharova, P.A. Kuznetsov, A.V. Baranenko, IOP Conference Series: Mater. Sci. Eng. **643**, 1 (2019)

[16] N. Khandelwall, A. Gupta, A. Tripathi, A. Pal, A. Jha, Our Heritage Journal **22**, 1 (2020)

E3S Web of Conferences 178, 01049 (2020)
HSTED-2020
https://doi.org/10.1051/e3sconf/202017801049

Theoretical aspects of optimization synchronous machine rotors

Timur Petrov[1,*], and *Alfred* Safin[1]

[1]Kazan State Power Engineering University, Kazan, Russia

Abstract. Permanent magnet synchronous electric machines are increasingly used in various drive technical complexes (oil industry, small generation, aviation industry, etc.). The presented optimization allows you to increase the energy performance of a synchronous machine without increasing the overall dimensions. Permanent magnet synchronous motors have become much more commonly operated in various drive technical complexes. For each drive complex, it is necessary to fulfil the requirements for the developed torque, cooling conditions and strength characteristics. The distribution of materials (topology) in the rotor casing of a synchronous machine (permanent magnets, iron, air, etc) determines the characteristics of permanent magnet synchronous electric machine.

1 Introduction

Topological optimization was initially carried out on the basis of a numerical method for solving partial differential equations, which was necessary to reduce mass and increase the strength characteristics of machine structures.

The proposed method allows you to transfer topological optimization to electromagnetic and thermal processes in a synchronous machine in order to increase the energy characteristics and reliability of the electric machine for a given mass and dimension indicators.

At the moment, there are no works on integrated topological optimization of rotors of synchronous electric machines with permanent magnets.

The algorithm of step-by-step topological optimization of rotors taking into account electromagnetic, thermal processes and strength calculation will allow designing prototypes of rotors with high energy characteristics for the preparation of conceptual and working design documentation.

The project aims to create a software package that will be used at electrical enterprises producing drive systems based on synchronous electric machines with permanent magnets.

The paper proposes a modern method of designing and optimizing synchronous motors with permanent magnets through the use of genetic algorithms. Operation of synchronous machines instead of asynchronous for electric drive sucker-rod pumping units will make it possible to obtain higher parameters of energy efficiency, and also synchronous motors have an extremely compact design. The structure of synchronous motors makes them much smaller and lighter than similar induction motors. A synchronous machine is up to 40 percent more compact than a similar induction motor. They correspond to a size that is two standard sizes smaller, which makes it possible for a more compact machine design. In addition, synchronous motors weigh 50 percent less and its inertia is 60 percent lower than that of a similar induction motor.

Designing an electric machine as a heuristic process does not guarantee a better solution. Methods are needed to complement the designer's experience and intuition in order to find the optimal (rational) solution. Design optimization is currently achieved by calculating the minimization of the functional of the variational problem. The proposed method allows you to transfer topological optimization to electromagnetic processes in synchronous motors to determine the directivity angle of the coercive force of the magnets used in the machine, the number of phases and turns of the windings, the size of the stator grooves and permanent magnets in order to increase the characteristics (power factor, reliability) of the electric machine at specified mass and size parameters.

2 Literature review

In Russia, the active use of topological optimization methods has been introduced relatively recently, and is mainly used to increase the specific strength of structures in the aerospace industry [1], optimization of turbine blades [2] or pistons of internal combustion engines [3]. The features of all these works are the topological optimization of only one material and the maximization of only one objective function (in most works of the Russian-speaking segment this is an increase in strength with a decrease in weight).

In foreign sources, work on topological optimization has been ongoing for a long time, and both strength parameters and the values of the electromagnetic and thermal fields are increasing.

The article [4] demonstrated the advantages of topological optimization of several materials compared

* Corresponding author: tobac15@mail.ru

to standard optimization (one material). The proof was the optimized design of the automobile engine for both optimization cases, and in the first case, a more rigid design with the same weight is obtained. The difference from our topic is that we will optimize synchronous motors with permanent magnets, and taking into account not one controlled parameter (weight in the article), but the electromagnetic, thermal and strength component, and this will allow us to obtain the most efficient motor design.

For example, an article is presented [5], in which the design of an electric motor without DC brushes is optimized (namely, the rotor configuration) based on the finite element method to increase specific power by increasing the electromagnetic field.

In [6], a two-dimensional model was considered for determining the electromagnetic field in the air gaps of synchronous motors with permanent magnets, and in scientific work [7] a method for designing permanent magnets that can already be used to obtain optimal designs is shown. In the article [8] a variant of topological optimization was presented through the use of a genetic algorithm, and in [9] optimization was achieved as a result of using the finite element method. In these works, optimization was necessary to increase the values of the electromagnetic field; thermal calculation was presented in [10], where an engine with the highest possible heat flux was designed. The design of engines can be made more efficient by using topological optimization based on sensitivity [11]. To increase the productivity of the genetic algorithm, optimal initial populations are necessary; work [12], which shows an algorithm for generating a population with high values of characteristics, can cope with this. The development of the genetic algorithm can also lie in the direction of adaptive grid changes right during optimization [13]; the use of the sequential linear programming method [14] is also an interesting solution.

In [15], an option was considered using the ON / OFF sensitivity method to optimize the design of the rotor. The following document [16] shows an interesting joint method for optimizing the design of the rotor, combining topological optimization and the limitations associated with the shape of the magnets.

It is worth noting that at the moment there are no works on complex topological optimization of permanent magnet synchronous motors, which allows taking into account the influence of the distribution of materials on the electromagnetic, thermal and strength characteristics of machines, in order to choose a rational design option.

3 Theoretical aspects of topological optimization

Topological optimization is one of the ways to find optimal solutions to structural problems for a particular structural model, including for electrical equipment designs.

This type of optimization is mainly used at the conceptual stage of design. The optimized function is the value of the potential energy of the system, as well as the movement of nodes. To solve the problems posed by this type of optimization, different methods of mathematical modeling are used. It is important to note that finding the exact solutions to topological optimization problems is most often difficult. And, therefore, in practice, algorithms for approximate numerical solutions are developed. In a broad sense, topological optimization can be classified as follows:

1) discrete element;
2) continuous.

The discrete-element type of optimization implies such an approach to design, in which there will be a ready-made set of possible discrete elements for the considered structural model. By changing the geometric parameters of any element from zero (the element disappears) to the maximum possible, the final structure will have different sizes and topologies. Continuous topological optimization is designing as a kind of continuous “void” or material with extremely low density values. By changing the distribution ratio, the "void" / material or density values within the maximum possible small values, the size value and the topology of the structural model of the object under consideration.

The distribution of materials in the rotor is one of the most important factors determining the magnetostatic field of the machine, thermal characteristics and determining the strength characteristics in electric machines.

In addition to commonly used materials such as steel sheet, permanent magnets (neodymium-ferrum-boron, samarium-cobalt), some recently developed magnetic materials, such as soft magnetic composites, amorphous and grain-oriented silicon steel can exhibit more high magnetic field parameters, for example, low losses and high values of saturation density, which leads to lower operating costs. Thus, we can say that the use of these types of magnets will allow the development of synchronous motors with the most optimal characteristics, both from a technical and economic point of view.

The production methods used, obviously, also have a fundamental role in the life cycle of developing synchronous motors, since they directly affect the quality of workmanship and actual performance.

It is important to note that in order to obtain the optimal characteristics of synchronous motors, the rotor can have such a complex configuration that it is difficult to manufacture, and this is a serious drawback.

Another disadvantage is that the characteristics of the materials in question will depend on the production methods. Optimum parameters can only be obtained with a full understanding of the theoretical aspects of magnets used for the design of synchronous machines, and accordingly this knowledge should affect production.

This must be taken into account when configuring the parameters of topological optimization.

Currently, the industrial production of permanent magnets with high energy performance is developing. Such magnets make it possible to create a large magnetic flux in small volumes, allowing one to significantly increase the ultimate power.

The most important requirements for engines in the drive mechanisms of rocking machines are:

- high energy-dynamic parameters;
- small dimensions and weight.

To achieve maximum values of the energy and dynamic parameters of synchronous motors with small dimensions, high-energy permanent magnets are used.

This allows:

- firstly, reduce the volume of permanent magnets;
- get the smallest moment of inertia of the rotating part of the motor rotor;
- thirdly, to obtain a greater magnetic flux, that is, large values of the starting, electromagnetic moment, specific power and efficiency.

Currently, permanent magnet synchronous motors use high-energy magnets based on samarium and cobalt and hard magnetic materials based on a neodymium-iron-boron alloy.

To use promising types of materials in the construction of an electric motor, it is important to emphasize that it is necessary to focus on the development of promising engine topologies based on new algorithms and topological optimization methods.

Thus, the rational design of the engine being designed should be acceptable in terms of manufacturability and energy performance.

The following algorithm of complex topological optimization is proposed (for example, a synchronous motor with permanent magnets and an internal rotor, the stator design and stator current are assumed to be given).

4 Calculation algorithm

Stage 1.

At this stage of topological optimization, the maximum torque on the shaft is determined.

The rotor design area, which is a cross section (axisymmetric task), is divided into a network of cells - finite elements with the material inherent in each cell (for example, air, steel, types of magnets with different directions of magnetization, which are designated 0, 1, 2, 3, 4 , 5, etc.).

Each topology is encoded with a unique code. Based on the genetic algorithm, the best solution is sought. Two initial samples are randomly selected. Next, several cells are selected for exchange by homogeneous crossing and, as a result, two new samples are created.

New samples inherit the best characteristics of previous samples, then the process repeats. A procedure has been proposed for cleaning materials to obtain homogeneous structures in the design area (for example, without air cavities in magnets), which makes it possible to simplify the manufacture of the prototype.

2 stage.

The level of thermal loads determines the ultimate impact on the material of permanent magnets, the insulation of the wires of the windings, bearing units, as well as other areas, limiting the life of the engine.

To reduce the heat load in the cross section of the prototype of the rotor obtained in stage 1, areas of ventilation ducts with dimensions are introduced to ensure the manufacturability of the manufacture of sheets of the rotor or its casting. For the same reason, the areas of the possible distribution of ventilation ducts must be symmetrical with respect to the radial guides and the minimum distance between the ventilation ducts and other materials is prescribed in the algorithm.

The first iteration sets the minimum possible size of ventilation ducts. The search for the best solution is based on a genetic algorithm, similar to stage 1. The thermal field of the 3D model is calculated (due to the uneven distribution of temperatures) of the electric machine based on the obtained topology, to determine the maximum temperatures in the rotor housing.

3 stage.

The obtained optimal solution of the thermal regime, taking into account the indicated limitations, is directed to the strength calculation, taking into account the maximum rotor speed and mechanical properties of the materials of the designed rotor. If the results are unsatisfactory, the prototype returns to stages 1 and 2.

Stage 4

The prototype of the first three stages is calculated torque on the shaft. If the received torque is not lower than 5% of the maximum torque obtained in stage 1, the optimization result is considered achieved. An unsatisfactory result is excluded from the list of possible solutions, and the calculation is repeated from stage 1.

A review of the scientific and technical literature, Internet resources, and data from manufacturers of electric machines shows that today there is no software package that implements complex topological optimization of the rotor of electric machines.

Due to the growing computational capabilities, direct numerical simulation by the finite element method today allows us to study in detail the various topologies of rotors with various types of materials (permanent magnets, steel, etc.), in order to determine the structure of the rotor that provides maximum torque, permissible thermal conditions and strength characteristics.

The creation of 3D prototypes of synchronous motors with specified parameters based on the results of integrated topological optimization, as well as full-scale models of electric motors and subsequent experimental studies allow us to create new engineering solutions with improved performance for various operating conditions.

The new ones are identification with a unique binary code of each element of the rotor of an electric motor, including various versions of the material used (materials of permanent magnets, steel grades, etc.) with specified properties, which allows you to calculate the necessary topology of an electric machine through an automated process, a procedure has been developed purification of materials to obtain homogeneous structures in the designed area (for example, without air cavities in magnets and steel), which allows to improve manufacturability prototype manufacturing.

The introduction into the topology of the areas of ventilation ducts and their calculation allows us to optimize the thermal conditions of the electric motor.

5 Modeling synchronous machines at the system level

Currently, developers of electric machines use finite element methods to optimize parameters such as torque, efficiency, and induced voltage. Using FEM tools, for example, ANSYS Maxwell, allow developers to make design changes to the geometry, properties of materials, electrical characteristics of power supplies, use mechanical load to optimize the magnetic system of an electric machine. Nevertheless, it is necessary to consider the interaction of an electric machine with the electrical and mechanical components of the entire system, which affects the behavior of the controller. To this end, the magnetic system of the machine design must be tested and optimized throughout the entire drive system, including: · Batteries and DC link capacitors. · Power electronics. · Mechanical components of the drive. · Controllers and software. ANSYS Simplorer provides a state-of-the-art platform for comprehensive drive research. To model the unsteady, dynamic behavior of a machine in a system model, it is necessary to represent it at the level of an equivalent circuit. It should be noted that direct non-stationary magnetic calculation of the finite element model under the control of the simulator also takes place and the solution provides high-precision results, including all the effects of a dynamically changing magnetic field. However, the cost of simulation time when calculating the dynamics of a control system with mechanical loads and a finite element engine model together can be prohibitive. On modern computing equipment, it may take several days to solve such a problem. The method of extracting equivalent circuits generates a descriptive model of a three-phase synchronous machine of the system level based on the results of magnetic finite element analysis, which allows to reduce the simulation time by the above method several times. ECE models cover basic non-linear magnetic saturation effects and groove effects. The resulting model is determined by a set of nonlinear differential equations describing the relationship between the phase currents of the machine and the magnetic flux.

Using the ECE model, a number of typical tasks can be solved:

- Harmonic analysis of voltage / current, moment;
- Dynamics of the control system;
- Short circuit protection.

The ECE technique was implemented for a three-phase electric machine, presented in the non-stationary model ANSYS Maxwell 2D. The method can work with six-phase machines and / or 2D / 3D models.

To extract the parameters of the ECE model, the procedure requires a non-stationary three-phase model of the ANSYS Maxwell 2D engine, as source data. The ANSYS Maxwell model must contain at least three windings and at least four windings for machines with a phase rotor (including an excitation winding). The existing axial symmetry is taken into account. A custom user interface tool is used, where the user enters data on the basic topology of the machine (number of poles, number of slots, maximum current, etc.) The ANSYS Distributed solve (DSO) feature for distributed solution of a parametric task within the local network will help to speed up the process significantly. computers or high-performance cluster. As a result of a series of calculations, a summary table is created of the dependence of the magnetic flux on the stator current, excitation current (for the phase rotor) and the position of the rotor.

6 Designed stand

The principle of operation of the developed genetic algorithm is as follows.

An example of calculating the electromagnetic moment in the Elcut program is presented for an engine with axial magnets.

In accordance with the current documentation for a permanent magnet synchronous motor, the motors are tested according to the following program:

1. Measurement of the resistance of the windings at constant current in a practically cold or heated state;
2. Test at reduced speeds under load;
3. Determination of idle speed characteristics;
4. Determination of losses and efficiency

The equipment that was used to create the test bench is presented:

1. The control cabinet (the element is composite, therefore it will be considered in more detail).
2. The load engine AIR90L2 2.2 kW.
3. The torque sensor M40-150.

M40 digital type torque sensors are designed to measure torque in a wide range of nominal values from 0.1 Nm to 300 kNm and can be used at rotational speeds up to 20,000 rpm.

Structurally, the sensors are made in the form of two separate components: a rotor and a stator, between which there are no brush contacts and bearings. The rotor incorporates an elastic strain element, slightly sensitive to the effect of bending moment, transverse and axial forces.

A microprocessor electronic module located on the rotor converts the strain gauge signals into a digital code, which is transmitted from the rotor to the stator via a non-contact telemetric communication channel. The digital encoded signal has a high noise immunity, provides high measurement accuracy, can be transmitted over considerable distances without distortion and loss of information.

The stator has a composite structure, which provides ease and simplicity of mounting the sensor on the test object.

M40 sensors are equipped with an optoelectronic speed sensor.

4. Compensation coupling MK-0.1.

MK disk couplings are designed to compensate for axial, radial, angular displacements, temperature deformations that occur during installation and during operation of torque sensors. MK couplings have significant axial and angular compliance, with high torsional stiffness. MK couplings are universal and can be used in various fields of mechanical engineering for transmitting torque between rotating shafts having misalignments and distortions of the axes.

5. Test engine valve.

In our case, we will use a synchronous permanent magnet motor with a power of 350 W BM1418 ZXF 350 W 48 V, efficiency> 75%, 40X40X30 cm, 12.0 kg., Rated speed: 2800 rpm.

6. Speed sensor LA8.

A suitable sensor is a prerequisite for a successful solution to the problem. The A5S Series speed sensors are based on the robust differential Hall effect. Their lower limit of 0 Hz allows you to control the machine up to a complete stop. Non-contact, wear-free, maintenance-free and insensitive to scattered magnetic fields and machine vibrations. Other sensors, such as MPU (magnetic induction) or static Hall sensors, unlike the sensors used on the differential Hall effect, are sensitive to interference and therefore not suitable for applications with strict requirements.

Each series of sizes is offered with different lengths and diameters, as well as with different types of connection, but all these sensors have the same functional characteristics.

7. DKC control cabinet rack.

DKC two-frame mounting racks are designed for installing network and telecommunication 19 "equipment inside office and industrial premises. It is most advisable to install racks in special rooms or halls where there are common ventilation and air filtration systems.

7 Optimization Research Results

As a result of optimization, the rotor sector (45 degrees) will look like this (Figure 1), where the filled squares are magnetic material, which will increase the engine torque (for each engine this value is different, in our case it was about 7%), with the same volume of magnets. But there are a few comments that still do not allow us to talk about the final optimization of the rotor:

1. It is necessary to optimize the thermal characteristics of the rotor (at least add an element such as air).

2. It is necessary to conduct a strength calculation of the resulting design of the rotor. These calculations will be carried out in the following articles.

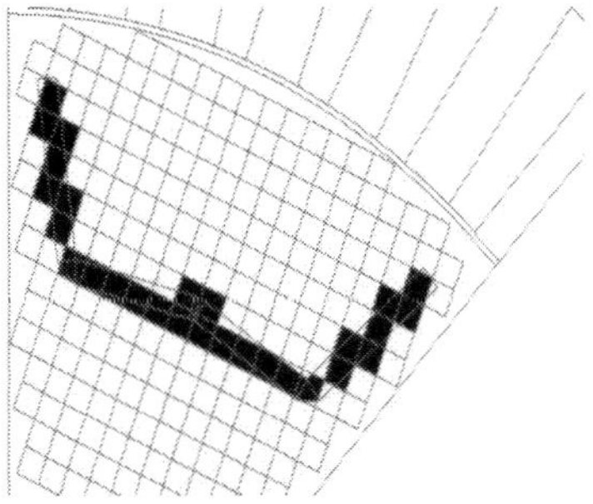

Fig. 1. Sector of the rotor after the optimization.

8 Conclusion

In this paper, the following results were obtained:

1. The analysis of modern scientific, technical, regulatory, methodological literature on the development and creation of synchronous motors with permanent magnets showed that they are currently being used, but there are still opportunities for the development of these machines, from the point of view of improving energy efficiency and optimization.

2. Currently, the industrial production of permanent magnets with high energy performance is developing. Such magnets make it possible to create a large magnetic flux in small volumes, allowing one to significantly increase the ultimate power of electric machines.

An analytical information search indicates that one of the most promising areas is the use of topological optimization for synchronous electric machines, as it is more conceptual and applied. The optimization of the topology of synchronous electric machines allows you to convert its geometric and shaped parameters, for example, the dimensions of the rotor grooves, which will lead to a harmonious distribution of the magnetic flux, which means less heating of the insulation and rotor windings.

Acknowledgements

The reported study was funded by RFBR, project number 19-37-90134.

References

[1] K.A. Bashin, R.A. Torsunov, S.V. Semenov, Methods of topological optimization of structures used in the aerospace industry, Bulletin of the PNRPU. Aerospace engineering **4**, 51 (2017)

[2] B.E. Vasiliev, L.A. Magerramova, Analysis of the possibility of using topological optimization in the design of uncooled turbine blades, Vestnik SGAU, 3-1 (2015)

[3] L.L. Myagkov, S.M. Sivachev, E.E. Strizhov, S.P. Chirsky, Topological optimization of a high-performance diesel piston, Dvigatelestroyeniye, 2, 3–10 (2018)

[4] C. Li, I.Y. Kim, Multi-material topology optimization for automotive design problems, Proceedings of the Institution of Mechanical Engineers, Part D: Journal of Automobile Engineering **232**, 14, 1950–1969 (2018)

[5] K.L. Shenoy, M.S. Kumar, Design topology and electromagnetic field analysis of Permanent Magnet Brushless DC motor for electric scooter application, International Conference on Electrical, Electronics, and Optimization Techniques, ICEEOT 2016, 1541–1545 (2016)

[6] T.I. Petrov, A.R. Safin, I.V. Ivshin, A.N. Tsvetkov, V.Y. Kornilov, The Prospects of Using a Synchronous Machine with Permanent Magnets in the Oil Industry, 14th International

Scientific-Technical Conference on Actual Problems of Electronic Instrument Engineering, APEIE 2018 – Proceedings, article no. 8546157, 336–338 (2018)

[7] D.N. Dyck, D.A. Lowther, Automated design of magnetic devices by optimizing material distribution, IEEE Trans. Magn. **32**, 3, 1188–1193 (1996)

[8] C.H. Im, H.K. Jun, Y.J. Kim, Hybrid genetic algorithm for electromagnetic topology optimization, IEEE Trans. Magn. **39**, 5, 2163–2169 (2003)

[9] S. Wang, D. Youn, H. Moon, J. Kang, Topology optimization of electromagnetic systems considering magnetization direction, IEEE Trans. Magn. **41**, 5, 1808–1811 (2005)

[10] H. Shim, S. Wang, K. Hameyer, Topology optimization of magneto thermal systems considering eddy current as Joule heat, IEEE Trans. Magn. **43**, 4, 1617–1620 (2007)

[11] D.H. Kim, J.K. Sykulski, D.A. Lowther, The implications of the use of composite material in electromagnetic device topology and shape optimization, IEEE Trans. Magn. **45**, 3, 1154–1157 (2009)

[12] J.S. Choi, J. Yoo, Structural topology optimization of magnetic actuators using Genetic algorithms and ON/OFF sensitivity, IEEE Trans. Magn. **45**, 5, 2276–2279 (2009)

[13] T. Labbe, B. Dehez, Convexity-oriented mapping method for the topology optimization of electromagnetic devices composed of iron and coils," IEEE Trans. Magn. **46**, 5, 1177–1185 (2010)

[14] J. Lee, N. Kikuchi, Structural topology optimization of electrical machinery to maximize stiffness with body force distribution, IEEE Trans. Magn. **46**, 10, 3790–3794 (2010)

[15] N. Takahashi, T. Yamada, D. Miyagi, Examination of optimal design of IPM motor using ON/OFF method, IEEE Trans. Magn. **46**, 8, 3149–3152 (2010)

[16] E.I. Gracheva, O.V. Naumov, A.N. Gorlov, Modelling Characteristics of Reliability Low-Voltage Switching Devices on the Basis of Random Checks on the Example of Contactors, Proceedings – 2019 1st International Conference on Control Systems, Mathematical Modelling, Automation and Energy Efficiency, SUMMA 2019, article no. 8947595, 641–643 (2019)

Comparative analysis of trend of change of Azerbaijan's energy sector functioning stability at the current development stage

Valeh Nasibov[1], and *Rana* Alizade[1]

[1]Azerbaijan Scientific – Research and Design–Prospecting Institute of Energetics, 94, Zardabi Str., Baku, Azerbaijan

Abstract. Methodological and practical aspects of the trend of the change of the Azerbaijan's energy sector functioning stability in 2012-2018, based on the methodology presented by the World Energy Congress (WEC), are analyzed and considered in the paper. The basis of the methodology for the assessment of power engineering stability is the use of an energy triangle, where the vertices are energy security, environmental sustainability and energy availability, where a set of indicators with their own weighting coefficients is used to evaluate each of the subsystems. The number of indicators, and, of course, the weighting coefficients are adjusted in time in order to fully take into account all the features of the power engineering. Recalculated The values of the Azerbaijan's energy sector functioning stability during the period under review, taking into account all available indicators, some of which were not taken into account in the calculation of the WEC for the assessment of the individual subsystems, have been recalculated, due to this fact the Azerbaijan's place in the ranking of countries has moved a few positions higher.

1 Introduction

The trends in the change of the Azerbaijan's energy sector stability index, determined by the methodology of the World Energy Congress, is considered in the paper. The dynamics of changes in the energy sector stability for 2010-2018 are analyzed, and potentials for each of the components of sustainable energy sector are determined.

The entire world is currently plunged into a deep social and economic crisis due to the rapid spread of a new COVID-19 coronavirus, which is destroying communities and livelihood sources.

Short-term economic recovery solutions must be in line with the medium-and long-term priorities set out in The United Nations' agenda for the period to 2030 and the Paris agreement, where energy transition is considered the main means of ensuring a secure environment while achieving a healthy, prosperous, fair and sustainable future.

Due to the widespread use of innovative technologies and the reduced cost of technologies for renewable energy sources transitions to sustainable energy sector take place in many countries around the world. Sustainable energy sector provides a balance of three often contradictory goals - energy security, energy availability, and environmental sustainability. One of the tools for assessing the level of energy sustainability is the application of the methodology for determination of the Energy Sustainability Index (ESI) developed by the World Energy Congress [1-8].

The ESI is a global initiative to create a system of indicators that characterize the energy industry, by means of which the components of a country's energy sector are assessed and the energy sustainability of various countries around the world are compared.

2 Method for determining the sustainable performance of energy

Energy sustainability is assessed using the efficiency of the energy sector on the one hand, and the peculiarities of the state's functioning on the other hand. The energy sector functioning should provide the economy with economically feasible energy of appropriate quality, without disturbing the ecological balance. To assess energy sector stability (Energy performance dimensions), three main components are included in the energy triangle of stability: Energy security, Energy equity, Environmental sustainability, and for the assessment of the state's functioning feature (Contextual performance dimensions) the following synthetic factors are used: Macroeconomic Environment, Governance, Stability for Investment and Innovation [1-8].

It should be noted that for the assessment of energy stability, the composition of indicators in each subsystem and their weighting coefficients changed in time, as shown in Table 1.

As can be seen from Table 1, until 2016, each of the Energy Performance subsystems (Energy Security, Energy Equity, Environmental Sustainability) was 25%, at that Country Context was 25% [1-3]. Since 2016, the

Table 1. Energy Trilemma Index structure and weighting.

2013	2016	2019
	ENERGY PERFORMANCE	
Energy Security 25%	**Energy Security 30%**	**Energy Security 30%**
Ratio of total energy production to consumption 3.6%	Diversity of primary energy supply 5%	Diversity of primary energy supply 6%
Diversity of electricity generation 3.6%	Energy consumption in relation to GDP growth 5%	Import dependence 6%
Distribution losses as a percentage of generation 3.6%	Import dependence 5%	Diversity of electricity generation 6%
Five year CAGR of the ratio of TPEC to GDP 3.6%	Diversity of electricity generation 5%	Energy storage 6%
Days of oil and oil product stocks 3.6%	Energy storage 5%	System stability and recovery capacity 6%
For importers – Net fuel imports as a percentage of GDP 3.6%	Preparedness (human factor) 5%	
For exporters – Fuel exports as a percentage of GDP 3.6%		
Energy Equity 25%	**Energy Equity 30%**	**Energy Equity 30%**
Affordability of retail gasoline 12.5 %	Access to electricity 5%	Access to electricity 6%
Affordability and quality of electricity relative to access 12.5%	Access to clean cooking 5%	Access to clean cooking 6%
	Quality of electricity supply 5%	Access to "modern" energy 6%
	Quality of supply in urban vs. rural areas 5%	Electricity prices 3%
	Electricity prices 3.3%	Gasoline and diesel prices 3%
	Gasoline and diesel prices 3.3%	Natural gas prices 3%
	Natural gas prices 3.3%	Affordability of electricity for residents 3%
Environmental Sustainability 25%	**Environmental Sustainability 30%**	**Environmental Sustainability 30%**
Total primary energy intensity 6.2%	Final energy intensity 5%	Final energy intensity 5%
CO2 intensity 6.2%	Efficiency of power generation and T&D 5%	Efficiency of power generation and T&D 4%
Effect of air and water pollution 6.2%	GHD emission trend 5%	Low carbon electricity generation 5%
CO2 grams/kWh from electricity generation 6.2%	Change in forest area 5%	GHG emissions trend 4%
	CO2 intensity 3.3.%	CO2 intensity 2%
	CO2 emission per capita 3.3%	CO2 emissions per capita 1%
	CO2 from electricity generation 3.3%	CH4 emissions per capita 1%
		PM2.5 mean annual exposure 4%
		PM10 mean annual exposure 4%

share of Energy Performance components has increased to 30%, and Country Context has decreased to 10%, respectively [4-8].

It should be noted that information on the indicators presented above is not available for all countries, therefore, for a comparative analysis of the energy sector of different countries, it has to be limited to a relatively narrow range of synthetic indicators (8 indicators in 2013-2015, 10 indicators in 2016-2018 and 9 indicators in 2019). While Energy Performance, according to Table 1, had to be evaluated using 12 indicators in 2013, 20 indicators in 2016, and 21 indicators in 2019.

The normalized value of each indicator, taking into account the degree of importance, is represented in the range from 0 to 10. When assessing energy sustainability, each country receives a balance rating, where the components of the energy sustainability index are as follows: Energy security, Energy equity, Environmental sustainability are estimated by literal notations A, B, C, D, with corresponding values for A - 10.00 – 8.01, B - 8.00 – 5.01, C - 5.00 – 2.51, D - 2.50 – 0.00.

Table 2. Setting Word's margins.

	2013	2014	2015	2016	2017	2018	2019
Energy Security	65.6	63.9	63.2	60.6	60.9	61.9	62.0
Energy Equity	83.5	83.9	84.2	84.1	83.2	82.6	83.0
Environmental Sustainability	63.8	60.1	59.6	58.8	61.1	63.3	63.0
Trilemma Score	71.0	69.3	69.0	67.8	68.4	69.3	69.3
Overall Rank	81 BBC	71 ABD	55 ABC	26 ABA	31 BBA	27 BBA	57 BBB
Key metrics	8	8	8	10	10	10	9

As can be seen from Table 2, according to the calculations of the WEC based on Key metrics for 2013-2019, the Energy Security value decreased by 5% from 65.6 to 62.0, the Energy Equity value almost did not change - 83 in 2019 against 83.5 in 2013, as well during this period the Environmental Sustainability value decreased slightly from 63.8 to 63. Such values of Azerbaijan's energy sector stability components are mainly related to the fact that not all indicators are used in the calculations to assess the energy components. Trends of change of some indicators for assessing Azerbaijan's energy sector sustainability are shown in Figure 1 [9-12].

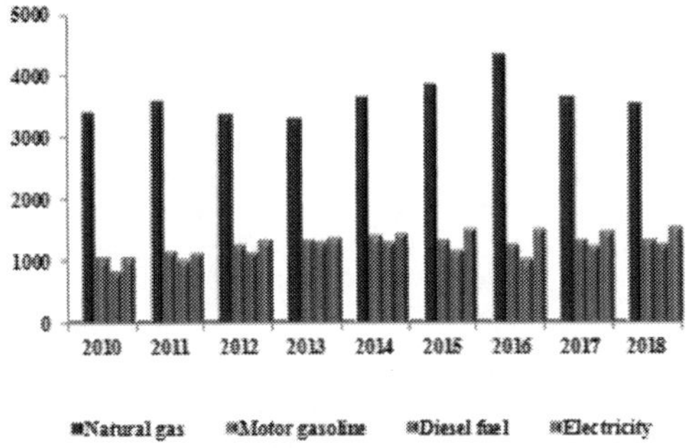

Fig. 1. Diversity of primary energy supply, thsd TOE.

As can be seen from Figure 1, Diversity of primary energy supply calculated according to Herfindahl index is about 0.21, that corresponds to high level of diversification (A) [13,14].

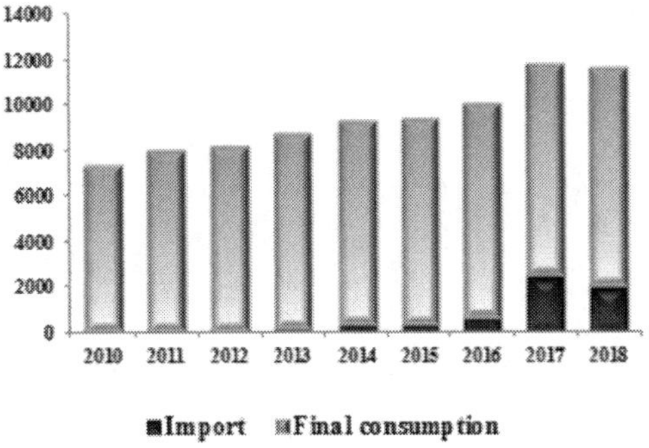

Fig. 2. Import dependence, thsd TOE.

The next indicator for assessing energy security is Import dependence. As can be seen from Figure 2, in recent years Azerbaijan has been importing a certain amount of energy resources. Since Azerbaijan is an exporter of energy resources, the purpose of energy import (natural gas) is not domestic consumption of the country, but the fulfillment of international obligations for the natural gas supply. For example, in 2019 the natural gas production in Azerbaijan was 19207. 1 mln. m3, of which 11076.1 mln. m3 was used for internal consumption, and exports, taking into account 1798.2 mln. m3 of imported natural gas, was 9911.8 mln. m3. Thus, this parameter for Azerbaijan corresponds to a very high level (A) [10-12].

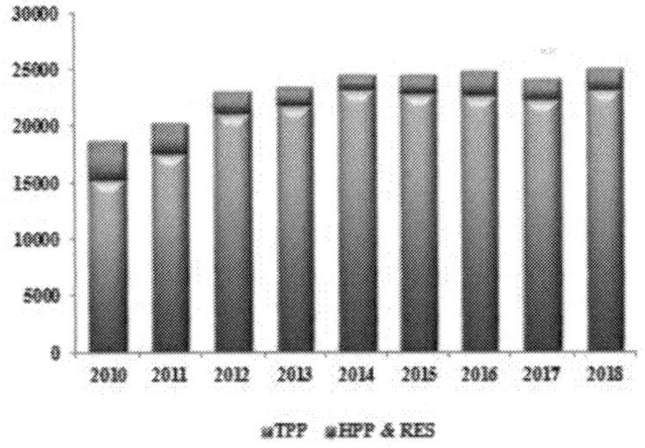

Fig. 3. Diversity of electricity generation, mln. kWh.

It should be noted that the share of HPPs and renewable energy sources in electric energy production varies in a quite large range (2010 - 18.4%, 2014 - 5.3%), as can be seen from Figure 3, at that in 2018 their share was 7.5%. Despite the low diversification of electric energy generation, different approach is needed to this parameter, since the fuel supply of thermal power plants, where the main share of electric energy (more than 90%) is produced, is provided at the expense of internal sources. The same approach to the assessment is required by the Energy storage indicator.

https://doi.org/10.1051/e3sconf/202017801050

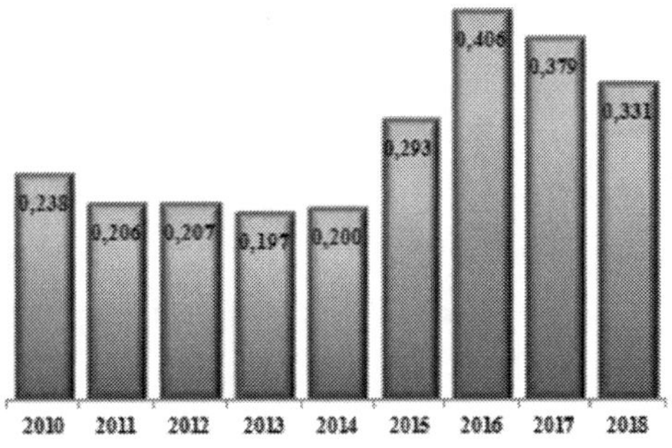

Fig. 4. Final energy intensity, TOE /1000 $.

As can be seen from Figure 4, starting from 2015 the Final energy intensity began to increase, reaching 0.406 TOE /1000 $ in 2016, which is related to the national currency devaluation. With the growth of Azerbaijan's GDP in the following years - 2017-2018, this indicator tends to decrease.

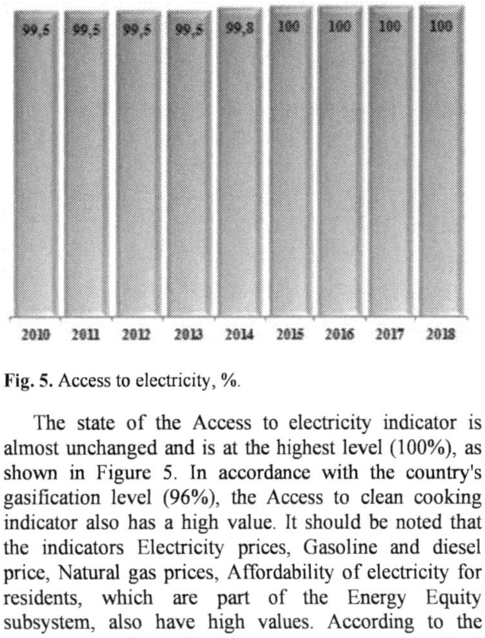

Fig. 5. Access to electricity, %.

The state of the Access to electricity indicator is almost unchanged and is at the highest level (100%), as shown in Figure 5. In accordance with the country's gasification level (96%), the Access to clean cooking indicator also has a high value. It should be noted that the indicators Electricity prices, Gasoline and diesel price, Natural gas prices, Affordability of electricity for residents, which are part of the Energy Equity subsystem, also have high values. According to the calculations of the World Economic Forum, in 2019 Azerbaijan took the second place in the world on the Energy Equity subsystem.

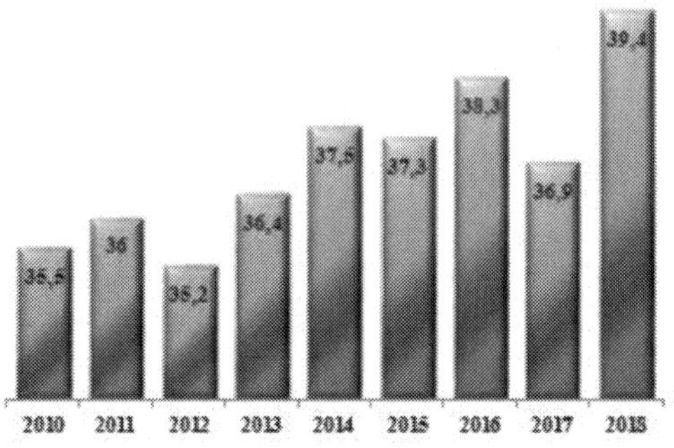

Fig. 6. Efficiency of power generation, %.

The Efficiency of power generation varies mainly depending on the volume of power production at Azerbaijan TPP, as shown in Figure 6, as it is the most powerful and less efficient power plant in the energy system. All other thermal power plants are modern with high efficiency (46-56 %) [10,15,16].

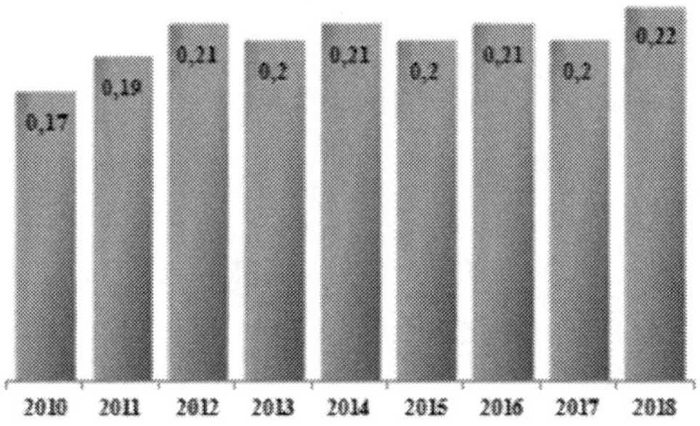

Fig. 7. CO_2 intensity, tonnes per 1000 $ GDP.

As shown in Figure 7, during 2010-2018 period the CO_2 intensity increased by about 30%, although this indicator is still about 18% less than the world average indicator (0.26). Another indicator for the assessment of the Environmental Sustainability subsystem is CO_2 emissions per capita, which is quite low in Azerbaijan. According to the European Commission, in 2018 the country released 34.7 mln. tons of CO_2 into the atmosphere, that is, only 3.5 tons per capita, which is below the established world norm of 4.9 tons [9,10,12].

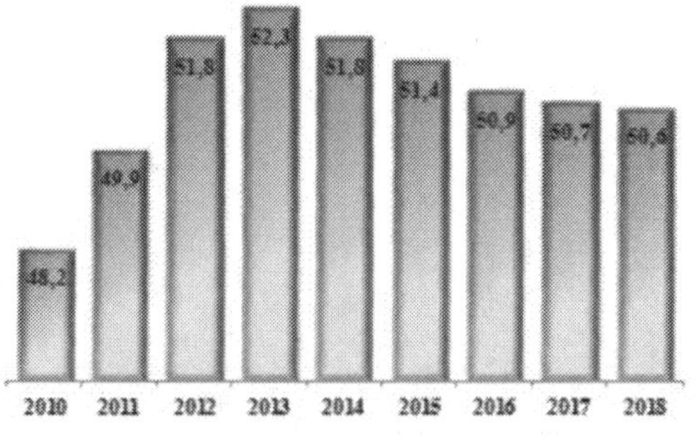

Fig. 8. GHG emissions trend, mln. tons.

The GHG emissions trend indicator for recent years (2013-2018) has a slight but steady decrease tendency. As shown in Figure 8, during the period under review this indicator decreased by about 4%.

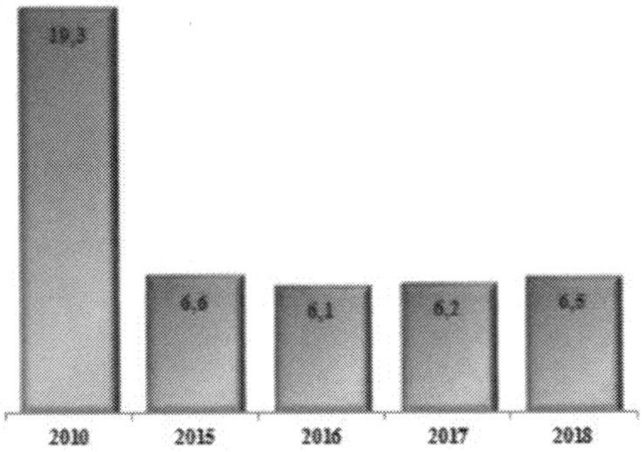

Fig. 9. PM mean annual exposure, thsd. ton.

The PM10 mean annual exposure indicator in Figure 9 has not practically changed in recent years, being at a relatively low level. The Energy Trilemma Index of Azerbaijan for 2019, constructed by the WEC with 9 key metrics and by us, taking into account 19 indicators, is presented in Figure 10 and 11. A comparative analysis of Azerbaijan's energy stability triangles shows that when all available indicators are fully taken into account, the subsystem values of energy sector have a significant difference. Energy Security increases to 78 against 62, Energy equity increases to 92 against 83 and Environmental sustainability increases to 64 against 63. At the same time, the Trilemma Score is equal to 78 against 69.3. With such indicators Overall Rank Azerbaijan could be at 40-45 place of rating [8,9].

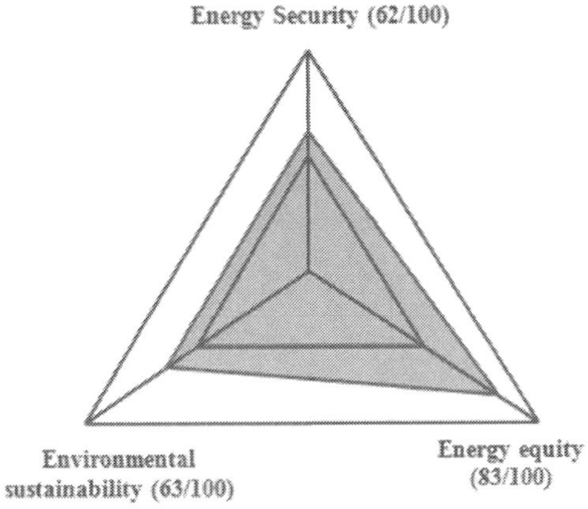

Fig. 10. Energy Trilemma Index of Azerbaijan (WEC).

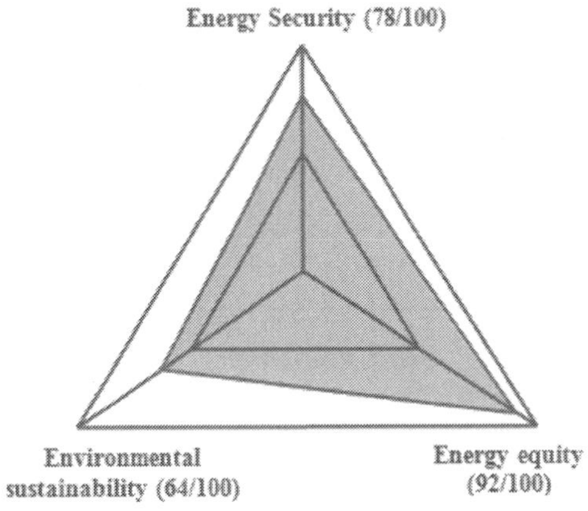

Fig. 11. Energy Trilemma Index of Azerbaijan.

3 Conclusions

1. The index of energy sustainability is calculated by the methods of World Energy Council on the basis of energy triangle, where the vertexes are Energy Security, Energy Equity and Environmental Sustainability.
2. With using all information about energy performance, the energy sustainability index rating of Azerbaijan will be at 40-45 places as against 57, by calculation of World Energy Council.
3. There are significant potentials for the improvement of status of Energy Security (78) and Environmental sustainability (64) subsystems in Energy Performance of Azerbaijan.

References

[1] World Energy Trilemma, Energy Sustainability Index [Electronic resource]. Available at: https://www.worldenergy.org/assets/images/imported/2013/09/2013-Energy-Sustainability-Index-VOL-2.pdf (2013)

[2] Energy Trilemma Index, Benchmarking the sustainability of national energy systems [Electronic resource]. Available at: http://www.wec-france.org/DocumentsPDF/Etudes_CME/Energy-Trilemma-Index-2014-ENG.pdf (2014)

[3] Energy Trilemma Index, Benchmarking the sustainability of national energy systems [Electronic resource]. Available at: http://www.wec-france.org/DocumentsPDF/Etudes_CME/2015_Energy_Trilemma_Index.pdf (2015)

[4] World Energy Trilemma Index [Electronic resource]. Available at: https://www.worldenergy.org/assets/downloads/Full-report_Energy-Trilemma-Index-2016.pdf (2016)

[5] The Global Energy Architecture Performance Index 2016: Methodological Addendum [Electronic resource]. Available at: http://www3.weforum.org/docs/WEF_Energy_Architecture_Performance_Index_2016.pdf (2016)

[6] World Energy Trilemma Index 2017 [Electronic resource]. Available at: https://www.worldenergy.org/assets/downloads/Energy-Trilemma-Index-2017-Report.pdf (2017)

[7] World Energy Trilemma Index 2018 [Electronic resource]. Available at: https://www.worldenergy.org/assets/downloads/World-Energy-Trilemma-Index-2018.pdf (2018)

[8] World Energy Trilemma Index 2019 [Electronic resource]. Available at: https://www.worldenergy.org/assets/downloads/WETrilemma_2019_Full_Report_v4_pages.pdf (2019)

[9] N.A. Yusifbeyli, V.Kh. Nasibov, Electroenergetics, electrotechnics, electromechanics + control **4**, 3, 13–23 (2013)

[10] Materials of State Statistical Committee of Azerbaijan Republic [Electronic resource]. Available at: https://www.stat.gov.az

[11] BP Statistical Review of World Energy 2019 [Electronic resource]. Available at: https://www.bp.com/content/dam/bp/business-sites/en/global/corporate/pdfs/energy-economics/statistical-review/bp-stats-review-2019-full-report.pdf (2019)

[12] World Energy Outlook 2018 [Electronic resource]. Available at: https://www.iea.org/reports/world-energy-outlook-2018 (2018)

[13] V.Kh. Nasibov, TECIS **51**, 30, 405–409 (2018)

[14] V.Kh. Nasibov, TECIS **51**, 30, 575–579 (2018)

[15] Clean energy for all Europeans package [Electronic resource]. Available at: https://ec.europa.eu/energy/topics/energy-strategy/clean-energy-all-europeans_en

[16] Renewables 2018 Global Status Report [Electronic resource]. Available at: https://www.ren21.net/wp-content/uploads/2019/08/Full-Report-2018.pdf (2018)

Reliability modeling of high-voltage power lines in a sharply continental climate

Shamil Rashidovich Khasanov[1,*], *Elena Ivanovna* Gracheva[1], *Muhayo Islomovna* Toshkhodzhaeva[2], *Shakhboz Tolibjonovich* Dadabaev[2], and *Dilafruz Saidulloevna* Mirkhalikova[2]

[1]Department Power supply of industrial enterprises, Kazan State Power Engineering University, Kazan, Russia
[2]Department of Power Supply and Automation, Khujand Polytechnic Institute of the Tajik Technical University Named after Academician M.S. Osimi, Khujand 735700, Tajikistan

Abstract. One of the important elements of the high-voltage transmission line is 110 kV, the damage of which leads to under-supply of electricity, i.e. reducing the reliability of the lines. Damage to the reinforcement from the applied load occurs due to material fatigue. The main causes of damage to the reinforcement are ambient temperatures and wind forces on this element. Periodic kinks of the wire are observed at the places of installation of the connecting and supporting clamps, vibration dampers, and cyclic lateral forces arise. The simultaneous impact of the above efforts leads to fatigue damage to the suspension systems. From the cyclic load, the nodes of rigid structures are destroyed, which bear the maximum load.

1 Introduction

Reliable and uninterrupted power supply to consumers mainly depends on the normal functioning of electric networks. The reliable operation of high-voltage power lines depends on not only operating conditions, technological conditions, operating modes of electric power networks, but also on the climatic conditions of the region. Abnormal cooling in the winter season, a sharp increase in temperature in the summer, increased wind, intense solar radiation, the presence of dust in the form of suspended particles; undoubtedly affect the operation of high-voltage power lines. As studies of high-voltage power lines 110 kV failures show, the main reason for the decrease in the reliability of high-voltage power lines 110, 220 kV is the influence of external factors, such as rain, snow, wind [1, 2 and 3]. The electric network refers to large systems that are characterized by a multiplicity of elements and the connections between them. To study large electric power systems, a systematic approach is used. The object of study is considered as a set of elements; therefore, their relationship is defined as an integral property of this set [4].

2 Methodology

Sag arrows and voltage across the wire vary with current load and ambient temperature. With increasing ambient temperature, the sag increases, therefore, the voltage on the wire decreases. With decreasing temperature, the opposite phenomenon occurs. When using the wires in the calculations, the values of the following temperatures are taken into account [5, 6]:

1. The highest temperature at which the wires extends the most and the sag reach the highest values.
2. The lowest temperature value at which the wires are reduced to the greatest degree and the voltage in them reaches the highest values.
3. The average annual temperature that meets the operating conditions.

To analyze the impact of natural and climatic loads on the 110 kV high voltage transmission line, it is necessary to determine the initial parameters of the line, which reflect the state of the high voltage cable under normal operating conditions. For conditions of sharply continental climate, the state of high-voltage transmission lines is accepted in the absence of wind and the temperature of the conductor material +25 °C.

When determining the initial mechanical loads, the total tension *Tt* is calculated, which consists of the vertical component *Vt* and the horizontal component Ht. There is a need to calculate the sag for all selected variation *ft* wire according to the formula 1. The mechanical stress in the conductors σ_t is determined from the equation of state [7, 8]:

$$f_t = \frac{\gamma_1 \cdot (l_{cal})^2}{8 \cdot \sigma_t} \tag{1}$$

$$\sigma_t - \frac{\gamma_1^2 \cdot (l_{cal})^2 \cdot E}{24(\sigma_t)^2} = \sigma_{dop} - \frac{\gamma_1^2 \cdot (l_{cal})^2 \cdot E}{24(\sigma_{dop})^2} - \alpha_{l.r} \cdot E \cdot (t - t_n) \tag{2}$$

* Corresponding author: shukrona14_01_2011@mail.ru

where t is the ambient temperature, *°C; t* = - 20 *°C* minimum temperature.

Components horizontal full tension of the conductor, H_t [7, 9 and 10]:

$$H_t = \frac{M_0 \cdot 10^{-3} \cdot (l_{cal})^2}{8 \cdot f_t} \tag{3}$$

Components vertical full tension of the conductor, V_t:

$$V_t = M_0 \cdot 10^{-3} \cdot l_{cal} \tag{4}$$

Full wire tension, T_t:

$$T_t = \sqrt{(H_t)^2 + (V_t)^2} \tag{5}$$

3 Modeling and simulation result

To calculate and build the dependencies of the sag arrows and tensions on the climatic loads, a software module was compiled in the Matlab / Simulink program [11-14]. The tension *Hυ, Vυ, Tυ* that occur in the wires and changes in the sag arrows, and the voltage on the wire when the ambient temperature changes t = 0, 10, 20, 25, 30, 40, 45 °C at a constant wind speed ϑ = 10 m/s are shown in figures 1 and 2 [7, 15 and 16].

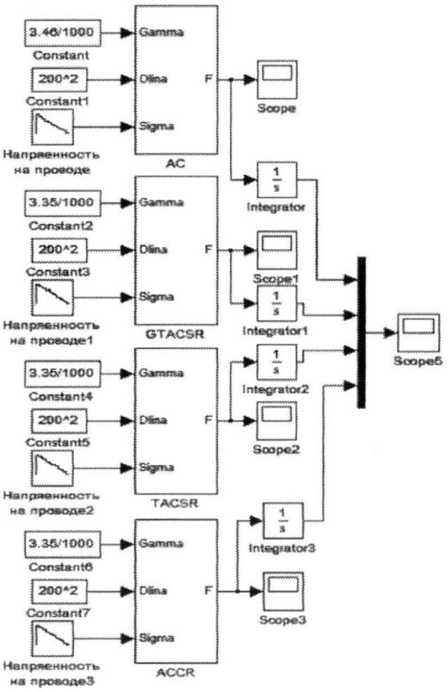

Fig. 1. Computer model for determining the sag of wires depending on temperature changes.

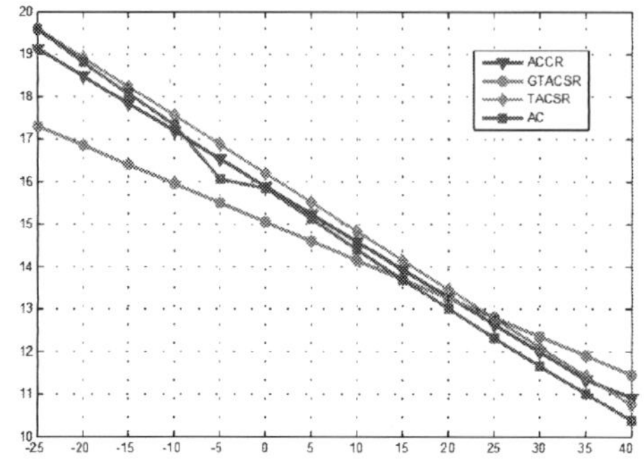

Fig. 2. The voltage on the wire depending from ambient temperature.

As can be seen from Figure 2, the voltage of all wires decreases linearly with increasing temperature. In the temperature range -25 ... +40 °C, the average decrease is 1.58 times. A characteristic of all wires is a temperature of +15 °C, at which the curves cross the permissible voltage. The tension on the GTACSR wire at a temperature of -25 °C is 10% less than that of the AC wire, and at a maximum temperature of 40 °C 10.3% more than that of the AC wire. A distinctive feature of GTACSR wires is a lower coefficient of temperature dependence, which is due to the presence of a gap between the conductors and the composite core [17-19].

The tension on the TACSR, ACCR, AC wires at the minimum temperature has the same value, but at maximum temperatures it is 10.3% higher than on the AC wires. This is due to the fact that the conductors are made of heat-resistant aluminum of the TAL type with the addition of zirconium, which makes it possible to increase the mechanical strength of the wire. TACSR and ACCR wires are 1.4 times stronger than permissible at minimum temperatures, and at maximum temperatures 1.3 times less than acceptable values. A distinctive feature of high-temperature wires is that at maximum temperatures they have more tension than traditional wires, therefore, the sag will be less than that of an AC wire. The change in the sag of the wires depending on the change in ambient temperature is shown in Figure 3

As can be seen from Figure 3, at the maximum temperature for all types of wires compared with the AC, the sag is greater, at maximum temperatures less than the wires of the AC. It should be noted that at maximum temperatures the sag of all wire grades is significantly less than the critical value. The dependence of the tension of the wires depending on the ambient temperature is shown in figure 4.

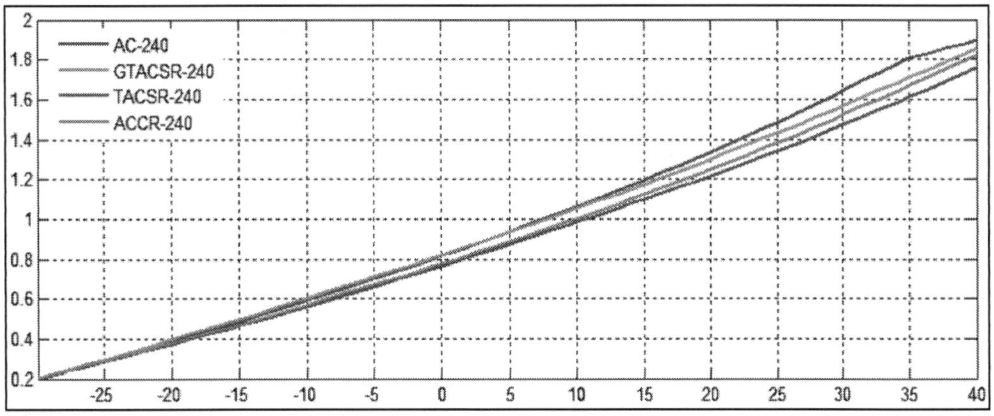

Fig. 3. Changing of wire sag depending on the ambient temperature.

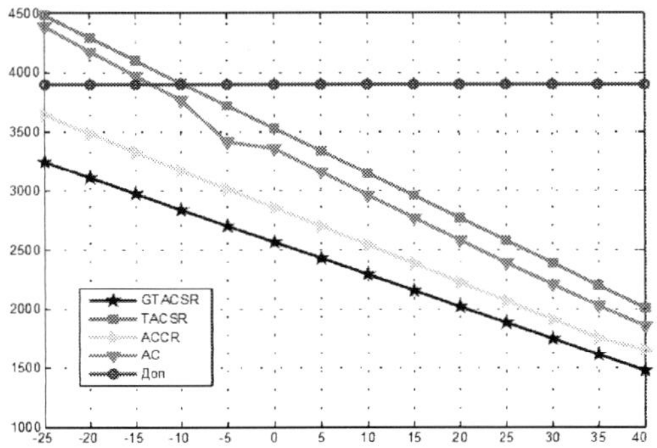

Fig. 4. Dependence of wire tension on ambient temperature.

As can be seen from Figure 4, the dependence of the full tension on the temperature on the wires is mainly due to the mechanical properties of the conductive materials. In the temperature range -25 .... + 10 °C, for TACSR and AC wires, the tension exceeds the permissible value, and for GTACSR and ACCR wires in the entire temperature range the tension is less than the allowable value.

The tension H_υ, V_υ, T_υ that occur in the conductors without ice, with a changing wind speed υ = 0, 10, 20, 30 m/s, at a certain ambient temperature t_g = + 10 °C, are calculated according to the equation 12. The value of the coefficient of non-uniformity depending on the magnitude of the wind speed α, the standard high-speed wind pressure at a height of 10 m from the earth's surface, taking into account the qn downwind region and the coefficient of increase of the high-speed wind pressure in height, are adopted according to [6, 7]. The region under consideration corresponds to the III district downwind, the repeat period is 10 years.

Sag arrows f_ϑ are determined from the expression

$$f_\vartheta = \frac{\gamma_4 \cdot (l_{cal})}{8 \cdot \sigma_\vartheta} \qquad (6)$$

The specific load on the wire γ_4 is calculated by the formula:

$$\gamma_4 = \sqrt{\gamma_1^2 + \left(\frac{\alpha \cdot C_x \cdot \vartheta^2 \cdot d_n}{1.6 \cdot 10^{-3} \cdot F}\right)^2} \qquad (7)$$

The voltage in the wire is calculated by the formula:

$$\sigma_\upsilon - \frac{\gamma_6^2 \times (l_{cal})^2 \times E}{24 \times (\sigma_\upsilon)^2} = \sigma_{dop} - \frac{\gamma_1^2 \times (l_{cal})^2 \times E}{24 \times (\sigma_{dop})^2} - \alpha_{l.r.} \times E \times (t_g - t_n) \qquad (8)$$

The horizontal components H_υ of the full tension of the wire are found from the expression:

$$H_\upsilon = \frac{\sqrt{(M_0 \cdot 10^{-3})^2 + (P_v)^2} \cdot (l_{cal})^2}{8 \cdot f_\upsilon} \qquad (9)$$

$$P_v = 6{,}25 \cdot 10^{-5} \cdot \alpha \cdot C_x \cdot \upsilon^2 \cdot d_n. \qquad (10)$$

The vertical components V_υ of the full tensile wire are calculated by the formula (3.21):

$$V_\upsilon = \sqrt{(M_0 \cdot 10^{-3} \cdot l_{cal})^2 + (P_v \cdot l_{cal})^2} \qquad (11)$$

Full conductor tension, T_υ:

$$T_\upsilon = \sqrt{(H_\upsilon)^2 + (V_\upsilon)^2} \qquad (12)$$

Graphs of changes in tension and sag arrows when changing wind speed at a certain air temperature t_g = 10 °C are presented in figure 5.

https://doi.org/10.1051/e3sconf/202017801051

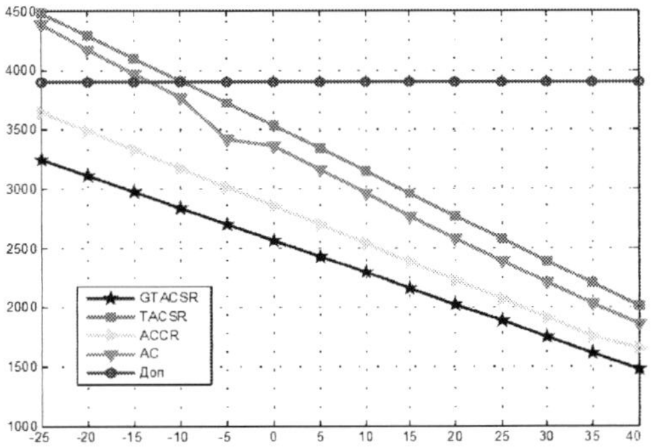

Fig. 5. Change of a arrow of a sag of wires depending from changes in wind speed.

The values of the changes in the sag depending on the wind speed are the same for GTACSR, TACSR and AC wires, but for the ACCR wire at the maximum wind speed, the change in the sag is 11% less than for AC wires. At the maximum value of wind speed, the value of the change in the arrow of the sag is less than permissible. This is due to the fact that in sharply continental climates there is practically no ice formation on the wires. The dependence of the tension of the wires depending on the wind speed is presented in Figure 6.

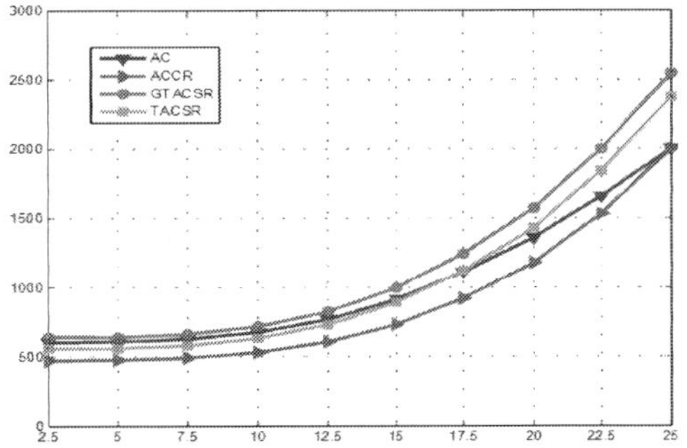

Fig. 6. Dependence of the tension of wires depending on wind speed.

From the above figures 5 and 6 and the calculations, we can conclude that the effect of wind pressure on the wires without ice is not a danger, therefore, even when the wind speed reaches 25 m/s, the tension on the wires does not exceed critical values, the overall dimensions of the lines are acceptable limits.

4 Conclusion

From the calculated data, it follows that the TACSR brand wire at maximum temperatures has the greatest value; therefore, the sag of such wires at maximum temperatures will be the smallest. Therefore, from the point of view of the influence of temperature on ensuring the high-voltage power transmission line reliability, the TACSR wire is optimal. The analysis of the results of the expected mechanical loads under the influence of natural factors showed that all the considered options for reconstruction of the 110 kV high-voltage power lines satisfy the specified degrees of reliability.

References

[1] E.I. Gracheva, O.V. Naumov, Vliyanie rezhima raboty na veroyatnostnye kharakteristiki elektricheskikh ustroystv (Operating Mode Influence on Probability Characteristics of Electric Devices), Journal of engineering and applied sciences **11**, 1, 2934–2938 (2016)

[2] E.I. Gracheva, O.V. Naumov, Estimation of Power Losses in Electric Devices of the Electrotechnical Complex, International Conference on Industrial Engineering, Applications and Manufacturing (ICIEAM) **6** (2019)

[3] D.S. Tolibjonovich, T.M. Islomovna, M.D. Saidulloevna, Modeling of starting transition processes of asynchronous motors with reduced voltage of the supply network, European Journal of Electrical Engineering **22**, 1, 23–28. DOI: 10.18280/ejee.220103 (2020)

[4] M.I. Toshkhodzhaeva, Sravnitel'nyy analiz mekhanicheskikh svoystv traditsionnykh i vysokotemperaturnykh provodov vlep – 110 kV (Comparative analysis of the mechanical properties of traditional and high-temperature vlep wires – 110 kV), Izvestiya Tul'skogo gosudarstvennogo universiteta, Tekhnicheskie nauki **3**, 169–175 (2017)

[5] M.I. Toshkhodzhaeva, A.A. Khodzhiev, Osobennosti diagnostirovaniya vlep-110 kv v usloviyakh rezko kontinental'nogo klimata, (Features of diagnosing high-voltage power lines of 110 kV in a sharply continental climate), Izvestiya Tul'skogo gosudarstvennogo universiteta, Tekhnicheskie nauki **2**, 364–369 (2019)

[6] O.S. Rakhimov, A.A. Khodzhiev, M.I. Toshkhodzhaeva, Povyshenie nadezhnosti vlep-110 kv na stadii proektirovaniya i ekspluatatsii (Improving the reliability of 110 kV high-voltage power lines at the design and operation stages), Elektrooborudovanie: ekspluatatsiya i remont **3**, 55–57 (2017)

[7] M. Amin, J. Stringer, The electric power grid: Today and tomorrow, MRS bulletin. **33**, 04, 399–407 (2008)

[8] W.C. Athas, et al., Low–power digital systems based on adiabatic–switching principles, IEEE Transactions on Very Large Scale Integration (VLSI) Systems **2**, 4, 398–407 (1994)

[9] P.P. Barker, R.W. De Mello, Determining the impact of distributed generation on power systems. I. Radial distribution systems, Power Engineering Society Summer Meeting, IEEE **3**, 1645–1656 (2000)

[10] L. Benini, A. Bogliolo, G. De Micheli, A survey of design techniques for system-level dynamic power management, IEEE transactions on very

large scale integration (VLSI) systems **8**, 3, 299–316 (2000)

[11] P. Gregorio, M. Ahmadi, M. Buehler, Design, control, and energetics of an electrically actuated legged robot, IEEE Transactions on Systems, Man, and Cybernetics, Part B (Cybernetics) **27**, 4, 626–634 (1997)

[12] J. Henkel, A low power hardware/software partitioning approach for core–based embedded systems, Proceedings of the 36th annual ACM/IEEE Design Automation Conference, Berlin: ACM, 122–127 (1999)

[13] F. Kiessling, et al., *Overhead power lines: planning, design, construction* (Springer, 321, 2014)

[14] E. Lakervi, E.J. Holmes, Electricity distribution network design, IET, 21, 130–142 (1995)

[15] W. Li, et al., *Reliability assessment of electric power systems using Monte Carlo methods* (Springer Science & Business Media, 540, 2013)

[16] R. Matsuoka, et al., Assessment of basic contamination withstand voltage characteristics of polymer insulators, IEEE transactions on power delivery **11**, 4, 1895–1900 (1996)

[17] D.C. Montgomery, E.A. Peck, G.G. *Vining, Introduction to linear regression analysis* (California, John Wiley & Sons, 821, 130, 2012)

[18] T.A. Reddy, *Applied data analysis and modeling for energy engineers and scientists* (Springer Science & Business Media, 320, 2011)

[19] J. Schlichting, Molybdenum disilicide as a component of modern high-temperature composites, High Temperatures – High Pressures **10**, 3, 241–269 (1978)

https://doi.org/10.1051/e3sconf/202017801052

Modelling of emergency modes with FACTS devices installed

Alexey Udaratin[1,*], *Kirill* Loginov[1], *Alexandr* Nemirovskiy[1], *Natalia* Rozhentsova[2] and *Elena* Gracheva[2]

[1]Vologda State University, Lenina str. 15, Vologda, Russia
[2]Kazan State Energy University, Krasnoselskaya str. 51, Kazan, Russia

Abstract. This paper considers installation of FACTS devices in a 500 kV line of substation (SS) Belozerskaya - SS Vologodskaya. Three FACTS devices in emergency mode were modelled: a static reactive power compensator (STATCOM), a static synchronous series compensator, and a unified power flow control system. The parameters of MATLAB Simulink blocks were shown. The obtained results are summarized in Table, which substantiates the choice of the best device for the Belozerskaya - Vologodskaya line in emergency mode.

1 Introduction

Electricity is an integral part of our modern world. It helps us to use a variety of electrical devices that make the life more comfortable.

However, the development of electrical power grids inevitably leads to two problems. Firstly, the resulting complex-closed and multi-level system cannot function without rigid control mechanisms (auto-control). Secondly, further development of the system requires involvement of new lands, which, given the introduction of private ownership of land, turns out to be very expensive. Under the current conditions, the Smart Grid ideology enabling controllability and increased transmission ability of grids is very promising. A number of experts claim that increasing the controllability of grids is a low-budget alternative to building new power lines. Actually, recently the transition to controlled grids has become an officially announced policy of the Russian network engineers. Active electro-technical network equipment (FACTS) is capable of flexibly change the characteristics of electric power transmission or conversion in order to optimize grid modes at once according to several criteria: throughput, technological loss level, stability, redistribution of power flows, power quality, etc. [1]

FACTS devices enable not only successful compensation of reactive power in the power grid, but also adequate response to abnormal conditions in it. Thanks to these parameters, flexible AC transmission systems are one of the best devices to be installed in line.

2 Parameters of power system units

The 500 kV power system of substation (SS) Belozerskaya - SS Vologodskaya was modelled in the MATLAB Simulink program. It allows one to use various blocks to simulate the process of electric power transmission.

A three-phase short circuit is simulated using the Three-phase Fault block. It implements an error (short circuit) between any phase and ground. When the external switching time mode is selected, the logic signal Simulink is used to control the failure operation. The data are shown in Table 1.

Table 1. Three-phase failure characteristics.

Parameter	Value
Switching time	[0.1 0.3 0.5] c
Failure resistance	200 Ohm
Ground resistance	0.001 Ohm
Shock absorber resistance	1 MOhm

The reactive power compensation is modelled using 3 blocks: Static Synchronous Compensator, Static Synchronous Series Compensator and United Power Flow Controller (Universal Energy Flow Control).

In MATLAB Simulink, these models are presented in two forms: Phasor model (a regular model that adjusts the main parameters of device) and Detailed mode (a model that allows one to configure devices for power compensation in more detail).

Static Synchronous Compensator (STATCOM) is one of the key FACTS devices. Based on a voltage converter, STATCOM regulates system voltage by absorbing or generating reactive power. Unlike a thyristor-based static variable compensator (SVC), the output current of the STATCOM (inductive or capacitive) can be controlled regardless of AC system voltage [2].

STATCOM parameters are shown in Table 2.

* Corresponding author: alexu79@mail.ru

Table 2. STATCOM parameters.

Parameter	Value
Rated voltage and rated frequency of the system	500 kV 50 Hz
Rated power of compensator	100 MVa
Rated resistance of compensator	[0.22/30, 0.22] r.u.
The initial value and phase of line current	[0, 0]
DC circuit rated voltage	40 kV
Equivalent capacitance of DC circuit	750 mF
Maximum change rate of reference voltage	10 r.u./s
Difference between STATCOM reference voltage and bus voltage	0.03
Gain of boost voltage regulator	[5 1000] r.u.
Gain of DC voltage regulator	[0.1e-3 20e-3] r.u.
Current gain	[0.3 10 0.22] r.u.

Static synchronous series compensator (SSSC), one of the key FACTS devices, consists of a voltage converter and a transformer connected in series with electric power transmission line. SSSC supplies a variable voltage quadratically to the line current, thereby simulating inductive or capacitive reactance. This emulated alternating reactance in series with the line may further affect the transmitted electric power. SSSC is used to suppress power fluctuations in electrical grid after a three-phase fault. [3]

The characteristics of SSSC are shown in Table 3.

Table 3. The characteristics of SSSC.

Parameter	Value
Rated voltage and rated frequency of the system	500 kV 50 Hz
Rated power of series compensator	[100, 0.1] MVa
Rated resistance of compensator	[0.22/30, 0.22] r.u.
The initial value and phase of line current	[0, 0]
DC circuit rated voltage	40 kV
Equivalent capacitance of DC circuit	750 mF
Maximum change rate of reference voltage	3 r.u./s
Difference between STATCOM reference voltage and bus voltage	0.03
Gain of boost voltage regulator	[0.03 1.5]/8 r.u.
Gain of DC voltage regulator	[0.1e-3 20e-3] r.u.

Unified power flow control (UPFC) systems have advantages in terms of static and dynamic operation of the power system. It also poses new challenges in the field of power electronics and power system design. The basic structure of UPFC consists of two voltage source inverters (VSI); where one converter is connected in parallel with the transmission line (STATCOM), and the other is connected in series with the transmission line (SSPK). [4]

UPFC parameters are shown in Table 4.

Table 4. UPFC parameters.

Parameter	Value
Rated voltage and rated frequency of the system	500 kV 50 Hz
Rated power of compensator	100 MVa
Rated resistance of static compensator	[0.22/30, 0.22] r.u.
The initial value and phase of line current	[0, 0]
Rated power of series compensator	[100, 0.1] MVa
Rated resistance of series compensator	[0.22/30, 0.22] r.u.
The initial value and phase of line current	[0, 0]
DC circuit rated voltage	40 kV
Equivalent capacitance of DC circuit	750 mF
Maximum change rate of STATCOM reference voltage	0.1 r.u./s
Difference between STATCOM reference voltage and bus voltage	0.03
Gain of boost voltage regulator	[5 1000] r.u.
Gain of DC voltage regulator	[0.1e-3 20e-3] r.u.
Current gain	[0.5 25]/3 r.u.
Gain of boost voltage regulator of series compensator	[0.025 1.5] r.u.

3 FACTS in emergency mode

3.1 Modelling of STATCOM in emergency mode

We simulate a three-phase short circuit using Three-Phase Fault. This is a three-phase fault unit that allows one to simulate various faults. The short circuit time is set so that it occurs at t=0.1-0.3 and t=0.5-1 seconds and disconnects the line for 0.15 seconds. Consider the graphs of three periods: the first short circuit at t=0.1-0.3 s, system recovery at t=0.2-0.4 s and the second short circuit at t=0.5-1 s.

Fig. 1 presents the power system, Fig. 2 presents voltage, and Fig.3 presents the active and reactive power during short circuit.

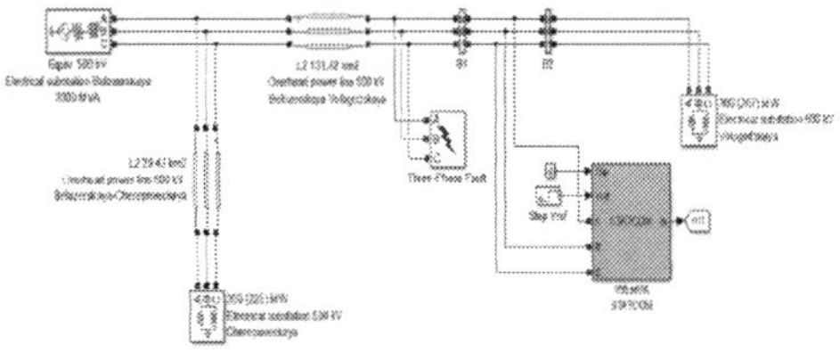

Fig. 1. Model of 500 kV SS Belozerskaya - SS Vologodskaya power system with STATCOM device with Three-Phase Fault block.

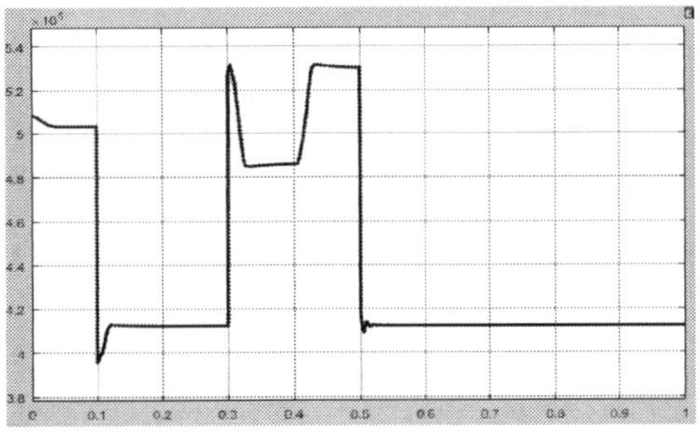

Fig. 2. Voltage on bus B1 with STATCOM device in emergency mode.

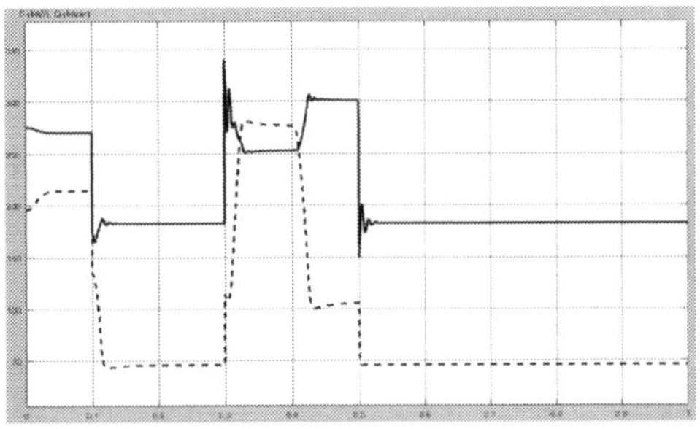

Fig. 3. Active and reactive power of line on B1 bus in the STATCOM emergency mode.

As it can be seen from graphs, after a short circuit at t=0.1-0.3 s and t=0.5-1 s, the line voltage decreases. When STATCOM is operating in normal mode, then at t=0.2-0.4 s, when the line voltage drops and power decreases, the reactive power increases to compensate losses, whereas for the emergency mode it decreases together with active power. When the system is restored to t=0.3, the network voltage instantly rises to 530 kV and then decreases to 490, along with it the active power drops. STATCOM at this moment is trying to compensate it by its reactive power, increasing it in the system and returning it to its normal value. [5]

3.2 Modelling of SSSC in emergency mode

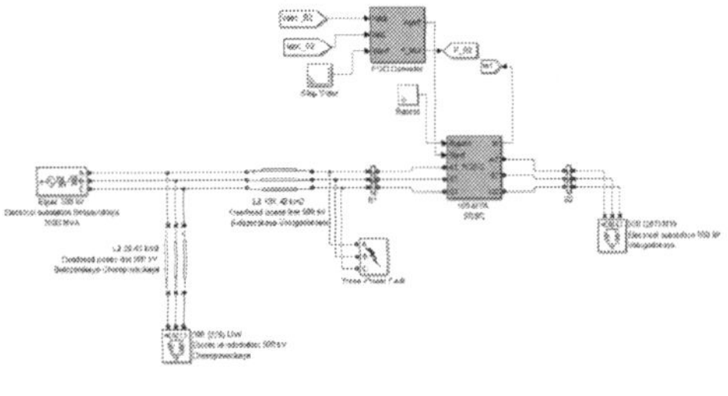

Fig. 4. Model of 500 kV SS Belozerskaya - SS Vologodskaya power system with SSSC device with Three-Phase Fault block.

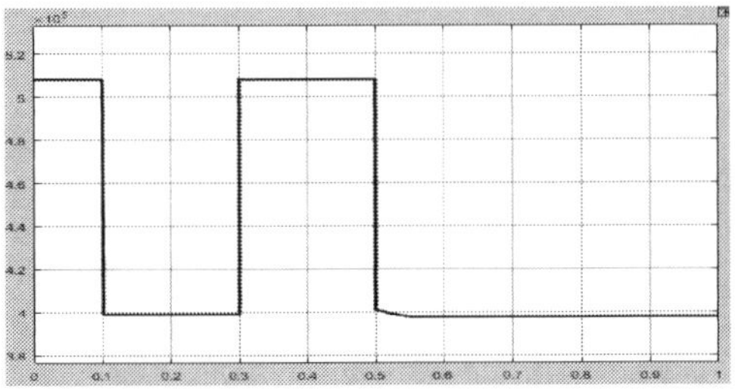

Fig. 5. Voltage on bus B1 with SSSC device in emergency mode.

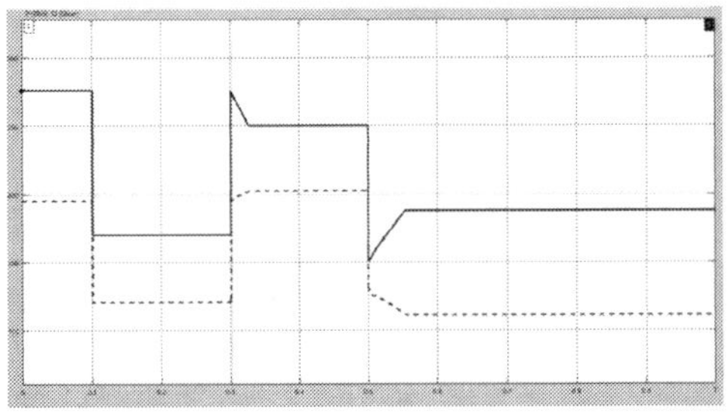

Fig. 6. Active and reactive power of line on B1 bus in the SSSC emergency mode.

Unlike STATCOM, the SSSC voltage practically does not change and drops only during a short circuit. The device operation during short circuit is not very different from its normal mode. During the first short circuit at t=0.1-0.3 s, both powers decrease together with voltage, while the reactive power in this section decreases less than with STATCOM. Then, at t=0.2-0.4 s, the series compensator decreases the active power and slightly increases the reactive one. However, unlike STATCOM and UPFC, the device continues to work as in normal mode, at t=0.5-1 s, reducing reactive power and increasing the active one. [6]

The power transmission line with SSSC in short circuit is shown in Fig. 4, and the voltage during short circuit is shown in Fig. 5. Active and reactive powers are shown in Fig. 6.

3.3 Modelling of UPFC in emergency mode

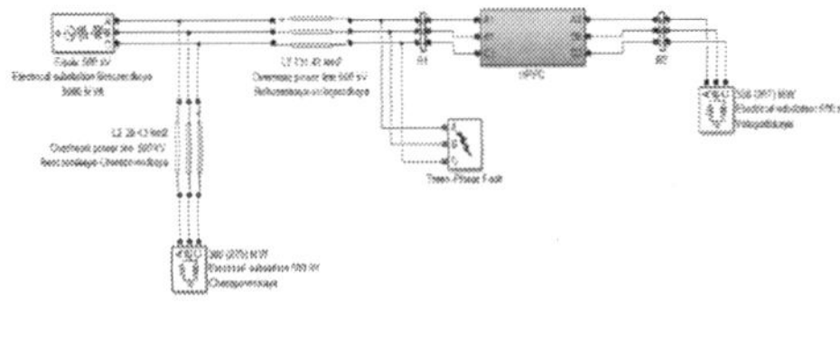

Fig. 7. Model of 500 kV SS Belozerskaya - SS Vologodskaya power system with UPFC device with Three-Phase Fault block.

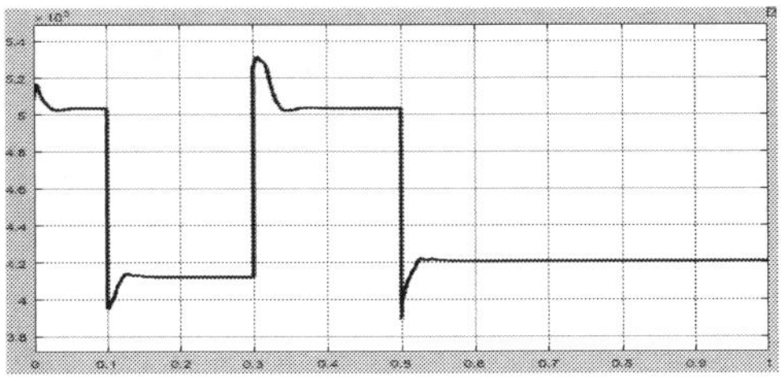

Fig. 8. Voltage on bus B1 with UPFC device in emergency mode.

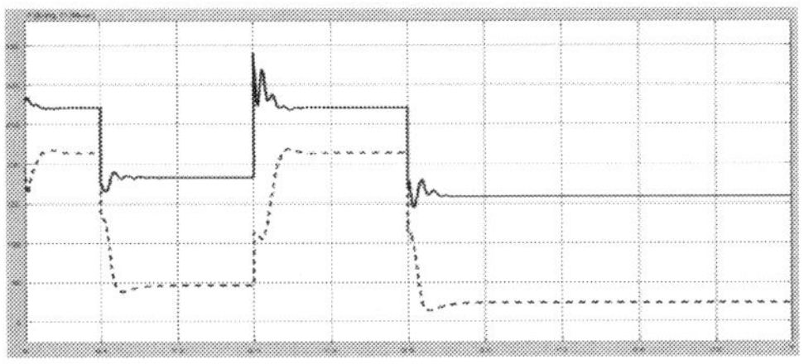

Fig. 9. Active and reactive power of line on B1 bus in the UPFC emergency mode.

The general UPFC operation principle during a short circuit is similar to the operation of static synchronous and static series compensators. It should be noted that the energy flow control system is most sensitive to voltage drops among the all devices. [7] At t=0.1-0.3 s, voltage decreases, and active and reactive power decreases. At t=0.2-0.4 s, an increase in reactive power occurs, without a decrease in active power. At t=0.5-1 s together with a decrease in voltage, the reactive power decreases along with the active one. [8]

The system is shown in Fig. 7. The voltage on bus B1 is shown in Fig. 8. Short-circuit capacities are shown in Fig. 9.

4 Conclusions

The performed modelling shows that FACTS devices act differently in emergency mode, in contrast to the normal mode. Using the data listed in Table 5, one can find out how the work could be continued and the reactive power could be compensated using compensations during abnormal mode. Successful power compensation consists of increasing active power and reducing reactive losses in power grid. [9] Two short circuits occur in this system, and a short recovery period. Using it, one can measure the active and reactive power, calculate the average values for each of the devices, and choose the most suitable one.

Table 5. Comparison of Q and P on bus B1 of FACTS devices with short circuits.

FACTS operation mode	Q, MVAr STATCOM	P, MW STATCOM	Q, MVAr SSSC	P, MW SSSC	Q, MVAr UPFC	P, MW UPFC
First short circuit (t=0.1-0.3 s)	50	175	125	175	49	175
The system after short circuit (t=0.2-0.4 s)	187.5	275	200	250	220	275
Second short circuit (t=0.5-1 s)	50	175	115	185	25	155
Average value	95.8	208.3	146.6	203.3	98	201.6

The table, created using experimental results, shows how much FACTS devices reduce reactive power and increase active power. The static synchronous compensator shows the best performance, with an active power of 208.3 MW and a reactive power of 95.8 MVAr.

References

[1] A. Udaratin, A. Alyunov, A. Krutikov, L.R. Mukhametova, O.O. Zaripov, I.V. Bochkarev, E3S Web of Conferences 124, 02020 (2019).

[2] A. Nemirovskiy, A. Kashin, V. Kosmach, Y. Titovec, I. Toptygin, D. Zaripova, IOP Conference Series: Earth and Environmental Science (EES) 337, 1, 012071 (2019).

[3] A. Ydaratin, T. Akhmetov, S. Makoev, IJCIET 10, 2, (2019).

[4] V. Zhukov, A. Shmelev, D. Mikheev, Russian Electrical Engineering 89, 5 (2018).

[5] M. Bashirov, A. Khismatullin, E. Sirotina, LNEE 641, 613-621 (2020).

[6] S. Tsyruk, S. Gamazin, Y. Ryzhkova, K. Charafeddine, Dynamics, 8601484 (2019).

[7] V. Kozlovsky, S. Petrovsky, V. Vakhnina, D. Skripnuk, ElConRus, 8657322, 569-573 (2019).

[8] E. Gracheva, O. Fedorov, Forecasting Reliability Electrotechnical Complexes of In-Plant Electric Power Supply Taking into Account Low-Voltage Electrical Apparatuses, ICIEAM 5 (2019).

[9] V. Ivanova, I. Ivanov, N. Rozhentsova, Justification of the set of technical and technological criteria when designing a combined power supply system for multi-story residential buildings, ICOECS 2019, 8949983 (2019).

Identification power line sections with increased electricity losses using sensors with Wi-Fi technology for data transmission

A.V. Vinogradov[1], *A.V.* Bukreev[2], *V.E.* Bolshev[1,*], *A.V.* Vinogradova[1], *M.O.* Ward[3], and *N.K.* Miftakhova[4]

[1]Federal Scientific Agroengineering Center VIM, 5 1st Institutsky Proezd, Moscow, 109428, Russia
[2]Orel State Agrarian University, 69 General Rodina St., Orel, 302019, Russia
[3]Damascus University, Baramkah, Damascus, 30621, Syria
[4]Kazan State Power Engineering University, 51 Krasnoselskaya St., Kazan, 420066, Russia

Abstract. The article presents portable timers-electricity meters (PTEM) which are devices for examining 0.4 kV power lines. There are two developed several versions of the devices: single-phase PTEM used to examine single-phase branch lines to consumers, and three-phase PTEM used for any power line sections. Also, the method to identify sections of power lines with increased electricity losses by means of these devices is presented. The paper considers the application of the three-phase PTEM with Wi-Fi technology to transmit data. Up to six independent three-phase PTEMs with Wi-Fi technology can be installed at different points of the power transmission line and transmit measured data in real time to the central unit. The use of these devices allows determining both technological and commercial losses of electricity in different power line sections and draw conclusions about whether these losses are overestimated. On the basis of this, measures are taken to reduce losses.

1 Introduction

Identification of the sections of 0.4 kV power transmission lines (PTL) with increased electricity losses is an urgent task. After solving this task electric grid companies can reallocate resources from the purchase of electric energy losses to reconstructing and modernizing electric networks. Significant forces of electric grid companies are aimed at identifying the sections with electricity losses, but nevertheless energy losses are still unreasonably high, especially commercial losses associated with unreasoned and non-contractual electricity consumption [1].

The peculiarity of this work is that it should be carried out mobilely without providing access to consumers' accounting devices at the first stage [2]. It should be possible to control electricity consumption in the power transmission lines and at the branch power lines to consumers, and compare data on consumption. In addition to the direct consumption of electric energy, it is rational to know the peak values of the consumed current and power, voltage changes at the measurement point. It will also make it possible to conduct a network survey, identify power line sections with insufficient transmission capability, deviations in power quality and reveal voltage unbalance in electrical network. In addition, it will be possible to verify the compliance of the power consumed by subscribers with the contractual conditions. It is desirable to organize all these measurements without disconnecting consumers and without attracting their special attention. There are existing devices that analyze the quality of electric energy and are able to control current and power in the network, for example, power quality analyzers. But they do not fully possess the necessary qualities as they are intended for several other purposes. They are also very expensive. Therefore, the purpose of this article is to develop technical means for organizing work to identify 0.4 kV power line sections with increased electricity losses.

2 Devices for examining 0.4 kV power lines

The device "portable timers-electricity meters " has been developed and allows performing the necessary functions of controlling the parameters of network operation modes and controlling electricity consumption. There are two versions of a device: single-phase (PTEM) [3] and three-phase (PTEM-3) [4]. These versions of the device have current and voltage sensors, but can operate only in a mode of current measurement when it is not possible to connect to voltage circuits. Taking into account the above tasks that a complex of technical means should identify energy losses, it should also allow autonomous continuous operation of the sensors and collecting measured data online by a unit for collecting and processing this data. Figure 1 shows the appearance of PTEM -3.

* Corresponding author: vadimbolshev@gmail.com

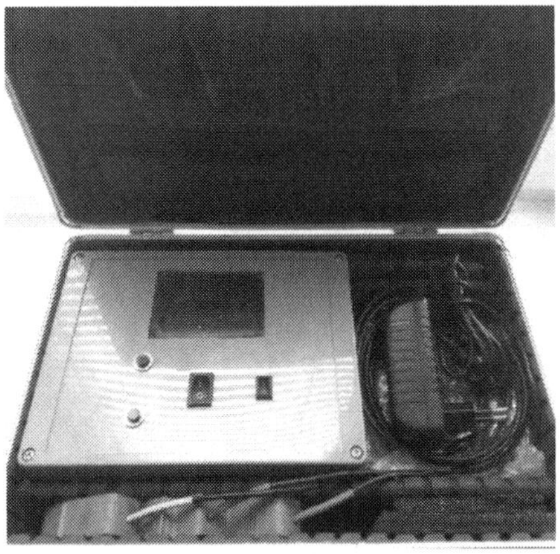

Fig. 1. The appearance of PTEM-3.

To identify power line sections with increased losses, it is necessary to install the simultaneous installation of several PTEM-3 devices at different sections of the power transmission line, to ensure their autonomous operation and then processing the received data [5]. Since the devices are synchronized in time, it is possible to carry out a comparison of these data and identify the sections where the energy losses are too high. It is also possible to analyze other modes of network operation modes such as a graph of the load, voltage, current and power consumption. Due to the fact that the data is written to a removable media, data processing does not require the device itself, that is, a new media can be installed and the device will continue to work at the same or a new control point of the network.

However, this version of the device does not provide the opportunity of online data transfer and, accordingly, the online analysis of information about power consumption modes and other network operation modes. This requires the ability to transfer data from devices using one of the cOmmunication technologies [6-9]. Therefore, the device was modernized in the following version - PTEM-3-W. This device contains current sensors with a current to voltage converter, voltage sensors, two mi-crocontrollers, TFT display, real-time clock unit, data recording unit, radio modules.

The central unit of the PTEM -3-W device (CU) is located either on the vehicle (4) or stationary, for example, on a 10/0.4 kV transformer substation. PTEM -3-W sensors (3) which register electric power consumption and other parameters of the network operation mode are installed at network control points, for example, at a transformer substation (1), at branch lines to consumers (2) and at separate sections of power lines. Data from sensors via communication channels are transmitted to the PTEM-3-W central unit. Figure 2 shows the appearance of PTEM-3-W. A sensor of PTEM-3-W is presented on the left, and the central unit is on the right.

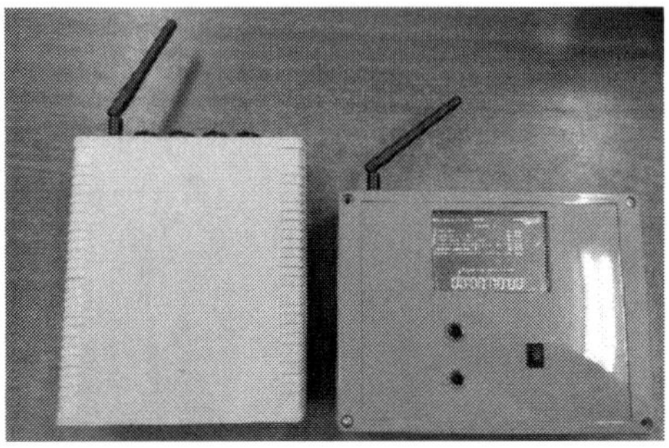

Fig. 2. Appearance of the sensor and the central unit of PTEM -3-W.

PTEM-3-W allows receiving data from sensors remotely via a Wi-Fi connection. Thanks to this, the device usage pattern is changing. The central unit of the device allows simultaneously working with six sensors that can be located at different points of an electrical network. These sensors transmit the measurement results to the central unit over the air. Figure 3 shows the utilization scheme of the device.

3 Method to identify power line sections with increased electricity losses

When using PTEM-3-W, the energy loss in the controlled power line section can be determined as the difference between the readings of the sensors and the readings of the consumers' metering devices.

For example, it is necessary to find out the loss of electricity in different power line sections shown in Figure 3. To that end, PTEM-3-W sensors will be set at the control points of the power transmission line and then the central unit will receive information from these sensors. The time of the control can be set from several minutes to several days or weeks. Also, if there is such an opportunity, it is rational to take readings from subscriber meters at the time of the beginning of measurements and at the time of their end. After receiving the data from the sensors, it will be processed. Technological losses in the section between PTEM -3-W 1 and PTEM -3-W 2 can be determined by equation (1):

$$\Delta W_{1-2} = W_1 - W_2 - W_3, \tag{1}$$

where W_1 is the amount of electricity consumed for a given period of time according to the readings of the PTEM sensor 1; W_2 is the amount of electricity consumed for a given period of time according to the readings of the PTEM 2 sensor; W_3 is the amount of electricity consumed for a given period of time according to the readings of the PTEM 3 sensor.

The actual total losses on the section can be calculated by equation (2) if the readings of the consumer's meter S1 is available:

$$\Delta W_{1-2\,fact} = W_1 - W_2 - W_{S1}, \tag{2}$$

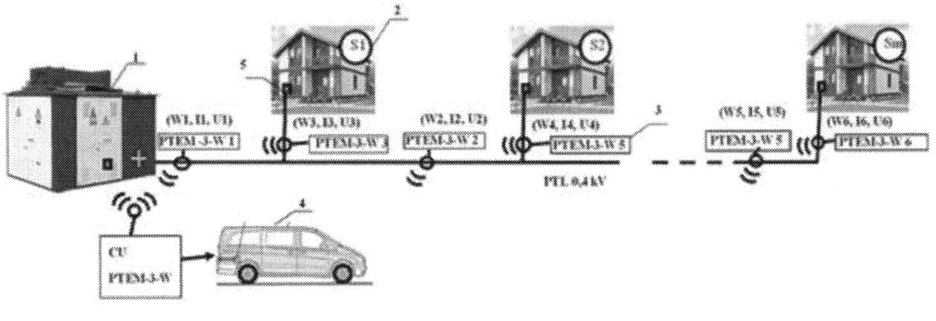

Fig. 3. PTEM -3-W utilization scheme to identify power line sections with high power losses.

where W_{S1} is the amount of electricity consumed for a given period of time according to the readings of consumer's meter S1.

Commercial losses can be calculated as follows (equation 3):

$$\Delta W_{1-2com.} = \Delta W_{1-2fact} - \Delta W_{1-2}, \qquad (3)$$

It should be should be taken into account that a small part of the technological losses will be included in the $\Delta W_{1\text{-}2com}$ losses, that is losses at the branch power line to consumer S1 (depending on the installation site of TEMP-3-W 3).

Similarly, it is possible to determine the losses in other sections of the power transmission lines. To identify sections with increased losses, it is necessary to understand what losses should be in these sections in the normal case which can be calculated by known methods [10, 11, 12]. Knowing the calculated values of technological losses in power line sections, they will be compared with the measured ones. Based on the received difference, a conclusion will be drawn on whether losses are overstated in this area or not. In any case, the calculated losses will differ from the measured ones since the calculation methods cannot sufficiently take into account all the parameters of power transmission lines (load changes constantly, voltage value at different sections is different and most often does not match the calculated value, etc.). However, the difference in the normal case will be negligible.

If there is a large difference between the measured and calculated losses, it is concluded that the losses in this section are overestimated and the reasons are analyzed.

Thus, it is possible to examine power transmission lines in terms of voltage losses on line sections, to construct a load graph for different sections, to determine peak values of current and power consumption and to estimate load and voltage unbalance. It allows verifying the adequacy of the transmission line capability, the power reserve of the transformer substation 10/0.4 kV, the compliance of the power quality with regulatory documents. Also, based on the measurement data, measures can be drawn up to modernize or reconstruct the examined power transmission lines.

4 Conclusion

As part of the study, the following results were obtained:

1. Technical means (portable timers-electricity meters in various versions) have been developed for examining 0.4 kV power transmission lines and, in particular, for solving the problem to identify power line sections with increased electricity losses.
2. It is most rational to use the PTEM version equipped with Wi-Fi data transmission technology to identify electric power losses in 0.4 kV power transmission lines since in this case several power transmission lines can be simultaneously monitored with the possibility of receiving information from sensors in real time.
3. The data obtained by PTEM allow calculating technological and commercial losses in the power transmission lines and draw conclusions about whether they are overestimated.

References

[1] A.V. Vinogradov, M.V. Borodin, Yu.A. Volchenkov, ZH.V. *Peshehonova, Sovershenstvovanie deyatel'nosti po energosberezheniyu i po osushchestvleniyu tekhnologicheskih prisoedinenij filiala OAO "MRSK CENTRA" – "ORYOLENERGO"* [Improvement of energy saving and technological connections for the branch of ORELENERGO, a branch of IDGC of Center] (Publishing House of Orel State Agrarian University, Oryol, Russia, 2015)

[2] A.V. Vinogradov, A.V. Vinogradova, V.E. Bolshev, M.O. Ward, N.V. Makhiyanova, L. V. Dolomaniuk, Justification for creating a mobile complex to assess electric energy loss in power transformers during the operation process, E3S Web of Conferences **124**. DOI:10.1051/e3sconf/201912402009 (2019)

[3] V.E. Bolshev, A.V. Vinogradov, A.V. Vinogradova, A.V. Bukreev, Tajmer-elektroschyotchik mobil'nyj portativnyj [Portable timers-electricity meter], RF patent No. 191056 (2019)

[4] A.V. Bukreev, A.V. Vinogradov, Tajmer-elektroschyotchik mobil'nyj portativnyj [Three-phase portable timers-electricity meter], RF patent No. 195905 (2020)

[5] A. Vinogradov, V. Bolshev, A. Vinogradova, M. Borodin, A. Bukreev, I. Golikov, *Mobile Measuring Complex for Conducting an Electric Network Survey (*In Handbook of Research on Energy-Saving Technologies for Environmentally-Friendly Agricultural Development, Hershey, PA: IGI Global, 243–267, 2020)

[6] M. Kuzlu, M. Pipattanasomporn, S. Rahman, Communication network requirements for major smart grid applications in HAN, NAN and WAN, Computer Networks **67**, 74–**88**. DOI: 10.1016/j.comnet.2014.03.029 (2014)

[7] S.S.S.R. Depuru, L. Wang, V. Devabhaktuni, N. Gudi, Smart meters for power grid: challenges, issues, advantages and status, 2011 IEEE/PES Power Systems Conference and Exposition, 1–7. DOI: 10.1109/PSCE.2011.5772451 (2011)

[8] E. Ancillotti, R. Bruno, M. Conti, The role of communication systems in smart grids: architectures, technical solutions and research challenges, Communication Technologies **36,** 1665–1697. DOI: 10.1016/j.comcom.2013.09.004 (2013)

[9] S. Galli, A. Scaglione, Z. Wang, For the grid and through the grid: The role of power line communications in the smart grid, Proceedings of the IEEE **99**, 6, 998–927. DOI: 10.1109/JPROC.2011.2109670 (2011)

[10] W.G. YU, Y.J. XIONG, X.F. ZHOU, G.L. ZHAO, N. Chen, Analysis on technical line losses of power grids and countermeasures to reduce line losses [J], Power System Technology **18**, 54–57 (2006)

[11] E.I. Gracheva, O.V. Naumov, Estimation of power losses in electric devices of the electrotechnical complex, International Conference on Industrial Engineering, Applications and Manufacturing (ICIEAM), 1–6. DOI: 10.1109/ICIEAM.2019.8742923 (2019)

[12] M. Stojkov, K. Trupinic, S. Nikolovski, Technical losses in power distribution network, MELECON 2006-2006 IEEE Mediterranean Electrotechnical Conference, 1048-1051. DOI: 10.1109/MELCON.2006.1653279 (2006)

DEVELOPMENT OF A NEW BURNER DEVICE BASED ON INJECTION-RADIATION METHOD FOR GAS FUEL COMBUSTION

A A Sinitsyn[1,*], *O Y* Nikiforov[1], *Timur* Akhmetov[2] and *D F* Karpov[1]

[1]Department of Heat and Gas and Water Supply, Vologda State University, Lenina str. 15, 160000, Russia
[2]Kazan State Power Engineering University, Kazan, Russia

Abstract. This paper presents the results of designing a promising burner device that combines injection and radiation methods for gaseous fuel combustion. Infrared radiation from the heated surface of nozzle of radiation burner provides intensive heat transfer in furnaces and boilers, specifies their high efficiency at low flue gas temperatures, and very low NOx emissions. Effective gas combustion is facilitated by additional heating of the initial fuel mixture during filtration through the hot walls of porous nozzle. The new design of radiation burner has high performance characteristics, provides effective combustion of gaseous fuel up to 99% and meets high environmental standards and requirements.

1 Introduction

Energy industry is one of the leading branches of human activity. Heat and light in houses, traffic flows and industry require a huge amount of energy. Currently, these needs are borne by mineral resources such as coal, natural gas, oil, reserves of which are limited and there are a number of works considering ways to increase the efficiency of their usage or other approaches for obtaining energy [1-5].

Today, the most economical and efficient type of fuel is natural gas. This is explained by its high calorific value, good transportability, minimum amount of harmful substances from combustion products, etc. The "blue fuel" releases heat during its burning in a mixture with an oxidizing agent. For this, various gas burner devices are used. In this connection promising is the development and construction of new devices and systems providing the most complete combustion of fuel, as well as reducing emissions of harmful substances into the air.

The aim of this work is to develop a new energy-efficient burner device with high performance characteristics, which meets environmental standards and requirements.

To achieve this goal, the following tasks were set:

- To conduct a review of currently used gas burner devices, to identify their main advantages and disadvantages;
- To develop a concept of a new burner device which doesn't have the identified disadvantages;
- To develop a device model;
- To perform computational and experimental studies;
- To give a technical assessment of the possible use of such devices;
- To draw conclusions on the performed work.

Many companies carried out research in the field of low-power burner development and fuel combustion, in particular NPO CKTI OJSC, ORGRES Firm, VNIIpromgaz, MosgazNIIproekt, Tomsk Polytechnic University. A substantial contribution to the development of such burner devices was made by R.B. Akhmedov, O.N. Bruhanov, A.S. Isserlin, M.I. Pevzner, R.I. Esterkin, V.A. Khristich, I.Ya. Seagal, K.F. Rider and others. The conducted research work allowed creating different types of burner devices, such as diffusion, blasting, injection, rotary, etc., which have received wide industrial implementation. In recent years, the main attention of international researcher groups has been directed to studying the processes of gasification and intensification of fuel combustion processes. For example, the works of S.E. Najim, A.C. Styles, N. Syred [6] and Nanjia Yu, Bo Zhao, Gonghan Li, Jue Wang [7] present experimental and simulation studies of fuel combustion intensification due to high-tech mixture formation.

Earlier studies were carried out on the efficiency of gaseous fuel burning under intensification by creating pulsations and exposure to magnetic field [8-10]. It is worth to note the developments [11,12], in which new methods for increasing the efficiency of fuel combustion and monitoring the pressure of the gas-air medium in fire equipment for various purposes are proposed.

At present burners are of various designs ensuring their vast usage and stable demand in the market. The performed patent studies required for development of new products [13-15] allowed us to analyze the existing technical solutions which received protection documents. A review of the selected patents for gas burner devices showed that they all have a number of disadvantages and the area of further research is free enough for the

* Corresponding author: sinitsyn.science@mail.ru

Table 1. Comparative characterization of burners available in the market.

Parameter	Ballu BIGH-4	Komfort Prometei	Bartolini Primavera I
Maximum thermal power, kW	4.5	2.3	4.2
Fuel type	propane/butane	propane/ natural butane	propane/butane
Nominal gas consumption, g/hour	207/327	180	100-300
Dimensions of device (WxHxD), mm	338x372x278	255x150x100	780x430x420
Weight	2.3	1.4	13
Producing country	China	Russia	Italy

introduction of new designs. The objective of this study is to reduce the loss of thermal energy, to increase the resistance of porous nozzle to thermal loads, to achieve more complete combustion of fuel, to increase the reliability of combustion by reducing the thermal conductivity of nozzle.

A comparison of devices currently available on the heating equipment market, which are analogues by the principle of operation, is given in table 1.

The review of gas burners showed that they have a maximum thermal power of 2.3-4.5 kW, while the nominal flow velocity can vary from 100 to 327 g/hour. All of them are quite compact, safe and easy to transport. They can operate both on liquefied and natural gas. They are mainly used for room heating, but all have German ceramic panels as a heating element.

A patent and literature search showed that there are analogues of burner devices based on the injection-radiation method for burning gaseous fuels, but they all have a number of disadvantages.

2 Concept of a new burner device

The purpose of this study is achieved by combining the injection and radiation components in the burner. The burner body has a standard version of an injection burner with a superstructure of integrated nozzle for burning and afterburning of fuel. The new device belongs to the thermal power systems, namely to gas flameless injection burners with refractory nozzles.

A feature of this device is a stacked nozzle made of tiles of heat-resistant ceramics. A combustion stabilizer is placed above the nozzle, which is made as a heat-resistant grating that acts as gas afterburner, flame stabilizer and additional emitter. These features result in a more complete combustion of fuel due to a combination of methods for burning and afterburning of the fuel-air mixture.

A sectional view of a burner device based on an injection-radiation method for burning gaseous fuel is presented in fig. 1.

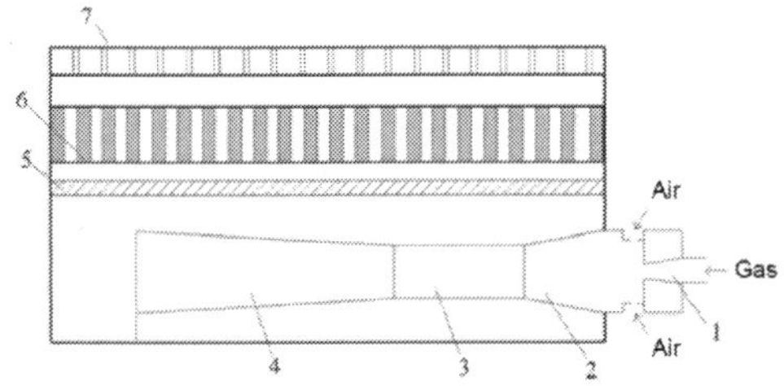

Fig. 1. Sectional view of a burner device based on injection-radiation method for burning gaseous fuels.

The operation principle of the device is as follows: gaseous fuel is supplied through the snout (1), while injecting the air necessary for combustion. In a mixer consisting of confuser (2), nose section (3), diffuser (4), fuel is mixed with air. The resulting air-fuel mixture through the distribution grid (5) enters the nozzle (6) and burns in the boundary areas of nozzle. The combustion stabilizer (7) contributes to fuel afterburning, the hot surface of nozzle and the combustion stabilizer create radiation flow and heat the surrounding objects. The nozzle, made of tiles of heat-resistant ceramics, provides optimal combustion conditions, has low thermal conductivity, prevents the spread of flame under the nozzle, even with wide limits of power control of the burner device.

The theoretical calculation of gas burners is very complicated and time-consuming, since it is associated with complex calculations of the processes of mixing, combustion and heat transfer, which should provide both high efficiency of gas fuel combustion, and the lowest possible concentration of harmful components in the combustion products. Due to the complexity of calculations, it is necessary to use a number of approximation data obtained from practice or particular experiments. This is demonstrated in [16,17], where the main stages of design and simulation of such devices were shown.

The design calculation of burner device was reduced to determination of dimensions of the main elements of gas burner: snout, nose section, mixer, confuser, diffuser and nozzle [18]. The design scheme is shown in fig. 2.

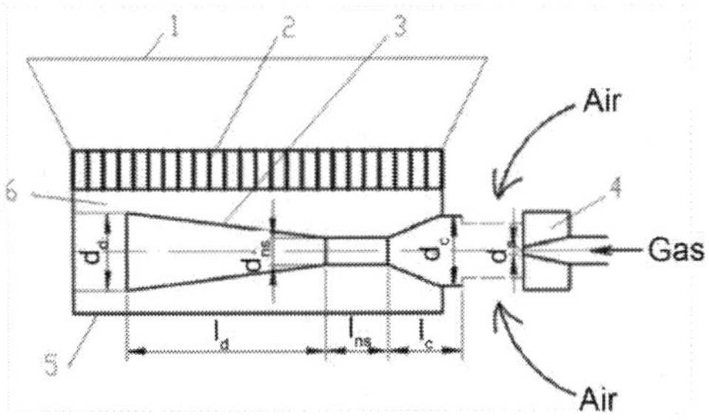

Fig. 2. Design scheme of an injection gas burner with a ceramic nozzle.

Fig. 2 conventionally shows the following elements: 1-reflector; 2-ceramic nozzle with holes; 3-injection mixer; 4-gas snout; 5-body; 6-channel head.

It should be noted that the ranges of stable operation of injection burners are limited by the possibility of flame lifting and flashback. Therefore, the calculation should be based on the velocity of gas-air mixture flow through the nozzle (0.11-0.16 m/s) and the size of fire holes in tiles (1.5-2 mm). So, the designed burner device has thermal load limits from 2.53 to 3.96 kW. The source data and the calculation results are presented in tables 2 and 3, respectively.

Table 2. The source data for design.

Name	Value	
Thermal load Q_g, W	2530	3960
Net volumetric calorific value Q_n, kJ/m³	35590	
Gas density ρ_0, kg/m³	0.74	
Gas pressure before snout P_g, Pa	3000	
Theoretically required amount of air V_g, m³/ m³	9.015	
Excess air ratio, α	1.05	

3 Results

The main design outputs are presented in table 3 [18] for two load parameters.

Table 3. Calculated data and design outputs.

Name	Value	
Total heat load Q_g, W	2530.00	3960.00
Gas consumption for a burner, V_g, m³/h	0.26	0.33
Diameter of snout cross section d_s, mm	1.03	1.29
Diameter of nose section of injection mixer d_{ns}, mm	12.37	15.48
Diameter of confuser d_c, mm	18.56	23.22
Diameter of diffusor d_d, mm	18.56	23.22
Length of confuser l_c, mm	18.56	23.22
Length of diffusor l_d, mm	61.86	77.39
Length of nose section of mixer l_{ns}, mm	30.93	38.69
The area of radiating surface of nozzle Fr, m²	0.06	0.09
Amount of tiles, pcs	16	25
Diameter of hole in tile, mm	2.0	
Amount of holes in tile, pcs	574	
Air-gas velocity, Wmix, m/s	0.11	0.16

Fig. 3 presents the optimal layout of a new gas burner device with dimensions of its main constituent elements.

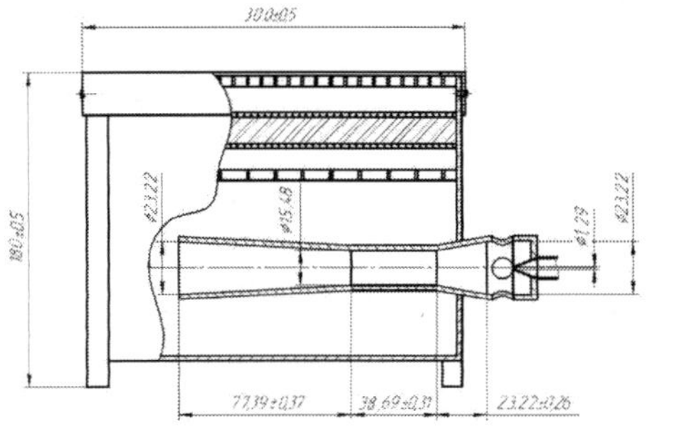

Fig. 3. Layout of a new gas burner with a heating capacity of 3.9 kW.

When developing a technical solution aimed at ensuring a proportional distribution of mixture at the outlet of the burner device, a three-dimensional model of the injection part of the burner was built. Geometry was developed in Autodesk Inventor graphic CAD system, which allows exporting geometry to ANSYS CFX.

Color maps of distributions of pressure and velocity in the calculated geometry are shown in Fig. 4. The simulation results show that the highest velocity is in the nose section. A further increase in the number of iterations does not affect the numerical calculation.

It should be noted that the burner has a symmetrical velocity profile resulting in a uniform temperature

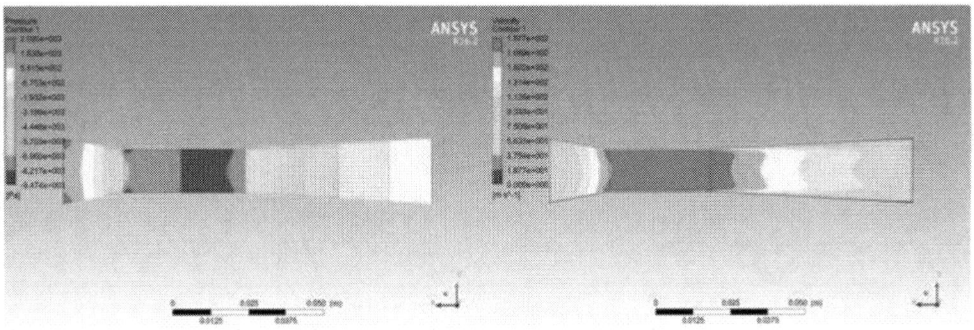

Fig. 4. Distribution of pressure (left) and velocity (right) of the air-gas mixture in the injection part of the burner in the ANSYS program.

profile. There is no explicit recirculation zone, which reduces hydraulic pressure losses. As it can be seen from figures, the design has a sufficiently high completeness of combustion and low velocities, due to which the temperature and concentration of nitrogen oxides at the outlet of burner device are reduced.

Three-dimensional modeling in the ANSYS CFX software package allowed developing technical solutions related to modernization and selection of optimal design of burners for domestic and power boilers.

4 Conclusions

The following results were obtained from the computational and experimental studies:

- The relevance of the use of gas burner devices is specified, the operation principle of such devices is described, and a review of manufacturers and patented technologies is carried out;
- Based on the review results, a comparison of devices was made, their advantages and disadvantages were identified, as well as the prospects of design;
- The concept of a new gas burner device was developed, which doesn't have the identified disadvantages of the existing devices;
- A new model of gas burner device was developed;
- A structural analysis was performed, which allowed us to determine the geometric parameters of the device;
- The injection part of burner was simulated in the ANSYS software according to the developed design.

It was found that infrared radiation from a heated surface of radiation burner nozzle provides intensive heat transfer in furnaces and boilers, their high efficiency at low flue gas temperatures, and very low NOx emission. Effective gas combustion is facilitated by additional heating of initial fuel mixture during filtration through the hot walls of the porous nozzle. The design of radiation burner has high operational properties, provides effective combustion of gaseous fuels up to 99% and meets the highest environmental standards and requirements. The burner was numerically simulated and appropriate presentations of physical processes were obtained.

References

[1] L.V. Larchenko, R.A. Kolesnikov, L. Mukhametova, Russian oil and gas industry as a sphere of international interests and economic cooperation, E3S Web of Conferences **161**, 01006 (2020).

[2] N. Politaeva, Y. Smyatskaya, R. Al Afif, C. Pfeifer, L. Mukhametova, Development of Full-Cycle Utilization of Chlorella sorokiniana Microalgae Biomass for Environmental and Food Purposes, Energies **13**, 10, 2648 (2020).

[3] V. Markov, A. Fedyukhin, I. Sultanguzin, B.G. Matisov, N.V. Nikitkov, V.V. Izrantsev, V.G. Knorring, G.A. Konrashkova, L.R. Mukhametova, Improvement of energy efficiency of motor fuel production at oil refinery, IOP Conference Series: Earth and Environmental Science 337, 1, 012080 (2019).

[4] I.A. Sultanguzin, E.V. Zhigulina, Y.V. Yavorovsky, I.D. Kalyakin, A.V. Govorin, A.V. Fedyukhin, A.A. Krolin, S.V. Guzhov, O. Derevianko, L. Mukhametova, Development strategy of the energy supply system for MPEI campus blocks based on green building, IOP Conference Series: Earth and Environmental Science 288, 1, 012071 (2019).

[5] E. Sukhareva, A. Fedyukhin, O. Derevianko, M. Egorov, L. Mukhametova, I. Akhmetova, Optimization of combined heat and power plant operating mode by means of underutilized equipment mothball, IOP Conference Series: Earth and Environmental Science 288, 1, 012122 (2019).

[6] S.E. Najim, A.C. Styles, N. Syred, Flame movement mechanisms and characteristics of gas fired cyclone combustors, Eighteen Symposium (International) on Combustion **18**, 1, 1949-1957 (1981).

[7] Nanjia Yu, Bo Zhao, Gonghan Li, Jue Wang, Experimental and simulation study of a Gaseous

oxygen, Gaseous hydrogen vortex cooling thrust chamber, Acta Astronautica **118**, 11-20 (2016).

[8] A.A. Sinitsyn, L. Mukhametova, Practical applicability of the method for measuring pressure of controlled medium on the example of a pulsating combustion boiler, IOP Conference Series: Earth and Environmental Science Proceedings of the Conference the international scientific conference Efficient waste treatment – 2018, EWT-2018, 012063 (2019).

[9] A.A. Sinitsyn, D. Rusakov. V.V. Izrantsev, S.A. Kravchenko, G.A. Konrashkova. Investigation of gas fuel combustion efficiency at magnetic field treatment, IOP Conference Series: Earth and Environmental Science Proceedings of the Conference the international scientific conference Efficient waste treatment – 2018, EWT-2018, 012082 (2019).

[10] A.N. Naimov, A.A. Sinitsyn A., F. Gazizov, A.P. Eperin, Yu.A. Rundygin, G.A. Agasiants, S.M. Galileev, T.R. Akhmetov, Mathematical modeling of heating temperature mode for a heat exchange system of the type «pipe in pipe», IOP Conference Series: Earth and Environmental Science Proceedings of the Conference the international scientific conference Efficient waste treatment – 2018, EWT-2018, 012068 (2019).

[11] A.A. Sinitsyn, Device for radiation-convective heating of working bodies, Patent of the Russian Federation, 129608, 27.06.2013, application no. 2012141775/06 dated 01.10.2012.

[12] D.A. Belyanskiy, A.A. Sinitsyn, A device for controlling the temperature of the working fluid, Patent of the Russian Federation, 119877 RU, Int. Cl.7 G01K 7/00, EnergoExpert LLC, pub. 27.08.2012.

[13] V.M. Shmelev, A.D. Margolin, Radiation burner, Patent of the Russian Federation, 2151957, MPK F23D14/12, declared 03.03.1999, publ. 27.06.2000.

[14] V.K. Baev, A.V. Potapkin, A.I. Trubitsin, Gas flameless burner, Patent of the Russian Federation, 2137040, MPK F23D14/14, declared 03.10.1998, dated 10.09.1999.

[15] A.F. Gekov, N.M. Lipovyi, A.N. Popov, F.N. Smus, B.E. Farfel, Radiation burner, Patent of the Russian Federation, 2094703, MPK F23D14/12, declared 18.06.1996, dated 27.10.1997.

[16] A.A. Sinitsyn, N.N. Boytsova, I.G. Akhmetova, S.O. Makoev, A. Asadpoori, O.A.Stepanov, N.V. Rydalina, E.O. Antonova, B.G.Aksenov, Study of operation of power-generating devices of gaseous fuels combustion, International Journal of Mechanical Engineering and Technology **10**, 2, 1480-1490 (2019).

[17] A.A. Sinitsyn, *Simulation of fire engineering processes in energy devices aimed at their optimization and improvement of reliability* (Life Science Journal, the Acta Zhengzhou University Oversea Version, Marsland Press: New York, 10, 2013)

[18] L.S. Berezina, A.A. Sinitsyn, *To the development of a new burner device of a combined type*, XII Annual scientific session of graduate students and young scientists: Materials of the interregional scientific conference: in 2 volumes, V. 1: Natural-technical direction (Vologda: VoGU, 258-260, 2018).

Calculation of pulse voltage stabilizer

Dmitry Cheremukhin[1,*]

[1]Far Eastern Federal University, Russia, Vladivostok

Abstract. This scientific article is devoted to one of the priority issues in electronics, namely the implementation of the engineering task of designing a power source. In my case, according to the terms of reference, which includes: the main parameters of the network, efficiency, a structural electrical circuit of a switching power supply was developed. In the process, the main functional units (blocks) were calculated and the choice of radio components from which they were built on the basis of: current values, reverse voltage, gain, operating frequency and other parameters of diodes, transistors, integrated circuits, Zener diodes, operational amplifiers was justified. In the "Methods" section, calculations will be made for the device blocks: a network rectifier, a push-pull transformer converter, a pre-amplifier circuit with a transformer input, an error signal comparison and amplification circuit, a sawtooth voltage generator, and a secondary power source. The information presented in the text is recommended to a narrow circle of specialists, priority in the field of power electronics, as well as in the field of communication systems.

1 Introduction

Switching voltage stabilizer - a voltage regulator in which the regulating element is in the cutoff mode (R_{max}) or under saturation (R_{min}), hence can be considered as the key [1,121]. Smooth change of voltage occurs due to the presence of integrating elements: the voltage rises as you accumulate them energy and decreases its impact in the load. This mode reduces energy loss and improves weight and overall dimensions, however, has its own characteristics.

Essential elements of a switching power supply are: integrator, which voltage can not change instantly, but gradually increases according to the accumulation of energy and then slowly drops as it returns to the load (for example a capacitor, to which included some non-zero resistance, which can be, for example, the internal resistance of the power source) and a device able to change the resistance to the passage of current from minimum to maximum and Vice versa.

2 Methods

This simulation was carried out at the Department of Electronics and Communications. During the calculations, the Mathcad program was used, as well as the development environment of electronic circuits Multisim.

3 Materials

Has been provided by the technical specification (requirements) regarding the network conditions for which it is planned to produce the product. Input the following:

Power supply voltage: Ups = 36 V; the Relative change in supply voltage: δUps = ± 20 V; Frequency mains current: F = 400 Hz; Type of mains supply: three phase with a "0" output star; Diagram of the network rectifier: three-phase half-wave; load Voltage: U load = 12 V; load Current max: I l max = 5 A; load Current min: I l min = 1 A; ambient Temperature: t = + 50 C; ripple Factor: Kr = 1.5 %; The switching frequency of the Converter: fs = 25 kHz; Diagram of converters: push-pull; efficiency: η = 92 %.

3.1 The calculation of the network rectifier

Depending on the number of phases supply rectification circuit may be a three-phase bridge, three phase half-wave, single-phase bridge.

In (Fig.1) shows a diagram of a network rectifier "three phase half-wave rectified".

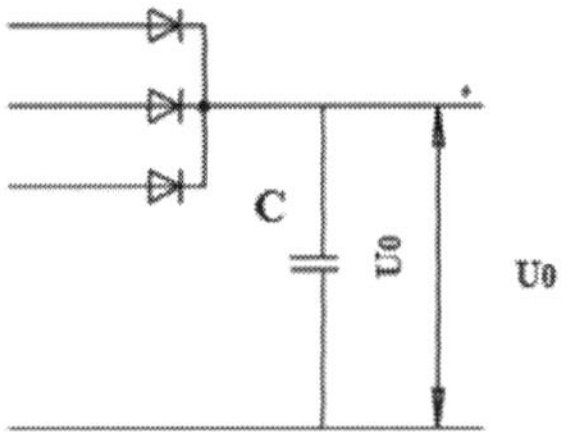

Fig. 1. Scheme of three-phase half-wave rectifier.

Calculate the main parameters:

* Corresponding author: dodgevipersrt15@mail.ru

The average value of the rectified voltage for three-phase half-wave circuits:

$$U_{0\ rated} = 0.95U_{ps}\sqrt{2} = 0.95 \cdot 36 \cdot \sqrt{2} = 47.9\ V$$

The maximum value of the rectified voltage:

$$U_{0max} = U_{0\ rated} \cdot \left(1 + \delta_{U_{ps}}\right) = 47.9 \cdot (1 + 0.2) = 57.46\ V$$

The minimum value of the rectified voltage:

$$U_{0min} = U_0 \cdot \left(1 - \delta_{U_{ps}}\right) = 47.9 \cdot (1 - 0.2) = 43.08\ V$$

The capacitance value:

$$C_1 = 1.3\frac{P_{load}}{\eta} = 1.3\frac{I_{load\ max} \cdot U_H}{\eta} = 1.3 \cdot \frac{5 \cdot 12}{0.92} = 84.15\ \mu F$$

Taken on a table e24: $C_1 = 100\ \mu F$.

The value of current-limiting (to limit charge current) resistance:

$$R_0 = \frac{U_{stabil\ max}}{I_{varib}} = \frac{60}{6} = 10\ Ohm$$

Taken on a table e24: $R_0 = 12\ Ohm$.

For selecting the rectifier diode, the required values:

The average current through the valve:

$$I_{1,2,3} = \frac{I_0}{3} = \frac{0.313}{3} = 0.104\ A$$

$$P_0 = I_0^2 * U = 0.9W$$

$$I_0 = \frac{P_0}{U_{0min}} = \frac{13}{43.08} = 0.313\ A$$

The magnitude of the reverse voltage to the idle valve:

$$U_{reverse} = U_{ps}\sqrt{2}\sqrt{3} = 36 \cdot \sqrt{2} \cdot \sqrt{3} = 87.19\ V$$

According to the data selected diode KD209B with the following parameters:

Table 1. Parameters of the Diode KD209B.

Diode	Ipr.cp, A
KD209B	0.5
U rev., V	I variable, A
600	6

3.2 Calculation of transformer push-pull converter

In this popular scheme (Figure 2), the performance is a combination of electric energy conversion and its regulation through pulse width modulation. Characteristic for them is the presence of a relatively strong smoothing LC filter [2, 63].

In half-bridge scheme of transformer push-pull converter, when the transistor VT1 is opened at time γ T/2, and VT2 is closed, energy transfer occurs from capacitor C1 through a transformer T to the cumulative LC choke and the load [3,51]. At the same time, the capacitance C2 is charged through the circuit: +U0, U1 winding, the capacitance C2 and –U0. In the next half cycle the process repeats: the S2 is open, C1 is charged. During the pause, the filter capacitor C is discharged to the load through the diode VD5. Capacitance C1, C2 form a capacitive divider.

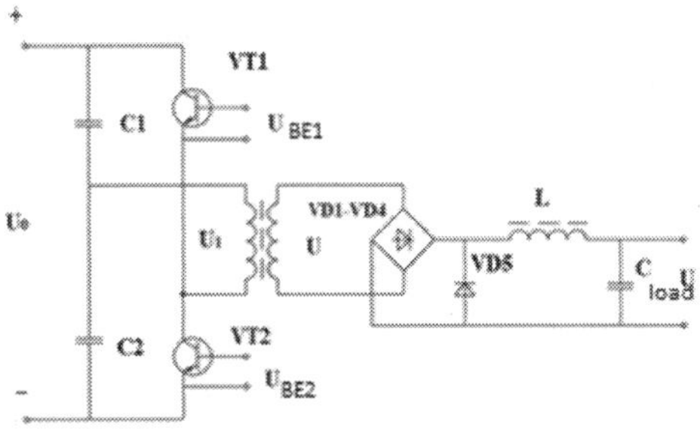

Fig. 2. Schematic diagram.

Capacity C_1 and C_2 form a capacitive divider:

$$U_1 = \frac{U_0}{2} = \frac{47.9}{2} = 23.95V$$

The fill factor of the pulse: $\gamma_{max} = 0.7$

The output voltage of an RF rectifier:

$$U'_{2\ min} = \frac{U_{load}}{\gamma_{max}} = \frac{12}{0.7} = 17.1V$$

$$U'_{2\ rated} = \frac{U'_{2\ min}}{\left(1 - \delta U_{ps\ max}\right)} = \frac{17.1}{(1 - 0.2)} = 21.4\ V$$

$$U'_{2\ max} = U'_{2\ rated} \cdot \left(1 + \delta U_{ps\ max}\right) = 21.4 * (1 + 0.2) = 25.6\ V$$

$$\gamma_{min} = \frac{U_{load}}{U'_{2\ max}} = \frac{12}{25.6} = 0.5$$

The voltage at the secondary winding of the transformer:

$$U_2 = U'_2 + U_{ddv} = 17.1 + 1 = 18.1V$$

where U_{ddv} - forward drop voltage of the diodes an HF bridge.

The value of the output inductor:

$$L = (5 \div 10) \cdot \frac{R_{load} \cdot (1 - \gamma_{min})}{2 \cdot f_s} = 7 \cdot \frac{12 \cdot (1 - 0.5)}{2 \cdot 25 \cdot 10^3} = 84\ \mu H$$

Taken on a number e24: $L = 100\ \mu H$.

Load resistance (R_{load}):

$$R_{load} = \frac{U_{load}}{I_{load\ min}} = \frac{12}{1} = 12\ Ohm$$

Taken on a number e24: $R_{load} = 12\ Ohm$.

The maximum current of the inductor:

$$I_{L\ max} = I_{load} + \Delta i_1 = 5 + 3.42 = 8.42\ A$$

Ripple current through the inductor:

$$\Delta i_1 = \frac{U'_2 \gamma_{max}}{2 \cdot L \cdot f_s} = \frac{17.1 \cdot 0.7}{2 \cdot 84 \cdot 10^{-6} \cdot 25 \cdot 10^3} = 2.85\ A$$

The capacity of the smoothing filter:
When $K_r = 2\%, get$:

$$C_{load} = \frac{(1 - \gamma_{min})}{8 \cdot L \cdot f_s^2 \cdot K_r} = \frac{(1 - 0.5)}{8 \cdot 84 \cdot 10^{-6} \cdot 25^2 \cdot 10^6 \cdot 0.02} = 50\ \mu F$$

Taken on a number e24: $C_{load} = 50\ \mu F$.
Average current of the diode VD5 and its voltage:

$$I_{a\ VD5} = I_{L\ max}(1 - \gamma_{min}) = 8.42 * (1 - 0.5) = 4.21\ A$$

$$U_{rev\ VD5} = U'_{2\ max} = 25.6\ V$$

The average value of the currents and reverse voltage for diodes HF bridge [4]:

$$I_{am} = \frac{I_{load}}{2} = \frac{5}{2} = 2.5\ A$$

$$U_{reverse} = U_{2\ max} = 19\ V$$

As the diode bridge is used KD213A with the following parameters:

Table 2. Parameters of the Diode bridge KD213A.

Diode bridge	$I_{direct.average}$, A
KD213A	10
$U_{rev.}$, V	I_{direct}
200	100
F, kHz	
100	

The maximum current value I_2 of the secondary winding:

$$I_2 = I_{L\ max} = 8.42\ A$$

For half-bridge circuits:
The collector current $I_{load\ max}$ and the max value of $U_{CE\ max}$ to pick VT:

$$I_{C\ max} = \frac{U_2 \cdot I_2}{U_1 \cdot \eta} = \frac{18.1 \cdot 8.42}{23.95 \cdot 0.92} = 6.87\ A$$

$$\eta_н = (0.9 \div 0.98)$$

$$U_{CE\ max} = U_{0\ max} = 57.46V$$

Selects the transistor VT1 – KT812A:

Table 3. Parameters of the transistor KT812A.

Type	H_{21e}	$U_{BE.\ saturation}$, V	
KT812A	30	2.2	
$U_{CE.\ saturation}$, V	f_{cutoff}, MHz	U_{CEmax}, V	I_{CEmax}, A
1.35	6.5	400	8

$q = (1.3 \div 2)$ – the coefficient of saturation,
The value of the base currents of the control:

$$I_{B2,3,4} = \frac{I_{C\ max} \cdot q}{h_{21\ min}} = \frac{6.87 \cdot 1.5}{30} = 0.34\ A$$

3.3 The calculation scheme of the pre-amplifier with transformer input

In (Fig.3) shows a diagram of a pre-amp.

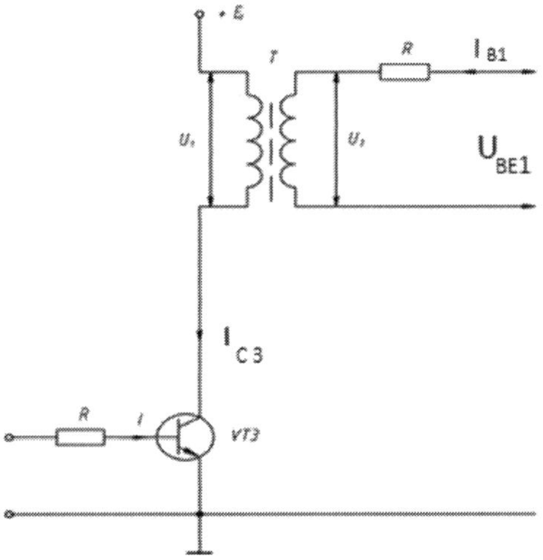

Fig. 3. Diagram of the preamplifier with transformer input.

$$U_2 = (3 \div 4) \cdot U_{BE\ saturation1} = 3.5 \cdot 2.2 = 7.7V$$

Balanced resistance is needed to limit the base current of the power transistor of the Converter (after replacing the transistor) due to the variation of parameters h21e, U_{BE} [5].
Balanced impedance:

$$R_b = \frac{U_2 - U_{BE\ saturation1}}{I_{b1}} = \frac{7.7 - 2.2}{0.27} = 20.4 Ohm$$

Is selected by row e24: $R_B = 22\ kOhm.$

$$I_{C3} = \frac{U_2 \cdot I_2}{U_1 \cdot \eta} = \frac{7.7 \cdot 0.34}{(15 - 1) \cdot 0.92} = 0.2\ A$$

$U_1 = E_1 - U_{CE\ saturation}$ – the amplitude of the voltage of the primary winding of the transformer;
Select transistor VT5-6:

Table 4. Parameters of the transistor KT630E.

Type	H_{21e},	$U_{BE\ saturat}$, V	$U_{CE\ saturat}$, V
KT630E (n-p-n)	300	0.85	0.11
f_{cutoff}, MHz	U_{CEmax}, V	I_{CEmax}, A	
300	60	1	

Current control I_{B3}:

$$I_{B3}=\frac{I_{C3}\cdot q}{h_{21}}=\frac{0.2\cdot 1.5}{300}=1\ mA$$

The resistance value R:

$$R=\frac{U^{1}_{out}-U_{BE\ saturation}}{I_{B3}}=\frac{14-0.85}{1\cdot 10^{-3}}=13\,kOhm$$

Is selected by row e24: $R = 15\ kOhm$.
,U_{1out} - voltage logical units on the I_{ps} output of the logical element "and" [6,124].

3.4 The calculation of the comparison circuit and the gain error signal

In (Fig.4) shows a diagram of the circuit built on the operational amplifier.

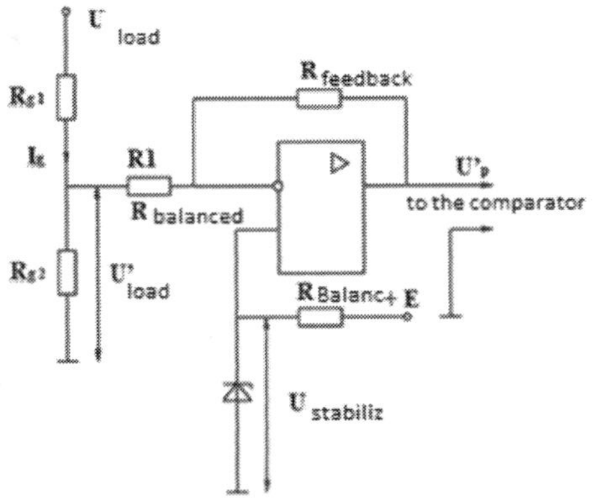

Fig. 4. Comparison Circuit and the gain error signal.

The voltage U_{st} on the stabilizer is selected from the conditions: $U_{st} < U_{in\ max}$, if $U_{in\ max}$ - is the maximum value of the input voltage of the operational amplifier [7,20].

Set the value U_{st} = 4.7 B, equating it to the voltage U'_{load}, t.e. $U'_{load} \approx U_{st}$ = 4.7 B.

The value of the current divider choose: I_d=1 mA.

Ballast resistance of reference voltage source:

$$R_h=\frac{E-U_{st}}{I_{st\ min}}=\frac{15-4.7}{3\cdot 10^{-3}}=3.4\,kOhm$$

Is selected by row e24: $R_b = 3.9\ kOhm$.
The resistance of the second divider:

$$R_{d2}=\frac{U'_{load}}{I_d}=\frac{4.7}{10^{-3}}=4.7\ kOhm$$

Is selected by row 24: $R_{d2} = 4.7\ kOhm$.
The division ratio of the divider:

$$K_d=\frac{U'_{load}}{U_{load}}=\frac{4.7}{12}=0.39$$

The resistance of the first divider:

$$R_{d1}=\frac{R_{d2}(1-K_d)}{K_d}=\frac{4.7\cdot 10^{3}\cdot(0.61)}{0.39}=7.4\,kOhm$$

Is selected by row (table) e24: $R_{d1} = 8.2\ kOhm$
As the operational amplifier selected chip LM392:

Table 5. Parameters of op-amp LM392.

Type	Type power supply V, U_{ps}	U_{in}, V	U_{out}, V
LM392	3-32	-0.3-32	U_{ps}-1.5
		0	0.4
I_{in}, nA	I_{out}, mA	K_u	Number of OG on the case
250	6	10^4	OG
250	20	$2\cdot 10^4$	Comparator

$$R_{feedback}=K\cdot R_1=30\cdot 10\cdot 10^{3}=300\ kOhm$$

Is selected by row e24: $R_{feedback} = 330\ kOhm$.
Select the Zener KC139A:

Table 6. Parameters of the Zener diode KC139A.

Model	$I_{st\ min}$, mA	$I_{st\ max}$, mA
KC139A	3	70
$U_{st\ Rated}$, V	$U_{st\ max}$, V	
3.6	4.3	

3.5 The calculation diagram of the sawtooth generator

The sawtooth generators find wide application in automation, television, communications technology, measurement technology, power supplies and other applications of electronics [8, 19].

Most often performed with an external control. The duration of the stroke is determined by the duration of the external control pulse of a rectangular shape [9]. If necessary, you can create generators running in sync mode or the oscillatory modex [10].

In the simplest case, when not required high linearity of the working area of the output voltage, they are executed according to the scheme of integrating RC-circuit with the reset of the charge during the return stroke of the saw. The formation of a sawtooth voltage based on the alternation in time of the processes of charge and discharge of the capacitor. The function of the element, creating a circuit to quickly discharge capacitance, performs a transistor in a key mode. Discharge of the capacitor and maintaining the voltage close to zero, occurs in the interval tp input signal when the transistor is in saturation (fully open). Linearly-varying voltage in the interval of the stroke is formed when the transistor is closed (nonconductive state).

In (Fig.5) shows a schematic of the sawtooth generator.

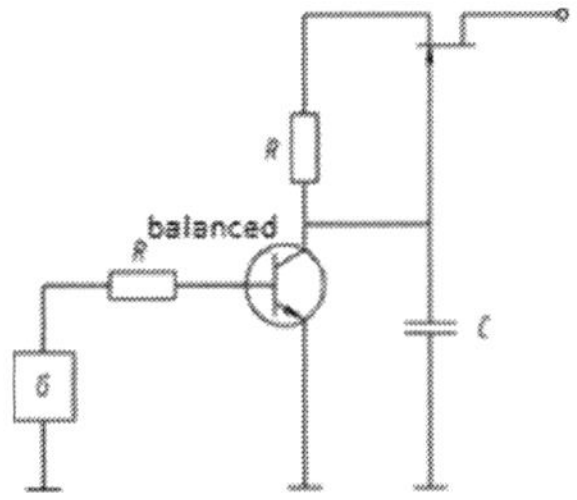

Fig. 5. Oscillator Circuit with a constant current charge.

The pulse period of the sawtooth is calculated according to the formula:

$$T = \frac{1}{f_{sw}} = \frac{1}{25000} = 40\,\mu s$$

The discharge period of the capacitor:

$$t_n = 0.02 \cdot 40 = 0.8\ \mu s$$

The period of charging of the capacitor:

$$t_p = T - t_n = 4 - 0.8 = 3.2\,\mu s$$

The calculation of the capacitance of the capacitor:

$$C_6 = \frac{I_c \cdot t_p}{U_g} = \frac{1 \cdot 10^{-3} \cdot 3.2 \cdot 10^{-6}}{3} = 1\ \text{nF}$$

Taken on a table e24: $C_6 = 1\ nF$

The value of the voltage of the generator is chosen equal to: $U_g = 3$ V.

The current value of the capacitor equal to: $I_{pow.s} = 1\ mA$

The calculation of the resistance of drain circuit:

$$R_9 = \frac{T}{c_6} = \frac{4 \cdot 10^{-6}}{1.06 \cdot 10^{-9}} = 3.77\ kOhm$$

Is selected by row e24: $R_{10} = 3.8\ kOhm$.
Selects the transistor VT1 – KT315G

Table 7. Parameters of the transistor KT315G.

Type	H_{21e},	$U_{BE.\ sat.}$, V
KT315G	350	1
$U_{CE.sat.}$, V	U_{kemax},V	I_{kemax},A
0.4	1	0.1

Select the VT2 transistor – KP302A

Table 8. Parameters of the transistor KP302A.

Type	$U_{GD\,max}$, V	$U_{GS\,max}$, V
KP302A	20	10
$U_{DS\,max}$, V	I_{max}, μA	
20	24	

The base current:

$$I_B = \frac{I_{ps}}{h_{21max}} = \frac{1 \cdot 10^{-3}}{350} = 2.85\ \mu A$$

Base resistance:

$$R_B = \frac{E - U_{BE\,sat}}{I_B} = \frac{15-1}{2.85 \cdot 10^{-6}} = 4.9\ MOhm$$

Is selected by row e24: $R_B = 5.6\ MOhm$.

4 Results

As a result, the circuit was simulated and substantially improved in the Multisim program using a separate power source for the control device, which allows controlling the processes occurring both in the network and in the load, regardless of the state of the power unit (RL = 0 or RL = ∞) [11], which simplifies the task of building a reliable control device.

Using the obtained calculations, it is planned to make a test model of the device. The results of this testing will be described in a future article.

5 Discussion

In this section, I want to discuss the merits of a switching voltage regulator relative to a linear regulator. Firstly, it has less energy loss for heating the regulating element, which increases the efficiency of the stabilizer and allows the use of a regulating element of lower power, and a radiator of smaller sizes and weights. You can also note a high efficiency, especially when working in a wide range of input voltages [12, 137]; small dimensions and mass (high specific power); fundamental possibility of galvanic isolation of input and output circuits.

6 Conclusion

Today, the scope of application of such devices has decreased in connection with reduction of power supply sources with transformerless input. Nevertheless, the use of switching regulators in a number of cases, it is economically more profitable than any other DC-DC Converter [13, 202].

In this paper we discuss the main provisions for the design of power supplies and electronic equipment. When developing a supply of funds was received for construction of circuits of power. Required the estimated coefficient characterizing their performance [14, 47]. On the technical task [15, 97] was calculated switching voltage regulator, developed in the software Multisim that will be done in further studies.

References

[1] V.P. Shuvalov, *Telecommunication systems* (M.: Radio and communication, 236, 2000)
[2] A.Kh. Shogenov, D.S. Strebkov, Ya.Kh. Shogenov; Ed. Analog, digital and power electronics: textbook (M.: FIZMATLIT, 215, 2017)
[3] E.M. Romash, *High-Frequency transistor converters* (M.: Radio Communications, 174, 1988).
[4] I. Hamidah, D.F. Ramadhan, The Performance Comparison Between Commercial Automatic Voltage Stabilizer And Programmable Automatic Voltage Stabilizer, Article (2020). ISBN: 1823-4690
[5] P. Bogomolov, I. Borzdyko, AC Voltage Stabilizer, Article. DOI: 10.30987/conference article_5c19e5e6ca1ba2.66670915 (2018)
[6] E.I. Brezhnev, *Effect Transistor apparatus of wide application: reference book* (M.: Radio and communication, 278, 1981)
[7] V.V. Durkin, *Circuitry of analog electronic devices, Basic concepts, feedbacks, the operation of the amplifying element in the circuit: study guide* (Novosibirsk: NSTU Publishing House, 165, 2017)
[8] A.V. Mitrofanov, *Pulsed sources of secondary power in consumer audio* (M.: Radio Communication, 204, 1987)
[9] R. Maharjan, S. Kamalasadan, Voltage Stability Index for Online Voltage Stability Assessment, Article. DOI: 10.1109/NAPS.2015.7335245 (2020)
[10] M.S. Rawat, S. Vadhera, Voltage Stability Assessment Techniques for Modern Power Systems, Article (2020). ISBN: 9781522585534
[11] V.K. Bityukov, D.S. Simachkov, *Sources of secondary power: textbook* (M.: Infra-Engineering, 2017)
[12] G. Sysoeva, *Power supply devices and systems of telecommunications: textbook* (VSL, 255, 2002)
[13] A.A. Bass, *Secondary power supply Sources with transformerless input* (M.: Radio Communication, 368, 1987)
[14] N.V. Sukhanova, Fundamentals of Electronics and Digital Circuitry: textbook (Voronezh: VGUIT, 247, 2017)
[15] E.T. Romancewe. *The Development and preparation of design documentation of electronic equipment*: a handbook (2nd ed. and additional, M.: Radio and communication, 114, 1989)

Assessment of feasibility of using the existing electric power infrastructure for charging electric and hybrid vehicles

Vyacheslav Rakov[1,*], *Oleg* Pikalev[1], *Timur* Akhmetov[2], *Tamara* Budavina[1] and *Peter* Smirnov[1]

[1]Vologda State University, Department of automobiles and automobile industry, 160000 Lenin st., 15, Russia
[2]Kazan State Power Engineering University, Kazan, Russia

Abstract. Many countries around the world are considering the possibility of a complete transition to electric cars. This raises questions about the capabilities of the existing infrastructure for electricity generation. This paper presents calculations on the example of the Russian power system. It is conventionally accepted that all cars with petrol engine will be run on electricity. To understand how much electricity consumption will change, a calculation technique has been proposed. As initial data, statistics on the production of electricity, petrol and diesel fuel over the past 20 years was used. As methodological approaches, a generalization and calculation were carried out, on the basis of which a theoretical experiment was conducted. The authors concluded that the increase in electricity consumption will be only 5%, which will not create an excessive load on the existing energy infrastructure.

1 Introduction

In many countries of the world, the possibility of converting a significant part of vehicles to run on electricity is being considered. It is planned to charge them using the capabilities of the existing power system. 2019 global annual growth in sales of electric vehicles worldwide exceeded 1 million units per year, while the annual growth rate over the past 5 years is about 30%. The governments of many countries and car manufacturers are gradually displacing cars with petrol and diesel engine and develop production of electric cars.

This paper considers the example of power system of Russia. In Russia, the number of electric vehicles already in use is insignificant: only about 6 thousand. However, government development programs contribute to their popularity. The largest number of electric buses among all European capitals of the world is in Moscow. The transport strategy of Russia for the period until 2030 involves increasing the environmental safety of transport, including by expanding the use of electric cars, which are environmentally-friendly vehicles. For example, since 2020, in Russia, customs duties on the import of electric vehicles are reset to zero, which clearly indicates a state stimulation of demand for them. The infrastructure of electric vehicle charging station network will also expand significantly in the coming years [1-2]. Along with an increase in the number of electric cars, the number of electric charging stations is also growing. According to the official website https://electromobili.ru, from 2018 to 2020, their number in Russia has grown 3 times and today is 161 pcs.

However, it should be noted that when planning the development of electric car fleet, it is necessary to assess the capabilities of the existing system for production and distribution of electricity. The influence of electric transport on power systems is considered in papers of D.Yu. Tsypulev [3-4]. Studies are also being conducted to optimize the placement of charging stations, taking into account the present infrastructure and the routes of electric vehicles [5]. However, after analyzing many scientific works, the authors did not find a sufficiently broad study of the considered issues. The present paper analyzes the possibility of adapting the electricity infrastructure to power a significant number of charging stations using the example of Russia.

For the period until 2030 the development dynamics of the power system of Russia implies a large-scale decrease in specific energy consumption in the economy and energy (by 2.1-2.3 times) with a significant increase (by 1.4-1.6 times) in the number of consumers and growth in electricity exports (by 1.1-1.2 times). Currently, the growth in energy consumption is not increasing, and even may slow down after 2020.

2 Materials and methods

Statistical information on the production of electricity, petrol and diesel fuel in Russia was used as the initial data for the study.

A characteristic of electricity production by years over the past 20 years was compiled based on the open source data of the Ministry of Energy of the Russian Federation, Figure 1. The plot shows that the annual electricity production in Russia is about 1,050 billion kW•h and has stabilized over the past year.

* Corresponding author: vyacheslav.rakov@mail.ru

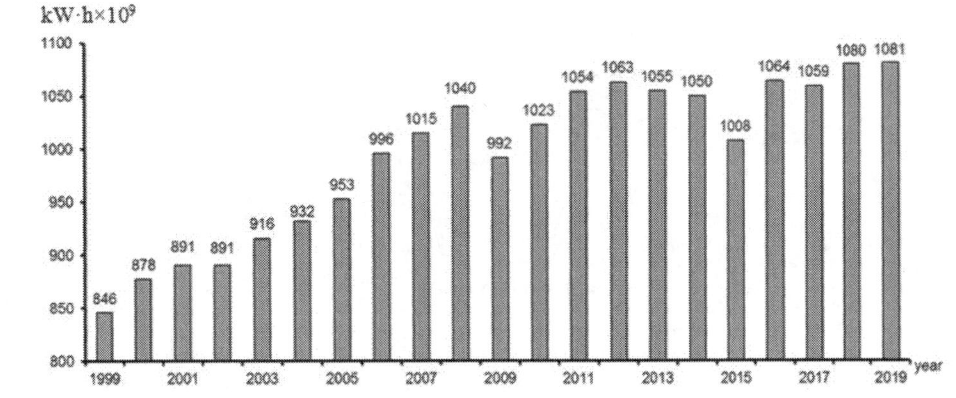

Fig. 1. The dynamics of electricity production over the past 20 years in Russia.

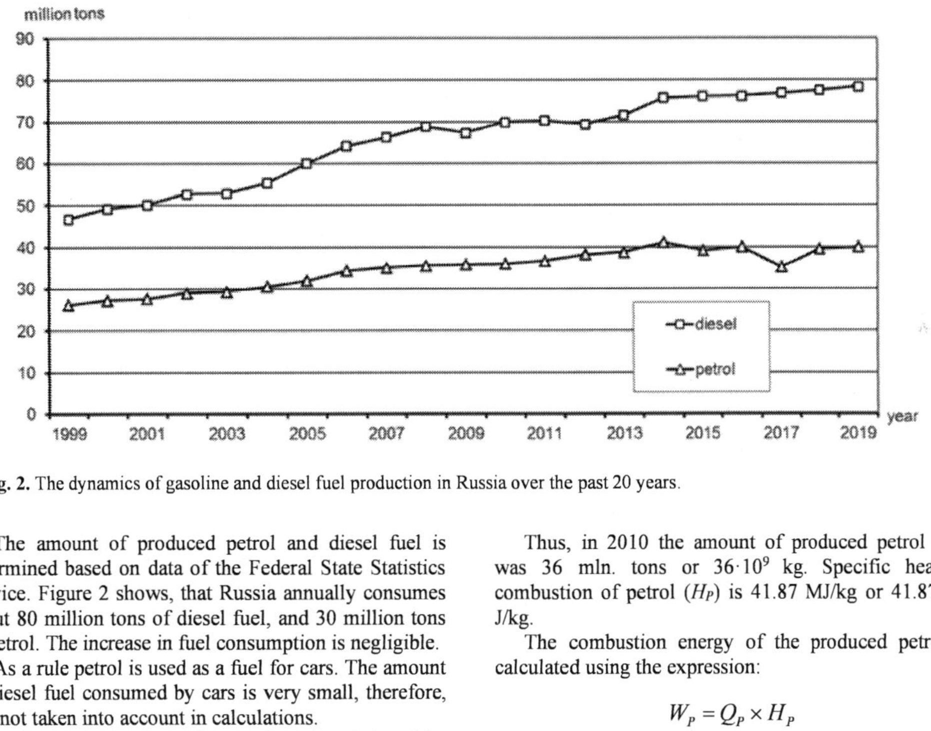

Fig. 2. The dynamics of gasoline and diesel fuel production in Russia over the past 20 years.

The amount of produced petrol and diesel fuel is determined based on data of the Federal State Statistics Service. Figure 2 shows, that Russia annually consumes about 80 million tons of diesel fuel, and 30 million tons of petrol. The increase in fuel consumption is negligible.

As a rule petrol is used as a fuel for cars. The amount of diesel fuel consumed by cars is very small, therefore, it is not taken into account in calculations.

We can compare the volumes of produced electricity and hydrocarbon fuel by reducing them to uniform terms [6]. For this, kW per hour and the energy of hydrocarbon combustion are reduced to MJ using the well-known relationship (1):

$$1\text{kW}\cdot\text{h} = 3.6\text{MJ} \tag{1}$$

For initial data on the produced electricity for the year 2010 we obtain:

$$W_e = 1023\cdot 10^9 \times 3.6 = 3683\cdot 10^9 \text{ MJ}$$

Thus, in 2010 the amount of produced petrol (Q_P) was 36 mln. tons or $36\cdot 10^9$ kg. Specific heat of combustion of petrol (H_P) is 41.87 MJ/kg or $41.87\cdot 10^6$ J/kg.

The combustion energy of the produced petrol is calculated using the expression:

$$W_P = Q_P \times H_P \tag{2}$$

For initial data on the produced petrol for the year 2010 we obtain:

$$W_P = 36\cdot 10^9 \times 41.87\cdot 10^6 = 1507\cdot 10^9\text{MJ}$$

The combustion energy of diesel fuel is calculated in a similar way. The amount of diesel fuel (Q_D) produced in 2010 was $69.9\cdot 10^9$ kg. The specific heat of combustion of diesel fuel (H_D) is $43.5\cdot 10^6$ J/kg.

When considering the initial data of 2010, the combustion energy of all diesel fuel will be:

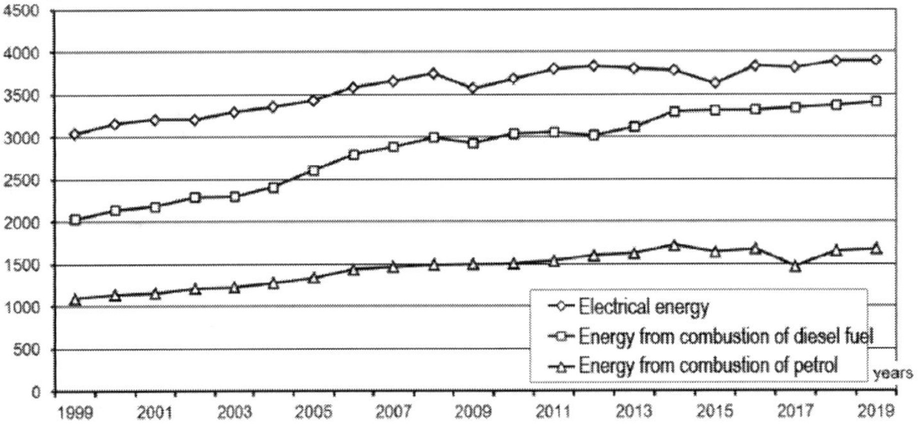

Fig. 3. The dynamics for electrical energy consumed in the Russian Federation, thermal energy from combustion of diesel fuel and petrol over 20 years, reduced to MJ.

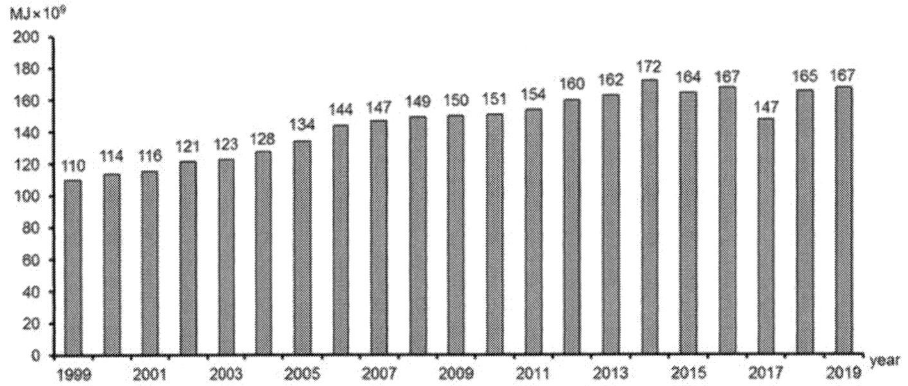

Fig. 4. The dynamics of useful transport work done by petrol cars over the past 20 years.

$$W_D = 69.9\cdot10^9 \times 43.5\cdot10^6 = 3041\cdot10^9\,\text{MJ}$$

Figure 3 presents comparison between the overall produced electrical energy and the energy of the produced petrol and diesel fuel in MJ. In uniform terms, it becomes clear that, on average, the energy of combustion of all produced petrol is about 3 times less than the overall produced electrical energy. But the energy of the produced diesel fuel in recent years is comparable to electrical energy.

However, not all fossil fuel energy is converted into useful work. Most of it is released into the atmosphere in the form of heat. The ratio between the thermal energy of fuel and the energy consumption for the transport process is determined by the operational efficiency of the vehicle [7].

The studies carried out at the Vologda State University [8] and by other scientific groups [9, 10] showed that when a car moves under real operation conditions, the fuel efficiency of internal combustion engine is very low. So in urban operation conditions, the average efficiency of cars with petrol engine approaches 5-6%, and the maximum value, as a rule, does not exceed 20%. Thus, taking the average value of the vehicle's efficiency equal to 10%, we can calculate the useful energy costs.

Energy costs for implementation of useful transport work W_{UW} can be determined by:

$$W_{UW} = 0.1\cdot W_i \tag{3}$$

For motor vehicles powered by petrol W_{UW} for the year 2010 is:

$$W_{UW} = 0.1\times1507\cdot10^9 = 151\cdot10^9\,\text{MJ}$$

Based on the performed calculations we constructed a plot of useful work carried out by all petrol vehicles, Figure 4.

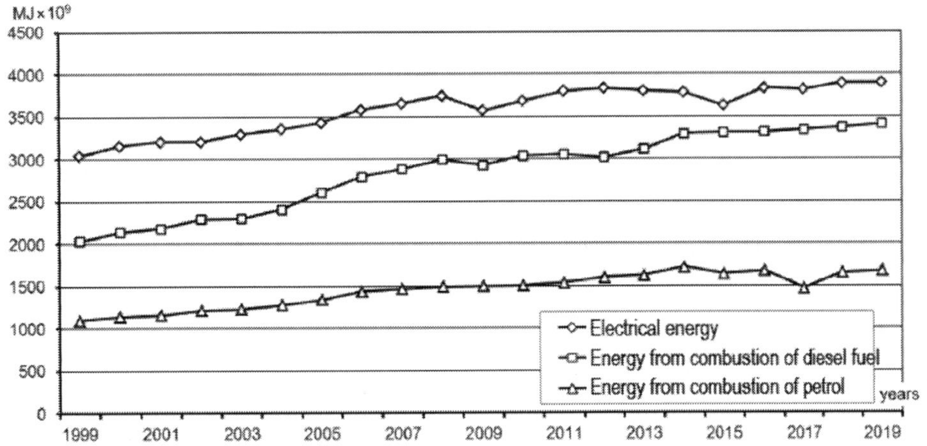

Fig. 5. Additional hypothetical costs of electrical energy in case of replacement of petrol cars in comparison with the amount of electricity produced by all power plants.

Now assume that the amount of transport work shown in figure 4 is performed by electric vehicles, the operational efficiency of which is approximately 0.65 [11-12]. With this in mind, the cost of electrical energy required to charge all electric vehicles is calculated by the formula:

$$W_{ev} = W_{tr} / 0.65 \tag{4}$$

Figure 5 shows how the total electricity production would correlate with additional hypothetical electricity costs for the charge of all electric vehicles.

Figure 5 shows that the increase in electricity consumption during the theoretical replacement of all petrol cars with electric cars will be about 5%, which should not cause a shortage of electricity in the power system of Russia.

3 Results

The presented analysis shows that if all petrol-powered cars are electrified, the load on the existing infrastructure for the production and distribution of electricity will increase by about 5%, which is not significant. Consequently, the wide-sweeping conversion of cars with petrol engine to run on electricity is quite achievable in the coming years. Nevertheless, it should be noted that the growth in electricity consumption can be uneven and will affect, first of all, large cities. Therefore, the infrastructure of charging stations should be developed step by step as far as the volume of electricity generation increases or it is saved through the introduction of energy-efficient technologies.

4 Conclusions

The calculations show that the wide-sweeping conversion of cars with petrol engine to run on electricity will not lead to a significant overload of the existing power system and can be carried out in the near future.

Wide-sweeping conversion to electricity of heavy vehicles and buses with diesel engines with the existing power system is impossible for 2 reasons. Firstly, there are no compact, energy-intensive and relatively cheap batteries of electrical energy, which will not lead to a significant loss of carrying capacity. The charging time of existing batteries also does not allow them to be used with significant daily runs. Secondly, the generation of electrical energy of the whole power system should almost double, which can't be achieved in a short time even if not take into account the need to modernize the existing distribution infrastructure.

The use of hybrid power plants in freight and passenger vehicles can become a more acceptable option for large cities. Moreover, it creates the possibility of a systematic development of power supply system with the final transition to electric cars in the future.

The presence of a transitional stage is also justified from an environmental perspective. The accumulated experience in operating vehicles with a hybrid engine shows that they have much higher environmental performance than conventional cars with internal combustion engines. In addition, with a limited daily mileage typical for operation in city, the use of plug-in hybrid vehicles is allowed, which further reduces the need for hydrocarbon fuel.

References

[1] D.I. Demidov, V.V. Pugachev, Forecast of global development of electric transport and infrastructure of electric charging stations [Prognoz global'nogo razvitiya elektrotransporta i infrastruktury elektricheskikh zapravochnykh stantsiy], News of the Orenburg State Agrarian University 5, 79, 173-178 (2019).

[2] V.E. Yutt, B.K. Ospanbekov, Promising directions for the development of charging stations for electric vehicles [Perspektivnyye napravleniya razvitiya zaryadnykh stantsiy dlya elektromobiley], Electronics and electrical equipment of transport 6, 10-12 (2013).

[3] D.Yu. Tsypulev, Influence of development of electric transport on power systems [Vliyaniye razvitiya elektrotransporta na energeticheskiye sistemy], Energy policy 1, 58-65 (2012).

[4] D.Yu. Tsypulev, Electric transport and its integration in intellectual networks of the future [Elektrotransport i ego integratsiya v intellektual'nyye seti budushchego], Alternative fuel vehicles 1, 31, 24-28.

[5] J. Dong, L. Changzheng, L. Zhenhong, Charging infrastructure planning for promoting battery electric vehicles: An activity-based approach using multiday travel data, Transportation Research Part C: Emerging Technologies 38, 44-55 (2014). DOI: 10.1016/j.trc.2013.11.001.

[6] I.K. Aleksandrov, V.A. Rakov, Analysis of prospects for the development of electrified road transport [Analiz perspektiv razvitiya ehlektrificirovannogo avtomobil'nogo transporta], Transport of the Russian Federation 2, 50-52 (2012).

[7] I.K. Aleksandrov, O.L. Belkov, V.A. Rakov, Assessment of the energy efficiency of ICE in unsteady mode [Otsenka energeticheskoy effektivnosti DVS v usloviyakh neustanovivshegosya rezhima raboty], Engineering Bulletin 6, 17-20 (2008).

[8] I.K. Aleksandrov, E.V. Nesgovorov, V.A. Rakov, Tractional Calculations for Vehicles with Adaptive Motors, Russian Engineering Research 30, 2, 122-124 (2010). DOI: 10.3103/S1068798X1002005X.

[9] V. Rakov, V.N. Ivanov, Y.G. Karpov, V.F. Melehin, R.A. Izmailov, V.K. Yun, D.A. Zaripova, Method for determining the basic energy characteristics of elements of a hybrid car engine, IOP Conference Series: Earth and Environmental Science 337, 1 (2019). DOI: 10.1088/1755-1315/337/1/012066.

[10] I.K. Aleksandrov, Energy efficiency of machine with partial energy recovery [Energeticheskiy KPD mashiny s chastichnoy rekuperatsiyey energii], Engineering Bulletin 9, 17-18 (2007).

[11] I.K. Aleksandrov, Determination of hybrid power installation parameters based on trolleybus driving cycle studies [Opredeleniye parametrov gibridnoy silovoy ustanovki na osnovanii issledovaniy yezdovogo tsikla trolleybusa], Alternative fuel vehicles 5, 41, 54-60 (2014).

[12] G. Pistoia, Electric and hybrid vehicles, Power sources, models, sustainability, infrastructure and the market (Oxford: The Netherlands Linacre House, Great Britain, 670, 2010).

E3S Web of Conferences **178**, 01057 (2020) https://doi.org/10.1051/e3sconf/202017801057
HSTED-2020

Analysis of influence of external atmospheric factors on the accuracy of fault location on overhead power lines

Olga Ahmedova[1], *Anatoliy* Soshinov[1], and *Natalia* Shevchenko[1]

[1]Kamyshin Institute of Technology (branch) Volgograd State Technical University, 403874 Russia, Volgograd region, Kamyshin, Lenina st. 6a

Abstract. Devices of fault location are widespread in electrical grids. In overhead power lines of voltages of 10 kV and higher, such devices are based on measurements of parameters in emergency mode. These devices can be divided into two main groups: designed to determine fault location of short circuits and of ground short circuit. In both cases, when detecting fault location, analyzed are not only the current and voltage parameters at the accident time, but also the parameters of overhead power line. When analyzing equivalent circuits of power lines, approximate tabular values of direct and quadrature parameters are used. It was revealed, that the line parameters are significantly affected by external atmospheric factors: ambient temperature, soil moisture, wind strength and direction, ice formation, etc. To accurately determine the fault location, it is necessary to evaluate the influence of these factors on the linear parameters of the overhead line. The paper presents analysis of the influence of changes in atmospheric factors on parameters of overhead power line. A methodology for calculating the operation setpoints of the device for fault location in power line is given.

1 Introduction

Uninterrupted and reliable transportation of electricity to consumers is one of the main tasks constantly solved by power engineers. To ensure this, electrical grids were created consisting of distribution substations and power lines connecting them.

When insulation of any place in a cable or extended overhead power transmission line (OHL) has a fault, the voltage applied to line creates a leakage or short circuit current through the disturbed section.

The insulation faults can be caused by various factors that either eliminate or continue their destructive influence.

Location of short circuit is especially important, since tripping of permanently faulted line is associated with a shortage of electricity and material damage to consumers. In these cases, accelerated fault search gives a large economic effect.

Devices for fault location (FL) primarily on overhead electrical lines of 10 kV and above, based on measuring emergency conditions, are widely used in electrical grids.

Now there are a wide variety of methods for fault location aroused from a variety of types and nature of faults, as well as structure and operation conditions of electric grids. They can be divided into two large groups: remote and topographic methods [1]. Remote FL methods consist in measuring the distance to the place of fault from the terminal or terminals of line. This group of methods, on one hand, satisfies the requirement on speed of fault location, but on the other hand, it has limited accuracy. Topographic FL methods detect the fault location on path of power transmission line, i.e. topographic location of fault. This group of methods has a high accuracy, but requires a lot of time. The classification of FL methods is presented in Fig. 1.

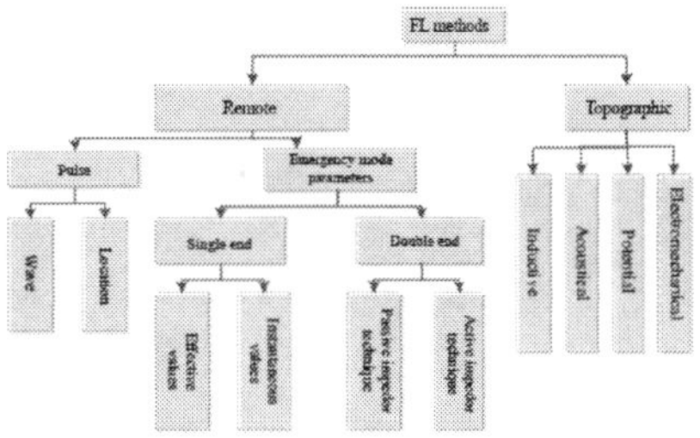

Fig. 1. Classification of FL methods.

To restore the normal operation mode of electric power systems, and reduce damage and costs, it is necessary to quickly and accurately detect the fault location on power lines. The requirements listed above are met by remote fault location methods, subject to increased calculation accuracy. As a rule, the calculation of power transmission line settings is based on short-circuit currents, which directly depend on the direct and quadrature parameters of the protected power line. Therefore, to ensure the necessary accuracy of calculations, it is necessary to analyze what external

* Corresponding author: alexu79@mail.ru

parameters affect the high-voltage power transmission line (HVL).

2 The method of calculation setpoints of FL device

Fig. 2 shows a diagram for fault location for a multi-circuit overhead power line.

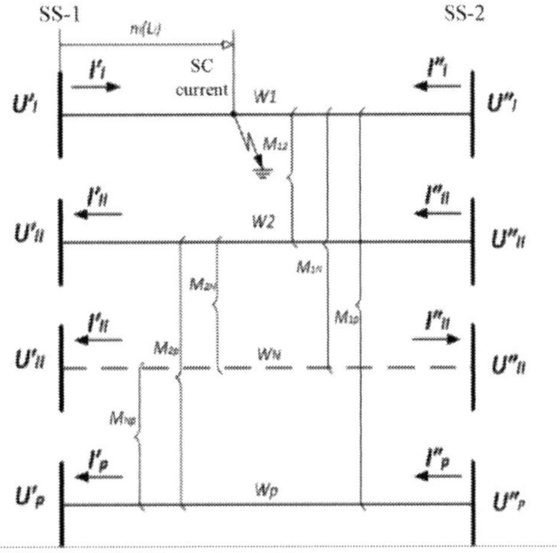

Fig. 2. Multi-circuit PL with mutual induction (n_I is the relative distance from SS-1 to short circuit location ($n_I=L_I/L_{OHL}$); L_I is the absolute distance from SS-1 to short circuit location, km; L_{OHL} is the length of OHL, km).

For the general case of homogeneous overhead lines electromagnetically coupled over the entire path (Fig. 2), the distance to SC location should be calculated [2] by solving a system of p linear algebraic equations (1) with p unknowns:

$$\begin{cases} z_1(I'_I + I''_I)n_I + Z_{12}I'_{II} + \dots + Z_{1P}I'_p = U''_I - U'_I + Z_I I''_I; \\ z_{21}(I'_I + I''_I)n_I + Z_{II}I'_{II} + \dots + Z_{2P}I'_p = U''_{II} - U'_{II} + Z_{21}I''_I; \\ \dots\dots\dots\dots\dots\dots\dots\dots\dots\dots\dots\dots\dots\dots ; \\ z_{p1}(I'_I + I''_I)n_I + Z_{p2}I'_{II} + \dots + Z_P I'_p = U''_p - U'_p + Z_{p1}I''_I. \end{cases} \quad (1)$$

where U'_I, U''_I, U'_{II}, U''_{II},... U'_p, U''_p are the zero sequence voltage measured at the terminals of overhead line W_I, - W_P;

I'_I, I''_I, I'_{II}, I''_{II},... I'_p, I''_p are the zero sequence currents measured at the terminals of the overhead line;

Z_I, Z_{II}, ... Z_p are the impedances of zero sequence of overhead lines W_I, W_{II}, ... W_P;

Z_{12}, Z_{21},... Zl_p, Zp_1, Z_{2p}, Z_{p2} are the mutual induction resistance between the lines $W_I - W_{II}$, $W_{II} - W_I$,... $W_I - W_P$, $W_P - W_I$, $W_{II} - W_P$, $W_P - W_{II}$.

Distance n_I for faulted WI OHL should be determined by the expression:

$$n_I = \frac{D_{nI}}{D} \quad (2)$$

where D is the determinant of the system of equations;

D_{nI} is the determinant obtained from determinant D by replacing a column composed of coefficients with unknown n_I by a column composed of free terms of the equation.

$$D = \begin{vmatrix} Z_I(I'_I + I''_I) & Z_{12} & \dots & Z_{1p} \\ Z_{2I}(I'_I + I''_I) & Z_{II} & \dots & Z_{2p} \\ \dots & \dots & \dots & \dots \\ Z_{p1}(I'_I + I''_I) & Z_{p2} & \dots & Z_p \end{vmatrix} \quad (3)$$

$$D_{nI} = \begin{vmatrix} U''_I - U'_I + Z_I I''_I & Z_{12} & \dots & Z_{1p} \\ U''_{II} - U'_{II} + Z_{21}I''_I & Z_{II} & \dots & Z_{2p} \\ \dots & \dots & \dots & \dots \\ U''_p - U'_p + Z_{p1}I''_I & Z_{p2} & \dots & Z_p \end{vmatrix} \quad (4)$$

To determine the short circuit location on WI overhead line, currents and voltages at the terminals of the damaged overhead line must be measured, as well as at all overhead lines having an electromagnetic connection with the damaged one.

The calculation formulas for a single-circuit overhead power line (Fig. 3) will take the following form:

$$l = nl = \frac{U'' - U' + Z_L I''}{Z_L(I' + I'')} L \quad (5)$$

where U',U'' and I',I'' are the voltages and currents at the terminals of the damaged overhead line, respectively; Z_L is the impedance of overhead line.

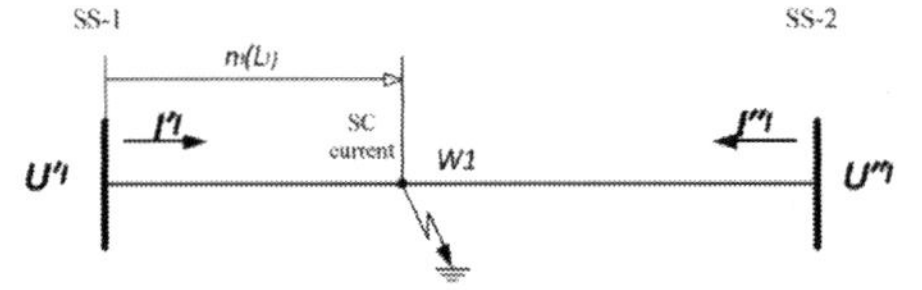

Fig. 3. Single-circuit PL.

The expression (5) shows that in order to accurately detect the OHL fault location, it is necessary to correctly determine the line resistance, take into account external factors that will directly affect it.

Setpoints of starting units (SU) of fault detectors should be calculated:

1) Starting units for symmetrical components of currents $I_{p.0}$, $I_{p.1}$, $I_{p.2}$;

2) Starting units for increments of symmetrical components of currents $\Delta I_{p.0}$, $\Delta I_{p.1}$, $\Delta I_{p.2}$;

3) Starting units for symmetrical components of currents, operating in conjunction with starting units for increments $\Delta I_{p.0}$ for ΔI, $\Delta I_{p.1}$ for ΔI, $\Delta I_{p.}$ for ΔI.

The main condition when choosing setpoints is to ensure the necessary sensitivity during short circuit within the network section where FL function is required. Additional (optional) conditions are elimination of false FL starts in load conditions, as well as minimization the number of unnecessary starts during short-circuit on external network elements in which the FL function is not provided.

Regulatory requirements for minimum sensitivity coefficient k_s have not been established. It is recommended to take k_s=2, which ensures reliable start-up of FL even with short-circuit, accompanied by significant transition resistance.

Pick-up current of the starting units according to current is determined based on ensuring the required sensitivity coefficient by the expression:

$$I_{p.n} = \frac{I_{SC.\min n}}{k_s} \quad (6)$$

where n is the sequence number: 1 – direct, 2 – reverse, 0 – zero; k_s is the required sensitivity coefficient; $I_{p.}$ is the pick-up current of the starting unit of direct, reverse or zero sequence; $I_{SC.min\ n}$. is the minimum value of the corresponding symmetrical component of the primary current at the point of switching on the device during short circuit in rated currents.

Three-phase short circuit current I_{SC} is calculated using the expression

$$I_{SC}^{(3)} = \frac{U_s}{\sqrt{3} \cdot \sqrt{r_e^2 + x_e^2}} \quad (7)$$

where U_s is the system voltage, V; r_e , x_e is the active and reactive resistances to the point of short circuit, reduced to the source voltage stage, Ohm.

From the expression (7) it follows that the active and reactive resistances of the overhead line are the variable and effective factors that affect the accuracy of fault location calculations. These parameters are affected by external climatic factors that are not taken into account in calculations.

The line impedance Z_L, Ohm, is calculated by the formula:

$$z_L = (r_0 + jx_0)L_L \quad (8)$$

where x_0 is the linear inductance of direct sequence, Ohm/km; r_0 is the linear resistance of direct sequence, Ohm/km; L_L is the line length, km.

As a rule, the linear parameters of power line are considered to be constant values, normalized in the reference literature. The given parameter data are calculated for normal external conditions (ambient temperature 20 °C, sunny, ground resistance under overhead lines is constant, etc.), although under actual conditions of overhead lines operation, climatic conditions vary significantly not only during the calendar year, but also in various climatic zones.

We analyze the influence of a change in atmospheric conditions on the linear parameters of an overhead power line.

Overhead power lines are characterized by the following parameters: specific resistance R, specific reactance X, specific conductance G and specific susceptance B. As a rule, these parameters are taken per unit length of the overhead line and the linear values, R0, X0, G0, B0, respectively, are used in the calculations. When analyzing power lines, tabular approximate values of direct and quadrature parameters are used in equivalent circuits, although solving problems in unsimplified form leads to significant refinements of known solutions.

3 Analysis of influence of changes in atmospheric factors on OHL parameters

3.1 Analysis of changes in HVL reactance

Weather factors directly affect the overlap of the OHL insulation, but only visually observed parameters are taken into account, and the effect of changes in soil resistance under HVL is not considered. This parameter is not taken into account in calculations, assuming that the soil has infinite conductivity and all current is concentrated on the surface. In reality, the soil has finite conductivity (Fig. 4) and the current penetrates to a certain depth, depending on the resistance of the soil and taking into account that the soil is homogeneous, decreases with distance from the wire deep into and to both sides.

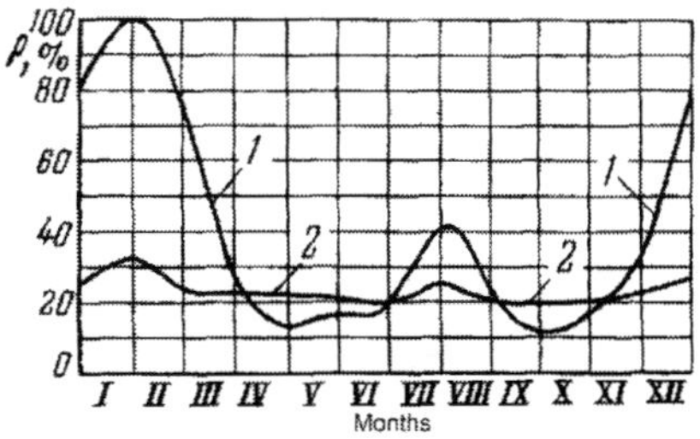

Fig. 4. Change in specific soil resistance over the year. 1 - at a depth of 0.7 m.; 2 – at a depth of 2.5 m.

Analysis of curves presented in Fig. 4, shows that a sharp decrease in soil resistance occurs with an increase in humidity of 15 - 20%, further humidification has a little effect on its resistance. Tabular resistances are given at a depth of about 0.7 m. However, Fig. 4 shows that in deeper layers of soil it decreases significantly and no longer has pronounced oscillations. This is due to the fact that soil is not subject to seasonal fluctuations (snow melting, overdrying in summer, freezing), but has a constant humidity value, therefore it is assumed that at a depth of 3 m and deeper the soil resistance will have a value of 15% of the tabular one [7].

A method for analyzing the effect of the finite conductivity of soil was proposed by Rüdenberg [9].

The wire reactance X_w is usually determined by the method of mirror images, i.e. the system "wire - image wire passing underground" is considered with a distance between the wires D_z such that the wire resistance of the two-wire line is equal to the wire resistance in the wire-soil system.

$$D_z = \frac{0.178}{\sqrt{f\gamma \cdot 10^{-9}}} \text{cm} \quad (9)$$

where f is the frequency, Hz; γ is the soil conductance, S.

Reactance of the wire of the wire-soil system is:

$$X_w = 29 f \lg \frac{D_z}{r_0} \cdot 10^{-4} \text{Ohm/km} \quad (10)$$

where r_0 is the wire radius, m.

From expressions (9), (10), it follows that the wire reactance depends on the soil resistance, which varies in a wide range under the influence of factors such as temperature and humidity [8].

Figure 5 shows the relationship between AS-120 wire reactance and the soil resistance.

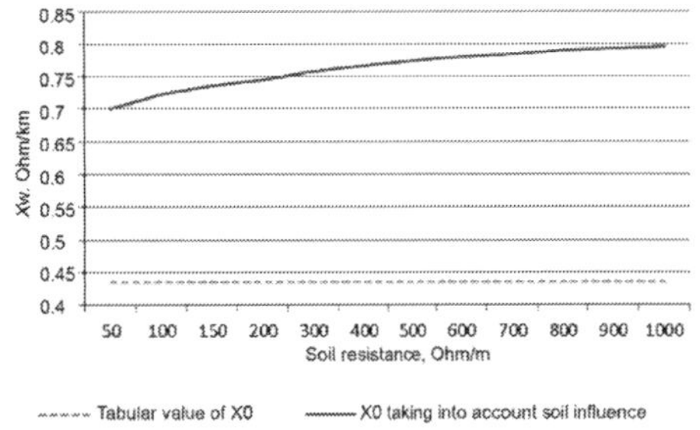

Fig. 5. Relationship between the reactance of 1 km wire and the soil resistance.

Therefore, if when determining the reactance of the overhead power line wire by the method of mirror images, the final soil conductivity under the overhead line is not taken into account, then the calculation error will be about 65%. The tabular value of reactance does not includes the soil conductivity, therefore it differs from the calculated one taking into account the finite conductivity of soil (10) almost by two times [8].

3.2 Analysis of the influence of weather conditions on the OHL active resistance

The data presented in the reference literature is designed for a wire temperature of 20 °C; they do not take into account seasonal changes in ambient temperature inherent for almost the majority of the Russian Federation.

The relationship between resistance and wire temperature is determined by the formula:

$$R_{0t} = R_{020}\left(1 + \alpha\left(t_w - 20°\right)\right) \quad (11)$$

where R_{020} is the tabular value of the specific resistance at wire temperature of 20° C; t_w is the wire temperature, °C; α is the temperature coefficient of electrical resistance, Ohm/deg.

The temperature of the overhead power line wires depends on the ambient cooling conditions (air temperature) and the current flowing through them. We analyze the relationship between AS-120 wire resistance and temperature $R_{0t} = f(t)$ in the indicated temperature range (Fig. 6).

The graph shows that increasing the wire temperature by 10 °C leads to an increase in wire resistance by 4%.

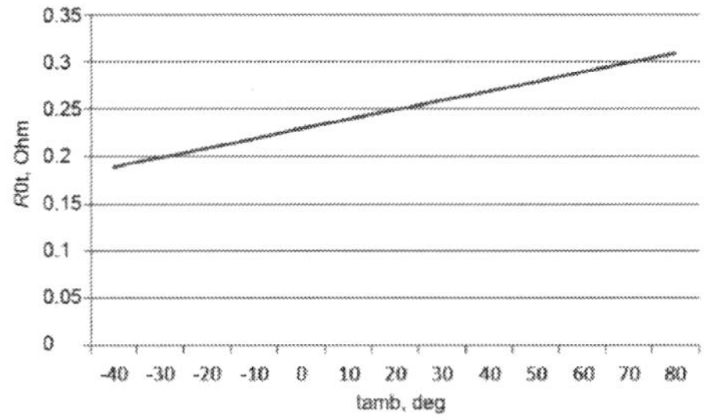

Fig. 6. Relationship between resistance of 1 km AS-120 wire and ambient temperature.

At current loads, which are limiting in terms of heating conditions, the wire temperature can reach + 70 °C, and at low ambient temperature and low loads up to - 50 °C, therefore, the resistivity can increase by 20% and decrease by 30%.

From the equilibrium condition, the steady-state wire temperature is determined for any OHL operation mode:

$$I^2 R_{020}\left(1 + \alpha\left(t_w - 20°\right)\right) = \sigma F\left(t_w - t_{\text{amb}}\right) \quad (12)$$

where I is the wire current, A; σ is the heat transfer coefficient equal to the amount of heat removed in 1 s from 1 cm^2 of the wire surface at a temperature difference between the wire and the environment of 1 °C, W/m^2 · deg; F is the wire cooling surface, cm^2; t_{amb} is the ambient temperature, °C.

Heat transfer coefficient:

$$\sigma = \sigma_L + \sigma_K = 2.8\left(100 + 0.6\left(T_w - 273\right)\right) \cdot 10^{-2} + 8.95\sqrt{\frac{p \cdot v}{d}} \quad (13)$$

where p is the air pressure, Pa; T_w is the wire temperature, K.

From expressions (12,13) it follows that the wire temperature is influenced not only by wind speed and current passing in line, but also by the ambient temperature.

Given expression (13), we find temperature of the overhead line wire from equation (11):

$$t_w = \frac{I^2 R_{020} + \sigma F t_{\text{amb}} - 20\sigma F}{\sigma F - I^2 R_{020}\alpha} + 20 \quad (14)$$

Based on expression (14), we construct the relationship between AS-120 wire temperature and the surrounding air temperature, wind speed, and current flowing through it (Fig. 7). Here we take into account atmospheric pressure equal to 1, and the ambient temperature in the range from + 40 ° C to -40 ° C, wind speed from 0.6 m/s, corresponding to motion of air masses only by heating up to 6 m/s. Since I_{add} is current passing through the overhead line at normal temperature

(+ 25 °C) and in the absence of wind, capable of heating the wire to the maximum permissible value (+ 70 °C), the influence of the current load was estimated in the range from 0 to I_{add}.

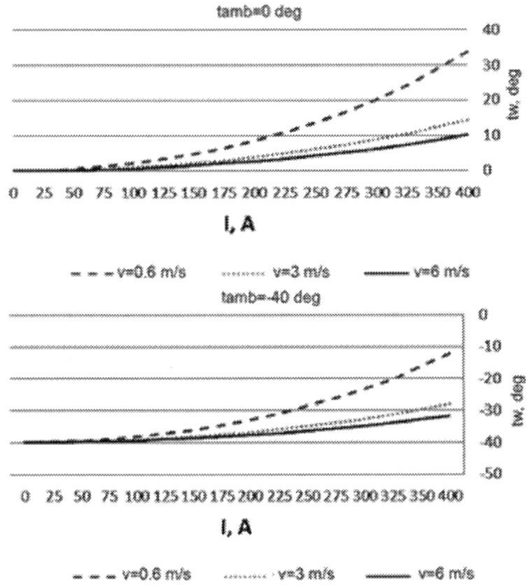

Fig. 7. Relationship between AS-120 wire temperature and the ambient temperature, wind speed, and current flowing through it.

Fig. 7 shows that at small and medium currents, passing through the overhead line and not high wind loads, significant changes in wire temperature occur mainly due to fluctuations in the ambient temperature. If the current load is more than 30% of the permissible line current and the wind speed is not high, then the current passing through the conductor has a noticeable effect on its heating. With increased wind speed, heat dissipation is significantly improved even for high flowing current.

Determine the wire impedance:

$$z_{0w} = \sqrt{R^2{}_w + (X_L - X_s)^2} \quad (15)$$

Based on expression (15), we plot the relationship between impedance of the AS-240 wire and the ambient temperature, taking into account the final soil conductivity (Fig. 8).

From Fig. 8 it follows that considering the ambient temperature in calculations significantly changes the line impedance. If we estimate the error at two extreme points (at a temperature of + 40 °C and - 40 °C), then it will be 8% and 23%, respectively. When taking into account the finite conductivity for the same ambient temperatures, the error will be 48% and 19%.

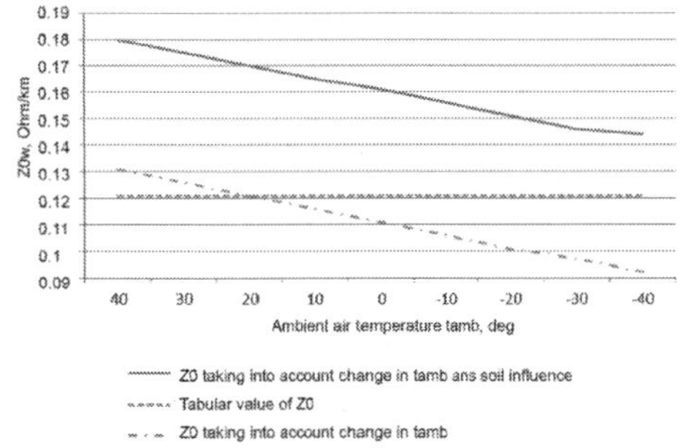

Fig. 8. Relationships between impedance of the AS-240 wire and the ambient temperature, taking into account the finite soil conductivity.

4 Conclusions

Based on the performed analysis, the following conclusions can be drawn:

1) The accuracy of fault location on overhead power line directly depends on the consideration of external climatic factors, since their influence cannot be foreseen analytically;

2) An increase in wire temperature by 10 °C results in an increase in wire resistance by 4%;

3) The tabular value of reactance does not take into account the soil conductivity, therefore it differs from the calculated one, which takes into account the soil finite conductivity by almost two times;

Therefore, fault location devices must be installed on overhead lines with sensors of wire temperature and soil moisture. Information from them at the time of accident will be transmitted to the computer unit, so the fault detection will be performed given the correct data of the power line parameters.

References

[1] A.I. Eisenfeld, Methods for determining short circuits on overhead power lines using fixing devices (Moscow: Energiya, 2014).

[2] A.I. Eisenfeld, G.M. Shalyt, Detection of places of short circuit on lines with branches, 2nd ed. (Moscow: Energoatomizdat, 1988).

[3] E.A. Arzhannikov, Application of remote principle in the conditions of ground faults for relay protection, automation and devices for fault location on power lines: Thesis for the Dr. tech. sciences (Ivanovo, 1996).

[4] E.A. Arzhannikov, V.Yu. Lukoyanov, M.Sh. Misrikhanov, ed. V.A. Shuina, Location of short circuit on high-voltage power lines (Moscow: Energoatomizdat, 2013).

[5] E.A. Arzhannikov, A.M. Chukhin, Methods and devices for fault location on power lines (Moscow: NTF Energopress, 1998, Biblioteka electromontera, issue 3).

[6] O.O. Akhmedova, Refined algorithm for calculating the active resistance of overhead power line

taking into account weather conditions, International Journal of Applied and Fundamental Research 12, 3, 387-398 (2016).

[7] O.O. Akhmedova, Analysis of linear parameters of overhead power line taking into account the final conductivity of soil, Successes in modern science and education 4, 4, 130-132 (2017).

[8] O.O. Akhmedova, A.N. Shilin, Study of the influence of ambient conditions on the parameters of overhead power lines for adjusting the settings of relay protection systems in real time, Energy and Resource Saving: Industry and Transport 2, 19, 32-36 (2017).

[9] Ollendorf. Erdstrome. (Berlin, 1928).

[10] Yu.Ya. Lyamets, V.A. Ilyin, N.V. Podshivalin, The software package for analysis of emergency processes and fault location to the power line, Elektrichestvo 12, 2-7 (1996).

[11] A.V. Zakabunin, V.V. Vlezkov, L. Klenovaya, D.A. Abramov, Features of choice of ASKUE depending on the technology of grid construction, Vestnik RGAZU 31, 36, 24 (2019).

[12] V.M. Rastorguev, Power cable monitoring and diagnostics, Vestnik RGAZU 31, 36, 43 (2019).

[13] A.L. Kulikov, M.D. Obalin, Development of software to support decision-making in the elimination of damage on power lines, News of higher educational institutions, Electromechanics 2 (2015).

[14] A.S. Butymov, A.S. Gaivoronsky, A.V. Purtov, A.L. Soloviev, A system for monitoring lightning discharges and fault location on overhead lines, Air lines 3, 37-42 (2012).

[15] A.L. Kulikov, M.D. Obalin, The use of adaptive procedures in algorithms for fault location on power lines, Industrial Energy 12, 35-39 (2013).

Analysis of the global energy resource market

Yulia Volotkovskaya[1], *Natalya* Volotkovskaya[2], *Alexander* Semenov[2,*], *Anastasia* Semkova[2], *Oleg* Fedorov[3], and *Oleg* Vladimirov[4]

[1]National Technical University "Dnieper Polytechnic", 49005, Dnipro, Ukraine
[2]North-Eastern Federal University n.a. M.K. Ammosov, Mirny Polytechnic Institute, 678170, Mirny, Russia
[3]Nizhny Novgorod State Technical University n.a. R.E. Alekseev, 603950, Nizhny Novgorod, Russia
[4]Kazan State Power Engineering University, Kazan, Russia

Abstract. The paper analyses the global demand for various energy resources in recent years. Nowadays, qualitative and quantitative changes in structure of distribution and consumption of energy resources take place. The paper has reviewed a number of papers predicting two-fold increase in the global energy production market by 2040. A number of other papers also predict the shift of energy demand from the USA and EU countries to the Asian market. The paper presents percentage distribution of world's energy requirements satisfied by the six types of resources among which oil, gas, and coil are still predominant. Some papers have stated that the demand for natural gas will continue to grow faster than the demand for oil or coil in the future. To confirm or deny this theory the authors have designed the mathematical model showing trends in demand for gas compared to the total demand for oil and coal; as a result, two different polynomial functions have been obtained with almost the same probability confirming this prediction. Trends in energy production and change of global energy demand across regions over the past twenty years have been studied. The studies show that, in general, energy production has been increased for the last decade. However, due to economic crises fluctuations, energy production in EU countries has abruptly decreased; it results in displacement of the EU countries from the market by the countries of the Middle East and the Near East. The paper also notes such important factors as decrease in solid fuel production in Great Britain; steady increasing importance of renewable energy sources for the last few years; reducing of the rate of nuclear energy development by some key countries; general lack of energy recourses in EU countries, etc.

1 Introduction

In the XXI century, the advent of new technologies in production results in qualitative and quantitative change in structure of distribution and consumption of energy recourses. In global economy, the majority of countries tend to reduce domestic energy consumption. They actively employ local kinds of conventional fuel that should be necessarily renewable one in industry [1]. During 2015-2019, increase in production of the main kinds of products was relatively small in most countries due to political instabilities. Besides, imposing sanctions on our country also influenced the structure and pattern of growth of production rate [2].

Over the next twenty years, in the world market, the bulk of the demand for energy resources is expected to shift from the USA and the EU countries to the Asian markets, which are rapidly developing now. According to this expectation, the global energy production market should almost double by 2040. This prediction is based upon the positive trend of economic indicators of the countries with developing economies. In general, they display a three-percent economic growth annually and are capable of doubling these indicators in the future (under favorable projections) [3]. However, these growth rates are compensated to a considerable extent by the improved energy efficiency. Thus, the demand for energy consumption in the world as a whole should not grow by 100 % but only by 30 %. At the same time, oils, gas, and coal will remain predominant sources of energy, while new renewable sources of energy alongside with nuclear and hydroelectric power will provide only a half of the expected 30 % of energy. Despite that, conventional sources of energy (oil, gas, and coal) will remain predominant. In 2040, they are expected to account for more than 75 % of supplies (in comparison with 81 % ones in 2018) (figure 1).

In 2018, six types of resources satisfied global energy requirements. The basis of demand was represented by hydrocarbons (81 %): oil (32 % of total consumption), gas (22 %), and coal (27 %). In the coming decades, this raw material, as earlier, will remain the basis of energy supply for the world's economy. Meanwhile, the proven oil and gas reserves will be sufficient at the current level of production for the next fifty years, as for the coal, its reserves will be sufficient for the next 150 years.

* Corresponding author: as.semenov@s-vfu.ru

 https://doi.org/10.1051/e3sconf/202017801058

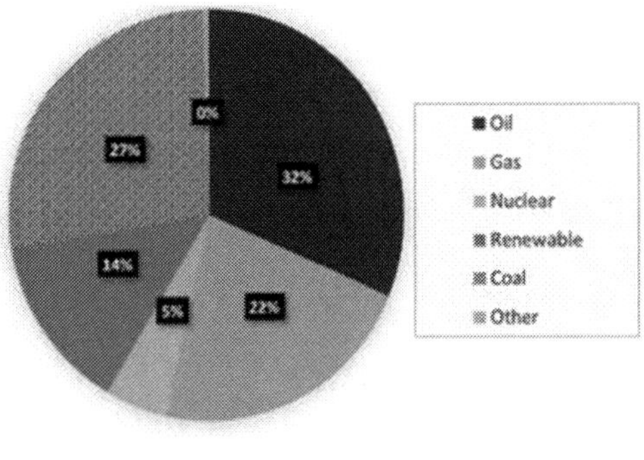

Fig. 1. Global requirement (demand) for various sources of energy in 2018.

2 Research results

Some scientific papers [4, 5] express the opinion that the demand for natural gas will continue to grow faster than the demand for oil and coal in the future. Larger volumes of liquefied natural gas production facilitates this process, as it is becoming available all over the world. In designing the mathematical model showing trends in demand for gas compared to the total demand for oil and coal, two different polynomial functions have been obtained with almost the same probability (97,96 % and 99,28 %), confirming the given forecast by 2040.

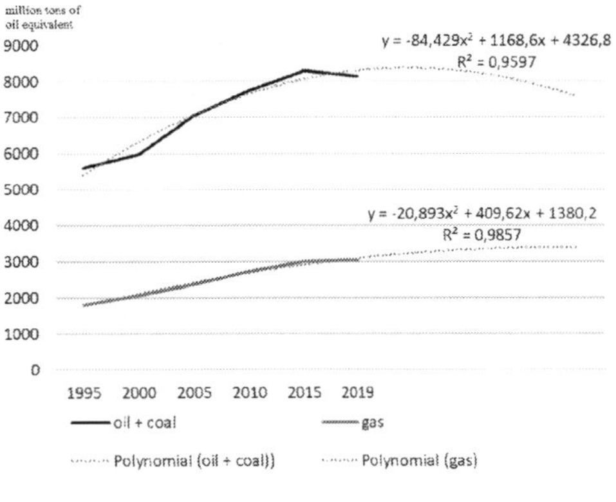

Fig. 2. Forecast of trends in demand for gas in comparison with the total demand for oil and coal.

Figure 2 shows that even without the data of the last drop in demand for oil (March-April 2020), which occurred due to a recession in the oil market caused by the Covid-19 pandemic in the world, according to the designed mathematical model $y = -84.429{\cdot}x^2 + 1168.8{\cdot}x + 4326.8$, minimum demand for oil and coal in 2020 was supposed to remain at the level of 2019 without a considerable growth. It is impossible to access the scope of the current drop of demand for oil within all the countries if only fluctuations of spot prices per a barrel of oil are taken into account. Nevertheless, after quarantine as early as in April 2020, China reduced its oil consumption by 17 % in comparison with the same period in 2019 [6].

Trends in energy production (supply) and energy requirement (demand) in definite regions of the world are shown in figure 3.

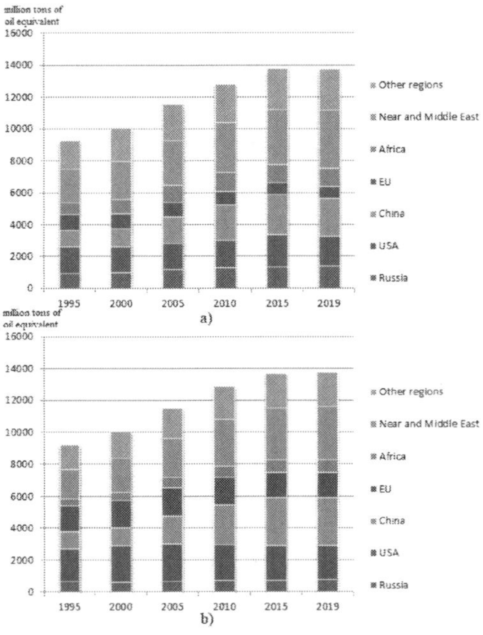

Fig. 3. Trend in energy production (a) and change of global energy demand (b) across the regions over the past ten years.

In 2018-2019, oil production increased by 0.5 % due to the countries of the Middle and Near East, at the same time the volume of production of these raw materials reduced in China and the USA [7]. In general, for the past ten years, increase in energy production has been evident. However, due to economic crisis fluctuations energy production in the EU countries has abruptly decreased. Countries of the Middle and Near East have displaced them. Meanwhile, gas production has increased approximately by 0.5 %, it became one of the lowest indicators of the growth rate for the last three decades [8]. Australia contributed to some revival in the development of the gas sector, as it began to operate new facilities for the production of shale gas, which facilitated the increase in volume of interregional supplies of this type of fuel by 4.8 % [9].

In the structure of energy consumption of the economies of developing countries, the share of hydrocarbon energy approaches 12 %, but in developed capitalist countries, this indicator approximates 20 % taking into consideration more intense use of nuclear energy and energy from renewable sources [10].

Increase in energy consumption in the developing countries mainly witnesses the development of transport infrastructure and an industrial sector as well as improvement of living conditions of the population. Nowadays their share accounts for about 60 % of the global demand [11].

 https://doi.org/10.1051/e3sconf/202017801058

In the course of cooperation with developed countries, which took place through the opening of production subsidiaries of transnational corporations on the territory of underdeveloped countries, these countries adopted modern energy-intensive technologies and moved to a higher level of energy development [12].

The data characterizing the production and the consumption of the main types of energy in the world are shown in table 1.

Table 1. Global production of main types of energy in 1995-2019, in mln. ktoe.

Types of Energy	1995	2000	2005	2010	2015	2019
Coal	4996	6001	7331	8660	9552	9594
Oil	1237	1212	1135	974	973	931
Gas	2116	2747	3701	4810	5552	5794
Nuclear	2332	2591	2768	2756	2570	2606
Renewable	2636	2836	3294	4200	5536	5939
including hydro	*2479*	*2618*	*2935*	*3445*	*3905*	*4061*
sunny	*33*	*55*	*141*	*408*	*1132*	*1330*
windmill	*130*	*164*	*226*	*360*	*517*	*571*
Other	40	52	58	68	80	82

Therefore, over the past ten years, the main increase in global demand for energy was provided China as well as India and Brazil (about 56 %). Since 2016, demand for primary energy in total consumption of these countries has exceeded 30% compared to 23 % in 2005. However, one should notice, that after stagnation caused by the pandemic of coronavirus at the beginning of 2020, the situation will change in the direction of a decrease in natural indicators and an increase in relative ones with regard to such countries as the USA, Russia, and EU countries. To compare: in the USA this indicator has increased by 21.3 % for 15 years; however, globally the demand decreased from 20.2 % in 2005 to 15.7 % in 2019 of total world energy consumption [13].

It should be also noted that Great Britain reduced solid fuel production (in 2006 – 11,4 mln. ktoe, in 2016 – 2,6 mln. ktoe) and closed the last mines in April 2017. As a result, this branch of industry – one of the major ones in the past – completed its more than two hundred-year-old journey (1815-2017) [14].

During the past several years, importance of electric energy generated by renewable sources is increasing. Its generation in 2006-2016 increased by 29 %. To compare: at the same period of time the production of basic types of hydrocarbon fuel increased by half – about by 15 %. Nowadays, the main sources of energy generation are nuclear power plants and large hydropower stations (with power more than 10-25 MW) [15–17].

Leaders on a scale of productivity of nuclear energy remain the USA (32,4 % of global generation) and France (15,4 %). Germany continues to adhere to the pace of drawdown of nuclear energy. Japan that had had 54 nuclear reactors decommissioned after the accident at NPP "Fukusima-1" (March 2011) did not re-established its facilities in their former extent. In 2015, they began disassembling of units that had been maintained for more than 40 years. However, annual purchase of coal and gas with 30 billion dollars complicated the country's balance of payment, and currently, a revival of the national nuclear industry has been outlined (till 2011 its share accounted for to 30 % of electric energy production). Japan will be able to increase production after completing complex works to improve safety of NPP according to the requirements of IAEA.

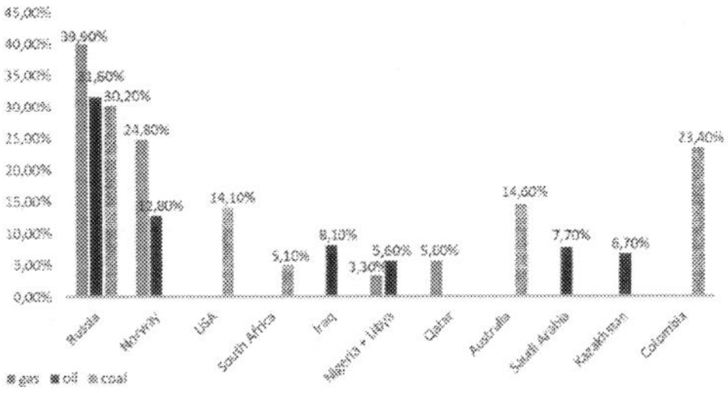

Fig. 4. Structure of import of hydrocarbon fuel in 2016 to the EU countries.

In the global energy economy, the greatest progress is made in the field of renewable sources (in 2016 – increase by 14,4 %). The leader of the segment is Holland. Last year its share accounted for about 50 % of energy generated in such a way, while the relative importance of solar power was 18 %. Nevertheless, the EU countries experience a general shortage of energy resources that is compensated by import from other countries. At the same time, Russia remains the main importer. The structure of import is shown in figure 4. It illustrates that the Russian Federation supplies the EU countries with all types of hydrocarbon fuel.

3 Conclusion

In general, over the past twenty years the efficiency of the used energy resources increased and energy consumption decreased. In 2005-2015, with the growth of GDP by 26,5 % (at constant prices) global energy consumption increased by only 19 %.

In the recent years, energy efficiency has increased amid falling consumed energy losses [18–20]. In the period from 2006 to 2010 with the growth of global GDP by 26,5 % (at constant prices) global energy consumption increased by only 17,8 %. Therefore, in 2006-2010 energy consumption rate of growth was more than 2 %, and the first half of the last decade was characterized by a steady lack of energy resources that was 130 million tons in average [20].

In 2011-2015 the situation started to change in reverse. Rates of growth of energy consumption of all types have decreased to 1 % for the recent three years, and the lack turned into excess. Thus, oil production exceeded the demand almost by 90 million tons per year back in 2018, while oil prices fell approximately by three times compared to the record levels in 2008 and 2012. At the beginning of 2020 before the beginning of the world crises, the exporters of fuel and energy products did not respond appropriately to stabilize the market in spite of the obvious production costs and losses of national budgets caused by the coronavirus pandemic.

References

[1] J.K. Parikh, Energy **23**, 69 (1998)

[2] V. Shatokha, Transactions of the Institutions of Mining and Metallurgy, Section C: Mineral Processing and ExtrActive Metallurgy **126**, 62 (2017)

[3] A.S. Ivanov, I.E. Matveev, Bureniye i neft' **12**, 4 (2017)

[4] O.V. Grigorash, M.M. Tarasov, Polythematic online scientific journal of Kuban State Agrarian University **118**, 949 (2016)

[5] A.S. Ivanov, I.E. Matveev, Russian Foreign Economic Bulletin **1**, 17 (2018)

[6] I.V. Kostenok, Y.V. Khomenko, A.V. Efimenko, Vestnik of the Plekhanov Russian University of Economics **1**, 93 (2020)

[7] A.N. Alekseev, A.V. Bogoviz, L.P. Goncharenko, S.A. Sybachin, International Journal of Energy Economics and Policy **9**, 95 (2019)

[8] A. Carlson, Energy Policy **31**, 951 (2003)

[9] V. Rajović, F. Kiss, N. Maravić, O. Bera, Energy Conversion and Management **118**, 96 (2016)

[10] L. Zhang, R. Long, H. Chen, X. Huang, Resources, Conservation and Recycling **136**, 248 (2018)

[11] V.S. Litvinenko, Natural Resources Research **29**, 1521 (2020)

[12] X. Hu, C.M. Iordan, F. Cherubini, Global Environmental Change **50**, 15 (2018)

[13] A. Kovalev, Yu. Tascheev, Scientific Bulletin of the Odessa National Economic University **9**, 62 (2018)

[14] V. Lir, Economy and Forecasting **14**, 107 (2016)

[15] N.S. Volotkovskaya, N.N. Kugusheva, A.S. Semenov, I.A. Yakushev, S.N. Pavlova, O.V. Kolosova, E3S Web of Conferences **124**, 04001 (2019)

[16] N.S. Volotkovskaya, A.S. Semenov, O.V. Fedorov, IOP Conference Series: Materials Science and Engineering **753**, 082033 (2020)

[17] N.V. Golubtsov, L.G. Efremov, R.G. Ismyatullin, Bulletin of the Chuvash University **1**, 247 (2013)

[18] D.D. Micu, I.V. Ivshin, E.I. Gracheva, O.V. Naumov, A.N. Gorlov, E3S Web of Conferences **124**, 02013 (2019)

[19] I.D. Elyakova, N.S. Buryanina, Y.I. Koroluk, A.L. Elyakov, L.I. Danilova, A.A. Khristoforov, E.E. Noeva, Journal of Advanced Research in Law and Economics **9**, 1222 (2019)

[20] I.D. Elyakova, A.L. Elyakov, T.A. Karataeva, A.A. Khristoforov, N.S. Buryanina, A.A. Fedorov, International Journal of Energy Economics and Policy **8**, 361 (2018)

[21] EU energy in figures. Statistical pocketbook 2018 [Electronic resource]. Available at: https://op.europa.eu/en/publication-detail/-/publication/99fc30eb-c06d-11e8-9893-01aa75ed71a1/language-en/format-PDF/source-78277087 (accessed: 07.05.2020)

Features of steam turbines diagnostics

Ilia Murmanskii[1,*], *Konstantin* Aronson[1], *Boris* Murmansky[1], *Andrei* Sosnovskii[1], *Vladimir* Novosyolov[1], and *Yuri* Brodov[1]

[1]Ural Federal University, Turbines & Engines Dep., 620002, Mira, 19, Ekaterinburg, Russia

Abstract. Enterprises of energy equipment and operational utilities set sights on diagnostic systems. This is necessary for state control and maintenance planning of steam turbines. It is useful for digitalization purposes too. So far, some mathematical systems are already used. Algorithms for flow part, heat expansion system, control system, vibration-based diagnostics and auxiliary equipment are already designed. We designed algorithms just in principle. Adapting them, for the PT-75/80-90 turbine we met with difficulties. Firstly, we should connect them to a single interface. Secondly, adaptation should include features of the equipment, its state (if not new), even operating conditions. Diagnostic signs for each turbine are the most important. We define them based on the operational data. When adapting the algorithms, we reconsider the signs list. We also estimate its coefficients of importance again. This requires experts to study designs, calculations, and modelling. We also analyzed a large amount of operation data at various power plants. To define the state we use tests. Adapting is based on the modes of a specific power station. Following this strategy, we adapt general algorithms for various turbines.

1 Introduction

At present, a number of power equipment manufacturers and large companies involved in its operation express an increased interest in developing diagnostic systems, for determining the actual condition of equipment and for planning the volumes and terms of its repair, based on this information. Over the last 20–25 years, a number of tasks have been tested of diagnosing steam turbines main and auxiliary equipment at individual power plants [1–9]. For these diagnostic systems, as a rule, the information was supplied by special sensors, and its implementation consisted of process parameters monitoring and of comparing the values of these parameters with standard characteristics. The developed diagnostic algorithms contained the necessary information about defects that occur in the equipment, as well as data for evaluating the manifestations of these defects. However, in most cases, the diagnosis was performed by experts on the basis of the process parameters monitoring data.

2 Methods and materials

The main difference between the diagnostic systems currently being created is the possibility of obtaining the information from standard measuring systems of automated process control systems and other (including commercial) information systems of thermal power plants. This allows one to diagnose equipment both in real time and post-operatively, solving, among other things, the forecast tasks. One of the most important issues of introducing diagnostic systems is the need to adapt the diagnosis and condition assessment algorithms developed by experts for specific equipment and its operating conditions. A number of specialists involved in the operation have an opinion about the possibility of extending the algorithms developed for one standard size of equipment to other standard sizes.

Before the diagnostic process, used data should be carefully validated. Three pre-diagnostic stages of validation are suggested for use. Firstly, the data, collected by the automated process control system (APCS), is checked according to availability of the sensor signal. Secondly, each sensor signal has a set range of values, which are determined by the technological process. Thirdly, as the diagnostic is performed not automatically, but by the expert, they have to look through the data by themselves to check if it looks realistic enough. After all, when the diagnostic algorithm is performed, the results are reviewed once again. If the data is too different from the nominal value or the previous one, the expert should check it once again.

Among the algorithms developed for diagnosis, a number of types can be distinguished that differ in the degree of universality (independence from specific conditions) and, on the contrary, in the degree of the diagnosis certainty. The first type of algorithms — calculated — implements diagnostic functions in the on-line mode within the automated process control system or in the postoperative mode. Such computational algorithms include diagnostic algorithms for a steam distribution and turbine regulation systems, for turbine auxiliary equipment (condensing unit, regenerative feed water heating system, hot water heating), for turbine flow path, etc. These algorithms describe the models of

* Corresponding author: i.b.murmansky@urfu.ru

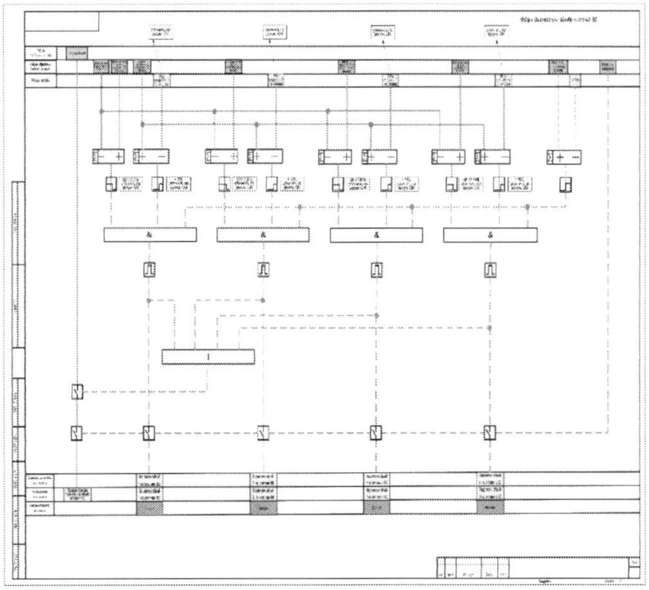

Fig. 1. Diagnostics of high pressure valves stem breakage.

processes occurring in equipment. To the second type of algorithms — the expert one — one can attribute expert systems that analyze, for example, the vibration state of rotating mechanisms, or the system of turbine thermal expansion, and the state of auxiliary equipment. As a rule, such algorithms describe the procedures for determining (evaluating) the signs of defects, and the probability of defects is calculated as the sum of the sign weighting coefficients or by Bayes formula. The third type of algorithms — the digital type — is represented by digital algorithms based on phenomenological relationships between the parameters of the process, the state of the equipment and other indicators, these relationships having been obtained over the test period of equipment operation. This type of algorithm is least related to the type of equipment, its features, and operating modes.

The need to adapt diagnostic systems to other kinds of equipment consists not only in the development of new diagnostic algorithms, but also in the formalization of discourse. This refers to the choice and justification of the normative state (prototype) [10], that is, to a certain model state, in comparison with which diagnostic signs are formed, as well as a numerical description of the signs of defects determined by the measurement results and required for diagnosing the sentences and opinions of experts. To develop a prototype, the operating modes of the equipment and its current normative characteristics must be taken into account; for example, when diagnosing the flow path, it is necessary to use the normative characteristics obtained from the results of periodic tests. As a prototype for diagnosing a turbine control system, its characteristics obtained from tests and adjustments after repair can be employed. When numerically describing defect signs, it is necessary, for example, to describe such expressions as turbine rotor increased vibration, reduced vacuum in the condenser, various steam pressure along the threads, etc. The complexity of this procedure lies in the fact that deviations of the indicators can be caused not only by the error of the instruments or of specific measurement scheme, but also by different values of the significant deviation of the same indicator in different diagnostic tasks.

Let us study the features of adaptation of each type of algorithms developed for a particular equipment and a particular thermal power plant to a different type of equipment and other conditions (schemes) of its operation.

3 Results

Calculation algorithms, for example diagnostic algorithms for a control system, as a rule, can be built into an automatic process control system, an automatic control system, or in a turbine regulating system. Figure 1, as an example, shows a block diagram of an algorithm for diagnosing a stem breakage in a turbine high-pressure control valve. The diagnostic sign of this defect is based on the fact that when the valve is open, the pressure behind it rises above the pressure in the control stage chamber. The diagnostic task is implemented as follows (for the 1st control valve as an example):

- the pressure in the control stage is compared on the adder with the pressure behind the 1st valve and the difference is applied to a threshold element that generates a discrete output signal at a predetermined difference value (for example, less than 0.5 MPa);
- similarly, the valve control signal is compared on the adder with the value corresponding to the start of the 1st valve opening (5%, as it is conventionally presumed

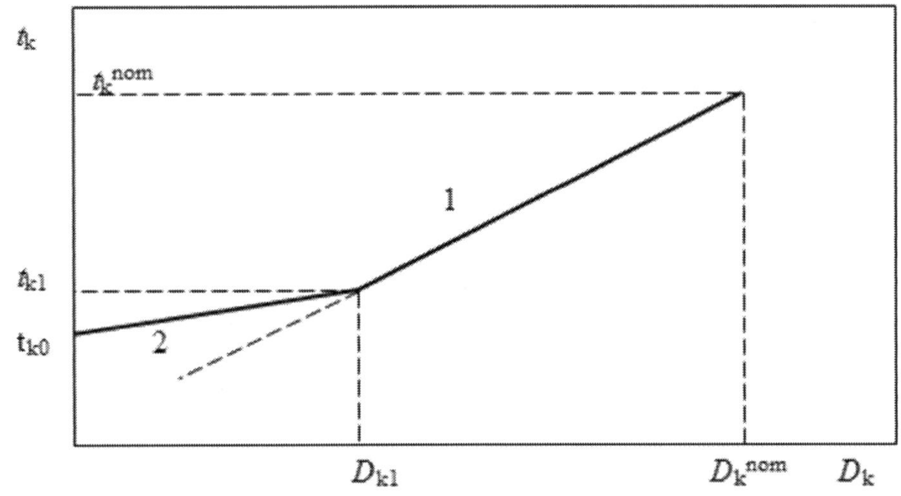

Fig. 2. Joint performance of the condenser and ejector: 1 — condenser performance without taking the ejector into account; 2 — joint condenser—ejector performance.

in Figure 1) and the difference is applied to a threshold element that generates a discrete output signal with a predetermined difference, for example, more than 10% (this means the unconditional opening of the 1st valve);

- the considered diagnosis of valve stem breakage is possible if there is steam flow to the turbine, while the turbine is operating under load; to do this, the current generator power value is compared on the adder with the minimum accepted value (for example, 5% of the rated power of the turbine), the difference is applied to a threshold element that generates a discrete output signal when it is positive;
- if all three of the indicated discrete signals are supplied to the "AND" logic element, then a discrete signal is generated at the output of the latter, indicating a possible valve stem break.

Algorithms of this type are developed for a specific design of the regulatory system; the functioning of these algorithms clearly depends on the availability of the necessary measurements (and usually requires the inclusion of additional sensors in the traditional volume, steam pressure sensors behind the valves in particular). When adapting algorithms for a different type turbine or the same turbine of another thermal power plant, but with the same type of steam distribution, it is necessary to adjust the settings of the calculation algorithm (moments in valve opening, pressure difference values, etc.), i.e., to change the description (formalization) of diagnostic signs. Moreover, a complete revision of the algorithms (calculation model) is not required. For algorithms of this type, the presented diagnosis almost uniquely determines the defect that has appeared. With another type of turbine steam distribution (throttle type), a complete revision of the diagnostic algorithm (model) is required.

Fig. 2 presents the diagnostic model, on the basis of which a postoperative calculation algorithm is developed for assessing the separate effect of increased air suction and tube contamination on turbine condenser steam pressure. This problem should be solved when the actual steam pressure in the condenser is above the standard value; the condition of the main ejectors of the condensing unit is satisfactory, the ejector performance corresponds to the rating characteristic. The minimum frequency for the task to be launched for a solution is once in 7–10 days [11]. The algorithm for solving this problem is based on the construction of a joint performance of the condenser and ejector. The joint condenser—ejector performance (the dependence of the vapor saturation temperature, t_k, on the steam flow rate in the condenser) consists of two sections, as shown in Fig. 2 [12]. Section 1 corresponds to the condenser performance at high steam flow rates into the condenser from D_{k1} to D_k^{nom}, where D_k^{nom} is the nominal steam flow rate, and $D_{k1} = (0.4 \ldots 0.6)D_k^{nom}$. In this range of D_k variation, the influence of the ejector at standard air suction does not affect the efficiency of the condenser; steam temperature and, accordingly, steam pressure are determined by the standard performance. Section 2 presents a joint condenser—ejector performance and describes the efficiency of the condenser at low steam flow rates. Sections 1 and 2 are linear functions.

Assuming that the steam pressure deviation in the condenser from the standard value is the sum of two pressure differences determined by increased air suction (ΔP_{air}) and by tube pollution (ΔP_{pol}) $\Delta P_k = \Delta P_{air} + \Delta P_{pol}$, so according to the calculation model [12], ΔP_{air} and then ΔP_{pol} are found.

When adapting this algorithm for a turbine of a different type and size, it is necessary to employ the regulatory characteristics of specific equipment, and depending on the specific measurement scheme, revise the algorithms for diagnostic signs evaluating, that is, change the procedures of their formalization. The solution to this problem for a cogeneration turbine equipped with a condenser having an integrated tube

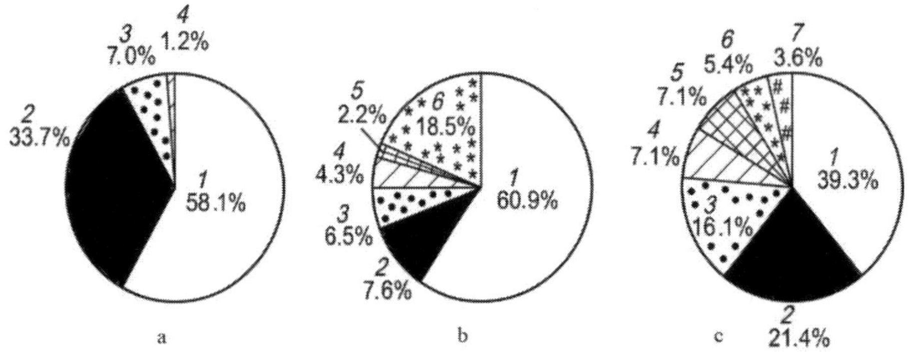

Fig. 3. Condenser damage distribution: a — distribution of condenser failure indications: 1 — increased condensate hardness, 2 — vacuum drop, 3 — si-phon breakdown, 4 — other; b — condenser failure defect distribution: 1 — condenser tube damage, 2 — valve damage, drain damage at al., 3 — tubes and tube sheets pollution, 4 — condenser housing defects, 5 — loss of tightness of damaged tubes plugs, 6 — other; c — damage causes distribution of the condenser tube systems: 1 — frictional wear of tubes or rolled joints of tubes with tube plate caused by various reasons, 2 — tube erosion, 3 — tube corrosion, 4 — poor quality of brass tubes, 5 — loss of zinc of the brass tubes, 6 — poor quality of rolling and of tube sheets during repair or manufacturing, 7 — tube breakage caused by blade fragments.

bundle, or for a condenser with a minimum steam consumption (when cogeneration diaphragm is completely closed) requires a change in the calculation model and a complete alteration of the algorithms.

Diagnostic algorithms of the first calculated (model) type are designed to identify specific defects and, in this aspect, provide maximum certainty of the analyzed defect.

The second type of algorithm (the expert one) presents the diagnosis of the presence of a defect in the form of probability. An algorithm of this type performs a procedure for assessing the presence or absence of signs characterizing a particular defect. The probability of a defect is calculated either as the sum of the weighting coefficients of the features present, or as a posteriori probability in the case of a Bayesian expert system.

The most important procedures for adapting these algorithms to various kinds of equipment are the analysis of the specific equipment damage at a specific thermal power plant and the formalization of expert knowledge regarding the assessment of a sign. An analysis of the damageability of equipment is of great importance, in particular, in determining the weight coefficients and probabilities of signs and a priori probabilities of hypotheses. As an example, Fig. 3 shows data on characteristic defects, the causes of defects and signs of their manifestation for turbine condensers, on the basis of which data the values of these probabilities are determined by the experts.

If the design of this equipment involves typical (previously used on other equipment) structural elements, it is advisable for diagnosis reliability increasing to evaluate (clarify) the probability of defects and signs occurrence based on statistics of their damageability.

Thus, to assess the probabilities of signs and defects when adapting expert-type diagnostic algorithms, one needs to formalize the knowledge of experts regarding a specific measurement scheme and design features of the equipment. The volume of equipment with measuring instruments is not always sufficient. In particular, when diagnosing a thermal expansion system, it is necessary to measure the following values [13]:

- absolute displacements of bearing housings to the left and right of the turbine axis,
- bevel slopes under bearing housings to the left and right of the turbine axis,
- lateral displacements of cylinder legs relative to bearing housings.

In the current industrial control systems, the absolute displacements of the bearing housings and the bevel slopes under the bearing housings are usually measured on one side, and the transverse displacements of the cylinder legs are not determined. It is necessary for each parameter used in the diagnosis to assign a value corresponding to its significant change. Moreover, a change in one and the same physical parameter in different tasks can be different. Thus, a significant temperature change for diagnosing thermal expansion system (a change in the temperature of the cylinder metal) that affects the cylinder displacement is 50 °C, while the circulation water temperature change for diagnosing the condensation unit is 0.5...1.0 ° C. In this case, the error of the device plays an important role; in automated control systems the aperture of addition to the database can be 2/3 of the error of the device.

The signs of defects encountered in the literature are formulated in a general way, and when developing algorithms for a specific unit, it is necessary to take into account its specific features: the number of bearings; the availability of steam extractions for industrial and heating purposes, their location relative to the turbine supports; the geometric characteristics of the cylinders, key joints, the possible values of the mutual displacement of the elements; the results of the thermal calculation determine the features of the algorithms for evaluating vibration signs and malfunctions, as well as a faulty system of turbine thermal expansions.

For example, the diagnostic sign of “vibration dependence on the operating mode” should be formalized for a specific unit with regard to its design and scheme for parameter measuring. The material of condensers tubes, and of low pressure feed water heaters, hot water heaters, the presence of an integrated tube bundle in the condenser, the number and type of heaters in the system of regenerative feed water heating must be taken into account when determining the list of diagnostic tasks for auxiliary equipment.

As an example, in [11] a diagnostic sign of misalignment of the steam turbine rotors is presented as “a change in vibration with a change in electric or thermal load and condenser pressure”. To create an algorithm for processing this feature, it is necessary to determine for each of the specified parameters a value, in case of exceeding which a change in the parameter is accepted. In addition, it is necessary to decide by what parameter the change in the heat load is estimated: by the pressure in the hot water heating extraction, by the opening of the diaphragm or the calculated parameter of the heat load, and if the turbine has several extractions, it is necessary to determine which rotor will be affected by the load change of each cogeneration extraction. For each parameter, it is necessary to describe a criterion — a limiting value at which it can be considered that the sign is present.

Thus, for expert-type algorithms, the adaptation of the diagnostic system to another type of turbine may consist in the adjustment of sign evaluation algorithms. When transferring these algorithms from one type of turbine to another, the list of defects, signs, and also algorithms for sign evaluating (formalization of signs) may change; however, the algorithm for calculating the probability of a defect remains unchanged. If there are not enough sensors and corresponding measurements to evaluate a particular sign, then the degree of certainty of the defect assessment (but not the probability, if the missing measurements are replaced by the expert answers) is reduced. In the ultimate case, that is in the absence of measurements (and expert answers), the probability of a defect coincides with its a priori probability for a system constructed by the Bayes theorem. Expert systems based on rigid logic, such as a “decision tree”, determine an unambiguous relationship between signs and defects. Such algorithms, which are expert in their form (developed by experts), in essence can be attributed to the calculation type, where the degree of certainty of the diagnosed defect is high.

The third type of algorithms — the digital one — involves the equipment malfunctions diagnosing by comparing the actual values of the process parameters with the model process generated on the basis of the data from the automatic process control system based on the results of a rather long-term operation of the turbine unit [14]. This approach requires minimal effort to adapt the algorithms developed for turbine unit of one type to another. However, in this case, the degree of certainty in the assessment of a specific defect is significantly reduced. So, during the operation of such a system at one of the thermal power plants, the staff received a notification about the defect of the turbine unit — a level decrease in the in the oil tank, which was caused by routine work due to the oil replacement in the turbine oil system. With digital algorithms even the participation of experts in the assessment of defects does not guarantee an increase in certainty in finding defects.

4 Discussion

Most of the diagnostic approaches [15-20] use the prescriptive documents of Russia Federation, equipment passport performances and results of regulated (periodic) tests. Expert systems are based on the huge volume of data analyzed. For particular diagnostic tasks, additional methods of results validation are used. For example, for calculations with a full turbine scheme including the heat-mass transfer, balances are being made up. All these parts of the diagnostic process allow us to accept the accuracy of the results.

5 Conclusions

Diagnostic features for steam turbine equipment is determined by wide range of various equipment and diagnostic tasks. In addition, each turbine is operating in various conditions and has design differences even comparing to same series turbines.

Authors designed algorithms for range of defects diagnostics and for estimation of state of steam turbine equipment: flow part efficiency; vibrational defects (over 30), heat expansion system defects (3 defects groups), regulation system (10 diagnostic tasks) and auxiliary equipment, which includes condensers, boilers, ejectors, etc. (15 diagnostic tasks). Some diagnostic tasks are operating on-line, others – in postoperative mode. Diagnostic tasks are intended for defects revealing and equipment state predicting.

Defects determination is embodied in various algorithm types: calculating, expert, digital types. Principles of algorithms adaptation to specific turbines and operating conditions are designed by authors. Adaptation is also provided for a range of new turbines. Some algorithms are already operating at thermal power plants for more than 10 years.

References

[1] I.A. Perminov, V.G. Orlik, Diagnostics of the technical condition of the flow path of high and medium pressure cylinders by operational measurements of steam pressure and temperature in the turbine, Electric Power Stations **6**, 38–41(2003)

[2] S.Sh. Rosenberg, L.A. Khomenok, A.A. Gordinsky, V.I. Ilyin, Diagnostics of the condition of radial clearances in the cylinder flow path during operation, CKTI Transactions **273**, 67–76 (1992)

[3] I.A. Kovalev, Development of algorithms for the functioning and recognition of defects for an

automatic vibration diagnostic system, CKTI Transactions **273**, 27–33 (1992)

[4] A.Z. Zile, M.N. Rudenko, S.B. Tomashevsky, V.G. Vasilyev, I.S. Khramov, The experience in implementing a system of vibrodiagnostic control of turbine units, Power Engineer **3**, 21–23 (1999)

[5] A.Z. Zile, A.A. Romashev, S.A. Lymar, et al., Automated system for vibration control and diagnostics of a turbine unit of T-250/300-240 type, Electric Power Stations **3**, 23–26 (1987)

[6] E.A. Don, D.V. Taraday, K.E. Buglaev, Diagnostic system for temperature expansion of turbine units, Electric Power Stations **9**, 42–47 (2012)

[7] L.S. Baran, Development of a system for comprehensive technical diagnostics of a condensing unit for a turbine of K-800-240-3 type, CKTI Transactions 279, 40–51 (1994)

[8] Yu.M. Brodov, K.E. Aronson, M.A. Nierenstein, Steam turbine condensing unit diagnostic system concept. Thermal Engineering **7**, 34–38 (1997)

[9] A.D. Trukhniy, N.A. Zroychikov, B.V. Lomakin, I.V. Sedov, Information and diagnostic system for monitoring hot water heaters for turbine of T-250/300-240 type, Thermal Engineering 1, 30–34 (1998)

[10] K.E. Aronson, B.E. Murmansky, Yu.M. Brodov, V.B. Novoselov, A.Yu. Sosnovsky, I.B. Murmansky, Forming of prototypes and features for diagnostics automation of steam turbine unit equipment. Thermal Engineering (2020)

[11] *Guidelines for steam turbine condensing units testing: MU 34-70-010-82* (M.: SPO Soyuztechenergo, 68, 1982)

[12] S.I. Khaet, K.E. Aronson, Yu.M. Brodov, A.G. Shempelev, Development and testing of system elements for monitoring the condition and diagnostics of a steam turbine condenser, Thermal Engineering **7**, 67–69 (2003)

[13] B.E. Murmansky, A.Yu. Sosnovsky, Yu.M. Brodov, Development of a module for monitoring and diagnostics of the system of steam turbine thermal expansions as part of modern automated control systems, Power Engineer **4**, 51–53 (2015)

[14] E.V. Uriev, *Fundamentals of reliability and technical diagnostics of turbomachinery: tutorial* (Ekaterinburg: USTU, 71, 1996)

[15] A. Brooking, P. Johns, F. Cox, ed. R. Forsyth, *Expert Systems, Principles and case studies* (M.: Radio & Communication, 191, 1987)

[16] Yu.M. Brodov, K.E. Aronson, B.E. Murmanskiy, I.B. Murmanskii, M.A. Nirenshtein, V.B. Novoselov, P.N. Plotnikov, A.Yu. Ryabchikov, Reliability of the equipment of steam turbine plants: a training manual (Ekaterinburg: Publishing House of the Ural Federal University, 144, 2017)

[17] Yu.M. Brodov, K.E. Aronson, A.Yu. Ryabchikov, M.A. Nirenstein, N.V. Zhelonkin, D.V. Brezgin, I.B. Murmanskii, *The current situation and trends in the design and operation of condensers of powerful steam turbines: a Training manual* (Ekaterinburg, publishing house of the Ural University, 104, 2019)

[18] Yu.M. Brodov, K.E. Aronson, A.Yu. Ryabchikov, M.A. Nirenshtein, I.B. Murmanskii, N.V. Zhelonkin, *The current level and trends in the design and operation of heaters for the regeneration system of steam turbines of TPPs and NPPs: textbook* (Yekaterinburg, Publishing House of the Ural University, 100, 2019)

[19] K.E. Aronson, S.N. Blinkov, V.I. Brezgin, Yu.M. Brodov, V.K. Kuptsov, I.D. Larionov, et al., *Heat exchangers of power plants: textbook for universities, third edition, revised and supplemented* (Yekaterinburg: UrFU, 828, 2014)

[20] Yu.M. Brodov, K.E. Aronson, Yu.M. Gofman, B.E. Murmansky, M.A. Nirenstein, A.Yu. Ryabchikov, P.N. Plotnikov, *Repair and maintenance of steam turbine equipment, reference book in 2 volumes* (Yekaterinburg, Publishing House of the Ural Federal University, 100, 2019)

E3S Web of Conferences 178, 01060 (2020) https://doi.org/10.1051/e3sconf/202017801060
HSTED-2020

MODELING OPERATION OF LIQUID METAL FUSES WHEN BREAKING OVERCURRENTS

A V Kuznetsov[1], *D S* Aleksandrov[1] and *Y P* Yurenkov[1,*]

[1]Ulyanovsk State Technical University

Abstract. This paper shows that successful switching of extremely high short-circuit currents I> 50 kA can be achieved by joint operation of a liquid-metal self-resetting current limiter and a circuit breaker connected in series. The type NFU-225 device from Mitsubishi was taken as an example. The time-current characteristic of joint operation of a liquid-metal self-resetting current-limiting device and a circuit breaker was compiled. However, further in the article physical processes occurring in a liquid-metal self-resetting current limiter with a complete transformation of fusible unit are considered. The result of work is modelling of operation of liquid-metal fuses when overcurrents are switched off based on the pilot studies obtained by the Japanese scientists. It is proposed to simulate the break process not at every time moment, but at specific time moments (reference points). At other time moments, current and voltage should be considered as approximately linearly changing characteristics. The work of current limiter can be represented by three stages: the pre-arc, the main arc and the final arc. If the current density is less than 1000 A/mm2, then the pre-arc operation stage of the current limiter includes the following sections for heating the fusible unit: primary heating to the melting temperature; melting and its transition to liquid state; secondary heating to evaporation temperature; evaporation of fusible unit.

1 Introduction

Implementation of new electrical equipment and promising circuit solutions, for example, by means of:

– Increasing the unit capacity of power transformers to Sn = 4 MVA;

– Switching of two or three power transformers to parallel operation;

– Using powerful autonomous energy systems (ship power systems);

– Advancement of powerful converters that provide conversion of alternating current to direct current,

depends on the presence of high current and current-limiting electrical devices which enable switching high short circuit (SC) currents.

Successful switching of extremely high short-circuit currents *I*> 50 kA can be achieved by joint operation of liquid-metal self-resetting current limiter (hereinafter referred to as current limiter) and a circuit breaker connected in series. A current limiter can be a structural part of a single electrical device, and in some cases it can be used separately, operating not with one circuit breaker, but with a group of linear circuit breakers. The combination current limiter - circuit breaker opens up the possibility of using conventional circuit breakers with a relatively low switching capacity. Limitation of short-circuit current is provided by a current limiter, and the circuit breaker completes the shutdown process, finally opening the circuit using its contact-arc damping system. Thus, the current limiter is actually an accompanying electrical apparatus to the main reusable device, the purpose of which is similar to that of a link of aM-type fuse.

The nature of joint operation of current limiter and circuit breaker is most easily understood using the combined time-current characteristics, obtained separately for the current limiter $t = f(i)$ and the circuit breaker (Fig. 1). As an example, consider an apparatus of the NFU – 225 type [1].

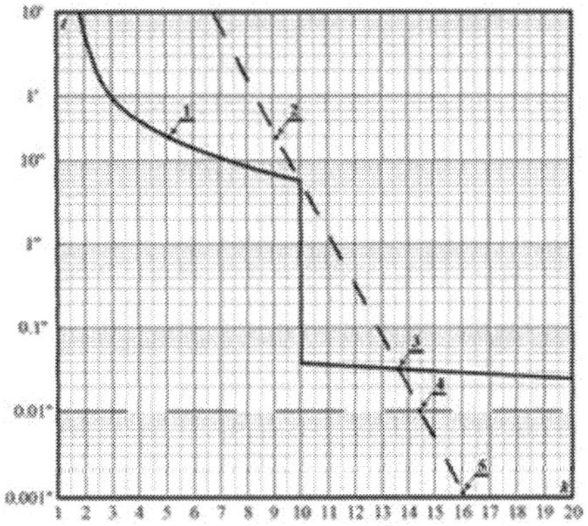

Fig. 1. Time-current characteristics of the circuit breaker (1) and current limiter (2): *t* is the response time; *k* is the relative short-circuit current by reference to the nominal current I_n (short circuit current multiplicity).

* Corresponding author: alxim2@mail.ru

In section 2-3, where the short circuit current multiplicity k is less than $15I_n$, the short circuit current is disconnected by a circuit breaker, and the current limiter can't operate fast enough. In section 3-5, where the short circuit current multiplicity $k>15I_n$, the short circuit current is breaking by the current limiter, and the circuit breaker "picks up" this process and completes it. The section 3-5 is divided into two sub-sections in terms of features of breaking the short circuit current. On the first subsection 3-4, the response time enables to reignite the arc at the location of fusible unit of the current limiter, since the possibility of "picking up" the short circuit current by a circuit breaker is limited by its relatively long operation. On the second subsection 4-5, the response time of current limiter is less than 10 ms, and the current passed through it is quite sufficient for reliable operation of the circuit breaker and completion of breaking after the first passing of current through zero.

The relevance of studying arc phenomena in current limiters is associated not only with the need to create them, but also with economic issues. In view of this, it becomes imperative to reduce the amount of expensive development work and testing due to wider use of analytical and numerical calculations.

The result of arc phenomena study should be a mathematical model that allows one to solve practical problems that arise not only for designers, but also for testers of electrical devices, designers of electrical systems. Creation of a mathematical model of arc phenomena in liquid-metal current-limiting devices is extremely complicated and the repeated attempts to obtain it [2-6] were unsuccessful.

Next, physical processes occurring in a current limiter with a complete conversion of a fusible unit of diameter d are considered under the following triggering conditions:

- Single-phase short-circuit currents in alternating current circuit;
- Limiting short-circuit currents with multiplicity k more than 100;
- Operation of circuit breaker is not taken into account.

Physical processes were studied with various degrees of depth in [2,7-10]. Therefore, there arises a task for comprehensive qualitative character of representation and generalization of physical phenomena in the current limiter at the main stages of work by summarizing the experimental research data presented in [1,4,11-16].

The analysis of the listed information sources shows that the work of current limiter can be represented in three stages: pre-arc, main arc and final arc.

2 Materials and Methods

Pre-arc operation stage.

Its time duration depends on the density of short-circuit current flowing through the lateral and uniform cross-section of the fusible unit with diameter d and length l_{fe} (Fig 2). An approximate boundary of overcurrent density in the fusible unit cross section is a value equal to j = 1000 A/mm^2 [17].

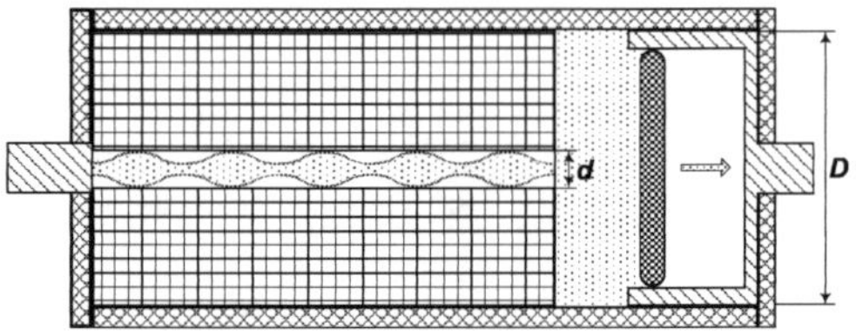

Fig. 2. The deforming impact of pinch effect on a fusible unit of diameter d and length l_{fe}: D is the diameter of chamber for pressure transfer from liquid metal fusible unit to the buffer chamber.

Consider a short-circuit current range with a current density *less* than 1000 A/mm^2.

If the current density is less than 1000 A/mm^2 [17], then the pre-arc stage of current limiter operation includes the following heating sections of the fusible unit (Fig.3):

- Primary heating to a melting point **1**;
- Melting and its transition to liquid state **2**;
- Secondary heating to evaporation temperature **3**;
- Evaporation of fusible unit **4**.

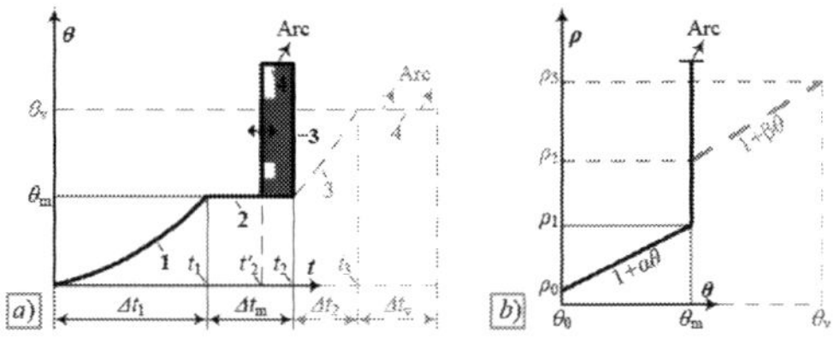

Fig. 3. Relationship between temperature of fusible unit $\theta = f(t)$ (a) specific resistance of metal and the heating temperature (b) $\rho = f(\theta)$.

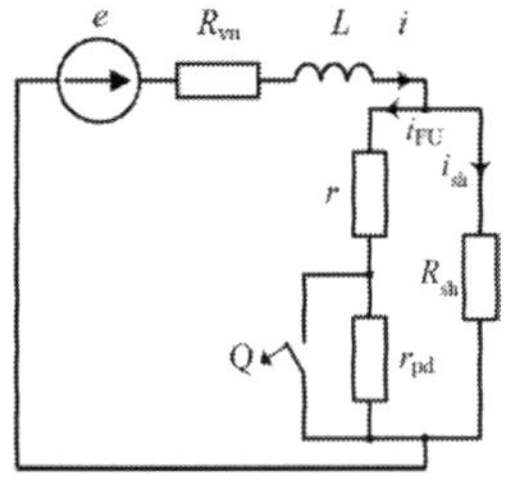

Fig. 4. Equivalent circuit for the main arc stage.

The primary heating section of fusible unit of current limiter **1** of Δt_1 duration, in which temperature varies from the initial temperature θ_{in} to the melting temperature θ_m.

The model of a homogeneous body [17] is adopted as a mathematical model of thermal process of the fusible unit. Heating of fusible unit is adiabatic, since the circuit break by a current limiter begins from current exceeding the maximum operating current of the electromagnetic breaker equal to $I_{av} \geq$ 10–12 I_n. In addition, due to the small temperature range of the primary heating section

θ_m-θ_0, the change in the fusible unit resistance r can be considered linear in the first approximation:

$$r_\theta = r_{\theta 0}\left(1 + \alpha\tau\right),$$

where r_θ is resistance at the current fuse temperature;

$r_{\theta 0}$ is resistance at ambient temperature θ_{amb};

α is the temperature coefficient of resistance of unmelted metal of fusible unit;

$\tau = \theta - \theta_{amb}$ is the difference between the current temperature over the ambient temperature, and the heat capacity C is constant and equal to heat capacity at the initial temperature of the fusible unit θ_0.

The melting section of the fusible unit **2** (Fig. 3, a) of duration Δt_m, at which temperature remains constant and equal to the melting temperature of metal θ_m = const.

After reaching the melting temperature θ_m, the fusible unit begins passing to a liquid state. At the melting temperature, the specific resistance of the molten metal will be much larger than the specific resistance of a solid metal. Consequently, the resistance of fusible unit consists of resistances of its two parts: the inner part, consisting of liquid metal with cross section s_2 and specific resistance ρ_2; and the outer part, which consists of a still hard metal with cross section s_1 and specific resistance ρ_1. The resistance behaviour during melting of fusible unit is shown in Fig. 3, b.

The melting time Δt_m can be determined by integrating a differential equation compiled based on the energy balance: the heat of melting and the energy released by the short-circuit current, without taking part of it into the associated elements of the current limiter.

The section of secondary heating of the fusible unit **3** (see Fig. 3) of duration Δt_2, in which the temperature varies from the melting temperature θ_m to the evaporation temperature of liquid metal θ_v.

For a short time Δt_2, the molten metal of the fusible unit continues to be heated by current. The specific resistance again increases linearly from ρ_1 to ρ_2 (see Fig. 3, b). As in the 1st heating section the model of a homogeneous body [17] is adopted as a mathematical model of thermal process of the fusible unit. Heating of fusible unit is also adiabatic. The change in resistance of fusible unit can be taken linear

$$r_\theta = r_{\theta m}\left[1 + \beta\left(\theta - \theta_m\right)\right],$$

where r_θ is the resistance at the current fuse temperature;

$r_{\theta m}$ is the resistance at temperature of the melted element θ_m;

β is the temperature coefficient of resistance of a fusible unit in a liquid state;

θ is the current temperature.

The heat capacity is assumed to be constant and equal to the average heat capacity $C = (C_{\theta m} + C_{\theta v})/2$ at the melting and boiling points of liquid metal.

The list of conditions for obtaining the instantaneous short-circuit current i_{t3} at time t_3 in general remains the same as that considered during analysis the processes of the primary heating section **1** and melting section **2**. The short-circuit current is set by the external inductive resistance of the circuit X_{ext} and does not depend on the resistance of the fuse element. If necessary, the resistance of the fusible unit r can be taken as the average value $r = (r_2 + r_4)/2$ = const during the secondary heating time Δt_2 to the boiling temperature θv.

If the relative short-circuit current is not less than fifty times greater than the nominal current of the current limiter I_n, then the instantaneous SC current i_{t3} calculated by this method and the time of its occurrence t_3 will be overestimated due to appearance of magneto-hydrodynamic instability of liquid metal already in the second section **2** of Δt_m duration (see Fig. 5, a). The current limiter in this case operates not as a result of evaporation of liquid metal, but actually due to destruction of integrity of fusible unit.

Upon occurrence of overcurrent with a density more than 1000 A/mm^2, the fusible unit of diameter d and length l_{fe} is heated to complete dilution. As a result of pinch effect (self-compression), it presents a chain of narrow straits and metal droplets connected in series and bounded by walls of channel with diameter d of the current limiter (Fig. 2). The longitudinal (tensile) component of force that arises in this case acts on the molten liquefied conductor in such a way that its wider part (drop) expands further, and the narrow part narrows further. This accelerates the rupture of circuit as a result of accelerated heating of straits and due to appearance of a breaking electrodynamic force in the places of narrowing the streamlines.

Figure 2 shows five drops and straits in order to clearly demonstrate the impact of pinch effect on a fusible unit in a liquid state. In fact, the size of one drop and strait adjacent to it does not exceed 0.5 mm [8]. Therefore, at least forty drops and straits may appear along the length of the fusible unit, for example, with a length l_{fe} = 2 cm.

So, if the current density is more than 1000 A/mm^2 [17], then due to the pinch effect, the pre-arc stage of the current limiter includes a smaller number of heating sections of fusible unit (Fig. 3, a):

- Primary heating to melting temperature **1**;
- Melting and its transition to a liquid state **2**.

Duration of stage **2** corresponds to the time range $(t_2 - t_1) \div (t'_2 - t_1)$ (see Fig. 3, a), since due to electrodynamic forces and accelerated heating of straits, short arcs can occur before the maximum time of the melting section t_2 and transition of fusible unit into a liquid state.

It is a very difficult task to take into account the impact of pinch effect on decreasing the instantaneous value of SC current i_{t3} and the time of its occurrence t_3. The simplest solution can be obtained by introducing the correction coefficient k'_{t3} or k_{t3}, taking into account the effect of self-compression of liquid metal of the fusible unit. The value of correction coefficient can be estimated from experimental data on the actually transmitted current i_{t3}. The actual values of arc appearance time t'_2 or t_2 are found by multiplying the calculated time of evaporation start t_3 by the correction coefficient k'_{t3} or k_{t3}, i.e., one should perform the following steps:

$$\begin{cases} t_2' = k'_{t_3} t_3 \\ or \\ t_2 = k_{t_3} t_3 \end{cases}.$$

Basing on consideration of thermal processes and assessment of their influence on electromagnetic processes, we can conclude that it is possible, as the first approximation, to construct a mathematical model based on two reference points with coordinates:

- First point – the process start $t_0 = 0 \rightarrow \begin{cases} i=0 \\ u=0 \end{cases}$.
- Second point – instantaneous voltage at time t'_2 or t_2 preceding arcing (the emergence of a chain of short arcs)

$$t'_2 \div t_2 \; or \; t_3 \rightarrow \begin{cases} i = i_{tr} \\ u = i_{tr} r_{t'_2}, i_{tr} r_{t_2} \; or \; i_{tr} r_{t_3} \end{cases}$$

The following notations are used in the formulas:

- i and u are the instantaneous SC current or the instantaneous current at the terminals of current limiter at the initial moment or at the time immediately preceding the beginning of arcing;
- i_{tr} is the transmitted current;
- $r_{t'2}$, r_{t2}, r_{t3} are the resistances of fusible unit at the corresponding times t'_2, t_2, t_3.

The presented discussion shows a prospect for constructing a simple mathematical model of the pre-arcing stage using two reference points by approximating the short-circuit current i and voltage u at the terminals of the current limiter by a straight line at a time interval of 0–(t'_2÷t_2) or time 0–t_3 depending on the presence or lack of pinch effect.

The necessary refinement of mathematical model is performed by transition from discrete to *continuous* representation of current and voltage by functional dependencies:

– $i = f(t)$, obtained for electrical short-circuited circuit without taking into account the resistance of fusible unit r;

– $u = f(t)$ voltages on current limiter terminals.

The main arc stage of the current limiter

Its time duration depends on the speed of switching the main part of external circuit current i to the branch of the shunt resistor R_{sh}.

Consider the range of SC currents with current density more than 1000 A/mm^2.

At the locations of narrow straits connected by drops, short arcs appear not simultaneously, but one after another over a very short period of time. Addition of every new arc increases the voltage across the arc gap in steps by a value not less than a value of the near-electrode voltage drop Ue = 20–25 V in each of them. As a result, before formation of an arc along the entire length of a fusible unit with a diameter d of length l_{fe} (see Fig 2), a large number of short arcs are formed in places of straits connected by drops with a total voltage of $U_\Sigma = \sum_1^n U_i$, where $U_i \approx$ 20– 25 V is voltage of one short arc. Due to the small size of narrow straits and drops of liquid metal, not exceeding fractions of a millimeter, and a sufficiently extended liquid metal element with a diameter d and length $l_{fe} \geq 1$–2 cm, an overvoltage of an order of one to two kilovolts $U_\Sigma \approx$ 1–2 kV appears. The time $t_{U\Sigma}$, when the overvoltage U_Σ appears, is negligibly little different from the time $t = t_3$ or $t = (t'_2 \div t_2)$ (see Fig. 3, a) due to the high growth rate of voltage on the contact gap of the current limiter, determined by the rate of introduction of elementary arcs. This implies the validity of the following expression

$$t_{U_\Sigma} \approx \begin{cases} (t_2' \div t_2) \\ or \\ t_3 \end{cases}.$$

The duration of increase of the total arc voltage in the range $0 < u_d < U_\Sigma$ does not exceed tens of microseconds. It can be estimated by the expression [1] $t = u_d/(U_m\varepsilon)$, where U_m is the amplitude of phase voltage of network, and $\varepsilon > 1000$ s^{-1} is the relative rate of increase of voltage across the arc. Therefore, limiting the overvoltages in current limiters with a complete conversion of fusible unit with diameter d and length l_{fe} provides a presence of a shunt resistor R_{sh} (Fig. 4). Acceptable overvoltage values can be obtained if R_{sh} of the shunt resistor is chosen so that [15]

$$U_{ph} / I_{Rsh} < R_{sh} < 3U_{ph} / I_{Rsh},$$

where U_{ph} is the acting value of the phase voltage of network;

I_{Rsh} is the acting value of the shunt current, not exceeding the current of maximum switching capacity of circuit breaker and the transmitted current i_{t3}.

The main arc stage of the current limiter is completed at time t_4, when the resistance of current limiter is actually equal to the resistance of the shunt resistor R_{sh}. The time t_4 can be determined by knowing the behaviour of external circuit current i, the current of current limiter i_{FU}, and the shunt current i_{sh}. This behaviour can be found if the behaviour of transition process in the circuit shown in Fig. 4 is known.

High-temperature plasma during the main arc stage of the current-limiting device evaporates part of material by heating the adjacent surface of ceramic sleeve. Thus, the diameter d of the fusible unit increases to a value of d_i, which during subsequent shutdowns will lead to an increase in the transmitted current i_{FU} and, accordingly, to an increase in the released energy. Starting from some n-th break of short circuit current, the diameter d_n of the fusible unit will increase so much that due to the increased transmitted current i_{t3}, the pressure inside the volume of the fusible unit will increase to an unacceptable value according to the strength conditions of the ceramic sleeve. This will lead to mechanical

 https://doi.org/10.1051/e3sconf/202017801060

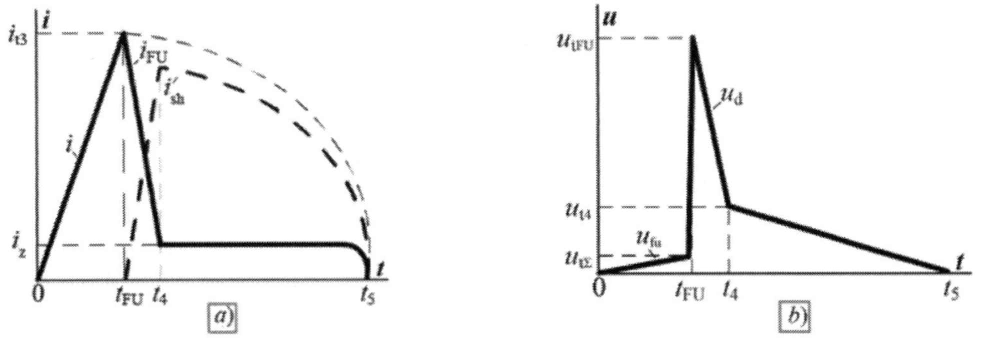

Fig. 5. Change in current of short-circuit at circuit break (a) and voltage at the terminals of current limiter (b).

destruction of the current limiter. Therefore, any liquid-metal current limiter has a specific resource in terms of the number of breaks of the limiting short-circuit currents.

The final arc stage of the current limiter

It occurs within the time interval t_4–t_5 (Fig. 5) and is characterized by burning of an arc in a current limiter at a significantly lower current i_z than the value of the transmitted current i_{FU}, when the relationship $i_z \ll i_{FU}$ is fulfilled. Therefore, the shunt current i_{sh} is actually equal to the current of external circuit i, i.e., $i_{sh} \approx$ i. Since the shutdown process is considered in an alternating current circuit, at time t_5 currents of external circuit, current limiter and shunt undergo a natural transition through zero.

3 Conclusions

1. A difficult problem of modeling pre-arc and arc processes in a liquid metal current limiter can be solved using a systematic approach, the initial stage of which is to qualitatively and fully consider arc phenomena.

2. The current level of knowledge does not allow constructing a generalized mathematical model of a liquid-metal current limiter suitable for calculating short-circuit current breaking process. Therefore, the formation of model should be carried out, focusing on the simplest structure of model, taking into account the basic physical processes and combining them with sufficient depth of their description.

3. It is proposed to simulate the shutdown process not at every time moment, but at specific time moments (reference points). At other time moments, current and voltage should be considered as approximately linearly changing characteristics.

4. The model of circuit break by a current limiter should be implemented as follows:

- Current, voltage and pressure should be considered;
- They should be determined not over the entire time period of the short-circuit current break, but only at the reference, specific time points. In the intervals between specific times, the currents and voltages are represented by straight lines;
- One should take into account the change in pressure inside the volume of the fusible unit during the main and final arc stages, caused not only by vapor of the liquid metal, but also by vapors of ceramic sleeve.

The model should be constructed, focusing, first of all, on the following specific control points of the general break process, delimiting its individual stages.

5. Further improvement of model should be carried out taking into account the nature of the influence of processes on successful operation of the current limiter in:

- Determining the limiting diameter of the fusible unit at which the current limiter can successfully turn off the limit value of the short circuit current;
- The impact of shock waves on the process of arc burning and on mechanical strength of elements of current limiter.

Acknowledgements

This work was funded by RFBR, project number 19-38-90307.

References

[1] S. Dzezhbicki, E. Valchuk, Current-limiting AC switches: trans. from Polish (Leningrad: Energoizdat, Leningrad branch, 116, 1982).

[2] P.A. Kulakov, O.Ya. Novikov, V.I. Prikhodchenko, V.V. Tanaev, High-current fuses, current limiters and switches with liquid metal contacts (Moscow: Informelectro, 49, 1984).

[3] V.V. Tanaev, A.F. Kuzmin, Features of arc burning in devices with liquid metal contacts, Electrotechnical industry 6, 5-6 (1982).

[4] S.L. Kuzhekov, B.N. Vasil'ev, N.N. Kurov, Calculation of dynamic processes occurring during the operation of liquid-metal self-resetting fuses, Proc. of Higher Educ. Institutions, Electromechanics 2, 131-134 (2012).

[5] S.L. Kuzhekov, B.N. Vasil'ev, N.N. Kurov, Performance evaluation of a self-resetting liquid metal

fuse, Proc. of Higher Educ. Institutions, Electromechanics 2, 134-136 (2012).

[6] A.V. Kuznetsov, Y.P. Yurenkov, Mathematical model of the short circuit process in the electrical network with self-recovering current-limiting device with a liquid metal, IOP Conf. Ser.: Mater. Sci. Eng. 643, ISEPC-2019 (St. Petersburg, Russia), 012083 (2019).

[7] L. Niemeyer, Evaporation dominated high current arcs in narrow channels, PAS-98, IEEE Trans 3, 950-958 (1978).

[8] J.De. Palma, J. Gelet, Observation of arcing inside a fuse under capacitor-discharge using 1 million frames per second X-ray imaging, 19th European Conf. on Power Electronics and Applications (Warsaw, Poland), 1-10 (2017).

[9] Zhuo Yang, Hailong he, Fei Yang, Yi Wu, Mingzhe Rong, Peng Zhao, Siyu Lv, Qing Wan, A Novel Topology of Liquid Metal Fault Current Limiter for MVDC Network Applications, IEEE Transactions on Power Delivery, 1 (2019).

[10] V.P. Mesheryakov, R.T. Sibatov, V.V. Samoilov, A.S. Topchiy, Calculation of the arc current and the time of its extinction in low-voltage circuit breakers, Electrical Engineering, Low voltage devices 2, 43-49 (2008).

[11] T. Itoh, T. Miyamoto, Y. Wada, Design considerations and the applications of permanent power fuse, Conference Paper NC72 103-5 IEEE Winter Meeting (1972).

[12] T. Itoh, T. Miyamoto, Y. Wada, T. Mori, H. Sasao, Design considerations on the PPF for a control center, IEEE Transactions on Power Apparatus and Systems 4, 1292 (1973).

[13] J.G. Murray, G.A. Bronner, Proposed method of rating fuses for pulsed service, IEEE Trans. Nucl. Sci. 4, 230-234 (1971).

[14] Resettable fuse: certificate of authorship, 1479972, kl. N 01 N 87/00 (1989).

[15] K.K. Namitokov, N.A. Ilina, I.G. Shklovsky, Devices to protect semiconductor devices (Moscow: Energoatomizdat, 279, 1988).

[16] H. He, et al., Study of Liquid Metal Fault Current Limiter for Medium-Voltage DC Power Systems, IEEE Trans. Compon. Packag. Manufacturing Technol. 8, 8, 1391-1400 (2018)

[17] R. Ryudenberg, ed. K.S. Demirchan, Operational modes of electric power systems and installations: trans. from German, 3rd publ. (Leningrad: Energiya, Leningrad branch, 578, 1980).

Improving the efficiency of electroosmotic drying of electric motors insulation

Alexandr Nemirovskiy[1], *Galina* Kichigina[1], *Irina* Sergievskaya[1], *Alexey* Udaratin[1], *Aleksandr* Alyunov[1] and *Yelena* Gracheva[2]

[1]Vologda State University, Lenina Street 15, Vologda, 160000 Russia
[2]Kazan State Power Engineering University, Kazan, Russia

Abstract. The actual problem of electric motors (EM) operation associated with wetting of windings insulation is considered. A new method of EM drying based on electrokinetic phenomenon of electroosmosis was developed at the Department of Electrical Equipment of Vologda State University. It was shown that water molecules can take part in formation of non-mobile hydrated positively charged hydrogen ions. In this case, hydrogen ions become bound, lose their mobility and ability to take part in the process of electroosmotic drying (EOD). EOD efficiency is drastically reduced. It was proved that it is possible to prevent ion hydration, increase their mobility, penetration ability and, thereby, increase the EOD efficiency by applying a pulsed voltage component to the EOD process with a duty factor of 4. As a result, a new EOD method was developed using a pulsed voltage component that increases the speed of EM insulation drying by three times.

1 Introduction

The number of failures of electric motors (EM) associated with wetting of windings insulation reaches 20-30% [1-3]. The environment of livestock complexes in agro-industrial production and at some industrial enterprises contains hydrogen sulfide, carbon dioxide, ammonia, etc. These gases, when combined with water vapour, form acids and alkalis, which bypass the EM insulation system and sharply reduce the resistance [4].

Moisture exposure to electrical insulation materials of EM insulation system result in hydrolysis of EM insulation, appearance of molds [5,6], deterioration of all electrical characteristics.

The appearance of coil faults in mesh EM winding takes place not only due to switching overvoltages when EM is turned on or its vibration during operation. The reason may be a winding break due to formation of galvanic pair in acidic compounds of moist aggressive environment.

A known effect is "steaming" wet insulation of EM when it is switched on [7]. In this case, the insulation resistance R sharply decreases and the probability of breakdown increases.

Thus, EM maintenance in a humid environment is complicated by the need for regular drying. Almost all known drying methods are thermal [7,8].

Any thermal drying method, to one degree or another, harms the EM insulation and reduces its "life". It is explained by warping, increased cracking, shrinkage, mechanical stress, destruction of molecular structures and thermal aging of insulation. Thermal drying methods are inconvenient, require dismantling and disassembly of EM, increased energy consumption and do not contribute to lower production costs [3, 9-11].

A new method of EM drying was developed, based on electrokinetic phenomena, one of which is electroosmosis [3,4,6]. Moisture is removed from the capillary EM insulation system using electric field [12] and a special electroosmotic drying device (DEOD) [3].

The main advantages of electroosmotic drying (EOD) are that it flows without increasing the EM insulation temperature. The energy consumption during EOD for drying EM with a power of 5 kW is incomparably lower than with thermal methods and amounts to 30-40 W•h. Moreover, EM does not need to be disassembled, which increases labour productivity, since it reduces labour costs by 2–3 times [3]. However, EOD applying is faced with the problem of reducing its duration and increasing efficiency.

The purpose of this study is to increase the efficiency of EM EOD by combining constant and pulsed electric fields.

2 Methodology foundations of research

Water is a strongly polar dielectric. Perfectly pure water weakly dissociates into H^+, OH^- ions; therefore, its electrical conductivity is insignificant and amounts to only 3.8·10-10 Ohm·m. Figure 1 shows a scheme of dissociation of water molecule into ions.

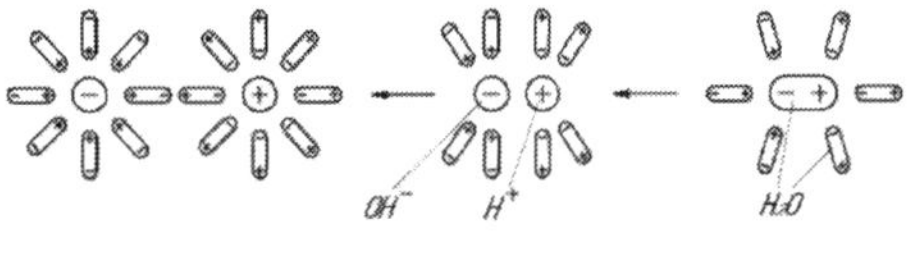

Fig. 1. A scheme of dissociation of water molecule with formation of ion complexes.

The theory of electrolytic dissociation explains the appearance of hydrated ions (ionic complexes), which are unstable compounds of water molecules (clusters). The existence of hydrated ions is proven. For example, a hydrogen ion always binds in solution with a single water molecule, forming a hydroxonium ion H_3O^-. H^+ ions with high penetration ability and mobility are the main driving force for fluid transfer during EOD. At the same time, there are few free H^+ ions in liquid, and for efficient EOD process its amount should be as great as possible [13].

The electrical model of EM insulation during EOD can be represented by an insulation equivalent circuit, by adding resistance R_e to electroosmosis current i_e and EMC of the double electric layer E_d (Fig.2). Connecting DEOD to the circuit induces currents i_∞, i_a, i_R, i_e.

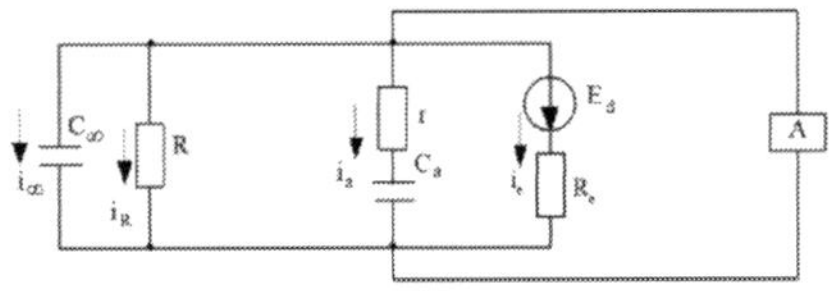

Fig. 2. Electrical insulation model for EOD: C is the geometric capacitance; R is the resistance of through and surface conductivities; r is the active losses associated with polarization of dielectric; C_a is the absorption capacity; Re is the resistance to electroosmosis current; i_∞, i_a, i_R, i_e are currents of geometric capacitance, through and surface conductivities, absorption, electroosmosis; E_d is the EMC of the double electric layer; A is the EOD device.

Currents of geometric capacitance i_∞ and absorption i_a do not have a drastic influence on EOD process, since they quickly decrease to insignificant values. For current i_e in the capillary of EM insulation one has:

$$i_e = \frac{\Delta P \varepsilon R^2 \xi}{4 \mu l}, \quad (1)$$

where ΔP is the pressure difference at the capillary ends;

R is the capillary radius;

ε is the dielectric constant of water;

μ is the dynamic viscosity coefficient of water;

ℓ is the capillary length;

ξ is the electrokinetic potential (zeta potential).

In expression (1) ΔP~ U, where U is the voltage at the output terminals of DEOD, $E_d \sim \xi$. i_e is induced by motion of water ions, which carry neutral H_2O molecules with them. As a result, an increase in i_e increases the EOD efficiency. Analysis of formula (1) shows that it is possible to actually increase i_e by acting on μ, U, ξ. This leads to the development of various ways to improve the EOD efficiency.

For through conductivity current i_R one can obtain [7]:

$$i_R = \frac{\pi R^2 \gamma U}{l^2}, \quad (2)$$

where γ is the specific electric conductivity of liquid.

The i_R current does not have a beneficial effect on the process of insulation dehydration. i_R shunts the electrodes and causes a decrease in DEOD output voltage, and hence the EOD efficiency. Thus, one must strive to reduce i_R. The surface conductivity current i_S caused by pollution and surface wetting of insulation is added to current i_R; therefore, for an effective EOD process, a clean and dry insulation surface is needed [7].

The criterion for EOD efficiency can be its duration τ. The faster drying takes place, the higher its efficiency is subject to other conditions (energy saving, low labour costs, etc.).

In [7], an engineering formula was obtained for determining the EOD duration as a function of initial R_1 and final R_2 insulation resistances, voltage at DEOD output U, and overall EM dimensions:

$$\tau = \frac{\delta l h}{\alpha \lambda U \left[h + (k+1) l \right]} ln \theta \left(R_1, R_2 \right), \quad (3)$$

where l, h are half width and height of EM groove;

δ is width of EM insulation groove;

λ, α are coefficients of electroosmosis and diffusion;

k is electric field averaging coefficient during EOD.

It follows from (3) that it is possible to decrease τ and increase the efficiency of EOD process by increasing U, however, the DEOD voltage for EM 0.38; 6-10 kV is not recommended to increase beyond certain limits due to sharp increase in i_R according to (2) and the risk of insulation breakdown. In addition, an increase in i_R shunts the electrodes, which results in a decrease in DEOD output voltage, and, consequently, the EOD efficiency. Thus, the possibility of intensification the EOD process due to a simple increase in U is quickly exhausted and the search for other ways is necessary.

An increase in the number of free H^+ ions in liquid will make it possible to increase the zeta potential ξ and EMC E_2 [13], which, according to expression (1) and fig. 1, increases current i_e. At the same time, the "harmful" current i_R somewhat increases, however, the positive effect of increasing i_e exceeds the negative impact of i_R growth on EOD efficiency. It is possible to increase the number of free H^+ ions by destroying the bound hydrated H^+ ions. An increase in the number of H^+ ions is equivalent to an increase in diffusion coefficient, which, according to (3), reduces τ.

The specifics of EM operating in agricultural and chemical industrial plants requires consideration of the influence of ammonia on the intensity of EOD process. Ammonia, combined with water, forms a hydrate of ammonium oxide NH_4OH:

 https://doi.org/10.1051/e3sconf/202017801061

$$NH_3 + H_2O \leftrightarrows NH_4^+ + OH^-.$$

NH_4^+ ions are also subject to hydration (fig. 3). The hydrated NH_4^+ ion not only becomes inactive, but also binds the neutral H_2O molecules. Thus, the EOD process in the ammonia medium slows down. Therefore, hydration of NH_4^+ ions must be prevented.

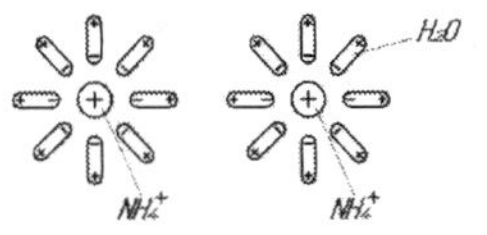

Fig. 3. Scheme of hydrated ammonia ions.

In this study, experimental samples were AIR EM of common industrial use with a power of 1.5 kW. Its insulation system includes Laviterm (groove box and groove wedge), film-asbestos millboard (interphase gaskets), PETV-2 (winding wire).

For statistical validity, three EMs were dried in each experiment (fig. 4). Drying was carried out by two DEOD devices developed at VoGU [3]. A constant voltage was generated at the output of the first DEOD. At the output of the second DEOD, a pulsed component was superimposed on DC voltage component, the parameters of which were determined from the expressions (4)-(12).

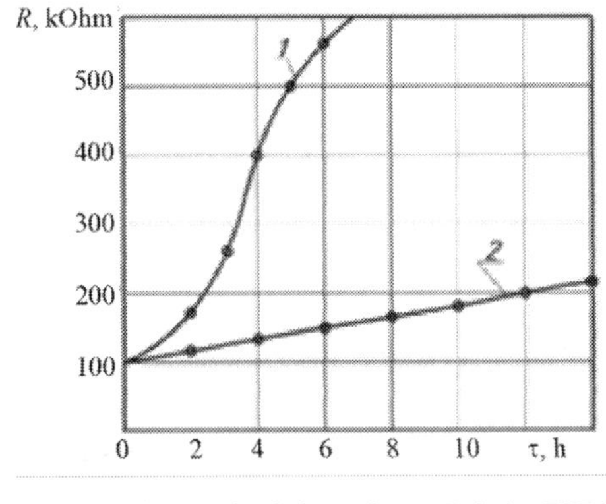

Fig. 4. Change in EM insulation resistance R during EOD by voltage of various shapes: 1 – Curve of EOD process when imposing a pulsed voltage component; 2 – The same during EOD under DC voltage.

The insulation resistance R was measured using the F4101 megaohmmeter. The positive pole from the DEOD output during drying was connected to EM windings connected to a star, and the negative pole to the casing. Preliminary humidification of EM windings was carried out in the KTV-0.4 moisture chamber at a temperature of T = 20 ± 2 °C and a relative air humidity of ω = 97 ± 3% without moisture condensation for 96 hours. To speed up the humidification process, the EM bearing shields were removed.

3 Results and discussion

It is possible to increase the penetration ability, mobility of ions and water molecules, and prevent their hydration by applying an additional pulsed electric field to the insulation system during EOD. Thus, we have developed a new approach for EOD. The essence of the phenomenon is explained from the standpoint of theoretical electrochemistry. H^+ ion or a dipole molecule H_2O located in a liquid medium has shells consisting of molecules or ions (figs. 1,3,5). The motion of the central ion (fig.5) towards the corresponding electrode is associated with destruction of shell and formation of a new one.

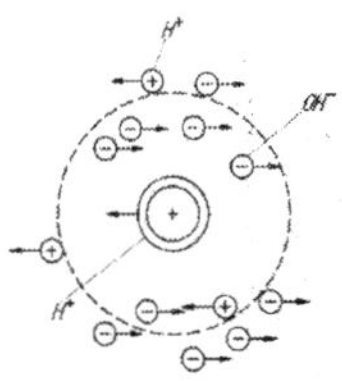

Fig. 5. Graphical representation of H^+ ion in the ionic shell in electric field during EOD.

The Debye – Onsager theory [14] suggests that behind the ion (hydrogen or ammonia) there will always be some excess charge of the opposite sign, and the resulting electric forces of attraction will inhibit its motion. So, when a short-term electrical pulse is supplied with sufficient strength to the insulation system, at certain parameters of pulse, the central ion will detach from the shell, and the new one will not have time to form. As a result, a non-mobile cluster will be destroyed, and its individual components will be freed.

The mechanism of destruction of ionic complexes in figs. 1,3,5 can be explained from the standpoint of energy. The energy supplied to the pulse insulation must be sufficient to destroy the clusters. Clusters have their own internal energy. Formation of clusters is associated with absorption of energy from water system, and their destruction is associated with energy release. The ions and molecules that make up the cluster oscillate in some way relative to their average positions. The amplitude and frequency of their oscillations is also determined by the internal energy of cluster. If the energy of electrical pulse applied to insulation resonates with the cluster energy, then the amplitudes of vibrations of its constituent ions and molecules will be so strong that the cluster will decay. Insufficient or excess pulse energies may not cause resonance and destruction of clusters. Moreover, excess energy of electrical pulses can lead to heating of EM insulation and increased energy consumption during EOD process. The Wien effect, the Debye Falkenhagen theory [14] suggest an increase in electrical conductivity of electrolyte solutions with increasing frequency of the applied electric field due to the destruction of ionic complexes and formation of free molecules and ions, and, therefore, confirm what was said.

https://doi.org/10.1051/e3sconf/202017801061

Application of pulsed voltage (fig.6) at the DEOD output terminals contributes to the destruction of ionic complexes (clusters) in figs. 1,3,5.

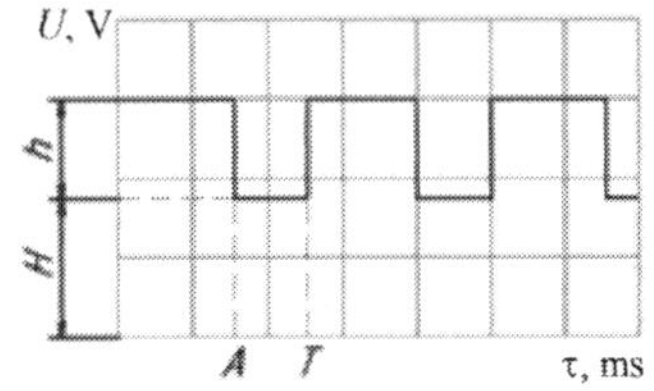

Fig. 6. EOD voltage curve with constant and pulsed components: H, h are amplitudes of constant U– and pulsed U~ voltages.

Pulsed voltage will also periodically weaken the level of space charge of the opposite sign at the electrodes, thereby weakening the electrode effect during EOD process. The positive electrode is the coil, and the negative is the EM housing. All this will positively affect the release of H^+ ions with high mobility. Motion of H^+ ions toward the EM shell will cause the motion of neutral water molecules due to its viscosity. Water will be displaced from the groove portion of EM insulation onto the surface of the stator iron and groove wedges. The insulation R will increase and after a while EM can be started. The work of EM will lead to its heating, evaporation of moisture from the surface of iron, grooved wedges and a further increase in insulation resistance.

When applying a pulsed component, it is prohibited to change the voltage polarity. Changing the electric field sign will lead to the transfer of H^+ ions in the opposite direction to the insulation of winding and EM will begin to moisten. To prevent the reverse transfer of moisture and re-wetting, the polarities of the constant and pulsed components of electric field must coincide.

In general, the optimum energy of voltage pulse during EOD process will be determined by a combination of parameters: shape, duty cycle, repetition rate, ratio of the amplitude of pulsed component to the maximum EOD voltage. Turning to the shape of pulse, we note that the steeper the front, the shorter the time for which a portion of energy is sent to the clusters. This facilitates the conditions for their decay. A rectangular pulse with a leading edge duration of 1 μs A is sufficiently steep and has a simplified engineering implementation in DEOD. The EM insulation system, as a DEOD load, has a predominantly capacitive nature. To reduce the distortion of waveform of DEOD output pulses one should not go beyond frequencies above 2000 Hz.

The expansion in a Fourier series of function that describes the change in voltage of EM insulation system during EOD process under the combined action of a constant and pulsed component has the form:

$$U(\tau) = H + \left(1-\frac{A}{T}\right)h - \frac{h}{\pi}\sin\frac{2\pi}{T}A\cos\frac{2\pi}{T}t - \frac{h}{\pi}\left(1-\cos\frac{2\pi}{T}A\right)\sin\frac{2\pi}{T}t + \ldots = H + \left(1-\frac{A}{T}\right)h + \frac{2h}{\pi}\sin\frac{\pi}{T}A\cos\left(\frac{2\pi}{T}t+\varphi\right) + \ldots, \quad (4)$$

where *H, h* are shown in fig.5.

Taking into account (4) and expression (2.87) from [15], the mobility of water molecules during EOD for voltage curve, shown in fig.5, will be determined as follows:

$$U = \frac{\varepsilon\xi}{\mu}[H + \left(1-\frac{A}{T}\right)h + \frac{\frac{2h}{\pi}\sin\frac{\pi A}{T}}{\sqrt{1+\frac{m^2d^2\omega^2}{\mu^2}}}\sin\left(\frac{2\pi}{T}t+\varphi\right)], \quad (5)$$

where ξ is the zeta-potential;

μ is the dynamic viscosity coefficient of water;

m, d are mass and diameter of liquid molecules;

ω is the oscillation frequency of AC voltage component.

Effective mobility is:

$$\vartheta_e = \frac{\varepsilon\xi}{\mu}\left[H + \left(1-\frac{A}{T}\right)h + + \frac{\frac{\sqrt{2}h}{\pi}\sin\frac{\pi A}{T}}{\sqrt{1+\frac{m^2d^2\omega^2}{\mu^2}}}\right]. \quad (6)$$

The mobility of water molecules $\vartheta_{e\sim}$ at alternating voltage is determined by expression (6), and by ϑ_{e-}at a constant voltage. Their ratio is:

$$\frac{\vartheta_{e\sim}}{\vartheta_{e-}} = 1 + \frac{h}{H}\left(1-\frac{A}{T}\right) + \frac{\frac{\sqrt{2}}{\pi}\left(sin\frac{\pi A}{T}\right)\frac{h}{H}}{\sqrt{1+\frac{4m^2d^2\pi^2}{T^2\mu^2}}}. \quad (7)$$

where $\vartheta_{e-} = \frac{\varepsilon\xi H}{\mu}$.

Consider the relationship between mobility of water molecule and the reciprocal of the duty cycle $X = \frac{A}{T}$. Expression (7) taking into account X is:

$$F(X)=\frac{\vartheta_{e\sim}}{\vartheta_{e-}}(X)=1+\frac{h}{H}(1-X)+ + \frac{\frac{\sqrt{2}}{\pi}\frac{h}{H}}{\sqrt{1+\frac{4m^2d^2\pi^2}{T^2\mu^2}}}\sin\pi X. \tag{8}$$

To maximize expression (8), we find the first and second derivatives:

$$\frac{\delta F}{\delta X}=-\frac{h}{H}+\frac{h\sqrt{2}}{H\sqrt{1+\frac{4m^2d^2\pi^2}{T^2\mu^2}}}\cos\pi X, \tag{9}$$

$$\frac{\delta^2 F}{\delta X^2}=-\frac{h\pi\sqrt{2}}{H\sqrt{1+\frac{4m^2d^2\pi^2}{T^2\mu^2}}}\sin\pi X, \tag{10}$$

As $\frac{\delta^2 F}{\delta X^2}<0$ for $0<X<1$, then any root of equation:

$$\frac{\delta F}{\delta X}=0 \tag{11}$$

At the interval]0;1[gives the function F (X) the largest value in this interval. Combining (9), (11), we obtain:

$$os\pi X=\frac{\sqrt{1+\frac{4m^2d^2\pi^2}{T^2\mu^2}}}{\sqrt{2}}. \tag{12}$$

In the first approximation $\sqrt{1+\frac{4m^2d^2\pi^2}{T^2\mu^2}}\approx 1$, so $cos\pi X=\frac{1}{\sqrt{2}}$. Hence X=1/4, and duty factor T/A=4.

The obtained pulse duty factor will ensure destruction of clusters, an increase in mobility of water molecules, and, consequently, will increase the efficiency of EOD process. The remaining set of pulse parameters was determined experimentally [7].

Figure 4 shows the curves of change in R of EM insulation groove during EOD by voltage of various shapes. Analyzing curves 1, 2, we see that EOD efficiency is several times higher for EOD device, at the output of which, in addition to the constant, a pulsed component is also generated. After 4 hours, the EM insulation R when using DEOD with a pulsed voltage component reached 400 kOhm. At the same time, the EM insulation R during EOD with only constant voltage was 130 kOhm.

4 Conclusions

1. Application of new method of cold drying of EM insulation eliminates thermal destruction, reduces energy consumption by 2-3 orders of magnitude and labour costs by 2-3 times.
2. Hydrated water molecules and ammonia ions become inactive, which reduces the EOD efficiency.
3. Imposition of additional pulsed electric field on EM insulation system during EOD increases the penetration ability, mobility of ions and water molecules and prevents their hydration.
4. A new EOD method was developed using a pulsed voltage component with a duty factor of 4. It increases the drying speed of EM insulation by three times.

References

[1] A.E. Nemirovskiy, A.I. Kashin, V.F. Kosmach, Y.F. Titovec, I.N. Toptygin, D.A. Zaripova, Innovative technology for dismantling the windings of electric motors using ultrasonic radiation, IOP Conference Series: Earth and Environmental Science 337, 012071 (2019).

[2] M. Rashevskaya, S. Yanchenko, S. Tsyruk, B. Kudrin, Assessing non-stationary power quality phenomena of induction motors, Proceedings of 18th International Conference on Computational Problems of Electrical Engineering, CPEE (2017).

[3] A.E. Nemirovsky, G.A. Kichigina, I.Yu. Sergievskaya, Electroosmotic drying and moisture protection of electrical equipment, Fedorov Readings 2017: materials of the XLVII International Scientific and Practical Conference with elements of scientific school (MPEI, Moscow, 2017).

[4] A.E. Nemirovsky, New ways of drying electrical equipment, Equipment in agriculture 6 (1989).

[5] Y.I. Gracheva, O.V. Naumov, Evaluation criteria of contact group technical state concerning electrical appliances, International Journal of pharmacy and Technology 4 (2016).

[6] V.V. Maslov, Moisture resistance of electrical insulation (Energiya, Moscow, 1973).

[7] A.E. Nemirovsky, Improving the efficiency of drying and moisture protection of insulation of electric motors used in agriculture, based on intensification of electroosmotic phenomena: thesis for the degree of doctor of technical sciences (SSAU, St. Petersburg-Pushkin, 1993).

[8] O.I. Khomutov, System of technical means and measures to increase the operational reliability of insulation of electric motors used in agricultural production: thesis for the degree of doctor of technical sciences (CHIMESH, Chelyabinsk, 1990).

[9] A. Alyunov, O. Vyatkina, V. Ivanov, Yu. Karpov, V. Melehin, A. Andreev, L. Mukhametova, Improving methods for identifying electric motor parameters in case of stator winding damage, IOP Conference Series: Earth and Environmental Science 337, 012070 (2019).

[10] A. Ydaratin, T. Akhmetov, S. Makoev, IJCIET 10, 2, (2019).

[11] I. Akhmetova, A. Tyfetylov, A. Tamakchi, G. Khadiyllina, O. Derevianko, Z. Syed, Improving the competitiveness of automobile engineering enterprises by advancement the internal control over its indicators, International Journal of Civil Engineering and Technology 9, 13, 1865.

[12] A.E. Nemirovsky, N.K. Moroz, K.P. Simakov, The method of electroosmotic drying of windings insulation of electrical machines, patent for invention, 2174280, 27 (2001).

[13] A.Ya. Fanin, Liquid motion in porous materials under the influence of electroosmosis, Automation and mining electrical engineering issues: Collection of papers, Don. UGI, 40 (Nedra, Moscow, 1969).

[14] B.P. Nikolsky, A new reference guide for chemist and technologist (MMIV, St. Petersburg, 2004).

[15] D.Ya. Vodenisov Study of influence of electroosmosis and ultrasound on isolation of ship electrical equipment: thesis for the degree of candidate of technical sciences (LPI, Gorky, 1981).

Universal Installation for the integrated utilization of flue gases and wastewater from thermal power plants

Sergey A. Glazyrin[1], *Mikhail G.* Zhumagulov[1,*], *Zhanar A.* Aydimbaeva[2], and *Abay M.* Dostiyarov[2]

[1]L.N. Gumilyov Eurasian National University, Dept. of Power engineering, 2 Satpayev str, 010003 Nur-Sultan, Kazakhstan
[2]Almaty University of Power Engineering and Telecommunications, Dept. of Power engineering, 050013 Baytursynuly, 126/1, Almaty c., Kazakstan

Abstract. For 30 years research has been carried out on the use of wastewater from thermal power plants and industrial boilers, as well as on the use and extraction of various components from flue gases such as carbon dioxide, sulfur and nitrogen. Technological solutions were developed and implemented in various productions at various times: - use of acid-forming components of flue gases for the regeneration of cation exchangers; - carbon dioxide from flue gases of 99.9% purity with "food" quality; - technical nitrogen of 95-99 purity from flue gases; - wastewater usage to increase the degree of sulfur oxides from flue gases. The article presents a technological solution for the integrated utilization of flue gases and wastewater from a thermal power plant with high-pressure boilers burning solid fuels.

1 Introduction

The main process of water treatment used in boilers is the technology using ion-exchange materials, which require periodic regeneration. In accordance with the increase in the production of desalted and softened water, more and more reagents are spent on the regeneration of ion exchangers.

For the regeneration of cation exchangers, solutions of sodium chloride or sulfuric acid are usually used. The costs of obtaining these reagents are increasing every year and it is getting difficult to obtain them, and sodium chloride is used in other industries, primarily for the production of caustic soda, chlorine and other substances.

Under these conditions, it is relevant to search for new low-cost and non-deficient reagents for the preparation of regeneration solutions and to develop technologies for water preparation and regeneration of ion exchangers that save already known reagents [1].

The process of using carbon dioxide in the production of electric welding at enterprises is improved year by year. When performing welding in a carbon dioxide environment, the quality of the weld is significantly higher than when using coated electrodes as the wire is used instead of the electrode. In addition, labor productivity during welding is increased by 2.5 - 4 times. Carbon dioxide is also required in the food industry

At the same time, the outgoing flue gases contain carbon dioxide, the production of which does not pose a great deal of difficulty for enterprises that have thermal power units [2].

In this regard, it is of interest to consider the possibility of using carbon dioxide for the preparation of regeneration solutions, which would make it possible to have the necessary amount of reagents of our own production.

For the preservation of heat and power equipment during idle time, the nitrogen preservation method is used, which is the simplest and most reliable way to protect against corrosion [3]. However, the method has not yet widely used in the energy sector due to the large capital expenditures on the construction of typical nitrogen-oxygen plants and the considerable costs of their maintenance. Therefore, obtaining cheap nitrogen at its own enterprise is also of current interest. In addition, the issue of using nitrogen to de-aerate the feedwater of boilers, in order to displace oxygen from water, in water-heating boilers, is being worked out, which, with a positive result, will allow us to avoid evacuation of the deaeration process.

Flue gases contain acidic components, which allows them to be used to neutralize alkaline waste water [4, 5]. If we technologically combine neutralization with the carbon dioxide production scheme, then, by taking off the flue gases after the neutralization unit, we get the gas almost completely purified from sulfur components.

Thus, the flue gases of industrial and power boilers are raw materials for the production of products used not only in the energy sector, but also in other industries.

2 Research methods

General scientific, empirical, theoretical methods and experiments on laboratory and industrial equipment were used in the study on the development of technology for

* Corresponding author: zhumagulov_mg@enu.kz

the integrated utilization of flue gases and wastewater from thermal power plants and boiler houses of industrial enterprises [6].

3 Results and discussions

For the extraction of individual components and their mixtures from exhaust gases, which are used in various industries, there are developments of installations for the production and use of these components, which have passed pilot tests for a long time, or are currently under operation. These are installations for the production of carbon dioxide, nitrogen, the preparation of regeneration solutions, the usage of acidic components of flue gases to neutralize alkaline waste water. Any enterprise with a boiler house needs: carbon dioxide – for electric welding; nitrogen – for preservation in TPPs or for grain storage at an industrial district boiler house; acid components – for preparing regeneration solutions. But installation of several separate installations in one enterprise is economically and technologically inefficient.

A universal installation was developed, which allows producing several types of products in one production using current developments [7]. Such an installation should not have high performance in obtaining individual components, but sufficient to meet the needs of not only the production itself, but also a small city. Part of the production, such as carbon dioxide and nitrogen, can be realized on the side. Low productivity is also explained by the fact that this installation cleans the flue gases, improving the environment, does not require a large staff, is mobile in operation, and should be planned separately at each enterprise, making the most of the existing space and equipment.

The following results were obtained based on the results of industrial tests

A flow chart for the preparation of regeneration solutions using exhaust flue gases, presented in Figure 1, has been developed. Studies have been conducted on the use of the solutions obtained in a water treatment unit for the regeneration of cation-exchange filters [8].

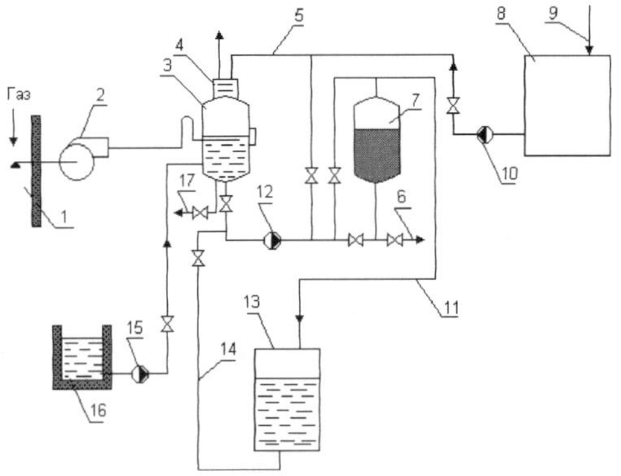

Fig. 1. Recovery scheme of using the regeneration solution for its reuse and disposal of flue gases (1 –flue-gas duct; 2 - fan; 3 - regeneration solution preparation tank; 4 - deaerator head; 5 - line for recirculation of the solution in the filter; 6 - supply of the regeneration solution to the filter; 7 - cation filter; 8 - collection tank of alkaline waste water from water treatment; 9 - alkaline water collection line during regeneration of anion-exchange filters; 10, 12 - pumps; 11 - discharge line of the regeneration solution for reuse; 13 - tank for receiving discharge of regeneration solution; 14 - supply line of spent regeneration solution for the preparation of a fresh solution; 15 - salt feeding pump; 16 - salt storage cell; 17 – drainage).

Based o the results of the study conducted we can note the following:

1. The regeneration solutions have the following composition at different pH values:

- at pH = 1 - H_2SO_4 to 0.1 g-mol / l and trace amounts of H_2SO_3;
- at pH = 2 - H_2SO_4 to 0.03 g-mol / l, H_2SO_3- 0.05 - 0.08 g-mol / l and trace amounts of HNO_3 and HNO_2.

Small amounts of V, Ba, Ti, and other heavy metals were detected in solutions, which in aqueous solution can be catalysts for the oxidation of SO_3^{2-} and SO_3^{2-} and HSO_3^- in SO_4^{2-} by the reaction

$$H_2SO_3 + ½ \cdot O_2 \rightarrow H_2SO_4 \qquad (1)$$

which explains the high concentration of H_2SO_4 and the practical absence of dissolved oxygen.

2. Using flue gases in heating boilers, TPP, it is possible to prepare regeneration solutions with pH = 1. These solutions effectively replace the salt solutions that are currently used for regeneration. The degree of regeneration of the solution with pH = 1 reaches 60 - 70%, which will save a large number of reagents.

3. The following is effective: regeneration by countercurrent, temperature is 20 - 40 °C, transmission rate of the solution 15 - 20 m / h. The ratio of the volume of the regeneration solution to the volume of sulfougol is 10 – 15

4. The regeneration solution can be reused after the new portion of the flue gas is saturated. The multiplicity of such use is 5-7, without precipitation and reducing the regeneration properties. This reduces the water consumption for regeneration. The reduction of water consumption is also facilitated by the fact that regeneration can be carried out without preliminary loosening, and 2–3 times less water is used for washing cationite after regeneration

5. High efficiency shows joint regeneration with a solution of NaCl with a concentration of 2% and a solution of flue gases. When this is carried out H - Na - cationization. At 30% salt saving, a high degree of regeneration of the cation exchanger is reached (up to 90–95%) and the content of the HCO_3^- ion decreases, which prevents alkaline corrosion in the boilers.

A universal installation for thermal power plants with high-pressure boilers was developed, shown in Figure 2 [9].

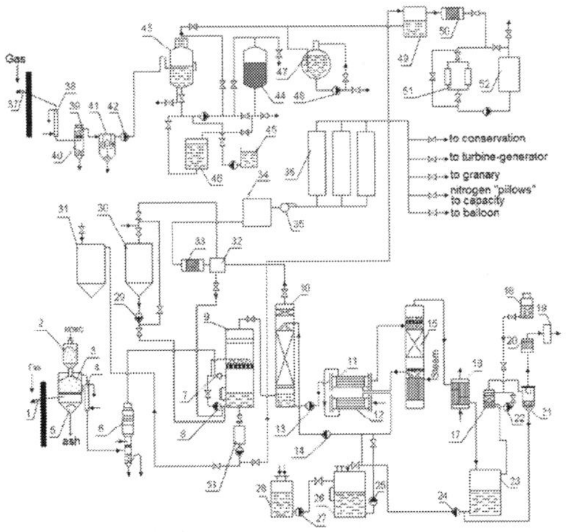

Fig. 2. Universal installation for thermal power plants with high pressure boilers (1 - high-temperature gas extraction unit; 2 - coke stock capacity; 3 - reactor; 4, 38 – tube-in-tube type heat exchanger; 5, 40 - ashpit; 6 - gas cooler; 7 - low-pressure ejector; 8 - circulation pump of soda solution; 9 - soda scrubber; 10 - absorber; 11 - heat exchanger MEA - MEA; 12 - heat exchanger MEA - water; 13, 14, 22, 24, 25, 27, 29, 48 - the pump; 15 - desorber; 16 - vapor-gas cooler; 17 - site washing of carbon dioxide with manganese-acid potassium; 18 - capacity of preparation of manganese-acid potassium; 19 - carbon dioxide gas tank; 20 - carbon dioxide moisture desiccant; 21 - droplet separator; 23 - condensate separator; 26 - tank for preparation of MEA working solution; 28 - storage capacity of a concentrated solution of MEA; 30 - tank for collecting acid waste water from the chemical water treatment; 31 - tank collection of alkaline waste water from the chemical water treatment; 32 - alkaline barrier filter; 33 – Nitrogen moisture dryer; 34 - nitrogen gas tank; 35 - compressor; 36 - receiver; 37 – low temperature gas extraction unit; 39 - filter for removal of mechanical suspension; 41 - bulk ash collector; 42 - fan; 43 - capacity of the preparation of the regeneration solution; 44 - sodium - cation filter; 45 - salt storage capacity; 46 - regeneration water collection tank; 47 - boiler drum; 49 - barrier soda filter; 50 - dehumidifier of moisture of flue gases used to prepare the regeneration solution;; 51 - node with absorbers loaded with activated carbon; 52 - vegetable storehouse with gas medium; 53 - intermediate tank).

After the high-temperature gas extraction unit 1, a reactor 3 is installed with a coke-2 storage capacity. The reactor produces low-temperature afterburning of oxygen residues from flue gases. Oxygen from flue gases is removed when gases pass through a layer of reducing agent (usually carbon). The interaction of carbon with an oxidizer proceeds with the participation of intermediate unstable formations. It is recommended to use low-ash fuel as a reducing agent, since a ash shell of smaller thickness forms on pieces of low-ash fuel, than on pieces of high-ash fuel. Combustion takes place at a temperature below the melting point of the ash, which does not affect the rate of burning of the particle

The gas extraction unit provides for isolation of the gas extraction pipe from the heating surface of the boiler in order to preserve the temperature of the gas to be extracted. In addition it should be noted that with different modes of operation of the boiler, the temperature of the gas entering the layer of reducing agent must not be lower than the calculated one (for coke it is 600 °C). Moreover, the sampling device must exclude additional oxygen suckers with flue gas entering the reactor 3 [10].

In order to sharply reduce the velocity of the gas, the area of the inlet of the probe should be 3-4 times as large as the area of the pipeline that takes gas from the furnace. With a decrease in speed and a change in the direction of motion, large dust particles "fall through" and do not fall into the probe.

The lower part of the reactor ends with an ash storage bin 5. The gases cleaned with 95 to 98% of the ash pass through a tube-in-tube heat exchanger, where they are pre-cooled, and sent to the gas cooler 6, in which they are freed from ash residues discharged through the bottom of the cooler periodic purging, as well as from the vast majority of sulfur components gases. Then the gas is sent to the carbon dioxide production unit.

Gas is taken into the nitrogen receiving unit from the upper part of the absorber and sent to the alkaline barrier filter 32, in which, due to the constantly flowing alkali from the tank 30 or directly from the alkali discharge pipeline from the water treatment plant, carbon dioxide remains are connected. After the barrier filter, the gas passes through a desiccant of nitrogen 33, which is filled with silica gel or zeolite, capable of absorbing moisture in the amount of 20% and 50% of its own weight, respectively. Nitrogen is sent to the gas tank 34, where, if necessary, the compressor 35 can be pumped out through the receivers 36 to the consumer. The alkali used in the filter 32 is periodically discharged to the suction of the soda-water circulation pump. The nitrogen obtained in the universal installation can be used for the conservation of power equipment, in grain storage facilities, in the turbogenerator, and pumped into cylinders [11].

At thermal power plants that have cation-resin and anion-exchange filters in make-up water preparation system, the project provides for a neutralization unit, in which two tanks 30 and 31 are included to collect waste alkaline water and waste acid after regeneration of filters [12]. In the neutralization unit, small tanks and pumps of various capacities are installed, which can be used in the universal installation, where one of the tanks (30) is designed to collect waste alkaline waters, and the second (31) - for waste acidic waters [13]. If necessary, the alkaline water is pumped into the soda scrubber by pump 29. After a short period of time, NaOH is transferred to Na_2CO_3 under the influence of carbon dioxide and subsequently used to trap residual sulfur from flue gases. The spent soda solution is discharged into the intermediate tank 53, where, if necessary, it is sent to an acid waste collection tank 31 or to a barrier soda filter 49.

For the second part of the installation, where oxygen removal is not required, low-temperature gas extraction by the probe 37 installed behind the exhauster along the gases from the boiler is carried out. Gases are sent to the heat exchanger 38 of the "pipe in pipe" type, where they are cooled with water, which reduces the volume of gases and allows the use of the fan 42, designed for lower temperatures and lower performance. When solid fuel is used at TPPs, a filter 39 is installed with anthracite loading to trap suspended particles and ash. An ashpit 40 is mounted in the lower part of the filter. A volumetric ash trap 41 is installed in front of the fan 42 to trap the remaining ash. To prepare the regeneration solution using the acidic components of the flue gases in the tank 43, the solution is recirculated by the pump. The source water from the recovery water collection tank 46 is removed by the recirculation pump and, after preparing the solution, the ready regeneration solution is supplied by the same pump for the regeneration of the sodium-cation filter 44. It is desirable to use one of the water treatment filters as the tank.

The gases leaving the deaerator head of the tank 43 do not contain sulfur components or contain them in a minimum quantity and are suitable for use as a source gas for preparing the wash solution when starting blocks from a cold state, as well as for washing copper deposits from the surface of the equipment. The rest of the gases are sent to a barrier soda filter 49 filled with a special solution. The purified gases pass through the dehumidifier of moisture 50 and enter the vegetable storehouse 52, or preliminarily pass the unit 51 with absorbers loaded with activated carbon to regulate the composition of the gas.

When the installation is placed at TPPs, the main units are mounted in the project room of the neutralization unit, at the water treatment plant, in the boiler room, in the area of smoke exhausters [14]. Compressors are mounted in a separate room, gas-holders and receivers - outdoors.

A gas extraction unit was designed with low-temperature oxygen removal in the reductant layer and ash removal.

For the selection of gas from the boiler, it is important to choose the installation location of the selected probe [15]. It should be taken into account that at different loads of the boiler operation, the temperature of the gas entering the layer of reducing agent should not be lower than the calculated one, for example, for coke - 600 ° C. In addition, the sampling device should exclude additional oxygen suckers with flue gas entering the reactor. The gas extraction unit (Figure 3) includes a bleed probe mounted on the end of the flue gas bleed pipeline.

The probe inlet should be 3–4 times the area of the extraction pipeline, in order to sharply reduce the velocity of the gas. With a decrease in the velocity and direction of gas movement, large dust particles "fall through" and do not fall into the probe. The inlet of the probe is directed "along the" selected flue gas. With the chosen dimensions and location of the probe, the amount of ash taken into the installation decreases by 20–30% already in the furnace at the entrance to the probe. The gas extraction pipeline from the furnace is mounted with a slope of at least 15% towards the reactor, which prevents its ash from being carried.

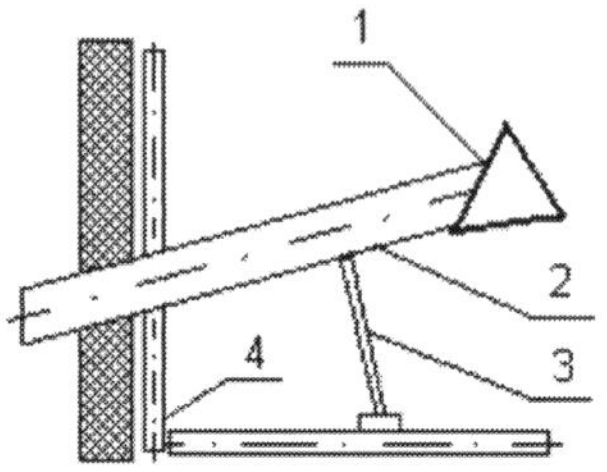

Fig. 3. Gas extraction unit (1 – bleed probe; 2 – gas bleed pipeline; 3 – support tube; 4 – heat surface).

When creating the first industrial unit, the ash from the flue gas was removed before it entered the reducing agent layer in the reactor. Various options were tested for ash collectors with bulk separation, with a layer of ceramics and with the installation of cyclone separation inside the ash collector. However, these ash collectors proved to be ineffective in removing ash (up to 75%), while at the same time their high hydraulic resistance (from 15.0 to 20.0 kPa) sharply reduced the gas flow rate for the installation [15]. The ash collectors installed before the reactor could not completely eliminate the drift of the semi-coke layer by fly ash from the flue gases. All ash traps installed in front of the reactor, in spite of their good thermal insulation, reduced the temperature entering the flue gas reactor by an additional 100–200 oC. In addition, when the reactor was heating up, the gas flow rate increased and, as a result, there was an increased drift of the semi-coke layer already in the first 2 to 3 hours after the unit was put into operation.

In the final installation, the main part of the ash was removed in a ash collector mounted under the reactor grate, and for the remaining ash, a volume separation ash collector was installed according to the scheme in the low temperature zone 110-120°C, which also had a positive effect on the ash removal process.

The rector operates at a temperature of 400–700°C and differs in shape, size, ash collectors under the grate, presence of a reducing agent hopper and a water cooling "jacket". The reactor of circular section (Figure 4) is made with a ratio of the height of the layer of reducing agent to the diameter of at least 2.

There is a repair hatch with a diameter of 500 mm and a hatch for "hiding" the hot layer of the semi-coke on the grate in its side wall. The thermal insulation of the walls of the reactor both inside and outside is provided. The reducing agent stock, designed for its additional loading on more than once every 2 months is installed above it. An ashpit is located under its grate. The housing of the ashpit and the reactor is made of a single pipe and is insulated from the outside and the inside. The combustion grate was made in two versions. First version was made of stainless steel sheet with a thickness of 10 mm, with holes of 8 mm diameter and thermal insulation protection from the reductant. However, it showed low reliability during operation. The removal of ash from the reactor was a significantly difficult: the "rubbling" was difficult due to the rough surface of the grid insulation, some of the holes were clogged with small pieces of semi-coke, preventing the ash from being removed. Second version of the grate was made of stainless steel tubes with a diameter of 16 mm, with a gap of 6 - 7 mm, which were fastened by welding to a corner welded to the reactor vessel.

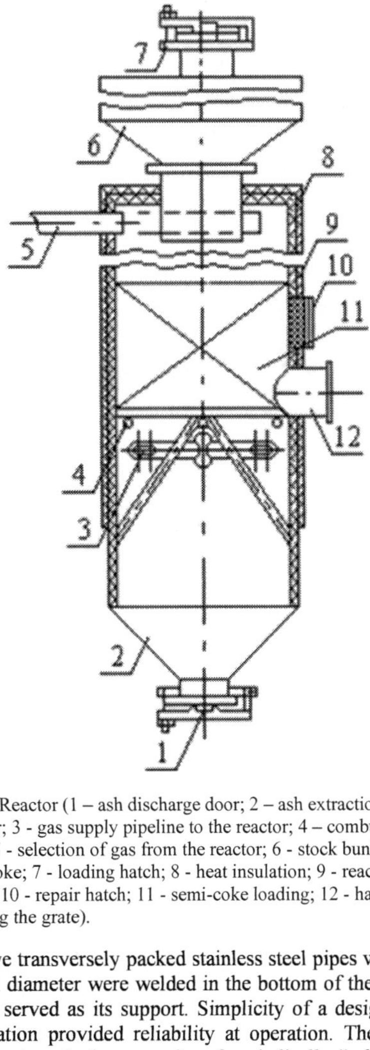

Fig. 4. Reactor (1 – ash discharge door; 2 – ash extraction retainer; 3 - gas supply pipeline to the reactor; 4 – combustion grate; 5 - selection of gas from the reactor; 6 - stock bunker for semi-coke; 7 - loading hatch; 8 - heat insulation; 9 - reactor vessel; 10 - repair hatch; 11 - semi-coke loading; 12 - hatch for cleaning the grate).

Five transversely packed stainless steel pipes with 50 mm in diameter were welded in the bottom of the grate, which served as its support. Simplicity of a design and installation provided reliability at operation. The main disadvantage is the necessity of periodically "rubbling" the reducing agent through the open hatch. The variant of the "swinging" grate was used to clean it from ash, i.e. the handle, mounted on the outside of the reactor, can be "pumped out" by opening the hatch cover. However, there is the problem of sealing grate sites in order to eliminate air suction in the reactor operating under vacuum. When choosing a design variant of the grate, it was taken into account that the frequency of "rubbling" of the layer of reducing agent above the grid is no more than 4–8 hours, the duration of the operation is 10–15 minutes.

4 Conclusions

The process flow diagram of the universal installation has been developed, which makes it possible to comprehensively utilize wastewater and flue gases at one thermal power plant, burning coal, and to receive several types of products, and to solve simultaneously several tasks, such as:

- regeneration of cation in make-up water preparation system byh regenerated solution prepared with the part of the washing water of cation-exchange filters saturated with acid-forming components of flue gases;
- production of carbon dioxide of food quality from the flue gases, which can be used both for the preparation of carbonated beverages and for welding, etc.
- production of technical nitrogen from flue gases of very high purity, which can be used for equipment conservation, for improving the quality of grain storage and other needs.

References

[1] S.A. Glazyrin, N.S. Glazyrina, B.G. Mukanova, Multiparametric Optimization of Water Treatment Process for Thermal Power Plants, ICCSAM 2015: 17th International Conference on Computer Science and Applied Mathematics, Venice, Italy, 13-14 April, 328–333 (2015)

[2] S.A. Glazyrin, A.I. Glazyrin, B.K. Aliyarov. Low-performance carbon dioxide plant. Scientific notes of Pavlodar State University (1998). ISBN: 1560-6120

[3] S.A. Glsazyrin, H. Hatjan, Analysis of metal protection methods for equipment of Astana-Energy JSC against corrosion, V International scientific conference: Actual problems of transport and energy, Astana, 438–441 (2017)

[4] N. Zlatov, S. Glazyrin, M. Zhumagulov, Z. Aidynbayeva, Wastewater treatment of the thermal power plants for desulfurization of flue gas, 2019 International Conference in Telecommunications, Informatics, Energy and Management (TIEM), Kavala, Greece, 12-14 September, 164–167 (2019). ISSN: 2367-8437

[5] S.A. Glazyrin, R.A. Musabekov, N.V. Zlatov, Zh.A. Aidymbayeva, Analysis of technologies for cleaning smoke gases from thermal power plants from sulfur oxides, XXIII Scientific Conference Energy, ecology, comfort, well-being, Sofia (Bulgaria), 29–32 (2018)

[6] Yu.N. Kolmogorov, A.P. Sergeev, D.A. Tarasov, S.P. Arapova, *Methods and means of scientific research: textbook* (Ekaterinburg: Publishing House Ural, 2017)

[7] S.A. Glsazyrin, N.S. Glsazyrina, Design of a universal installation for the disposal of flue gas components of boiler houses with low pressure boilers, International Scientific Conference

Problems and Achievements in Industrial Energy, Yekaterinburg (Russia), 74–77 (2010)

[8] S.A. Glsazyrin, Investigation of the properties of cation exchanger during repeated regeneration by a solution prepared on the basis of waste wash water using flue gas components, Herald of PSU (Pavlodar), 2, 58–63 (2008)

[9] S.A. Glsazyrin, Universal installation for the use of flue gas components at thermal power plants with high pressure boilers, Kaz. GOSINTI, order no. 111 (1998)

[10] S.A. Glsazyrin, Design of a flue gas sampling unit for boilers to use their components. International Scientific Conference Current State and Prospects for the Development of Energy, Tashkent (Uzbekistan). 18-20 December, 134–136 (2006)

[11] S.A. Glsazyrin, A. Kyrykbay, K.Sh. Erzhanov, Analysis of existing methods for reducing and cleaning sulfur and nitrogen oxide emissions. International Scientific Conference of Young Scientists "XII Satpayev Readings", Pavlodar **21**, 70–76 (2012)

[12] S.A. Glsazyrin, A.T. Seitbekova, The process of improving the water treatment process in the boiler, VII International Scientific Conference: Actual problems of transport and energy, Astana. 504–511 (2019)

[13] S.A. Glsazyrin, N.S. Glsazyrina, Problems of operation of existing design schemes for water treatment for supercritical boilers, Herald of PSU, Pavlodar, 2, 73–82 (2017)

[14] S.A. Glsazyrin, A.I. Glsazyrin, Installation for the production of nitrogen from flue gases, Kaz. GOSINTI, Order no. 102,103, 104 (1998)

[15] S.A. Glsazyrin, Options for the integrated use of waste TPP, International scientific conference: Quality of education: management, credit system of education, achievements, problems, 213–215 (2006)

Fuel Saving And Electric Power Quality Improvement In Autonomous Power Supply Systems By Using Energy Storages

Kamil Bakhteev[1,*], *Alexander* Fedotov[1], and *Rinat* Misbakhov[1]

[1]Engineering center, Computer modeling, and engineering in the field of power engineering and power engineering, Kazan State Power Engineering University, 420066 Kazan, Russia

Abstract. In autonomous power supply systems for diesel power stations located in remote areas, the main component of costs is the cost of imported fuel. The use of energy storage systems reduces fuel consumption for an autonomous power station. Reducing the northern import of liquid fuel is a relevant task for Russia. The organization of the charge-discharge process of an energy system storage according to the criterion of the minimum daily fuel consumption depends on both the load schedule and the type of flow rate characteristic of the primary engine. When making design and investment decisions on the feasibility of using alternative technical solutions for the modernization of sources in isolated energy supply systems, a comparative assessment of their economic efficiency is necessary. The paper studies the choice of energy storage system power in terms of fuel economy, reducing the installed capacity of a power station for electric power supply systems.

1 Introduction

The use of energy storage systems (ESS) may be appropriate for diesel-generator units, primarily diesel fuel, gasoline, fuel oil, as their operation is associated with the need for constant delivery of fuel and its seasonal storage, with the need to work in intermittent modes following the load of consumers, which significantly increases fuel consumption, and, consequently, current costs [1-3]. To assess the economic efficiency of the designed facility, it is necessary to compare the capital (investment) costs of creating an electrotechnical complex (ETC) and the current costs of its operation. As a rule, for a quick project assessment at the stage of feasibility studies, the static method is used and the project payback period is calculated. The combination of ESS with an autonomous generator assumes solving a number of issues related to the optimization of the parameters of the created "generator-ESS" ETC. Battery Energy Storage Systems (BESS) are capable of damping power imbalances for short and long periods of time, storing large amounts of energy. As an example, it is possible to note that the General Electric Corporation and the California-based energy company Southern California Edison announced the implementation of the first project in the world of a hybrid power station in which a gas turbine unit (GTU) works in conjunction with ESS [4-6]. Such an application will provide a solution to such problems as:

- increasing the limits of the dynamic stability of GTU with small values of the moments of inertia during accidents in the grid;

Research is made with financial support of the Ministry of Education and Science of the Russian Federation within implementation of the federal special program «Research and development in the priority directions of scientific and technological complex of Russia for 2014-2020», the agreement on granting a subsidy № 075-15-2019-057, unique identifier of applied scientific research (project) RFMEFI57418X0188. All research articles should have a funding acknowledgment statement included.

- stable operation of the auxiliary system and the GTU excitation system due to the maintenance of the required voltage level at the GTU clamps with significant voltage fluctuations in the network during autonomous GTU operation.

But, as you know, power units - electric generator drives for autonomous small thermal power plants, apart from GTU, can be diesel-generator units (DGU), gas piston installation (GPI). The determining criteria for the selection of power units for use in an autonomous power supply system (APSS) are fuel consumption issues, the level of operating costs, and the payback period of power station equipment. In addition, when choosing power units, factors such as ease of operation, the level of maintenance and repair, and also the place where the power units are to be repaired should be taken into account. These issues are primarily associated with costs and problems that may arise during the operation of an APSS. The tasks of ensuring the stabilization of the GPU rotation speed and optimal control of the ESS charge-discharge process should be worked out.

* Corresponding author: kam.rav@bk.ru

2 The cycle of charge-discharge of the battery without taking into account energy losses

One of the problems that can be solved by applying the ESS based on electric batteries is fuel economy. Other problems solved by ESS are presented in [7–9]. The organization of the charge-discharge process of an ESS according to the criterion of the minimum daily fuel consumption depends on both the load schedule and the type of flow rate characteristic of the primary engine. Fig. 1 shows the relevant data for various types of engines, covering almost all types of flow characteristics.

Consumption characteristics 1 and 3 (Figure 1) are qualitatively the same. They are distinguished only by the fact that for diesel power stations the passport values of fuel consumption for a 25% load are given.

From Fig. 1 it is seen that a decrease in the generated power leads to an increase in specific fuel consumption. Thus, it is possible to provide a fuel economy in the "generator-ESS" ETC if the load schedule is substantially irregular. Then, during the minimum load hours, the BESS is charged, and during maximum load hours, it transfers its energy to the grid, Fig. 2a, graph 3, corresponding to the battery charge at constant current.

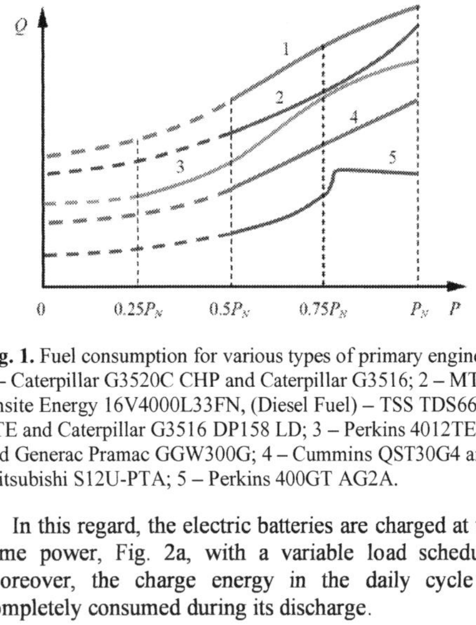

Fig. 1. Fuel consumption for various types of primary engines: 1 – Caterpillar G3520C CHP and Caterpillar G3516; 2 – MTU Onsite Energy 16V4000L33FN, (Diesel Fuel) – TSS TDS6606 LTE and Caterpillar G3516 DP158 LD; 3 – Perkins 4012TESI and Generac Pramac GGW300G; 4 – Cummins QST30G4 and Mitsubishi S12U-PTA; 5 – Perkins 400GT AG2A.

In this regard, the electric batteries are charged at the same power, Fig. 2a, with a variable load schedule. Moreover, the charge energy in the daily cycle is completely consumed during its discharge.

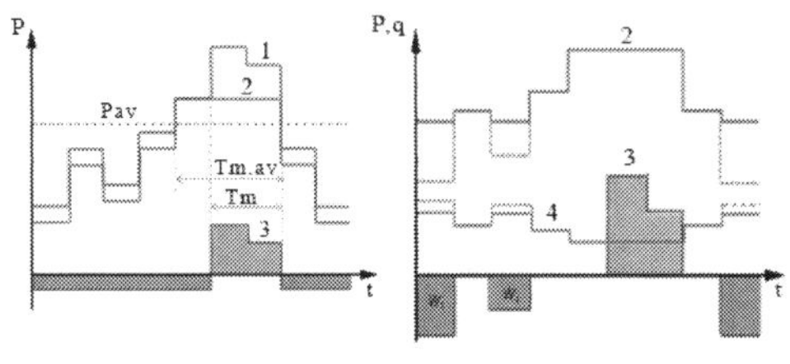

Fig. 2. Daily power and specific fuel consumption profiles: a - load power profiles (1), generator (2), ESS (3); b - power profiles and specific fuel consumption (4).

When optimizing the operation of a power station, the values of power losses at the first stage of development of the technique are not used. In this regard, a simpler method can be applied: either prescriptively assign the maximum duration, as is done in Fig. 2a, or calculate the average load on the daily time interval (Pav, Fig. 2a) and its part that is above the average, be attributed to the maximum load. This will highlight the interval of the duration of the maximum load Tm.av, Fig. 2a. In Fig. 2b shows an optimized BESS charge graph. It is formed on the basis of the laws of increasing specific fuel consumption with decreasing load. In order to reduce the daily fuel consumption, it is necessary to organize the recharging of the BESS in such a way as to maximize the GPU power generation during the day.

For definiteness, we assume that the characteristic of the specific fuel consumption of a power station (1) has the form [10]

$$Q_0 = Q_{nL} + bP^{\mu} \tag{1}$$

where Q_{nL} is the idle fuel consumption; b, μ are the approximation parameters, and $0 < \mu \leq 1$.

In [8, 9], the derivation of the formula is described which shows that for the optimal operating mode of the installation with a constant load, we obtain fuel consumption (2)

$$Q_{\text{опт}} = Q_{nL}T + (1{,}4)^{\mu} q P_{L1}^{\mu} T \tag{2}$$

In the case when ESS is not used, we have (3)

$$Q = Q_{nL}T + q\left(P_{L1}^{\mu} * 0{,}9T + 5^{\mu} P_{L1}^{\mu} * 0{,}1T\right) = Q_{nL}T + \left(0{,}9 + 5^{\mu}\right) q P_{L1}^{\mu} T. \tag{3}$$

The specific monetary effect of the use of ESS depends on the specific values of the coefficients of the expenditure characteristic.

3 Comparative economic efficiency of the energy storage system in combination with traditional sources of generation

When making design and investment decisions on the feasibility of using alternative technical solutions for the modernization of sources in isolated energy supply systems, a comparative assessment of their economic efficiency is necessary. As a basic variant of power supply:

option 1 was adopted - Power supply using Caterpillar 3512 DGU with a rated power of 1020 kW [13].

The following two technically feasible options have been adopted as alternatives:

option 2 - Power supply using a Caterpillar 3412 DGU with a rated power of 720 kW [4] in combination with a Liotech LT-LFP 700P ESS with a total power of 329 kW and a battery capacity of 3500 Ah, consisting from 5 modules in every 147 batteries [13; 14];

option 3 - Power supply with the use of Caterpillar G3516 GPU, with a rated power of 1030 kW in combination with Liotech LT-LFP 700P ESS with a total capacity of 329 kW and a battery capacity of 3500 Ah, consisting from 5 modules in every 147 batteries [13; 14];

A GPU with a rated power of 1030 kW was selected for greater clarity when comparing options, which is shown in Table 1 (in reality, it would be possible to install a GPU of lower power Caterpillar G3512 at 725 kW and, therefore, cheaper).

To assess economic efficiency, it is necessary to have data on capital investments in the scheme under consideration, exchange rates, technical specifications and the number of hours of use of the equipment, as well as on the consumption and cost of the fuel used.

The recommended GPU life cycle is 30–40 years [15; 16], DGU - 20-30 years, the battery life depends on various factors.

The capital investments necessary for the implementation of each of the options are shown in Table 1. Construction and installation work includes the cost of building the foundation of the generating installation, access roads, installation and commissioning, rental of construction equipment, and payment of wages to builders.

Table 1. Models and manufactures of fault indicators.

Parameter Name	Opt. 1	Opt. 2		Opt. 3	
	DGU	***DGU***	***BESS***	***GPI***	***BESS***
Rated power, kW	1020	720	329	1030 / 725	329
Capital expenditures for equipment (Keq), thousand euros	309.4	135.8	908.9	319.6/ 260.1	908.9
Capital expenditures for construction and installation works (Kci), thousand euros	24.7	10.8	72.7	25.5/ 20.8	72.7
Capital costs for transport and logistics services (Ktl), thousand euros	18.5	8.1	54.5	19.7/ 15.6	54.5
Total costs in ETC (Kes = Keq + Kci + Ktl), thousand euros	352.7	154.8	1036.2	364.4/ 296.5	1036.2

As can be seen from Table 1, the largest amount of capital investment corresponds to option 3 using GPI in combination with ESS. This is explained by the significant costs of both the purchase of the installation itself and the ESS included in the ETC as a BESS, as well as the costs of design and construction, work. Due to the high cost of ESS, the total capital costs in options 2 and 3 are significantly higher than in option 1 without ESS.

Maintenance and repair costs are calculated individually for each type of equipment. For DGU and GPI, the maintenance costs are determined depending on the number of operating hours of the equipment, in addition, the costs associated with changing the oil, filters, checking the condition of the fastening and electrical connections, checking the instrumentation and indicators on the control panel, etc. must be taken into account. [15, 16, 17].

The next step is the assessment of current costs for different operating modes of the installations. Table 2 presents the technical specifications from manufacturers for fuel consumption Q for DGU and GPI, as well as specific fuel consumption q.

 https://doi.org/10.1051/e3sconf/202017801063

Table 2. Fuel Consumption for DGU and GPI.

The degree of utilization of installations from the nominal,%	DGU (1020 kW)		DGU (720 kW)		GPI (1030 kW)	
	Q, l/hour	*q, l/kWh*	*Q, l/hour*	*q, l/kWh*	*Q, m³/hour*	*q, m³/kWh*
100	271.6	0.272	171.8	0.264	351.0	0.351
75	209.6	0.279	128.9	0.264	316.0	0.421
50	147.6	0.295	86.0	0.265	282.0	0.564
35	110.3	0.315	63.0	0.277	218.3	0.624
20	72.4	0.362	46.8	0.360	155.8	0.779
5	33.9	0.679	26.4	0.812	93.3	1.867

The table shows that with a decrease in the load on the generator in all three cases, the specific fuel consumption begins to increase, therefore, it is advisable to load the generator throughout the day from 50 to 100%.

For reference: the isolated power system of Kamchatka is 50% energy-efficient (at peak loads). This means that for a significant part of the time, electrical units operate in a mode with increased fuel consumption. In the no-load operation, fuel consumption is reduced by 70-80% of the "peak load" mode.

The number of hours of using the maximum load of consumers shows how many hours a year a consumer could work with the maximum load for a given annual power consumption (C^n_{year}) [14]: for oil refining h^n_{max} = 6000 - 8000 hour / year, for non-ferrous metallurgy or oil production h^n_{max} = 7000 - 7500 h / year, for ferrous metallurgy h^n_{max} = 6500 h / year, for machine building h^n_{max} = 3500 - 5000 h / year, for light industry h^n_{max} = 3000 - 4500 h / year.

With such a number of hours of work per year, the main component in the total costs is fuel costs. In general, the regions of the "northern delivery" have an extremely high price for imported fuel: it varies from 0.72 to 1.44 thousand euro/ton (for comparison: the cost of diesel fuel for consumers in the central part of Russia is approximately 0.66 thousand euros/ton) [18]. The share of the transport component in the price of fuel at the final consumer reaches 30-90% [16; 19]. In Yakutia, the average price of diesel fuel on the "northern delivery" in 2016 was 0.92 thousand euro/ton [20, 21]. The price of Arctic diesel fuel consumed by DGU, taking into account delivery to remote rural settlements - 1.14 euro/liter [22]. The cost of associated petroleum gas (APG) production is up to 3.60 euro, the price of such gas on the market is set at no more than 7.21 euro [23].

To calculate and compare fuel costs for DGU and GPI, it is necessary to bring them into a comparable form, bringing everything to one unit of measurement. Rostekhnadzor established its average fuel density coefficients for converting liters to tons, including diesel fuel - 830 kg/m3. For GPI, it is necessary to convert m3 into tons, taking into account the fact that APG is used for GPI, which consists of 22 components of the gas mixture, among which the main four are methane, ethane, propane, butane, their content, and density of natural gas and for APG, are presented in Table 3.

Table 3. The composition of APG oil field.

Gas mixture components	Component designation	Density according to GOST 2939—63, kg/m3	Natural gas, content in%	APG, content in%
Methane	CH_4	0.656	94.3442	61.7452
Ethane	C_2H_6	1.26	2.9114	7.7166
Propane	C_3H_8	1.8641	0.4312	17.5915
Butane	C_4H_{10}	2.48	0.0719	4.8729
Other gases	C_5H_{12}, C_6H_{14} C_6H_6, C_7H_{16} and 14 more components	0.98	2.2413	8.0738

Thus, based on Table 3, knowing the composition of the gas mixture and the density of its components, we determine according to the mixing rule the average density of the mixture (4):

$$P_n = \frac{(P_1V_1 + P_2V_2 + \text{L } P_nV_n)}{100} \quad (4)$$

where P1, P2 ... Pn is the density of gas fuel components, kg/m³; V1, V2 ... Vn - content of components, volume in percent.

Thus, according to formula (4), the average APG density will be 0.9463 kg/m³. Thus, using the obtained average fuel density, we translate everything into tons of standard fuel, the total fuel costs per year are calculated by the formula (5) and are summarized in Table 4 at T_{max}=2628 h and T_{min}=6132 h for DGU and GPI:

$$C_T = Q_T \cdot F_T, euro \quad (5)$$

where Q_T is the annual fuel consumption, ton; F_T is fuel cost, euro/ton.

Table 4. Fuel costs per year.

Name	DGU (1020 kW)	DGU (720 kW)	GPI (1030 kW)
F_T, euro/ton	1380	1380	13.66
Q_T, ton	1 325.24	1110.93	2 474.92
C_t, thousand euro	1 829.6	1 533.7	33.8

This table does not take into account the costs of replacing oil, spare parts, maintenance, etc., however, they are not significant (about 10% of all costs), and the main share (about 90%) of all costs are fuel costs.

From Table 4 it can be seen that the highest annual fuel costs are observed when using DGU (option 1) when using an ESS and choosing a smaller DGU (option 2), fuel costs are reduced by 295.9 thousand euro while ensuring that the same load under the same conditions during the operation of the GPI (option 3) will require 33.8 thousand euro. This difference in fuel costs is due to the fact that APG is used for GPI, which is released during the production and preparation of oil. DGU, as a rule, operate at moderate and low installed capacity utilization factors, i.e., when equipment is not loaded enough and seasonal and daily energy consumption schedules typical for our country are insufficient, while DGU minimum fuel consumption is provided only at rated load. In addition, the GPI has several advantages over DGU:

GPI oil consumption: 330 g/MWh oil change after 2000 moto hour (220 l) - significantly lower than that of DGU, for which the flow rate is from 400 g/MWh. and a replacement period of 600 moto hour;

The term for an overhaul of the DGU requires overhaul through 30-45 moto hour, and GPI on average - after 60-80 moto hour;

An additional advantage of the GPI is the possibility of heating the premises: along with electric energy, you can get 1000 kW of heat every hour for heating.

Since the compared options differ significantly in terms of capital and current costs, the most preferred option is selected based on the calculation of the payback period of additional capital investments, which is determined by formula (6) as the ratio of their size to savings from lower operating costs:

$$T = \frac{K_2 - K_1}{C_1 - C_2} \tag{6}$$

where K1, K2 - investment in the compared options; C1, C2 - operating costs for the same options.

The value of T shows the period of time during which additional capital costs are covered by savings on operating costs. By combining the options and comparing the calculated values with each other, the optimal option is determined with the minimum value of the calculation period.

The additional costs in option 2 with ESS will pay off due to fuel economy for 3 years. For option 3, the cost of annual gas consumption does not exceed 34 thousand euro, which makes this option uncontested in the presence of APG.

When choosing an ESS to ensure the stable operation of the GPI, it is important to coordinate the power of the ESS with the dynamic characteristics of the GPI. The power of the ESS will determine the dynamic mode, and the capacity will determine the graphs of the load.

4 Conclusions

The use of ESS in combination with a generating unit allows you to save fuel by optimizing the operating mode of the drive motor. For maximum fuel economy in the daily cycle, it is necessary to have preliminary information on the load schedule. This implies the advisability of forming a research direction in terms of automatically storing power profiles and optimizing the operation of ESS. Given the high cost of ESS, it is necessary to develop a line of the nominal power of ESS with a clear link to the characteristics of load schedules.

To make a decision on investing in a project and using a DGU, a DGU with ESS or GPI with ESS, it is necessary to consider the values of all the above parameters. GPI is advisable to apply with a significant need for electrical energy. Fuel for them is very cheap if it is possible to use APG. Despite significant capital investments in the application of GPI, its reliable operation during operation in combination with ESS provides a gain in the total accumulated costs from possible emergency situations, which allows us to recognize this option of energy supply as the most effective. Operating costs for GPI with ESS are significantly lower.

The possible reduction in the installed capacity of both DGU and GPI concomitant with the ESS installation should be evaluated comprehensively, comparing investments with current operating costs: more powerful power installations usually have a lower specific fuel consumption in the load range of 40-60% than those operating with a load 60 –100% less powerful power installations.

When choosing ESS to ensure the stable operation of the GPI, it is important to coordinate the power of the ESS with the dynamic characteristics of the GPI. power of the ESS will determine the dynamic mode, and its capacity will determine the graphs of the load. The introduction of ESS will contribute to the differentiation of power supply schemes and a general increase in energy efficiency.

Acknowledgements

Research is made with financial support of the Ministry of Education and Science of the Russian Federation within implementation of the federal special program «Research and development in the priority directions of scientific and technological complex of Russia for 2014-2020», the agreement on granting a subsidy № 075-15-2019-057, unique identifier of applied scientific research (project) RFMEFI57418X0188. All research articles should have a funding acknowledgment statement included.

References

[1] Storage in the electric power industry [Electronic resource], Energy Bulletin, Analytical Center under the Government of the Russian Federation **60** (2018). Available at: http://ac.gov.ru/files/ publication/a/16882.pdf

[2] The accumulation and storage of energy [Electronic resource], Supercapacitors and lithium-ion capacitors. Available at: http://www.socomec.ru/v2/range-dynamic-energy-storage_ru

[3] Energy storage based on lithium-ion batteries [Electronic resource]. Available at: http://liotech.ru/userfiles/10121131_liotech_broshure.indd.pdf

[4] The future of energy: lithium-ion hybrid gas turbine power plants [Electronic resource]. Available at: https://aftershock.news/?q=node/672325&full

[5] Gas power plant in combination with energy storage [Electronic resource]. Available at: http://renen.ru/gas-power-plant-in-combination-with-an-energy-storage-unit

[6] Distributed Energy of Russia 2010-2015 [Electronic resource], Market for gas turbine plants for small and medium power plants Available at: http://infoline.spb.ru/m/catalog/products/issledovaniya-rynkov/raspredelennaya-energetika-2010-2015-rynok-gazoturbinnykh-ustanovok-dlya-elektrostantsiy-maloy-i-sre

[7] K. Bakhteev, A. Fedotov, N. Chernova, R. Misbakhov, Methodological approaches to the choice of energy storage and optimization of their parameters to improve the electric power quality in various types of electric power systems, Proceedings of the 10th International Scientific Symposium on Electrical Power Engineering, Elektroenergetika, 488–493 (2019)

[8] K. Bakhteev, A. Fedotov, R. Misbakhov, The improving efficiency of electric receivers on the industrial enterprises in case of short-term power outages, Proceedings of the 2019 20th International Scientific Conference on Electric Power Engineering, 347–352. DOI: 10.1109/EPE.2019.8778181 (2019)

[9] A. Fedotov, R. Misbakhov, E. Fedotov, K. Bakhteev, Influence of voltage dips on the stability of excitation of synchronous machines, Proceedings of the 2019 20th International Scientific. 330–335. DOI: 10.1109/EPE.2019.8778167 (2019)

[10] I.M. Markovich, Energy Modes of energy systems (372, 1969)

[11] A. Fedotov, R. Misbakhov, N. Chernova, K. Bakhteev, IOP Conf. Series: Materials Science and Engineering, Utilization of electrochemical energy storage devices in autonomous power supply systems**,** 643. DOI: 10.1088/1757-899X/643/1/012105 (2019)

[12] A. Fedotov, R. Misbakhov, K. Bakhteev, N. Chernova, FarEastCon, The Use of Electrochemical Energy Storages to Limit the Effects of Short-Term Power Outages. DOI: 10.1109/FarEastCon.2019.8934810 (2019)

[13] Production of energy facilities for various industries: Genmotors plant [Electronic resource]. Available at: https://www.energo-motors.com

[14] Lithium-ion (lithium-iron-phosphate) batteries (Russia) [Electronic resource]. Available at: http://www.pea.ru/docs/equipment/batteries/lioteh

[15] A.S. Platonov, V.V. Pikhletsky, Modern Science: Actual Problems of Theory and Practice: Series "Natural and Technical Sciences" [Electronic resource], Performance evaluation of mini-CHP based on gas-piston units, 11-12 (2013). Available at:http://www.nauteh-journal.ru/index.php/----etn13-11/1059-a.

[16] An example of calculating the payback of mini-thermal power plants [Electronic resource]. Available at: http://www.esist.ru/info/okupaemost

[17] I.G. Karapetyan, Handbook for the design of electrical networks [Electronic resource]. Available at: https://info.wikireading.ru/244128

[18] Dynamics of retail prices for diesel fuel (Russia, rubles) for the Republic of Tatarstan [Electronic resource]. Available at: https://news.yandex.ru/quotes/11119/20010.html?1

[19] N.M. Parnikov, Increasing the energy efficiency of decentralized power supply complexes, as an example: the Republic of Sakha (Yakutia) (24, 2009)

[20] D.N. Karamov, The relevance of the use of renewable energy sources in decentralized settlements of Russia by the example of the Lensky district of the Republic of Sakha (Yakutia) [Electronic resource], ISTU Bulletin **11,** 82, 279–283 (2013). Available at: https://elibrary.ru/item.asp?id=20886279

[21] V.R. Kiushkina, Decentralized power supply of the regions of Yakutia using wind energy (24, 2005)

[22] Sakhaneftegazsbyt JSC: the source contains information on wholesale prices for petroleum products from March 1, (2016) [Electronic resource]. Available at: http://www.sngs.ykt.ru

[23] Associated Petroleum Gas: Refining and Use or Disposal [Electronic resource]. Available at: https://pronedra.ru/gas/2017/03/03/pererabotka-poputnogo-neftyanogo-gaza

A method for construction of an energy-efficient ice floating pier in the Arctic using hardened ice

G L Kagan[1,*], *L R* Mukhametova[2] and *A Y* Velsovskij[1]

[1]Department of Highways, Vologda State University, Lenina ul. 15, 160000, Russia
[2]Kazan State Power Engineering University, Kazan, Russia

Abstract. This paper presents studies on increasing the ice strength using various additives. It is indicated that addition of wood fiber (artificial composition) and vegetable fiber (natural composition - frozen peat) to ice is an energy-efficient method to increase its strength. This enhances the possibilities of using ice during building and construction works in the Arctic. As an example, the authors proposed a floating ice pier design.

1 Introduction

One of the directions of the Russian state program "Socio-economic development of the Arctic zone of the Russian Federation" is to develop coastal infrastructure during reclaiming of the Arctic zone. This work should have a fast, but at the same time rational, resource-saving approach, so it is important to use proprietary technologies that make it possible to obtain energy-efficient constructions at relatively low costs. These constructions have to solve the problems of defense, national security and industrial development, including the placement of energy facilities. One of such technologies is the design and method of construction an energy-efficient ice floating pier in the Arctic using hardened ice.

Therefore, the purpose of this study was to develop a technological platform - a floating ice pier using hardened ice. The purpose of this pier is mooring and safe parking of ships and vessels, as well as other offshore facilities, loading and unloading operations along the entire length of the pier using pneumatic wheels and caterpillar tracks cranes of various carrying capacities, placement of various technological and power equipment. It should meet the following criteria:

- Local materials should be used to the maximum extent possible during its production;
- The cost of production and operation should be lower than that of analogues of floating piers or stationary piers under comparable conditions;
- It should provide an ability of quick erection on site;
- It should provide year-round use.

2 Materials and methods

In the Arctic, the main local building material is ice. Ice is an accessible and very widespread substance in nature. Today this material finds its application in construction. In the northern regions of Russia, ice is used to strengthen the roadbed in the winter ("winter roads"), however it is necessary to solve the problem of reducing the coefficient of adhesion of wheel to road.

Ice is also used in the construction of ice crossings for cars and trains. An interesting historical fact is the construction of the largest railway ice crossing through Lake Baikal in 1904, which connected the eastern and western branches of the Great Siberian Route.

Regular ice has a limited usefulness even at low temperatures, due to its low strength characteristics, in particular tensile strength, and the brittle nature of fracture. For hardening of ice, various additions can be used that act as reinforcing materials, and the result is a building material - ice concrete.

Studies on improving the strength characteristics of ice using various additives were carried out by G.L. Kagan in the laboratory of the Arctic and Antarctic Research Institute [1] as well as by other authors [2-7]. Sawdust, fiberglass, cotton and wood fibers, and other materials were used as additives. The efficiency of the obtained compositions from hardened ice depends both on freezing of ice crystals with reinforcing fibers, as well as on the total number of fibers and their strength. The studies [8] showed that the most effective additive to ice providing increased strength characteristics is wood fiber. It provides an increase in tensile strength, including quadruple bending, and eliminates the brittle nature of fracture. When the ultimate compressive loads are reached, ice deforms as a plastic material without visible fracture. Wood fiber is a technological product used in the pulp and paper industry. However, for the regions of the Arctic and the Far North, it is not the most accessible material.

Peat is one of the most widespread materials in the north of the Russian Federation. Owing to the cold climate and high humidity within a long period, northern

* Corresponding author: v2u@yandex.ru

https://doi.org/10.1051/e3sconf/202017801064

Table 1. Elastic characteristics of ice.

Characteristics	Temperature, °C	Elastic characteristics
Modulus of elasticity, MPa: a) determination by static method b) determination by dynamic method Shear modulus, MPa: a) determination by static method b) determination by dynamic method Poisson's ratio	– (3-27) – (3-20) – (0-15) – (0-15) – (5-15)	(3-7) 10^3 (7-10) 10^3 (1-2) 10^3 (2.3-3.5) 10^3 0.33-0.37

Table 2. Ice strength.

Deformation type	Temperature, °C	Tensile strength range, MPa
Compression	– (6-11)	2.5-6.0
Tensile	– (0-18)	0.7-1.5
Bending	– (3-15)	0.5-2.5
Shear	– (0-15)	0.5-1.2

peat has a more porous structure and differs significantly from peat that forms in the south of our country. In such peat, plant residues decompose more weakly, preserving its fibrous structure, since they almost immediately fall into a zone with low oxygen content and, accordingly, low oxidative ability. Peat with high content of strong (poorly decomposed) fibers can be effectively used as a reinforcing element in various compositions. The lower is the degree of peat decomposition, the stronger are fibers, and the better is the quality of the resulting composition. Peat is the most promising reinforcing material for ice hardening, both in terms of its availability (local material), and low cost.

Studies to determine the complex of mechanical characteristics of ice with peat were carried out by G. Kagan during development of methodology for calculating the bearing capacity of a frozen peat cover at the base of fishing roads. These roads were the main transport network for development of oil fields in Western Siberia. During the tests, peat of high bogs was used with a minimum decomposition degree of plant residues. The degree of peat decomposition in the upper horizons of deposits was 5-10 percent. The studies were carried out in the laboratory of the Tyumen Industrial Institute [1,9]. When freezing, peat, like mineral soils, dramatically changes its properties. Frozen peat is a multicomponent system, the main elements of which are mineral particles, ice, plant fibers, water and gases. The gas content remains unchanged when freezing peat and is 4-5%. When water is freezing it is transferred to its solid phase - ice and cementation of plant fibers by it provide a sufficiently high monolithicity and strength of frozen peat. The resulting structure makes it similar to artificial composite materials, in particular reinforced plastics. Thus, frozen peat is a natural composition in which ice serves as a binder and plant fibers as a reinforcing material. The work of such a material is determined by the properties of the constituent elements and their interactions.

Plant fibers in peat are oriented parallel to the freezing surface. The length of fibers varies in a wide range - from fractions of a millimeter to several centimeters, and the tensile strength varies from 16 to 86 MPa. Another component of frozen peat - ice is a crystalline body. The existence of ice in nature is limited to a period of time with a negative temperature close to the melting point. The lowest temperature of natural ice differs from its melting point by only a few tens of degrees. Therefore, ice to a large extent exhibits rheological properties (creep, relaxation), and its resistance to external loads is determined by the time of their action. So, if the load action time is very short (instantaneous load) and plastic deformation does not have time to develop, ice deforms as an elastic body. In this case, ice destructs as a brittle material. The elastic characteristics of ice are given in table 1 [10].

The strength of ice is determined by the nature of ice formation, structure, temperature, mode and type of loading. The range of changes in the strength characteristics determined in the conditions of fast loading is shown in table 2 [10].

The development of creep deformation in ice is observed at almost any stress. It is known that if the stress of ice under load does not exceed a certain limit, then over time the creep rate is fixed and ice does not break. According to experimental data, the ultimate creep of ice at shear is 0.2 MPa at a temperature of -1.8 °C. Upon deformation in the steady-state creep stage, ice is considered as a viscous liquid.

Plant fibers have a higher modulus of elasticity than ice, so when they are used together, the tensile strength of frozen peat is greater than that of ice. Thus, the mechanical properties of the ice - peat composition and their resistance to loads are determined by the joint work of ice and plant fibers.

There are few experimental data on mechanical properties of frozen peat. The elastic properties of frozen peat are manifested only under dynamic loads. So, the elastic constants can be defined using dynamic test methods. Using this method, the elastic modulus was determined for the ice-plant fiber composition at a temperature of –5 ° C, and are presented in table 3 [7].

As it can be seen, with an increase in the content of wood fiber the elastic modulus decreases. Thus, the elastic modulus of a composition with a 7% content of wood fiber decreases by 25% as compared to the elastic modulus of pure ice.

Table 3. The elastic modulus E the ice–plant fiber composition.

Content of wood fiber to the total sample mass, %	Average value of E, MPa
0	8700
3	8300
7	6500

The results of these studies showed that the strength characteristics of frozen peat are similar to those with addition of wood fiber. The materials of these studies are given in [9].

Thus, ice with fiber content can be considered as a composite material, the filler of which is fibers, and ice serves as a binder. If wood fiber is used in this composition, it represents an artificial composition, and if plant fibers, it represent the natural one, that is, frozen peat.

3 Results and Discussion

As noted above, peat soil is widespread in areas adjacent to the Arctic zone and in the Far North, and it can be considered where as local material. The use of hardened ice in the Arctic opens up opportunities for the use of more heavy-duty energy-efficient engineering structures there.

The structural design of a floating ice pier proposed by the authors [11] is an example of such an energy-efficient structure. It involves construction of a floating ice pier in the Arctic, with optimal cost and using mainly local building materials. The pier design enables its long operation period involving mooring, parking and handling of ships at different times of year.

The essence of the reported technical solution is as follows. During formation of an ice sheet with a thickness that ensures the pier construction, its bearing mass of ice with the wood or plant fibers additives is frozen, and in the central part cavities are formed to ensure circulation through them of air of negative temperatures.

Usually hardened ice is obtained using layer-by-layer distribution of fibers from various reinforcing materials on a previously prepared surface, followed by filling them with a layer of water and cooling to hardening.

The hardened ice can be produced using various devices, the general principle of which is a layer-by-layer deposition of pulp, a water mixture and a 3-6% fibrous material on a frozen surface, followed by forced cooling to hardening. The number of pulp freezing cycles depends on the required thickness of the structure. Hardened ice does not contain air inclusions and areas with increased internal stress, which positively affects its strength [12].

The freezing of ice mass is carried out using formwork installed along its outer contour, which includes layers of heat- and waterproofing.

The cavity can be formed using formwork of pipes made of various materials - metal, composite, flexible shells, etc. The use of high-strength materials, such as metal, provides increased rigidity and increased structural strength of the floating pier.

The cavities in the pier body provide its necessary buoyancy and thermal stability of the frozen mass when air of negative temperatures circulates through them. In the winter period, cooling is carried out due to circulation of external air in the cavities. If necessary, in warm period, artificial cooling of air supplied to the cavity is used.

This allows one to freeze the pier mass with a height greater than the height of the natural frozen cover, but at the same time thawing from the bottom is excluded. Thus, these piers can be used as both temporary and permanent structures.

The figure shows a cross section of a floating flax pier. It is an artificially frozen mass of ice with addition of fibers 1, inside of which cavities 2 are formed using pipes 3 as formwork and a device along the contour of the mass of protective layer 4, including formwork during freezing of the ice-fiber mass, heat and waterproofing.

The structure can be of any dimensions, which is

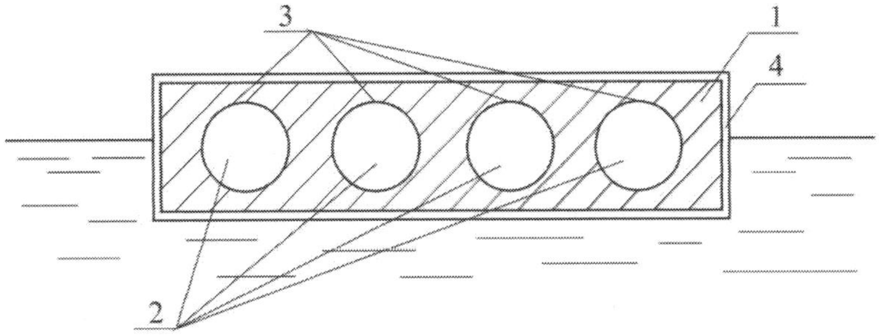

Fig. 1. Cross section of a floating ice pier: 1 - artificially frozen massif of ice with addition of fibers; 2 - cavities inside pipes; 3 - pipes; 4 - formwork for freezing the ice-fiber array, heat- and waterproofing.

selected according to the customer requirements. For comparison of cost indicators we considered dimensions similar to the three sequentially connected sections of the PM-61M metal floating pier connected with a 17-meter access bridge (length of 125 m, side height of 2.2 m, hull width of 8.2 m), which was ordered by a shipyard in Murmansk in April 2020. The estimated cost of delivery, assembly, installation and unfastening of the analogue, PM-61M metal pier, is currently 130 million rubles. The cost of the proposed construction is 31.5 million rubles, comprising 16 million rubles for metal pipes of 1.6 diameters and inserts; 3.5 million rubles for in-line axial fans and equipment for cooling in the summer period, 0.5 million rubles for peat, 7 million rubles for formwork and insulation; 4.5 million rubles for installation and unfastening. The service life of such a structure is unlimited; periodic replacement or repair of fans and cooling systems is required in accordance with their service life.

4 Conclusions

In the difficult natural conditions of the Far North regions, the use of traditional piers, floating piers and pier facilities is expensive and ineffective due to remoteness from construction bases. The proposed platform - a pier made of hardened ice can be produced on site in a short time and from local materials, and it requires only several fans and a cooling unit in the summer. Therefore, this solution can be proposed for the quick and cost-effective development of the northern territories, and the technology of hardened ice will find its application in construction of modern ice structures, including piers, roads, runways and other facilities as a part of creation a new innovative infrastructure in the Russian Arctic region.

References

[1] G.L. Kagan, Strength characteristics of frozen peat and methodology for their study in relation to the calculation of roads on the frozen cover of swamps, thesis for the degree of candidate of geological and mineralogical sciences (Tyumen, 143, 1974).

[2] I.S. Peschansky, Glaciology and ice engineering (Leningrad: Gidrometeoizdat, 464, 1967).

[3] V.V. Lavrov, Deformation and strength of ice (Leningrad: Gidrometeoizdat, 208, 1969).

[4] V.G. Zanegin, N.G. Khrapatyi, Experimental studies of ice strength, Transactions of the Leningrad Polytechnic Institute 361, 85-87 (1978).

[5] N.A. Tsytovich, Mechanics of frozen soils (Moscow: Vyschaya Shkola, 448, 1973).

[6] R.L. Kobl, U.D. Kingeri, Artificial hardening, reinforcement of ice, in the book: Ice and snow, Properties, processes, use (Moscow: Mir, 480, 1966).

[7] N.F. Savko, N.M. Tupitsin, Yu.V. Yushkov, et al., Construction of roadbed on a frozen peat base, Automobile Roads 10, 4-5 (1972).

[8] G.L. Kagan, I.S. Peschansky, Z.I. Schweinstein, Yu.L. Nazintsev, Study of strength properties of a new ice-fiber material, Transactions of coordination meetings on hydraulic engineering, XXIII (Moscow: Publishing House Energiya, 1965).

[9] S.S. Vyalov, G.L. Kagan, A.N. Voevoda, V.I. Muravlenko, Construction of commercial structures on frozen peat (Moscow: Nedra, 144, 1980).

[10] K.F. Voitkovsky, Mechanical properties of ice (Moscow: Publishing house of USSR Academy of Sciences, 100, 1960).

[11] G.L. Kagan, A.Yu. Velsovskij, The method of construction an ice pier in the Arctic, Patent 2715034 of the Russian Federation, MPK7 E02B 17/00, B63B 35/44, applicant and patent holder Vologda State University, 2019104118, declared 14.02.2019, publ. 21.02.2020, bull. no. 6, 5.

[12] V.I. Moiseev, A.V. Panyushkin, N.K Vasil'ev, The method of formation of hardened ice, Certificate of authorship USSR, SU 1 249 277 A1, MPK F25C 1/12. applicant Leningrad Institute of Railway Engineers named after V.N. Obraztsov, 3911768, declared 07.01.1985, publ. 07.08.1986.

https://doi.org/10.1051/e3sconf/202017801065

Optimize the cost of paying for electricity in the water supply system by using accumulating tanks

Alexei Kapanski[1,*], *Nadezhda* Hruntovich[1], *Siarhei* Bakhur[1], *Larisa* Markaryants[2], and *Leonid* Dolomanyak[3]

[1]Sukhoi State Technical University of Gomel, Prospect Octiabria, 48, 246746, Gomel, Republic of Belarus
[2]Federal State Budgetary Educational Institution of Higher Education «Moscow State Linguistic University» (MSLU), 119034, 38 Ostozhenka St. Moscow
[3]Kazan State Power Engineering University, str. Krasnoselskaya, 51, 420066, Kazan, Russia

Abstract. The article considers the method of optimizing the pumps in the water supply system of the first water rise station in the zones of the day, where there is a different system of payment for electric energy. To assess the regulatory capacity of pumps on a temporary parameter, the authors of the article propose to use existing water tanks in the water supply system, which act as a buffer, smoothing the unevenness of water consumption. The studies have revealed that the comprehensive optimization of pumps on the criterion of minimizing specific electricity consumption by lifting water and optimizing the operation of pumps in the zones of the day allows to significantly improve energy efficiency and reduce the cost of extraction and transportation of water to the consumer. In the article, the authors examine an algorithm that allows us to assess the economic potential of pump regulation in the real-world conditions of the system.

1 Introduction

Implementation of a set of organizational and technical measures aimed at reducing energy consumption is always a pressing task in the current conditions of the water and sewerage industry. Active introduction of modern energy-saving technologies, development of the system of assessment and forecasting of energy efficiency indicators, development of effective ways and methods to identify hidden reserves of energy savings leads to a reduction in energy intensity of products [1]. The importance of state control in the implementation of planned energy-saving measures is explained by the need to improve the system of tariff regulation of water utilities, which includes a unit energy component of which the share of which reaches 25% (Fig. 1) [2].

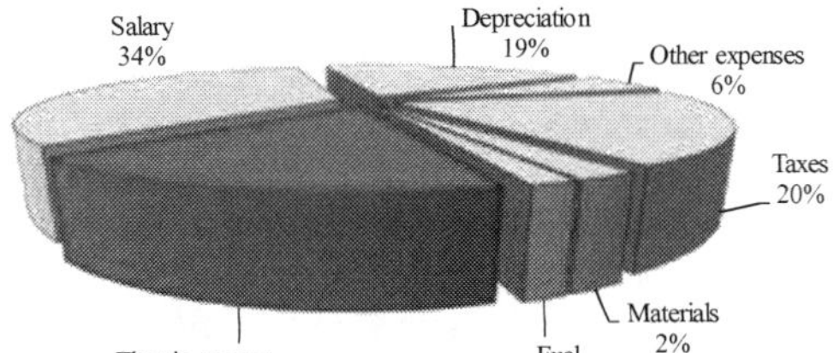

Fig. 1. Cost structure for water production and transportation.

Tariff regulation of drinking water contributes to the implementation of technical measures aimed at improving the energy efficiency of water utilities, as a result of which the monetary costs of extracting and transporting water to the consumer are reduced. Such measures include the introduction of modern energy-efficient equipment, regulation of the operating modes of pumping stations of the first and second rise, optimization of pressure schedules, reduction of water losses during transportation, construction of water reuse facilities, etc.

On the other hand, when planning the cost of water production, factors determining the change in the cost of energy resources should be taken into account: price change indices; payment for active electric power in the zones of the day [3, 4]. Under the current operating conditions of water utilities, the development of measures aimed at regulating the operating modes of pumping units to optimize electrical load schedules according to the criterion of minimum payment for energy carriers is relevant. The article presents the results of a study on the example of the water intake of one of the water intakes of the Republic of Belarus.

2 Electricity payment system

In the Republic of Belarus, the electricity tariff for industrial consumers with connected capacity above 750 kVA is determined by the formula:

$$C_p = a \cdot P_{\max} + b \cdot W, \qquad (1)$$

where a – basic rate of a two-part tariff for electric capacity; $P_{\max}$ – the actual value of the largest half-hour combined active power for the billing period; b – additional rate of two-part tariff for electricity.

* Corresponding author: kapanski@mail.ru

Thus, the consumer pays for energy not only for consumed electricity, but for participating in the maximum electrical loads of the energy system [1]. With the availability of automated metering systems that allow recording the values of electric loads with 30-minute discretization, the consumer has the opportunity to switch to a differentiated payment system in which the tariff for electricity is determined by the formula:

$$C_d = a \cdot k_a \cdot P_{\max} + b \cdot \left(k_n \cdot W_n + k_{hp} \cdot W_{hp} + k_p \cdot W_p\right), \quad (2)$$

where k_a – decreasing coefficient to the basic rate of the two-part tariff, set at 0.5; k_n, k_{hp}, k_p – respectively night, half-peak and peak tariff coefficients; W_n, W_{hp}, W_p – energy costs in the night, peak and half-peak zones.

The rate coefficient is $k_{hp} = 1$. The coefficient k_n, k_p determined on the basis of the calendar number of days and established time zones (night t_n, peak t_p and half-peak t_{hp}):

$$k_n = 1 - \frac{a \cdot (1 - k_a) \cdot (4 \cdot t_p - t_n)}{b \cdot d \cdot (t_n^2 - t_p^2)}; \quad (3)$$

$$k_p = 1 + \frac{a \cdot (1 - k_a) \cdot (4 \cdot t_n - t_p)}{b \cdot d \cdot (t_n^2 - t_p^2)}, \quad (4)$$

where d – calendar number of days in the billing period.

The energy system has identified the following time zones: from 23:00 to 6:00 h - night zone; from 6:00 to 8:00 h and from 11:00 to 23:00 h the half-peak zone and from 8:00 to 11:00 h the peak zone.

In addition, the evening peak load $P_{e.\max}$ (from 18:00 to 21:00) is highlighted, which should not exceed the morning maximum $P_{m.\max}$ (from 08:00 to 11:00) in the billing period. Otherwise, the payment for electricity is not differentiated by the zones of the day and is calculated by the formula 1.

Thus, the choice of a system of payment for electricity and the possibility of gradation by day zone increases the motivation of energy personnel to regulate electric loads in order to reduce payment for electricity [2, 3, 4].

Analyzing the formula 2, it can be noted that the main ways to reduce the cost of electricity are:

- reduction of total energy costs by increasing the energy efficiency of equipment;
- reducing the maximum electrical load due to the optimal distribution during the work shift and the optimal distribution of electrical loads in time zones [5].

The results of comparing the average daily payment for electricity under real operating conditions of the water utility in the city of Zhlobin (Republic of Belarus) are shown in the Table 1.

Later, when comparing the economic efficiency of the transition from a two-part tariff to a differentiated tariff, the formula was used:

$$\Delta C = \left(C_p - C_d\right) / C_p \cdot 100\,\%. \quad (5)$$

Fig. 2 shows a graph of the electrical load of the studied water intake, indicating tariff zones.

Table 1. Comparison of the average daily cost of electricity for various payment systems.

Index	Designation	Units rev.	Value	
Rate type	–		№1	№2
Base rate	a	\$/kW	0.43	0.43
Additional rate	b	\$/kWh	0.11	0.11
Maximum power	$P_{\max}$	kW	152.0	152.0
Tariff coefficients	k_a	–	–	0.5
	k_n	–	–	0.76
	k_{hp}	–	–	1
	k_p	–	–	2.2
Electricity consumption by tariff zones	W_n	kWh	–	384
	W_{hp}	kWh	–	1902
	W_p	kWh	–	380
Total electricity consumption	W	kWh	2666	2666
Payment for maximum power	C_P	\$	64.8	32.4
Electricity payment	C_W	\$	311.5	349.5
Total	C	\$	376.3	381.9
In Table 1 tariff № 1 is two-part; tariff № 2 – two-part differential				

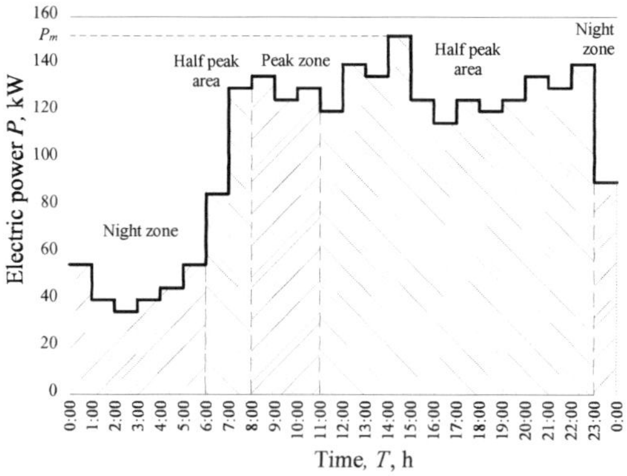

Fig. 2. Chart of daily electric power water intake.

Of particular interest is the change in the cost of electricity with a shift in the graph of electrical load [10, 11, 12, 13]. For the studied water utility, in Fig. 3, the boundaries of economic efficiency are marked with the shift of the electric load 2 hours ahead and 3 hours ago. For the studied water utility, in Fig. 3, the boundaries of economic efficiency are marked with the shift of the electric load 2 hours ahead and 3 hours ago.

In real conditions, changing the electrical load by shifting the production cycle is not possible, since the operating modes of the equipment are determined by the needs of the population and industry for water supply.

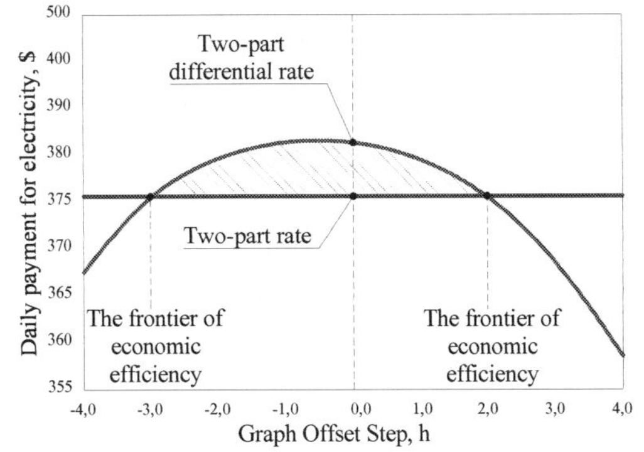

Fig. 3. Change in the cost of electricity at various tariffs and the shift of the electrical load.

Under existing conditions, the transition to a differentiated payment system is impractical $\Delta C = -1.5$ %. In this connection, a search for new methods for regulating the schedule of electric load is required.

3 Storage tanks as a tool for regulating electrical load

To regulate the supply of water to the city and preserve the fire reserve of water, three reinforced concrete control tanks are provided for the studied water intake. The capacity of each tank is 6000 m^3. The tanks are tied with the following pipelines: water supply from the deferrization station to the tanks; water supply from the tank to the pumping station of the 2nd lift; overflow pipe to prevent overfilling of the tank; full discharge pipe. All pipelines are equipped with shut-off and control valves.

The water level in the tanks ranges from $h_{\min} = 2.0$ m to $h_{\max} = 3.75$ m (Fig. 4).

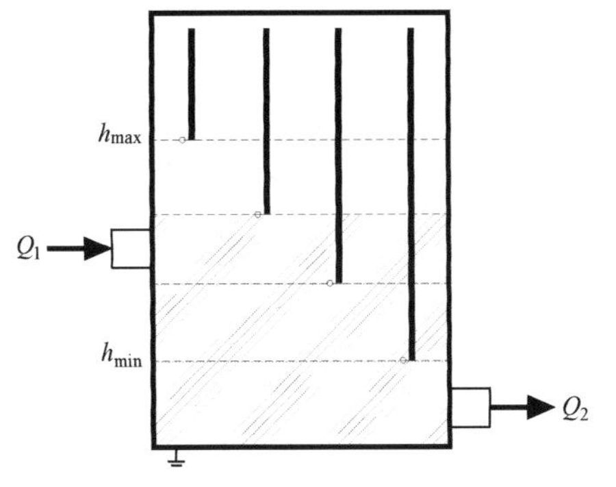

Fig. 4. Accumulation tank with location levels for water level sensors.

The minimum level is due to the need for a fire reserve, when the maximum level is reached, water enters the overflow pipe. To control the water level in each tank, sensors are installed (measuring pressure transducer), the signal from which is output to the pumping station of the second rise. Water level control is carried out by the pumping unit operator. Consider the ability to control the water level in the tank [6].

It is necessary to build a system that maintains a given level of water in the tank h_0. We assume that water is pumped into the tank continuously. To control the water level h, we can change the volume of water Q_1 raised from the wells. Thus, the water level h is an adjustable quantity. The change in the water level in the tank depends on the difference in the volumes of water raised by Q_1 and supplied to the network Q_2 and the area of the tank. The area of all tanks is $S = 3600$ m^2.

Suppose that at time $t = 0$ the water level in the tank is equal to a predetermined value, and the input Q_2 and output Q_1 volume are equal to each other, so that the water level does not change. This mode determines the nominal water level in the tank. In the calculations, we assume the nominal level equal to $h_0 = 3.0$ m. Then the control system model can be described by an equation that determines the change in water consumption:

$$h(t) = \frac{1}{S}\int_0^t (Q_1(t) - Q_2(t))dt. \tag{6}$$

The resulting equation can be represented in differential form:

$$\frac{dh(t)}{dt} = \frac{1}{S}(Q_1(t) - Q_2(t)). \tag{7}$$

Target function minimizing the cost of cash for electricity when using a differentiated tariff:

$$\begin{aligned} C_d &= a \cdot k_a \cdot w_{sp} \cdot Q_{h.\max} + \\ &+ b \cdot (k_n \cdot W_n + k_{hp} \cdot W_{hp} + k_p \cdot W_p) \rightarrow \min \end{aligned} \tag{8}$$

where w_{sp} – specific consumption of electricity for rising water, kWh / m^3; $Q_{h.\max}$ – maximum hourly water consumption, m^3 / h.

To implement the system, it is necessary to fulfill a number of conditions. Firstly, the daily volume of water rise should not be lower than the volume supplied to the pipeline network [7].

The system of linear constraints takes the form in the optimization function takes the form:

$$\begin{cases} Q_{1.d} \geq Q_{2.d}; \\ Q_{1.h} \leq V_{t.\max}; \\ Q_{1.h} \leq Q_{h.wf}; \\ Q_h \geq 0; \\ P_{m.\max} \geq P_{e.\max}; \\ h \geq h_{\min}; \\ h \leq h_{\max}, \end{cases} \tag{9}$$

where $Q_{1.d}$, $Q_{2.d}$ – daily volume of raised and supplied water, m^3;

$V_{t.\max}$ – total tank volume, m^3;

$Q_{1.h}$ – hourly flow rate, m^3/h;

https://doi.org/10.1051/e3sconf/202017801065

$Q_{h.wf}$ – hourly maximum volume of water production, m^3/h;

h_{max}, h_{max}, h – minimum, maximum and actual water level, m.

Secondly, the hourly rise in water cannot be greater than the maximum volume of the tank. Thirdly, it is necessary to fulfill the condition in which the minimum value of the 30-minute power of the morning maximum will be less than the maximum value of the 30-minute power of the evening maximum of loads $P_{e.max}$.

Also, in the system of linear restrictions, it is necessary to include the boundaries of the change in the water level in the tank.

In Fig. 5 shows the results of optimizing the schedule of electric load minimizing the cost of paying for electricity.

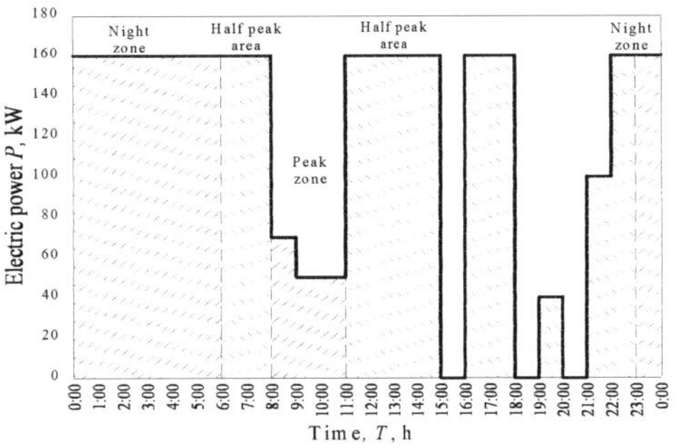

Fig. 5. Chart of daily electric power water intake after optimization.

The economic efficiency after the measures was ΔC = 10.2%, which is a very significant indicator.

4 Conclusions

Maneuvering the load schedules does not directly lead to a decrease in power consumption, however, due to a reduction in the cost of purchasing electricity, the electric power component of the cost is reduced. When using a differentiated payment system by day zones, it becomes possible to significantly increase the economic efficiency of water utilities. In the peak zone, where the maximum payment for electricity, shutdown of the well pumps is supposed, while the consumer will be provided with water filled in the tanks. The implementation of such an event in practical conditions allows reducing the cost of paying for electricity by more than 10%.

References

[1] N.V. Hruntovich, A.A. Kapanski, D. Baczynski, G.V. Vagapov, O.V. Fedorov, Optimization of a variable frequency drive pump working on a water tower, EDP Sciences, E3S Web of Conferences **124**, 05060 (2019)

[2] A.A. Kapansky, Energy Efficiency of the Technological Systems of Water Supply and Drainage and Methods of Its Evaluation, Izv. higher studies. institutions and energy. about the CIS, Energy **5**, 436–451 (2016)

[3] O.V. Fedorov, Expeditious forecasting of power consumption, 2017 International Conference on Industrial Engineering, Applications and Manufacturing, ICIEAM, IEEE (2017)

[4] Y.I. Gracheva, O.V. Naumov, Estimation of Power Losses in Electric Devices of the Electrotechnical Complex, International Conference on Industrial Engineering, Applications and Manufacturing, ICIEAM, 6 (2019)

[5] A.A. Shpiganovich, K.A. Pushnitsa, E.V. Churkina, O.V. Fedorov, Features of the operation of power supply systems of ferrous metallurgy enterprises, Ferrous metals, 5, 56–61 (2017)

[6] Y.I. Gracheva, A.N. Alimova, Calculating Methods and Comparative Analysis of Losses of Active and Electric Energy in Low Voltage Devices, International Ural Conference on Electrical Power Engineering, UralCon, 361–367 (2019)

[7] A. Khosravi, N. Saeid, C. Doug, Construction of optimal prediction intervals for load forecasting problems, IEEE Transactions on Power Systems **25.3**, 1496–1503 (2010)

[8] X. Liu, et al., Real-time household load priority scheduling algorithm based on prediction of renewable source availability, IEEE Transactions on Consumer Electronics **58.2**, 318–326 (2012)

[9] A. Khosravi, N. Saeid, C. Doug, *Load forecasting and neural networks: A prediction interval-based perspective, Computational intelligence in power engineering* (Springer, Berlin, Heidelberg, 131–150, 2010)

[10] F. Danitz, et al., Use based allocation methods for payment of electricity transmission systems, Proceedings, IEEE, International Conference on Power System Technology **2** (2002)

[11] M. Assili, M. Hossein, D.B. Javidi, R. Ghazi, An improved mechanism for capacity payment based on system dynamics modeling for investment planning in competitive electricity environment, Energy Policy **36.10**, 3703–3713 (2008)

[12] F. Zhao, et al., Payment cost minimization auction for deregulated electricity markets with transmission capacity constraints, IEEE Transactions on Power Systems **23.2**, 532–544 (2008)

[13] F. Zhao, et al., Bid cost minimization versus payment cost minimization: A game theoretic study of electricity auctions, IEEE Transactions on Power Systems **25.1**, 181–194 (2010)

[14] Philipp-Bastian Brutscher, Payment Matters?-An Exploratory Study into the Pre-Payment Electricity Metering (2011)

[15] M. Misiti, et al., Optimized clusters for disaggregated electricity load forecasting, Revstat **8.2**, 105–124 (2010)

The study of structural fields of daily gas consumption of the balance groups of the regional gas supply system

Deniz Moroz[1], *Nadzeya* Hruntovich[2], *Nikolay* Hruntovich[2], *Svetlana* Jhukovets[2], *Aleksey* Kapansky[2], and *Leonid* Dolomanyak[3]

[1]State Enterprise "Research Institute Belgiprotopgaz", Domashevskij lane,11a, Minsk, Republic of Belarus
[2]Sukhoi State Technical University of Gomel, Prospect Octiabria, 48, 246746, Gomel, Republic of Belarus
[3]Kazan State Power Engineering University, str. Krasnoselskaya, 51, 420066, Kazan, Russia

Abstract. The study of the balance sheet structure of the regional gas supply system is necessary to solve a number of operational tasks. An information database has been created for conducting research. It includes information on daily gas consumption and the actual average daily outdoor temperature for a six-year period for more than 539 elements of the balance structure of the regional gas supply system. The most significant factors determining gas consumption were identified for 7 enlarged balance groups. It´s investigated the dynamics of factors in time. The article presents the types of daily fields "gas flow – average daily outdoor temperature" for enlarged groups of the balance structure of the regional gas supply system. Four temperature areas were identified and substantiated in the daily fields. The assessment of qualitative and quantitative indicators characterizing the temperature areas is presented in the article.

1 Introduction

A systematic approach to the study of properties, patterns of development and functioning of energy objects in the 60s of the last century was laid by academician L. A. Melentyev and continued its development in the works of his students and followers around the world [1, 2]. A systematic approach involves a comprehensive review of the system, taking into account its external and internal relationships, all possible research purposes or management criteria, various restrictions and consequences of decisions. The meaning and content of the system approach proposed by academician L. A. Melentyev for energy systems, which includes the gas supply system, is clarified by the allocation of four components [3, 4]: the study and accounting of external relations of the system under consideration; a hierarchical view of the internal structure of complex objects and their management processes; accounting for uncertainties caused by incomplete source information, multicriteria and other factors; the use of mathematical modeling methods and information and computing tools to describe the object under study [5, 6, 7, 8]. Effective management of the modern regional gas supply system (RGS) should be based and developed on the principles of the system analysis, which primarily involves the study of the properties, patterns of functioning of enlarged balance groups and the gas supply system as a whole [9, 10, 11, 12]. For these purposes, it is proposed to use the structural fields of daily gas consumption from the average daily outdoor temperature (Bday-tday) and a system of indicators that characterize temperature zones over an annual time interval. The conducted research allowed us to form approaches to the construction of structural fields, which consist in rationing the daily values of gas consumption for all enlarged groups of the balance structure relative to the average annual values. The average daily outdoor temperature was converted to Kelvin degrees and correlated with the average annual values, which made it possible to exclude negative temperatures and further compare both types of structural fields and their characteristics for different years, both for the balance group and between different enlarged balance groups.

2 Research results

The enlarged balance structure of gas consumption for the studied RGS can be represented by seven groups: «Energetics», «Agriculture and cooperation», «Municipal, population», «Ministry of housing and communal services», «Industry», «Other non-industrial purposes», «Other no affiliation». The contribution of each enlarged group to the total gas consumption of the region is shown in figure 1. The largest gas consumption is accounted for the group «Energetics» – 50.2%, «Industry» – 22.6%, «Municipal, population» – 13.2%, «Ministry of housing and communal services». The «Industry» group includes 6 subgroups of various departmental affiliation, and in total, the «Industry» group includes 106 balance elements – industrial enterprises [1, 2].

To construct the structural fields «B_{day}-t_{day}», an information database was created for 2012–2017, containing information on daily gas consumption and average daily outdoor temperature for 539 elements of

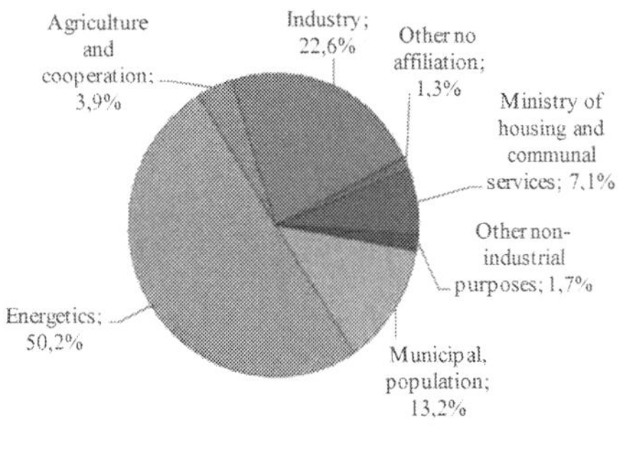

Fig. 1. Structure of gas consumption by enlarged groups RGS.

An assessment of the degree of closeness of the relationship between the «B_{day}-t_{day}» for the enlarged balance groups and the RGS as a whole was made (table 1). The most stable and high degree of closeness of the relationship between the average daily temperature and daily gas consumption has the RGS as a whole: R (t_{day}–B_{day}) is in the range -0.92 ÷ -0.94. The RGS shows properties characteristic of large technical systems: inertia and emergence [3, 4]. A high correlation coefficient R (t_{day}-B_{day}) is observed among the groups «Energetics», «Ministry of housing and communal services», «Municipal, population», which is mainly associated with the operation of the heating system. The «Industry» group has the lowest correlation coefficient R=-0.46÷-0.64, which indicates the need to search for and justify additional factors that determine the gas consumption of end consumers.

Table 1. Results of the assessment of the degree of closeness of the relationship between daily gas consumption and outdoor temperature for the enlarged balance groups of the RGS for 2012-2017.

Enlarged groups of the balance structure of the RGS	Correlation coefficients between the average daily outdoor temperature and daily gas consumption, R (t_{day}–B_{day})					
	2012	2013	2014	2015	2016	2017
Submission (RGS in general), including:	-0.94	-0.92	-0.93	-0.92	-0.94	-0.94
Group «Municipal, population»	-0.93	-0.94	-0.87	-0.89	-0.92	-0.91
Group «Energetics»	-0.93	-0.94	-0.68	-0.91	-0.92	-0.91
Group «Industry»	-0.80	-0.59	-0.61	-0.53	-0.52	-0.49
Group «Other non-industrial purposes»	-0.61	-0.59	-0.51	-0.46	-0.63	-0.64
Group «Agriculture and cooperation»	-0.84	-0.78	-0.87	-0.71	-0.80	-0.66
Group «Other no affiliation»	-0.85	-0.88	-0.79	-0.58	-0.85	-0.82
Group «Ministry of housing and communal services»	-0.96	-0.95	-0.92	-0.92	-0.93	-0.93

Structural fields for 2012-2017 were constructed for the enlarged balance groups and for the RGS as a whole and their type was determined (figure 2). There are two main types of «B_{day}–t_{day}» fields:

a) a field with a clear separation into two temperature regions, determined by the state of the heating system («on» or «off») and the overlap of these areas in the zone of unstable, uncertain state of the RGS (from 0°C to 13.6°C). This type of field is characteristic for RGS and for the following groups «Municipal, population», «Energetics», «Ministry of housing and communal services».

b) a field without a pronounced delamination into temperature regions, with a dependence of the growth of daily gas consumption when the outdoor temperature decreases with a significant vertical spread of B_{day} at the same value of the outdoor temperature. This type of field is typical for such groups as «Other non-industrial purposes», «Agriculture, cooperation», «Other no affiliation», «Industry».

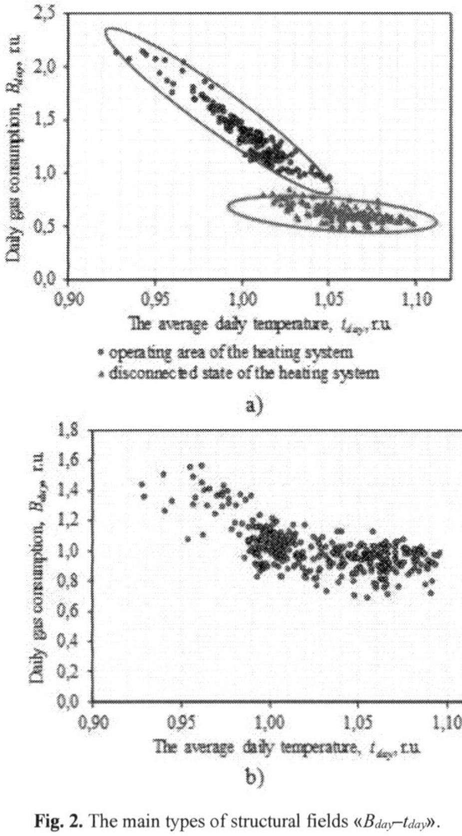

Fig. 2. The main types of structural fields «B_{day}–t_{day}».

Four temperature ranges are proposed for the analysis of the structural fields «B_{day}–t_{day}» (figure 3): first temperature range 0,9÷0,97 r.u. that corresponds to the range of daily average temperature from -27,3°C to -8,2°C; the second temperature range 0,97÷1 r.u. that corresponds to the range of daily average temperature from -8,2°C to 0 °C; the third temperature range

1.0÷1.05 r.u., that corresponds to the range of daily average temperature from 0°C to +13,6°C; a fourth temperature range is more than 1.05 r.u., that corresponds to an average daily temperature of more than +13,6°C.

The choice of average daily temperature ranges is due to several reasons:

- the first and second temperature range refers to the winter period with stable values of negative temperature, which implies that the heating system is switched on. The allocation of the second temperature range t_{day}=0.97÷1 for the winter period is due to the fact that it includes up to 60% of the day of the heating period;
- the third range t_{day}=1÷1.05 is selected separately for analysis, because in this range the greatest uncertainty appears in the modes of operation of the RGS. In this range two states of the heating system are possible: «enabled» and «disabled». This range includes the days of March, April, and October;
- the fourth range t_{day} > 1.05 combines a day with a stable positive temperature value and a disabled state of the heating system.

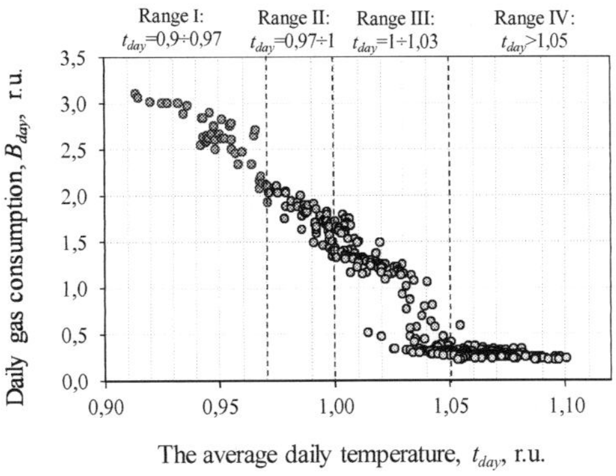

Fig. 3. Temperature ranges for the analysis of the structural fields «B_{day}–t_{day}».

Structural fields «B_{day}–t_{day}» were evaluated by the following parameters: change in outdoor temperature, t_{day}, expressed in absolute (°C) and relative (r.u.) units; average daily gas consumption in the temperature range, $B_{day.avg}$, r.u.; minimum daily gas consumption in the temperature range, $B_{day.min}$, r.u.; maximum daily gas consumption in the temperature range, $B_{day.max}$, r.u.; number of days in the temperature range, N, day; standard deviation of gas consumption in the range, σ, r.u.; coefficient of variation for gas in the temperature range, V, %; relative change in temperature in the range, Δt, r.u.; sensitivity coefficient of relative change in gas consumption ΔB with a relative change in outdoor temperature Δt, K_s, r.u.

The greatest interest for research is K_s, which is calculated as the ratio of the difference between the maximum and minimum gas consumption ΔB in the range to the difference between the maximum and minimum outdoor temperature Δt:

$$K_s = \frac{\left(B_{day.\max} - B_{day.\min}\right)}{\left(t_{day.\max} - t_{day.\min}\right)} = \frac{\Delta B}{\Delta t}. \quad (1)$$

Table 2 shows the average K_s values for 2012-2017.

Table 2. Results of assessing the degree of closeness of the relationship between daily gas consumption and outdoor temperature for the enlarged balance groups of the RGS for 2012-2017.

Large balance group	Sensitivity coefficient, $K_{s,avg}$ 2012÷2017 y. over the ranges t_{day}, r.u.			
	Range I t_{day}=0,9÷0,97	Range II t_{day}=0,97÷1	Range III t_{day}=1÷1,05	Range IV t_{day} >1,05
RGS	7,01	19,66	26,36	9,73
«Municipal, population»	10,08	26,24	42,37	12,90
«Industry»	4,60	14,74	15,87	6,10
«Other no affiliation»	7,82	21,69	27,27	9,05
«Agriculture and cooperation»	9,05	26,73	27,33	10,37
«Other non-industrial purposes»	9,07	25,50	36,32	13,75
«Energetics»	8,74	25,80	37,79	12,97
«Ministry of housing and communal services»	12,22	30,90	54,56	14,39

Evaluation of indicators that characterize the structural fields «B_{day}-t_{day}»:

– the first range corresponding to the lowest outdoor temperature includes the least number of days in the annual time interval. In 2015, this range included 7 days, and in 2012 - 42 days. This range has a lower sensitivity coefficient for all groups: K_s, I<< Ks, II << Ks, III. This indicates a decrease in the influence of regulation in heating systems and the reached limit values of heating loads;

– the second range with a temperature t_{day}=0.97÷1 or from -8.2°C to 0°C has a higher K_s compared to the first and fourth regions, which indicates the regulatory ability to consume gas in the RGS;

– the third range with the temperature t_{day} =1÷1.05 (from 0 to 13.6 °C) is unstable in terms of modes for such groups as «Municipal, population» and «Ministry of housing and communal services», the RGS as a whole, since the state of the heating system is ambiguous in this range. The Ks for this range is several times higher than the K_s of the fourth and first ranges. In this range the daily field of the B_{day} is divided into two areas with low gas consumption (the heating system is not included) and with a high 1.5 to 2 fold increase in the average daily gas consumption (the heating system is switched on).

– the fourth range, corresponding to the highest outdoor temperature, combines from 53% to 46% of the number of days of their annual value. For this range, the sensitivity coefficient of the K_s is several times less than K_s of the third and second ranges: K_s, IV<< Ks, II<< K_s, III.

It should be noted that for the «Industry» group, the fourth and the third range K_s are the lowest of all enlarged RGS groups, which indicates a lower degree of influence of outdoor temperature. The daily gas consumption regimes of the «Industry» group are formed at the expense of technological consumption for output.

3 Conclusions

For 7 enlarged balance groups and RGS as a whole the field types are defined as «B_{day}–t_{day}»: a field with a clear stratification into two temperature regions, defined by the condition of the heating system (enabled or disabled) and the overlap of these areas in the area of the unstable state of RGS; field without a pronounced stratification into temperature regions, with a dependence of the growth of daily gas consumption with a decrease in the temperature of the outdoor air with a significant vertical dispersion of B_{day} when the outdoor temperature is the same. 4 temperature ranges and a system of indicators for analyzing gas consumption modes are proposed. The proposed K_s allows us to evaluate the gas regulation capacity in each temperature range, as well as to evaluate individual features of gas supply modes for enlarged balance groups.

References

[1] L. Melentiev, *Selected works, Methodology of system researches in power engineering* (Moscow, Nauka, 289, 1995)

[2] Ed. N. Voropay, *System research in the energy sector: retrospective of scientific directions of the SEI – ISEM* (Novosibirsk, Nauka, 686, 2010)

[3] N. Ilkevich, *Multilevel modeling of the development of gas supply systems* (Novosibirsk, Nauka, 217, 2014)

[4] N. Ilkevich, T. Dzyubina, Zh. Kalinina, S. Okuneva, *Simulation of the equilibrium of the flows of the cost of supply and demand on natural gas, Pipeline systems of power engineering: methodical and applied problems of mathematical modeling* (Novosibirsk, Nauka, 318–331, 2015)

[5] M. Shevchenko, Development of private models for monitoring the regional gas supply system. Technological audit and production reserves **6**, 2 (26), 40–46 (2015)

[6] G. Galustov, S. Brovchenko, S. Meleshkin, *Mathematical modeling and forecasting in technical systems: Educational* (Taganrog, Publishing house of TTI JUFU, 30, 2008)

[7] F. Peregudov, *Introduction to system analysis* (Moscow, Vysshaja shkola. 367, 1989)

[8] N.V. Hruntovich, A.A. Kapanski, D. Baczynski, G.V. Vagapov, O.V. Fedorov, Optimization of a variable frequency drive pump working on a water tower, EDP Sciences, E3S Web of Conferences **124**, 05060 (2019)

[9] O.V. Fedorov, Expeditious forecasting of power consumption, 2017 International Conference on Industrial Engineering, Applications and Manufacturing (ICIEAM), IEEE (2017)

[10] Y.I. Gracheva, O.V. Naumov, Estimation of Power Losses in Electric Devices of the Electrotechnical Complex, International Conference on Industrial Engineering, Applications and Manufacturing, ICIEAM, 6 (2019)

[11] A.A. Shpiganovich, K.A. Pushnitsa, E.V. Churkina, O.V. Fedorov, Features of the operation of power supply systems of ferrous metallurgy enterprises, Ferrous metals **5**, 56–61 (2017)

[12] Y.I. Gracheva, A.N. Alimova, Calculating Methods and Comparative Analysis of Losses of Active and Electric Energy in Low Voltage Devices, International Ural Conference on Electrical Power Engineering (UralCon), 361–367 (2019)

[13] N. Gruntovich, D. Moroz, V. Shalonik, About necessity of study the structure and regularities of the functioning of regional gas supply systems, Energeticheskaja strategija, 1, 42–46 (2017)

[14] D. Moroz, A. Kapanskij, Development of models of gas consumption by structural elements of the regional gas supply system, Energy saving and efficiency in technical systems, Materials of the IV International Scientific and Technical Conference of Students, Young Scientists and Specialists, Tambov State Technical University, 416–418 (2017)

[15] D.R. Moroz, N.V. Hruntovich, Regional Gas Supply System Considered from the Standpoint of System Analysis and Regularities of its Functioning. Energetika, Proc. CIS Higher Educ. Inst. and Power Eng. Assoc. **61**, 4, 359–371. DOI: 10.21122/1029-7448-2018-61-4-359-371 (2018)

[16] R. Kantjukov, M. Sukharev, V. Meshalkin, R. Gimranov, A. Popov, I. Ryzhenkov, Forecast gas consumption – the basis for making rational decisions on the structure and technological parameters in the design and reconstruction of the gas supply system, Electronic scientific journal Oil and Gas Business **1**, 201–221 (2015)

Simulation modeling balancing current in the traction network of a substation group

Valerij Li[1], and *Yuliya* Konstantinova[1,*]

[1]FESTU, Institute of Power Engineering, 680021 Serysheva st. 47, Khabarovsk, Russia

Abstract. This article is devoted to assessing the transit of electricity from an external electrical power supply system through the traction network of a railway district that includes nine traction substations. Substation transformers has star-delta-11 connection. The paper presents a simulation model of the railway district implemented in the MATLAB/Simulink software package. Various options of loading the external power supply system are modeled including into account the symmetrical load of the external network and the single-phase load of traction substations. The values of the voltage difference on the traction windings of power transformers of adjacent substations that occurs at different loads, the balancing currents that this difference creates and real power losses in the inter-substation areas are calculated.

1 Introduction

At present, the volume of traffic on the Far Eastern Railway of the Russian Railways branch is increasing and their implementation requires the strengthening of the traction power supply system, which, as a rule, leads to a decrease in the resistance of the traction network and helps to increase the flow of electricity from the external power supply system to the traction system. When the traction network works together with the external power supply system, a balancing current (BC) occurs in the traction network. This current flowing through the elements of the network creates additional energy losses. Thus, the energy indicators of the electrified railway are reduced, and Russian Railways Company incurs economic damage from imperfections in the interfacing of external and traction power supply systems. Some disadvantages in the joint operation of the external power supply system with the traction network are caused by the simultaneous electrification of the railway and settlements, industrial and agricultural loads in the regions [1-3].

There are the following causes of the equalizing current:

• Several identical traction substations (TS) are connected to the same line,

• A low power transformer is installed on one of the TSs,

• A double transformation is applied at one of the TSs (transformation from 220 to 110 kV by autotransformer and from 110 to 27.5 kV by transformer),

• Adjacent TSs are powered by networks of various voltage classes,

• The TSs group receives power from a power system with a voltage class of 220 (110) kV, and one of the TSs receives power from a distribution substation with a voltage of 110 (220) kV,

• The third winding of the traction transformer TS is supplying a powerful load.

Three-phase transformers in which the star-delta-11 connection supplies two inter-substation areas (ISA) each with a separate phase. As a result, the operating rate for the phases, are significantly different. The third phase of the external power supply network is underloaded and to reduce asymmetry when each subsequent substation is connecting to the network, the underloaded phase is alternated. Thus, as a rule, the same ISA is simultaneously powered by the leading phase of one of the adjacent substations and the lagging phase second substation. Voltage losses in the traction windings of transformers that supply ISA with the leading and lagging phase are determined by the formulas:

$$\Delta\bar{U}_{lead} = \frac{2}{3}\cdot I_{r1}\cdot X_1'\cdot sin\varphi_{r1} - \frac{1}{3}\cdot I_{l1}\cdot X_1'\cdot sin\left(60-\varphi_{l1}\right) \quad (1)$$

$$\Delta\bar{U}_{lagg} = \frac{2}{3}\cdot I_{l2}\cdot X_2'\cdot sin\varphi_{l2} + \frac{1}{3}\cdot I_{r2}\cdot X_{r2}'\cdot sin\left(120-\varphi_{r2}\right), \quad (2)$$

where I_{r1}, I_{l1} is the magnitude of current phasor on the right and left ISA for the first substation, A; φ_{r1}, φ_{l1} is the power-factor angle on the right and left ISA for the first substation, electrical degrees; I_{r2}, I_{l2} is the magnitude of current phasor on the right and left ISA for the second substation, A; φ_{r2}, φ_{l2} is the power-factor angle on the right and left ISA for the second substation, electrical degrees; X'_1, X'_2 is the equivalent inductive reactance of the external power supply system and the transformer for the first and second substation, Ohm. As can be seen from the formulas, even with equal loads in the right and

* Corresponding author: kamfes@mail.ru

left ISA, voltage losses in the lagging phase will be higher than in the leading phase.

The purpose of the research is to evaluate the flow of electric energy amount from an external electrical power supply system through an AC traction network on a railway district. The objectives of the research are, firstly, a creating a simulation model of the power supply system of an electrified railway district, secondly, simulating the flow of electricity with various load cases of the power supply system; thirdly, assessing the effect of power-factor correction units and series capacitor banks on the flow of electricity through the traction network; fourthly, assessing the effectiveness of measures to reduce balancing current.

2 Power supply system simulation

The considered electrified railway district includes 9 traction substations (Fig.1). Substations No. 1–2, 5–9 are connected to a network voltage of 220 kV, substation No. 4 receives power from a network voltage of 110 kV. Substation No. 1 receives electricity mainly from the adjacent power supply system (S1), substation No. 2 is connected to buses 220 kV of the system-forming substation with a higher voltage of 500 kV (S2). Substations No. 3 and 4 receive electricity from power supply sources S3-S5. A group of substations consisting of No. 5–9 is connected to a double-circuit line 220kV which connects power supply sources S5 and S6. The data on the voltages supplying the inter-substation areas are presented in table 1.

Table 1. The data of the inter-substation areas voltages.

ISA	Phase voltage	Type phase voltage for the substation left/ right
1-2	U_B	lagging / leading
2-3	U_C	lagging/lagging
3-4	U_B	leading /lagging
4 -5	U_A	leading /lagging
5– 6	U_C	leading /leading
6 –7	U_A	lagging / leading
7– 8	U_B	lagging / leading
8 – 9	U_C	lagging/lagging

Power-factor correction units (PFCU) are installed at traction substations No. 1-3, 5-7 and 9. Series capacitor banks (SCB) are connected between the rail and the output of the transformer winding 27.5 kV at substations No. 4 and 8. This study used simulation in the MATLAB / Simulink software package [4,5].

The following blocks were used when creating the model:

• The three-phase source block as power supply sources and system-forming substations (S1-S6),

• The pi section line as overhead high-voltage line section,

• The three-phase transformer as a transformer of the external network (220/110) and traction substation (220 (110) / 27.5),

• The series RLC branch as section of the traction network (overhead line 25 kV and a rail circuit),

• The three-phase series RLC load as symmetric three-phase load (L1, L2),

• The series RLC load as point traction load (40 MVA).

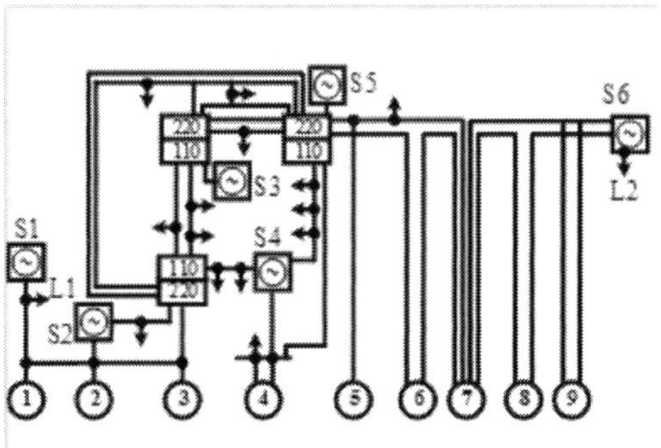

Fig. 1. Electrical power supply system of the electrified railway district.

2.1 Simulation options

The following options for loading the external network were simulated:

• Only symmetrical three-phase load in the network (Fig.1 indicated by blocks L1 and L2) at rated currents in the wire on sections S1-1 and S5-5, when the PFCUs and SCBs are disconnected (op1),

• Only symmetrical three-phase load in the network at rated currents in the wire on sections S1-1 and S5-5, when and SCBs are enabled the PFCUs are disconnected (op2),

• Only symmetrical three-phase load in the network at rated currents in the wire on sections S1-1 and S5-5, when and SCBs are disconnected the PFCUs are enabled (op3),

• Only symmetrical three-phase load in the network at half of rated currents in the wire on sections S1-1 and S5-5 (the PFCUs and SCBs are disconnected) (op4),

• Symmetrical three-phase load in the network at half of rated currents in the wire on sections S1-1, S5-5 and the traction load for all ISAs except for the ISA under consideration (the PFCUs and SCBs are disconnected) (op5).

2.2 Simulation results

The results of simulation the balancing current that occurs in the traction network with the above-described load variants of the external system are presented in table 2.

The flow of electric energy from the external electrical power supply system through the traction network is minimum value at the ISAs 2-3 and 3-4, which can be explained by the location of these substations at the near the power supply sources. The maximum value of the BC for all the simulation options is at the area inter substation No. 6 and 7, which is also due to the most remote location of them from the power supply sources.

Table 2. Results of calculating the balancing current.

ISA	Balancing current, A				
	op1	op2	op3	op4	op5
1-2	120.9	121.0	119.2	60.9	89.3
2-3	2.5	3.6	3.0	9.9	9.4
3-4	16.2	34.9	10.7	11.8	31.5
4-5	129.0	158.0	110.4	62.9	111.6
5-6	139.8	139.9	140.1	60.4	69.1
6-7	167.1	168.5	175.7	73.9	87.2
7-8	121.6	139.2	131.2	52.6	70.7
8-9	131.8	139.1	126.4	57.0	47.9

The resistance of the two adjacent inter-substation areas decreases when the SCB is switched on at the substation located in the middle. At the same time, the balancing current increases by 11-31% at areas that receive power from the TS No. 8. An increase in surge current amounted to 50% for the area located to the right of substation No. 4 and almost 400% for the ISA posted to the left.

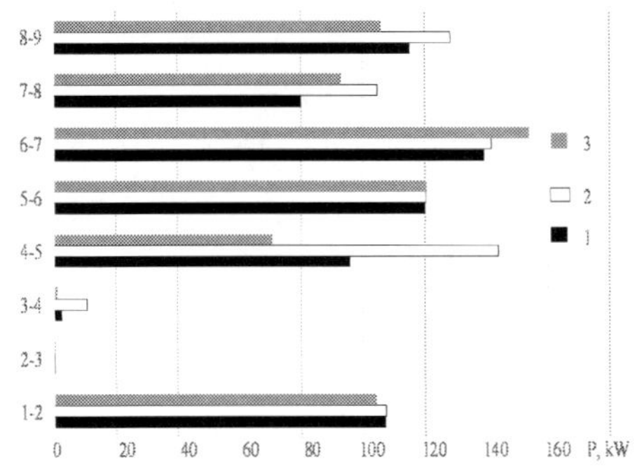

Fig. 2. Bar chart of the real power losses for options 1-3.

The use of PFCUs mainly reduced the balancing current at the inter-substation areas 1-2 (1.4%), 3-4 (34%), 4-5 (14%) and 8-9 (4%), however, overcompensation caused current a BC increase at the ISAs 2-3 (17%), 5-6 (0.2%), 6-7 (5%) and 7-8 (8%). The real power losses of the options considered are presented in Fig. 2. The application of the SCB s leads to a general increase in real power losses at the areas by 15%, while losses when turning on PFCUs are reduced by 1.4%.

Real power losses increase at all inter-substation areas (fig.3), when the external network load combines a symmetrical three-phase and traction load (options 5), the exception is the inter-substation area 8-9 because substations supply this ISA by lagging phases. In this case, the voltage losses caused by the flow of the traction load of adjacent ISAs have the effect of equalizing the voltage on the buses of these substations (the same effect is at ISA 2-3).

The voltage difference at ISA 1-2, which is supplied simultaneously by the leading and lagging phase, increases, because the voltage loss is higher at TS No.1. Real power losses are significantly increased at ISA 4-5, because load losses at TS No. 5 are higher (lagging phase) than at TS No. 4 (leading phase). Real power losses in the inter-substation area 3-4 are small compared to the others, since this zone is the longest in the section, its length is 62.7 km (the shortest zone between substations 4 and 5 is 38.2 km long). Based on the calculation results, it is recommended to take measures to reduce the balancing current in the traction network in the ISA 1-2, 4-5, 5-6, 6-7, 7-8 and 8-9.

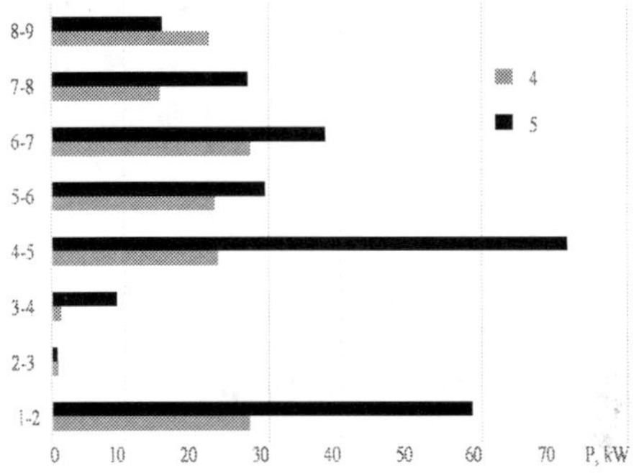

Fig. 3. Bar chart of the real power lose for options 4 and 5.

3 Modeling of measures to reduce balancing current

The following measures to reduce BC are simulated:

- relocation of the substation`s No. 4 neutral section,
- grid reconnection of traction substations transformers,
- application of a booster transformer at ISAs 1-2, 4-5, 5-6, 6-7, 7-8 and 8-9.

3.1 Relocation of the neutral section

The simulation of the TS No. 4 neutral section relocation 12 km closer to TS No. 3 showed a decrease in the balancing current at the ISA 4-5 by 19 A, while increasing the BC at the ISA 3-4 is 2 A. The total power loss in two ISAs decreased 8 -10%, depending on the option, losses on the entire district from this event decreased by 2-3%.

3.2 Grid reconnection of traction substations transformers

As mentioned above the third phase of the external power supply network is underloaded and to reduce asymmetry when each subsequent substation is connecting to the grid, the underloaded phase is alternated. In the considered electrified railway district, the alternation of the underloaded phase is as follows: C-A-A-C-B-B-C-A-A. The study simulated the work of the power supply system when changing grid reconnection of transformers at traction substations No. 3 -5 and 9, which led to the alternation of the underloaded phase C-A-B-C-A-B-C-A-C. Simulation of the grid reconnection of traction substations transformers revealed a BC decrease when the network supplies only a symmetric three-phase load (options 1 and 4) by 1-2%. Balancing current decreased at the ISA 3-4 by 28 A but increased at other areas by 2 to 39 A, which led to a rise in total losses by 13% when the network supplies a symmetric three-phase and traction load.

3.3 Application of a booster transformer

The most effective way to reduce the flow of electricity from the external electrical power supply system through the traction network is to equalize the voltage on the 27.5 kV buses of adjacent TSs. In Fig. 4 shows a bar chart value geometric $|\Delta\bar{U}|$ and algebraic $|\bar{U}1|$-$|\bar{U}2|$ of the ISAs voltage difference have been calculated for option 4 and 5.

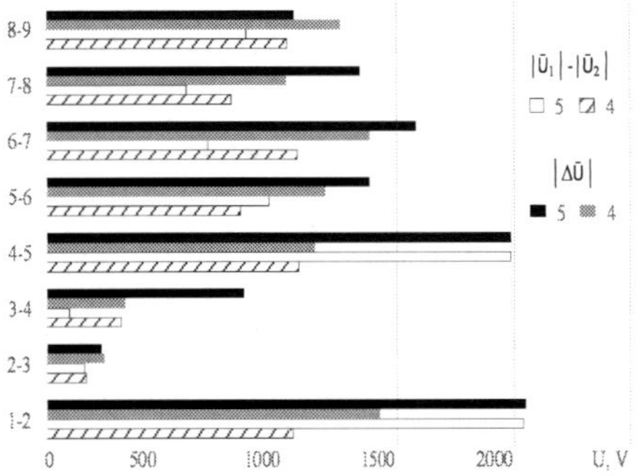

Fig. 4. Bar chart value geometric $|\Delta\bar{U}|$ and algebraic $|\bar{U}1|$-$|\bar{U}2|$ of the voltage difference for option 4 and 5..

As can be seen from the bar chart, these differences usually differ by an average of 20-30%, so equalization of the voltage on-load tap-changer will be ineffective [1, 6]. An automatic balancing current reduction device is proposed [1], which consists of a block of voltage addition and contains an input transformer, an rectifier [7, 8], a voltage inverter, a rectifier filter, an output filter [9] and a voltage-boosting transformer [3]. The additional voltage is formed as follows. The input transformer steps down traction network voltage to the rated voltage of the rectifier. Then, after rectification and filtering, a DC voltage is applied to the input of the inverter. Meanwhile, the control unit receives data from phasor measurements units [10] of the adjacent TSs and calculates the angle and magnitude of the additional voltage. Next, control signals are sent to the inverter to generate a voltage with pulse width modulation [11]. After filtering the sinusoid with a frequency of 50 Hz and with an angle of additional voltage, it is fed to the primary winding of the booster transformer. Thus, an additional voltage is formed on the secondary winding of the BT:

$$\Delta\bar{U}_b = -\Delta\bar{U} = -\left(\bar{U}_1 - \bar{U}_2\right) = \frac{\bar{U}_{in} \cdot K_r \cdot H \cdot K_{T2}}{K_{T1}}, \quad (3)$$

where $\bar{U}_1$, $\bar{U}_2$ is the voltage phasors on the buses of adjacent substations, kV; $\bar{U}_{in}$ is the input voltage, kV; K_{T1} is the input step-down transformer ratio; K_r is the ratio between the rectified and the alternating voltage; H is switching functions of PWM inverter, representing closed states of inverter semiconductors; K_{T2} is booster transformer ratio.

The use of a device with a booster transformer reduced the losses of electricity traction network by 97%.

4 Conclusions

Firstly, it was found that the use of series capacitor banks significantly increases the flow of electricity from an external power supply system through the traction network. Secondly, the traction load of other inter-substation areas flows through the inter-substation areas, on which there are no electric locomotives. Thirdly, the balancing current rises due to the one inter-substation area is powered by both the leading and the lagging phase at the same time. Finally, the most effective measure to reduce the flow from the external network to the traction network is to install booster devices.

References

[1] Y.A. Konstantinova, V.N. Li, A.M. Konstantinov, International Scientific Siberian Transport Forum TransSiberia 2019: VIII International Scientific Siberian Transport Forum **1115**, 15–24 (2020)

[2] Y.A. Konstantinova, V.N. Li, E.Y. Tryapkin, 2018 International Multi-Conference on Industrial Engineering and Modern Technologies (FarEastCon), IEEE Press, 1–5 (2018)

[3] A.B. Kosarev, B.I. Kosarev, Russian Electrical Engineering **89**, 9, 531–535 (2018)

[4] U.J. Shenoy, K.G. Sheshadri, K. Parthasarathy, H.P. Khincha, D. Thukaram, 2004 IEEE Region 10 Conference TENCON, IEEE Press, Chiang Mai **3**, 508–511 (2004)

[5] G.B. Worku, A.B. Kebede, Global Energy Interconnection **1**, 2, 187–196 (2018)

[6] X. Zhang, R. Kang, M. McCulloch, A. Papachristodoulou, Sustainable Energy, Grids and Networks **5**, 27–38 (2016)

[7] S.S. Moolamkunnam, J. Mathew, S. Joseph, International Journal of Innovative Research in Electrical, Electronics, Instrumentation and Control Engineering **7**, 4, 50–58 (2019)

[8] N.A. Samsudin, D. Ishak, A.B Ahmad, Engineering Science and Technology **21**, 2, 189–200 (2018)

[9] M. Pomalis, R.C. Leborgne, A.R. Herrera-Orozco, A.S. Bretas, Electric Power Systems Research **176**, 1–9 (2019)

[10] S. Affijulla, P. Tripathy, Protection Applications, IEEE Transactions on Smart Grid **9**, 2, 1250–1260 (2016)

[11] D.G. Holmes, Power Converters Principles and Practice, John Wiley & Sons, Inc., 155–214 (2003)

 https://doi.org/10.1051/e3sconf/202017801068

Structure of many-level adaptive automatic voltage regulation system

I.O. Golikov[1], *A.V.* Vinogradov[2], *V.E.* Bolshev[2,*], *A.V.* Vinogradova[2], *M.* Jasinski[3], and *R.R.* Gibadullin[4]

[1]Orel State Agrarian University, 69 General Rodina St., Orel, 302019, Russia
[2]Federal Scientific Agroengineering Center VIM, 5 1st Institutsky Proezd, Moscow, 109428, Russia
[3]Wroclaw University of Science and Technology, 27 Wybrzeże Wyspiańskiego st., Wrocław, 50-370, Poland
[4]Kazan State Power Engineering University, 51 Krasnoselskaya St., Kazan, 420066, Russia

Abstract. This article describes the features of voltage regulation in electrical networks of 35, 110, 220 kV. The structural diagram of the 35/10/0.4 kV network is presented. The paper also describes the adaptive automatic voltage regulation system which allows regulating the voltage taking into account the actual voltage values at the consumers' inputs. The structural diagram of the adaptive automatic voltage regulation system in the 0.4 kV electrical network using a boost transformer as an additional means of voltage regulation is given. The system is based on voltage sensors installed in different parts of an eletcrical network sending information on voltage values to to the processing unit which generates a signal for voltage regulating supplied to the executive device and the working body whuch, in turn, change the on-load tap-changer position of a transformer. The paper justifies the need for the enhancement of the adaptive automatic voltage regulation system for different voltage classes wich allows controlling a voltage change at different power supply system levels and regulating voltage level in accordance with this change. For this problem the multi-level adaptive automatic voltage regulation systemis proposed. The system allows regulating the voltage not only in the 0.4 kV network but also in networks of higher voltage classes. The proposed system can be integrated into the structure of intelligent electrical networks.

1 Introduction

When organizing intelligent electrical networks [1-3], the different elements must be considered for e.g.

- load and renewable energy sources production forecasting [4-5],
- power control [6-7],
- smart metering [8-9],
- power quality [10-11],
- security issues [12-13],
- economic dispatch [14-15].

Not indicated but urgent issue is the implementation of adaptive automatic voltage regulation in them. Currently, voltage regulation is automated mainly at substations of 35, 110, 220 kV using counter regulation by load regulation devices (on-load tap-changers) [16-17]. In networks of 0.4 ÷ 10 kV, the regulation by on-load tap-changers is hardly used.

A voltage change in the 35 ÷ 110 kV network causes a change in voltage in the 6 ÷ 10 kV distribution network and in the internal 0.4 kV networks of consumers. Obviously, following this change in the electric network parameters, voltage regulation at different voltage levels of the electric network will be required. Figure 1 shows the structural diagram of the 35/10/0.4 kV network.

2 Structure of adaptive automatic voltage regulation system

The use of an adaptive automatic voltage regulation system in rural 10/0.4 kV electric networks are considered in works [18-19]. There are the developed methods and technical means for an adaptive automatic

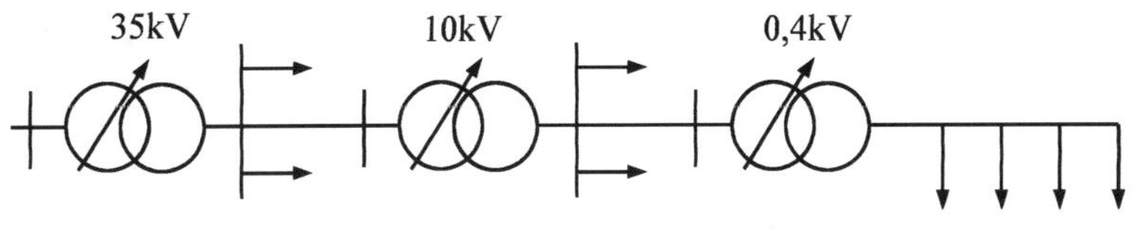

Fig. 1. Structural diagram of the 35/10/0.4 kV network.

* Corresponding author: vadimbolshev@gmail.com

 https://doi.org/10.1051/e3sconf/202017801068

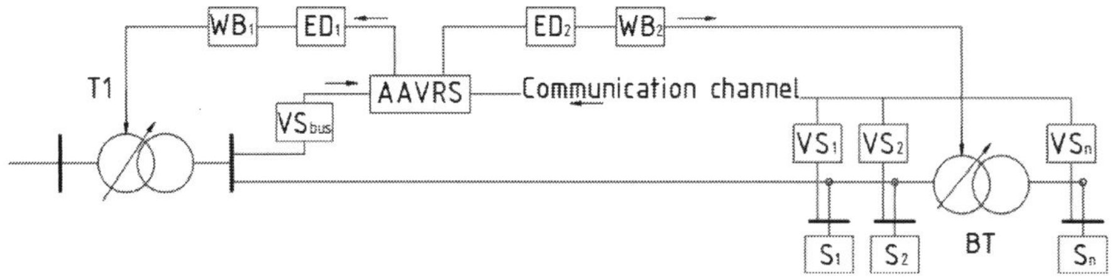

Fig. 2. Structural diagram of AAVRS using a boost transformer as an additional means of voltage regulation.

voltage regulation. According to them, voltage regulation is supposed to be carried out centrally by the on-load tap-changer of a transformer or a stabilizer installed at the beginning of the 0.4 kV power transmission line. Also, regulation options are provided using additional technical means, for example, boost transformers.

Figure 2 presents the structural diagram of the adaptive automatic voltage regulation system (AAVRS) using a boost transformer as an additional means of voltage regulation. Information about the voltage in different parts of the network by sensors VS1, VS2, VSn, VSbus is sent to the AAVRS processing unit. It generates a signal for voltage regulating supplied to the executive device ED1. The executive device ED1 acts on the working body WB1 which changes the on-load tap-changer position of the transformer T1. If it is necessary, the regulation signal is supplied to the executive device ED2 and the working body WB21, which affect the voltage boost transformer BT. Sensors VS1, VS2 are installed at the inputs of consumers S1, S2. The necessary value of voltage regulation is determined to take into account the voltage values measured by all voltage sensors [18]. This system is capable of responding to changes in voltage in a controlled electrical network by adjusting the voltage in a possible range taking into account the characteristics of a transformer on-load tap-changer and a boost transformer.

3 Structure of many-level Adaptive Automatic Voltage Regulation System

The enhancement of the adaptive automatic voltage regulation system for different voltage classes will allow controlling a voltage change at different power supply system levels and regulating voltage level in accordance with this change. Figure 3 presents the structural diagram of the many-level AAVRS.

It is proposed to install AAVRS at different voltage levels of an electrical network. So, for example, the AAVRS processing unit installed at the 0.4 kV side controls the voltage deviation at the 0.4 kV electrical network.

For example, if it is impossible to regulate the voltage both by means of the on-load tap-changer of the 10/0.4 kV transformer and by means of additional technical means, the AAVRS-4 processing block transmits a signal to the AAVRS-2 processing block about the need for voltage regulation. The AAVRS-2 analyzes the possibility of voltage regulation taking into account the data coming from all other 0.4 kV AAVRS connected to the 10 kV network of 10/0.4 kV transformer substation (AAVRS-3, AAVRS-5). The voltage value at the high side of all connected transformers 10/0.4 kV is also analyzed. If possible, voltage is regulated at the substation 35/10 kV using

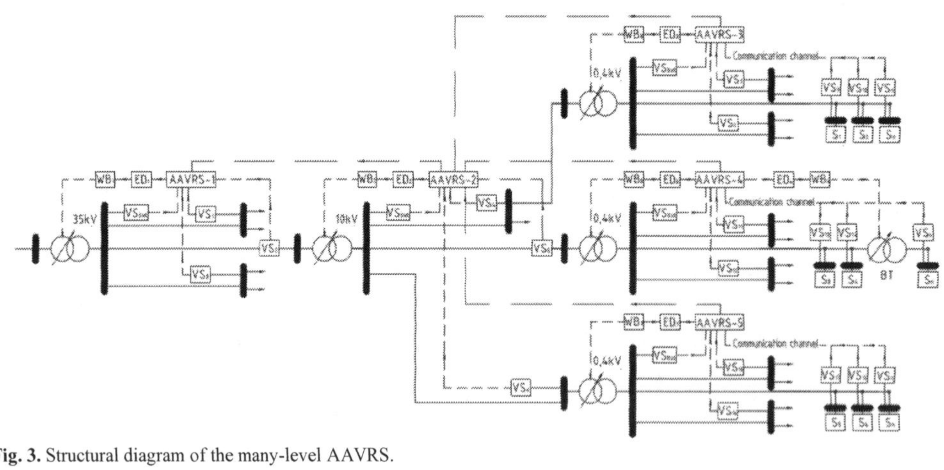

Fig. 3. Structural diagram of the many-level AAVRS.

AAVRS-2. Similarly, regulation is performed using AAVRS-1 at 110/35 kV transformer substation.

The advantage of this system is its extensibility and it can adapt to electric network operation modes. Such a structure allows regulation taking into account the actual voltage values, that is, the voltage losses that occur in all network elements [20].

The integration of many-level AAVRS into intelligent electric networks can be carried out using the IEC-104, IEC-61850 data protocols and existing data transmission channels and communication tools [21-22]. It can use channels of other systems such as control systems for power quality and power supply reliability [23], smart metering systems [24-25].

4 Conclusion

The development of intelligent electric networks also affects the issues of ensuring a stable voltage level at the consumers' inputs. To do this, adaptive automatic voltage regulation systems must be used. The system allows voltage regulation based on the measuring actual voltage values at different electrical network points including at the consumers' inputs. The proposed structure of the many-level adaptive automatic voltage regulation system makes it possible to solve the problem.

References

[1] A. Arif, Z. Wang, Service restoration in resilient power distribution systems with networked microgrid, Proceedings of the 2016 IEEE Power and Energy Society General Meeting (PESGM), 1–5. DOI: 10.1109/PESGM.2016.7741533 (2016)

[2] J.A. Momoh, Smart grid design for efficient and flexible power networks operation and control, Power Systems Conference and Exposition, PSCE '09, 1–8. DOI: 10.1109/PSCE.2009.4840074 (2009)

[3] A. Ipakchi, F. Albuyeh, Grid of the future, IEEE Power and Energy Magazine **7**, 52–62. DOI: 10.1109/MPE.2008.931384 (2009)

[4] Y. Yang, W. Li, T.A. Gulliver, S. Li, Bayesian Deep Learning-Based Probabilistic Load Forecasting in Smart Grids, IEEE Transactions on Industrial Informatics **16**, 7, 4703–4713. DOI: 10.1109/TII.2019.2942353

[5] V. Suresh, P. Janik, J. Rezmer, Z. Leonowicz, Forecasting Solar PV Output Using Convolutional Neural Networks with a Sliding Window Algorithm, Energies **13**, 3, 723, DOI: 10.3390/en13030723

[6] M. Beccali, L. Bellia, F. Fragliasso, M. Bonomolo, G. Zizzo, G. Spada, Assessing the lighting systems flexibility for reducing and managing the power peaks in smart grids, Applied Energy **268**, 114924. DOI: 10.1109/TSG.2019.2942850

[7] M. Jalali, V. Kekatos, N. Gatsis, D. Deka, Designing Reactive Power Control Rules for Smart Inverters Using Support Vector Machines, IEEE Transactions of Smart Grid **11**, 2, 1759–1770. DOI: 10.1109/TSG.2019.2942850

[8] D.L. Donaldson, D. Jayaweera, Effective solar prosumer identification using net smart meter data, International Journal of Electrical Power & Energy Systems **118**, 105823. DOI: 10.1016/j.ijepes.2020.105823

[9] Y. Chawla, A. Kowalska-Pyzalska, P.D. Silveira, Marketing and communications channels for diffusion of electricity smart meters in Portugal, Telematics and Informatics **50**, 101385. DOI:10.1016/j.tele.2020.101385

[10] M. Jasiński, et al., Influence of Measurement Aggregation Algorithms on Power Quality Assessment and Correlation Analysis in Electrical Power Network with PV Power Plant, Energies **12**, 18, 3547. DOI: 10.3390/en12183547

[11] W. Grycan, B. Brusilowicz, M. Kupaj, Photovoltaic farm impact on parameters of power quality and the current legislation, Solar Energy **165**. DOI: 10.1016/j.solener.2018.03.016

[12] A.W. Altaher, A.H. Hussein, Intelligent security system detects the hidden objects in the smart grid, Indones. J. Electr. Eng. Comput. Sci. **19**, 1, 188. DOI: 10.11591/ijeecs.v19.i1

[13] F. Wei, Z. Wan, H. He, Cyber-Attack Recovery Strategy for Smart Grid Based on Deep Reinforcement Learning, IEEE Transactions of Smart Grid **11**, 3, 2476–2486. DOI: 10.1109/TSG.2019.2956161

[14] T. Sikorski, et al., A Case Study on Distributed Energy Resources and Energy-Storage Systems in a Virtual Power Plant Concept: Economic Aspects, Energies **12**, 23, 4447. DOI: 10.3390/en12234447

[15] H. Li, Z. Wang, G. Chen, Z.Y. Dong, Distributed Robust Algorithm for Economic Dispatch in Smart Grids Over General Unbalanced Directed Networks, IEEE Transaction of Industrial Informatics **16**, 7, 4322–4332. DOI: 10.1109/TII.2019.2945601

[16] C. Vournas, C. Lambrou, M. Glavic, T. Van Cutsem, An integrated autonomous protection system against voltage instability based on Load Tap Changers, In 2010 IREP Symposium Bulk Power System Dynamics and Control – VIII (IREP), 1–14. DOI: 10.1109/IREP.2010.5563286 (2010)

[17] C. Vournas, M. Karystianos, Load tap changers in emergency and preventive voltage stability control, IEEE Transactions on Power Systems **19**, 1, 492–498. DOI: 10.1109/TPWRS.2003.818728 (2004)

[18] I.O. Golikov, A.V. Vinogradov, *Adaptivnoe avtomaticheskoe regulirovanie napryazheniya v sel'skih elektricheskih setyah 0,38 kV* [Adaptive

automatic voltage regulation in rural electric networks 0.38 kV] (Publishing House of Orel State Agrarian University, Oryol, Russia, 2016)

[19] A. Vinogradov, A. Vinogradova, I.O. Golikov, V. Bolshev, Adaptive Automatic Voltage Regulation in Rural 0.38 kV Electrical Networks, International Journal of Emerging Electric Power Systems **20**, 3. DOI: 10.1515/ijeeps-2018-0269 (2019)

[20] E. Gracheva, A. Alimova, Calculation Methods and Comparative Analysis of Losses of Active and Electric Energy in Low Voltage Devices, In 2019 International Ural Conference on Electrical Power Engineering (UralCon), 361–367. DOI: 10.1109/URALCON.2019.8877627 (2019)

[21] M. Kuzlu, M. Pipattanasomporn, S. Rahman, Communication network requirements for major smart grid applications in HAN, NAN and WAN, Computer Networks **67**, 74–88. DOI: 10.1016/j.comnet.2014.03.029 (2014)

[22] E. Ancillotti, R. Bruno, M. Conti, The role of communication systems in smart grids: architectures, technical solutions and research challenges, Communication Technologies **36,** 1665–1697. DOI: 10.1016/j.comcom.2013.09.004 (2013)

[23] A. Vinogradov, V. Bolshev, A. Vinogradova, T. Kudinova, M. Borodin, A. Selesneva, N. Sorokin, A System for Monitoring the Number and Duration of Power Outages and Power Quality in 0.38 kV Electrical Networks, In P. Vasant, I. Zelinka, G.W. Weber, (eds) Intelligent Computing & Optimization, ICO 2018, Advances in Intelligent Systems and Computing **866,** 1–10. DOI: 10.1007/978-3-030-00979-3_1 (2019)

[24] Y. Kabalci, A survey on smart metering and smart grid communication, Renewable and Sustainable Energy Reviews **57**, 302–318. DOI: 10.1016/j.rser.2015.12.114 (2016)

[25] R.R. Mohassel, A. Fung, F. Mohammadi, K. Raahemifar, A survey on Advanced Metering Infrastructure, Electrical Power and Energy Systems **63**, 473–484. DOI: 10.1016/j.ijepes.2014.06.025 (2014)

Steam reheater with helical tube bundle for wet steam turbine

Mikle Egorov[1,2,*], *Alexander* Ivanov[1], *Ivan* Kovalenko[2], *Irina* Krectunova[2], *Nadezhda* Litvinova[2], and *Elena* Popova[1,3]

[1]Peter the Great St. Petersburg Polytechnic University, Russia, 195251 Saint-Petersburg, Polytechnicheskaya 29
[2]St. Petersburg State University of Aerospace Instrumentation, Russia, 190000 St. Petersburg, Bolshaya Morskaya str. 67
[3]Concern Avrora Scientific and Production Association JSC, Russia, 194021 Saint-Petersburg, Karbysheva str. 15

Abstract. Since steam heat exchangers, used at steam cycle of Russian nuclear power stations, were designed while the knowledge about the separation and the heat exchange processes was limited, deviations between its empirical and theoretical characteristics occur. This limitation also determined application of heating pipes with simple straight shape rather than curved. This study explores a steam heat exchanger with helical heating pipes. It was shown that the model may work stably within the range of parameters, simulating work conditions of the moisture separator and steam reheater at Leningrad nuclear power plant. The experiment included processing of pure water steam as well as mixture of steam and nitrogen. It was obtained a relationship between empirical the heat transfer coefficient and the steam mass flow rate. It was noted that presence of incondensable gas does not affect significantly the heat transfer from the coils, processing high pressure steam.

1 Introduction

Identification of the sections of 0.4 kV power Moisture separators and steam reheaters (MSR) are widely used at Russian nuclear power plants (NPP) with VVER, RBMK reactor types [1–3]. These installations process steam between the high pressure stage of the turbine the low pressure stage. MSR improve thermal parameters of steam at all modes of the reactor and even in the mode with natural circulation of the coolant [4, 5]. Application of such equipment pursues several goals such as to improve the efficiency of the steam cycle and to avoid decay of turbine parts. The erosion process may occur when steam, leaving the high pressure stage, becomes saturated with moisture that may erode blades of the following turbine stages [6–8].

Furthermore, water droplets continuously contacting with warm surface of heating pipes induce cyclic stress which results in pipe ruptures. Notably, that pipes of horizontal helical shape are more prone to breaks in contrast to that of vertical and straight shape. This circumstance determined limited usage of MSR with helical coils and widespread of models with straight heating elements.

At present, different constructions of MSR are in use. Since theories of steam separation and heat exchange were not well developed at the time when MSR were engineered, various deviations between simulated and real characteristics of MSR can be observed. For instance:

- uneven profile of the outgoing steam from the separator which results in moisture transfer to reheater and consequently into nonattainment of the designed temperature value [9];
- appearance of overheated and overcooled zones within the heating pipes bundle because of uneven flow spread and different length of steam supply pipes. This circumstance creates undesirable tensions, and generates condensation areas, initiating vibration, respectively. Condensation from a two phase flow is more complex than that from a single phase because it hinders heat transfer between mediums as shown in [10]. Therefore, complex models like Hammouda are used instead of ordinary heat exchange formulas for the heat transfer coefficient estimation of the pipe bundle [11]. Recent studies show satisfactory precision of the approach [12–14].

In addition to these deviations, other disadvantages were discovered:

- high friction loss along the steam duct due to arrangement of moisture separator above MSR (which is common for VVER–440 project);
- appearance of incondensable gases (ICG) in heating pipes diminishes heat exchange between the heating steam and the pipe surface [15–16];
- vertical arrangement of heating pipes in the MSR bundle (which is typical for VVER–1000 project) has lower heat transfer coefficient in contrast to the horizontal arrangement. Furthermore, many researches were conducted within the last decades aiming on heat transfer intensification [17–19].

In order to improve disadvantages of existing constructions, several modernizations were made. For instance, after redesigning of separating bundles of the MSR model "SPP–500–1" at the Leningrad NPP, located near Saint-Petersburg, Russia, several goals were achieved. On first, uneven moisture distribution was

* Corresponding author: mikhail.yu.egorov@gmail.com

eliminated and on second, the moisture concentration value reached its calculated range [20–21].

Aiming to overcome disadvantages, which are inherent to MSR with vertical arrangement of heating pipes, two experimental heat exchangers were designed. First unit, which consisted of two full-scale helical tubes and shell, justified that horizontal coils may work steady.

1.1 Research objectives

Since the first unit showed satisfactory results, it was suggested to increase the number of heat exchanging pipes in order to simulate construction of the MSR heating section.

It was also suggested to expand experimental objectives as follows:

- justify multicoil unit's ability to perform stably;
- estimate the heat transfer coefficient at nominal mode and its deviation from the simulated coefficient;
- study how presence of ICG in steam (0,02% by mass) affects heat transfer and hydrodynamic stability of the unit (simulating work conditions, occurring at the NPP with single loop reactor type).

2 Materials and methods

The multicoil experimental unit consist of 10 double-twisted helical tubes, mounted inside cylindrical shell. A 3d model of the unit, which was designed by authors, illustrated at the Fig. 1.

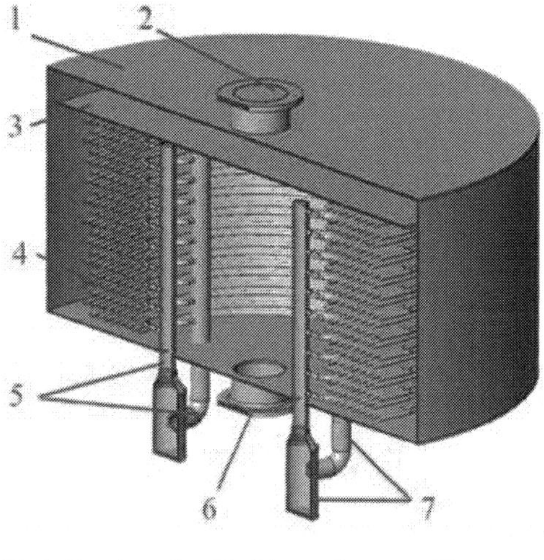

Fig. 1. Experimental multicoil heat exchanger model cross section: 1 – shell; 2 – LP steam inlet flange; 3 – sheet; 4 – coil bundle; 5 – HP steam inlet channel; 6 – LP seam outlet flange; 7 – HP steam-condensate outlet channel.

The unit has two steam loops: low pressure (LP) and high pressure (HP). From the inlet nozzle, the HP vapor is distributed between two vertical steam channels and then it is spread among 10 helical pipes. Steam transfers its heat to pipe's walls and condenses partially. The steam and condensate mixture flows towards one of two vertical condensate channel; then it fills up the condensate tank and drains after cooling.

When the LP steam reaches the unit's inlet, it flows through the gap between flat round sheet and cylindrical shell. Then, it goes around several helical coils and warms up. At last, reheated steam is supplied back.

Each heating pipe is plain and circular in cross section with diameter Ø18 mm, wall thickness of 2 mm and length of 8.5 m. Tubes are twisted in flat spirals (Archimedes' spirals) with step of 20 mm. Its dimensions are as follows: about Ø1,0 m (internal) and Ø1,7 m (external). Coils has square layout with inline pattern in the front section. The total surface of 20 spirals is F ~ 46.2 m2.

Unit's cylindrical shell with diameter is 2100 mm is covered by insulation made from clay bricks, placed with a small gap between, and surrounded by external metal cylindrical shell. All unit parts were made of carbon steel alloy 'steel 20'.

During the experiment, the unit's performance was tested within the range of parameters as shown in the Table 1 below.

Table 1. Experimental unit's parameters range.

Parameter	LP	HP
Steam flow rate G, kg/s	1.9 – 5.3	0.18 – 0.38
Incoming flow temperature t_{in},	134 – 170	232 – 301
Pressure P, MPa	0.18 – 0.27	2.62 – 4.05

After every change of the mass flow rate, the empirical heat transfer coefficient K_{emp}, W/m^2 K was estimated using following equation:

$$K_{emp} = \frac{G^{hp}\left(r + C_P\left(t_{in}^{hp} - t_{out}^{hp}\right)\right)}{F\Delta\bar{t}}$$

where

G^{hp} – condensate flow rate, kg/s;

r – enthalpy of vaporization, J/kg [22];

C_P – isobaric heat capacity, J/kg K;

t_{in}^{hp} – tubeside inlet temperature, K;

t_{out}^{hp} – tubeside outlet temperature, K;

F – coil surface area, m^2;

$\Delta\bar{t}$ – logarithmic mean temperature difference, K, which can be founded using following equation [23]:

$$\Delta\bar{t} = \frac{\left(t_{in}^{hp} - t_{in}^{lp}\right) - \left(t_{out}^{hp} - t_{out}^{lp}\right)}{\ln\frac{\left(t_{in}^{hp} - t_{in}^{lp}\right)}{\left(t_{out}^{hp} - t_{out}^{lp}\right)}}$$

where:

t_{in}^{lp} – shellside inlet temperature, K;

t_{out}^{lp} – shellside outlet temperature, K.

The theoretical linear heat transfer coefficient K_{theor}, W/m^{2}K can be obtained from the relationship:

$$K_{Theor} = \frac{1}{\frac{1}{\alpha_1 d_1} + \frac{1}{2\lambda_{shell}} \ln\frac{d_2}{d_1} + \frac{1}{\alpha_2 d_2}}$$

where:

d_1 – external diameter of the coil, m;

d_2 – internal diameter of the coil, m;

α_1 – heat transfer coefficient of the pipe wall (in case the flow is crossing the heating pipes bundle), W/m^{2}K.

The coefficient α_1 can be computed from the following equation:

$$\alpha_1 = 0.2\ \mathrm{Re}^{0.64}\ \mathrm{Pr}^{0.35}$$

where Re – the Reynolds number which can be computed from the following relationship:

$$\mathrm{Re} = \frac{\bar{v} d_1}{\nu}$$

where $\bar{v}$ – average velocity of the HP flow, m/s;

d_1 – external diameter of the pipe, m;

ν – kinematic viscosity of the fluid, m^2/s;

Pr – the Prandtl number, given as:

$$\mathrm{Pr} = \frac{C_p \mu}{\lambda}$$

where:

μ – dynamic viscosity, Pa s ;

λ – thermal conductivity, W/m K.

In case condensation occurs inside heating pipes, the heat transfer coefficient α_2, W/m^{2}K can be determined using the equation $\alpha_2 = 0{,}02\, A\, C_z\, q^{-0{,}5}\, l^{0{,}3}\, d_2^{0{,}2}$.

3 Results

1. The empirical heat transfer coefficient K_{emp} , W/m^2 K was estimated while the steam mass flow of both: HP and LP was changed in a wide range. The result of the estimation can be seen from the Fig. 2.

As it can be noted from the Fig. 2, the heat transfer coefficient rises steadily with increase of steam flow mass rate. Though the heat transfer coefficients of the 10-coils model and that 2-coil model follow similar pattern, the gap between coefficients of models is ~30%.

A relationship between the heat transfer coefficient, estimated experimentally, and the flow mass rate of the 10-coil model can be expressed using following equation:

$$K_{emp} = 4{,}546(\gamma W)^2 + 76{,}411.$$

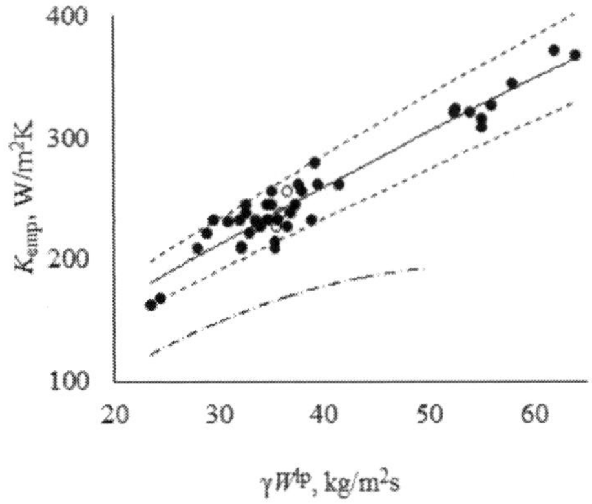

Fig. 2. Empirical heat transfer coefficient by the flow mass rate of low pressure steam γW:
■ – experiment with pure vapor;
□ – experiment with mixture of vapor and nitrogen;
▬ – approximation (10-coils model);
▬ · ▪ – approximation (2-coils model)..

Analyzing the equation (1) it can be noted that the heat transfer coefficient rises steadily with increase of steam flow mass rate. Though the heat transfer coefficients of the 10-coils model and that 2-coil model follow similar pattern, the gap between coefficients of models is ~30%.

2. It is observed a 18% difference between theoretical and empirical values of the heat transfer coefficient as shown at the Fig. 3.

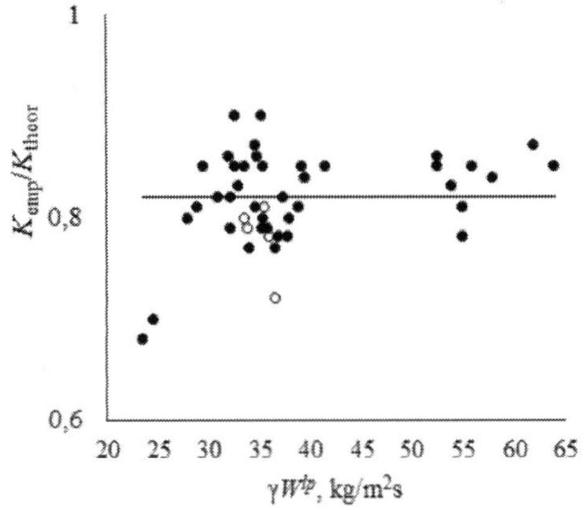

Fig. 3. The empirical to theoretical heat transfer coefficient ratio by the low pressure steam mass flow rate:
■ - experiment with pure vapor;
□ - experiment with mixture of vapor and nitrogen;
▬ – approximation (10-coils model).

3. Observing Fig. 2 and Fig. 3 it can be seen that results of experiment with pure steam and that with the nitrogen does not differ significantly.

4 Discussion

A remarkable deviation between heat transfer coefficients of 10-coil model and that of 2-coil model, which can be seen at the Fig. 2, caused by:

- better precision while 10-coil model manufacturing;
- more steady flow which multicoil bundle assure by contrast to that of 2-coil model.

As it can be inferred from the Fig. 2, there is a positive relationship between the empirical heat transfer coefficient and HP mass flow rate value. The densest dots cluster on the left of the figure shows boundaries of the nominal mode. The cluster center indicates the coefficient value in nominal mode which is $K_{emp} = 250$ W/m^2 K when $\gamma W^{\text{ap}} = 37{,}5$ kg/m^2.

Theoretical values of linear heat transfer coefficient from inner side are larger than that from outer side. Thus, overall heat transfer coefficient of the model mainly depends from factors, affecting condensation such as: position coils in model's space, medium velocity, inner and outer diameter of the pipe; overall length of coil and the number of loops, inner surface roughness and other [24].

When experiment was completed, empirical heat transfer coefficients were benchmarked with its theoretically computed values. As it can be seen from the Fig. 3, empirical values lag behind theoretical because of:

- presence of baffles, partly obscuring coil surface but needed for coils fixation;
- poor heat exchange of coils, that located the top and in the bottom of the bundle, due to imperfect steam distribution which is determined by the shell and the plate geometry.

Absence of impact of incondensable gas on the heat transfer coefficient, which can be observed at the Fig. 2 and Fig. 3, can be explained by high pressure of HP steam. This finding matches with [25] though discharge of incondensable gases from the model is suggested in [16].

5 Conclusion

In result of the study it was obtained:

- the multicoil model may work stably within wide range of parameters and modes, which corresponding to that of moisture separator and reheater MSR, installed at the Leningrad NPP;
- 3D-model of the experimental unit;
- the empirical heat transfer coefficient lags behind theoretical. In nominal mode its value is 250 W/m^2 K;
- incondensable gas does not hinder significantly the heat transfer from the coil surface.

References

[1] G. Li, Ch. Yan, J. Wang, Annals of Nuclear Energy **73** (2014)

[2] Y. Liu, Zh. Qu, Energy Procedia **105** (2017)

[3] W. Xinjun, Ch. Peng, X. Jinwei, Zh. Zijie, Journal of Mechanical Science and Technology **29**, 8 (2015)

[4] V. Bezlepkin, I. Kukhtevich, S. Svetlov, S. Alekseev, A. Solodovnikov, A. Molchanov, Y. Krylov, A. Blagoveshchenskij, O. Krektunov, Tyazheloe Mashinostroenie **8** (2004)

[5] A. Blagoveshchenskii, S. Bor, M. Konovich, V. Mityukov, V. Sokolov, B. Shumskii, S. Vygovskii, A. Pinegin, L. Bogachek, V. Bai, N. Ignat'ev, Thermal Engineering **51**, 2 (2004)

[6] N.I. Kolev, *Multiphase Flow Dynamics 5* (Berlin, Springer, 2011)

[7] L. Liu, B. Bai, Nuclear Engineering and Design **298** (2016)

[8] Zh. Huang, L. Qianfeng, Q. Benke, B. Hanliang, Ch. Feng, Annals of Nuclear Energy **92** (2016)

[9] V. Legkostupova, A. Sudakov, St. Petersburg Polytechnic University Journal of Engineering Science and Technology **23**, 3 (2017)

[10] N. Agafonova, I. Paramonova, Thermal Engineering **60**, 3 (2013)

[11] S. Alekseev, Y. Ilyukhin, V. Kukhtevich, I. Paramonova, S. Svetlov, V. Sidorov, High Temperature **37**, 4 (1999)

[12] N. Agafonova, I. Paramonova, Thermal Engineering **61**, 8 (2014)

[13] N.D. Agafonova, I.L. Paramonova, Journal of Engineering Physics and Thermophysics **89**, 4 (2016)

[14] M. Gorpinyak, A. Solodov, Thermal Engineering **66**, 6 (2019)

[15] G. Bychkova, V. Melikhov, O. Melikhov, A. Bel'skii, Thermal Engineering **52**, 3 (2005)

[16] V. Sharapov, M. Malikov, Power Eng.: Research, Equipment, Tech., 5–6 (2007)

[17] S. Sapozhnikov, V. Mityakov, A. Mityakov, A. Babich, E. Zainullina, Technical Physics Letters **45**, **4** (2019)

[18] S. Sapozhnikov, V. Mityakov, V. Seroshtanov, A. Gusakov, J. Phys.: Conf. Ser. **1421**, 012064 (2019)

[19] G. Lepesh, A. Lepesh, S. Luneva, J. Technico-tehnologicheskie problemy servisa**4**, 38 (2016)

[20] V. Legkostupova, A. Sudakov, St. Petersburg Polytechnic Univ. J. of Eng. Sci. and Tech. **4**, 254 (2016)

[21] V. Legkostupova, A. Sudakov, Thermal Eng. **62**, 3 (2015)

[22] A. Alexandrov, B. Grogor'ev, *Tables of thermophysical properties of water and steam* (Moscow, MEI, 1999)

[23] P.L. Kirillov, A.V. Zhukov, N.I. Loginov, *Reference book on thermohydraulic calculations in nuclear power industry* (Moscow, IzdAT, 2013)

[24] D.I. Volkov, Energomashinostroenie **6** (1969)

[25] A. Tugov, E. Ivanov, Y. Radin, R. Shuvalov, V. Guseva, Thermal Engineering **53**, 7 (2006)

Improving the reliability of electric motors during startup using electroosmotic drying of insulation

Alexandr Nemirovskiy[1], *Irina* Sergievskaya[1], *Regina* Salikhova[2] and *Alexey* Udaratin[1]

[1]Vologda State University, Department of Electrical Equipment, Vologda, Russia
[2]Kazan State Power Engineering University, Kazan, Russia

Abstract. The article considers the actual problem of electric motors (EM) operation associated with wetting of winding insulation. The condition of wet insulation is getting worse due to aggressive environment of premises of agro-industrial complex, and some workshops in industry. It is noted that together with humidification, the effect of insulation "steaming" occurs after EM is put into operation. In this case, the insulation resistance R sharply decreases and its breakdown may occur, after which a major overhaul of EM is required. A new method for EM drying based on the electrokinetic phenomenon of electroosmosis has been developed. Under the influence of electric field forces, moisture is extruded from insulation capillaries onto the surface of EM winding and stator iron. R increases, and EM becomes ready to turn on. Under further operation, EM is heated and moisture evaporates into the environment. It is shown how electroosmotic drying (EOD) and the developed device (DEOD) help to get rid of the effect of insulation "steaming" and thereby ensure trouble-free EM operation.

1 Introduction

Currently, one of the topical issues of electric motors (EM) operation is wetting of windings insulation. The number of EM failures associated with insulation wetting reaches 20–30% [1,2,3].

The aggressive environments of livestock complexes in agricultural and individual workshops in industrial production further deteriorate the condition of wet insulation. Hydrogen sulfide, carbon dioxide, ammonia and other gases form acids and alkali when combined with water. Conductive bridges appear in insulation, which reduce the insulation resistance R [4].

Moisture changes the dielectric properties of insulation, and worsens almost all its electrical characteristics: dielectric strength, resistance R, dielectric loss tangent, etc. Breakdown can occur not only through the insulation thickness, but also on a wet surface. Moisture causes chemical changes: hydrolysis of insulation and formation of molds [5].

Humidified EM in an aggressive environment is a galvanic cell. One of the plates of this element is a copper winding, the other is stator iron. A mixture of various acids formed in a humid aggressive environment acts as electrolyte. Hydrogen sulfide and carbonic acids are obtained by interaction of moisture with hydrogen sulfide and carbon dioxide, other acids can be formed by hydrolysis and decomposition of insulation in water. In this case, chemical reactions occur that are typical for a galvanic pair, and the copper wire of winding is destroyed. The latter circumstance leads to a break in windings and turn-to-turn short-circuit.

Humidification leads to the emergence of the insulation "steaming" effect after EM setting in operation [6]. EM switching-on is accompanied by heating of wet insulation. Moisture in pores and capillaries expands and additionally shunts the dielectric. The insulation resistance sharply drops, and a breakdown may occur. The smaller is R before EM turning on, the stronger the "steaming" effect is, and the more probable the EM failure is. If the accident is avoided, then with further EM operation the insulation will dry up.

Thus, EM must be periodically dried. Today, there are a significant number of different ways of EM drying. As a rule, they all are thermal: light bulbs, ovens, blowers, welding transformers, etc. [6,7].

Thermal methods of EM drying can cause warping, thermal degradation, local overheating and thermal aging of insulation. These methods are implemented with dismantling and disassembly of EM at the operation fields. They require special equipment, significant labor costs and increased energy consumption [8-11]. In this regard, there is a need to develop and apply new drying methods that would not have disadvantages of thermal drying.

A new method of drying EM has been developed at the Department of Electrical Equipment of the Vologda State University (VoGU), based on the electrokinetic phenomenon of electroosmosis [12-15]. The essence of electroosmosis is the motion of fluid in capillary systems under the influence of electric field [13]. The materials used for EM windings insulation are capillary systems, therefore, the electroosmotic effect of moisture removal can be realized in EM insulation [17].

Electro-osmotic drying (EOD) is cold and does not heat the EM insulation. The energy consumption during EOD is 2 to 3 orders of magnitude less than that of traditional thermal drying methods. EOD does not require to disassemble and dismantle EM from its workplace, which reduces labor costs by 2 to 3 times [13-15].

The purpose of research is to effectively use EOD not only for EM drying, but also for eliminating the insulation "steaming" effect when EM is switched on.

2 Materials and methods

The serial AIR electric motors of common industrial use of 1.5 kW power were used as experimental samples. The insulation system is as follows: groove box and groove wedge are made of Laviterm, phase-to-phase gaskets are made of film-asbestos millboard, winding wire is made of PETV-2.

Similar EM samples with insulation system additionally encapsulated with organosiloxane rubber to enhance moisture resistance were also used in studies [6].

For statistical reliability, three EM were used in each experiment. In order to smooth the insulation "steaming" effect, an EOD device (DEOD) developed and manufactured at VoGU was used [15]. Measurements of insulation resistance R were made using the F4101 megaometer. A positive plus from DEOD output was connected to EM windings connected in a star, and the negative was connected to the case. Preliminarily, the EM windings were moistened in a KTV-0.4 moisture chamber at a temperature of T = 20 ± 20°C and a relative air humidity of ω = 97 ± 3% without moisture condensation for 96 hours. To speed up the humidification process, the EM bearing shields were removed.

3 Results and discussion

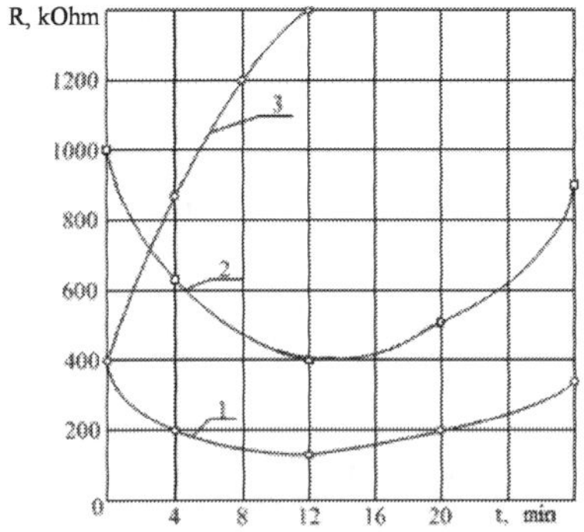

Fig. 1. Relationship between insulation resistance and operation duration of wet electric motors after startup: 1,2 - conventional startup of electric motors of general industrial design and encapsulated ones; 3 - startup of electric motors of general industrial design after preliminary processing of insulation by electric field from EOD device.

Figure 1 shows the curves of change in insulation R after switching on the electric motor. Curves 1,2 correspond to EM insulation, which were not pre-treated by electric field before startup (DEOD was not used). Curves 1, 2 show a rather strong manifestation of the "steaming" effect: the insulation resistance of non-encapsulated EM decreases to almost 100 kOhm, which is significantly less than the allowable 500-1000 kOhm. Obtaining experimental data for curve 1 was accompanied by such a pronounced "steaming" effect that one of EMs failed due to a short circuit in the groove part of insulation. The insulation R of this EM apparently decreased below 100 kOhm, which could not be recorded during measurements. In encapsulated EMs, the "steaming" effect is less pronounced (curve 2), which can be attributed to a high initial resistance and a more moisture-resistant electrical insulation design. Curve 3 presents the EM startup without insulation "steaming". The insulation system of this EM batch was characterized by initial R = 250 kOhm and was subjected to a 15-minute exposure to electric field from DEOD before starting. As a result, R increased to 400 kOhm. After switching on the EM in the network, after 12 minutes, the resistance reached 1.5 MOhm due to self-feeding effect, and the increase in R occurred continuously without jumps. Thus, "steaming" was excluded.

To explain the experimental results (curves 1, 2, 3 in Fig. 1) we use the theory of contact drying of capillary-porous bodies [16].

Figs. 2 (a), 3 (a) show idealized versions of the structure of EM insulation dielectric with an extensive network of microcapillaries partially filled with water (1-1′, 2-2′ – through capillaries). When EM is switched on, the current-carrying part is heated, the moisture in contact with it partially evaporates and condenses on the walls of capillaries, forming conductive bridges (Fig. 2, b). Heating of air in microcavities leads to its expansion and causes moisture motion in capillaries. Water is redistributed and another possible option for appearance of a conductive bridge is shown in Fig. 3 (b). Thus, "steaming" is associated with appearance of shunting fluid bridges. Redistribution of moisture, and consequently, a decrease in R, is observed until the insulation warms up throughout the volume and evaporation of moisture from the surface begins. From this moment an increase in R begins.

Physical processes occurring in the insulation system of EM during EOD enable quite simple and reliable eliminating the undesirable effect of "steaming". For this, immediately before startup, voltage from DEOD is supplied to EM insulation system (plus to the windings, minus to the case). Exposure by external electric field causes water to move toward the negative electrode and frees the near-anode regions of capillaries, which are located near the current-carrying parts of EM. Fig. 4 (a) shows the EM insulation dielectric in humidified condition before DEOD connection. The same dielectric after exposure to electric field is shown in Fig. 4 (b). The near-anode regions are dehydrated, there is no contact of water with current-carrying parts in microcapillaries,

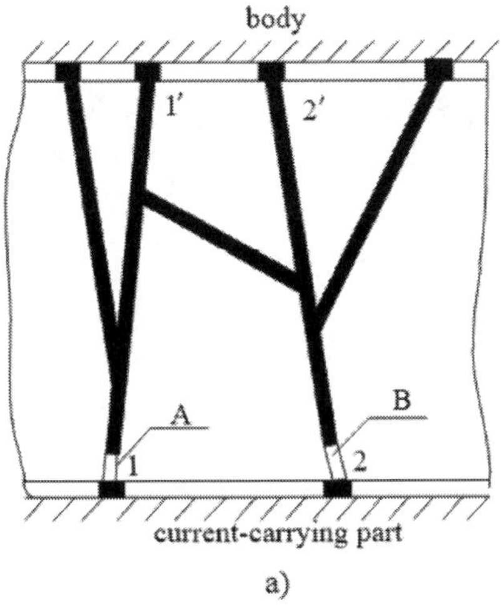

Fig. 2. Structure of a humidified dielectric of electric motor insulation (1st option): a) over- humidified condition; b) moisture condensation on the walls of capillaries after starting the electric motor; A, B are air inclusions.

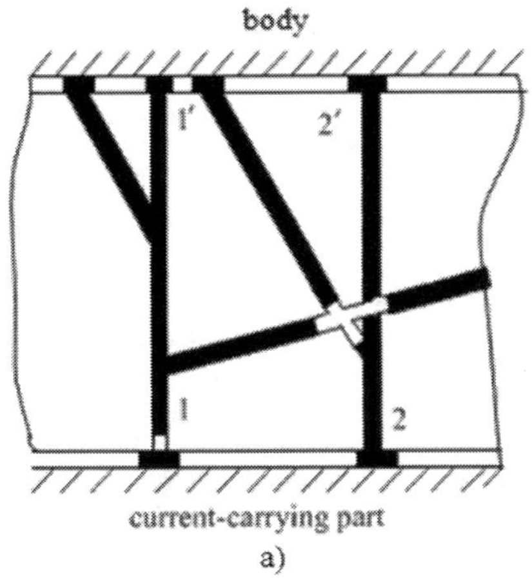

Fig. 3. Structure of a humidified dielectric of electric motor insulation (2nd option): a) over- humidified condition g; b) redistribution of moisture in capillaries after starting the electric motor.

there will be no water evaporation and through conductive bridges are not formed. After EM switching-on, the heat released in windings heats the air in the near-anode regions. The volume of air inclusions that appeared after DEOD connection (Fig. 4, b) increases when heated, and continues to displace moisture from the center to the periphery, resistance of the EM insulation system increases.

It should be noted that after DEOD disconnecting, moisture in EM insulation system can be redistributed. As a result, the initial unfavorable state of insulation can be restored (Fig. 2,3). Therefore, to reliably eliminate "steaming", one needs to turn on EM immediately after DEOD disconnecting. If EM insulation is dried more than enough up to 1.5-2 MOhm, then one should not be aware of "steaming" and there is no requirement for EM switching-on immediately after drying. The method of EM starting without insulation "steaming" expands the practical use of electroosmosis and DEOD to increase the operational reliability of EM.

Implementation of EOD as part of the developed devices [13,15] into the practice of EM operation allowed us not only to eliminate the disadvantages of traditional thermal drying methods, but also to remove the effect of insulation "steaming". Consequently, this very significantly increased the level of accident-free operation of EM.

4 Conclusions

1. The developed new method for cold drying of EM winding insulation excludes thermal degradation,

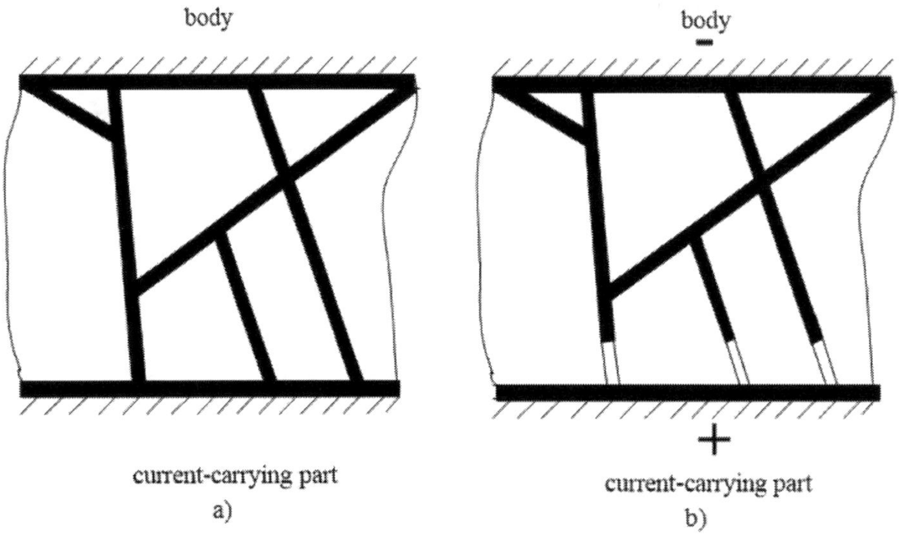

Fig. 4. To the issue of "steaming" of an electric motor insulation: a) insulation condition before DEOD connection; b) the same after connection of DEOD, which eliminates "steaming".

reduces power consumption by 2–3 orders of magnitude and labor costs by 2–3 times, increases the EM operational reliability by eliminating the "steaming" effect during startup.

2. Connecting the humidified EM to DEOD for 15 minutes will eliminate the "steaming" effect and possible breakdown of EM insulation during startup. A further increase in insulation resistance occurs due to self-drying by EM heating during startup and operation.

3. In order to reliably eliminate the insulation "steaming" effect, it is necessary to turn on EM immediately after DEOD disconnecting. If EM insulation is dried more than enough up to 1.5-2 MOhm, then one should not be aware of "steaming" and there is no requirements for EM switching-on immediately after drying.

References

[1] A.E. Nemirovsky, G.A. Kichigina, I.Yu. Sergievskaya, Calculation of heating of electric motor windings during electroosmotic drying of insulation and low temperature,s Fedorov Readings 2017: materials of the XLVII International Scientific and Practical Conference with elements of scientific school, MPEI, Moscow (2017).

[2] A. Nemirovskiy, A. Kashin, V. Kosmach, Y. Titovec, I. Toptygin, D. Zaripova, Innovative technology for dismantling the windings of electric motors using ultrasonic radiation, IOP Conference Series: Earth and Environmental Science 337, 012071 (2019).

[3] M. Rashevskaya, S. Yanchenko, S. Tsyruk, B. Kudrin, Assessing non-stationary power quality phenomena of induction motors, Proceedings of 18th International Conference on Computational Problems of Electrical Engineering, CPEE (2017).

[4] S.A. Mikoyan, Justification of parameters of convective-electroosmotic drying of electric motors and startup protection equipment in agricultural production, thesis for the degree of candidate of technical sciences (SSAU, St. Petersburg-Pushkin, 1991).

[5] V.V. Maslov, Moisture resistance of electrical insulation (Energy, Moscow, 1973).

[6] A.E. Nemirovsky, Improving the efficiency of drying and moisture protection of insulation of electric motors used in agriculture, based on intensification of electroosmotic phenomena, thesis for the degree of doctor of technical sciences (SSAU, St. Petersburg-Pushkin, 1993).

[7] O.I. Khomutov, System of technical means and measures to increase the operational reliability of insulation of electric motors used in agricultural production, thesis for the degree of doctor of technical sciences (CHIMESH, Chelyabinsk, 1990).

[8] A.E. Nemirovsky, G.A. Kichigina, I.Yu. Sergievskaya, Electroosmotic drying and moisture protection of electrical equipment, Fedorov readings 2017: materials of the XLVII International Scientific and Practical Conference with elements of scientific school, MPEI, Moscow (2017).

[9] A. Udaratin, A. Alyunov, A. Krutikov, L.R. Mukhametova, O.O. Zaripov, I.V. Bochkarev, Efficiency study of the reactive shunt compensation device in power lines, E3S Web of Conferences 124, 02020 (2019).

[10] A.N. Alyunov, O.S. Vyatkina, L.R. Mukhametova, V.P. Markovskiy, R.M. Mustafina, Automation of calculations of power transformers windings parameters taking into account the position of

the voltage regulator, E3S Web of Conferences 124, 02016 (2019).

[11] A.N. Alyunov, O.S. Vyatkina, I.G. Akhmetova, R.D. Pentiuc, K.E. Sakipov, Issues on optimization of operating modes of power transformers, E3S Web of Conferences 124, 02015 (2019).

[12] A.E. Nemirovsky, N.K. Moroz, K.P. Simakov, The method of electroosmotic drying of windings insulation of electrical machines, Invention patent, 2250550, bull. no. 11 (2005).

[13] S.V. Kirillov, Electroosmotic drying of transformers, Lokomotiv 1 (2014).

[14] A.E. Nemirovsky, I.Yu. Sergievskaya, Experience in the operation of moisture protection devices for windings insulation of electrical equipment at oil and gas enterprises, Actual problems of the state and development of the oil and gas complex of Russia: 7th All-Russian Scientific and Technical Conference, Gubkin Russian State University of Oil and Gas (2007).

[15] A.E. Nemirovsky, I.Yu. Sergievskaya, Experience in the operation of electroosmotic drying devices for windings insulation of electrical equipment at oil and gas enterprises, Actual problems of the state and development of the oil and gas complex of Russia: 7th All-Russian Scientific and Technical Conference, Gubkin Russian State University of Oil and Gas (2007).

[16] A.V. Lykov, Theory of Drying (Moscow, Energy, 1968).

[17] A.Ya. Fanin, Fluid motion in porous materials under the influence of electroosmosis, Issues on automation and mining electrical engineering: Collection of papers, Don. UGI, no. 40 (Nedra, Moscow, 1969).

Increasing the productivity of a plant for moisture evaporation from oil sludge by application of efficient method for technological parameters calculation

Alisa Surikova[1,*], *Elena* Sytsyanko[1], *Ekaterina* Solov'eva[1] and *Liliya* Mukhametova[2]

[1]Vologda State University, Lenina str. 15, 160000, Russia
[2]Kazan State Power Engineering University, Kazan, Russia

Abstract. The paper presents the method of heat calculation of recovery stages for the plant of moisture evaporation from oil sludge. It enables to increase the plant power and reduce fuel consumption per one unit of evaporated moisture in heating boiler.

1 Introduction

At the modern stage of chemistry and petrochemistry development, the processes of management of waste, including oil and sludge, are under strict complex requirements. These include reducing energy and resource consumption, increasing their technical and economic efficiency, ensuring environmental safety according to the world standards [1,2]. The existing technologies for oil sludge disposal often do not respond to the stated requirements, and have a negative impact on the environment by their own waste, which have to be buried.

Over the past decades, various technological solutions for disposal of oil sludge have been proposed. However, despite the achieved results, these technologies need significant improvement due to annual enforcement of laws related to environmental protection and increasing requirements for their performance.

The issues of processing and using the resource potential of chemical, petrochemical, oil production waste and heavy oil residues (oil sludge) and assessing their environmental impact were considered in works of many Soviet and Russian scientists, such as: R.Kh. Fassakhov, M.Yu. Malikova, N.S. Yamanina, A.A. Desyatkin, T.A. Kulagina, A.F. Nadein, L.K. Altunina, L.I. Swarovskaya, A.Yu. Evdokimov, N.T. Elovshin, I.V. Efremov, A.G. Ishkov, V.M. Ferdman, A.B. Dmitrevsky, N.L. Guryleva, V.G. Maksimovich, V.A. Argunov, O.G. Fetisov, B.E. Prusenko, G.L. Handel, N.V. Lavrov, R.V. Surzhikova, M.S. Shpinkova, T.Sh. Aksanov, G.G. Gladyshev, N.M. Ismailov and others.

But despite the obvious urgency of the problem of oil sludge disposal, such productions are still not as widespread as required. This is explained by the fact that such plants, both Russian and international, have a high cost with low productivity and efficiency. In recent times oil sludge separation units (BRNSh) are the most promising in the field of oil-containing waste recovery [3]. Such plants are based on evaporation of moisture from oil sludge waste, they also can be potentially improved and achieve higher energy efficiency.

In the present work, a method of thermal calculation of a plant for moisture evaporation from petroleum products is proposed. Using it, technological parameters are identified that significantly affect the performance and thermal efficiency of the BRNSh-3 unit (the number "3" means the volume of oil sludge (m3) loaded in plant) [4].

2 Technological scheme of BRNSh plant

To determine the potential for improving and increasing efficiency of the oil sludge separation unit (BRNSh) an efficient design scheme is proposed [3]. Using variable approaches the proposed calculation procedure allows one to introduce new technological modes and thereby increase the thermal efficiency and productivity of the considered plant.

Consider the technological scheme and operation principle of BRNSh.

Figure 1 shows the technological thermal scheme of plant for moisture evaporation from oil sludge with additional recovery heat exchangers.

The technological scheme of plant consists of the following stages: WHB is water heating boiler; ES is evaporation stage (main), heated by water from boiler; RS is recovery evaporation stage; OR is recovery heater of oil sludge; WP is vacuum water ring pump; S is separator; K is calorifer. The plant includes the following elements: circulation pump for water and oil sludge; lines of heating and cooling water, steam and condensate [5].

The plant has the following operation stages:

1. The preheated oil sludge in equal volumes is loaded into the main and recovery evaporation stages.

* Corresponding author: alisa-surikova@bk.ru

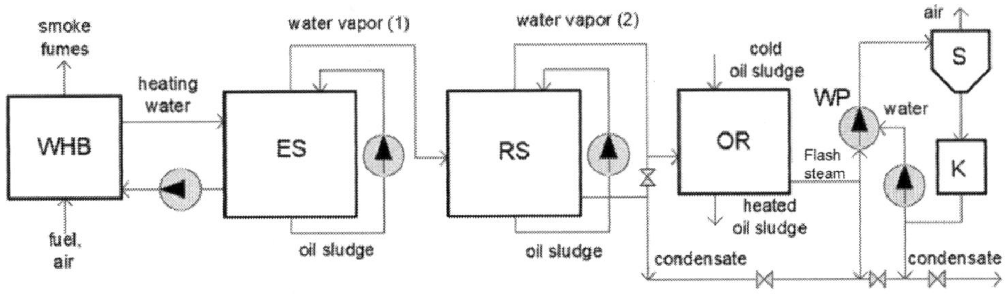

Fig. 1. Thermal scheme of plant for moisture evaporation from oil sludge (BRNSh) with recovery heat exchangers.

Oil sludge is heated to reduce the time of water evaporation at the main stage.

2. After the plant gets into action, the process of evaporation of moisture from oil sludge starts. The primary steam directed to heat the recovery steam stage (RS) is condensed inside the heating pipes. In this case, the condensation heat is spent to moisture evaporation in RS and to cover the heat losses.

3. The secondary steam leaving RS is sent to heat the recovery heater OR, where the unheated oil sludge is located.

4. The condensate of heating steam, which goes out of recovery stage and heater, is discharged into a container, from which it is drained as necessary.

5. Non-condensed steam and non-condensing gases from the RS recovery stage are sent through the valve through the secondary steam line to the OR.

6. The steam from OR recovery stage is sent to the WP vacuum pump, where it condenses under the influence of cooling water.

7. The resulting mixture of water and gases is sent to separator C having a discharge into the environment.

8. After removing gases from separator, the remaining heated water from separator C is fed to the heater K, where it is cooled and then sent to a vacuum pump.

It is advisable to compose the heating surface RS from horizontally arranged pipes with a slight slope for condensate drainage. Inside the pipes, the heating steam condenses, and oil sludge is heated using free convection [6].

Further operation of the heat-recovery evaporation stage CP is considered in more detail (Fig. 2).

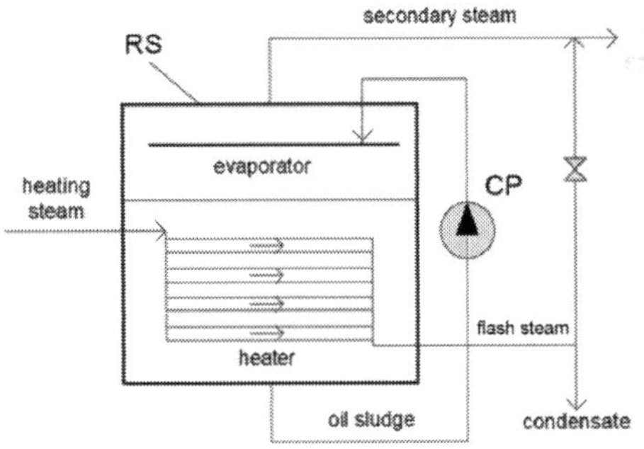

Fig. 2. Schematic diagram of the heat-recovery evaporation stage.

Heated oil sludge is fed into the CP tank. After evaporation process starts in the main evaporation stage, the primary steam leaving it with temperature t1 and pressure p1 enters the heating pipes of heat-recovery stage, where it condenses at about the same temperature and pressure. Some reduction in steam pressure due to hydraulic losses is not taken into account. After heating of recovery stage begins, the oil sludge is heated to the evaporation temperature t2, which is determined by pressure p2 in the tank of recovery stage, after which the process of evaporation of moisture from the oil sludge begins.

Evaporation in the main and heat-recovery stages, connected in series along the steam flow, takes place under vacuum created by the vacuum ring water pump WP. The heated oil sludge is taken from the lower part of tank of evaporation stages by central circulation pumps and fed to evaporator shelves located in the upper part of tank, which serve to increase the mass transfer surface of oil sludge. With partial evaporation of moisture on the shelves of evaporator, the oil sludge is cooled by several degrees, and flows back to the lower part of the stage tanks [7].

 https://doi.org/10.1051/e3sconf/202017801071

3 Operation principle of the plant

To introduce the necessary parameters into the calculation model, it is necessary to consider the plant operating modes, when evaporation stages operate as evaporators.

Denote the absolute pressure at evaporator stages by p_1 and p_2. The corresponding temperatures of steam formed in them are t_1, and t_2. The maximum temperature head between the heating steam and oil sludge in the recovery stage is $\Delta t_{max}=t_1-t_2$ [8].

The average temperature head is $\Delta t=\zeta\cdot\Delta t_{max}$. Here the coefficient $\zeta \le 1$ depends on hydraulic losses during heating steam movement and on flow rate of oil sludge pumped by the circulation pump in the recovery stage and affecting the average oil sludge temperature in the tank [9].

The smaller the hydraulic losses are and the greater the oil sludge flow rate is, the closer is the coefficient ζ to unity. To minimize hydraulic losses, it is necessary to use a circuit with parallel steam motion inside the heating pipes.

Steam flow rates G_1, G_2 (kg/s) are generated in evaporator stages in the set operation mode, and steam flow rate G_1 is pre-set according to the procedure [3]. Assume heat of condensation (evaporation) of water steam r_1 and r_2 at pressures p_1 and p_2 are in pressure range of 0.006÷0.06 MPa at condensation temperatures of 36.2÷85.9 °C. The plant condensation heat varies from 416 to 2293 kJ/kg.

Suppose $Q_1=G_1\cdot r_1$ is the heat flux (W), which is released during the complete condensation of the primary steam with the flow rate G_1 inside the pipes of the recovery stage heater; $Q_2=G_2\cdot r_2\ r_1$ is the heat flux (W), which is spent for formation of secondary steam G_2, and is released during its condensation:

The heat balance of the recovery stage is written in the form:

$$Q_2=\eta\cdot Q_1, \qquad (1)$$

where $\eta \le 1$ is the thermal efficiency of recovery stage. Thermal insulation of the outer surface of the stage housing provides a heat balance with efficiency of $\eta \ge 0.99$.

If coefficients $\eta=1$ and r=const, then for the set mode the condition is fulfilled: $Q_1=Q_2$; $G_1=G_2$ and the total efficiency of plant is determined by the formula $G_\Sigma=2\cdot G_1$. In this case the productivity will increase by about 2 times, the fuel consumption burned in boiler does not change, and the specific fuel consumption (per 1 kg of evaporated moisture) will decrease by 2 times.

The heat flux through the heater RS surface is equal to:

$$Q_1=\frac{Q_2}{\eta}=k\cdot\Delta t\cdot F=k\cdot\zeta\cdot\Delta t_{max} \qquad (2)$$

where k is the heat transfer coefficient from heating steam to oil sludge, (W/(m²·K));

Δt and Δt_{max} are the average and maximum temperature heads, (°C) ($\Delta t_{max}=t_1-t_2$);

ζ is the coefficient which relates Δt and t_{max} ($\Delta t=\zeta\cdot t_{max}$);

F is the surface area of heater determined from outer pipe diameter, (m²).

If we neglect the hydraulic resistance during the heating steam motion and the change in condensation temperature, then the coefficient ζ can be calculated by the expression:

$$\zeta=\left(1+\frac{0.5\cdot k\cdot F}{W}\right)^{-1},$$

where $W=c_o\cdot G_o$; c_o is the mass heat capacity of oil sludge , (J/(kg·K));

G_o is the mass flow rate of oil sludge pumped by circulation pump (kg/s).

The latter formula shows that with an infinite increase in W, the coefficient ζ tends to unity, which will require infinite energy expenditures for pumping oil sludge [10].

Figure 3 shows the relationship between coefficient ζ and $\omega=W/(k\cdot F)$.

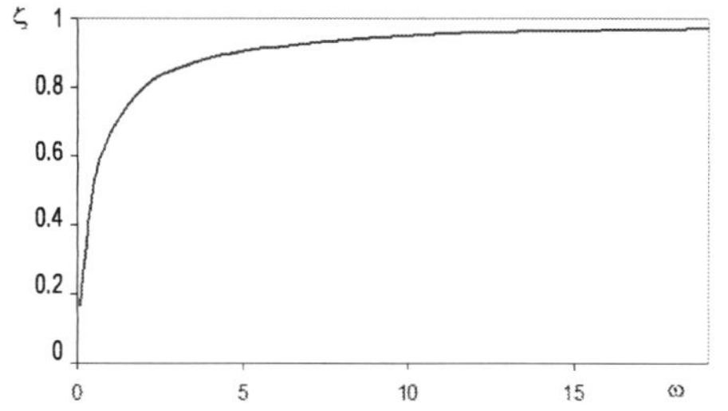

Fig. 3. Relationship between coefficient ζ and $\omega=W/(k\cdot F)$.

Fig.3 shows that for $\omega=5$ ($W=5\cdot k\cdot F$) the coefficient $\zeta \approx 0.9$. When ω is increased from 5 to 10, the coefficient ζ increases from 0.9 till 0.95, i.e. for only 5.5 %, while electricity consumption N for pumping oil sludge increases by about 2^{3}=8 times (as $N \sim W^3$). When ω is decreased from 5 to 2 the coefficient ζ decreases from 0.9 to 0.8, i.e. by 12.5%, while electricity consumption for pumping oil sludge decreases by 15.6 times. Decreasing of ζ requires the corresponding increase in heater surface, i.e. its price. Optimal values of ω and ζ should be determined from technical and economic estimates. One can assume that reasonable value of ζ is in the range 0.8÷0.95, which corresponds to ω=2÷10; and reasonable flow rate of oil sludge is determined using the formula $G_o=(2\div10)\cdot k\cdot F/c_o$[11].

The heat transfer coefficient from heating steam to oil sludge k in the recovery stage RS is determined by expression, as for a flat wall [4]:

$$k = \frac{1}{\frac{1}{\alpha_1} + \frac{\delta_c}{\lambda_c} + \frac{1}{\alpha_2}}, \quad (3)$$

where α_1 is the heat exchange coefficient from condensing steam to the inner surface of pipes;

α_2 is the heat exchange coefficient from outer surface of pipes to oil sludge;

λ_{pw} is thermal conductivity coefficient of pipe wall;

δ_{pw} is thickness of pipe walls.

Note that the thermal resistance of heat exchange on the inner surface of pipe $1/\alpha_1$ and pipe wall δ_{pw}/λ_{pw} is many times less than the thermal resistance of heat exchange from the outer surface of pipe $1/\alpha_2$. Therefore, in a first approximation, we can take $k \approx \alpha_2$. In this case, the average temperatures on the outer and inner surfaces of pipes in the first approximation are equal to $t_{pw2} \approx t_{pw1}=t_1 - \delta t_w$, where t_1 is the temperature of condensation (saturation) of the primary steam; δt_w value is a fraction of a degree.

Since coefficient α_1 is approximately two orders of magnitude higher than coefficient α_2 and weakly affects the heat transfer coefficient k, so the minimal heat exchange coefficient α_1 can be determined by the Nusselt formula for calculating heat exchange during condensation of stationary steam inside horizontal pipes [12]:

$$\alpha_1 = 0.65 \cdot \left(\frac{\lambda_w^3 \cdot \rho_w^2 \cdot g \cdot r}{\mu_w \cdot \delta t_w \cdot d_{in}} \right)^{0.25}, \quad (4)$$

where thermo-physical parameters of condensate (water) λ_w, ρ_w, μ_w are chosen for the condensation temperature

t_1 (λ_w) is thermal conductivity coefficient;

ρ_w is density;

μ_w is dynamic coefficient of water viscosity;

r is heat of steam condensation at temperature t_s;

$\delta t_w = t_1 - t_{pw1}$ is the difference between temperatures of steam condensation and the inner surface of heating pipe; d_{in} is the pipe inner diameter.

All values in (4) are presented in SI units.

The heat exchange coefficient to oil sludge α_2 can be calculated using the well-known formula for calculating heat exchange for free convection of liquid near horizontal pipes [13]:

$$\alpha_2 = 0.5 \cdot \lambda_o \left(\frac{g \cdot \beta_o \cdot \delta t_o}{\nu_o^2 \cdot d_{out}} \right)^{0.25} \cdot \left(\frac{\Pr_o}{\Pr_{c2}} \right)^{0.25} \quad (5)$$

where λ_o, β_o, ν_o are coefficients of thermal conductivity, thermal volumetric expansion, kinematic viscosity of oil sludge for its average temperature t_o;

d_{out} is the outer pipe diameter;

Pr_o, Pr_{pw2} are the Prandtl numbers of oil sludge at temperatures t_o and t_{pw2};

$\delta t_o = t_{pw2} - t_o$ is the difference between temperatures of outer surface of pipes and average temperature of oil sludge.

The average temperature of oil sludge at the recovery stage is determined by the expression

$$t_o = t_2 + \Delta t_{\max} (1 - \zeta) = t_1 - \zeta \cdot (t_1 - t_2).$$

Temperature t_o is approximately 1÷2 °C more than evaporation temperature t_2.

The thermal conductivity and viscosity coefficients, as the main thermo-physical properties of oil sludge, affect the coefficient α_2 and k to a large extent. Thermo-physical properties of oil sludge, affecting heat exchange, are weakly dependent on moisture content, if it is less than 30%.

Coefficient α_2 is significantly affected by temperature difference δt_o. It is known that δt_o is equal to the average temperature head of the considered stage: $\delta t_o \approx \Delta t = t_1 - t_o$. Expression (5) at $k \approx \alpha_2 \sim \Delta t^{0.25}$ and density of heat flux from heating flux to $q = k \cdot \Delta t$ shows that q is proportional to $t^{1.25}$.

The values of k and q are determined as follows: assume $\delta t_o \approx \Delta t$, using expression (5) the coefficient α_2 is calculated in the first approximation, so the heat flux density is $q \approx \alpha_2 \cdot \Delta t$. Since $q = \alpha_1 \cdot \delta t_w$, the heat flux on the inner surface of pipe δt_w is found from equation (4):

$$\delta t_w = \left(\frac{q}{0.65} \right)^{\frac{4}{3}} \left(\frac{\mu_w \cdot d_{in}}{\lambda_w^3 \cdot \rho_w^2 \cdot g \cdot r} \right)^{\frac{1}{3}}. \quad (6)$$

From the known δt_w, the heat exchange coefficient α_1 on the inner surface of pipes is calculated from expression (4). Further, expression (3) is used to defined k more accurately and to determine heat flux density $q = k \cdot \Delta t$. The value δt_o is found from expression: $\delta t_o = \Delta t - q(1/\alpha_1 + \delta_{pw}/\lambda_{pw})$. Using formula (5) one can calculate coefficient α_2, and determine $q = \alpha_2 \cdot \delta t_o$.

Expressions (6) and (4) are used to find α_1; expression (3) is used to calculate the heat transfer coefficient k.

4 Computation model

To perform the calculation, one should take into account the production characteristics of the plant:

- Temperature of the heating steam t_1=75 °C,
- Steam pressure 0.039 MPa.
- Average temperature of oil sludge t_o=55 °C.
- Average temperature difference between steam and oil sludge $\Delta t = t_1 - t_o$=20 °C.
- Heating pipes of heat-recovery stage are horizontal steel pipes Du-40 with outer diameter d_{out}=0.048 m, and inner diameter d_{in}=0.040 m,
- Thermal conductivity of pipes λ_{pw}=51.5 W/(m·K).

Thermo-physical properties of oil sludge are taken according to [5] (transformer oil):

- Heat exchange coefficient α_2 on the outer surface of heating pipes calculated by (5) for $\delta t_o \approx 20$ °C, α_2=90.9 W/(m^2·K),
- Value $q \approx \alpha_2 \cdot \Delta t$=1818 W/m^2. Value δt_w, calculated by (6), is $\delta t_w \cong 0.1$ °C,
- Heat exchange coefficient α_1, calculated by (4), is α_1=27980 W/(m^2·K),
- Heat transfer coefficient, calculated by (4), is k=90.1 W/(m^{2}K),
- $k \approx \alpha_2$, and $\delta t_o \approx \Delta t$.

During design calculation inside the evaporator stages of recovery stage for the temperatures of the primary and secondary steam t_1 and t_2, coefficient ζ, and flow rate of the primary steam G_1 from the main stage in evaporation mode, the pressures are taken as p_1 and p_2.

Heat flux transmitted during condensation of primary steam is $Q_1=G_1 \cdot r_1$. Expressions (3-6) are used to calculate the heat transfer coefficient k, and formula (2) is used to determine the necessary area F of heater surface of the recovery stage [14].

The pipes surface area is $F=\pi \cdot d_{out} \cdot l \cdot n$, where l is length of one pipe; n is the amount of pipes.

If to assume G_1 =0.031 kg/s, then for r_1=2320 kJ/kg the value Q_1 is equal to 72.7 kW.

For Δt=20 °C and k=90.1 W/(m^2·K) he necessary area of heater surface of the recovery stage is F=40.3 m^2.

As heating pipes we choose horizontal pipes Du-40 of length l=2.5 m. For the outer pipe diameter d_{out}=0.048 m the design amount of pipes is $n \approx 112$.

Steam of temperature t_2, re-entering from the recovery stage RS inflows for heating to OR, where it heats cold oil sludge of mass M of initial temperature $t_{cold} \approx 25$ °C. This heater is a tank facing out, in the bottom of which heating pipes are located. These pipes are immersed in oil sludge, where the secondary steam condenses, and part of it, non-condensed flash steam, is pumped out by a vacuum pump. During OR operation oil sludge is heated approximately to the condensation temperature of the secondary steam t_2 [15].

The condition for plant operation is $M=M_1 + M_2$, where M_1 and M_2 are the oil sludge masses loaded in the main and recovery evaporation stages. Masses M_1 and M_2 are selected from the condition that the evaporation process of the main and utilization stages ends simultaneously. The mass ratio of M_1 and M_2 can be determined from the heat balance equation:

$$M_1 \cdot \varphi \cdot r_1 \cdot \eta = M_2 \cdot \varphi \cdot r_2 \cdot \eta$$

It is seen that $M_2/M_1=\eta \cdot r_1/r_2$. At temperatures t_1=75 °C and t_2=55 °C evaporation heat is equal to: r_1=2320 kJ/kg; r_2=2370 kJ/kg. If assume η=0.98, then $M_2/M_1 \cong 0.96$.

Part of heat of the secondary steam from RS is spent on heating the oil sludge in recovery stage. For the temperature of cold oil sludge t_{cold} and the temperature of the heated oil sludge $t_o \approx t_2$, the amount of heat spent for its heating will be: $Q_o=c_o \cdot M \cdot (t_o - t_{cold})$. The amount of heat released during condensation of the secondary steam $Q_s=M_2 \cdot \varphi \cdot r_2$, where φ is the initial moisture content of oil sludge.

Taking $M_2 \approx M_1$, $M \approx 2M_2$; φ=0.2; c_o=2.5 kJ/(kg·K); r_2=2370 kJ/kg; t_{cold}=20 °C; t_o=55 °C, the ratio of Q_o and Q_s is obtained:

$$Q_o / Q_s \cong 2c_o \cdot (t_o - t_{cold}) / (\varphi \cdot r_2) \approx 0.25$$

The calculation showed that only a small part of the secondary steam heat can be used to heat the oil sludge. To achieve higher efficiency of this process, it is necessary to calculate the heating surface area of the recovery heater in accordance with the area of the recovery stage [16].

5 Conclusions

This work describes a general thermal scheme and a method for thermal calculation of a plant for moisture evaporation from oil products. The authors proposed a thermal calculation for the existing oil sludge separation unit BRNSh-3. As a result of the performed analysis, the technological shortcomings were identified and recommendations were developed to improve thermal efficiency and productivity of the plant. The first recommendation is the need to increase the parameter of thermal power transmitted from the heating water to oil sludge up to the nominal heat output of water boiler. The second recommendation is to use the heat of steam generated during moisture evaporation from oil sludge using a recovery evaporation stage.

References

[1] Y. Smyatskaya, N. Politaeva, L. Mukhametova, Physico-chemical studies of sorption materials based on biomass waste, E3S Web of Conferences **161**, 02003 (2020).

[2] N. Politaeva, V. Prokhorov, E.I. Lezhnev, V.A. Polyanskii, B.G. Matisov, L.R. Mukhametova, I.A. Mukhin, A new approach for recycling of spent activated sludge, IOP Conference Series: Earth and Environmental Science **337**, 1, 012077 (2019).

[3] F-R. Ahmadun A. Pendashteh, L.C. Abdullah, D.R.A. Biak, S.S. Madaeni, Z.Z. Abidin, Review of technologies for oil and gas produced water treatment, Journal of Hazardous Materials **170**, 530-551 (2009).

[4] A. Al-Futaisi, A. Jamrah, B. Yaghi, R. Taha, Assessment of alternative management techniques of tank bottom petroleum sludge in Oman, Journal of Hazardous Materials **141**, 557-564 (2007).

[5] A.A. Al-Shamrani, A. James, H. Xiao, Separation of oil from water by dissolved air flotation, Colloids and Surfaces A: Physicochemical and Engineering Aspects **209**, 15-26 (2002).

[6] T. Appleton, R. Colder, S. Kingman, I. Lowndes, A. Read, Microwave technology for energy efficient processing of waste, Applied Energy **81**, 85-113 (2005).

[7] A. Surikova, E. Sytsyanko, V.F. Kosmach, Yu.F. Titovec, T.R. Akhmetov, Thermal calculation of the plant for the moisture evaporation from petroleum products, IOP Conf. Series: Earth and Environmental Science **337** (2019).

[8] P.C. Blokker, Spreading and Evaporation of Petroleum Products on Water, Proceedings of the Fourth International Harbour Conference, Antwerp, Belgium, 911-919 (1964).

[9] M. Bobra, *A Study of the Evaporation of Petroleum Oils*, Manuscript Report Number EE-135 (Environment Canada, Ottawa, ON., 1992).

[10] M. Bobra, E.J. Tennyson, Photooxidation of Petroleum, Proceedings of the Twelfth Arctic Marine Oilspill Program Technical Seminar, Environment Canada, Ottawa, Ontario, 129-147 (1989).

[11] P.J. Brandvik, L.-G. Faksness, Weathering Processes in Arctic Oil Spills: Meso-scale Experiments with Different Ice Conditions, Cold Regions Science and Technology, 160-166 (2009).

[12] P.W.M. Brighton, Evaporation from a Plane Liquid Surface into a Turbulent Boundary Layer, Journal of Fluid Mechanics **159**, 323-345 (1985).

[13] P.W.M. Brighton, Further Verification of a Theory For Mass and Heat Transfer From Evaporating Pools, Journal of Hazardous Materials **23**, 215-234 (1990).

[14] W. Brutsaert, *Evaporation into the Atmosphere* (Reidel Publishing Company, Dordrecht, Holland, 299, 1982).

[15] J.N. Butler, *Transfer of Petroleum Residues from Sea to Air: Evaporative Weathering* (Marine Pollutant Transfer, Ed H.L. Windom and R.A. Duce, Lexington Books, Toronto, 201-212, 1976).

[16] P.J. Drivas, Calculation of Evaporative Emissions from Multicomponent Liquid Spills, Environmental Science and Technology **16**, 726-728 (1982).

Restoring normal network operation mode by automatically disabling network ATS

A.V. Vinogradova[1], *A.I.* Psarev[2], *A.V.* Vinogradov[1], *V.E.* Bolshev[1,*], *M.* Jasinski[3], and *R.R.* Gibadullin[4]

[1]Federal Scientific Agroengineering Center VIM, 5 1st Institutsky Proezd, Moscow, 109428, Russia
[2]Orel State Agrarian University, 69 General Rodina St., Orel, 302019, Russia
[3]Wroclaw University of Science and Technology, 27 Wybrzeże Wyspiańskiego st., Wrocław, 50-370, Poland
[4]Kazan State Power Engineering University, 51 Krasnoselskaya St., Kazan, 420066, Russia

Abstract. Automatic sectionalizing and redundancy of power transmission lines is one of the conditions for network intellectualization and can significantly reduce the power supply interruption time for rural consumers. However, both sectionalizing and redundancy of power transmission lines in 0.4 kV electric networks are used extremely limitedly since there are no sufficiently effective methods and technical means for their implementation. This article presents a method to automatically deactivate the automatic load transfer switch when restoring the normal network operation mode. The paper considers the implementation of the method on the example of damage occurring in the network where power is supplied to consumers from two power sources using an automatic load transfer switch and two sectionalizing units. A device is developed which allows implementing the developed method of automatic load transfer switch deactivation. The device consists of sectionalizing unit, automatic load transfer switch, voltage presence sensor, power direction sensor, memory element, repeater element, element AND and 3 elements NOT. The work of the device is described. The proposed method for automatically disconnecting the automatic load transfer switch in order to return the consumers' power supply circuit to its initial state solves the problem of eliminating unreasonable power supply interruptions for consumers.

1 Introduction

Improving power supply reliability and power quality is the main task to be solved when creating intelligent electrical networks (Smart Grids) [1-4]. Network intellectualization, first of all, requires the development of new automation methods to minimize power supply interruptions to consumers [5]. Automatic sectionalizing and redundancy of power transmission lines (PTL) is one of the conditions for network intellectualization and can significantly reduce the power supply interruption time for rural consumers [6-10]. However, both sectionalizing and redundancy of power lines in 0.4 kV electric networks are used extremely limitedly since there are no sufficiently effective methods and technical means for their implementation [11-13]. At the same time, technical means for solving this problem were proposed in [14]. These are universal sectionalizing units (SU) that combine the functions of network sectionalizing and redundancy (SUATS). Such a device can operate both in the mode of a sectionalizing units (SU) and the units of automatic load transfer switch (ATS) depending on the specified mode. One of the disadvantages of network redundancy is the lack of ways to automatically turn off the ATS to restore a power supply system back to a normal operation mode. This need arises when the damage at the redundant PTL section is repaired and power is supplied to it from the main power source. The study aims to develop a method for automatically disconnecting a network ATS when restoring a normal network operation mode.

2 Materials and methods

Figure 1 shows a diagram of power supply to consumers from two power sources using the means of network sectionalizing and redundancy.

In normal operation mode, power supply to consumers S1 and S2 (Figure 1 shows one consumer, but in reality, it can be consumer groups connected to a particular PTL section) is provided from the transformer substation (TS) TS1. The universal sectionalizing SUATS1 is turned on and works in the sectionalizing unit (SU) mode. SUATS2 operating in the mode of network ATS unit is disabled. Power supply to consumers S3 and S4 is provided from TS2. SUATS3 is turned on and operates in the SU mode.

If damage occurs to the section 1 of the power lines 1 (sec1PTL1), then the protective device installed on TS 1 will disable this section. SUATS1 will turn off since the voltage at sec1PTL1 will disappear. SUATS2 will turn on and supply power to sec2PTL1, thus supplying consumer S2 from TS2. After repairing the damage at the sec1PTL1, it is first necessary to turn off SUATS2

* Corresponding author: vadimbolshev@gmail.com

 https://doi.org/10.1051/e3sconf/202017801072

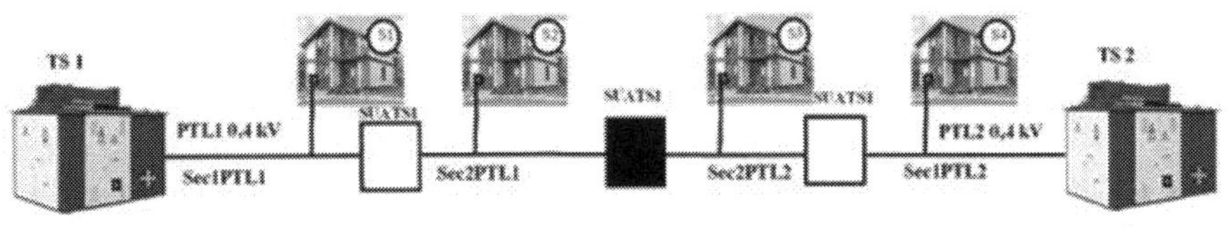

Fig. 1. Diagram of power supply to consumers from two power sources using the means of network sectionalizing and redundancy.

and then turn on SUATS1 to restore the normal power supply circuit. Consumer S2 will again lose power for a short time. This will lead to unreasonable damage from power supply interruption especially if the technological process of consumer S2 is sensitive to short-term power outages. Therefore, it is necessary to develop a method for automatically returning the power supply circuit to normal operation mode without power supply interruption to consumer S2. This method was previously developed for 10 kV electric networks [15], but it is not suitable for 0.4 kV networks. Analyzing sequences of changes in such parameters of 0.4 kV electric network operating modes as voltage and power flow direction is used for a new method.

3 Results and Discussion

The proposed method for disconnecting ATS when restoring the normal operation mode of a network making possible not to disconnect consumers of the reserved PTL section is illustrated in Figure 2. Figure 2 shows the structural diagram of a device that implements this method.

The essence of the method is that after turning on the ATS3 to supply power to consumer S13 of the reserved PTL section, the direction of power flow through it is controlled. If the power flow is directed towards the reserved section of the power transmission line, this

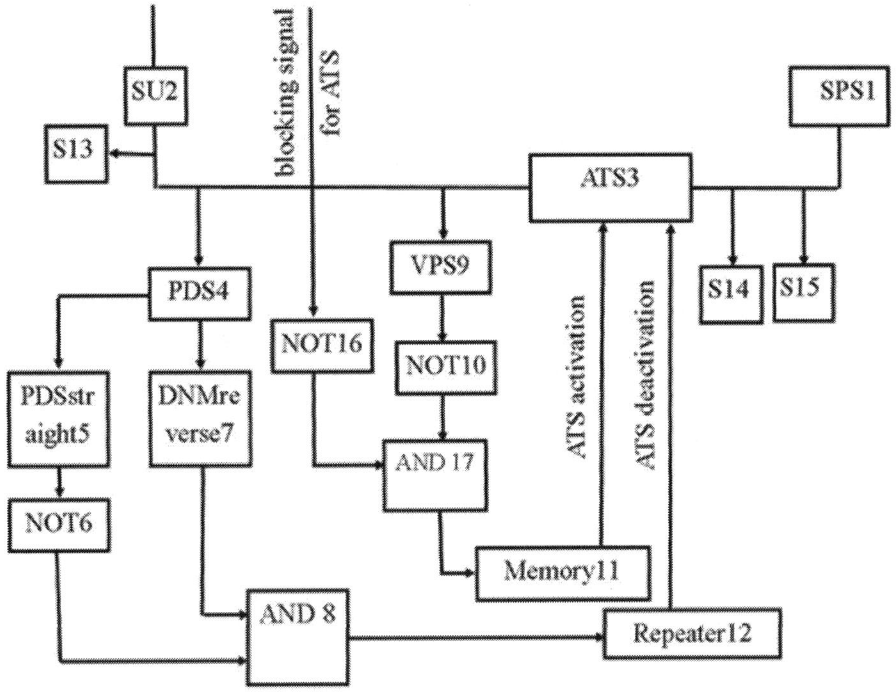

Fig. 2. Structural diagram of a device that implements the method of automatic ATS deactivation.

means that the consumer of this section is supplied by the standby power source (SPS1) and the ATS should not be disconnected. If the power flow changes direction, it is concluded that the power of the consumers of the reserved section from the main source is restored, that is, the sectionalizing unit SU2 is turned on. In this case, the ATS is turned off and the power supply circuit returns to its original state without power supply interruption to consumers connected to the reserved section of the power transmission line [16].

A device for implementing the method works as follows. In normal operation mode of the network, consumer S13 is supplied from the main power source through the sectionalizing unit SU2. Consumers S14 and S15 are supplied from standby power supply SPS1. ATS3 is turned off. There is a signal at the output of the voltage presence sensor VPS9, which monitors the phase voltages in the reserved section. Therefore, there is no signal at the outputs of the elements NOT10 and Memory 11. After the voltage disappears in the reserved PTL section for any reason, for example, due to damage to the main power source, a signal from the output of VPS9 element will disappear, a signal from the output of NOT10 element will appear, which will be remembered by Memory 11 element and ATS will be turned on. After turning on the ATS using power direction sensors PDS4, PDSstraight5 and DNMreverse7 control the direction of power flow through it. A signal from the PDSstraight5 output appears when the power is directed from the reserving section to the reserved one of the power transmission line. A signal from the DNMreverse7 output appears when the power is directed from the reserved section to the reserving one of the power transmission line. If there is a signal from the output of PDSstraight5, there is no a signal from the output of the element NOT6 and accordingly no a signal at the input of the element AND 8. There is no a signal from the output of DNMreverse7, the ATS is not turned off.

When the sectionalizing unit SU2 is switched on after the restoration of the main power source, the direction of the power flow will change. Since the power flow changes direction, a signal will appear at the output of DNMreverse7 element and will be fed to the input of AND8. At the second input of AND8 element, the signal will also appear since there will be no signal at the output of PDSstraight5 and there is a signal at the output of element I8. AND8 will give a signal to turn off the ATS and reset the signal from Memory 11. To eliminate the coincidence of the duration of these signals, the circuit provides for the installation of Repeater 12 element. It increases the duration of the signal sending to ATS deactivation. After the deactivation, ATS returns to its original position. It should be noted that the proposed method requires a number of conditions. Firstly, the reserved and reserving power lines must be phased to prevent interphase short circuits when the SU and the ATS are on together. The second condition is that the primary and standby power sources must be able to work in parallel. And the third condition is that the flow separation point should be on the side of the standby power source concerning the ATS unit.

4 Conclusion

The use of technical means of sectionalizing and redundancy of 0.4 kV power transmission lines requires the development of new methods to minimize losses from lack of electricity supply to consumers, including those associated with short-term power supply interruptions necessary for switching in electric networks. The proposed method for automatically disconnecting the network ATS in order to return the consumers' power supply circuit to its initial state solves the problem of eliminating unreasonable power supply interruptions for consumers.

References

[1] J.A. Momoh, Smart grid design for efficient and flexible power networks operation and control, Power Systems Conference and Exposition, PSCE '09, 1–8. DOI:10.1109/PSCE.2009.4840074 (2009)

[2] A. Arif, Z. Wang, Service restoration in resilient power distribution systems with networked microgrid, Proceedings of the 2016 IEEE Power and Energy Society General Meeting (PESGM), 1–5. DOI: 10.1109/PESGM.2016.7741533 (2016)

[3] A. Ipakchi, F. Albuyeh, Grid of the future, IEEE Power and Energy Magazine **7**, 52–62. DOI: 10.1109/MPE.2008.931384 (2009)

[4] M. Jasinski, T. Sikorski, P. Kostvla, K. Borkowski, Global power quality indices for assessment of multipoint Power quality measurements, 10th International Conference on Electronics, Computers and Artificial Intelligence (ECAI), 1–6. DOI:10.1109/ecai.2018.8679053 (2018)

[5] M. Wang, *Reliability Analysis with Various Transfer Switch Technologies in Open-ring Distribution Systems* (Doctoral dissertation, McGill University Libraries, 2013)

[6] S.A. Tsyruk, S.I. Gamazin, Y.N. Ryzhkova, K.F. Charafeddine, Determination of Source Fault Using Fast Acting Automatic Transfer Switch, In 2018 Dynamics of Systems, Mechanisms and Machines (Dynamics), 1–4. DOI: 10.1109/Dynamics.2018.8601484 (2018)

[7] P. Ilyushin, K. Suslov, Operation of Automatic Transfer Switches in the Networks with Distributed Generation, In 2019 IEEE Milan PowerTech, 1–6. DOI: 10.1109/PTC.2019.8810450 (2019)

[8] H. Falaghi, M.R. Haghifam, C. Singh, Ant colony optimization-based method for placement of sectionalizing switches in distribution networks using a fuzzy multiobjective approach, IEEE Transactions on Power Delivery **24**, 1, 268–276. DOI: 10.1109/TPWRD.2008.2005656 (2008)

[9] F.T. Asr, A. Kazemi, Determining optimum location of automated switches in distribution network, IEEE Canada Electric Power Conference, 1–6. DOI:

10.1109/EPC.2008.4763358 (2008)

[10] A. Abiri-Jahromi, M. Fotuhi-Firuzabad, M. Parvania, M. Mosleh, Optimized sectionalizing switch placement strategy in distribution systems, IEEE Transactions on Power Delivery **27**, 1, 362–370. DOI: 10.1109/TPWRD.2011.2171060 (2011)

[11] A. Vinogradov, A. Vasiliev, V. Bolshev, A. Semenov, M. Borodin, *Time factor for determination of power supply system efficiency of rural consumers. Handbook of Research on Renewable Energy and Electric Resources for Sustainable Rural Development* (Hershey, PA: IGI Global, 394–420, 2018). DOI:10.4018/978-1-5225-3867-7.ch017

[12] A. Vinogradov, A. Vasiliev, V. Bolshev, A. Vinogradova, T. Kudinova, N. Sorokin, N. Hruntovich, *Methods of Reducing the Power Supply Outage Time of Rural Consumers, Renewable Energy and Power Supply Challenges for Rural Regions* (Hershey, PA: IGI Global, 370–392, 2019). DOI: 10.4018/978-1-5225-9179-5.ch015

[13] A. Vinogradov, A. Vinogradova, V. Bolshev, A.I. Psarev, Sectionalizing and Redundancy of the 0.38 kV Ring Electrical Network: Mathematical Modeling Schematic Solutions, International Journal of Energy Optimization and Engineering (IJEOE) **8**, 4, 15–38. DOI: 10.4018/IJEOE.2019100102 (2019)

[14] A.V. Vinogradov, A.V. Vinogradova, A.I. Psarev, R.P. Belikov, Metodika vybora mest ustanovki i kolichestva universal'nyh sekcioniruyushchih punktov s funkciej AVR v elektricheskoj seti 0,38 kV po kriteriyu obespecheniya chuvstvitel'nosti zashchitnyh apparatov [Methodology for selecting installation locations and the number of universal sectioning points with an ABP function in an 0.38 kV power network according to the criterion of ensuring the sensitivity of protective devices], Vesti vysshih uchebnyh zavedenij CHernozem'ya [News higher educational institutions of the Black Earth Region] **3**, 57, 38–49 (2019)

[15] V.G. Vasiliev, A.V. Vinogradov, S.M. Astakhov, M.A. Kolesov, Sposob vosstanovleniya pitaniya potrebitelej rezerviruemogo uchastka linii po iskhodnoj skheme [A way to restore power to consumers of a reserved section of the line according to the original scheme], RF patent No. 2313878 (2006)

[16] V.E. Bolshev, A.V. Vinogradov, A.V. Vinogradova, A.I. Psarev, Sposob i ustrojstvo otklyucheniya AVR pri vosstanovlenii normal'nogo rezhima raboty seti [The method and device for disconnecting the ATS when restoring the normal operating mode of the network], RF patent No. 2687053 (2019)

Methodology for assessing technological efficiency and safety at the enterprise for production of reinforced concrete structures

Anton Sinitsyn[1,*], *Oleg* Stratunov[1], *Timur*. Akhmetov[2], and *Nikolay* Monarkin[1]

[1]Department of Heat and Gas and Water Supply, Vologda State University, Lenina str. 15, 160000, Russia
[2]Kazan State Power Engineering University, Kazan, Russia

Abstract. This paper considers the main results of instrumental analysis of technological efficiency and safety, and suggestions for improving the working conditions of workers engaged in steaming of reinforced concrete products in steam curing chambers operating on saturated steam. This work is a part of comprehensive energy study of enterprise to find measures that increase the energy potential of not only the enterprise itself, but also reduce the cost of primary and auxiliary products by reducing costs in its production. An optimized scheme for delivery, transporting and consumption of water vapour was implemented in the real sector of the economy in Vologda (Russia).

1 Introduction

Precast reinforced concrete factories are the enterprises where sanitary and hygienic working conditions and safety precautions are not only an important criterion for labor productivity, but they also ensure health preservation of each employee. In order to reduce the time for dismantling reinforced concrete structures and putting them under load, builders have always sought to accelerate the concrete hardening. This issue has gained particular relevance in manufacturing of concrete and reinforced concrete products at factory, as enterprises are interested in maximizing the use of production facilities and in reducing the time for product manufacturing.

Nowadays, heat treatment is the most commonly used way to accelerate the concrete hardening. It allows obtaining products with specified delivery strength in a short time, which can be transported to the construction site and mounted in buildings and structures. At factories, heat treatment is carried out by steaming products in steam curing chambers. During steaming, the molded products are kept in a chamber in saturated steam until concrete reaches the specified strength. In steam curing chamber one should create not only a favourable temperature for accelerated hardening (in the range of 60-100 °C), but also an optimal medium humidity, which contributes to preservation of moisture in concrete for its further hardening even after steaming.

The effectiveness of steaming and other types of heat treatment is determined by the choice of rational treatment mode in full accordance with the accepted composition of concrete, the characteristics of components, the cement peculiarity, the dimensions and configuration of products, the initial strength of concrete at the time of processing, etc.

Humidity conditions of hardening are especially important for formation of concrete structure; therefore, in many cases, preference should be given to the heat-moisture treatment of reinforced concrete products. In this case, a particularly important task is to develop a technological flow chart of product processing ensuring operation safety and labour protection.

The main activity of employees of enterprises for the production of reinforced concrete structures is carried out in strict accordance with the requirements of the Model Instructions for Labour Protection for the molder of reinforced concrete products and structures, as well as the Rules for the Construction and Safe Operation of steam and hot water pipelines.

It should be noted that processing modes are assigned according to instructions and recommendations with the obligatory experimental verification and refinement of parameters of modes or experimental-computational methods, which are characterized by inefficient use of thermal energy in calibration experiments, significant costs and labour r costs for their implementation. Experimental verification is carried out, as a rule, on model samples that differ from reinforced concrete products by the kinetics of heat and mass transfer processes. In practice, the assigned modes often do not ensure the design energy consumption and product quality, therefore, it is necessary to carry out mode adjustment at manufacturing sites, to increase production areas for concrete curing, which leads to an increase in production cost.

However, from the point of view of saving energy resources, reducing costs and laboriousness the most promising are the calculation and experimental methods for assigning modes. Such a method was proposed by E.I. Shmitko and A.A. Fedin, the method for calculation the limiting temperature gradients in reinforced concrete

* Corresponding author: sinitsyn.science@mail.ru

products during electrothermal treatment was developed by S.V. Fedosov, V.I. Bobylev, A.M. Ibragimov and A.M. Sokolov. M.S. Bibik and V.V. Babitsky developed a technique and a computer program for assigning of energy-saving thermos regimes of steam curing treatment (SCT) of concrete and reinforced concrete products in pit steam curing chambers, taking into account the minimum cost.

The scientific group of K. Maekawa, T. Ishida, T. Kishi, A.K. Schindler, K.J. Folliard modeled structural concrete [1] and hydration of binders [2]. The development of an express-test for monitoring the heat release of concrete mixtures in laboratory and field conditions is shown in works of W. Kejin, G. Zhi, J. Jim Grove, M. Ruiz, R. Rasmussen [3]. Today calculation and experimental methods are applied in scientific, practical and technological research [4-6], in particular for considering SCT modes to increase the efficiency of works on production of reinforced concrete products which remain relevant.

2 Problem statement and the applied methods

This paper presents the main results of determining the technological conditions for the steam curing treatment of reinforced concrete products in pit steam curing chambers of the joint venture enterprise "Vologda Plant of Reinforced Concrete Products and Construction Details" of JSC Russian Railways.

The company specializes in the production of concrete products for construction of load-bearing and enclosing structures of buildings and constructions, as well as temporary auxiliary structures.

The testing ground of steam curing treatment of reinforced concrete product was used to compare the actual data on steam consumption for concrete production (t/h) with the calculated values obtained using the engineering method [7]. These guidelines for construction and thermotechnical calculation of the pit steam curing chamber are used in scientific and educational process of the REC "Problems of the modern technological environment", which is shown in [8-9].

Further the methods used in this work are being discussed. The methods for analyzing a technical object were based on the principles of systematic approach. In this work, the technical system was understood as the interconnection of its main elements - the source, the consumer complex, and distribution network. The structure of technical system was determined by the composition of its basic elements and the ways of their connections. The set of all possible system states was depended on the number of elements and was determined by the levels of connections between them, as well as by the functions of technical system. At the same time, the decomposition method was used to solve the complex technical problem of analyzing the efficiency of the enterprise steam supply system. It is reduced to dividing the system into subsystems and individual elements for the purpose of their detailed study with their subsequent synthesis. The modelling method was of greatest importance due to specifics of the technical object design. The modelling cycle includes the steps of creating a technical object model, model study, model transformation, and transition from model to technical object. At this stage of modelling, correspondence rules were established expressing the relationship between properties of a real technical object and properties of mathematical objects. Similar methods are clearly discussed in [10–13].

Experimental work is the most important part of these scientific studies. To achieve the set goals, a hypothesis was developed, a program for experimental work was created, conditions for implementation of the experimental work procedure were provided, ways and methods of fixing the course and results of experiment were developed, experimental tools were prepared, and experimental data were processed and analyzed.

3 Mathematical model of heat losses in a steaming chamber

The main mathematical expressions used in this work are listed below. Heat losses through building enclosures (kJ/period) are calculated by the general formula:

$$Q_{loss} = 3.6 \cdot S_{enc} \cdot \alpha_{ext} \cdot \left(t_{exts} - t_{amb}\right) \cdot \tau, \qquad (1)$$

where S_{enc} is the surface area of the chamber building enclosures, m^2;

α_{ext} is the heat exchange coefficient from external surfaces, W/($m^2 \cdot$°C);

t_{exts} is the average temperature of external surface of the building enclosures, °C;

t_{amb} is the ambient temperature, °C;

τ is the operation time of installation, h.

During the chamber operation the temperatures of external surface of walls, the bottom and the cap are often not equal, and these temperatures are different in the period of rise and isothermal holding, so heat losses are determined by the formula:

$$Q_{loss} = Q_c + Q_w{}^h + Q_w{}^l + Q_b \qquad (2)$$

or

$$\begin{aligned} Q_{loss} = 3.6 \cdot \{ & S_c \cdot \left[\alpha_{exts}{}^{cr} \left(t_{exts}{}^{cr} - t_{amb}\right) \cdot \tau_r + \alpha_{exts}{}^{ch} \cdot \left(t_{exts}{}^{ch} - t_{amb}\right) \cdot \tau_h\right] + \\ & + S_w{}^a \cdot \left[\alpha_{exts}{}^{wr} \cdot \left(t_{exts}{}^{wr} - t_w\right) \cdot \tau_r + \alpha_{exts}{}^{wh} \cdot \left(t_{exts}{}^{wh} - t_w\right) \cdot \tau_h\right] + \\ & + S_w{}^b \cdot \left[k^{rw} \cdot \left(t_c{}^r - t_{soil}\right) \cdot \tau_r + k^{hw} \cdot \left(t_c - t_{soil}\right) \cdot \tau_h\right] + \\ & + S_b \cdot \left[k^{rb} \cdot \left(t_c{}^r - t_{soil}\right) \cdot \tau_r + k^{hb} \cdot \left(t_c - t_{soil}\right) \cdot \tau_h\right] \}, \end{aligned} \qquad (3)$$

where Q_c, $Q_w{}^a$, $Q_w{}^b$, Q_b are the losses of thermal energy through the surfaces of the cap, walls above and below the zero mark, and the bottom, respectively, kJ/period;

S_c, $S_w{}^a$, $S_w{}^b$, S_b, are the surface areas of the cap, walls above and below the zero mark, the bottom, respectively, m^2;

τ_r, τ_h are the periods of temperature rise and holding, respectively;

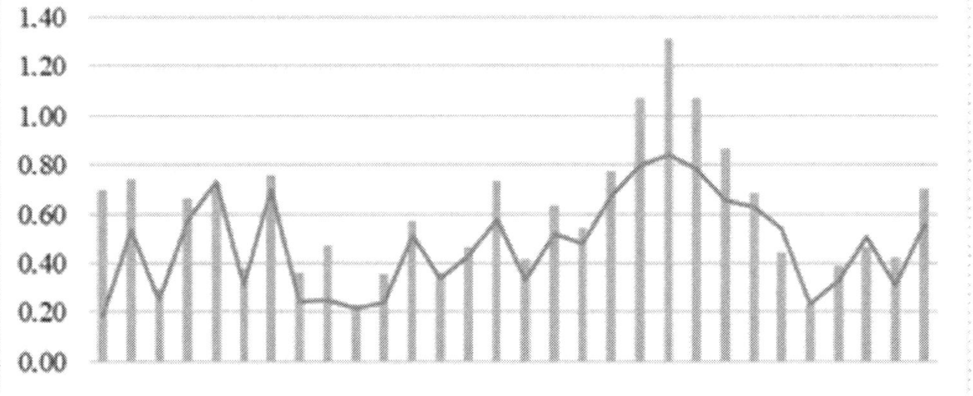

Fig. 1. Comparison of actual specific steam consumption (columns) for concrete production (t/h) with calculated data (line) by the method [1].

t_{exts}^{wr},t_{exts}^{wh} are the average temperatures of the external surface of walls during the periods of temperature rise and holding, respectively, °C;

t_{exts}^{cr},t_{exts}^{ch} are the average temperatures of the cap enclosure during the periods of temperature rise and holding, respectively, °C;

t_c^r,t_c is the surface temperature below zero mark during the periods of temperature rise and holding, respectively, °C;

t_{soil} is the soil temperature, °C;

k^{rw}, k^{rb} are the heat transfer coefficients from the internal of the chamber to the soil during temperature rise, W/m²·°C.

k^{hw}, k^{hb} are the heat transfer coefficients from the internal of the chamber to the soil during temperature holding, W/m²·°C.

α_{exts}^{cr}, α_{exts}^{wr} are heat exchange coefficients from the surface of cap and walls above the zero mark during the temperature rise period W/m²·°C.

$$\alpha_{exts}^{wr} = 9.8 + 0.07\left(t_{exts}^{wr} - t_w\right), \quad (4)$$

where t_{exts}^{wr} is average temperature on the surface of walls during the period of temperature rise;

α_{exts}^{wh}, α_{exts}^{ch} are the heat exchange coefficients from the surface of cap and walls that are above the zero mark and the cap during the holding period, W/m²·°C. For example, for walls:

$$t_{exts}^{wh} = \frac{\left(t_{exts1} + t_{exts2}\right)}{2}, \quad (5)$$

α_{exts}^{cr} is defined in a similar way.

α_{exts}^{ch}, α_{exts}^{wh} are the heat exchange coefficients from the surface of cap and walls that are above the zero mark during the temperature holding period, W/m²·°C For example, for walls:

$$\alpha_{exts}^{wh} = 9.8 + 0.07\left(t_{exts}^{wh} - t_{amb}\right), \quad (6)$$

where t_{exts}^{wh} is the temperature on the outer surface of walls during isothermal holding, °C ($t_{exts}^{wh}= t_{exts2}$);

t_{amb} is the ambient temperature in the workshop, °C;

Heat released by cement during thermal moisture treatment, kJ/kg, is determined by the formula:

$$Q_{ecem} = 1.85 \cdot Q_{cem28} \cdot \left(Wat / Cem\right)^{0.44} \cdot \left(1 - v\right), \quad (7)$$

where Q_{cem28} is the heat dissipation of cement at 28 days hardening under normal conditions, kJ/kg;

Wat is water consumption per 1 m³ of concrete, kg;

Cem is cement mass in 1 m³ of concrete, kg;

v is the value depending on degree hours (n) of concrete hardening.

If the number of degree hours (n) does not exceed 375, then $v = e^{-0.0015n}$ and (7) takes the form:

$$Q_{ecem} = 1.85 \cdot Q_{cem28} \cdot \left(Wat / Cem\right)^{0.44} \cdot \left(1 - e^{-0.0015n}\right), \quad (8)$$

and if 375 < n < 2000, then $v = 0.666 \cdot e^{-0.0004n}$, (7) takes the form:

$$Q_{ecem} = 1.85 \cdot Q_{cem28} \cdot \left(Wat / Cem\right)^{0.44} \cdot \left(1 - 0.666\, e^{-0.0004n}\right). \quad (9)$$

4 The results of computational and experimental studies

The engineering calculation made it possible to determine the heat energy consumption for the heat treatment of materials and products. In particular steam consumption was calculated for loading concrete and overall dimensions of molds for manufacturing of concrete and reinforced concrete products, as well as the internal dimensions of the pit steam curing chamber and its penetration into the ground, taking into account design features of walls, bottoms and caps of chamber.

A plot comparing the actual specific steam consumption for concrete production, t/h, with the

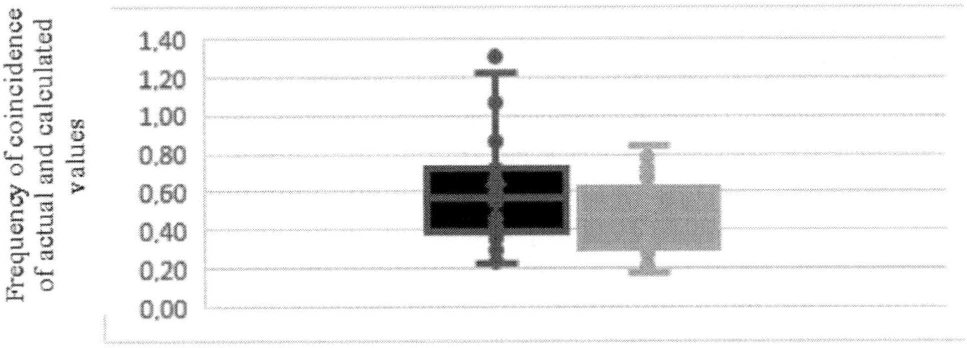

Fig. 2. The indicator diagram for frequency of coincidence of actual (dark color) and calculated (gray) specific steam consumption for concrete production, t/h.

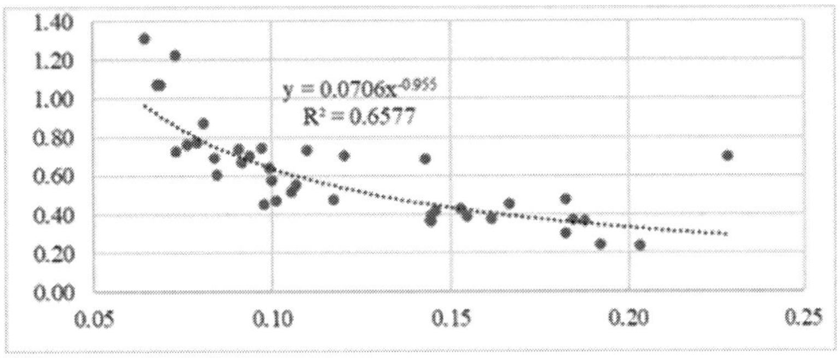

Fig. 3. Relationship between specific steam consumption (t/h) and the load of the steam chamber.

calculated data according to the methodology [7] is presented in Fig. 1.

Based on data from Fig. 1, a conclusion is drawn on the adequacy of mathematical model and engineering methodology to the actual data on consumption of water vapour for production obtained from the metering device (the computational line practically repeats the dynamics of values from the meter). Indicator diagram 2 shows that the range of the most frequently encountered values of actual specific steam consumption for concrete production is 0.38-0.75 t/h.

The calculations showed a decrease in steam consumption with an increase in the degree of loading of the steam curing chamber, which can be described by a power-law function with a correlation accuracy of 0.66.

A similar analysis was performed for the reinforced concrete products during its heat and moisture treatment.

Analysis of data on reinforced concrete products also allow us to confirm the compliance of mathematical model and engineering methodology with actual data on water vapour consumption for production. The range of the most frequent values of the actual specific steam consumption for concrete production is 0.35-0.45 t/h. A decrease in steam consumption with an increase in the degree of loading of the steam curing chamber is also confirmed, with the correlation accuracy of 0.75.

These calculations made it possible to formulate the basic technological indicators of the process of steam curing treatment of reinforced concrete products, as well as to give recommendations on the technological safety of work associated with the use of saturated steam, as well as sanitary and hygienic requirements for production processes and technological equipment of precast concrete enterprises.

Recommendations were made on the implementation of production processes in accordance with national safety requirements and the use of production equipment.

It was also noted that pit-type curing chambers should be equipped with hydraulic locks or other special design, ensuring the tightness of connection of cap with the chamber to prevent the release of steam, while the working pressure of the medium in the chamber should not exceed that indicated in the regime flow chart. Chambers located indoors must be equipped with a steam extinguishing system or exhaust ventilation, while all chambers (of testing ground and reinforced concrete unit (RCU)) must be equipped with devices that control the temperature of medium in it.

The results of determining the costs of thermal energy and water vapour for production needs allowed us to compare the calculated and actual costs of the enterprise production line. For example, Table 1 shows a

Table 1. Comparative analysis of actual data of commercial metering unit and calculation results for the heat load of consumption

Date	Outside temperature, ◦C	Technological unit of the enterprise	Actual data	Calc. data	Deviation
1	2	3	4	5	6
February 28, 2019.	+1	Polygon	3.83	2.94	23%
		RCU	8.42	9.43	12%
March 02, 2019	-8◦C	Polygon	3.85	3.12	19%
		RCU	7.58	7.26	4%

comparative analysis of the actual data of commercial metering unit and the results of calculating the heat load of consumption for Feb., 28 2019 and March, 02 2019 (days with comparable technological loads).

Data comparison shows that in general, the calculated and actual values coincide within the deviation of no more than 23% for the studied loading modes of steam consumption system. So the production conforms to the recommended technological modes.

Studying the issue of assessing the efficiency of power use in steam curing treatment of concrete, it was found that the efficiency of steam curing chambers depends on many external and internal factors. The thermal efficiency of precast concrete production can be significantly increased by reducing the heat loss associated with unsatisfactory condition of the steam curing chambers, heat networks, valves and means for controlling the steam flow.

Analysis of distribution of thermal energy consumption in the steam curing chambers of Plant of Reinforced Concrete Products and Construction Details LLC showed that the largest share of thermal energy is used to heat the building enclosures (54%) and to make up for heat loss through the building enclosure (23%).

5 Conclusions

The work presents a comprehensive study of the enterprise to find measures that not only increase the energy potential of the main production, but also reduce the cost of production by reducing production expenditures. An optimized scheme for supply, transporting and consuming water vapour was implemented in the real sector of the economy of Vologda.

Based on the survey results, measures were developed and proposed to improve the efficiency of the existing steam supply system of the enterprise, its digital GIS version was created for the operational planning of energy-saving measures.

The relevance of the work is associated with the need of enterprise to increase the energy efficiency of its own production, the results of which can become the basis for making subsequent technical solutions and additional measures to save energy and reduce the cost of basic and related products.

Acknowledgements

The team of authors sincerely thanks the employees of the energy sector of the company for the production of reinforced concrete structures of the joint venture "Vologda Plant of Reinforced Concrete Products and Construction Details" of JSC Russian Railways (Vologda) for the materials provided and the invaluable assistance in conducting an instrumental study of the enterprise's steam supply system.

References

[1] W. Kejin, G. Zhi, J. Jim Grove, M. Ruiz, R. Rasmussen, *Developing a Simple and Rapid Test for Monitoring the Heat Evolution of Concrete Mixtures for Both Laboratory and Field Applications: Final Report* (Center for Transportation Research and Education Iowa State University, Ames, 75, 2006).

[2] K. Maekawa, T. Ishida, T. Kishi, *Multi-scale modeling of structural concrete* (London and New York: Taylor & Francis, 655, 2009).

[3] A.K. Schindler, K.J. Folliard, Heat of hydration models for cementitious materials, ACI Materials Journal **102**, 24-33 (2005).

[4] N. Politaeva, Y. Smyatskaya, R. Al Afif, C. Pfeifer, L. Mukhametova, Development of Full-Cycle Utilization of Chlorella sorokiniana Microalgae Biomass for Environmental and Food Purposes, Energies **13**, 10, 2648 (2020).

[5] V. Markov, A. Fedyukhin, I. Sultanguzin, B.G. Matisov, N.V. Nikitkov, V.V. Izrantsev, V.G. Knorring, G.A. Konrashkova, L.R. Mukhametova, Improvement of energy efficiency of motor fuel production at oil refinery, IOP Conference Series: Earth and Environmental Science **337**, 1, 012080 (2019).

[6] A.A. Kalyutik, D.V. Grigorieva, A.V. Fedyukhin, O.V. Derevianko, I.G. Akhmetova, Reduction of the technological minimum by bypass method with environmental impact estimation on the example of energy blocks with T-100 and T-250/300-240 turbines, IOP Conference Series: Earth and Environmental Science **288**, 1, 012094 (2019).

[7] S.I. Koryukin, A.A. Sinitsyn, *Fundamentals of designing heat supply systems: a textbook for course works and diploma projects* (2013).

[8] N. Sovetova, A. Sinitsyn, E. Tritenko, O. Derevianko, L. Mukhametova, A. Fedukhin, S. Makoev, A. Kalyutik, Regional management of district heating, IOP Conference Series: Earth and Environmental Science Collection of materials International Scientific and Practical Conference. Institute of Physics and IOP Publishing Limited (2019).

[9] A. Sinitsyn, L. Mukhametova, Practical applicability of the method for measuring pressure of controlled medium on the example of a pulsating combustion boiler, IOP Conference Series: Earth and Environmental Science Proceedings of the Conference the international scientific conference Efficient waste treatment – 2018, EWT-2018 (2019).

[10] D.A. Cauich-López, L.F. Barrera Payan, A.M.N. Abdelhalim, E. Socolova, A.A. Sinitsyn, I.G. Akhmetova, Small-scale steam generation from local solid waste, International Journal of Civil Engineering and Technology **10**, 2 (2019).

[11] A.A. Sinitsyn, Experience of small innovative enterprises foundation in the Vologda region, Middle East Journal of Scientific Research **16**, 10 (2013).

[12] A.A. Sinitsyn, Application of small innovative business model in case of research and education center Teploenergetika, Problems and trends of economy and management in the modern world Proceedings of the International Conference (2012).

[13] A.A. Kalyutik, D.V. Grigorieva, A.V. Fedyukhin, O.V. Derevianko, I.G. Akhmetova, Reduction of the technological minimum by bypass method with environmental impact estimation on the example of energy blocks with T-100 and T-250/300-240 turbines, IOP Conference Series: Earth and Environmental Science **288**, 1, 012094 (2019).

ANALYSIS OF THE CONDITION AND MONITORING OF EMERGENCY POWER OUTAGES IN 10 kV OVERHEAD LINES OF THE ORYOL REGION

Igor Nikolaevich Fomin[1*], *Alexander Vladimirovich* Vinogradov[2], *Roman Pavlovich* Belikov[3], *Tatyana Anatolyevna* Kudinova[4], and *Marat Firdenatovich* Nizamiev[5]

[1]Orel State Agrarian University named after N.V. Parakhin, Orel, Russia
[2]Leading Researcher,Federal State Budgetary Scientific Institution "Federal Scientific Agroengineering Center VIM", Moscow, Russia
[3]Orel State Agrarian University named after N.V. Parakhin, Orel, Russia
[4]Orel State Agrarian University named after N.V. Parakhin, Orel, Russia
[5]Kazan State Power Engineering University, Kazan, Russia

Abstract. Efficiency improvement of modern production is associated with uninterrupted power supply to consumers. Power supply reliability to consumers is ensured by reliable operation of power lines. The number of outages depends on their condition and operation quality. An increase in the number of outages affects the further degradation of power line state, significantly reducing the resource of lines and equipment; and it takes a lot of time to find faults and eliminate them. Monitoring of power outages in overhead power lines will reduce the time of emergency elimination.1

1 Introduction

The need to implement power outage monitoring was predetermined by the analysis of the condition and fault rate of 10 kV radial overhead lines in the Oryol region. Analysis of the condition and fault rate of overhead lines in the Oryol region includes: analysis of the length of lines, technical condition of 10 kV overhead lines, indicators and main causes of line breakdowns. These indicators characterize the dynamics of the state of overhead lines and make it possible to forecast their faults [1, 2]. Frequent outages of overhead power lines, including emergency ones, are associated with large time expenditures, financial resources and power undersupply to consumers. Implementation of power outage monitoring in 10 kV overhead lines will allow attending personnel to eliminate faults as soon as possible and also to prevent them [3-6].

2 Theoretical research

The overhead lines of 10 kV in the electric networks of the Oryol region today are made mainly of wire grade AS - 35. The cross-section of this wire is two steps lower than currently recommended for trunk sections under the condition of providing mechanical strength. This fact alone is one of the reasons for the increased number of line faults.

Another reason is related to the fact that the main part of the 10 kV overhead lines (60%) was built before 1990 and therefore their service life has already exceeded the standards. Lines in unsatisfactory condition are dismantled (not in full, many of them continue to operate), while new lines are not fully put into service, which leads to the total length reduction of electric networks. Over the past 5 years, it decreased by 570.1 km (by 4.13% of the length of 2015), that is, by an average of 142.5 km per year (0.8%). At the same time, the length of some overhead lines is much higher than the recommended one and in many cases is more than 25 km, and sometimes more than 50 km. It also affects the number of line outages, as it reduces their reliability and operation quality. It is practically impossible to provide the required voltage level for the consumer with such a line length and wire cross-section [7-9].

In percentage terms, the length of lines less than 25 km is about 75%, from 25 km to 50 km is 21-22%, more than 50 km is 2.4%.

Analysis of the number of power outages shows that this indicator is growing every year due to an increase in the number of planned and emergency outages (Fig. 1.).

Over the 5 years, the number of planned outages increased by 284 outages, and the number of emergency outages increased by 492 ones. And if in 2014 emergency outages amounted 67.4% of the number of planned outages, then in 2017 they made up 81.1%.

* Corresponding author: igor.fom-in@yandex.ru

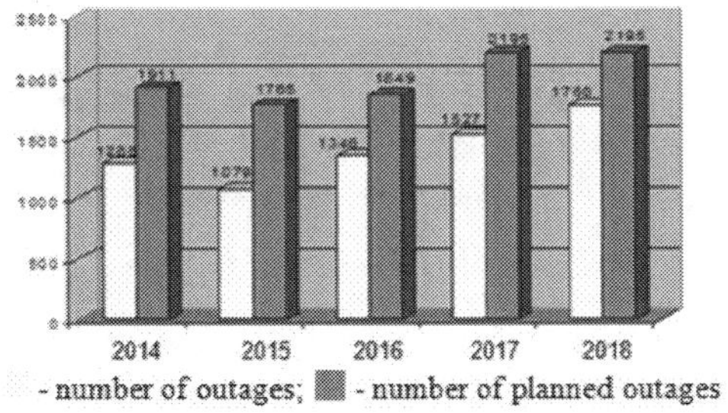

Fig. 1. Number of power outages in 10 kV overhead lines in the Oryol region.

Planned outages are determined by the repair work (the number of repairs has increased, therefore, there is a need in them). At the same time, the number of emergency (unplanned) outages is growing; it means that it will take some time to find out the cause, to locate and to eliminate the fault (in case if the power outage was not cleared by the automatic reclosing (AR) or manually), so the duration of power supply interruptions and the power undersupply to consumers is growing [10, 11].

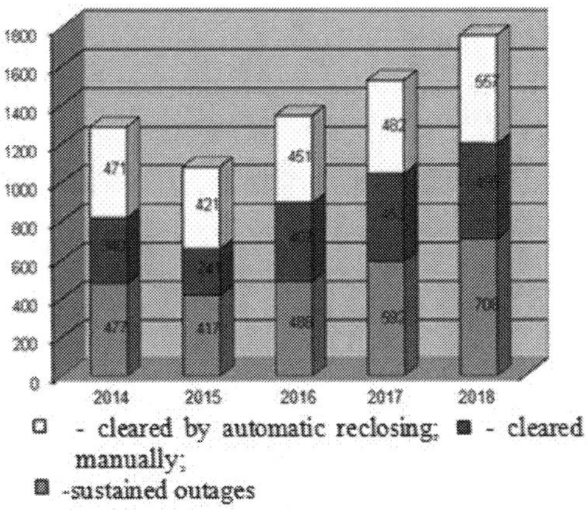

Fig. 2. Number of emergency outages in 10 kV overhead lines in the Oryol region.

The ranking of emergency outages, in turn, shows that sustained outages make up the majority. The same component is growing most rapidly. The efficiency of automatic reclosing, instead of theoretical 70%, in practice is 56.5% in 10 kV networks. This is due to the fact that most drives of automatic reclosing are obsolete models [12, 13].

Due to the increased rate of faults in 10 kV overhead lines, over the specified period of time, the power undersupply to consumers has significantly increased. The increase in the number of outages affects the further degradation of power line state, significantly reducing the resource of lines and equipment. The large length of the lines, in turn, increases the power undersupply due to the fact that it takes a lot of time to locate and to eliminate the faults.

In order to minimize these negative developments, an algorithm for remote monitoring of emergency outages in overhead lines was proposed; it is based on the analysis of various operation options of line switches.

The algorithm is a program of work that the means of remote monitoring of emergency outages must fulfill, regardless of how these means are implemented: on relay elements, microcircuits, or using a programmable logic matrix (PLM). Correctly composed work algorithm and its implementation determine the correct operation of the system as a whole. In this case, the algorithm (Fig. 3) implements the developed methods for monitoring emergency outages of circuit breakers in 10 kV overhead lines [4, 14, 19, 25].

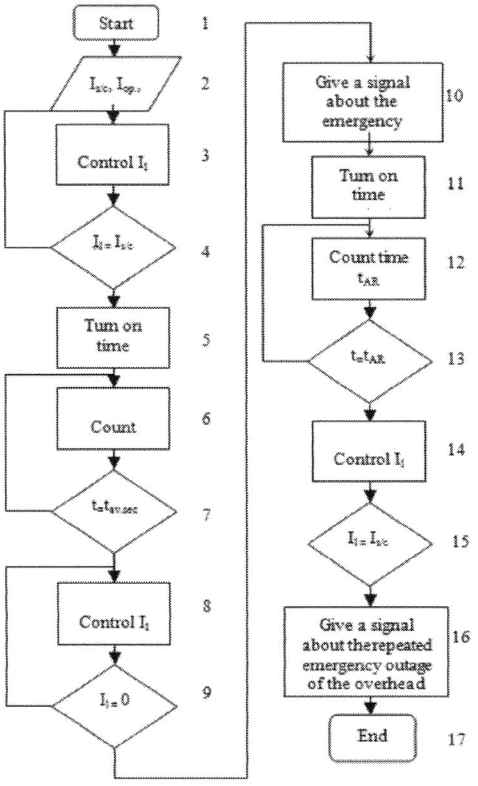

Fig. 3. Algorithm for monitoring emergency outages of circuit breakers in 10 kV overhead lines.

The beginning of the algorithm provides for the presence of embedded data on the values of the operating current and the minimum short circuit current, the rise rate of the short circuit current, the holding time of the automatic reclosing of the switch. The beginning of the algorithm (Fig. 3) provides monitoring of the appearance of the short-circuit current in 10 kV overhead lines, not only in its magnitude, but also in the rise rate. When the conditions laid down in block 4 of the algorithm are fulfilled, it is concluded that a short circuit has occurred and the time counter is started, counting down the time equal to the protection response time of the controlled circuit breaker in overhead lines. This is ensured by the operation of blocks 6 and 7. At the end of the countdown using blocks 8 and 9, the current is controlled [15, 16].

When the condition laid down in block 9 is fulfilled, block 10 gives a signal about the emergency outage of the overhead line switch. And then, the time counter of block 11and the exposure time of the automatic reclosing of this switch will turn on. This is ensured by the operation of blocks 12 and 13. At the time the countdown ends, the current is controlled by blocks 14 and 15. When the condition laid down in block 15 is fulfilled, block 16 gives a signal about the repeated emergency outage of the overhead line switch [17, 18, 20, 21].

3 Technical implementation

The proposed algorithm is implemented in an information processing device, made on the basis of a personal computer, with subsequent software-based processing of the received data. [21-24].

The device consists of a block for connecting current sensors and a block for converting and analyzing the input signal. This device has passed laboratory tests on the basis of FSBEI HE "Oryol State Agrarian University named after N.V. Parakhin" and proved its effectiveness [22, 24]. Currently, an agreement has been reached on the installation of an experimental emergency outage monitoring device in the Oryol Power Distribution Zone.

4 Conclusions

1. Analysis of the state of 10 kV overhead lines in the Oryol region made it possible to rank the outages. Ranking of emergency outages, in turn, showed that the majority of these outages were sustained ones.
2. In order to reduce the time for eliminating faults in overhead lines, an algorithm for remote monitoring of emergency outages was proposed.
3. Implementation of monitoring of emergency outages will reduce the time to eliminate the fault. It will lead to increased reliability of power supply to consumers through the adoption of the necessary decisions by attending personnel.

References

[1] I.N. Fomin, R.P. Belikov, Algorithm for Remote Monitoring of the Head Switch Line, Bulletin of Orel State Agrarian University **61**, 4, 72–77 (2016)

[2] L.D. Surov, I.N. Fomin, Monitoring of Successful Automatic Reclosing Of Sectionalizing Circuit-Breaker Line Ring Network, Bulletin of Orel State Agrarian University **22**, 1, 23–25 (2010)

[3] L.D. Surov, I.N. Fomin ,Monitoring Failure of Automatic Reclosing of the Sectionalizing Circuit-breaker, ELEKTRO, Electrical engineering, electricity, electrical engineering, 6, 46–48 (2009)

[4] L.D. Surov, I.N. Fomin, The Statefulness of the Head Switch Line Ring Network, Bulletin of Orel State Agrarian University **29**, 2, 112-117 (2011)

[5] M.V. Borodin, *Increase of Efficiency of Functioning of Power Supply Systems by Monitoring Power Quality* (Dissertation ... of candidate of technical Sciences: 05.09.03 / Lipetsk State Technical University, Lipetsk, 177, 2013)

[6] M.V. Borodin, A.V. Vinogradov, *Increase of Efficiency of Functioning of Power Supply Systems by Monitoring Power Quality, Monograph* (Orel: Publishing house of Orel SAU, 185, 2014)

[7] S.M. Astakhov, N.S. Sorokin, E.A. Semenov, Investigation of the Reliability of Information on the Occurrence of Short Circuits, Bulletin of Orel State Agrarian University **22**, 1, 25–28 (2010)

[8] A.V. Vinogradov, A.N. Vasilyev, A.E. Semenov, A.N. Sinyakov, V.E. Bolshev, The Analysis of Time of Interruptions in Power Supply of Rural Consumers and Methods of its Reduction Due to Monitoring of a Technical Condition of Power Lines, Bulletin of All-Russian Institute of Farm Electrification **2**, 27, 3–11 (2017)

[9] N.S. Sorokin, N.N. Chaadaeva, *The method of Estimating the Reliability of Power Supply Systems* (Bulletin of Construction and Architecture, the Collection of Scientific Papers, Orel, 127–130, 2016)

[10] A.V. Vinogradov, A.E. Semenov, A.N. Sinyakov, Analysis of Recovery Time of Power Supply of Rural Consumers in Case of Failure in Power Lines, Innovation in Agriculture: Problems and Prospects **1**, 13, 12–22 (2017)

[11] A.V. Vinogradov, A.V. Vinogradova, A.E. Semenov, A.N. Sinyakov, The analysis of Idle Time of the Electric Equipment Involved in Agricultural Technological Processes After Restoration Of Electric Power Supply, Agrotechnics and Power Supply **2**, 15, 26–31 (2017)

[12] A.V. Vinogradova, A.V. Vinogradov, V.A. Konstantinov, Application of Partitioning of Transmission Lines 0,38 kV to Improve the Reliability of Electric Power Supply of Rural Consumers, Innovation in Agriculture: Problems and Prospects **2**, 10, 12–26 (2016)

[13] A.V. Vinogradova, S.V. Blinov, The Improving of the Reliability of Transmission Lines of 0.38 kV With Their Partitioning, News of Higher Educational Institutions of Chernozem Region, 3, 8–12 (2014).

[14] L.D. Surov, I.N. Fomin, A way of Monitoring Successful and Unsuccessful Automatic Reclosing of the Switches in the Partitioned Line Ring Network, Patent of Russia No. 2305356, Bull., 24 (2007)

[15] A.V. Chernyshov, *Identification of Defects of Insulation of Overhead Transmission Lines With Voltage of 10 kV With the Formation of Artificial Ground Fault* (Physics and Modern Technology in Agriculture, Proceedings of the IX International Youth Scientific-Practical Conference, Orel: Publishing house of Orel State Agrarian University, 270–273, 2017)

[16] B.N. Meshkov, V.A. Chernyshov, Diagnostics of Damage of Overhead Power Lines in the Distribution Network Complex 10 kV, Innovations in Agriculture **2**, 12, 87–91 (2015)

[17] L.D. Surov, V.V. Filippov, The Monitoring of Disconnection and Failure of Automatic Reclosing of the Main Switch of Line, The News of Volgograd State Technical University **11**, 8 (135), 88–91 (2014)

[18] L.D. Surov, I.N. Fomin, N.V. Makhianova, Method of Monitoring of Disconnection and Failure of Automatic Reclosing Sectionalizing Circuit-Breaker Line Ring Network, Patent of Russia No. 2304338, Bull. 22 (2007)

[19] L.D.Surov, I.N. Fomin, N.V. Makhianova, R.V. Lebedev, Method of Monitoring State Changes of the Head Switch Line Ring Network, Patent of Russia No. 2410817, Bull. **3** (2011)

[20] L.D. Surov, V.V. Filippov, Determination of the Damaged Section of the Ring Network Line, Bulletin of the Volgograd State Technical University **8**, 187, 105–110 (2016)

[21] N.S. Sorokin, Using of the Integrated Circuit FT232R Chip AD7495AR and in a Block of Digital Processing in the System to Monitoring the Operation of Electrical Networks 6-35 kV, Bulletin of Orel State Agrarian University **60**, 3, 3, 87–93 (2016)

[22] Y.A. Kuznetsov, L.D. Surov, I.N. Fomin, Remote Monitoring of Automatic Reclosing Sectionalizing Circuit-breakers, Machinery and Equipment for the Village, 7, 29–33 (2015)

[23] N.S. Sorokin, Improving the Operation Efficiency of Electrical Networks with Voltage of 6-35 kV by Monitoring the Parameters of the Electrical Network, National University of Life and Environmental Sciences of Ukraine. Series: Technics and Energy of the Agro-industrial Complex, 209-2, 92–95 (2015)

[24] N.S. Sorokin, Connection Unit for Alarm Detection Sensor in Distribution Networks 6-35 kV, Bulletin of Orel State Agrarian University **29**, 2, 118–119, (2011)

[25] V.A. Chernyshov, E.A. Pechagin, *The Remote Monitoring of Overhead Transmission Lines with Insulated Neutral* (In the book: Energy Conservation and Efficiency in Technical Systems, Proceedings of the IV International Scientific-Technical Conference of Students, Young Scientists and Specialists, Tambov State Technical University, 355–356, (2011)

Environmental study of use of coal-water slurry fuel as an alternative to traditional fuels

Anton Sinitsyn[1,*], *Irina* Akhmetova[2], *Ekaterina* Filippova[1] and *I A* Vladimirov[3]

[1]Department of Heat and Gas and Water Supply, Vologda State University, Lenina str. 15, 160000, Russia
[2]Kazan State Power Engineering University, Kazan, Russia
[3]Peter the Great Saint-Petersburg Polytechnic University, Saint-Petersburg, Russia

Abstract. In this work we consider coal-water slurry fuels as an alternative to the traditional types of fuels, in particular, coal. Assessment is performed using a software tool for calculating the combustion process of coal-water slurry fuels. The obtained results will allow one to perform computer-aided environmental calculations during CWS combustion in experimental and industrial plants.

1 Introduction

Although the main public attention is focused on electrification of power sources and personal vehicles, liquid fuel still has a 100 times higher energy capacity than the best modern lithium-ion batteries, and this difference is predicted to last for more than a dozen years. Given this fact, investments in improving heat-generating devices and internal combustion engines continue to be relevant, especially owing to the fact that its efficiency increases together with significant reduction of harmful emissions to environment.

Currently, the problems of sustainable development of economic systems at the level of enterprises, regions, individual industries and the national economy as a whole are the subject of scientific research [1-5]. The modern conditions are characterized by acceleration the process of change and increasing uncertainty, both within economic systems and in the external environment, so solving problems of sustainable development are highly relevant. However, to date, the tools and mechanisms for ensuring sustainable growth remain insufficiently studied. This is due to variety of approaches for determination the content of the concept of sustainable development of an enterprise. Some works of Yu.P. Melentyev, V.M. Brodyansky, A.P. Merenkova and K.N. Trubetskoy are devoted to assessment the efficiency of the enterprise with transition to new energy sources and identification of the hidden reserves. However, most researchers studying the problems of energy conservation consider only factors providing energy conservation, affecting the development of preconditions for energy saving. Promising studies related to the use of new energy carriers are presented in the works of Delyagin G.N., Morozov A.G., M. J. Rini and others [6-8].

The authors of the present article believe that development of preconditions for energy conservation is only a necessary condition for energy conservation, while a sufficient condition for saving energy resources is to achieve an effective balance between the volume of energy and energy needs of the enterprise through the implementation of effective factors. The example is implementation of energy-saving technologies in modern energy systems of the real sector of the economy [9,10].

Currently, the world is rethinking the role and place of coal in ensuring the energy and economic security of countries. Moreover, an increase in the share of coal in the fuel balance is a stabilizing factor in protecting against the occurrence of deep energy crises. However, environmental problems arising from the use of coal fuel require the development and implementation of new environmentally friendly coal technologies [11]. In this regard, the use of coal in the form of coal-water slurry (CWS) is promising. The introduction of the CWS provides conservation of energy and material resources, as well as the environment. In addition, the use of CWS is the most effective and environmentally friendly method for disposal of thin coal sludge from coal mining and coal processing enterprises.

The focus was made on devices operating on vibrational (pulsating) combustion of solid, liquid, and gaseous fuels. The prospects of such studies are also emphasized by results of applying magnetic field to the fuel flow before its combustion.

The high price of traditional energy sources (gas and heating oil) makes one to look for ways to replace them with cheaper ones, which will reduce the cost of heat. One of the most promising products for such substitution is coal-water slurry fuel - a mixture of crushed coal (fractions of 1-70 microns), water and reagent. It is characterized by good calorific value and high degree of combustion.

Studying the practical experience in the production and use of CWS at the Amaltea LLC (Russia), special attention should be paid to the problem of increasing the environmental effect of replacing traditional fuel with a

* Corresponding author: sinitsyn.science@mail.ru

modernized one. The set task is to study the advantages and disadvantages of using water-carbon emulsion in real boilers based on the previously developed software tool for calculating the environmental indicators of the coal-water slurry combustion.

2 Subject of study

The subject of study is coal-water slurry fuel. To obtain coal-water slurry fuel, coal of gas grades is mainly used, which is characterized by high content of volatile substances. It is delivered to an open platform, and is fed into the receiving hopper of crusher using the front loader. Ball, roll and hammer mills are used for coal grinding. Almost each of these devices is characterized by the complexity and bulkiness of the equipment used, as well as by high energy costs. This requires implementation of new and more efficient grinders in the preparation of coal-water slurry fuels.

In general case, CWS is a complex heterogeneous structure consisting of finely dispersed coal and water, with addition of various additives and plasticizers that keep solid particles in suspension [11]. This structure complicates the creation and development of methods for combustion of slurry fuels, which should provide satisfactory indicators of combustion and emissions of nitrogen and sulfur oxides.

Theoretical and experimental studies on ignition of coal-water fuel particles prepared from various coals were carried out in [11-13]. The processes of pyrolysis, lighting, and combustion of water-carbon fuels were studied. Particles were suspended on quartz filaments. Heat was supplied by radiation. For ignition, air was introduced into the cuvette. The particle diameter ranged from 200 to 700 μm. The study identified six characteristic times of the process:

1. The beginning of thermal destruction and release of volatiles.
2. The beginning of the process of increasing the particle volume.
3. Completion of the process of increasing particle volume.
4. Completion of thermal decomposition and combustion of volatiles.
5. The beginning of burning of particles with charring of the agglomerate and its mechanical destruction.
6. Complete burnout of coke residue.

It was previously established that water has a significant influence on the ignition of CWS particles. As a result of moisture evaporation, a highly porous carbon skeleton with high thermal resistance is formed at the surface. In this case, the temperature of a water-coal particle rises much faster than that of conventional coal fuel. Water vapour resulting from evaporation is filtered through a porous skeleton. Accordingly, a zone of increased pressure is formed in the particle immediately near the evaporation front.

Water in any fuel is usually the ballast, which should be avoided. An alternative to dry grinding is wet grinding while homogenizing the resulting suspension. The use of wet grinding for grinding sludge solves several technological problems:

- Instead of a dry explosive powder, a liquid suspension is obtained, pumped by the pumps and evenly fed into the boiler furnace. The dosed supply of dry dust is a more complex process, and for boilers of both small and medium power it is even more expensive;
- The need for preliminary drying of sludge with the cost of evaporation is higher than just water evaporation;
- The overall cleanliness of fuel preparation area improves.

During wet grinding process, water is usually added to obtain a homogeneous suspension, although any waste, flushing from fuel oil tanks, etc. would be preferable. Given the moisture content of sludge from 25% and higher, the amount of added moisture is not more than 15%.

The presence of additional moisture will certainly affect the fuel balance of boiler. However, as numerous calculations show, such an increase in humidity is more economically effective than a set of measures for dewatering sludges and solving issues of their dosed supply. This is especially true for small and medium-sized energy facilities [12].

The environmental reasons for using CWS are fairly obvious: burning CWS should lead to reduction in formation of nitrogen oxides (NOx) and to more convenient measures for the capture of sulfur oxides (SOx). Nevertheless, the results on the completeness of burning coal in the form of CWS are widely known [11–16], but there are no real practical data on emissions from burning coal-water fuel both in large boilers and in boilers of small and medium power in the public domain.

In order to promote CWS on the market, Amaltea LLC (Moscow, Russia) together with Effective Energy Technologies GmbH (Vienna, Austria) created an industrial CWS preparation plant. A series of combustions of CWS from various types of raw materials was carried out using this equipment. The aim of works was to carry out measurements on the volume of emissions from CWS combustion according to the CE standards and to study some boiler operation modes.

This work presents the results of implementation of a previously developed mathematical model in a form of software and comparison of environmental friendliness of combustion of coal and CWS.

3 Environmental calculations using the developed program

The new software tool for computer-aided calculation of parameters of coal-water slurry fuel combustion process will simplify and speed up the calculations of the percentage composition of the resulting fuel; of the thermal characteristics of the source and obtained fuels with their subsequent comparison; of the amount of pollutants emitted with flue gases into the atmosphere when burning CWS; and of calorimetric temperature of fuel combustion. Automation of calculations avoids possible errors in calculations.

Table 1. Initial data for environmental calculations.

Parameter	Designation	Coal	CWS
1	2	3	4
Rated power of boiler, Gcal/h	Q_{rat}	0.4	0.4
Boiler efficiency,%	η	81	85
Heat losses with chemical incomplete combustion,%	q_3	2	0
Heat losses with physical incomplete combustion,%	q_4	7	0.7
Calorific value, kcal/kg	Q_c	5 453	4 112
The same, MJ/kg		22.8	17.2

The program has a form of file created in Microsoft Office. The software consists of three interrelated sections: "Composition and thermo-technical characteristics of fuels", "Calculation of pollutant emissions during CWS combustion", "Calculation of calorimetric temperature of CWS combustion", and applications containing the necessary data in the form of tables and plots.

In order to quantitatively evaluate the economic and environmental effect of CWS usage, a calculation and comparison of fuel consumption and the amount of pollutants emitted into the atmosphere with flue gases during combustion of coal and CWS was carried out. The comparison was carried out under the following conditions: coal was burned in the KVr-0.4 boiler with a nominal heating capacity of 0.4 Gcal/h with an efficiency of 81% (accepted from its technical characteristics). The combustion of coal-water slurry fuel is carried out in a boiler having a similar heat output and an efficiency of 85% (accepted averaged over the published results of previous studies).

The following initial data were used for calculations (table 1).

The obtained calculation results are shown in table 2.

Fig. 3 presents the calculation results in graphical form.

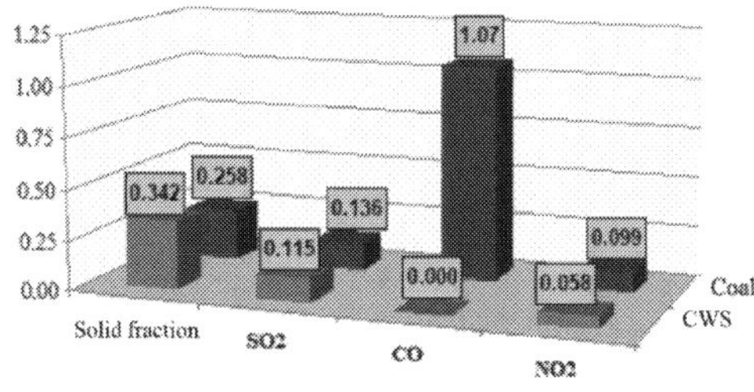

Fig. 3. Emissions of pollutants from combustion of coal and CWS.

Analysis of table 2 shows the following results:

1) Fuel consumption of boiler when using CWS increases by 21%;

2) Emissions of solid fractions increased by 33%;

3) Emissions of sulfur oxide decreased by 15%;

4) There are no emissions of carbon oxides when using CWS;

5) Emissions of nitrogen oxide decreased by 41%.

When using CWS the required fuel consumption for a given boiler capacity increases compared to that of coal. However due to more complete combustion of CWS coal particles, the total fuel consumption will be less by 10%. This is confirmed by the following calculation (table 3).

To sum up, we can conclude that the use of coal-water slurry fuel as an alternative to traditional types of fuels will reduce the cost of a unit of generated heat energy, make benefit from the products previously

Table 2. The calculated fuel consumption and the amount of pollutant emissions.

Index	Designation	Coal	CWS	Reduction in %
1	2	3	4	5
Fuel consumption per boiler, kg/h	B	90.6	114.4	-
The same g/s		25.2	31.8	-
Emissions of solid fraction, g/s	M_{sol}	0.258	0.342	-
Emissions of sulfur oxides, g/s	M_{SO2}	0.136	0.115	15
Emissions of carbon oxides, g/s	M_{CO}	1.07	0.0	100
Emissions of nitrogen oxides, g/s	M_{NO2}	0.099	0.058	41

Table 3. The economic effect of CWS usage.

Fuel type	The required consumption, kg/s	Efficiency of use, %	Total consumption, kg/s
1	2	3	4
Coal (grate firing)	90.6	70	129.4
Coal-water slurry fuels	114.4	98	116.7

disposed as waste, and save the environment.

4 Conclusions

The following conclusions can be made:

1) In general, the accumulated results of the previous authors and the present studies show the possibility of expanding the market for consumption of water-coal fuel, primarily in small and medium heat generation, and of eliminating the typical administrative and technical difficulties.

2) The use of compact devices for CWS preparation, the co-combustion of water-coal fuel and supporting fuel, accelerates its implementation, reduces the capital costs for implementation and operating costs associated with environmental protection;

3) Switching from grate firing of coal to the use of CWS results in an increase in the required fuel consumption of 21%, but due to more complete combustion of CWS coal particles, the total fuel consumption will be less by 10%.

Acknowledgements

The authors are grateful to Amaltea LLC for the information provided on the CWS production cycle.

References

[1] N. Politaeva, Y. Smyatskaya, R. Al Afif, C. Pfeifer, L. Mukhametova, Development of Full-Cycle Utilization of Chlorella sorokiniana Microalgae Biomass for Environmental and Food Purposes, Energies **13**, 10, 2648 (2020).

[2] L.V. Larchenko, R.A. Kolesnikov, L. Mukhametova, Russian oil and gas industry as a sphere of international interests and economic cooperation, E3S Web of Conferences **161**, 01006 (2020).

[3] V. Markov, A. Fedyukhin, I. Sultanguzin, B.G. Matisov, N.V. Nikitkov, V.V. Izrantsev, V.G. Knorring, G.A. Konrashkova, L.R. Mukhametova, Improvement of energy efficiency of motor fuel production at oil refinery, IOP Conference Series: Earth and Environmental Science **337**, 1, 012080 (2019).

[4] I.A. Sultanguzin, E.V. Zhigulina, Y.V. Yavorovsky, I.D. Kalyakin, A.V. Govorin, A.V. Fedyukhin, A.A. Krolin, S.V. Guzhov, O. Derevianko, L. Mukhametova, Development strategy of the energy supply system for MPEI campus blocks based on green building, IOP Conference Series: Earth and Environmental Science **288**, 1, 012071 (2019).

[5] A.A. Kalyutik, D.V. Grigorieva, A.V. Fedyukhin, O.V. Derevianko, I.G. Akhmetova, Reduction of the technological minimum by bypass method with environmental impact estimation on the example of energy blocks with T-100 and T-250/300-240 turbines, IOP Conference Series: Earth and Environmental Science **288**, 1, 012094 (2019).

[6] G.N. Delyagin, A.P. Petrakov, G.S. Golovin, E.G. Gorlov, Water dispersed systems based on brown coal as energy and process fuel, Russian Chemical Journal, 6, 72-77 (1997).

[7] M.J. Rini, R.C. Flesh, J.G. McGowan, Large-scale combustion applications of beneficiated fuels, Conference on coal and slurry technologies, Coal and Slurry Technologie Association, 65-76 (1990).

[8] K.V. Thumbimuthu, Environmental benefits of coal liquids fuels, Canada Center for mineral and Energy Technology, CANMET, 23 (1989).

[9] R. Kuusik, T. Kaljuvee, H. Veskimae, Yu. Roundygine, A. Keltman, Reactivity of Oil Shale Ashes towards Sulfur Dioxide, 3, Recurrent Use of Ash for Flue Gas Purification, Oil Shale **16**, 4, 303313 (1999).

[10] N. Sovetova, A. Sinitsyn, E. Tritenko, O. Derevianko, L. Mukhametova, A. Fedukhin, S. Makoev, A. Kalyutik, Regional management of district heating, IOP Conference Series: Earth and Environmental Science Collection of materials International Scientific and Practical Conference, Institute of Physics and IOP Publishing Limited (2019)

[11] Lee Sunggyu, G. Speight James, K. Loyalka Sudarshan, *Handbook of alternative fuel technologies* (CKC Press, New York, 2007).

[12] S.I. Mosin, A.G. Morozov, T.N. Delyagin, Russian experience in introducing industrial technology for the production of coal-water fuel, News of heat supply, 9 (2008).

[13] A.A. Sinitsyn, The study of vibrational characteristics of operation of boiler units Journal of Instrument Engineering 7, 51 (2008).

[14] A.A. Sinitsyn, D. Rusakov, V.V. Izrantsev, S.A. Kravchenko, G.A. Konrashkova, Investigation of gas fuel combustion efficiency at magnetic field treatment, IOP Conference Series: Earth and Environmental Science Proceedings of the Conference the international scientific conference Efficient waste treatment – 2018, EWT-2018, 012082 (2019).

[15] A.A. Sinitsyn, The calculation of the characteristics and of the geometrical sizes of power devices, European Applied Sciences **4** (2013).

[16] E.N. Filippova, A.A. Sinitsyn Development of software tool for calculating the combustion process of coal-water slurry fuel, Intelligent Information Technologies and Intelligent Business, INFOS-2019, Materials of the Tenth International Scientific and Technical Conference (2019).

Determination of the required power for bus hybrid engine

Vyacheslav Rakov[1,*], *Timur* Akhmetov[3], *Alexander* Capustin[2] and *Anatoly* Vostrov[1]

[1]Vologda State University, Department of automobiles and automobile industry, 160000 Lenin st., 15, Russia
[2]Saint-Petersburg State University of Architecture and Civil Engineering, 190005, St. Petersburg, 2nd Krasnoarmeyskaya st., 4, Russia
[3]Kazan State Power Engineering University, Kazan, Russia

Abstract. This article presents a methodology for determining the required engine power of hybrid city buses. The vehicle's driving cycle and its main technical characteristics were used as the initial data. The calculated change in power on the driving wheels is an intermediate result and is used to analyze the chains of energy transfer from gasoline engine to driving wheels. In this approach, a sequential type of circuit in a hybrid drive is used. A bus weighing 4 tons was considered as an example, and the calculations showed that the maximum power of the internal combustion engine should be 15.2 kW.

1 Introduction

City buses operate under conditions of frequent acceleration and braking, so the operation mode of the internal combustion engine (ICE) and transmission are constantly changing. Electromechanical hybrid engines fit very well for such working conditions. ICE in such engines is not directly connected to the driving wheels and can operate in the optimal stationary mode, producing energy for traction motors [1,2]. Therefore, when designing hybrid engines, it is especially important to determine the optimal characteristics of ICE, ensuring its minimum specific fuel consumption, low cost and sufficient performance.

To select the required power of ICE hybrid bus engines, a simplified traction calculation is often used, which does not take into account energy losses during energy transfer and power balance when consuming and refilling energy during motion. Thereby, often after manufacturing an experimental car, it turns out that the selected engine operates at low or, conversely, at high loads. This leads to a decrease in efficiency of engine, a lack of power when driving, or an increase in specific fuel consumption. For this reason, many hybrid bus projects have not been implemented, despite the large amount of money spent. Thus, a methodology for calculating the main parameters of a hybrid engine of a vehicle, and, first of all, its main source of energy - an internal combustion engine is needed.

Studies on the adaptation of ICE to unsteady operating conditions were presented by us earlier [3]. Also, some components of method are considered in [4,5]. A number of authors address the issues of modeling hybrid engines for vehicles [6,7].

The presented methodology is used to determine the maximum power of ICE operating in combination with an electric drive. It takes into account mass and aerodynamic characteristics of vehicle, the specified dynamic driving conditions and the characteristics of the drive elements themselves (transmission, electric motor, energy storage, power converter, generator). This methodology includes traction and energy calculation of drive. Both sections are related with each other, which enables to perform a comparative analysis and select the optimal ICE characteristics for a given vehicle, as a function of specified driving cycle. The methodology for calculating the maximum ICE power was modeled and became a part of the algorithm of the GSU-AVTO software [5], which made it possible to carry out theoretical modeling and obtain the necessary characteristics without the cost for making experimental running models.

2 Materials and methods

The basis for calculation the maximum power of ICE hybrid bus engine is the driving cycle. It is a set of data on the change in vehicle speed and longitudinal angle of the road plane as a function of time ($v=f(t)$; $\alpha=f(t)$). These data sets can be obtained in experiments by recording the change in parameters when moving along a given route or can be set theoretically. Thus, to perform calculation, it is advisable to use the average motion cycle. This paper presents an example of calculation based on the low-speed part of the standard WLTC driving cycle, which has been operating in Europe since 2018 (figure 1) [8]. This cycle consists of a series of accelerations to a speed of not more than 56 km/h and braking, including a complete stop. The average speed is 18.9 km/h, and speed-up during acceleration and braking are also suitable for light buses. For given operating conditions of bus, it is possible to set other types of driving cycles. So, some researchers use bus cycle models for a particular city [9,10].

Using the known traction calculation formulas, the resistance forces acting on bus with a hybrid engine were

* Corresponding author: vyacheslav.rakov@mail.ru

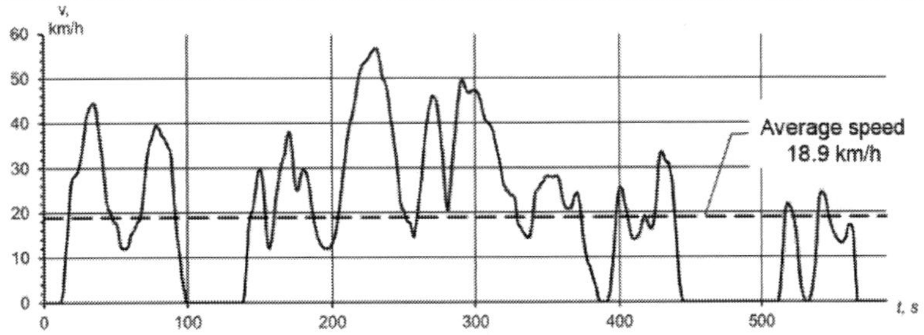

Fig. 1. Low-speed part of the standard test cycle WLTC.

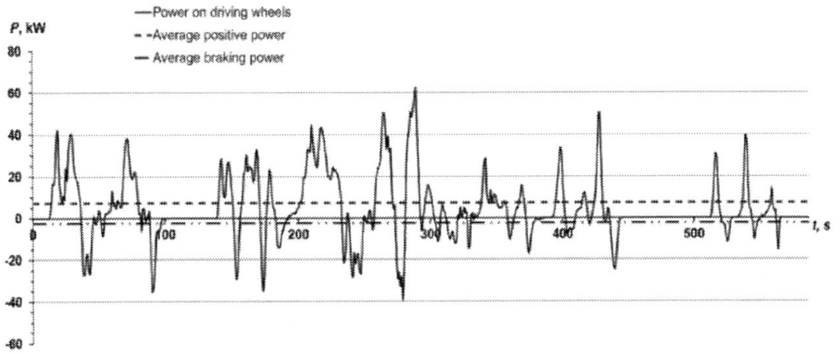

Fig. 2. Change in power on the bus driving wheels.

determined at each i-th point: N_{Ki} is the rolling resistance force; N_{vi} is the air resistance force; N_{Ii} is the inertia resistance force; N_{Pi} is the resistance to lift.

The power of traction force on bus driving wheels will be equal to the sum of all powers of resistance forces acting at a given time moment:

$$N_{Ti} = N_{Ki} + N_{vi} + N_{Ii} + N_{Pi}, \,[kW] \quad (1)$$

Thus, after calculating all the resistance forces, one can find the characteristic of power change of the driving wheels. Figure 2 shows the calculation results for power of the driving wheels of a bus weighing 4 tons when moving according to the standard cycle shown in figure 1.

Further, the obtained change in power of the bus driving wheels is used to analyze energy flows in engine.

The average power of traction force over the entire driving section $N_{POS.AVER}$ (figure 2) shows the average value of only positive traction power over the entire calculated section, n is the number of points of the test section. In fact, it is an average power of the hybrid drive without taking into account the energy returned during the recovery (formula 2):

$$N_{POS.AVER} = \frac{\sum_{1}^{n} N_{POS.AVERi}}{n}, \,[kW] \quad (2)$$

$N_{BRAK.AVER}$ (figure 2) is the average value of only negative values of traction power over the entire calculated section. In fact, this is the average power returned back to the energy storage:

$$N_{BRAK.AVER} = \frac{\sum_{1}^{n} N_{BRAK.AVERi}}{n}, \,[kW] \quad (3)$$

Thus, based on expressions 1-3, the ICE power in general terms will be calculated by:

$$N_{ICE} = \frac{N_{POS.AVER}}{K \times \eta_{DRIVE}} - N_{BRAK.AVER} \times \eta_P, \,[kW] \quad (4)$$

where N_{ICE} is the required ICE power, taking into account losses in drive, kW; K is the coefficient of efficient operating mode of ICE for this section (determined experimentally [3,4]); η_{DRIVE} is the drive efficiency; η_R is the efficiency of recovery on the route.

 https://doi.org/10.1051/e3sconf/202017801076

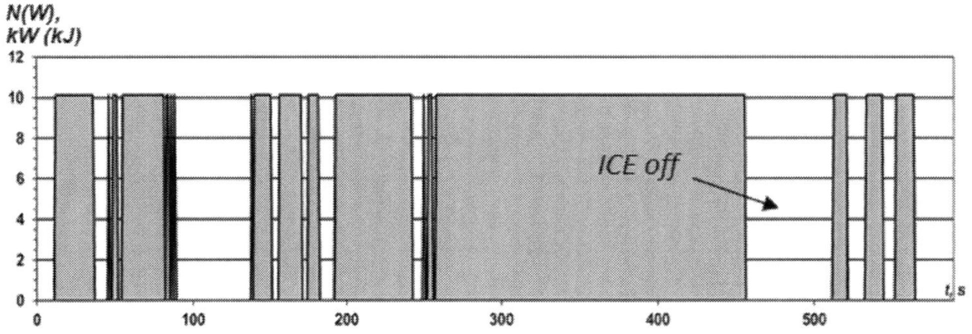

Fig. 3. The developed power of an internal combustion engine of a hybrid power plant.

Further, the power flows between ICE and driving wheels were determined, taking into account the features of the applied typical scheme of a hybrid engine.

A sequential energy transfer scheme is often used in hybrid bus engines [5]. The mechanical connection between the driving wheels and the internal combustion engine in this scheme is completely absent. The advantages and disadvantages of such a scheme were considered by the authors earlier [1]. In this case energy will be transmitted along the following chains:

- From ICE to the generator, then to the traction motor and to the driving wheels (N_1);
- From ICE to the electric generator, then to the high-voltage battery (HVB), then to the traction motor and to the driving axle (N_2);
- During recovery, energy will be transmitted from the driving wheels through the generator to HVB, and further spent on motion again (N_2).

The energy transfer chains in a hybrid engine were described in detail by the authors in [4, 11].

Therefore, the required ICE power is made up of all 3 energy transfer chains:

$$N_{ICE} = N_1 + N_2 - N_3, \left[kW\right] \quad (5)$$

Figure 3 shows the ICE operating mode, from which it follows that it operates in quasi-stationary mode. The developed ICE power in this case is about 10.5 kW. While calculating N_{ICE} value, it is necessary to take into account the optimal ICE loading mode, which is usually equal to 75% of the maximum power. Thus, the maximum power of ICE will be more by 25%. One should also take into account power consumed by all on-board devices, which can reach 2 kW. Therefore, taking into account the safety factor K_P and on-board energy consumers N_C, the final ICE power will be calculated using the following expression:

$$N_{ICE}^{MAX} = K_P \times N_{ICE} + N_C \quad (6)$$

3 Results

For a hybrid engine of a bus weighing 4 tons when driving in urban conditions, the maximum ICE power should be 15.2 kW. In this case, ICE operates in a quasi-stationary mode corresponding to the minimum specific fuel consumption. Peak power values resulting from accelerations will be compensated by the energy of buffer storage of electrical energy of sufficient power. The overall balance of electrical energy will always remain constant. In this mode, the hybrid engine of bus will almost always work in the mode of minimum specific fuel consumption. And when there is no load on the driving wheels, ICE will generate energy to charge the buffer drive or will be completely off.

It should be noted that the methodology presented by the authors will work only with a sufficient capacity of high-voltage battery providing a long supply of energy and peak power surges on the bus driving wheels. If necessary, one can use a smaller battery, but the maximum ICE power in this case will be higher. Evaluation of the required battery capacity of a hybrid engine will be considered by the authors in their further papers.

4 Conclusions

Thus, the methodology presented by the authors allows one to determine the required ICE power of a bus hybrid engine, at which it will perform the specified transport work with minimum specific fuel consumption.

This methodology can be useful for conducting theoretical research and engineering calculations during design of energy-efficient buses with hybrid engines.

References

[1] A. Kapustin, V.A. Rakov, Hybrid cars [Gibridnye avtomobili] (Vologda: Vologda State University, 96, 2016).

[2] G. Pistoia, Electric and hybrid vehicles, Power sources, models, sustainability, infrastructure and

the market (Oxford: The Netherlands Linacre House, Great Britain, 670, 2010).

[3] I.K. Aleksandrov, E.V. Nesgovorov, V.A. Rakov, Tractional Calculations for Vehicles with Adaptive Motors, Russian Engineering Research 30, 2, 122-124 (2010). DOI: 10.3103/S1068798X1002005X.

[4] V. Rakov, V.N. Ivanov, Y.G. Karpov, V.F. Melehin, R.A. Izmailov, V.K. Yun, D.A. Zaripova, Method for determining the basic energy characteristics of elements of a hybrid car engine, IOP Conference Series: Earth and Environmental Science 337, 1 (2019). DOI: 10.1088/1755-1315/337/1/012066.

[5] I.K. Aleksandrov, Determination of hybrid power installation parameters based on trolleybus driving cycle studies [Opredeleniye parametrov gibridnoy silovoy ustanovki na osnovanii issledovaniy yezdovogo tsikla trolleybusa], Alternative fuel vehicles 5, 41, 54-60 (2014).

[6] M.V. Yutt, Principles of building a combined energy installation of a minibus for inter-city transportation [Printsipy postroyeniya kombinirovannoy energoustanovki mikroavtobusa dlya vnutrigorodskikh perevozok], Electronics and electrical equipment of transport 1, 14-16 (2008).

[7] T.J. Boehme, B. Becker, M. Ruben-Weck, M. Rothschuh, A. Boldt, C. Rollinger, R. Butz, H. Rabba, Optimal design strategies for different hybrid powertrain configurations assessed with European drive cycles, SAE world congress, Technical Paper, 1751 (2013). DOI: 10.4271/2013-01-1751.

[8] M. Tutuianu, P. Bonnel, B. Ciuffo, et al., Development of the World-wide harmonized Light duty Test Cycle (WLTC) and a possible pathway for its introduction in the European legislation, Transportation Research Part D: Transport and Environment 40, 61-75 (2015).

[9] H. Wi, J. Park, J. Lee, W. Kim, et al., Development of a City Bus Driving Cycle in Seoul Based on the Actual Patterns of Urban Bus Driving, SAE Technical Paper (2009). DOI: 10.4271/2009-01-2914.

[10] B.H. Wang, Y.G. Luo, J.W. Zhang, Simulation of city bus performance based on actual urban driving cycle in China, Int. J. Automot. Technol. 9, 501-507 (2008). DOI: 10.1007/s12239-008-0060-3.

[11] V.A. Rakov, A.V. Smirnov, Determination of the required power of internal combustion engines of hybrid engines of vehicles [Opredeleniye neobkhodimoy moshchnosti DVS gibridnykh silovykh ustanovok transportnykh sredstv], Engineering Bulletin 4, 29-31 (2010).

To the issue of assessing the efficiency of distributed generation in low-voltage electric networks

A.N. Alyunov[1], *A.Yu* Belyanin[1], *A.E.* Nemirovsky[1] and *D.A.* Zaripova[2]

[1]Vologda State University, Vologda, Russia
[2]Kazan State Power Engineering University, Kazan, Russia

Abstract. This article considers an urgent task of using autonomous sources for power supply systems of industrial enterprises of small and medium-sized production. The efficiency of choosing the time of their switching-on in peak hours of power system in order to reduce the cost of electricity is shown on the example of diesel generators.

1 Introduction

At present the problem of increasing electricity tariffs for centralized electricity supply is brewing in some areas of industry. Statistical analysis over the past few years shows that the cost of electricity to the end consumer is increasing. The following factors have an influence on this trend:

- Transmission of electricity over long distances results in severe losses;
- Obsolete electrical equipment is still in operation;
- Electric networks have a lack of flexibility and mobility.

This problem can be solved by switching from centralized power supply to distributed generation. The main advantage of this process is production of electricity directly at the place of consumption, which guarantees optimal efficiency during its transportation [1-4].

The aim of the study is to assess the efficiency of distributed generation in low-voltage electric networks for production facilities.

Currently, in the Russian Federation, the primary consumers of electricity are industrial enterprises whose installed power range from several kilowatts to ten megawatts.

Today, the following categories of distributed generation can be considered:

1. Isolated generating plants.
2. Combined heat and power plants.
3. Diesel, gas turbine and gas piston power plants.

At first glance, these categories of distributed generation have more disadvantages than advantages for small production facilities.

To obtain the optimal saving effect from the use of a diesel generator when powering an object, the necessary condition is to come up to coincidence of hours the generator is turned on with hours of peak loads in power system and with hours of power take-off. Studies of electricity consumption schedules show that it is possible to reduce the cost of electricity and power in the specified hours [5].

For enterprises of small and medium-sized production it is inefficient to use large isolated generating plants, due to the lack of territory for their construction and significant costs for their construction and maintenance. In addition, many enterprises do not have gas distribution stations, substations and main gas pipelines in the immediate vicinity. One of the advantages of diesel power plants is the possibility of their widespread use, since diesel fuel is much more affordable than other types of fuel.

The comparative characteristics of diesel and gas power plants are presented in table 1.

Diesel generators are oriented to longer work in comparison with gasoline generators. Gasoline generators can operate continuously for only a few hours and have power limitations.

Undoubtedly, to obtain the greatest economic effect, it is more expedient to switch to autonomous power supply using gas generators. However, due to significant one-time investments related to the commissioning of gas power plants for small and medium-sized productions, it is more rational to use diesel generators.

For any organization, the expenditures for purchasing electricity comprise a significant share in the cost of produced goods and services. All subjects of the Russian Federation according to the procedure for settlements for electricity can be divided into two groups:

- Zones of wholesale market in which the price of electricity is formed under competition;
- Non-price zones without competition, in which the price for electricity purchase is approved by the local executive body with authority in the field of tariff regulation.

It should be noted that the bulk of the electricity price is formed by the prices of transmission services, sales

Table 1. Advantages and disadvantages of diesel and gas power plants.

Power source	Advantages	Disadvantages
Gas generators	1) The cost of a unit of generated electricity is much lower than that of gasoline and diesel generators; 2) Low noise during operation; 3) High environmental performance.	1) Gas is explosive fuel, the probability of ignition of which is much higher than that of diesel fuel; accordingly, it is necessary to observe special conditions of operation, storage, transportation of fuel tanks; 2) The relatively large dimensions of the installation; 3) Obtaining special permission from the relevant authorities to connect the unit to the gas pipeline; 4) Temperature conditions for storage of gas fuel.
Diesel generators	1) Low operation costs; 2) A relatively high efficiency, within 50%; 3) Low noise during operation; 4) The ability to work in various weather conditions; 5) Compactness; 6) Quick start, stable performance.	1) High cost of primary fuel; 2) High requirements to the quality of diesel fuel; 3) Load factor of diesel generator should be at least 40%; 3) High power diesel generators require additional installation of a ventilation system.

allowances and purchase prices of electricity in the wholesale electricity market [6], [7].

Currently, six price categories can be distinguished with different procedures for the formation of the final price for electricity. Practice shows that the right choice of price category can reduce the cost of purchasing electricity by up to 30 percent.

Using a diesel generator at the enterprise and turning it on during peak hours of power system allows one to get an additional effect by reducing expenditures for power purchase [8], [9].

2 Materials and methods

Consider the idea of implementation of own generation in a low-voltage electric network using the example of electricity consumed by a manufacturing enterprise for 1 month (December) [10].

The greatest economic effect can be achieved by reducing the payments for power. For this, it is necessary to turn on the diesel generator during peak hours (peak load monitoring hours), thereby reducing the amount of power paid for in the wholesale electricity market. For this purpose, it is necessary to simultaneously solve the problem of choosing the optimal power and strategy of the diesel generator [10].

Electricity consumption should be considered only during hours of power take-off, that is, during peak hours on working days of month.

For clarity, we present them in the average monthly load schedule during power take-off hours (Fig. 1).

The methods of mathematical statistics were used to study the X criterion, i.e. power at peak hours on working days of month.

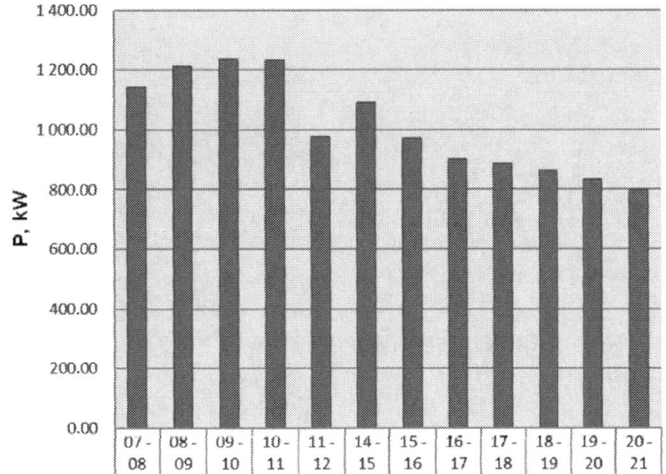

	07-08	08-09	09-10	10-11	11-12	14-15	15-16	16-17	17-18	18-19	19-20	20-21
P, kW	1 143	1 213	1 234	1 233	975.9	1 090	969.3	899.1	886.8	863.8	832.8	799.3

Fig. 1. Monthly average load schedule during power take-off hours (December)

The average value (analog of mathematical expectation) of the investigated criterion is:

$$\overline{x} = \frac{\sum_{k=1}^{i}(X_k \cdot m_k)}{n} = 1011.52\ kW. \quad (1)$$

The variance is

$$D = 96097.76\ kW^2 \quad (2)$$

The sample standard deviation is the most accurate variation indicator, which indicates by what value the entire set of random variables may differ from the average value (mathematical expectation):

$$\sigma = \sqrt{96097.76} = 310\ kW. \quad (3)$$

Based on the study, we determine the possible number of operation hours of the diesel generator, provided that the costs of the diesel generator will be covered by the cost of compensated power. The amount of paid power will be lower, and additional benefits will be achieved by reducing the amount of purchased electricity [10]:

$$t = \frac{C_W + C_N}{C_D \cdot N_D} \approx 12\ h, \quad (4)$$

where C_W is the purchase price of electricity in the wholesale electricity and power market, \$/kW; C_N is the system utilization charge of electrical networks of the selected tariff, \$/kW; C_D is the price of kWh of electricity generated by a diesel generator, \$/kW·h; N_D is the number of working days in a month.

Practice shows that to obtain additional savings, it is necessary to partially reschedule the work of production equipment from peak hours to another time, if the technological process enables this change.

3 Results and discussion

Thus, we find that the optimal number of operation hours of a diesel generator per day (in December according to the original schedule) is 12 hours. For the considered situation, by switching on the generator at the indicated hours, we are guaranteed to reduce both the amount of power paid in the wholesale market and the network power by the value of the nominal power of the diesel generator.

Based on the data obtained in the study, it can be argued that the diesel generator power should be selected from the condition:

$$\sigma \le P \le 2.5 \cdot P_{MIN}, kW,$$

where P_{MIN} is the minimum power consumption, kW; σ is the standard deviation from average power, kW

It is impractical to set power over $2.5 \cdot P_{MIN}$, since the load factor of a diesel generator in some hours will be less than 40%, which will lead to accelerated wear of the internal mechanisms of unit.

We obtain that

$$310 \le P \le 2.5 \cdot 407.45\ kW.$$

According to the range of standard power we accept P=400 kW. Given the generator's operating strategy for 12 hours a day during peak hours (from 8 to 12 hours and from 15 to 21 hours, inclusive), it is possible to determine the economic effect of using a diesel generator when reducing the power supply. However, turning on the unit for all 12 hours is irrational, therefore, it is necessary to determine the most probable reporting hours. Since peak hours are published at the end of the reporting period, the most probable hours should be determined based on statistical data over the past periods of time.

We denote H as the peak load hour on working days of the considered December, obtained as a result of measurements and determine the numerical characteristics of the considered criterion H:

$$mean\ \overline{h} = \frac{1}{2}\sum_{i=1}^{21} h_i,$$

$$variance\ D(H) = \frac{1}{n}\sum_{i=1}^{21}\left(h_i - \overline{h}_i\right)^2 = 7.8.$$

We denote Y as the peak hour on working days of another month (February) obtained as a result of observations. The determined numerical characteristics of the investigated characteristic Y are:

$$mean\ \overline{y} = \frac{1}{n}\sum_{j=1}^{20} y_j = 10,$$

$$variance\ D(Y) = \frac{1}{n}\sum_{j=1}^{20}\left(y_j - \overline{y}\right)^2 = 3.55.$$

We use the Pearson's chi-squared test, as one of the most frequently used in studies, to test the hypothesis that the peak hour of the load is the same in different months. This fact allows us to judge that the populations from which the samples are extracted are the same, so they have the same distribution law. The description of the experimental data using the well-known distribution law makes it possible to present the probability density function and the distribution function in an analytical form. This is useful in finding the probability of falling into a certain interval of the real peak load hours. [11].

Table 2 shows peak hour data. h_i in December and y_i in February, as well as the number of working days in which this hour took place.

Further we calculate the empirical value of chi-square:

$$\chi_E^2 = M \cdot N \sum_{i=1}^{l} \frac{\left(\frac{h_i}{M} - \frac{y_i}{N}\right)}{h_i + y_i} \approx 11.44$$

Table 2. Statistics of peak hours.

Peak hour	h_i	y_i
8	0	2
9	3	9
10	5	3
11	10	5
17	0	1
18	3	0
Total	$M = \sum_{i=1}^{6} h_i = 21$	$N = \sum_{j=1}^{6} y_i = 20$

Using the chi-square table for the number of degrees of freedom $k=l-1=6-1=5$, we find the probability that, due to random reasons (rather than sample homogeneity), the value χ^2 will be not less than empirical value [11].

The found probability is quite small, therefore, the hypothesis of uniformity of samples, indicating the same choice of peak hour of power system every month using the Pearson criterion chi-square, is rejected. In addition, the spread of obtained numerical characteristics speaks in favor of different distribution of peak hour of power system in the indicated months. Studies based on probabilistic and statistical methods show that the choice of optimal hour is irrational.

Therefore, it is necessary to determine the time interval for the strategy of own generation. When choosing the optimal solution, other power characteristics in low-voltage electric networks can be taken into account, for example reactive power [12].

References

[1] K.S. Sambaiah, T. Jayabarathi, Optimal reconfiguration and renewable distributed generation allocation in electric distribution systems, International Journal of Ambient Energy (2019). DOI: 10.1080/01430750.2019.1583604.

[2] S. Kansal, B. Tyagi, V. Kumar, Cost–benefit analysis for optimal distributed generation placement in distribution systems, International Journal of Ambient Energy 38 (2017). DOI: 10.1080/01430750.2015.1031407.

[3] E. Sukhareva, A. Fedyukhin, O. Derevianko, M. Egorov, L. Mukhametova, I. Akhmetova, Optimization of combined heat and power plant operating mode by means of underutilized equipment mothball, IOP Conference Series: Earth and Environmental Science 288, 1, 012122 (2019).

[4] A.A. Sinitsyn, N.N. Boytsova, I.G. Akhmetova, S.O. Makoev, A. Asadpoori, O.A. Stepanov, N.V. Rydalina, E.O. Antonova, B.G. Aksenov, Study of operation of power-generating devices of gaseous fuels combustion, International Journal of Mechanical Engineering and Technology 10, 2, 1480-1490 (2019).

[5] A. Jabbary, H. Shayeghi, Smart imposing of operational limits in optimising the technical and economic indexes of a distribution network in the presence of distributed generation sources, International Journal of Ambient Energy (2018). DOI: 10.1080/01430750.2018.1492451.

[6] J. Salehi, M. R.J. Oskuee, Optimal planning of distributed generation from the DisCo and the DGO viewpoints considering the uncertainties in future demand and electricity price, International Journal of Ambient Energy 39 (2018).

[7] N. John, V. Janamala, J. Rodrigues, Impact of Variable Distributed Generation on Distribution System Voltage Stability, International Conference on Data Science and Communication (IconDSC), Bangalore, India, 1-5 (2019). DOI: 10.1109/IconDSC.2019.8817028.

[8] F.R.A. Bukit, N. Mubarakah, D.M. Nasution, Minimizing The Electrical Power Distribution Cost By Optimizing the Distributed Generation Placement, 3rd International Conference on Electrical, Telecommunication and Computer Engineering (ELTICOM), Medan, Indonesia, 60-65 (2019). DOI: 10.1109/ELTICOM47379.2019.8943868.

[9] V.V. Dubaylova, N.D. Mukhlynin, O.J. Polyakova, Distributed Generation Control in Energy Market Conditions Using Advanced Model Predictive Control, 16th Conference on Electrical Machines, Drives and Power Systems (ELMA), Varna, Bulgaria, 1-4 (2019). DOI: 10.1109/ELMA.2019.8771696.

[10] A.N. Alyunov, V.A. Babarushkin, IT service for energy saving in power supply systems, Energy-efficient technologies in a modern institution, International Energy Forum (2013).

[11] O.I. Mikryukova, S.V. Ivanova, A.Yu. Belyanina, Nonparametric criteria for testing statistical hypotheses (Vologda State University, Vologda, 2017).

[12] D. Matvoz, R. Leskovec, M. Maksić, Optimized reactive power characteristics for distributed generation sources in the low voltage network, IEEE Manchester PowerTech, Manchester, 1-6 (2017). DOI: 10.1109/PTC.2017.7981070.

https://doi.org/10.1051/e3sconf/202017801078

Features of electroosmosis usage for drying and moisture protection of electrical equipment

Alexandr Nemirovskiy[1], *Galina* Kichigina[1], *Regina* Salikhova[2] and *Alexandr* Alyunov[1]

[1]Vologda State University, Lenina Street 15, Vologda, 160000 Russia
[2]Kazan State Power Engineering University, Kazan, Russia

Abstract. The article describes a new method for drying electric motors and transformers based on electrokinetic phenomenon of electroosmosis. The process of electroosmotic motion of moisture in insulation capillaries is described. The advantages of electroosmotic drying associated with a significant reduction in energy and labor costs are presented.

1 Introduction

One of the problems arising during operation of electrical equipment in Russia and other countries is wetting of winding insulation. The annual number of failures of electric motors (EM) due to wetting of windings insulation in the agricultural sector and some other industries reaches 20-30% [1, 2].

The development of specialized EM for wet environment of agricultural premises and mines carried hope for the possibility of solving the problems of insulation wetting. However, the experience in operating such EMs has cast doubt on their ability to withstand long-term exposure to a humid environment. The idea of sealing EM enclosures of special designs could not be fully realized. Rubber gaskets and seals at places where the shaft goes out of the housing were not able to provide sealing at change of temperature regime. The engine "breathed" and moist air was sucked in. When a winding cooled down, the moisture condensed, which caused hydrolysis and reduction of the insulation resistance R [3, 4]. Calculations showed that in EM of a 160 kW capacity and a cavity volume of 0.05 m^3, 1.4 g of moisture accumulate during one day of inactivity [5]. So, the EM insulation R in short-term operating conditions in an environment with 100% air humidity decreased to 100 kOhm [6].

A similar situation is typical for transformers [7, 8], although the number of failures is less, but it is considerable, especially in terms of economic damage. So, according to William Bartley, the head of the working group for damage analysis of power transformers of the International Association of Engineering Insurers (IMIA), the damage caused by failure of only one transformer due to insulation wetting amounted to \$ 175,000 [9]. The main source of transformer wetting is atmospheric moisture penetrating under the influence of pressure gradient, especially through weakened seals. According to Vanina B.V. damage to power transformers with voltage of 110-500 kV in operation over a period of 12-25 years is largely due to wetting of windings, as well as penetration of atmospheric moisture through high-voltage bushings (130 cases) [10].

There are quite a few ways to dry EM, but all of them are limited to thermal ones: electric lamps, ovens, blowers, welding transformers, etc. [6, 3]. Transformers are also dried mainly by thermal methods: induction heating, short circuit current, direct current, zero sequence current, circulating hot oil, heating the tank with infrared radiation, blowing by hot air, in a drying chamber, etc.

Thermal drying methods for EM and transformers can cause warping, thermal degradation, local overheating and thermal aging of insulation. These methods are often associated with disassembling and dismantling of EM and transformers from their workplaces, as well as with a presence of special equipment and increased energy consumption [11, 12].

The known methods of moisture protection are mostly thermal. They are associated with energy consumption and heating of windings of electrical equipment (EE). Moisture protection can be implemented by sealing (encapsulation) of EE windings [3]. However, in the first case there are drawbacks inherent to EOD. In the second case, complete sealing, as a rule, cannot be achieved and the windings continue to become wet.

At the Department of Electrical Equipment of the Vologda State University, new methods of drying and moisture protection of EE have been developed, based on the electrokinetic phenomenon of electroosmosis [13-20].

Electroosmosis was discovered in 1809 by F.F. Reuss and consists in motion of liquid in capillary systems induced by electric field. Modern concept of electroosmosis mechanism is based on the idea of the existence of a double electric layer (DEL) at the interface between liquid and solid phases. A phase with higher dielectric constant (water) is charged positively, and with lower one (dielectric) is charged negatively. Due to extremely high dielectric constant of water, a wide variety of dielectrics get a negative charge in it. The reason for DEL formation may be different. At the interface between two phases, there is a possibility for absorption of ions from the surrounding solution, which are formed there due to electrolytic dissociation. According to another theory, at a certain temperature in the ionic lattice on the surface and inside the solid phase of dielectric, there are some free places. The charge will be negative if the cation vacancies are in excess, and positive if there is an excess of anion vacancies.

The physics of DEL is shown most consistently in [21]. DEL at the dielectric-liquid interface is represented as two parallel surfaces (Fig. 1), where the dielectric surface is negatively charged, and the dense layer of counterions of liquid phase has a positive charge.

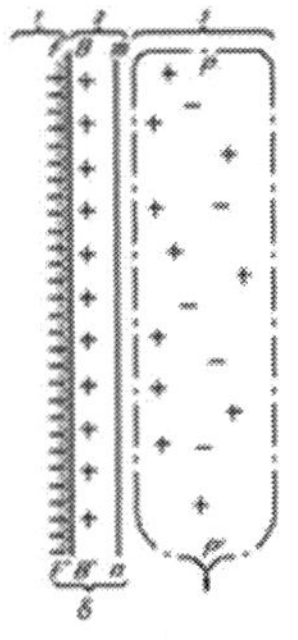

Fig. 1. Structure of a double electric layer [21]: *BB'* is the dense layer of counterions; *FF'* is the dielectric surface; *mn* is the sliding surface of the diffusion part of liquid; *PP'* is the diffusion layer of counterions; 1 is the solid phase; 2 is the adsorption (fixed) liquid layer of thickness δ; 3 is the moving liquid layer.

Counterions in liquid phase are under the influence of two mutually opposite forces. On one hand, they are electrostatically attracted by dielectric surface, on the other hand, the thermal motion of liquid particles, in which counterions also participate, tends to uniformly disperse them in the thickness of liquid. As a result of interaction of these oppositely directed forces, the counterions in liquid form a diffuse ionic atmosphere near the solid surface of adsorbent. Concentration of counterions is greatest near the surface *FF'*, and then gradually decreases. If liquid adjacent to the solid surface *FF'* is divided into a series of thin layers f_1K_1, f_2K_2, f_3K_3, then each of them will be charged with counterions to a potential, which decreases when this layer is moved from the solid wall (FC curve in Fig. 2). The line CQ indicates the boundary of the diffuse layer, and the distance F'C corresponds to its thickness.

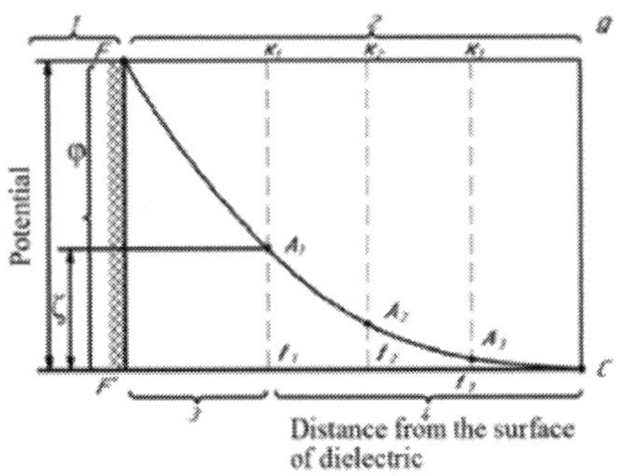

Fig. 2. Potential distribution in the double electric layer and in the diffuse part of liquid [21]: CQ is the boundary of diffuse layer of counterions; 1 - solid phase (dielectric); 2 - liquid phase (water); 3 - adsorption layer; 4 - moving liquid layer.

When liquid moves along a solid surface, liquid phase slides not along the boundary between the solid surface and liquid, but along another surface, slightly shifted toward the liquid phase (surface *mn* in Fig. 1). The thinnest adsorption layer of liquid *B'Bmn* directly adheres to the solid surface *FF'* and is held by it very firmly. During general liquid motion, this layer and the counterions inside it ("bounded" counterions) remain motionless. Only counterions that are behind the sliding surface *mn* ("free" counterions) can flow out together with liquid. The fixed adsorption layer *B'Bmn* contains only a certain part of counterions. This part only lowers the potential of dielectric solid surface, but does not completely compensate its charge. In this case, as a result of interaction between positive and negative charges in a fixed layer of thickness δ, some potential that is a fraction of total potential of a solid surface remains not neutralized. Thus, at the boundary *mn* (Fig. 1), i.e. on the sliding surface, a special difference (jump) in potentials is created, which is called the electrokinetic or zeta potential ζ. The ζ potential is equal to the potential difference between the solid surface together with adsorption layer of liquid on one hand, and its moving part on the other. Total drop of potential from its value on the surface *FF'* to zero at point *C* corresponds to the maximum potential difference between the solid wall of dielectric and all counterions (Fig. 2) .This maximum potential difference is s thermodynamic potential φ, but the electrokinetic potential ζ plays a major role in the electrokinetic phenomena.

When electric field is applied to a capillary with liquid, the excess ions of the same sign in the outer DEL part begin to move to the oppositely charged pole (Fig. 3, a). Ions of the inner part of DEL, located directly on the solid wall, and ions of the opposite sign in the adsorption layer do not participate in liquid motion. In the middle part of the capillary far from the wall, there are equal numbers of ions of both signs.

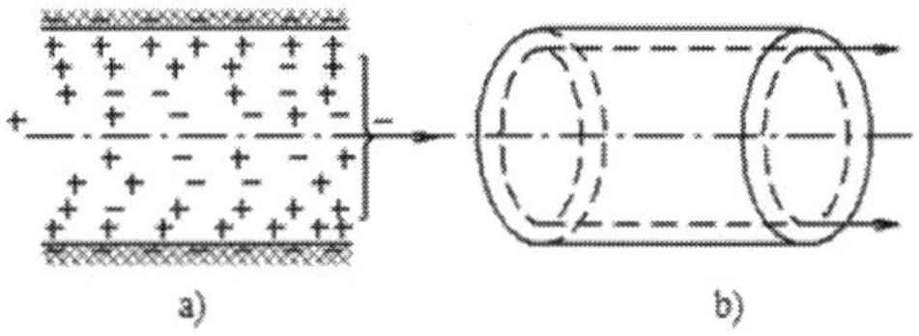

Fig. 3. Capillary model of electroosmosis: a) charge structure; a) motion of a liquid volume.

Therefore, in electric field ions of different signs move in opposite directions with velocities corresponding to their mobilities and potential gradient. H^+ cations are more mobile than hydroxyl groups of OH^- anions; therefore, the motion of ions in aqueous medium occurs mainly from the positive to the negative pole. Thus, a directed flow of excess ions of diffuse layer is created near the wall. In a round-shaped capillary, such an ion flow has a cylindrical shape and moves to the oppositely charged pole (Fig. 3, b). During its motion, the flow carries away the rest mass of liquid in the capillary, which is facilitated by friction and molecular cohesion.

Note that the latter circumstance is most significant when water is removed from the capillary, because water weakly dissociates into ions and dehydration only due to motion of ions is insignificant. At the same time, the described representation of electroosmotic liquid transfer indicates that the more ions of the same sign are in the diffuse part of DEL, i.e. the larger ζ is, the stronger is the resulting action of the moving near-wall layer of ions and the faster the entire liquid mass in the capillary should move.

If the potential difference U is applied to the capillary ends, and electric field strength is parallel to the axis of the capillary tube, then the electric force acting on the surface unit of the charged layer (cylinder) of liquid can be determined as:

$$F = \frac{\varepsilon\zeta U}{4\pi\delta l}, \tag{1}$$

where l is the capillary length;

δ is the thickness of liquid adsorption layer;

ε is the dielectric constant of liquid.

The moisture transfer mechanism in a single direct circular capillary in Fig. 3 can be transferred to any heterocapillary system, which consists of capillaries of various sections and directions. The materials used to insulate EE windings are capillary systems. The pores and capillaries that are found in electrical insulating materials (EIM) can be of a variety of shapes and sizes, open-type and blind. They can connect to the environment directly or using a branched system. According to the classification of Rebinder P.A. [22], the forms of moisture connection with material are divided into three groups: chemical, physico-chemical, physico-mechanical. Of these, only the third form of connection is exposed to electroosmosis. Thus, under the influence of electric field at EOD, all moisture from EIM cannot be removed. But on the other hand, water is displaced, that is in the most dangerous condition for electrical properties of insulation.

Moisture from windings can be removed by applying a potential difference between the current-carrying parts and the housing, i.e. by creating electric field in insulation, which causes electroosmotic motion of water towards the negative electrode. As a rule, EM windings are connected into a star, positive potential is applied to them, and negative potential is applied to the housing [11, 12]. Under such connection of the electroosmotric drying device (EOD), the sequence of moisture motion in slot insulation is shown in Fig. 4.

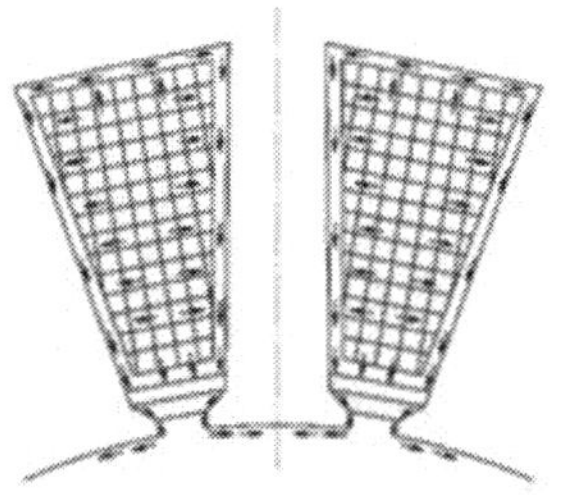

Fig. 4. Geometrical model of electroosmosis for EM slot insulation (arrows indicate the motion of moisture at EOD).

Here, a cross section of two stator slots is shown, made along a layer that insulates the individual plates of the magnetic circuit. The arrows indicate the direction of moisture motion. Moisture from the slot parts of insulation is displaced to the slot walls. Further, under the influence of excess pressure moisture is displaced into the stator cavity through the slots and the end parts of the slots. Our calculations show that this excess pressure in EOD process is six times greater than the atmospheric one. Moisture is removed from inter-turn parts of insulation, under the action of diffusion gradient. From the interphase parts of insulation that are not directly exposed to electric field during EOD according to the "star-housing" scheme [20], water moves to the slot parts also under the action of diffusion gradient and the process is repeated. When EM is turned on and heated, moisture evaporates into the environment.

EOD does not heat the insulation, drying is cold, and energy consumption is 2-3 orders of magnitude lower than during traditional methods of EE thermal drying. During EOD there is no need to disassemble EE or to dismantle it from its workplace, which reduces labor costs by 2-3 times [11].

Electroosmotic moisture protection (EOMP) is based on electroosmosis like EOD, but it is used more rarely than thermal methods of moisture protection. When potential is applied to the winding of inactive EM relative to the housing, electric field is created in the insulation system, which prevents moisture penetration into the winding, thus a preventive moisture support is implemented [11]. EOMP has all the advantages typical for EOD.

The process of EOD and EOMP for transformers is more complicated than that for EM because of complexity of electrical insulation design, the presence of oil and the lack of close contact of insulation between the winding conductors and the housing [19]. Potential difference from EOD and EOMP devices is supplied on the winding conductors and the transformer housing. Therefore, EOD and EOMP for transformers require special consideration [23].

2 Conclusions

1. The electrokinetic phenomenon of electroosmosis is applicable for drying and moisture protection of insulation of EE windings.

2. The use of EOD does not reduce the service life and the resource of EE electrical insulation, which is typical for thermal drying methods.

3. The EOD is cold, does not require disassembling and dismantling of EE from its workplace. It requires minimal energy and labor consumption in comparison with thermal drying methods.

References

[1] A.E. Nemirovskiy, G.A. Kichigina, I.Yu. Sergievskaya, Calculation of heating of windings of electric motors during electroosmotic drying of insulation under low temperatures, Fedorov Readings 2017: Materials of XLVII international Scientific and Practical Conference with Elements of a Scientific School, MPEI, Moscow (2017).

[2] A.E. Nemirovskiy, A.I. Kashin, V.F. Kosmach, Y.F. Titovec, I.N. Toptygin, D.A. Zaripova, Innovative technology for dismantling the windings of electric motors using ultrasonic radiation, IOP Conference Series: Earth and Environmental Science 337, 012071 (2019).

[3] A.E. Nemirovskiy, Improving the efficiency of drying and moisture protection of insulation of electric motors used in agriculture, based on intensification of electroosmotic phenomena, thesis for the D.Sc. in engineering (SSAU, St. Petersburg-Pushkin, 1993).

[4] M. Rashevskaya, S. Yanchenko, S. Tsyruk, B. Kudrin, Assessing non-stationary power quality phenomena of induction motors, Proceedings of 18th International Conference on Computational Problems of Electrical Engineering, CPEE (2017).

[5] L.G. Prischep, F. Bulte, Condensation accumulation in a sealed electric motor, Mechanization and electrification of agriculture 6 (1981).

[6] O.I. Khomutov, System of technical means and measures to increase the operational reliability of insulation of electric motors used in agricultural production, thesis for the D.Sc. in engineering (CHIMESH, Chelyabinsk, 1990).

[7] A.N. Alyunov, O.S. Vyatkina, L.R. Mukhametova, V.P. Markovskiy, R.M. Mustafina, Automation of calculations of power transformers windings parameters taking into account the position of the voltage regulator, E3S Web of Conferences 124, 02016 (2019).

[8] A.N. Alyunov, O.S. Vyatkina, I.G. Akhmetova, R.D. Pentiuc, K.E. Sakipov, Issues on optimization of operating modes of power transformers, E3S Web of Conferences 124, 02015 (2019).

[9] W.H. Bartley, Analysis of Transformer Failures, Energy and Management 1, 58 (2011).

[10] B.V. Vanin, Yu.N. L'vov, M.Yu. L'vov, et al., Damages to power transformers with voltage of 110-500 kV in operation, Electric stations 9 (2001).

[11] A.E. Nemirovskiy, G.A. Kichigina, I.Yu. Sergievskaya, Electroosmotic drying and moisture protection of electrical equipment, Fedorov Readings 2017: Materials of XLVII international Scientific and Practical Conference with Elements of a Scientific School, MPEI, Moscow (2017).

[12] A.E. Nemirovskiy, Calculation of parameters of electroosmotic drying of windings of electric motors, Technique in agriculture 2-3 (1992).

[13] A.E. Nemirovskiy, N.K. Moroz, K.P. Simakov, Method of electroosmotic drying of insulation of windings of electrical machines, Patent for invention, 2174280, bull. no. 27 (2001).

[14] A.E. Nemirovskiy, N.K. Moroz, K.P. Simakov, T.A. Nemirovskaya, Method of drying the solid insulation of transformer windings in a sealed tank, Patent for invention, 2174281, bull. no. 27 (2001).

[15] A.E. Nemirovskiy, N.K. Moroz, K.P. Simakov, Method of electroosmotic drying of paper cable insulation, Patent for invention, 2254631, bull. no. 17 (2005).

[16] A.E. Nemirovskiy, N.K. Moroz, K.P. Simakov, Method of electroosmotic drying of insulation of windings of electric machines, Patent for invention, 2250550, bull. no. 11 (2005).

[17] A.E. Nemirovskiy, N.K. Moroz, V.G. Bugakov, A.V. Busyrev, Method of drying of insulation of windings of electric machines, Certificate of authorship, 1705972, bull. no. 2 (1992).

[18] S.V. Kirillov, Electroosmotic drying of transformers, Lokomotiv 1 (2014).

[19] A.E. Nemirovskiy, I.Yu. Sergievskaya, Operational experience of moisture protection devices for insulation of windings of electrical equipment at enterprises of oil and gas complex, Actual problems of the state and development of the oil and gas complex of Russia: 7th All-Russian Scientific and Technological Conference, Gubkin Russian State University of Oil and Gas (2007).

[20] A.E. Nemirovskiy, I.Yu. Sergievskaya, Operational experience of electroosmotic drying devices for insulation of windings of electrical equipment at oil and gas enterprises, Actual problems of the state and development of the oil and gas complex of Russia: 7th All-Russian Scientific and Technological Conference, Gubkin Russian State University of Oil and Gas (2007).

[21] A.Ya. Fanin, Liquid motion in porous materials under the influence of electroosmosis, Issues on automation and mining electrical engineering: Collection of papers Don. UGI, no. 40 (Nedra, Moscow, 1969).

[22] V.V. Maslov, Moisture resistance of electrical insulation (Energy, Moscow, 1973).

[23] A. Udaratin, A. Alyunov, A. Krutikov, L.R. Mukhametova, O.O, Zaripov, I.V. Bochkarev, Efficiency

study of the reactive shunt compensation device in power lines, E3S Web of Conferences 124, 02020 (2019).

E3S Web of Conferences **178**, 01079 (2020)
HSTED-2020
https://doi.org/10.1051/e3sconf/202017801079

ELECTRON SPECTROSCOPY OF CdMeTe NANOSTRUCTURES CREATED ON CdTe SURFACE UNDER ION BOMBARDMENT

Ergashov Yokub[1,*], and *Umirzakov* Boltakhodja[1]

[1]Tashkent State Technical University, 100095, 2-University street, Tashkent, Uzbekistan

Abstract. The electronic structure of CdMeTe nanostructures was studied by a variety of techniques of secondary and photoelectron spectroscopy. Particularly, for the first time it has been shown that in the conductivity band of the CdZnTe films there are maximums peaks of free electronic states with energies 3.5 and 4.4 eV relative E_v.

1 Introduction

Multi-layer thin-film structures based on semiconductor compounds A^2B^6 are widely used to design diverse optoelectronic devices, including solar cells. We have investigated crystalline structure of CdTe, CdZnTe and HgCdTe films, as well as their optical properties and how they change under different influences [1-6]. It was shown that properties of these films mainly depend on a synthesis technique, on their thickness and properties of a near-contact (transitive) layer formed on the boundary film-substrate. In recent years a techniques of low-energy ion implantation is frequently used for nano-sized structures on the surface and in near-surface region of semiconductor and dielectric films [7-10]. Particularly, the films of $Cd_{0.5}Ba_{0.5}Te$ were produced by the Ba^+ ion implantation into CdTe in combination with annealing [10]. It was established that the state densities of valence electrons and the forbidden band width of the CdBaTe nanofilms for d≤25-30 Å depend on film thickness.

This paper reports the energy band parameters and the results of the studyof the density of electronic states in the valence band and in the conductivity band for three-component nanofilms of $Cd_{1-x}Me_xTe$ produced by Ba and Zn ion implantation into CdTe.

2 Experimental methods

Single crystalline CdTe(100) of the n-type with the sizes 10×10×1 mm^3 were polished and treated with an etchant. The investigation was carried out in vacuum (10^{-6} bar) by Auger-electron spectroscopy (AES), ultraviolet photoelectron spectroscopy (UPS), electron energy loss spectroscopy (EELS) and by measurements of energy dependencies of the coefficients of secondary electron emission (SEE) and intensities of light transmitting through the sample. Morphology and degree of surface roughness were studied by scanning electron microscopy (SEM) and atomic force microscopy (AFM). The profiles of atom distributions in depth were defined by AES in combination with ion etching.Ion bombardment was made perpendicular to the target surface under vacuum (10^{-6} bar). Before ion implantation the CdTe samples were degassed by heating up to T=1100 K in combination with mild etching with the Ar^+ ions under vacuum (10^{-7} bar). By varying energy of the Ba^+ and Zn^+ ions within the range 0.5-5 keV the three-component nanoscale structures with the thickness 20-50 nm were produced. The optimal modes of ion implantation and annealing for the nanocrystals (NC) and nanofilms (NF) of CdMeTe/CdTe(100) are presented in Table 1.

Table 1. Optimal modes of formation of three-component nanostructures of CdMeTe/CdTe (100).

Ions	E_0, keV	D, cm^{-2}	T_0, K	Structure	Approximate composition	Film thicknessd, Å
Ba	1	10^{15}	900	NC	$Cd_{0.6}Ba_{0.4}Tc$	35-40
		$6 \cdot 10^{16}$	1000	NF	$Cd_{0.5}Ba_{0.5}Te$	35-40
	3	10^{15}	900	NC	$Cd_{0.7}Ba_{0.3}Te$	50-60
		$8 \cdot 10^{16}$	1000	NF	$Cd_{0.5}Ba_{0.5}Te$	50-60
Zn	1	10^{15}	850	NC	$Cd_{0.6}Zn_{0.4}Te$	40-45
	1	$8 \cdot 10^{16}$	800	NF	$Cd_{0.7}Zn_{0.3}Te$	-

* Corresponding author: yergashev@rambler.ru

3 Results and discussion

The photoelectron spectra of pure CdTe(100) and CdTe(100) with the $Cd_{0.6}Zn_{0.4}Te$ film which were measured for hν=10.8 eV are presented in Fig.1. The spectral structure, i.e. the density of valence electron states of matrix and film, greatly differ from each other. The peaks corresponding to E_1, E_2 and E_3 in the CdTe spectrum can be associated with hybridization of the 5p and 5s electrons of Te with the 5s and 4d ones of Cd. In the case of a three-component film, along with the Cd and Te electrons the 3d and 4s electrons of Zn will participate in electron hybridization owing to which the re-distribution of the density of valence electron states takes place. Besides these peaks, in the spectra there are peaks associated with the surface states (E_{ss}) and impurity levels (E_{imp}). The spectrum width ΔE of film is 0.2-0.3 eV less than that of matrix.

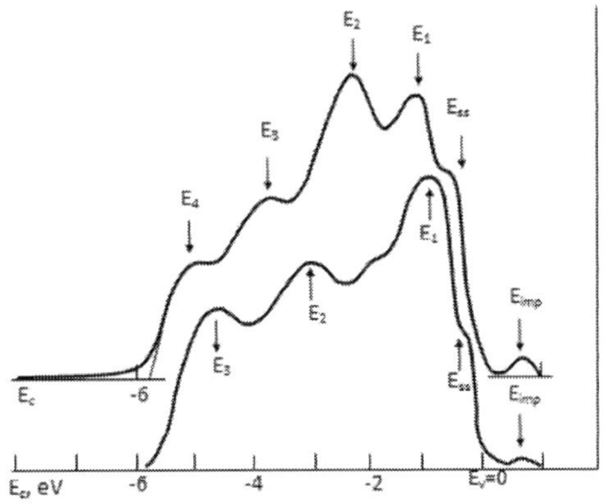

Fig. 1. The photoelectron spectra of pure CdTe (100) and with $Cd_{0.6}Zn_{0.4}Te$ nanofilm for hν=10.8 eV.

The spectra $N(E_2)$ measured for E_p=5 eV for pure CdTe and for CdTe with a submonolayer Cs film (θ≈0.2) are presented in Fig.2. E_2 is the energy of secondary electrons. Zero on the E_2 axis corresponds to a vacuum level of the CdTe film. From Fig. 3 it is seen that after Cs deposition the emission of true-secondary electrons into vacuum increases and the spectrum start sifts to lower energy by ~2.6 eV, i.e. the potential barrier (electron affinity χ) decreases by 2.6 eV. In this part of the spectrum there are peculiarities for energy E_{c1}=-1 and E_{c2}=-1.9 eV (relative to E_v for energy 3.5 and 4.4 eV). The positions of these peculiarities correspond to maximums of the density of electronic states in the conductivity band. By increasing the thickness of the Cs film up to 0.5-0.6 in the monolayer it is possible to decrease χby ~3-3.5 eV. However, beginning with θ~0.4-0.5 of monolayers, in the spectrum there are peculiarities associated with the Cs atoms.

The dependences of true-secondary electrons δ, elastically reflected ones χ and dR/dE_p on primary electron energyE_p for CdTe(111) within the range of E_p=0-20 eV are presented in Fig.3. The peculiarities of the curve $R(E_p)$, i.e. maximums of the curves $dR/dE_p(E_p)$, correspond to the electron transitions from the valence band maximums into the conductivity zone maximums [11]. Analysis of the $dR/dE_p(E_p)$ curve (spectra of elastically reflected electrons) for photoelectrons (Fig.1) and true-secondary electrons showed that the maximum E_1' corresponds to the electron transition from E_v to E_c (approximately equal to E_g); E_2' to the transition from E_1 to E_{c1}; E_3' to the transition from E_v to E_2; and E_4' to the transition from E_2 to E_{c2}. The maximums $E_6'$and E_7' can be probably associated with the electron transition from the maximums of the density of the valent electron states into the maximum of free states in vacuum. The position E_3'corresponds to an initial immediate increase in δ, which is due to the electron transition from the top of the valent zone into vacuum. For E_p=11.2 eV there is also an immediate decrease in R (maximum on the curve dR/dE_1). However, for this energy E_p there is no immediate increase in δ. Owing to the above statements it is possible to suppose that the valence electrons move not to vacuum but to a maximum of the conductivity band. For the width of the forbidden band of CdTe to be defined more accurately, the energy dependence of the intensity I of light transmitting through CdTe was measured within the range hν=0.6-2 eV [10]. From Fig.4 it is seen that for CdTe$E_g$is ~1.45 eV.

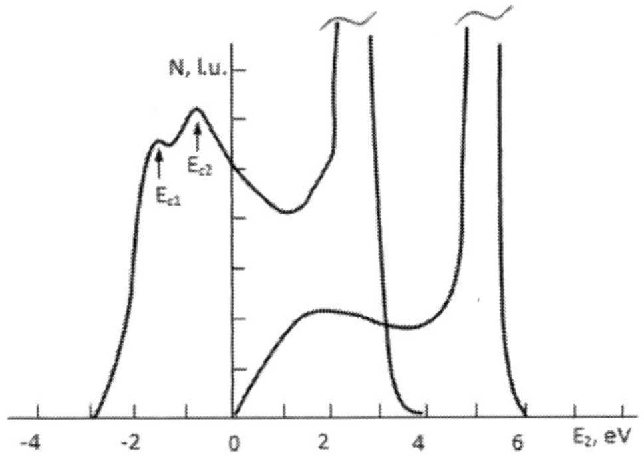

Fig. 2. The spectra $N(E_2)$ measured for E_p=5 eV for pure CdTe and CdTe with the submonolayer(θ≈0.2) film of Cs.

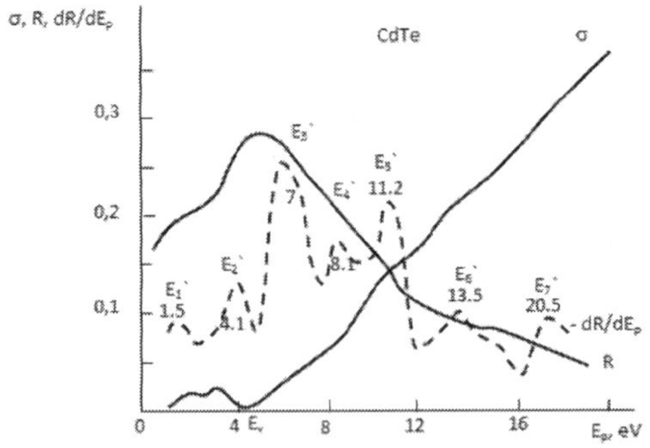

Fig. 3. The dependences of δ, Rand - dR/dE_p on energy of primary electrons for CdTe(111).

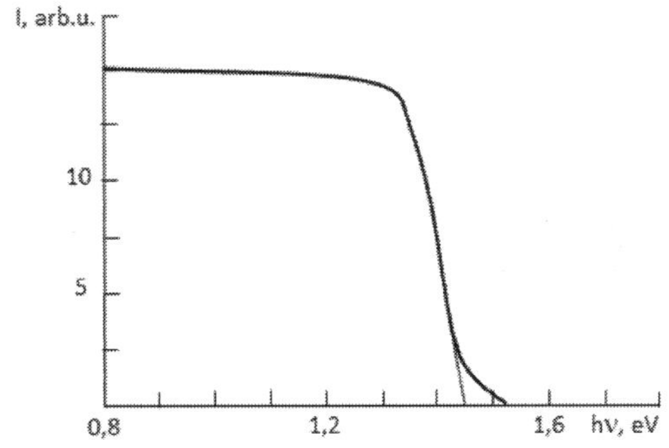

Fig. 4. The energy dependence of the intensity I of light transmitting through CdTe.

4 Conclusion

Thus, in this paper the nano-crystalline phases and nanofilms of CdBaTe and CdZnTe were formed and their band-energy parameters and the densities of occupied and free electronic states have been studied by the technique of low-energy ion implantation on the surface of single crystals CdTe(111).

References

[1] C.H. Lee, S.W. Park, Lee Jaesun, Y.M. Moon, et al., J. Electron. Mater. **27**, 6, 668 (1998)
[2] A.V. Brodovoy, S.G. Bunchuk, Z.F. Tcibriy, FTT **53**, 3, 524–526
[3] P.N. Tkachuk, FTT **42**, 11, 1961–1963 (2000)
[4] K.D. Glinchuk, A.P. Medvid, A.M. Mychko, et al., FTP **47**, 4, 435–441 (2013)
[5] N.I. Klyuy, V.B. Lozinskiy, A.N. Lukyanov, V.A. Morojenko, R.K. Savkina, et al., JTF **82**, 8, 83–86 (2012)
[6] M.K. Herndon, A. Gupta, V.I. Kaydanov, R.T. Collins, Appl.Phys. Lett. **75**, 3503 (1999)
[7] S.A. Muzafarova, Sh.A. Mirsagatov, J. Janabergenov, FTT **49**, 6, 1111–1116 (2007)
[8] B.E. Umirzakova, D.A. Tashmukhamedovab, M.K. Ruzibaevaa, F.G. Djurabekova, S.B. Danaev, NuclearInstr.AndMehodsinPhys. Research Section B: Beam Inerac. WithMaterials and Atoms 326, 322–325 (2014)
[9] Y.S. Ergashov, D.A. Tashmukhamedova, B.E. Umirzakov, Journal of Surface Investigation **11**, 2, 480–484 (2017)
[10] Z.E. Mukhtarov, Z.A. Isakhanov, B.E. Umirzakov, T. Kodirov, E.S. Ergashev, Technical Physics **60**, 12, 1880–1884 (2015)

Quality management of electric energy and compensation of reactive power in the electric power system by means of stimulating tariffs

Vladislav Chikin[1,*], *Anatoly* Kuznetsov [1], and *Diana* Rebrovskaya[1]

[1]Ulyanovsk State Technical University, 32, Severny Venetz str., 432027 Ulyanovsk, Russia

Abstract. The article noted that feed-in tariffs are effective enough to attract consumers to participate in the process of improving power quality and reactive power compensation. It is shown that the application of premiums and discounts must not modify the revenues of the budget of ESO. Otherwise there is a contradiction to the principles of state regulation of tariffs for services of natural monopolies. In this case, legal barriers do not allow to use the model in practice. The proposed criteria that must be considered in the development and approval of the scale of discounts and allowances to ensure the legal purity of the incentive rate, to avoid its cancellation, as it was in 2000. Then, the proposed mechanism of discounts and allowances with the support of the Antimonopoly service and the Ministry of justice of the Russian Federation was cancelled as violating the principles of state regulation of tariffs in the conditions of the natural monopoly position of the network organization.

1 Introduction

Stimulating tariffs are widely used in world practice to attract electricity consumers to participate in the process of improving the quality of electricity and compensating for reactive power. The incentive rate C_{stim} for the consumer is the sum of the main C_{bas} and incentive part. The main part is the same for all consumers of the electric grid organization, receiving power at the same voltage level, and includes the average costs associated with the production and transmission of electricity [1].

The stimulating part is not the same for all consumers, it represents a fee for the distortion of power quality indicators (PKE) or for violation of the mode of reactive power consumption and is formed by surcharges and discounts to the main part of the tariff [2,3].

2 Relevance, scientific significance

For certain values of the incentive part of the tariff ΔC_{ND}, the installation and operation of technical means improving the quality of electric power and improving the mode of consumption of reactive power in a customer's network can be beneficial and stimulated to actions aimed at their use [4]. Unfortunately, the application of incentive tariffs is currently facing legal barriers. Attempts to introduce incentive tariffs in the form of discounts and surcharges fail. The lack of scientific knowledge and approaches for making organizational decisions currently does not allow the use of incentive tariffs.

3 Formulation of the problem

Despite the saturation of the market, stimulating tariffs are not acquired by technical means and are not used by electricity consumers. The quality of electrical energy (QE) remains low. The damage from low QE in Russia is estimated at a minimum of about $ 25 billion per year [5]. Reactive power flows are increasing. Nevertheless, the need to develop and implement a scale of discounts and allowances in the Russian Federation are actively discussed in the technical literature [6]. The development of a scale of discounts and allowances is associated with the need to formulate requirements, conditions and restrictions that should remove the legal barriers to their use in practice. Without this, it is not possible to interest consumers in the use of technical means of improving the quality of electricity and compensation of reactive power.

4 Theoretical part

The mathematical expression for the incentive tariff in the case of the use of allowances ΔC_{ND} in one form of the record has the form

Equations should be centred and should be numbered with the number on the right-hand side.

$$C_{stim} = C_{bas} + \Delta C_{ND} \quad (1)$$

Or in another form

* Corresponding author: chikin_vladislav@mail.ru

$$C_{stim} = C_{bas}\left(1+\frac{\Delta C_{ND}}{C_{bas}}\right) \quad (2)$$

At the same time (1+ $\Delta C_{ND}/C_{bas}$) a can be called a raising factor to the tariff.

If a discount ΔC_{CK} is applied to a consumer, then the expressions for the incentive tariff take the form:

$$C_{stim} = C_{bas} - \Delta C_{CK} \quad (3)$$

or:

$$C_{stim} = C_{bas}\left(1-\frac{\Delta C_{CK}}{C_{bas}}\right) \quad (4)$$

where (1- $\Delta C_{ND}/C_{bas}$) can be called a reduction factor.

The effectiveness of the incentive tariff should be considered on the model. To this end, it is proposed to consider the incentive tariff model, which consists of three interrelated elements. One of them is the energy supplying organization (ESO). The second is a consumer with a load that distorts the power quality indicator or a consumer that violates the mode of consumption of reactive power. We call this consumer culprit violation of electricity consumption (CVC). The third element is a consumer who is forced to perceive the consequences of a violation of the mode of electricity consumption by a consumer CVC. Consequences may be associated with low power quality, limited ability to increase power consumption due to network load reactive power, which limits the bandwidth of the supply network and so on. Using legal terminology we will call this consumer "victim" (PVS). The consumer of the VC seems to be a kind of equivalent generalized consumer, which unites all the consumers of the network organization except the consumer CVC.

It is obvious that there are a lot of parameters of the mode of electricity consumption, which may not meet the established requirements. Among them are PKE, indicators of consumption of reactive power and energy. In the model, the consumer CVC is considered as a violator of one of the parameters of the mode of electricity consumption. If the violation of the mode of power consumption occurs by several parameters, it is necessary to consider the model for each of these parameters separately. Suppose this parameter is non-sinusoidal [7, 8].

In this case, CVC violates the requirements for the PKE not only in its network, but also in the network of the energy supplying organization, which, due to circumstances beyond its control, has to supply poor quality electricity to the consumer VS. This consumer is harmed. The incentive tariff model assumes a formalized way of compensating for damage to a consumer PVS as the culprit for the distortion of the PKE - the consumer CVC, but not directly, indirectly through the ECO. Damages occur in two stages.

At the first stage, the damage to the consumer CVC will compensate the ECO in the form of a discount to the tariff. At the same time, the income of the ECO will be reduced, which it could have received if it were not a consumer CVC, distorting the PKE in the network of the energy supplying organization. From a legal point of view, the right of the ECO to supply high-quality electricity to the consumer of the CVC is violated. As a result, the energy supplying organization does not receive a portion of the income that it could receive under normal conditions of civilian traffic. Lost are referred to as lost profits incomes in accordance with Art. 15, p. 2 of the Civil Code of the Russian Federation. The loss of profit for the ECO is measured by the amount of compensation for damage ΔC_{CK} to the consumer PVS.

At the second stage, the consumer CVC compensates to the ECO the lost profit in the amount of the damage paid to the consumer VS. Consumer CVC provides a premium to the tariff ΔC_{ND}. The energy supplying organization at the same time compensates for its expenses for indemnification of damage to the PVS. The premium ΔCND is presented as a formalized way to compensate for the loss of benefits to the ECO on the one hand, and as compensation for damage to the consumer VS on the other hand. The role of the energy supplying organization in this model is to pay damages ΔC_{CK} to the consumer of the PVS and recover damages ΔC_{ND} from the consumer CVC in the form of the loss of profit resulting from this.

The model assumes the participation of the energy supplying organization in the process of collecting and paying for damage, although this activity is not the main activity for it. Nevertheless, the supply of poor-quality electricity to the consumer PVS for the energy supplying organization affects its interests. She is forced to supply poor quality electricity and is responsible for this.

It is obvious that the model provides for the responsibility of the offender in the form of a surcharge to the tariff and implies the inevitability of its application. The premium is a stimulating factor. It has quite a strong influence on the consumer in terms of the installation of compensating and filter-compensating devices (FCD). Without it, it is hardly possible to influence the consumer's decision to install and operate such devices.

5 Research results, practical significance

Successful operation of the model is possible only under certain conditions and requirements. The requirements relate to the correctness of the application of discounts and surcharges by the state regulatory authority in case of formalized compensation to the consumer PVS at the expense of the consumer CVC. The discount is a refund. Surcharge is for lost profits. Discounts and allowances, one way or another, change the income part of the ECO budget. When using discounts, the revenue side decreases. When applying allowances, the revenue side increases. In this case, it is necessary to provide guarantees to the energy supplying organization not to have difficulties associated with the main activity and to keep the budget revenues unchanged, due to the basic component of the incentive tariff C_{bas}. Prevent its unjustified reduction, on the one hand, and increase, on the other.

This requirement is dictated by the principle of state regulation of tariffs for the services of natural monopolies. Revenues of the network organization as a natural monopoly should be economically justified. The tariff for the transmission of electricity, included in the main, basic part C_{bas} of the incentive tariff, is subject to state regulation and approved by government bodies. This is the Federal and Regional Tariff Service (FST and RST). When using the incentive part, the change in income of the ECO is unacceptable. Otherwise, a legal inconsistency arises.

The latter is quite possible in the conditions of market relations, when officials who make decisions about the size of discounts and premiums do not have clear and definite criteria for assessing decisions made. They must ensure the accuracy and transparency of information about the absence of additional income from the network organization. Incomes in excess of revenues due to the approved tariff for transmission of electricity. In this case, consumers will not have distrust and doubts about the fairness of the use of allowances. They will be ready to accept and maintain the mechanism for applying discounts and surcharges.

This criterion decisions can be the criterion of economic feasibility of the revenue of the ECO, which was first proposed in [9].

For the model, it represents

$$k_{ECO} = 1 + \frac{D^{*}_{ECO} + \Delta C_{ND} - \Delta C_{CK} - D^{*}_{ECO}}{D^{*}_{ECO}} \to 1 \quad (5)$$

where D^{*}_{ECO} - the value of the income of the ECO related to the supply of electricity to consumers CVC and VS, which is formed by charging fees associated with the supply of electricity without discounts and premiums; ΔC_{ND} and ΔC_{CK} - the value of the allowance to the consumer CVC and the amount of the discount to the consumer PVS.

In general case ΔC_{CK} in this model represents

$$\Delta C_{CK} = \frac{\sum_{i=1}^{n-1}\left(\Delta C_{CK,i} \cdot W_{\mathrm{PVS},\,i}\right)}{W_{\mathrm{PVS}}} \quad (6)$$

where $\Delta C_{CK,i}$ - the discount for everyone i-consumer in the ECO except the culprit; $W_{PVS,i}$ - electricity consumed by each consumer in an ECO, except for the culprit; n - the number of consumers in an ECO; W_{PVS} - electricity consumed by the equivalent consumer PVS.

$$W_{\mathrm{PVS}} = \sum_{i=1}^{n-1} W_{\mathrm{PVS},\,i} \quad (7)$$

When k_{ECO}=1, the income of the energy supplying organization does not depend on the application of discounts and surcharges. The principle of economic feasibility of income is not violated.

The criterion for evaluating the decision made (5) implies the interdependence of the amount of the allowance and the discount. For example, if a surcharge is defined, the discount cannot be selected without considering its value. The communication between their values is as follows.

$$\Delta C_{ND} W_{\mathrm{CVC}} = \Delta C_{CK} W_{\mathrm{PVS}} \quad (8)$$

where W_{CVC} - electricity is consumed by the CVC.

$$\frac{\Delta C_{ND}}{\Delta C_{CK}} = \frac{W_{\mathrm{PVS}}}{W_{\mathrm{CVC}}} \quad (9)$$

If take

$$W_{\mathrm{ECO}} = W_{\mathrm{CVC}} + W_{\mathrm{PVS}} \quad (10)$$

$$W^{*}_{\mathrm{CVC}} = \frac{W_{\mathrm{CVC}}}{W_{\mathrm{ECO}}} \quad (11)$$

$$W^{*}_{\mathrm{PVS}} = \frac{W_{\mathrm{ECO}} - W_{\mathrm{CVC}}}{W_{\mathrm{ECO}}} = 1 - \frac{W_{\mathrm{CVC}}}{W_{\mathrm{ECO}}} = 1 - W^{*}_{\mathrm{CVC}} \quad (12)$$

Condition (8) translates to

$$\frac{\Delta C_{ND}}{\Delta C_{CK}} = \frac{1 - W^{*}_{\mathrm{CVC}}}{W^{*}_{\mathrm{CVC}}} \quad (13)$$

or

$$\Delta C_{ND} = \Delta C_{CK}\left(\frac{1 - W^{*}_{\mathrm{CVC}}}{W^{*}_{\mathrm{CVC}}}\right) = \Delta C_{CK}\left(\frac{1}{W^{*}_{\mathrm{CVC}}} - 1\right) \quad (14)$$

In expressions (8) - (14): W_{ECO} - electricity consumed by all consumers ECO; W^{*}_{CVC} - represents the proportion of electricity consumption by the consumer of security products; W^{*}_{PSV} - represents the share of electricity consumption by the consumer PVS.

6 Conclusion

Condition (14) must be considered when developing and approving a scale of discounts and premiums. If this is not done, legal barriers arise that will not allow the mechanism to be used to control the quality of electricity and the process of compensation and reactive power.

An attempt to introduce a stimulating tariff model at the state level took place [2, 3]. However, she failed. The implemented mechanism for applying discounts and surcharges caused great resistance from consumers, and with the support of the Anti-monopoly Service and the Ministry of Justice of the Russian Federation was canceled as violating the principles of state regulation of tariffs in the context of the natural monopoly position of the grid organization [10].

Acknowledgements

The study was carried out with the financial support of the Russian Foundation for Basic Research and the Ulyanovsk Region in the framework of the research project № 18-48-730025.

References

[1] A.V. Kuznetsov, The use of incentive tariffs to improve the quality of electricity and reduce reactive power flows in the electric power system, Elektrika **12**, 12–17 (2012)

[2] Rules for applying discounts and surcharges to tariffs for the quality of electricity, Industrial energy 8, 49–52 (1991)

[3] Rules for applying discounts and surcharges to tariffs for electric energy for the consump-tion and generation of reactive energy, Entered into force on 1 December 1997), Industrial energy **10**, 43–52 (1998)

[4] A.V. Kuznetsov, The organizational and legal aspects of the management of power con-sumption modes, Power Stations **12**, 52–56 (2003)

[5] L.A. Dobrusin, Investments in the power industry of Russia and the program of increasing their efficiency, Report, VI All-Russian Energy Forum Fuel and Energy Complex of Russia in the XXI Century, Moscow, the State Kremlin Palace (1-4 April 2008)

[6] V.E. Vorotnitsky, Energy Efficiency and Reactive Power Compensation in Electrical Net-works. Problems and solutions [Electronic resoruce], Energy Council **1**, 47 (2017). Available at: http://www.energosovet.ru/ (accessed: 21.04.19)

[7] GOST 32144-2013, Standards of quality of electric energy in general-purpose power supply systems [Electronic resource], Electronic fund of legal and regulatory and technical documentation. Available at: http://docs.cntd.ru/document/1200104301 (accessed: 25.04.19)

[8] G.Ya. Vagin, A.A. Sevostyanov, *Indicators of the quality of electricity in power supply systems: studies, Po-dobe* (Nizheoro. state Tech. Unto them. R.E. Alek-seeva, N.Novgorod, 188, 2017)

[9] A.V. Kuznetsov, L.T. Nagazinnik, *Improving the management efficiency of electrical energy consumption modes* (M.: Energoatomizdat, 105, 2006)

[10] Regarding the application of the provisions of the Instructions on the procedure for pay-ments for electric and thermal energy [Electronic resource], Electronic Fund of legal and regulatory and technical documentation. Available at: http://docs.cntd.ru/document/gost-r-7-0-5-2008 (accessed: 21.04.19)

E3S Web of Conferences **178**, 01081 (2020) https://doi.org/10.1051/e3sconf/202017801081
HSTED-2020

Technical and economic efficiency of proactive diagnostic systems for power transformers

A.N. Alyunov[1], *O.S.* Vyatkina[1], *A.V.* Udaratin[1] and *D.A.* Zaripova[2]

[1]Vologda State University, Vologda, Russia
[2]Kazan State Power Engineering University, Kazan, Russia

Abstract. The expediency of using systems of proactive diagnostics of power transformers is substantiated. The diagnostic system expands the possibilities of analyzing the transformer condition in operating mode. It allows one to minimize the expected damage from interruptions in power supply and equipment failure. The results can be used in world practice during construction of various control systems to identify developing transformer defects caused by windings deformations.

1 Introduction

The strategy of international scientific and technological development comprises transition to advanced digital, intelligent manufacturing technologies, robotic systems, new materials and design methods, creation of systems for processing large volumes of data, machine learning and artificial intelligence.

Under the current conditions, it is possible to bring the properties of protections installed on transformers closer to ideal. It is achieved by implementation of proactive actions of relay protection by identifying potential damage at early stage of its development. This significantly reduces the risk of sudden failures in transformers, and, therefore, the risk of sudden disturbances in power system.

Internal short-circuit in transformers of 110 - 500 kV are caused, first of all, by damage of on-load tap-changers, high-voltage bushings and windings. The most severe consequences occur during development of the following defects [1]:

- Reduction of electric strength of oil channel of high-voltage sealed bushings due to deposition of sediment on the inner surface of porcelain and on the surface of internal insulation;
- Reduction of electric strength of paper-oil insulation of high-voltage unsealed bushings due to wetting and pollution;
- Pollution, moisture and wear of insulation of transformer windings;
- Burnout of winding insulation and turns of windings due to prolonged non-shutdown of through short-circuit current on the low voltage side of transformer;
- Errors during installation, operation and repairs.

The causes, behavior, and consequences of damage to transformer windings are well understood, however, traditional protection can detect damage when 65-70% of the winding turns are closed, which is clearly not enough. Moreover, various damages, including deformation of windings, may remain unnoticed, which leads to the further progress of emergency situations and conversion of winding faults to multiphase short circuits. The disadvantage of the existing methods is the large turn-off time of winding circuits and uneven sensitivity to symmetric and asymmetric short circuits.

It should be noted that most of these defects could be timely detected using methods and means of technical diagnostics [2, 3].

There are methods for diagnostics of power transformers according to the experience of idling and short circuit, requiring disconnection of transformer from the network. There are systems for transformer monitoring that calculate an integrated coefficient of current technical condition of transformer. Registration of currents and voltages of the primary and secondary windings of transformer in such modules is used to calculate only short circuit resistance. These approaches do not consider other parameters (short circuit losses, current and idle losses), so they have a decreased accuracy of damage diagnosis [4–6].

The system of diagnostics of transformers and other equipment should have full information, technical, regulatory support, as well as a decision-making strategy on the possibility and expediency of its further operation or the need for repair.

The main guidance documents regulate the list of transformer tests during commissioning and operation, the maximum permissible values of the controlled parameters and deadline monitoring. In recent times, there appeared a need for a significant enhancement of the controlled parameters of power transformers. Along with the traditional criteria for insulation evaluating (resistance and the tangent of angle of dielectric loss of insulation, breakdown voltage of oil, etc.) which have a long history of practical experience, additional indicators

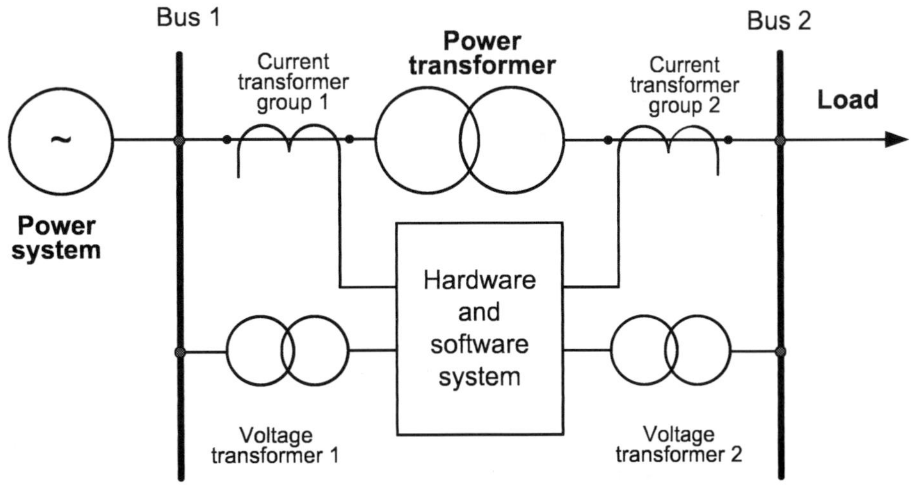

Fig. 1. Connection diagram of hardware and software complex for power transformer diagnostics.

are introduced. These include: chromatographic analysis of gases dissolved in oil, measurements of polymerization degree, control of furan content in oil, measurements of short circuit resistance, thermal imaging control.

Experience shows that although traditional tests are necessary and are the basis for assessment of transformer condition, they do not always enable to detect defects at an early stage and provide timely information on the development of processes that lead to a decrease in reliability and performance of equipment. Therefore, the use of additional controlled parameters is objectively justified and is developing both in Russia and other countries. At the same time, the system of regulatory parameters for assessing the condition of power transformers in operation is also being improved [1, 10-12].

Unfortunately it is impossible to accurately predict the degree of wear of transformer windings insulation by analyzing the impact of operational factors. This is explained by a rather complicated network of chemical reactions in insulation, leading to degradation of cellulose, and a vast amount of reasons that affect their development.

The aim of this study is to increase the accuracy of assessing the condition of transformer according to parameters of its normal operation mode . It is implemented by monitoring a number of the most important diagnostic indicators of transformer without disconnecting it from the network. This can be achieved by collecting and monitoring additional parameters and generating a warning signal in case when the monitored parameter exceeds the permissible values.

2 materials and methods

To achieve this goal it is necessary to solve a number of tasks. In this framework, the magnitudes and phase angles of voltages and currents of all transformer windings are measured in two different load conditions according to the circuit shown in the figure 1.

The measured values are used to determine the resistance and short circuit loss, current and idle loss by solving the equations of the power transformer state, compiled on the basis of equivalent circuit of transformer. The technical condition and operability of transformer are assessed judging from excess of the obtained values over the corresponding reference values. By choosing transformer operation modes with different loads, the measured currents of windings and voltage drops are recorded. Further the first harmonic component of the registered currents and voltages of windings is derived, their symmetric components are determined, which are used in subsequent calculations. The zero phase-sequence resistance is determined by the parameters of asymmetric mode. This value is compared with the reference value, and the technical condition of electrical circuits connecting the transformer windings to the ground is determined by deviation of this value from the reference one. The active and inductive resistances of each of windings are determined and are used to assess the technical state of winding insulation and the state of current-carrying parts of windings. The transformer transformation coefficient is determined and is used to specify the technical condition of transformer voltage regulation device.

An improved method for determining the parameters of direct and zero sequence of power transformers will differ from the existing ones in that it will enable

 https://doi.org/10.1051/e3sconf/202017801081

Table 1. Characteristics of the hardware and software complex.

№	Parameter	Value
1	Number of measuring channels	Not less than 16
2	Accuracy class	0.2S
3	Sampling frequency	250 kHz
4	Transformer voltage	6-110 kV
5	Supply voltage	80 -240 V
6	Operation temperature	-40 - +65 °C
	Interfaces	RS485, USB, Ethernet, 4-20 mA
7	Power consumption, W	Determined by the research results
8	Mass, g	Determined by the research results
9	Dimensions LxWxH, mm	Determined by the research results

determining the parameters of a multi-winding (including two- and three-winding) transformer for an arbitrary position of regulator under load, different groups and connection schemes of windings based on data from digital registration of processes.

The algorithm for proactive control of the degree of deformation of windings of power transformers based on data of registration of emergency processes is also different in that the development of a defect can be established in the operating mode and thereby prevents occurrence of a sudden accident.

Early detection of potential damage will allow one to gain a sufficient margin of time to localize the damage without using an emergency shutdown.

The hardware and software complex for diagnosing internal damage in a power transformer is proposed to connect to measurement cores of current and voltage transformers with accuracy class 0.2 or 0.5. The changes in equivalent circuit parameters should be monitored in normal conditions. This is sufficient to predict the development of winding faults and assess the windings deformation. The characteristics of the hardware and software complex are given in the table 1.

3 Results and discussions

The economic effect of the study will be in reduction of damage from technological failures caused by sudden damage to the transformers, and, therefore, in reduction of risk of sudden disturbances in power system and interruptions in power supply to consumers.

We estimate the approximate damage from an emergency shutdown of a transformer with a rated power of S_B = 63 MV·A due to damage to its windings.

Complete economic losses (disbenefit) arising from damage to the power transformer include damage from equipment damage and damage from interruptions in power supply to consumers:

$$D=D_{EQ}+D_{US}, \$ \quad (1)$$

where D_{EQ} is the losses from equipment damage, $.

D_{US} is the losses from interruptions in power supply to consumers, $.

We estimate the approximate damage from an emergency shutdown of transformer with rated power S_B=63 MV·A due to damage to its windings.

The average specific damage per unit transformer power, according to [7], is d=8990 $/MV·A.

According to data of Table 4 [7] the damage rate of transformer windings k of the total number of damage is:

$$k=\frac{N}{N_\Sigma}\cdot 100\%,$$

where N is the amount of internal damage to transformers;

N_Σ is the total damage.

$$k=\frac{24}{94}\cdot 100\%=25.5\%.$$

Then the expected losses from the emergency shutdown of transformer with a rated power of S_B = 63 MV·A due to damage to its windings will be:

$$D=k\cdot d\cdot S_B=0.255\cdot 8990\cdot 63=144420 \$.$$

Analysis of the final data of Table 1 [7] shows that the share of losses caused by equipment damage is 56.95%, and the share of losses from interruptions in power supply accounts for 43.05%.

Taking into account formula (1), we obtain:

$$D_{EQ}=82250 \$;$$

$$D_{US}=62170 \$$$

According to [1,8,9], the damageability of new transformers is 0.2% (every 500th transformer). For the 30-years transformers this value is 1% (every 100th transformer). According to [10], for transformers and autotransformers with voltage of 110–500 kV and power of 63 MV·A or more, operating in electric networks, the specific damage, accompanied by the occurrence of internal short circuits, is 0.45% per year or every 222nd transformer . In this case, damage to the windings occurs in transformers with any service life [1].

Installing the proposed hardware and software complex on 222 transformers will prevent the annual expected losses valued at $ 144,420. Analysis of market offers of leading manufacturers of microprocessor devices showed that the estimated cost of one set of hardware and software kit will be $ 1270, so the investment in 222 transformers should be $ 281,940.

4 Conclusions

The presented results promote opportunities of analysis the condition of transformers in operating mode. They can be used in construction of various monitoring systems to identify developing transformer defects caused by windings damage. The implementation of algorithm for monitoring the degree of windings deformation in the protection against internal damage allows one to prevent the approaching moment of damage and prevent occurrence of sudden accidents.

The proposed diagnostic system enables comprehensive assessment of power transformers condition, make decisions on their further operation and implement the technical policy of transition to repair according to the actual equipment condition. Calculations show that the discounted payback period for capital investments in this system will be about two years.

References

[1] B.V. Anin, Yu.N. L'vov, M.Yu. L'vov, B.N. Neklepaev, K.M. Antipov, A.S. Surba, M.I. Chitinsky, Power stations 9 (2001).

[2] S. Miyazaki, Yo. Yamamasu, H. Sakaki, Method and device for diagnosing soundness in transformer, Patent JP2013061310A, publication of 2013-04-04.

[3] Z. Hui-Min, Three-phase three-winding transformer fault area diagnostic method Publication, Patent CN104076227A, Publication of 2014-10-01.

[4] M. Mostafaei, J. Faiz, P. Venikar, M. Sudhakar, Ballal Turn-to-turn fault monitoring methods in electrical power transformers-State of the art, International Transactions on Electrical Energy Systems 28, 2 (2018). DOI: 10.1002/etep.2644.

[5] E. Kornatowski, S. Banaszak, IEEE Transactions 29, 1398-1406 (2014). DOI: 10.1109/TPWRD.2013.2289944.

[6] F. Haghjoo, M. Mostafaei, H. Mohammadi, IEEE Transactions on Power Delivery, 1-1 (2017). DOI: 10.1109/TPWRD.2017.2688419.

[7] W. Bartley, Power Engineering and Management 1, 58 (2011).

[8] B.V. Vanin, Yu.N. Lvov, M.Yu. Lvov, et al., Damage to 110-500 kV power transformers in operation, Power stations 9 (2001).

[9] V.S. Bogomolov, N.N. Khublarov, M.Yu. A.O. Lvov, Problems of autotransformers operation in systems with fault currents upgrowth, CIGRE 12 (2000).

[10] M.Yu. Lvov, Power transformers of 110 kV and higher, Diagnostics will determine the future, Electrical News 6, 24 (2003).

[11] A.N Alyunov, O.S. Vyatkina, L.R. Mukhametova, V.P. Markovskiy, R.M. Mustafina, E3S Web of Conferences 124, 02016 (2019). DOI: 10.1051/e3sconf/201912402016.

[12] A.E. Nemirovskiy, A.I. Kashin, V.F. Kosmach, Yu.F. Titovec, I.N. Toptygin, D.A. Zaripova, Innovative technology for dismantling the windings of electric motors using ultrasonic radiation, IOP Conference Series: Earth and Environmental Science 337 (2019). DOI: 10.1088/1755-1315/337/1/012071.

Software product to reduce commercial energy losses

Maksim Vladimirovich Borodin[1,*], *Tatyana Anatolievna* Kudinova[2], and *Zumeyra Munirovna* Shakurova[3]

[1]Orel State Agrarian University named after N.V. Parakhin, Orel, Russia
[2]Orel State Agrarian University named after N.V. Parakhin, Orel, Russia
[3]Kazan State Power Engineering University, Kazan, Russia

Abstract. The personnel of electric grid companies spend considerable amount of time searching for non-contractual and non-meteredelectricity consumption. The raids carried out by personnel of electric grid companies to detect electricity theft are not always effective, as they are carried out selectively and often do not take into account many factors that could affect the number of detected cases of electricity theft. The article proposes a software product that allows, on the basis ofconsumers' registration, who have already stolen electricity, have arrears in payment for consumed electricity, made uncontrolled consumption, carry out a process of technological connection or have non-meter accounting, to send personnel conducting a raid to identify electricity theft. It also makes possible to generate the necessary acts, certificates, etc.in real time, which allows staff to reduce the time to fill out paper versions and transfer the necessary documents to the police. The software product allows to increase the efficiency of raids by electric grid companies, by reducing the time of inspections. In turn, the implementation of the proposed software product will allow electric grid companies to significantly reduce energy losses by increasing the territoriality of raids.

1 Introduction

The authors [1, 9, 14, 15] indicate that, at present, in Russia, the imperfection of the regulatory framework in the field of electricity metering leads to non-meteredand non-contractual use of electricity, as well as legal disputes between the consumer and the power supply organization. The work carried out by the state and power supply companies in this direction is not sufficiently effective.

In order to reduce the number of non-metered and non-contractual energy consumption, it is necessary to improve the regulatory framework, make the necessary amendments and changes to it to implement tools aimed at reducing the number of non-metered and non-contractual electricity consumptionmore efficiently. Also, in order to reduce the number of non-metered and non-contractual electricity consumption, electricity grid companies themselves must use modern tools aimed at reducing electricity losses.

Literary sources [2-8,10-13,16-20] indicate that a lot of time is spent by the electric grid company personnel on identifying electricity theft by consumers. Raids carried out by electric grid company personnel to detect electricity theft are not always effective, as they are carried out selectively and often the balance of consumption by feeder, settlement, etc. is not taken into account.

2 Research methodology

In the electric grid company ofPJSC "IDGC(Interregional Distribution Grid Company) of Center", the rules of procedure for WG BP 7 / 03-06 / 2018 "Work on the identification and exclusion of non-contractual, non-meteredelectricity consumption", regulating the algorithm for conducting raids by personnelare developed, but they are not entirely effective from our point of view.So, in clause 6.1 of the procedure rules of the WG BP 7 / 03-06 / 2018 "Work on the identification and exclusion of non-contractual, non-metered electricity consumption" it is indicated that SAP has the CIRMS (corporate information resource management system) functionality (SAP IS-U "Route sheet"). But the CIRMS functionality does not take into account non-meter consumers (that is, consumers who do not have a meter, but they pay according to the standard) and consumers found to be stealing electricity. It is necessary to take into account non-meterconsumers, since they can consume electricity above the established norm. At the same time, the personnel performing the raids do not have devices that can measure electricity consumption. This is also a huge drawback, when performing raids.

All of the above can be changed if a software product that can be installed on a tablet would be developed; the developed program would allow personnel, performing the raids, to show first consumers who have already stolen electricity, have arrears in payment for electricity

* Corresponding author: maksimka-borodin@yandex.ru

consumed, made non-contractual consumption, carry out a process of technological connection or have non-meter accounting at the inspected site (feeder, settlement). In this case, the personnel, performing the raid, in the first place, would go to them with checks. If there are non-meter consumers on the feeder, it is imperative to establish devices that control consumption. While personnel are carrying out raids, metering devices that control consumption can measure power consumption of non-meter consumers. If employees have identified theft from the above mentioned consumers, then they enter the necessary data into the program, which subsequently generates the necessary acts, certificates, etc. in real time and sends them for printing.

3 Results and Discussion

To implement the above mentioned idea, a software product algorithm has been developed. It provides for the identification of consumers who have already committed the electricity theft, have arrears in payment for consumed electricity, made non-contractual consumption, carry out the process of technological connection or have non-meter accounting. It also provides for the formation of the necessary acts, certificates, etc.in real time. The algorithm of the software product is shown in Figure 1.

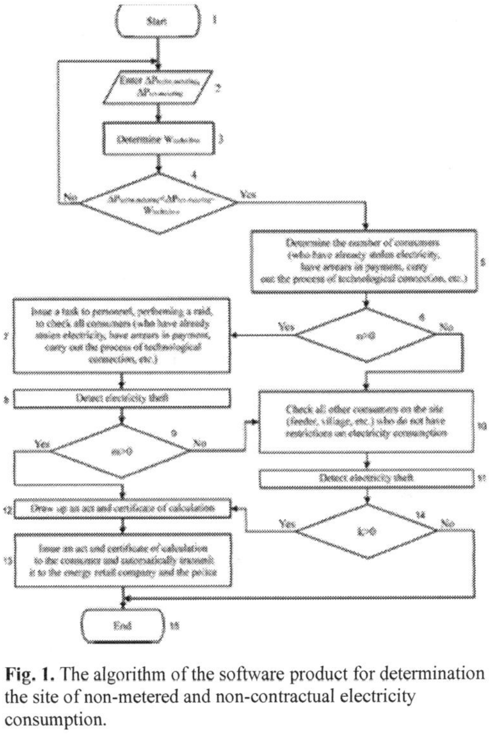

Fig. 1. The algorithm of the software product for determination the site of non-metered and non-contractual electricity consumption.

In the diagram: T1 - power transformer of the main poLet's consider the operation of the presented algorithm in cases where personnel are going to perform a raid to detect the electricity theft in a certain area (feeder, village).In this case, the beginning of the algorithm involves entering data on the amount of electricity passed through the site, the feeder, in the village ($\Delta P_{techn.\ metering}$) and the total amount of electricity ($\Delta P_{revenue\ metering}$) received by consumers. The indicated data can be selected for every hour, day, month or year. Unit 3 determines commercial losses of electricity (ΔW) and technological losses of electricity ($W_{techn.\ loss}$). Unit 4 compares the commercial losses of electricity in the area (feeder, village) with the planned indicators. When the values of commercial losses of electricity deviate from the norm, unit 4 sends a signal to unit 5, which determines the number of consumers who have already stolen electricity, have arrears in payment for consumed electricity, made non-contractual consumption, carry out the process of technological connection or have non-meter accounting. If such consumers are detected, unit 6 sends a signal to unit 7, which is responsible for issuing tasks to the personnel performing the raid to conduct the inspection. Unit 7 sends a signal to unit 8, which is responsible for detectingelectricity theft.If the electricity theft was detected during a raid by consumers who have already stolen electricity, have arrears in payment for consumed electricity, made non-contractual consumption, carry out the process of technological connection or have non-meter accounting, then unit 9 determines the number of detected cases of electricity theft. After that, from unit 9, the signal goes to unit 12, which is responsible for drawing up the act and the certificate of calculation. Then the signal goes to unit 13, which is responsible for issuing the act and the certificate of calculation to the consumer on the spot and automatically transmits all the drawn up and filled in documents to the energy retail company and the police. If there are no consumers at the site (feeder, village) who have already stolen electricity, have arrears in payment for consumed electricity, made non-contractual consumption, carry out the process of technological connection or have non-meter accounting and new cases of electricity theft of the above mentioned consumers have not been revealed, a signal from units 6 and 9 goes to unit 10, which is responsible for checking all other consumers available on the site (feeder, village). Next, the signal is transmitted to unit 11, which is responsible for identifying new electricity thefts. If such a theft is detected, then the signal of unit 11 goes to unit 14, which determines the number of new cases of electricity thefts by "new consumers". If the electricity theft by consumers ("new consumers") was detected during the raid, then from unit 14 the signal goes to unit 12, which is responsible for drawing up the act and the certificate of calculation. Then the signal goes to unit 13, which is responsible for issuing the act and the certificate of calculation to the consumer on the spot and automatically transmits all the drawn up and filled in documents to the energy retail company and the police. In that case, if during the raid new cases of electricity theft were not detected, the signal from unit 14 is sent to unit 15.

The number of consumers who have already stolen electricity, have arrears in payment for consumed electricity, made non-contractual consumption, carry out the process of technological connection or have non-meter accounting, is indicated by the variable n, while the number of new detected cases of electricity theft by the above mentioned consumers is indicated by the variable m. The number of new cases of electricity theft by consumers not specified above is indicated by the variable k.

To implement the software product in electric grid companies, it will need to be integrated with the SAP program.This integration will make it possible to obtain initial data on the amount of electricity consumed by consumers on the site (feeder, village), on consumers who have already stolen electricity, have arrears in payment for consumed electricity, made non-contractual consumption, carry out the process of technological connection or have non-meter accounting, etc.. If necessary, the program can show existing regulations aimed at conducting checks and raids to identify the electricity theft.

The increase in spot raids to detect electricity thefts will allow the electric grid companies to increase the territoriality of inspections. At the same time, the implementation of the software product in electric grid companies will significantly reduce the time for performing raids to detect electricity thefts.

4 Conclusions

Reducing the number of electricity thefts in electric grid companies is an urgent task. To reduce non-metered and non-contractual energy consumption, an algorithm that can be implemented as a software product has been developed. The presented software product allows, based on the registration of consumers who have already stolen electricity, have arrears in payment for consumed electricity, made non-contractual consumption, carry out the process of technological connection or have non-meter accounting, send to themthe personnel,performing a raid, to identify electricity theft. It also makes possiblein real time to generate the necessary acts, certificates, etc., that allows personnel to reduce the time to fill out paper versions of documents.

The software product allows to increase the efficiency of raids, performed by electric grid companies, by reducing the time of inspections. In turn, the implementation of the program will allow electric grid companies to significantly reduce electricity losses by increasing the territoriality of raids.

References

[1] A.S. Semenov, Lower Economic Losses in Electric Networks, Applied Sciences and technologies in the United States and Europe: common challenges and scientific findings Proceedings of the 3th International scientific conference, ed. Ludwig Siebenberg, 139–141 (2013)

[2] F. Jamil, On the Electricity Shortage, Price and Electricity Theft Nexus, Energy Policy **54**, 267–272 (2013)

[3] A. Vinogradov, V. Bolshev, A. Vinogradova, T. Kudinova, M.Borodin, A. Selesneva, N. Sorokin, A System for Monitoring the Number and Duration of Power Outages and Power Quality in 0.38 kV Electrical Networks, In: Vasant P., Zelinka I., Weber GW. (eds) Intelligent Computing & Optimization, ICO 2018, Advances in Intelligent Systems and Computing **866**, 1–10 (2019)

[4] S.M. Astakhov, R.P. Belikov, State and Ways to Improve the Performance of Distribution Networks in the Agro-industrial Complex, Bulletin of OrelSAU **29**, 2, 106–108 (2011)

[5] M.V. Borodin, N.S. Uryupin, Justification of Energy-efficient Cross-section of Power Lines with Voltage Deviations of ± 5%, ± 10%, Bulletin of Nizhny Novgorod Institute of Engineering and Economics **2**, 81, 20–29 (2018)

[6] M.V. Borodin, Analysis of Electric Power Losses in Electric Networks of Orlovsky Distribution Zone of Orelenergo branch of IDGC of Center, PJSC, Agrotechnics and Power Supply **4**, 17, 46–53 (2017)

[7] A.I. Psarev, Analysis of Electric Power Losses in Electric Networks of the Branch of IDGC of Center – Orelenergo PJSC, Orlovsky Distribution Zone, Innovative Development of the University Library: Management and Marketing, Proceedings of the II Scientific and Practical Conference, 65–68 (2018)

[8] A.E. Semenov, Analysis of Electric Power Losses in Electric Networks of the Branch of IDGC of Center – Orelenergo, PJSC Livensky Distribution Zone, Innovative Development of the University Library: Management and Marketing, Proceedings of the II Scientific and Practical Conference, 68–71 (2018)

[9] D.V. Lukovnikov, Unaccounted-forand non-metered power consumption, Proceedings of Bratsk State University. Series: Natural and Engineering Science **1**, 43–47 (2017)

[10] M.V. Borodin, Yu.A. Volchenkov, A.V. Vinogradov, Development of Measures to Reduce Electricity Losses in the Branch of IDGC of Center – Orelenergo, Bulletin of DonAgrarian Science **4**, 32, 27-34 (2015)

[11] M.V. Astiazara, D.A.C. Barone, Application of an Artificial Immune System to Predict Electrical Energy Fraud and Theft, ICEIS 2012 – Proceedings of the 14th International Conference on Enterprise Information Systems, 265–271 (2012)

[12] R.R. Shaikhutdinov, E.I. Gracheva, The Use of Modern Electrical Equipment to Reduce Electrical Energy Losses in Distribution Networks, Science and Education **3**, 47, 29–31 (2019)

[13] M.V. Borodin, R.P. Belikov, N.V. Makhiyanova, Improving Power quality Through the Calculation of Voltage Losses, Bulletin of Don Agrarian Science **3**, 47, 35–40 (2019)

[14] S.V. Krivonogov, Analysis of Electricity Theft in the Domestic Sector, Bulletin of Nizhny Novgorod

Institute of Engineering and Economics **10**, 41, 82–88 (2014).

[15] O.A. Chausovsky, Modern Methods ofElectricity Theft and the Fight Against Them, Industrial Energy **1**, 54–56 (2016)

[16] N.S. Sorokin, Improving the OperationEfficiency of Electrical Networks with Voltage of 6-35 kV by Monitoring the Parameters of the Electrical Network, National University of Life and Environmental Sciences of Ukraine, Series: Technics and Energy of the Agro-industrial Complex, 209-2, 92-95 (2015)

[17] K. Yevgeniy, Problems in Reimbursing Actual Damage in Cases of electricity theft, Ecology, Environment and Conservation **21**, AS59–AS62 (2015)

[18] D.D. Micu, I.V. Ivshin, E.I. Gracheva, O.V. Naumov, A.N. Gorlov, Influence of Resistance of Contact Units of Switching Devices on the Electric Power Losses in Shop Networks of Low Voltage, E3S Web of Conferences 2019, 02013 (2019)

[19] A.V. Vinogradov, A.V. Vinogradova, V.E. Bolshev, M.O. Ward, N.V. Makhiyanova, L.V. Dolomaniuk, Justification for Creating a Mobile Complex to Assess Electric Energy Loss in Power Transformers During the Operation Process, E3S Web of Conferences 2019, 02009 (2019)

[20] E.I. Gracheva, O.V. Naumov, Estimation of Power Losses in Electric Devices of the Electrotechnical Complex, 2019 International Conference on Industrial Engineering, Applications and Manufacturing, ICIEAM 2019, 8742923 (2019)

https://doi.org/10.1051/e3sconf/202017801083

Identification power line sections with increased electricity losses using sensors with Wi-Fi technology for data transmission

Igor Nikolaevich Fomin[1,*], *Roman Pavlovich* Belikov[2], *Tatyana Anatolyevna* Kudinova[3], and *Alexei Nikolaevich* Tsvetkov[4]

[1]Orel State Agrarian University named after N.V. Parakhin, Orel, Russia
[2]Orel State Agrarian University named after N.V. Parakhin, Orel, Russia
[3]Orel State Agrarian University named after N.V. Parakhin, Orel, Russia
[4]Kazan State Power Engineering University, ul. Krasnoselskaya 51, Kazan, Russia

Abstract. Modern highly mechanized and electrified agriculture places high demands on power supply reliability and uninterrupted work. To increase the reliability of power supply to agricultural consumers, in some cases, taking into account the configuration of electric distribution networks and the availability of responsible consumers, a conditionally closed ring network is created. Interruptions in power supply lead to downtime of agricultural production, a decrease in the volume of output, damage to the main technological equipment [1, 2]. In this regard, there is a need to make informed decisions on the choice of ways to increase the reliability of uninterrupted power supply due to the reservation of various elements of the power supply system, improving the organization of maintenance, and the operational diagnostics of faulty elements.

1 Introduction

To increase the reliability of power supply, power lines are partitioned. In addition, electricity is reserved, i.e. radial sectioned lines can be connected by a point of automatic transfer switch (ATS) with the formation of a conditionally closed ring network [1]. However, despite the technical support of the regulatory levels of power supply reliability, equipment failure of the power supply system may occur, such as a spurious tripping of the sectionalizing circuit-breaker in the power line. The shutdown may be caused by any malfunction or non-selective protection action. In order to timely receive information about the spurious tripping of the sectionalizing circuit-breaker in such a network, a new method of remote control has been developed.

2 Theoretical research

Analysis of the statistical data on the change in the inrush current in the ring network line at characteristic time intervals [2] gave full visibility of the processes and was the source of the necessary material for the development of remote monitoring methods [3,4,5]. Investigations of the authors [6] revealed the signs that allow remote monitoring of the false shutdown of the sectioning switch. Based on this, a new method has been developed to control the spurious tripping of the sectionalizing circuit- breaker in the ring network line.

According to this method, a change in current is monitored at the beginning of the main power supply line, and if a drop in the operating current to a value determined by the load of the line connected after the sectionalizing circuit-breaker is detected, and there is no short-circuit current inrush at this time, then the countdown equal to the delay time of the switching-on of the point switch of the ATS is started. At the end of the countdown of this time, the appearance of a current surge at the beginning of the backup power supply line is controlled, and if a current surge occurs, by a value determined by the disconnected load of the main power supply line and no short circuit current occurs, then the fact of a spurious tripping of the sectionalizing circuit-breaker is established [7].

To implement such control, Figure 1 shows a circuit of a conditionally closed ring network and a block diagram. The scheme works as follows. In normal mode, the monitored sectionalizing circuit-breaker Q2 is turned on and consumers S1, S2 are powered from the buses of the power transformer of the main power source T1 (see Fig. 1).

At the output of CT 1 and CT 12 there is a certain value of the output signal (Fig. 2, diagram 1 and diagram 12), due to operating currents, but insufficient for tripping of SCCS 2, OSC 3 and SINGLE-VIBRATOR 13. Presence of the output signal from the element NOT 4, coming to the element MEMORY 11, and from the element NOT 5, coming to the first input of the circuit element AND 7, does not lead to their operation. The circuit does not start.

* Corresponding author: igor.fom-in@yandex.ru

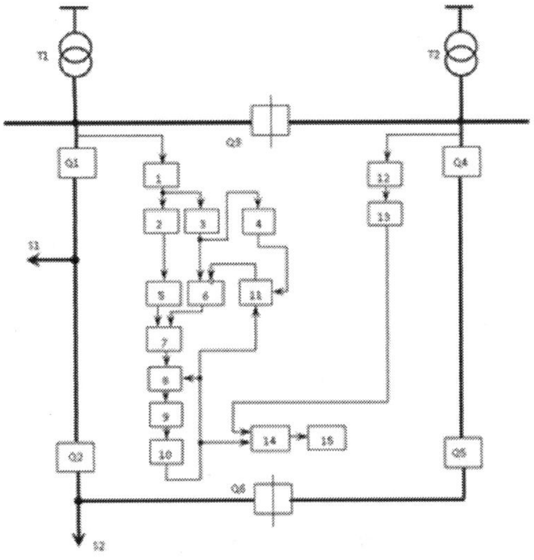

Fig. 1. Diagram of a conditionally closed ring network and a structural diagram of remote control.

In the diagram: T1 - power transformer of the main power source; T2 - power transformer of the reserve power source; Q1 - head circuit-breaker of the main power supply line; Q2 - sectionalizing circuit-breaker of the main power supply line; Q3 - substation sectional circuit-breaker; Q4 - head circuit-breaker of the reserve power supply; Q5 - sectionalizing circuit-breaker of the reserve power supply line; Q6 – circuit-breaker of the point ATS; S1, S2 - line load of the main power supply; 1- current transformer (CT), 2- short circuit current sensor (SCCS), 3 - operating current sensor (OCS); 4 - element NOT; 5 - element NOT; 6 - the element INHIBIT; 7 - element AND; 8 - element MEMORY; 9 - element DELAY; 10 - element REPEATER; 11 - element MEMORY; 12 - CT; 13 - SINGLE-VIBRATOR; 14-element AND; 15 - recording device (RD).

If the sectionalizing circuit-breaker Q2 is switched off false, at the beginning of the line a drop in the operating current will be observed (Fig. 2, time point t1). At the output of CT 1 (Fig. 2 diagram 1), the output signal will decrease by a value determined by the line load connected after the sectionalizing circuit-breaker Q2, which is fed to the input SCCS 2 and OSC 3. The value of the output signal CT 1 will not be sufficient for tripping SCCS 2, therefore, a signal does not appear at its output (Fig. 2, diagram 2), which would be input to the element NOT 5. The signal will not disappear from the output of the element NOT 5 (Fig. 2, diagram 5).

At the output of the OSC 4, which is triggered when the operating current drops to a value determined by the line load switched on after the sectionalizing circuit-breaker, a signal will appear (Fig. 2 diagram 4), which will be input to the element INHIBIT 6 and element NOT 4. The signal will disappear from the output of element NOT 4 (Fig. 2 diagram 4). At the output of element INHIBIT 6, a signal will appear (Fig. 2, diagram 6) and will go to the second input of element AND 7. The simultaneous presence of two input signals at the input of element AND 7 will lead to the appearance of its output signal (Fig. 2, diagram 7), which will go to input of element MEMORY 8. This signal will be remembered by element MEMORY 8 (Fig. 2 diagram 8) and will go to the input of element DELAY 9.

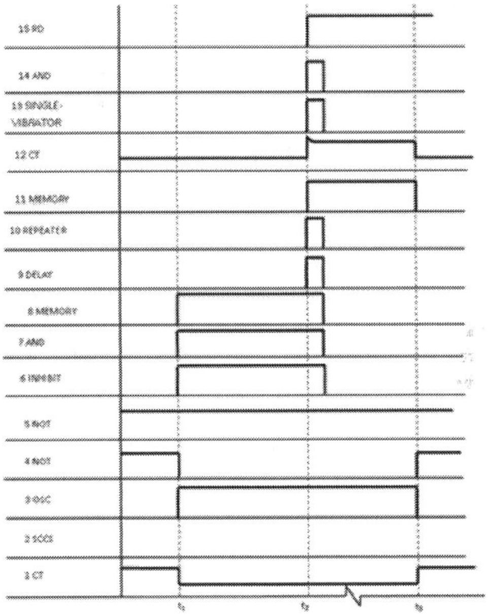

Fig. 2. The diagrams of the output signals of the structural diagram: t_1 –time point of the spurious tripping of the sectionalizing circuit-breaker S2; t_2 –time point of turning on the point switch of the ATS; t_3 – time point of normal operation mode restoration.

From the output of element 14, the signal appears after a time equal to the delay time of the switching-on of the point switch of the ATS (Fig. 2 time interval is t_1-t_2). After counting the delay time of the switching-on of the point switch of the ATS, the signal from the output of element DELAY 9 (Fig. 2 diagram 9) will be input to element REPEATER 10. Element 10 will give a single impulse (Fig. 2 diagram 10), which will go to the input of the element MEMORY 11, to the input of the element AND 14 and "reset" the element MEMORY 8. This signal will be remembered by the element MEMORY 11, from the output of which the signal (Fig. 2, diagram 11) will go to the inhibitory input of the element 6. At the output of the element INHIBIT 6, the signal will disappear (Fig. 2, diagram 6) entering the second input of the element AND 7. In this case at the output of element AND 7, the signal will also disappear (Fig. 2, diagram 7). If the circuit-breaker of the ATS point is turned on at the same time, then the output of the CT 12 will have a signal (Fig. 2 diagram 12), increased by a value determined by the load of consumers S2. The magnitude of this signal is sufficient for operation of the SINGLE-VIBRATOR 13. The output signal from the SINGLE-VIBRATOR will go to the second input of element AND 14. The presence of two input signals at

AND 14 will cause the signal to appear at its output (Fig. 2, diagram 14). The output signal from AND 14, having entered the RD 15, will provide information in it about the spurious tripping of the sectionalizing circuit-breaker.

When normal operation mode is restored (Fig. 2, time point t_3), the controlled sectionalizing circuit-breaker Q2 will turn on and consumers S2 will be powered by the buses of the power transformer of the main power source T1 (see Fig. 1). At the output of CT 1 and CT 12 there will be a certain value of the output signal (Fig. 2, diagram 1 and diagram 12), due to operating currents, but insufficient for tripping of SCCS 2 and SINGLE-VIBRATOR 13. At the output of OSC 3, which is triggered when the operating current drops to a value determined by the line load connected after the sectionalizing circuit-breaker, the signal (Fig. 2, diagram 3) which was input to element INHIBIT 6 and element NOT 4 will disappear. A signal which will "reset" the element MEMORY 11 will appear at the output NOT 4 (Fig. 2 diagram 4). The circuit will return to the initial state of control [8].

3 Technical implementation

The proposed remote control method is implemented in a device for monitoring the spurious tripping of a sectionalizing circuit-breaker in a ring network line based on an analog-to-digital converter, with subsequent processing of the data by software on a computer.

Structurally, it consists of two blocks [9, 10]. One of the blocks is the current sensor connection block. The second block is necessary for converting and analyzing the input signal, as well as for generating and transmitting digital data to a computer for further analysis (information processing block) [11].

The algorithm of the above-described structural diagram of the remote method, implemented in the device, while simulating the situations of spurious tripping of the sectionalizing circuit-breaker during laboratory studies, has proved its operability. During the experiments, the remote monitoring software [12-14] correctly recognized the simulated situation of a spurious tripping of the circuit-breaker during each test.

Thus, the developed device makes it possible to receive timely information about the spurious tripping of the sectionalizing circuit-breaker in the ring network line.

4 Conclusions

1. The proposed method of remote control makes it possible to receive timely information about the spurious tripping of the sectionalizing circuit-breaker in the ring network line.
2. Laboratory tests fully confirmed the performance of the developed technical means.
3. The implementation of the proposed method will increase the reliability of power supply to consumers by taking the necessary decisions based on the information received by operational personnel.

References

[1] L.D. Surov, I.N. Fomin, Control Failure of Automatic Reclosing of the Sectionalizing Circuit-breaker, ELEKTRO, Electrical engineering, electricity, electrical engineering **6**, 46–48 (2009)

[2] L.D. Surov, V.V. Filippov, The control of Disconnection and Failure of Automatic Reclosing of the Main Switch of Line, Bulletin of Volgograd State Technical University **11**, 8, 88-91 (2014)

[3] L.D. Surov, I.N. Fomin, V.F. Shumarin, The Control Method of the Shutdown and Successful Automatic Reclosing of the Sectionalizing Circuit-Breaker in the Ring Network Line, Applicant and patent holder FSEE HPE Orel State Agrarian University, Patent No. 2337453 Russian Federation IPC H02 J 13/00, No. 2007139420/09, declare 23.10.2007, publ. 27.10.2008, bul. no. **30**, 7 (2007)

[4] L.D. Surov, I.N. Fomin, V.V. Grekov A Method for Controlling of a sectionalizing circuit-breaker in Case of a Failure to Open a Circuit-breaker of a Point of Automatic Transfer Switch Which has been Switched on for a Stable Short-circuit in a Ring Network, Applicant and patent holder FSEE HPE Orel State Agrarian University, Patent No. 2371826 Russian Federation IPC H02J 13/00, No. 2008138127/09, declare 24.09.2008, publ. 27.10.2009, bull. **30**, 6 (2008)

[5] L.D. Surov, I.N. Fomin, N.V. Makhianova, Method of Control of Disconnection and Failure of Automatic Reclosing of the Sectionalizing Circuit-Breaker in the Ring Network Line, Patent of Russia No. 2304338, bull. **22** (2007)

[6] I.N. Fomin, R.P. Belikov, Algorithm for Remote Control of the Head Switch Line, The Bulletin of Orel State Agrarian University, **61**, 4, 72–77 (2016)

[7] L.D. Surov, I.N. Fomin, V.V. Grekov, A method for Monitoring the False Tripping of a Sectioning Switch in a Ring Network Line, Applicant and patent holder FSEE HPE Orel State Agrarian University, Patent No. 2378754 Russian Federation IPC H02J 13/00, No. 2008148640/09, declare 09.12.2008, publ. 10.01.2010, bull. **1**, 6 (2010)

[8] L.D. Surov, V.V. Filippov, Determination of the Damaged Section of the Ring Network Line, Bulletin of the Volgograd State Technical University **8**, 187, 105-110 (2016)

[9] N.S. Sorokin, Using of the Integrated Circuit FT232R chip AD7495AR and in a Block of Digital Processing in the System to Control The Operation of Electrical Networks 6-35 kV, Bulletin of Orel State Agrarian University **60**, 3, 87–93 (2016)

[10] N.S. Sorokin, Improving the Efficiency of

Functioning of Electric Networks with Voltage of 6-35 kV by Monitoring the Parameters of the Electric Network, Scientific Bulletin of the National University of Bioresources and Environmental Management Ukraine. Series: Technique and Power Engineering of Agro-Industrial Complex **209**, 2, 92–95 (2015)

[11] V.A. Chernyshov, E.A. Pechagin, The Remote Control of Overhead Transmission Lines with Insulated Neutral, Energy Conservation and Efficiency in Technical Systems, Proceedings of the IV International Scientific-Technical Conference of Students, Young Scientists and Specialists. Tambov State Technical University, 355–356 (2017)

[12] I.N. Fomin, R. Belikov, V. Zelyukin, E.V. Mikhailova, Improvement of Power Supply Reliability by Means of Remote Control of the Automatic Repeated Switching-on of Sectionalizing Circuit-breakers, E3S Web of Conferences, 01042 (2019)

[13] O.V. Naumov, Estimation Power Losses in Electric Devices of the Electrotechnical Complex, International Conference on Industrial Engineering, Applications and Manufacturing (ICIEAM), 6 coauth (2019)

[14] O.V. Naumov, A.N. Gorlov, Influence of Value Resistance Contact Units of Switching Devices on Losses of the Electric Power in Low Voltage Shop Networks, IOP Conference, Materials Science and Engineering, 6 coauth (2019)

Some aspects of use of organic containing waste for electrical and thermal energy generation

I A Vladimirov[1,*], *Liliya* Mukhametova[2], and *M V* Yamashkin[1]

[1]Peter the Great Saint-Petersburg Polytechnic University
[2]Kazan State Power Engineering University, Kazan, Russia

Abstract. This article is devoted to the problem of energy generation from the non-processed part of waste, both industrial and municipal. Some aspects of use of gasification products of organic-containing wastes and landfill gas are considered. The presented comparison shows that landfill and generator gases can be attributed to the group of medium-calorific gases used both as boiler fuel and for production of liquid synthetic fuels. It should also be noted that generator and landfill gases contain a significant proportion of inert gases (from 22% to 60%), which is significantly higher than that of natural gases. Inert gases in landfill and generator gases are different. In the first case it is CO_2, in the second it is N_2. Unlike solid fuel boilers, in a boiler designed to burn gaseous fuels (natural gas), synthetic gases can also be burned without significant additional reconstruction. Due to the difference in gas calorific value, the ratio of fuel to air (with a constant coefficient of excess air α) will change to obtain the necessary thermal power, however, the volume of flue gases will differ slightly. Therefore, the flow sections of the boiler furnace can be saved.

1 Introduction

Municipal solid waste (MSW) and industrial waste management is currently a global problem. Waste must be sorted. Some of its components, such as metal, plastic, ceramics, are recyclable and reusable [1], food waste can be biodegradable. A significant part of waste (about 30% by volume of MSW) is not recyclable, the organic-containing part of such waste can be subjected to thermal utilization in order to generate thermal and electric energy [2-5]. The leaders in the field of MSW processing are Japan, where 75% of MSW are processed, USA (46%), Switzerland (53%) [6], Denmark (46.3%), the Netherlands (51.7%), Germany (66.1%), Sweden (48 %), France (39.5%) [7].

As a rule, organic-containing waste is a solid loose mass of various fractional compositions. The main types of organ-containing waste considered in this article include: municipal solid waste, logging debris, wood processing waste, agricultural waste: straw and biomass of cultivated plants. Therefore, during direct combustion, a solid fuel boiler is required. The efficient use of such a boiler is accentuated by the fact that solid fuel boilers are designed and calculated for a specific type of fuel with the known thermophysical properties [8]. However, the thermophysical properties of organic-containing waste are unstable, so the efficiency of such a boiler will significantly change. In some cases it cannot be used, for example, for slag with low melting temperature and upon slagging of the screen surfaces. At the same time, gas boilers do not have such a drawback [9]. To use organic-containing waste in the energy recovery cycle, it is necessary to use gaseous products obtained from waste by gasification, pyrolysis, or by collecting landfill gas. Consider some aspects of gasification of organic-containing waste and combustion of generator gas.

2 Materials and methods

To determine the higher Q_h and lower Q_l^w calorific values of fuel the formulas of D.I. Mendeleev are used:

$$Q_{l(h)} = 0.01 \cdot (Q_{l(h)H_2} \cdot H_2 + Q_{l(h)CO} \cdot CO + \\ + Q_{l(h)H_2S} \cdot H_2S + \Sigma Q_{l(h)C_nH_m} \cdot C_nH_m) \quad (1)$$

where $Q_{l(h)}$ is the lower (higher) heat of combustion of generator gas components, (MJ/m^3); H_2, CO, H_2S, C_nH_m is the content of generator gas components, %. When substituting the known calorific values of gases, the formula is reduced to:

- for higher calorific value of gas:

$$Q_h = 126.4 \cdot CO + 127.8 \cdot H_2 + \\ +397.8 \cdot CH_4 + 630.6 \cdot C_2H_4 + \\ +699.7 \cdot C_2H_6 + 972.2 \cdot C_3H_8 + 1285.9 \cdot C_4H_{10} \quad (2)$$

- for lower calorific value of gas:

$$Q_l^w = 126.4 \cdot CO + 107.8 \cdot H_2 + \\ +358.6 \cdot CH_4 + +591.2 \cdot C_2H_4 + \\ +638.1 \cdot C_2H_6 + 913.4 \cdot C_3H_8 + 1187.6 \cdot C_4H_{10} \quad (3)$$

* Corresponding author: vladimirov@nil-teplo.ru

Table 1. Comparison of thermophysical characteristics of fuel gases.

Gas	Gas content, volumetric,%							ρ, kg/m³	Q_l^w, MJ/m³
	Combustion				Inert gases		O_2		
	H_2	CO	CH_4	C_nH_m	CO_2	N_2			
Biomethane [10]			95	3		1.5			36
Landfill gas: [11]									
high calorific			70	-	30	-	-	1.098	25.102
low calorific			33	-	60	-	-	1.48	11.834
Semi-coke [12]	23.5	6.5	51.5	4.5	7.0	5.5	1.5	0.764	24.65
Coke [12]	57	6.0	24.0	3.0	3.0	7.0	-	0.324	17.6
Shale [12]	24.7	10.0	16.2	5.0	16.4	26.8	0.7	1.040	13.85
Generator gas*	18.1	24.4	1.6	-	10.3	45.4	0.2	1.09	4.5
Generator gas **	13.0	27.6	0.6	-	6.0	53.2	0.2	1.141	5.15
Generator gas ***	37.6	39.2	-	-	3.02	19.9	-	0.52	8.25
Blast-furnace [12]	3.0	30	-	-	9.0	58.0	-	1.283	4.10
Natural [12] (Korobovskiy)	-	-	81.5	14.8	-	3.2	0.5	0.901	41.45
Natural [12] (Igrimskiy)	-	-	95.7	2.8	-	1.3	0.2	0.741	36.47

The theoretically required air consumption for mixture burning (m³/m) is:

$$V_0 = 0.0476 \cdot \big(0.5 \cdot CO + 0.5 \cdot H_2 + 2 \cdot CH_4 + \\ + \Sigma\left(m + \frac{n}{4}\right) \cdot C_m H_n + 1.5 \cdot H_2S - O_2\big) \tag{4}$$

When changing the compositions of gas and blast, the theoretical flow rate of blast for the complete combustion of 1 m³ of gas is calculated by the formula:

$$V = \frac{100}{K} \cdot V_{0,O_2} \tag{5}$$

where K is the oxygen content in blast, %. K = 21% for air blast;

$$V_{0,O_2} = 0.5 \cdot (H_2O + Co) + 1.5 \cdot H_2S + \\ + \Sigma\big((n+m) \cdot C_n H_m - O_2\big) \cdot 0.01 \tag{6}$$

The volume of combustion products, m³/m³ is determined by the formula:

$$V_{RO_2} = 0.01 \cdot \left(CO + CH_4 + \Sigma C_m H_n + H_2S + CO_2\right) \tag{7}$$

Mass fraction of water in wet gas, % is:

$$W_g = \frac{\rho_{H_2O} \cdot H_2O}{\rho_{CO_2} \cdot CO_2 + \rho_{CH_4} \cdot CH_4 + \rho_{H_2O} \cdot H_2O} \tag{8}$$

And the volume of water vapour is:

$$V_{ws} = 0.01 \cdot (H_2 + 2 \cdot CH_4 + \Sigma \frac{n}{2} \cdot (C_m H_n + \\ + H_2S + 0.124 \cdot W_g) + 0.016 \cdot V_0 \cdot \alpha_{air} \tag{9}$$

The temperature of mixture combustion is

$$T_{combustion} = \frac{Q_l^w + h_{mix} + h_{air}}{V_{mix} \cdot C_{mix}^{`}} \tag{10}$$

or a detailed expression

$$T_{combustion} = \frac{Q_l^w + h_{mix} + h_{air}}{V_{CO_2} \cdot C_{CO_2}^{`} + V_{N_2} \cdot C_{N_2}^{`} + V_{H_2O} \cdot C_{H_2O}^{`} + V_{air} C_{air}^{`}} \tag{11}$$

where h_{mix} and h_{air} is the heat of mixture and air, respectively, kJ/kg, the values of which under normal conditions can be neglected.

The heat capacity of generator gas is determined based on the principle of additivity, i.e. the products of heat capacities of individual components by their mass fractions are summed up.

3 Results and discussion

The thermophysical properties of generator gas depend on the gasification method and blast composition. Using the listed formulas, the properties of landfill gas of various calorific values were calculated (boundary values were adopted depending on the MSW composition) and the generator gas obtained by various gasification technologies, taking into account various blast composition.

For clarity, the properties of generator and landfill gases are given in comparison with fossil natural gas of various fields, as well as other synthetic gases. The comparison results are shown in table 1.

The comparison shows that landfill and generator gases can be attributed to the group of medium-calorific gases used both as boiler fuel and for production of liquid synthetic fuels. It should also be noted that generator and landfill gases contain a significant proportion of inert gases (from 22% to 60%), which is significantly higher than that of natural gases. This fact explains its relatively low calorific value. It should be also noted that the composition of landfill gas is variable and depends on the morphological composition of

Table 2. The amount of flue gases produced during combustion of fuels with various heat of combustion for α=1.2.

Parameter	Natural gas	Generator gas
Heat of combustion Q_i^w, MJ/kg	35.3	4.96
Fuel consumption, m^3/s	0.283	7.2
Air consumption, m^3/s	7.218	5.155
Consumption of flue gases, m^3/s	7.845	7.800

Table 3. Temperature of generator gas combustion.

Gas mixture composition	$T_{combustion}$, K
CH_4 - 40%, CO_2 – 60%	2498
CH_4 - 70%, CO_2 – 30%	2833
CH_4 – 33%, CO_2 – 60%, H_2O – 7%	1975
CH_4 – 63%, CO_2 – 30%, H_2O – 7%	2241

municipal solid waste located at the landfill. For unsorted waste, biogas includes [11]:

- combustible component methane (CH_4) in the amount from 33 to 70%,
- inert gas carbon dioxide (CO_2) in the amount from 60 to 30%,
- impurities, their amount is insignificant, but hydrogen sulfide H_2S is presented in it,
- water vapor, the maximum content of which in landfill gas reaches 7%.

Inert gases in landfill and generator gases are different. In the first case, it is CO_2, in the second it is N_2. Dilution of combustion gases with inert (ballast) gases impairs their flammability. The type of inert gas can significantly affect the flammability limits, and for CO_2, this effect is much stronger than for N_2. The concentration limits of methane ignition (the combustible component of biogas) in a mixture with CO_2 are in the range from 6 to 30% [15]. There, for comparison, a similar dependence is given for a mixture of methane with inert gas (N_2); for such a mixture, the concentration limits are wider and amount to 6–42% [15].

At the lowest possible concentration of CH_4 in biogas, ignition of biogas is problematic, as it is in a very narrow range of air concentrations. In such cases, a quite common technique that improves flammability is the addition of high-calorific gas from an external source (natural gas, propane, butane, etc.).

3.1 Volume of flue gases

Among the aspects of burning synthetic gases, it is important to note that, unlike solid fuel boilers, synthetic gases can also be burned in a boiler designed to burn gaseous fuels without significant additional reconstruction. Of course, due to the difference in the calorific value of gas, the ratio of fuel to air (at a constant coefficient of excess air α) will change to obtain the necessary thermal power, however, the volume of flue gases will differ slightly. Therefore, the flow sections of the boiler furnace can be saved. The results of calculating the volume of flue gases during combustion of natural and generator gases at α = 1.2 are shown in table 2.

The presented data allow us to conclude that the generator gas, as well as biogas, can be used in various combustion devices, for both industrial (steam and hot water boilers, process furnaces, drying plants, heat generators, heating devices, etc.), and domestic (heating stoves, hot water devices, etc.) purposes without significant reconstruction. Due to changes in gas and air flow rates, only burner devices should be replaced. For efficient and reliable burning of generator gas and optimization of furnace devices, it is advisable to study the propagation velocity of the combustion front, the length of flame and the completeness of combustion in burners of various types for various biogas composition and flow rates of blast air.

Thus, it is possible to convert the existing mini-CHPPs to work on generator or landfill (in case of a mini-CHPP located near the MSW landfill) gas. For such a transfer, an add-on to a mini-CHP plant with a gas generator and a generator gas purification system (if necessary) will be required. The boiler room (with the exception of burners), turbine equipment, or a gas piston unit, as well as flue gas removal and purification system can be left unchanged.

3.2 Generation of nitrogen oxides

The formulas listed above were used to calculate the combustion temperature of generator gas as a function of gas mixture composition. The calculation results are shown in table 3.

It should be noted that the calculation results differ from the Russian literature data [15], which show that when burning the $2CH_4$+CO_2 mixture in air, the flame temperature is 2025 K. Therefore, experimental confirmation is required to correctly determine the combustion temperature using the proposed formulas.

At the same time, when burning natural gas, the temperature in the combustion zone reaches 2200–2500 K [16]. The combustion temperatures for natural and generator gases are comparable, the methods for reducing the formation of nitrogen oxides during their combustion can be the same [17].

The presence of SO_X in the combustion products is also not desirable in the thermotechnical sense, since it prevents deep utilization of heat of combustion products [18]. When combustion products are cooled below the dew point, sulfuric acid H_2SO_4 is formed from sulfur oxides, which causes intense corrosion of convective surfaces (the speed of which can reach 1 mm per year). Therefore, in case of energetic use of biogas, it is advisable to clean it from H_2S. The presence of sulfur oxides in the combustion products can be eliminated by preliminary purification of generator and landfill gases from H_2S, for example, using special absorbers (methanol, amines, activated carbon, aluminum oxide, etc.). If H_2S content in gas is high, the installations catching it can be supplemented with sections for production of elemental sulfur and sulfuric acid. However, this solution requires a separate feasibility study to calculate the minimum H_2S content in gas at which the sulfur or sulfuric acid production unit will be profitable.

4 Conclusion

The organic-containing waste can be used to produce electrical and thermal energy both through direct burning, and with their preliminary gasification. This is one of the most effective ways to handle non-recyclable waste.

Direct burning must be carried out using specially designed combustion devices, the design of which depends on the morphological composition and thermophysical properties of organic-containing waste.

The preliminary gasification of waste has some advantages:

1. The possibility of utilization gasification products in a binary cycle, or in gas-piston units. It enables applying the principle of cogeneration in energetic use of organic-containing waste with greater efficiency.

2. When operation of an existing boiler unit is converted from natural to generator or landfill gas, it is necessary to replace the burner devices. The flow sections of the furnace remain the same due to the almost unchanged volume of flue gases.

The simplest way to use non-recyclable waste for energy production is to perform a preliminary sorting already at the stage of their formation (that is, directly at the factory or by citizens). If waste sorting is provided by law, and all the materials that make up waste are pre-marked, the production of energy from waste becomes more economical by reducing the cost of sorting. The use of non-processed organic-containing waste for energy production should be considered when developing long-term plans and strategies for energy supply in the regions, as well as waste management as a priority measure.

References

[1] H. Sözer, H. Sözen, Waste capacity and its environmental impact of a residential district during its life cycle, Energy Reports 6, 286-296 (November 2020).

[2] N. Politaeva, Y. Smyatskaya, R. Al Afif, C. Pfeifer, L. Mukhametova, Development of Full-Cycle Utilization of Chlorella sorokiniana Microalgae Biomass for Environmental and Food Purposes, Energies 13, 10, 2648 (2020).

[3] A.A. Sinitsyn, N.N. Boytsova, I.G. Akhmetova, S.O. Makoev, A. Asadpoori, O.A. Stepanov, N.V. Rydalina, E.O. Antonova, B.G. Aksenov, Study of operation of power-generating devices of gaseous fuels combustion, International Journal of Mechanical Engineering and Technology 10, 2, 1480-1490 (2019).

[4] A. Fedyukhin, I. Sultanguzin, A. Gyul'Maliev, V. Sergeev, Biomass pyrolysis and gasification comprehensive modeling for effective power generation at combined cycle power plant, Eurasian Chemico-Technological Journal 19, 3, 245-253 (2017)

[5] L.E. Bell, Cooling, heating, generating power, and recovering waste heat with thermoelectric systems, Science 321, 5895, 1457-1461 (12 September 2008).

[6] H. Ruohomaa, N. Ivanova, From solid waste management towards the circular economy and digital driven symbiosis, IOP Conference Series: Earth and Environmental Science 337, 1, 012032 (18 November 2019).

[7] Eurostat database [Electrical resource]. Available at: http://ec.europa.eu/eurostat/data/database (accessed: 12.04.2020).

[8] K. Jesionek, M. Kantorek, H. Karcz, Main problems concerning co-firing biomass mixture with hard coal in pulverized-fuel boilers, 27th Annual International Pittsburgh Coal Conference 2010, PCC 2010 2, 1300-1308 (2010).

[9] A. Szkarowski, S. Janta-Lipińska, T. Dąbrowski, Research on co-combustion of gas and oil fuels, Rocznik Ochrona Srodowiska 20, 1515-15299 (2018).

[10] E.S. Pantchava, V.A. Pozharnov, L.V. Zysin, et al., Biomass energy conversion, Russian experience, Heat and Power Engineering 5, 33-38 (1996).

[11] V. Maslikov, E. Negulyaeva, A. Cheremisin, A. Chusov, D. Molodtsov, A. Stroganov, Experimental assessment of biogas and leachate emissions from MSW landfills for energy reclamation with the aim of sustainable urban development, Solid State Phenomena 871, 199-207 (2016).

[12] Industrial heat power engineering and heat engineering: reference book (Moscow: Energoatomizdat, 551, 1983).

[13] V.V. Sergeev, Gasification of plant biomass (SPb.: Publishing house of the Polytechnic, University, 242, 2016).

[14] A.N. Brattsev, V.A. Kuznetsov, V.E. Popov, A.A. Ufimcev, Arc Gasification of Biomass: Example of Wood Residue, High temperature 49, 2, 244-248 (2011).

[15] A.S. Isserlin, Foundations of burning gas fuel (Leningrad: Nedra, 356, 1987).

[16] H.K. Kayadelen, Effect of natural gas components on its flame temperature, equilibrium combustion products and thermodynamic properties,

Journal of Natural Gas Science and Engineering 45, 456-473 (2017).

[17] M.H. Waldner, R. Halter, A. Sigg, B. Brosch, H.J. Gehrmann, M. Keunecke, Energy from Waste – Clean, efficient, renewable: Transitions in combustion efficiency and NOx control, Waste Management 33, 2, 317-326 (February 2013).

[18] R. Saidur, E.A. Abdelaziz, A. Demirbas, M.S. Hossain, S. Mekhilef, A review on biomass as a fuel for boilers, Renewable and Sustainable Energy Reviews 15, 5, 2262-2289 (June 2011).

The experiment results of using the competency-based approach in the occupational safety management field to reduce microtraumas in the energy sales company

Elena Bakiko[1,*], *Vitaly* Serdyuk[1], *Vladimir* Kuleshov[1], and *Ivan* Ignatovich[1]

[1]Omsk State Technical University, Department of Industrial ecology and safety, 644050 Omsk, Russia

Abstract. Reducing injuries problems in the electrical energy industry require the modern tools search and application for managing professional risks. The authors suggest the labor protection services specialists' professional competence improving in organizations as one of these tools. Conducting experiment of the labor protection specialists' professional competence improving impact assessment on the labor protection management effectiveness of was performed on the example of the energy sales company. The experiment purpose was the electrical injuries level decrease. The competency-based approach testing in the labor protection field is based on the researches results of more than 1000 organizations, where the assessment of more than 2000 labor protection specialists' professional competence state was performed. The results obtained showed the occupational safety specialists professional competence improving effectiveness not only in electrical injury reducing by 1.5 times over 2 years, but also in the occupational safety management system effectiveness improving in the company as a whole. The authors forecast the further positive company's competence approach impact in the labor protection field in dynamics. The experiment results are of practical interest to the labor protection services heads, personnel management services, leading top managers, employers.

1 Introduction

International policy in the occupational safety management field, the occupational safety specialists associations activities are focused on reducing occupational diseases, occupational injuries, stress, as well as managing occupational risks [1-2].

The labor protection management effectiveness improving is one of the most relevant research areas [3-7]. The necessity to find ways to reduce occupational injuries is noted in the researches of many foreign authors [8-10]. Research results show that a high injuries level leads to large capital losses in the organization [11-12]. The fatal injuries consequence in the world was about 2.34 million dead people [13-14].

One of the tools for reducing injuries is the directorate and employees involvement in occupational safety management [15-17]. An effective management tool is the analysis and accounting of the negative production risks impact on employees [18].

International practice and the best practices using in reducing injuries shows the effectiveness of accounting not only for microtraumas, but also for near misses (minor discrepancies). The system organization for microtraumas, incidents recording and their analysis is a task for occupational safety specialists (safety managers). The authors' research shows that the effectiveness of such work depends on their qualifications and professional competence [19-20].

Since 2015, when Vladimir Putin at the G-20 Antalya Summit declared Russia's commitment to the Vision Zero concept, the directorate of the energy sales company under study has set this concept principles implementing task to the labor protection service.

The management's commitment to this concept and the labor protection service work in the energy sales company showed a positive result - a reduction in injuries in the period 2014-2017. However, the electrical microtraumas state remained at an unacceptable level. In 2017 the company used the competency-based approach as a tool for microtraumas reducing at the authors' suggestion. It is aimed at the labor protection service specialists' professional competence level increasing, which contributes to the effective labor protection management in the company.

2 Methods

We used empirical research methods (observation, comparison, measurement, experiment), statistical analysis and synthesis.

Experiment results forecasting is based on the author's researches on the occupational safety specialists professional competence level impact on the industrial injuries level, conducted on the basis of more than 1000

* Corresponding author: bakiko@mail.ru

organizations, including energy sales companies, which employed more than 2000 occupational safety specialists from 18 regions of Russia for the period 2014-2019 [19, 21].

3 Results

The consistently high energy sales company employees electrical injuries level associated with their work makes the management continuously search for methods and means to effectively prevent these undesirable and adverse events [22].

The initial injuries analysis showed that in the period from 2007 to 2019, 95 accidents of various severity occurred in the company under study (Fig. 1, 2).

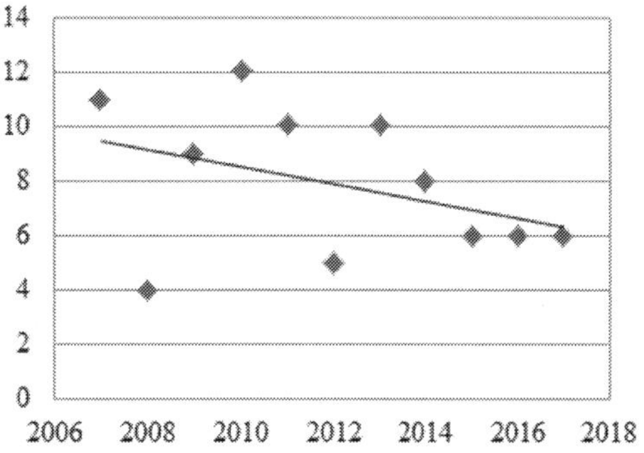

Fig. 1. Accidents number dynamics in an energy sales company.

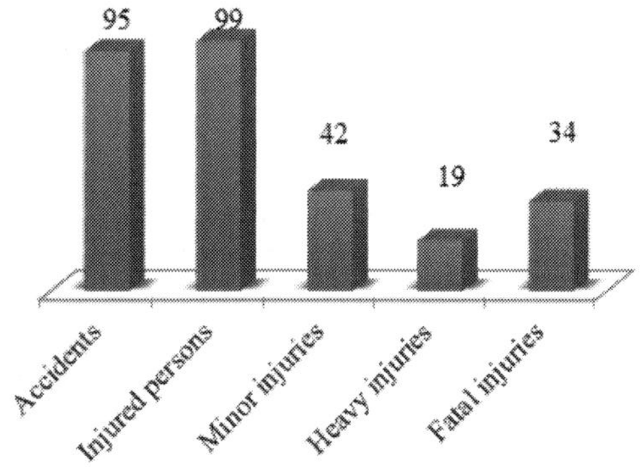

Fig. 2. Injury rate in the energy sales company for the period 2007 – 2019.

At the experiment initial stage the labor protection service specialists' professional competence level assessment of studied energy sales company in the Omsk region was performed. By the expert method the authors rated it as an average level. Service specialists had higher education, but not specialized education, professional retraining, periodically passed advanced training courses and had work experience of 3 years or more [19].

The authors suggested the company to use a competency-based approach to improve the labor protection management effectiveness, which is associated with labor protection specialist professional competence level increasing. Therefore, in 2017, the company recruited labor protection specialist with higher level of professional competence– a higher educational institution graduate (Omsk State Technical University) with a master's degree in the field of Technosphere safety with the profile "Labor Safety" without work experience. He was assigned the accounting organizing task of not only accidents and occupational diseases, but also the accounting and analysis of microtraumas, circumstances and causes of their occurrence.

According to the authors' forecast – as a result of the occupational safety specialist professional competence level increasing in the studied company, the procedures implementation for microtraumatism accounting and analysis will be expertly organized, reducing microtraumatism rate by at least 1.5 - 2 times in the dynamics over 5 years.

The authors' expert opinion, supported by the experience of testing the proposed approach in other organizations in the region, was convincing for the management. For example, according to the OJSC "Russian Railways" activities report in the sustainable development field for 2017 labor protection measures, including measures for recording and registering microtraumas, allowed for a 3-fold reduction in injuries over 7 years.

The experiment next stage showed that a specialist of a higher professional competence level, in accordance with the task set by the management, considered the possibility of using the "Accidents pyramid model" for analyzing injuries in the company. It properly illustrates the relationship between accidents, incidents and near misses in the organization [23].

The analysis showed that in the context of the company, the top layer is "accidents, including fatalities". They are subject to investigation and registration. Corrective measures are being developed for prevention and preventing similar cases.

The middle layer is "incidents with less severe consequences", namely cases of disability, which are also recorded. This reduction in injuries is provided by the existing average labor protection service specialists professional competence level (the injuries level in 2015-2017 in Fig 1).

The lower layer is "incidents with minimal consequences (discrepancies)", including electrical microtrauma, which can pass unnoticed, so it is not taken into account.

During the experiment, a specialist with an increased professional competence level ensured the electronic "Incidents and near misses register" introduction in a number of company structural divisions for 3 months. It is important that the recordings were made in a non-personalized manner.

Due to active departments' heads involvement to the microtraumas accounting process the following intermediate result was obtained:

1. One-third of top managers considered these labor costs non-production;

2. One-third of top managers stated obtaining positive results in their units;

3. The company's management positively considered the issue of journal mandatory and regular maintenance in all structural company divisions in the short term. After a month of logging "undervalued" improper operations monitoring increased. It has become obvious that this is an effective way to predict the number and type of microtraumas and accidents; differentiated work state assessment to reduce microtraumas in departments; underestimated detection;

4. Labor protection specialists have noted an increase in the volume of information about the true and reliable improper operations causes. This allowed them to develop effective measures to reduce the number of electrical microtraumas opportunely and in a speedy fashion.

5. The "Accidents pyramid " for 2019 was built in the ratio of 4: 41:1912, from which it follows that for 1 accident in the company there are 10 incidents, and for 1 incident there are 478 near misses.

4 Discussion

At the final stage of the experiment, after two years, the following result was obtained:

1. Reducing the number of near misses, microtraumas, accidents (Fig. 3). Reducing the accidents consequences severity;

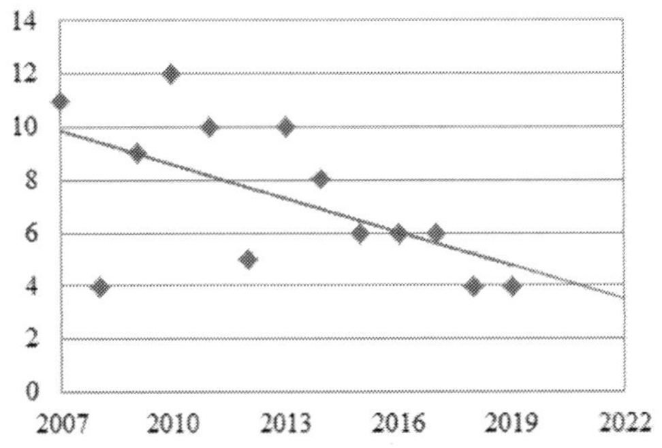

Fig. 3. Reduction of injuries as the occupational safety specialist competence improving result.

2. Company socio-economic losses reduction for reasons related to accidents and microtraumas;

3. Company occupational health and safety management system effectiveness improving;

4. Increasing the motivation of directorate and divisions direct line managers in managing professional risks;

5. Analytical electronic resource emergence for quantitative and qualitative safety improper operations assessment in divisions;

6. Increase in the complexity associated with maintaining the register has led to a decrease in the complexity of preparing weekly (and monthly) job inspections reports. All the necessary information has been pre-processed and structured. In 2020, an automatic service was developed in the company's electronic document flow structure whereby the heads of departments will be able to record all incidents on a daily basis. At the same time, the inspections results reports in divisions are automatically generated;

7. Improving the accuracy of the information provided. The workplaces planned regular inspections results formal filling percentage has decreased;

8. Employee training effectiveness improving on labor protection issues in the company by analyzing the identified reliable improper operations causes;

9. Continuing and improving the working group effectiveness on microtraumas reducing in the company, according to the authors' forecasts, will lead to a continuing injuries reducing trend;

10. Emergence of interest in this solution applying experience from other energy sales companies at the regional level.

Thus, the competency-based approach using by the directorate of the studied energy sales company in the labor protection field to reduce injuries and microtraumas showed a positive result in the period 2017-2019. The labor protection specialists professional competence improving allowed reducing the number and severity of accidents in the company by at least 1.5 times in the first 2 years. The company is expected to continue positive dynamics in the short term.

References

[1] J. Verbeek, I. Ivanov, Essential occupational safety and health interventions for low-and middle-income countries: an overview of the evidence, Safety and health at work 4, 2, 77-83 (2013).

[2] A.M. Elin, Labor Protection and Economics, Working conditions control as a managerial influence factor 4, 17, 4-8 (2014).

[3] D.C. Darabont, C. Bejinariu, I. Ionita, M.A. Bernevig-Sava, C. Baciu, E.R. Baciu, Considerations on improving occupational health and safety performance in companies using ISO 45001 standard, Environmental Engineering & Management Journal (EEMJ) 17, 11 (2018).

[4] A. Górny, Assessment and management of risk in improving the OHS management system, System Safety: Human-Technical Facility-Environment 1, 1, 105-111 (2019).

[5] T. Jordan, The ISO 45001: 2018 Implementation Handbook: Guidance on Building an Occupational Health and Safety Management System, Quality Progress 52, 1, 54-54 (2019).

[6] L. Morgado, F.J.G. Silva, L.M. Fonseca, Mapping Occupational Health and Safety Management Systems in Portugal: outlook for ISO 45001: 2018 adoption. Procedia Manufacturing 38, 755-764 (2019).

[7] H. Lingard, Occupational health and safety in the construction industry, Construction management and economics 31, 6, 505-514 (2013).

[8] M.A. Rodrigues, A. Sá, D. Masi, A. Oliveira, G. Boustras, S. Leka, F. Guldenmund, Occupational Health & Safety (OHS) management practices in micro-

and small-sized enterprises: The case of the Portuguese waste management sector, Safety Science 129, 104794 (2020).

[9] P. Swuste, C. van Gulijk, J. Groeneweg, F. Guldenmund, W. Zwaard, S. Lemkowitz, Occupational safety and safety management between 1988 and 2010: review of safety literature in English and Dutch language scientific literature. Safety science 121, 303-318 (2020).

[10] R.C. Sinclair, T.R. Cunningham, P.A. Schulte, A model for occupational safety and health intervention diffusion to small businesses, American journal of industrial medicine 56, 12, 1442-1451 (2013).

[11] P. Yakovlev, R.S. Sobel, Occupational safety and profit maximization: Friends or foes?, The Journal of Socio-Economics 39, 3, 429-435 (2010).

[12] E.V. Bakiko, V.S. Serdyuk, E.V. Yakovleva, Omsk Scientific Bulletin. Series Society. History. Modernity, Socio-economic impact of adverse working conditions and their differentiation 4, 4, 149-156 (2019).

[13] X.B. Liu, H.B. Liu, D.D. Han, Current situation and analysis of occupational safety and health in the world, Mod. Occup. Saf. 3, 91-93 (2016).

[14] H. Chen, C. Hou, L. Zhang, S. Li, Comparative study on the strands of research on the governance model of international occupational safety and health issues, Safety science 122, 104513 (2020).

[15] M. Simard, A. Marchand, The behaviour of first-line supervisors in accident prevention and effectiveness in occupational safety, Safety science 17, 3, 169-185 (1994).

[16] B. Fernández-Muñiz, J.M. Montes-Peón, C.J. Vázquez-Ordás, Safety climate in OHSAS 18001-certified organisations: Antecedents and consequences of safety behavior, Accident Analysis & Prevention 45, 745-758 (2012).

[17] A. Skład, Assessing the impact of processes on the Occupational Safety and Health Management System's effectiveness using the fuzzy cognitive maps approach, Safety science 117, 71-80 (2019).

[18] M. Kim, I. Lee, Y. Jung, International project risk management for nuclear power plant (NPP) construction: Featuring comparative analysis with fossil and gas power plants, Sustainability 9, 3, 469 (2017).

[19] E.V. Bakiko, V.S. Serdyuk, V.V. Fadeeva, Omsk Scientific Bulletin, Series Society, History. Modernity, Organizational and economic aspect of labor protection specialist professional competence formation 5, 2, 160-168 (2020).

[20] S.P. Voroshilov, N.N. Novikov, G.Z. Feinburg, Occupational health and safety in construction, Fundamentals of the methodology for assessing the level of professional risk of an employee due to the level of his professional competence 5 (2011).

[21] V.V. Kuleshov, E.V. Bakiko, V.S. Serdyuk, Development of Additional Education to Increase the Level of Competence of Specialists in the Field of Technosphere Safety, International Scientific Conference Far East Con (ISCFEC 2020), 2744-2749 (2020).

[22] A.C. Ahmad, I.N.M. Zin, M.K. Othman, N.H. Muhamad, Hazard Identification, Risk Assessment and Risk Control (HIRARC) Accidents at Power Plant, EDP Sciences, MATEC Web of Conferences 66, 00105 (2016).

[23] S. Yusupova, Application of international indicators for the analysis of industrial injuries at the Zhezkazgan processing plant, 1, LLP Kazakhmys Corporation, The Works Of The University 1, 78-81 (2016).

E3S Web of Conferences **178**, 01086 (2020) https://doi.org/10.1051/e3sconf/202017801086
HSTED-2020

Assessment of efficiency of diesel generators use in distributed energy industry

Alexandr Alyunov[1,*], *Olga* Vyatkina[1], *Ivan* Smirnov[1], *Alexandr* Nemirovskiy[1] and *Elena* Gracheva[2]

[1]Vologda State University, Lenina str. 15, Vologda, Russia
[2]Kazan State Power Engineering University, Kazan, Russia

Abstract. The existing methods for calculation electricity tariffs for enterprises are analyzed numerically. The efficiency of diesel generators use for power supply was estimated using a real object as an example. A methodology has been developed for choosing the optimal price category, based on the schedule for consumption of electrical energy of a particular enterprise, taking into account introduction of its own generation in the 0.4 kV network. It is shown how to get the maximum benefit using one of the most affordable and autonomous power sources, taking into account the peculiarities of tariff setting for electricity.

1 Introduction

In modern electric power industry, there is a problem of constant growth of electricity tariffs, under centralized electricity supply. Analysis of statistical data shows that the cost of electricity has grown more than five times for the final consumer, and it likely to grow further.

The introduction of own generation at industrial facilities is the most optimal solutions to the problem, both from practical and from economic points of view [1], [2].

As it is known, the main consumers of electricity are industrial enterprises, and a significant share of consumption is accounted by small and medium-sized enterprises, the installed capacities of which are in the range from several kilowatts to ten megawatts [3] - [5].

We evaluate the efficiency of distributed generation for small and medium-sized enterprises. For such enterprises it is inefficient to use large block stations, due to the lack of territory for their erection, as well as high costs of their construction and maintenance. Also, most enterprises do not have gas distribution stations, substations and main gas pipelines near them. One of the advantages of diesel power plants is the possibility of their widespread use, since diesel fuel is much more affordable than other types of fuel. We take into consideration this category of own generation [5] - [8].

The cost of a unit of electricity generated by a diesel generator is higher than the cost of a unit of electricity purchased from the network, for example, in the night. This is due to the relatively high cost of diesel fuel. Thus, the continuous use of diesel generators for autonomous power supply is disadvantageous. However, switching it on for certain periods of time, for example, during peak hours or hours of power take-off, can significantly reduce the cost of purchasing electricity.

The purpose of this work is:

1. To numerically analyze the existing methods for calculating electricity tariffs for enterprises (price categories).
2. To evaluate the efficiency of using diesel generators for power supply using an example of a real object.
3. To develop a methodology for selecting the optimal price category based on the schedule for consumption of electrical energy of a particular enterprise, taking into account the introduction of its own generation in a 0.4 kV network.

2 Materials and methods

We consider the choice of optimal price category and assessment of efficiency of using diesel generators in a 0.4 kV network using the example of a woodworking enterprise.

The initial data provided by the enterprise include:

- Volumes of electricity by feeder at hourly intervals for the billing period (we take into consideration the volumes of electricity for December 2019 [9]).
- The voltage side on which electricity is being purchased (for this enterprise).

In addition to the initial data the components of the maximum levels of unregulated prices and monthly reports on peak hours are required for the calculation of all price categories.

The price category is calculated according to the algorithm described in [10], with a detailed description of each component of the electricity price: the purchase price of electricity in the wholesale market; electricity transmission services; sales allowance. The need to account for all three components is due to the fact that they in total show approximately 98% of the final cost of

* Corresponding author: alyunov@mail.ru

electricity for the consumer provided by the guaranteeing supplier. The calculation does not include value added tax. The results of the cost calculation are presented in table 1.

The use of a diesel generator at the enterprise, when it is switched on during peak hours (hours for peak load monitoring), allows one to obtain a significant economic effect due to reduction in power cost. This is possible when calculating for the 3-6 price categories.

Thus, the task is reduced to the choice of optimal power and strategy of the diesel generator operation.

The correct choice of a diesel generator allows one to save significant money, due to different nominal flow characteristics.

We compare the cost of 1 kW•h of electricity generated by diesel generators of equal power, but from different manufacturers.

The cost of 1 kW•h of energy generated by a diesel generator is determined by the formula (1):

$$S_D = \frac{Q_{DF} \cdot S_{DF}}{P_W}, \$ / \text{kW·h}, \quad (1)$$

where Q_{DF} is the diesel fuel consumption, l/h;

S_{DF} is the diesel fuel price, $/l;

P_W is the rated power of the diesel generator, kW;

S_D is the cost of 1 kW·h of electricity generated by a diesel generator, $/kW·h.

The calculations presented in [10] show that the cost of 1 kW·h of electricity generated by a diesel generator is in the range of 8–9 $/kW·h. For the first price category, this value is higher than the cost of electricity purchased from the network, therefore, the use of a diesel generator at the enterprise is disadvantageous. For the second price category, this value is higher if electricity is purchased at half-peak, night and day hours, but during peak hours load, the cost of electricity purchased from the network becomes higher than the cost of electricity generated by a diesel generator. Therefore, for the second price category, it is advantageous to turn on the diesel generator only during peak hours. If the calculations are carried out for the 3–6 price categories, then, as indicated above, the cost of electricity is determined for each hour separately, therefore, it is almost impossible to choose the hour at which the diesel generator must be put into operation in order to get a profit.

The greatest economic effect can be achieved by reducing the cost of power. To achieve this goal, it is necessary to simultaneously solve the problem of choosing the optimal power and strategy of the diesel generator.

Based on the studies presented in [3] - [8], we derive an algorithm for choosing the optimal price category taking into account the use of a diesel generator:

1. Initial data is entered: the volume of electricity at hourly intervals for the billing period, the voltage side, the amount of cash (that is, the amount that the customer is ready to allocate for the purchase of a diesel generator).

2. Calculation of the cost of electricity is performed for a price category without taking into account the diesel generator.

3. A load schedule is constructed during power take-off hours for the working days of the month.

4. For peak hours of working days of month, the X indicator is studied (the power at the i-th hour of the j-th working day), obtained as a result of measurements using mathematical statistics methods. The standard deviation of mean power (δ) and minimum power consumption (*Pmin*) are determined.

5. The capacity of diesel generator (P_W) is selected based on the amount of money allocated for its purchase in accordance with the condition (2):

$$\delta \le P_W \le 2.5 \cdot Pmin, \text{kW}, \quad (2)$$

where $Pmin$ is the minimum power consumption, kW;

δ is the standard deviation from average power, kW.

To determine the operation strategy of the diesel generator, the possible number of operation hours of the diesel generator per day (*t*) is calculated. Most often, *t* is equal to the number of peak hours.

In case the funds are insufficient to purchase the diesel generator of the required capacity, P_W can be determined from the condition (3).

$$P_W \le \delta, \text{kW}, \quad (3)$$

Further the strategy of diesel generator is determined from the conditions of optimal reduction of the network power and power being paid in the wholesale electricity market.

6. The calculation of the cost of electricity by price category is carried out taking into account the work of diesel generator according to the developed strategy.

7. The efficiency of using one or another version of the diesel generator is evaluated.

3 Results and discussions

Using the developed algorithm, we calculate the cost of electricity for a woodworking enterprise.

According to conditions (2) - (3), we choose the power of diesel generator. We consider two possible options.

Option No. 1: Amperos AD 400-T400 diesel generator.

Option No. 2: TSS AD-200S-T400-1RKM11 diesel generator.

The possible amount of operation hours of the diesel generator per day (t), taking into account data [5], is 12.

The strategy for the December operation of a 200 kW diesel generator (option No. 2) during peak hours is 5 hours a day: from 8 am to 11 am and at 3 pm.

The strategy for the December operation of a 400 kW diesel generator (option No. 1) during peak hours is 12 hours a day: from 8 am to 12 am and from 3 pm to 9 pm.

The results of calculating the cost of electricity are presented in table 1.

Table 1. Total prices by price categories for December 2019 when using diesel generators.

Price category (PC)	Cost without diesel generator, $/month	Cost with the Amperos AD 400-T400 diesel generator, $/month	Cost with TSS AD-200S-T400-1RKM11 diesel generator, $/month
1. PC1	53 690	55 000	56360
2. PC2/2	59 717	60 356	60376
3. PC 2/3	65 892	65 613	64480
4. PC3	51 402	51 592	52041
5. PC4	50 018	48 273	46527
6. PC5 (1.1)	51 296	51 493	51685
7. PC5 (0.9)	51 181	51 381	51585
8. PC5 (ideal)	50 968	51 180	51409
9. PC6 (1.1)	49 755	48 026	46306
10. PC6 (0.9)	49 641	47 914	46203
11. PC6 (ideal)	49 585	47 862	46160
Minimum cost	49 585	47 862	46160
Profit	0	1722	3381
T, year	-	0.81	0.75

The payback period (*T*) of both generators is approximately 10 months. However, the use of a 400 kW diesel generator is preferable, since monthly income is more important. The disadvantage of this solution is the high initial capital investment.

4 Conclusions

Analysis of the obtained results showed that the right choice of price category allows one to reduce energy costs by 30%, and the introduction of own generation can further reduce these costs by 5-10%.

References

[1] J.S. Savier, D. Das, Impact of Network Reconfiguration on Loss Allocation of Radial Distribution Systems, IEEE Trans. Power Del. 22, 4 (2007). DOI: 10.1109/TPWRD.2007.905370.

[2] V.V. Dubaylova, N.D. Mukhlynin, O.J. Polyakova, Distributed Generation Control in Energy Market Conditions Using Advanced Model Predictive Control 16th Conference on Electrical Machines, Drives and Power Systems (ELMA), Varna, Bulgaria, 1-4 (2019). DOI: 10.1109/ELMA.2019.8771696.

[3] S.A. Eroshenko, A.I. Khalyasmaa, The improvement of distributed generation integration efficiency, 11th IEEE International Conference on Compatibility, Power Electronics and Power Engineering, CPE-POWERENG (2017). DOI: 10.1109/CPE.2017.7915151.

[4] F.R.A. Bukit, N. Mubarakah, D.M. Nasution, Minimizing The Electrical Power Distribution Cost By Optimizing the Distributed Generation Placement, 3rd International Conference on Electrical, Telecommunication and Computer Engineering (ELTICOM), Medan, Indonesia, 60-65 (2019). DOI: 10.1109/ELTICOM47379.2019.8943868.

[5] A.N. Alyunov, O.S. Vyatkina, I.G. Akhmetova, R.D. Pentiuc, K.E. Sakipov, Issues on optimization of operating modes of power transformers, E3S Web of Conferences 124, 02015 (2019) DOI: 10.1051/e3sconf/201912402015.

[6] A. Nemirovskiy, A. Kashin, V. Kosmach, Y. Titovec, I. Toptygin, D. Zaripova, Innovative technology for dismantling the windings of electric motors using ultrasonic radiation, IOP Conference Series: Earth and Environmental Science (EES) 337, 1, 012071 (2019). DOI: 10.1088/1755-1315/337/1/012071.

[7] A.N. Egorov, Ya.S. Kharitonov, V.M. Khubieva, I.A. Yakushev, N.V. Golubtsov, Z.M. Shakurova, The impact of higher voltage and current harmonics on the operation of electrical systems, E3S Web of Conferences 124, 05067 (2019) DOI: https://doi.org/10.1051/e3sconf/201912405067

[8] Y.I. Gracheva, O.V. Naumov, Estimation of Power Losses in Electric Devices of the Electrotechnical Complex, International Conference on Industrial Engineering, Applications and Manufacturing, ICIEAM (2019). DOI: 10.1109/ICIEAM.2019.8742923.

[9] Trading System Administrator JSC: Components of limiting levels of unregulated prices [Electrical resource]. Available at: http:// www.atsenergo.ru.

[10] A.N. Alyunov, Web-service Online Electric [Electrical resource]. Available at: http://online-electric.ru.

The labor protection specialist competence influence on the professional risk management state

Elena Bakiko[1,*], *Vitaly* Serdyuk[1], *Svetlana* Yanchij[1], *Ivan* Ignatovich[1] and *Ekaterina* Bardina[1]

[1]Omsk State Technical University, Department of Industrial ecology and safety, 644050 Omsk, Russia

Abstract. The results of the authors' research in more than 600 organizations have shown the labor protection specialist competence influence on the occupational risk management system state and the overall labor protection management system. It is revealed that the leading role in the labor conditions and safety improving strategy choosing and formation, as well as in the scientifically based method of professional risk assessment belongs to a labor protection specialist with a certain competence. The paper describes 16 occupational safety specialists' professional competence levels and the main factors that determine their formation. The possibilities and abilities of specialists' with different professional competence levels to assess professional risks are considered and presented in tabular form. About 100 data of different approaches and methods for assessing occupational risks are given, from which only a competent occupational safety specialist can choose the best method for his organization. His decision influences not only the assessing and managing professional risks procedure costs, determined by the particular method choosing, but also the overall occupational safety management system effectiveness in the organization. The conducted research has shown a significant occupational safety competent specialists role in ensuring the professional risks management in the organization and allowed to link his capabilities in assessing professional risks with his professional competence level. This information can motivate the employer to pay attention to the labor protection specialists' role in ensuring the industrial safety and take care of their timely skills improvement.

1 Introduction

The risk-based approach to occupational health management in Russia is not just a motto of the future, but an existing reality. International experience shows that the production efficiency among other issues is aligned with occupational risks identification and management [1, 2]. The existence and functioning of a modern occupational safety management system is a necessity for the employer and the regulatory authorities close attention subject. Its effectiveness depends directly on the professional risk management procedures integration.

The professional risk management system ensures workplace working conditions improvement, which leads to savings on non-production costs [3, 4].The successful workplace safety organization based on the occupational risk levels reduction is obvious [2, 5-7].

The leading role in the professional risk assessment and management organization is assigned to occupational safety specialists who do not always possess the necessary competence. Sometimes they are not sufficiently trained and are not actually able to organize an effective occupational risk management system in the company, because they do not have sufficient knowledge for this, which does not allow for an acceptable occupational safety management system functioning level. This circumstance testifies to the relevance of the working conditions and labor protection state dependence problem in the organization on the labor protection specialists' competence level studied by the authors.

2 Methods

The research used factor and statistical analysis methods, synthesis, method of principal components

The professional risk management system state studies in more than 600 organizations employing more than 1,000 occupational safety specialists in 18 regions of Russia, Western Siberia and the Khanty-Mansi Autonomous Area-YUGRA for the period 2017-2019 were conducted.

3 Results

The professional risk management system implementation in Russian organizations and its integration into occupational health and safety management systems, quality management systems and occupational safety and health management systems is an urgent task related to a number of methodological and

* Corresponding author: bakiko@mail.ru

competency-based problems. Appropriate tools are needed to solve these problems and minimize risks.

In addition to the technical and technological factor of reducing occupational risks, it is necessary to take into account the human and organizational factors [8-10]. Competence potential as a tool for reducing the accidents number and severity is an axiom [11-14]. At this point the occupational safety specialists' professional competence becomes one of the main tools for effective occupational risks management.

The accumulated experience using in the professional risks assessing and managing in international companies has led to the practice of simple step-by-step procedures by safety managers in organizations. However, in the Russian legislation in the professional risk management methodology and its practice application field not all is unambiguously. Labor protection specialists face problems that are difficult or impossible to solve at the current level of their professional competence.

Therefore, the occupational safety specialists' assessment by the degree of readiness and ability to implement their competencies in the professional risks identification, assessment and management field, and their basic knowledge state can give employers and top managers the necessary tools to improve the professional risk management system effectiveness.

The competent occupational safety specialist role affects the organization strategy choosing and formation for working conditions and safety improving on the ground of scientifically based approach to professional risks managing.

The ambiguity of a professional risk assessment method choosing that is suitable for this industry and production activities specifics are explained by the following circumstances.

Firstly, there is no unified methodology in Russia, and at the same time a significant number of different methodological approaches to professional risk management are proposed. In such event, the employer has the right to choose an effective method for his production, but labor protection specialists have their first difficulties here. Today in Russia there are about 100 methods and approaches to professional risk assessment prepared and proposed for employers [15].

Professional risks are assessed:

– by direct and indirect method;

– using a qualitative, quantitative, or semi-quantitative method;

-by normative, scientific and methodological, expert or international method;

– by the individual risks forecasting method, group risks forecasting method or the whole organization risks;

- by the individual method with or without account taken of employees personal data.

It is difficult for most occupational safety specialists to study them and choose the best one for their organization.

Secondly, there are still conceptual problems with risk assessment. The various methodology conceptual aspects increase the methods analyzing complexity and complexity of choosing the optimal one for implementation in production.

For example, risks are assessed in the occupational medicine context (employees health deterioration risk), in the safety context (industrial risks), in the labor protection context (occupational risks), in the social insurance fund context (organizational professional risk classes), in the effective management context (risk management).

At this stage, an insufficiently competent occupational safety specialist is aware of the lack of their basic professional knowledge and is not able to make the right decision that the employer needs [16].

This leads to a third group of problems – the competence group. The personality influence studies as the main factor in preventing accidents [13, 17] prove that effective professional risks management requires not only highly qualified managers, but also highly qualified labor protection specialists. Therefore, in addition to the methodological nature problems, according to the authors, there are problems associated with the low organizations labor protection specialists professional competence, on which the direct and leading participation in the effective professional risk management system organization for certain production depends.

According to the "Occupational Safety Specialist" professional standard occupational safety specialist of 6th qualification level should identify, analyze and manage professional risks in the organization; occupational safety specialist of 7th qualification level should evaluate the occupational safety and occupational risk management systems effectiveness, develop corrective actions, develop and improve the effectiveness of these systems. Thus, this professional standard also creates a need in high labor protection professional competence level specialists for organization.

The authors' research has shown that the occupational safety specialists' professional competence level in the organization in the field of determining occupational risks will depend on:

- labor protection specialist basic education;

-professional development, including identification and methodology of professional risk assessment;

– work experience as a labor protection specialist in the organization, including experience in the field of professional risk assessment;

– corporate standards and procedures for professional risk management systems;

- motivation of heads and managers at all levels in the company.

It is revealed that the labor protection specialists professional competence level, their work experience, ability to train and motivate organization's employees – often determine the methodology choosing for assessing professional risks and the effectiveness of risk management. Also, the research conducted by the authors has shown that the leading factor in the professional competence formation is basic education.

The using of the author's 16-level of occupational safety specialists professional competence scale for managing occupational risks by the basic education factor is shown in table 1. The conducted research allowed defining and describing the labor protection specialists' professional competence levels without taking into account the industry specifics of enterprises, the number of employees in the organization, the region where the enterprise is located, etc. (Table 1).

Table 1. Occupational safety specialists' professional competence levels.

Level	Labor protection specialists professional competence main factors formation description
1	Higher education in "Technosphere safety" (hereinafter referred to as occupational safety) (postgraduate, doctoral studies), labor protection specialist independent certification (hereinafter referred to as certification) and experience as a labor protection specialist for at least 3 years
2	Higher education in occupational safety (postgraduate studies) and certification
3	Higher education in occupational safety (postgraduate course)
4	Higher education in occupational safety (master's degree) and certification
5	Higher education in occupational safety (master's degree)
6	Higher education in occupational safety and certification
7	Higher education in occupational safety
8	Higher education in the sciences, additional professional education (retraining) in occupational safety and certification
9	Higher education in the sciences and additional professional education (retraining) in labor safety
10	Higher education in the sciences and advanced training
11	Higher education in the humanities, additional professional education (retraining) in occupational safety and certification
12	Higher education in the humanities, additional professional education (retraining) in occupational safety
13	Higher education in the humanities and advanced training
14	Specialised secondary education, additional professional education (retraining) in occupational safety and experience as a labor protection specialist for more than 5 years
15	Specialised secondary education, additional professional education (retraining) in occupational safety
16	Specialised secondary education and advanced training

4 Discussion

As a result of researches conducted on the basis of Omsk State Technical University, it can be claimed that specialists in labor protection with a bachelor's degree will find it difficult to assess and manage professional risks in the organization. They will face both methodological and practical difficulties. For specialists with a master's degree, this task can be solved. However, practical experience in the organization is necessary. The occupational safety specialists' professional competence levels that will allow them to solve these tasks are presented in Table 2.

Table 2. Ability to manage professional risks.

<table>
<tr><th>Professional competence levels</th><th>Levels' estima tion</th><th>The analysis of possibilities and difficulties in occupational risks assessment</th><th>Ability to manage professional risks</th></tr>
<tr><td>1</td><td rowspan="5">High</td><td rowspan="2">It will not cause any difficulties</td><td rowspan="2">The specialist is able to manage professional risks</td></tr>
<tr><td>2</td></tr>
<tr><td>3</td><td>It will not cause any difficulties. Practical experience in the organization will be needed</td><td>The specialist is able to manage professional risks by having experience</td></tr>
<tr><td>4</td><td>Risk assessment is possible</td><td>The specialist is able to manage professional risks in certain circumstances</td></tr>
<tr><td>5</td><td>Risk assessment is possible. It will cause difficulties of practical nature</td><td rowspan="3">The specialist is able to manage professional risks in certain circumstances, but difficulties occur</td></tr>
<tr><td>6</td><td rowspan="4">Medium</td><td>Risk assessment is possible. It will cause difficulties of methodological nature</td></tr>
<tr><td>7</td><td>Risk assessment is possible. It will cause difficulties of methodological and practical nature</td></tr>
<tr><td>8</td><td rowspan="2">Only formal risk assessment can be executed (according to sample)</td><td rowspan="9">The specialist is not able to manage professional risks</td></tr>
<tr><td>9</td></tr>
<tr><td>10</td><td rowspan="4">Low</td><td rowspan="4">Risk assessment is impossible</td></tr>
<tr><td>11</td></tr>
<tr><td>12</td></tr>
<tr><td>13</td></tr>
<tr><td>14</td><td colspan="2" rowspan="3">The specialist is incompetent in this field</td></tr>
<tr><td>15</td></tr>
<tr><td>16</td></tr>
</table>

Thus, the occupational safety specialists' professional competence directly affects the effectiveness of occupational safety and occupational risk management systems in the organization. Without higher education in "Technosphere safety", it is not possible to choose a scientifically based methodology and assess professional risks. Only professional retraining cannot give students the necessary competencies for this. In addition, the occupational safety specialist' qualification level, the form of training (full-time, part-time) and the training duration play a significant role.

The authors' research also showed:

– occupational safety specialists professional competence level is reduced by one step without regular and effective professional development;

– work experience from one year to 3 years increases the existing level by one step;

– work experience of more than 5 years simultaneously with regular and effective advanced training increases the professional competence level by two levels.

Thus, if the labor protection services specialists are assigned the assessment, analysis and effective risk management tasks, then the methodological complexity will consist in the scientifically-based choosing of appropriate methodology. There is a need to choose the optimal and effective method from a large list of authors' approaches recommended by various sources and international approaches for a particular organization. Therefore, the labor protection specialist professional competence should be of a high level.

Therefore, the labor protection services competence potential objective assessment and their professional competence improvement necessary range identification can serve as a tool for employers and top managers to improve the occupational risk management system effectiveness and the overall labor protection management system effectiveness in organizations.

References

[1] R. Ramesh, D.M. Prabu, S. Magibalan, P. Senthilkumar, Hazard Identification and Risk Assessment in Automotive Industry, International Journal of ChemTech Research 10, 4, 352-358 (2017).

[2] A.C. Ahmad, I.N.M. Zin, M.K. Othman, N.H. Muhamad, The 4th International Building Control Conference 2016 (IBCC 2016), Hazard Identification, Risk Assessment and Risk Control (HIRARC) Accidents at Power Plant, 00105, 66 (2016).

[3] M. Cioni, M. Savioli, Safety at the workplace: accidents and illnesses, Work, Employment and Society 30, 5, 858-875 (2016).

[4] E.V. Bakiko, V.S. Serdyuk, E.V. Yakovlev, Socio-economic impact of adverse working conditions and their differentiation, Omsk Scientific Bulletin, Series Society, History, Modernity 4, 4, 149-156 (2019).

[5] P. Carayon, P. Hancock, N. Leveson, I Noy, L. Sznelwar, G. van. Hootegem, Advancing a sociotechnical systems approach to workplace safety – developing the conceptual framework, Ergonomics 58, 4, 548-564 (2015).

[6] D.A. Hofmann, M.J. Burke, D. Zohar, 100 years of occupational safety research: From basic protections and work analysis to a multilevel view of workplace safety and risk, Journal of Applied Psychology 102, 3, 375-388 (2017).

[7] A.M. Lay, R. Saunders, M. Lifshen, F.C. Breslin, A.D. Lamontagne, E. Tompa, P.M. Smith, The relationship between occupational health and safety vulnerability and workplace injury, Safety Science 94, 85-93 (2017).

[8] M. Dağdeviren, İ. Yüksel, Developing a fuzzy analytic hierarchy process (AHP) model for behavior-based safety management, Information Sciences 178, 6, 1717-1733 (2008).

[9] J.E. Skogdalen, J.E. Vinnem, Quantitative risk analysis offshore – Human and organizational factors , Reliability Engineering & System Safety 96, 4, 468-479 (2011).

[10] M.J. Djapan, D.P. Tadic, I.D. Macuzic, P.D. Dragojovic, A new fuzzy model for determining risk level on the workplaces in manufacturing small and medium enterprises, Proceedings of the Institution of Mechanical Engineers, Part O: Journal of Risk and Reliability 229, 5, 456-468 (2015).

[11] A. Badri, B. Boudreau-Trudel, A.S. Souissi, Occupational health and safety in the industry 4.0 era: A cause for major concern?, Safety Science 109, 403-411 (2018).

[12] G.C. Gauchard, J.M. Mur, C. Touron, L. Benamghar, D. Dehaene, P. Perrin, N. Chau, Determinants of accident proneness: a case–control study in railway workers, Occupational Medicine 56, 3, 187-190 (2006).

[13] S.P. Voroshilov, N.N. Novikov, G.Z. Feinburg, Fundamentals of the methodology for assessing the level of professional risk of an employee due to the level of his professional competence, Occupational health and safety in construction 5 (2011).

[14] A.M. Elin, Working conditions control as a managerial influence factor, Labor Protection and Economics 4, 17, 4-8 (2014).

[15] V.S. Serdyuk, A.M. Dobrenko, O.A. Tsorina, E.V. Bakiko, S.V. Yanchij, Mathematical models for estimating production risks, Journal of Physics: Conference Series, Mechanical Science and Technology Update (MSTU-2018) 1050, 012077 (2018).

[16] E.V. Bakiko, V.S. Serdyuk, V.V. Fadeeva, Organizational and economic aspect of labor protection specialist professional competence formation, Omsk Scientific Bulletin, Series Society, History, Modernity 5, 2, 160-168 (2020).

[17] E. Visser, Y.J. Pijl, R.P. Stolk, J. Neeleman, J.G.M. Rosmalen, Accident proneness, does it exist? A review and meta-analysis, Accident Analysis & Prevention 39, 3, 556-564 (2007).

AUTHOR INDEX

AUTHOR INDEX

AUTHOR INDEX